Atomic Masses (Based on carbon-12)

Element	Symbol	Atomic number	Atomic mass	Element	Symbol	Atomic number	Atomic mass
Actinium	Ac	89	(227)[a]	Mercury	Hg	80	200.59
Aluminum	Al	13	26.98	Molybdenum	Mo	42	95.94
Americium	Am	95	(243)	Neodymium	Nd	60	144.24
Antimony	Sb	51	121.75	Neon	Ne	10	20.183
Argon	Ar	18	39.948	Neptunium	Np	93	(237)
Arsenic	As	33	74.92	Nickel	Ni	28	58.71
Astatine	At	85	(210)	Niobium	Nb	41	92.91
Barium	Ba	56	137.34	Nitrogen	N	7	14.007
Berkelium	Bk	97	(249)	Nobelium	No	102	(253)
Beryllium	Be	4	9.012	Osmium	Os	76	190.2
Bismuth	Bi	83	208.98	Oxygen	O	8	15.9994
Boron	B	5	10.81	Palladium	Pd	46	106.4
Bromine	Br	35	79.909	Phosphorus	P	15	30.974
Cadmium	Cd	48	112.40	Platinum	Pt	78	195.09
Calcium	Ca	20	40.08	Plutonium	Pu	94	(242)
Californium	Cf	98	(251)	Polonium	Po	84	(210)
Carbon	C	6	12.011	Potassium	K	19	39.102
Cerium	Ce	58	140.12	Praseodymium	Pr	59	140.91
Cesium	Cs	55	132.91	Promethium	Pm	61	(147)
Chlorine	Cl	17	35.453	Protactinium	Pa	91	(231)
Chromium	Cr	24	52.00	Radium	Ra	88	(226)
Cobalt	Co	27	58.93	Radon	Rn	86	(222)
Copper	Cu	29	63.54	Rhenium	Re	75	186.23
Curium	Cm	96	(247)	Rhodium	Rh	45	102.91
Dysprosium	Dy	66	162.50	Rubidium	Rb	37	85.47
Einsteinium	Es	99	(254)	Ruthenium	Ru	44	101.1
Erbium	Er	68	167.26	Samarium	Sm	62	150.35
Europium	Eu	63	151.96	Scandium	Sc	21	44.96
Fermium	Fm	100	(253)	Selenium	Se	34	78.96
Fluorine	F	9	19.00	Silicon	Si	14	28.09
Francium	Fr	87	(223)	Silver	Ag	47	107.870
Gadolinium	Gd	64	157.25	Sodium	Na	11	22.9898
Gallium	Ga	31	69.72	Strontium	Sr	38	87.62
Germanium	Ge	32	72.59	Sulfur	S	16	32.064
Gold	Au	79	196.97	Tantalum	Ta	73	180.95
Hafnium	Hf	72	178.49	Technetium	Tc	43	(99)
Helium	He	2	4.003	Tellurium	Te	52	127.60
Holmium	Ho	67	164.93	Terbium	Tb	65	158.92
Hydrogen	H	1	1.0080	Thallium	Tl	81	204.37
Indium	In	49	114.82	Thorium	Th	90	232.04
Iodine	I	53	126.90	Thulium	Tm	69	168.93
Iridium	Ir	77	192.2	Tin	Sn	50	118.69
Iron	Fe	26	55.85	Titanium	Ti	22	47.90
Krypton	Kr	36	83.80	Tungsten	W	74	183.85
Lanthanum	La	57	138.91	Uranium	U	92	238.03
Lawrencium	Lw	103	(257)	Vanadium	V	23	50.94
Lead	Pb	82	207.19	Xenon	Xe	54	131.30
Lithium	Li	3	6.939	Ytterbium	Yb	70	173.04
Lutetium	Lu	71	174.97	Yttrium	Y	39	88.91
Magnesium	Mg	12	24.312	Zinc	Zn	30	65.37
Manganese	Mn	25	54.94	Zirconium	Zr	40	91.22
Mendelevium	Md	101	(256)				

[a] Values in parentheses are for the most stable isotope.

Chemistry

Chemistry

Henry I. Abrash
California State University, Northridge

Kenneth I. Hardcastle
California State University, Northridge

Glencoe Publishing Co., Inc.
Encino, California
Collier Macmillan Publishers
London

Copyright © 1981 by Glencoe Publishing Co., Inc.

Printed in the United States of America

All rights reserved. No part of this book may be
reproduced or transmitted in any form or by any
means, electronic or mechanical, including photocopying,
recording, or by any information storage and retrieval
system, without permission in writing from the Publisher.

Glencoe Publishing Co., Inc.
17337 Ventura Boulevard
Encino, California 91316
Collier Macmillan Canada, Ltd.

Library of Congress Catalog Card Number: 78–71729

ISBN 0–02–471100–4
ISBN 0–02–978480–X (International Edition)

1 2 3 4 5 6 7 8 9 10 85 84 83 82 81

The symmetry of molecules and atomic arrangements in the
solid state is manifested in the outward form and symmetry
of the object we see. For example, the atomic
arrangement on the cover represents the structure
of diamond—perfect tetrahedral bonding between
atoms, and six-sided channels through the crystal.

to Barbara and Ingrid,
Gladys, Elizabeth, Kenneth,
and Geoffrey

Contents in Brief

chapter 1   Introductory Concepts   1
chapter 2   Stoichiometry   29
chapter 3   The Behavior of Gases   47
chapter 4   Atomic Structure: The Constituent Parts   79
chapter 5   Quantum Mechanics   101
chapter 6   The Electronic Structures of Ions and Molecules   135
chapter 7   The Symbolic Representation of Ions and Molecules   157
chapter 8   Chemical Thermodynamics   189
chapter 9   Solids, Liquids, and Changes of State   217
chapter 10   Physical Properties of Solutions   257
chapter 11   Ionic Solutions, Acids, and Bases   283
chapter 12   Chemical Kinetics   303
chapter 13   Chemical Equilibrium   329
chapter 14   Aqueous Equilibria   351
chapter 15   Electrochemical Reactions   379
chapter 16   Molecular Spectroscopy   405
chapter 17   Hydrogen, Oxygen, and Water   421
chapter 18   The Metallic Elements: Periodic Groups IA, IIA, and IIIA   449
chapter 19   Carbon and the Group IVA Elements   473
chapter 20   The Nonmetallic Elements and Noble Gases: Periodic Groups VA–0   505
chapter 21   Transition Metals, Lanthanides, and Actinides   539
chapter 22   Nuclear Properties and Radioactivity   573
chapter 23   Biochemistry   599

Appendices   629
Glossary   649
Answers to Selected Exercises   661
Index   671

Contents

Preface xxiii

chapter 1

Introductory Concepts 1

1.1 The Scientific Method 1

1.2 The Nature of Substances 3
Pure Substances and Mixtures 3
Chemical Reactions 5
Elements and Compounds 5

1.3 Mass Relationships for Chemical Reactions 8
Mass 8
SI Units 9
The Law of Conservation of Mass 10
The Law of Definite Proportions 11

1.4 Dalton's Atomic Theory 12
The Quantization of Mass 12
Applications of Dalton's Theory 13
Isotopes 13
Formulas and Atomic Mass 15
Atomic Mass Units 15
The Law of Multiple Proportions 16

1.5 Molecular Motion and Heat 16
Force and Momentum 17
Work and Energy 18
The Law of Conservation of Energy 19
Heat 19

Special Resource Section: Newton's Laws of Motion 21

Summary 22
Vocabulary List 24
Exercises 24

chapter 2   Stoichiometry   29

- 2.1 **Formulas**   29
  - Molecular Formulas   29
  - Molecular Mass   30
  - Determination of Molecular Formulas   30
  - Simplest Formulas   31
  - The Mole   32
  - Determination of Simplest Formulas   34
  - Determination of Compositions from Formulas   35
- 2.2 **Chemical Equations**   36
- 2.3 **Stoichiometric Calculations for Reactions**   38
  - Mass Calculations   38
  - Limiting Reactants and Theoretical Yields   40

Summary   43
Vocabulary List   43
Exercises   43

chapter 3   The Behavior of Gases   47

- 3.1 **Gas Laws**   47
  - Pressure   47
  - Boyle's Law   49
  - Absolute Temperature   52
  - Charles's Law and Gay-Lussac's Law   53
  - Standard Temperature and Pressure   55
  - Dalton's Law of Partial Pressures   55
- 3.2 **Volume Relationships in Gas Reactions**   57
  - The Experiments of Gay-Lussac   57
  - Avogadro's Hypothesis   58
  - The Ideal Gas Law   60
  - Determining Molecular Mass from Gas Density   61
  - Gas Volume Stoichiometry   62
- 3.3 **The Kinetic-Molecular Theory of Gases**   63
  - Brownian Motion   63
  - The Kinetic-Molecular Theory   64
  - Qualitative Applications of the Kinetic-Molecular Theory   65
- Special Resource Section: The Derivation of the Ideal Gas Law   65
  - Graham's Law of Diffusion   67
  - Random Molecular Motion   68
  - Nonideality   69

Summary   71
Vocabulary List   72
Exercises   72

chapter 4

Atomic Structure: The Constituent Parts 79

- 4.1 **Properties of Elements** 79
  - The Periodic Table 79
  - Atomic Number 81
  - The Structure of the Periodic Table 82
- 4.2 **Chemistry and Electricity** 83
  - Charge 83
  - Electrostatic Forces 83
  - Magnetism 84
  - Chemical and Electrical Phenomena 86
- 4.3 **The Electron** 89
- 4.4 **The Nucleus** 92
  - Rutherford's Experiments 92
  - Isotopes and Nuclear Mass 94
  - Protons and Neutrons 95

Summary 97
Vocabulary List 97
Exercises 98

chapter 5

Quantum Mechanics 101

- 5.1 **The Nature of Light** 101
  - Wave Phenomena 101
  - Quantum Theory 105
- 5.2 **The Quantized Atom** 106
  - Atomic Emission Spectra 106
  - x-Ray Emission and Atomic Number 108
- Special Resource Section: The Details of the Bohr Model 108
- 5.3 **Wave–Particle Duality** 112
  - Particles and Waves 112
  - Standing Waves and Quantization 114
  - The Heisenberg Uncertainty Principle 115
  - The Schrödinger Wave Equation 115
  - Atomic Orbitals of Hydrogen 116
  - Higher-Energy Hydrogen Orbitals 117
  - Quantum Numbers 118
  - Electron Spin 121
- 5.4 **Electron Configuration and the Periodic Table** 122
  - Polyelectron Atoms 122
  - Hydrogenlike Orbitals 123
  - The Pauli Exclusion Principle 123
  - Electron Configuration 123
  - Hund's Rules 125
  - Electron Configuration and Chemical Similarity 128

CONTENTS

Summary 129
Vocabulary List 130
Exercises 130

chapter 6

The Electronic Structures of Ions and Molecules 135

6.1 **Ions** 136
  Ionization Potential and Electron Affinity 136
  Ionic Compounds 139
  The Stability of Ionic Compounds 140
  Stable Ions and Electron Configuration 142

6.2 **Covalent Bonds** 142
  The Properties of Covalent Substances 142
  Molecular Quantum Mechanics 144
  Covalent Bonding and Electron Configuration; He_2 146
  Diatomic Molecules of Second-Row Elements 146

6.3 **Polar Covalent Compounds** 148
  Properties of Polar Compounds 148
  Polar Covalent Bonds 149
  Electronegativity 150
  Molecular Shape and Molecular Polarity 151

Summary 152
Vocabulary List 152
Exercises 153

chapter 7

The Symbolic Representation of Ions and Molecules 157

7.1 **Structural Formulas** 157
7.2 **The Lewis Octet Rule** 159
  Applications of the Octet Rule 160
  Isomerism 161
  Nonoctet Structures 163
7.3 **The Shapes of Covalent Molecules** 164
  Hybrid Orbital Method 165
  The Valence-Shell Electron-Pair Repulsion Method (VSEPR) 168
7.4 **Resonance** 169
7.5 **Oxidation–Reduction** 171
  Oxidation State 171
  Oxidation States and Chemical Similarity 174
  Redox Reactions 174
  Net Ionic Equations 175
  Balancing Redox Equations 176

CONTENTS

7.6 **Chemical Nomenclature** 180
Summary 184
Vocabulary List 185
Exercises 185

chapter 8 — Chemical Thermodynamics 189

8.1 **Energy and Enthalpy** 190
Thermodynamic Systems 190
Work and Energy 191
Calorimetry 194
Enthalpies of Reaction 196
Enthalpies of Formation 199
Hess's Law 199

8.2 **Entropy** 202
Spontaneous Change 202
The Second Law of Thermodynamics 204
Entropy and Spontaneous Change 205

8.3 **Chemical Stability** 205
Entropy and Chemical Reactions 205
Gibbs Free Energy 206
Free Energy and Work 208

Summary 211
Vocabulary List 212
Exercises 212

chapter 9 — Solids, Liquids, and Changes of State 217

9.1 **The Regularity of Crystal Structures** 217
Crystal Symmetry 218
Crystal Lattices 220
Crystal Systems and Bravais Lattices 221

9.2 **x-Ray Crystallography** 221
Bragg's Law 223
Determination of Crystal Structures 225

9.3 **Some Typical Unit Cells** 226
Close-Packed Arrangements 226
Free Space and Cubic Closest Packing 229
Body-Centered Cubic Lattice 231
Diamond Structure 232
Ionic Lattices 232
Polymorphism 233

9.4 **Crystal Properties** 233
Lattice Forces 233

Crystal Imperfections 235
Amorphous Solids and Crystallinity 235
9.5 **The Disorder of Liquids** 236
9.6 **Changes of State** 237
Heating Curves 237
Melting 237
The Thermodynamics of Melting 238
Supercooling 239
Pressure Effects on Melting 240
Le Chatelier's Principle 241
Evaporation 241
Equilibrium Vapor Pressure 241
Special Resource Section: **Temperature Effects on Vapor Pressure** 242
The Thermodynamics of Evaporation 245
Boiling 245
Phase Diagrams 246
Critical Temperature and Pressure 248
9.7 **Liquid Crystals** 248

Summary 249

Vocabulary List 250

Exercises 251

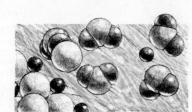

chapter 10 **Physical Properties of Solutions** 257

10.1 **Definitions** 257
Concentration 258
10.2 **Solubility** 260
Saturated Solutions 260
The Thermodynamics of Solubility 261
Supersaturated Solutions 263
10.3 **Phase Changes and Solutions** 263
Vapor Pressure 263
Distillation 264
Nonideality 266
Vapor Pressures of Saturated Solutions 267
Colligative Properties 267
Henry's Law 274
10.4 **Colloids** 276

Summary 279

Vocabulary List 280

Exercises 280

chapter 11 Ionic Solutions, Acids, and Bases 283

- 11.1 **Electrolytes** 283
  - The Arrhenius Theory 283
  - Ions in Solution 285
  - Weak Electrolytes 286
- 11.2 **Acids and Bases** 287
  - Arrhenius Definitions 287
  - Neutralization 288
  - Brønsted–Lowry Definitions 289
  - Lewis Definitions 292
  - Polyprotic Acids 292
  - Amphoterism 293
- 11.3 **Stoichiometry in Solutions** 293
  - Titration 293
  - Standardization 294
  - Normality 295

Summary 298

Vocabulary List 298

Exercises 299

chapter 12 Chemical Kinetics 303

- 12.1 **The Measurement of Reaction Rates** 304
- 12.2 **Reaction Mechanisms** 306
  - The Effects of Concentration 306
  - Determining Kinetic Orders 308
  - Mechanisms of Reactions 310
  - Heterogeneous Reactions 313
- 12.3 **Catalysis** 313
  - Homogeneous Catalysis 313
  - The Theory of Homogeneous Catalysis 314
  - Heterogeneous Catalysis 315
- 12.4 **Collision Theory** 316
  - Temperature Effects 316
  - Collision Theory 318
- Special Resource Section: Absolute Rate Theory 321

Summary 322

Vocabulary List 323

Exercises 324

chapter 13   Chemical Equilibrium   329

13.1 The Law of Mass Action   329
- General Principles   329
- Equilibrium in the Gas Phase   330
- Equilibrium Calculations   333
- Other Forms of the Mass-Action Law   335
- Nonideality   337

13.2 Equilibrium, Thermodynamics, and Kinetics   339
- The Thermodynamics of Equilibrium   339
- The Effect of Temperature on Equilibrium   340
- Kinetics and Equilibrium   342

Summary   344
Vocabulary List   345
Exercises   345

chapter 14   Aqueous Equilibria   351

14.1 Acid–Base Equilibria   351
- The Autoprotolysis of Water   351
- Hydrogen-Ion Concentration; pH   352
- Strong Acids   354
- Weak Acids   355
- Bases   356
- Buffer Solutions   357
- Indicators   361
- Titration Curves   362
- Titrations of Polyprotic Acids   365

14.2 Solubility Equilibria   367
- Solubilities of Salts; Solubility Products   367
- The Common-Ion Effect   368
- The Effect of pH on Solubility   369
- The Solubilities of Amphoteric Hydroxides   370
- Complex Ions and Solubility   371

Special Resource Section: Simultaneous Equilibria   371

Summary   372
Vocabulary List   374
Exercises   374

CONTENTS xvii

chapter 15 Electrochemical Reactions 379

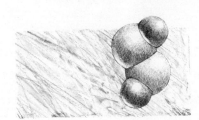

15.1 Reduction Potentials 379
   Galvanic Cells   379
   Cell Notation   381
   Reaction Potentials   382
   Half-Cell Potentials   383
   Concentration Effects   386
   Concentration Cells   388
   Cells as Practical Energy Sources   389

15.2 Electrolysis 392
   Decomposition Potentials   392
   Concentration Effects in Electrolysis   394
   Competing Electrolyses   394
   Commercial Electrolysis   395
   Coulometric Titrations   395
   Electrolytic Corrosion   397

Summary 398
Vocabulary List 399
Exercises 399

chapter 16 Molecular Spectroscopy 405

16.1 The Measurement of Spectra 405
   Spectrophotometers   407
   The Beer–Lambert Law   407
   Selection Rules   408

16.2 Rotational–Vibrational Spectra 409
   Molecular Rotations and Microwaves   409
   Molecular Vibrations and Infrared Radiation   410

16.3 Electronic Spectra 413
   Ultraviolet–Visible Spectra   413
   Fluorescence and Phosphorescence   414

16.4 Nuclear Magnetic Resonance 415

Summary 417
Vocabulary List 418
Exercises 418

CONTENTS

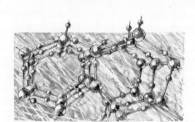

chapter 17 Hydrogen, Oxygen, and Water 421

17.1 **Hydrogen** 421
   The Preparation of H_2 422
   Properties 423
   Hydrogen Isotopes 425
17.2 **Compounds of Hydrogen** 426
   Ionic Hydrides 427
   Molecular Compounds 428
   Metallic Compounds 428
   Polymeric and Electron-Deficient Hydrides 429
17.3 **Hydrogen Bonds** 429
17.4 **Molecular Oxygen** 432
   Properties 432
   Preparation 432
   Allotropes of Oxygen 433
17.5 **Compounds of Oxygen** 434
   Oxides 434
   Hydroxides and Oxyacids 436
   Peroxides and Superoxides 437
17.6 **Water** 439
   Properties 439
   Compounds of Water 442

Summary 443
Vocabulary List 445
Exercises 445

chapter 18 The Metallic Elements: Periodic Groups IA, IIA, and IIIA 449

18.1 **The Group IA Elements (Li, Na, K, Rb, Cs, and Fr)** 449
   Occurrence 449
   Preparation and Properties 450
   Reactions 452
   Compounds 453
18.2 **The Group IIA Elements (Be, Mg, Ca, Sr, Ba, and Ra)** 455
   Properties and Occurrence 456
   Preparation 456
   The Stability of the +2 Oxidation State 457
   Carbonates and Bicarbonates 458
   Other Compounds 460
18.3 **The Group IIIA Elements (B, Al, Ga, In, and Tl)** 462
   Occurrence 462

Properties and Preparation 463
Compounds 465

Summary 467
Vocabulary List 469
Exercises 469

chapter 19 Carbon and the Group IVA Elements 473

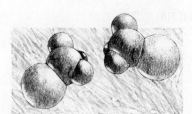

19.1 **Atomic and Molecular Carbon** 474
Allotropes 474
The Isotopes of Carbon 475

19.2 **Compounds of Carbon** 476
Carbides 476
Oxides of Carbon 476
Hydrogen Cyanide and Cyanide Ion 478

19.3 **Organic Chemistry** 479
Functional Groups 479
Alkanes 480
Enantiomers 481
Alkenes and Alkynes 485
Cyclic Compounds 487
Aromatic Compounds 487
Structure Proof 490

19.4 **Silicon, Germanium, Tin, and Lead** 491
Properties of the Metals 491
Preparation 493
Compounds of Silicon 494
Compounds of Germanium, Tin, and Lead 496

Summary 498
Vocabulary List 499
Exercises 499

chapter 20 The Nonmetallic Elements and Noble Gases: Periodic Groups VA–0 505

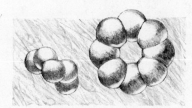

20.1 **Group VA Elements (N, P, As, Sb, and Bi)** 505
Occurrence 507
The Nitrogen Cycle and Nitrogen Fixation 507
The Haber Process 508
Preparation of the Group VA Elements 509
Properties and Uses of the Elements 509
Reactions and Compounds of Nitrogen 511
Reactions and Compounds of P, As, Sb, and Bi 515

20.2 **Group VIA Elements (O, S, Se, Te, and Po)** 518
 Occurrence 519
 Properties 519
 Reactions and Compounds of Sulfur 521
20.3 **Group VIIA Elements (F, Cl, Br, I, and At)** 524
 Occurrence 525
 Properties and Preparation of the Elements 525
 Halides 527
 Oxides, Oxyacids, and Oxyanions 529
20.4 **The Noble Gases (He, Ne, Ar, Kr, Xe, and Rn)** 531
 Occurrence 531
 The Discovery of the Noble Gases 531
 Uses 532
 Noble Gas Compounds 532
 Liquid Helium 534

Summary 534
Vocabulary List 536
Exercises 536

chapter 21 Transition Metals, Lanthanides, and Actinides 539

21.1 **General Considerations** 539
 Electron Configurations 539
 Metallic Properties 542
 The Lanthanide Contraction 542
21.2 **Coordination Compounds** 543
 The Work of Alfred Werner 543
 Geometric Isomers 544
 Chelates 548
 Optical Isomers 550
 The Valence-Bond Description of Coordination Complexes 550
 Special Resource Section: Crystal-Field Theory 552
21.3 **Properties of the Elements** 556
 Groups IIIB, IVB, and VB 556
 The Chromium and Manganese Subgroups 557
 Iron, Cobalt, Nickel, and the Noble Metals 559
 Groups IB and IIB 563
 The Lanthanides (or Rare Earth Elements) and the Actinides 565

Summary 568
Vocabulary List 569
Exercises 569

CONTENTS xxi

chapter 22   Nuclear Properties and Radioactivity   573

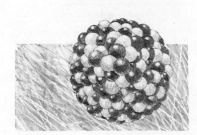

- 22.1 **Radioactivity**   573
  - The Discovery of Radioactivity   573
  - The Measurement of Radioactivity   574
- 22.2 **Nuclear Properties**   576
  - Nuclear Binding Energy   576
  - Nuclear Stability   578
  - Theories of Nuclear Binding   579
  - Nuclear Spins   580
- 22.3 **Nuclear Transformations**   580
  - Alpha Decay   581
  - Beta Decay and Neutrinos   581
  - Positron Emission   581
  - Electron Capture   582
  - Gamma Rays   582
  - The Kinetics of Radioactive Decay   582
  - Natural Radioactivity   585
  - Induced Radioactivity   586
  - Fission   589
  - Nuclear Reactors   590
  - Fusion   591
- 22.4 **The Uses of Radioactivity**   592
  - Analytic Chemistry   592
  - Geology   592
  - Archeological Dating   593
  - Medical and Biological Applications   593

Summary   594
Vocabulary List   595
Exercises   595

chapter 23   Biochemistry   599

- 23.1 **Carbohydrates and Lipids**   600
- 23.2 **Proteins**   602
- 23.3 **Enzymes**   607
- 23.4 **Nucleic Acids**   609
- 23.5 **Protein Synthesis**   612
- 23.6 **Bioenergetics**   615
- 23.7 **Photosynthesis**   619

Summary   620
Vocabulary List   621
Exercises   622

Appendices 629

Appendix A. Mathematics 629
A.1 Exponentials 629
A.2 Operations with Exponents 629
A.3 Significant Figures 630
A.4 Logarithms 631
A.5 Natural Logarithms 632
A.6 Mathematical Operations with Logarithms 633
A.7 Quadratic Equations 633
A.8 Linear Equations 634
A.9 Dimensional Analysis 634

Appendix B. Logarithms 636
Appendix C. Commonly Used Units Derived from Basic SI Units 638
Appendix D. Conversion Factors 639
Appendix E. Fundamental Physical Constants 640
Appendix F. Nomenclature of Inorganic Complexes (Coordination Compounds) 641
F.1 Formulas 641
F.2 Nomenclature 641

Appendix G. Vapor Pressure of Water from 0 to 100°C 643
Appendix H. Equilibrium Constants (Room Temperature) 644
H.1 Ionization Constants for Acids 644
H.2 Ionization Constants for Bases 645
H.3 Solubility-Product Constants 645
H.4 Dissociation Constants for Complex Ions 645

Appendix I. Standard Reduction Potentials 646

Glossary 649
Answers to Selected Exercises 661
Index 671

Preface

During our teaching careers we have grappled with all the usual problems of teaching general chemistry—problems that may be all too familiar to you. While some of our students have strong backgrounds in physics and mathematics, many others know no physics and have only a rudimentary grasp of algebra. How do we teach modern chemistry to the second group of students without depriving the better prepared students of the stimulation they need and deserve? How do we coordinate a lecture course devoting several weeks to atomic structure, quantum mechanics, and chemical bonding with a laboratory that is ill-equipped to deal with these topics? What balance should there be between theoretical and descriptive chemistry? How can we condense a steadily expanding body of subject matter to fit into a rigid academic year?

This book developed from our efforts to teach as accurate, balanced, up-to-date, and accessible a course as possible in the face of these challenges. We set definite goals: to emphasize the experimental nature of chemistry, to present the physical basis for molecular behavior, to show the conceptual framework chemists use to organize their observations, and to give students a foundation for more advanced chemistry courses.

Our approach to these goals is pragmatic, and to accomplish them we have selected those techniques that work best in the classroom. Our students have taught us not to stint on specific facts and detailed discussions; they have taught us to avoid generalities and cursory digests of theories. Our discussions of such key topics as quantum mechanics, crystal structure, thermodynamics, kinetics, and equilibrium reflect our belief in the benefits of going directly to the heart of the matter and the futility of a superficial brush with a subject. We have been careful, also, to develop these topics logically and with the student's needs in mind. We believe that our approach makes realistic demands on both student and instructor and that its rewards are real.

Organization

A discussion of the scientific method; the fundamental definitions of matter (pure substances, mixtures, elements, and compounds); units; Dalton's theory; and the physical definitions of energy, heat, and temperature opens the book. The principles of Dalton's theory are applied to stoichiometry in Chapter 2. Chapter 3 covers the behavior of gases and the applications of the ideal gas law to such problems as finding molecular masses from gas densities and gas–volume relationships in chemical reactions. Chapters 4 through 7 discuss atomic and molecular structure and redox chemistry. Chemical thermodynamics is introduced in Chapter 8, and this chapter forms the basis for the discussions of changes of state, solutions, kinetics, equilibrium, and electrochemistry provided in Chapters 9 through 15. Spectroscopy is covered in Chapter 16, which is an interlude before the survey of descriptive chemistry in Chapters 17 through 21. Organic chemistry is not treated as a distinct topic but is incorporated into the discussion of carbon chemistry in Chapter 19. Nuclear chemistry and biochemistry are covered in Chapters 22 and 23, the final two chapters of the book.

Features

Readability. Because no textbook can be effective without being read and understood, this book has been written, rewritten, and rewritten again with the student in mind. We have tried to write in a clear, straightforward style, avoid unnecessary jargon, and clearly define all technical concepts and terms. The terms are highlighted in boldface type where they are introduced, and they are supported by examples. We have used our classroom experience to anticipate sources of frequent misconceptions among students and have been especially careful to illuminate these areas.

Coordination with the Laboratory. A major goal in organizing this book was to achieve the best possible coordination between lecture and laboratory schedules. As shown by student polls, nothing is more damaging to morale than covering one set of concepts in the lecture and then moving to the laboratory to perform totally unrelated experiments for which the theory has not been presented.

To obtain lecture–laboratory coordination, we delay introducing atomic structure, quantum mechanics, and molecular structure until we have presented stoichiometry and the gas laws. There are few suitable (in other words, inexpensive) experiments dealing with atomic and molecular structure, but by introducing stoichiometry and gas laws early in the course, we can tap a large enough reserve of experiments relating to these topics to carry student and instructor through the lean period of atomic and molecular structure.

Emphasis on Empirical Observations. To emphasize the experimental nature of chemistry, we present observations before introducing the laws that generalize these observations or the theories that explain them. Concepts are illustrated with real examples, not with hypothetical reactions between imaginary substances, and we use real data in these examples. We do not define a quantity without also discussing, however briefly, the experimental technique for measuring it. In Chapter 6, for example, we describe the differences in physical properties of ionic and covalent substances before giving theoretical descriptions of the types of bonds involved.

Chapter Sequence. In addition to developing a chapter sequence to match a laboratory schedule, we have been careful to arrange the chapters so that a major concept introduced in one chapter is utilized in the chapters that immediately follow. Thermodynamics, for example, is introduced in Chapter 8, and then is used frequently in subsequent chapters in the discussions of changes of state, solutions, kinetics, equilibrium, and electrochemistry.

Development within the Chapters. Each chapter moves from simple to complex material and from experimental observations to theoretical explanations—in other words, from the concrete to the formal.

Descriptive Chemistry. The last third of the book is devoted to descriptive chemistry. The primary goals of these chapters, other than to convey specific information, are to review and apply the theoretical principles presented earlier and to illustrate the practical problems, both industrial and ecological, we face today.

Rigor. The level of rigor required in this text is moderate and is comparable to that of most other general chemistry textbooks currently available. We have not assumed any background in chemistry on the part of the reader, but the discussions in this book do require a firm grasp of algebra and understanding of logarithms and trigonometric functions. No calculus is needed, but calculus is not ignored completely; its role in obtaining integrated rate equations is mentioned, but the students are not required to perform these operations.

Any modern treatment of chemistry requires some reference to mechanics, electrostatics, magnetism, and light. Because most beginning chemistry students, we have found, do not have the necessary background in physics, we devote sections of Chapters 1, 4, and 5 to these topics. We have found that a brief, coherent presentation of physical topics in the main text improves student performance in a way that is not achieved by presenting them as a series of ad hoc definitions and equations or by relegating them to appendices.

Exercises. We believe that student practice is an essential part of learning, and so we have provided abundant carefully thought out exercises. Most chapters offer at least forty exercises, and some have more than sixty. About half the exercises (all those with numbers shown in color) are answered in the back of the text, and many are paired so that an unanswered exercise follows a similar answered one. This arrangement is a valuable resource for individual study. The exercises cover a wide range of difficulty. The simpler exercises appear first, and the last few exercises in each chapter (those set off by a rule) illustrate advanced topics to superior students.

Worked-Out Examples. All major mathematical and symbolic operations and the techniques of equation balancing are illustrated by detailed, step-by-step solutions of sample exercises.

Supplementary Teaching Aids. A detailed summary and a list of key terms appear at the end of each chapter. A complete glossary is also included at the end of the book. Discussions of historical, social, and ecological issues, set off throughout the text, place chemistry in a broader perspective.

Key data tables are included in the chapters, and complete tables are included in the appendices for easy reference. Appendices also cover mathematical operations, significant figures, units and dimensional analysis, and nomenclature.

Supplemental Materials. A variety of supplemental material has been especially written for use with this text. An *Instructor's Manual*, by the authors of this text, contains model lecture and laboratory schedules, additional chapter information, sample examination questions, and suggested demonstrations and teaching aids. A *Solutions Manual*, also by the authors, contains detailed solutions and explanations for each exercise in the text. *Laboratory Experiments in General Chemistry*, by Dorothy Barnes and John Chandler (both of the University of Massachusetts, Amherst), and a *Student Study Guide*, by Charles Millner (of California State Polytechnic University, Pomona) and David I. Miller (of California State University, Northridge), also complement the material presented in this book.

Acknowledgments

The authors wish to acknowledge the help of many people in the preparation of this text. Marion Hansen's excellent copy editing skills and attention to chemical accuracy without a doubt make this a better book. Al Burkhardt created the attractive design, and Mary Burkhardt is responsible for the finely rendered artwork that appears on the cover and on all chapter opening pages. The drawings that illuminate the text were provided by Scientific Illustrators. Phyllis Niklas did a superb job of managing all the production work for the entire book.

People too numerous to mention contributed suggestions that were incorporated into the final version of the text. We especially want to thank the following individuals for their extensive reviews of our manuscript:

Sheldon Clare, University of Pittsburgh, Johnstown
Fred Decker, University of Connecticut
James Hill, California State University, Sacramento
Paul Hunter, Michigan State University
Richard Lungstrom, American River College
Benjamin Naylor, California State University, San Jose
Gary Pfeiffer, Ohio University
Don Roach, Miami-Dade Community College
Sami Talhouk, Pennsylvania State University
Max Taylor, Bradley University
Dale Warren, Western Michigan University
Stephen Webber, University of Texas, Austin

Chemistry

chapter 1

Introductory Concepts

Molecular hydrogen, H₂, the simplest molecule

The impact of chemistry on the modern world, both its benefits and the problems it creates, may not be as obvious as other forms of technology, but it pervades our whole society. Perhaps our use of synthetic fibers does not generate the excitement of a trip to the moon, yet it has a far greater influence on our lives. For most people, the prospect of nuclear warfare is far more frightening than chemical pollution, even though the latter could turn out to be a greater threat to our well being. We must learn more about the materials around us, those occurring naturally and those we make ourselves, and know how to use them in the best way. This is the task of the modern chemist.

It is hard to give a concise, precise, and accurate definition of chemistry. Chemistry does involve the study of matter, but physicists, engineers, biologists, geologists, and physicians are all closely involved with the nature of matter. Instead of attempting a definition now, let us start with the idea that chemistry is what chemists do. The rest of this book is devoted to exploring how chemists view matter, the sorts of experiments they perform, and the theoretical framework used to organize the information gained from these experiments. We will also see how chemistry is applied to such diverse problems as predicting the yield of energy from a chemical reaction, extracting metals from ores, and describing how living systems work.

1.1 The Scientific Method

There is no prescription for scientific progress. The development of science is the trial-and-error result of the efforts of many individuals working in many different ways but sharing certain basic assumptions and definitions. The first assumption is that quantitative measurement is possible—that scientists, whatever their differences in training or theoretical commitment, can agree on reliable procedures for measuring quantities within mutually recognized limits of uncertainty. The second important assumption is that there are relationships between observable facts, and that there is more to the universe than a random collection of independent phenomena and events. Without this assumption we could never have made so much theoretical and technical progress. It is impossible to remember all the specific observations dealing with chemistry, but it is relatively easy to grasp a few general relationships between these observations and use them for reliable predictions.

A **scientific law** is a general relationship based on experience. A scientific law is proposed when all the available information fits the law and when the scientist who proposes the law is intuitively confident that all future observations will also follow the general relationship. Being a mental creation of a scientist, a scientific law has neither the inevitability nor inviolability popularly associated with "natural laws" (which have no place in science). If a future observation should disagree with a currently accepted law, the law—not the observation—would have to be modified or even abandoned. A law is accepted or discarded according to our ability to make valid predictions based upon it.

Scientific laws are not explanations. Two persons can agree on the validity of a law and disagree about its causes. For example, a sixteenth-century astronomer who thought that the sun rotated around the earth had a serious theoretical disagreement with one who believed that night and day were due to revolution of the earth around its own axis. Yet both agreed that the sun would rise in the east and set in the west every 24 hours—a scientific law based on experience.

As the ordering process of science leads to more extensive and daring generalizations, the generalizations may involve hypothetical relationships between causes and effects. These proposed cause-and-effect relationships are **theories**. At the time a theory is developed, its proposed causes are hypothetical rather than directly observed. The scientist who proposes these hypothetical causes uses logic to predict the behavior of the hypothetical model under observable conditions. If the predicted behavior of the model matches observed behavior, the theory is satisfactory.

The scientific method contains a self-testing feature. A valid theory must be consistent not only with past observation but also with all future observations. If a single observed fact disagrees with a previously accepted theory, it is the theory (not the fact) that must be reconsidered and modified. Like a law, a theory is judged by its predictive reliability. As the modifications required of a theory increase, the appeal of the theory diminishes. What first seemed to be an explanation of facts begins to look like the rationalization of them if we are forced to continually invent new hypotheses to explain away predictive failures. When an old theory sags under the weight of its modifications, it is time to replace it by a new, fundamentally different theory.

There are many cases in the history of science in which two conflicting theories competed with each other. These theoretical conflicts cannot be decided by data alone because we can usually explain any set of observations in several logical ways. The relative predictive value of the two theories might offer a criterion, but the theory that predicts incorrectly can be modified to correct its failure and the controversy can continue.

The conflict is often settled by the criterion that the theory with the simpler set of hypotheses is preferable. This idea can be misused. Simple does not mean traditional or popular, and the idea of simplicity is not a proper means for maintaining the theoretical status quo. Sometimes there is a tendency to judge the simplicity of a theory by its symbolic devices (for example, pictures of molecular structures) rather than its fundamental hypotheses, and this is also wrong. A solid line in a molecular structure may be easier to draw than a dotted line, but that does not mean that the bond represented by a dotted line is more complex.

There is no board of scientists that decides between conflicting theories. Instead, there is an informal but effective mechanism for rejecting theories. Theories are the working tools of science, and an accepted theory is one that is being used to make predictions. A needlessly complex and unwieldy theory, through its inherent inefficiency, eventually leads us up a blind alley. When scientific progress stalls due to the flaws in its theoretical assumptions, we are forced to reexamine these assumptions and develop new theories. The effectiveness of a new theory is proven if it opens our eyes to natural relationships that had previously escaped our attention.

1.2 The Nature of Substances

One major goal of chemistry is to describe the properties of the many different forms of matter we encounter. It does not take any special training to see the difference between water and steel, feel the difference between sand and clay, taste the difference between salt and sugar, or smell the difference between ammonia and camphor, but a knowledge of chemistry allows us to make much more precise and subtle distinctions between materials. Many of these distinctions involve careful measurements of properties such as densities, boiling points, electric and magnetic properties, the way matter interacts with light, and the amount of heat generated during chemical changes.

A second task facing chemists is to understand the relationship between the properties of materials and their internal structures. In this chapter we will present the idea that all matter is composed of fundamental building blocks called *atoms* and *molecules*. The observable properties of a particular material depend on the structures of the atoms and molecules within it and the ways in which these building blocks interact with each other.

Pure Substances and Mixtures

Most of the materials that we encounter in our daily lives, such as air, milk, gasoline, and steel, are mixtures. **Mixtures** contain two or more distinct substances that can be physically separated from each other. Some mixtures, such as sand mixed with gravel, are **heterogeneous**—in other words, the different components of the mixture occupy distinct regions within the sample. You would have no trouble distinguishing the chunks of gravel from the grains of sand, and the mixture could be separated by a simple mechanical procedure such as sifting the sand from the gravel. Other mixtures, for example a mixture of alcohol and water, are **homogeneous**—the mixture has the same composition throughout the entire sample. There is no portion of the homogeneous alcohol-water mixture that is either pure water or pure alcohol.

There are many different physical methods that can be used to separate mixtures, including homogeneous ones, into their pure components. For example, we can take advantage of the fact that alcohol evaporates more readily than water to separate an alcohol-water mixture by selective evaporation (a process called distillation). Other methods include: selective melting of one component of a solid mixture, leaving the other in its solid form; selective absorption of one component of a mixture on a solid surface; selectively dissolving one component in a certain liquid. This is just a partial list, and we will present details of different separation techniques later in the book.

A **pure substance** is matter that cannot be separated into other substances by physical means. Among common pure substances are oxygen, water, table salt,

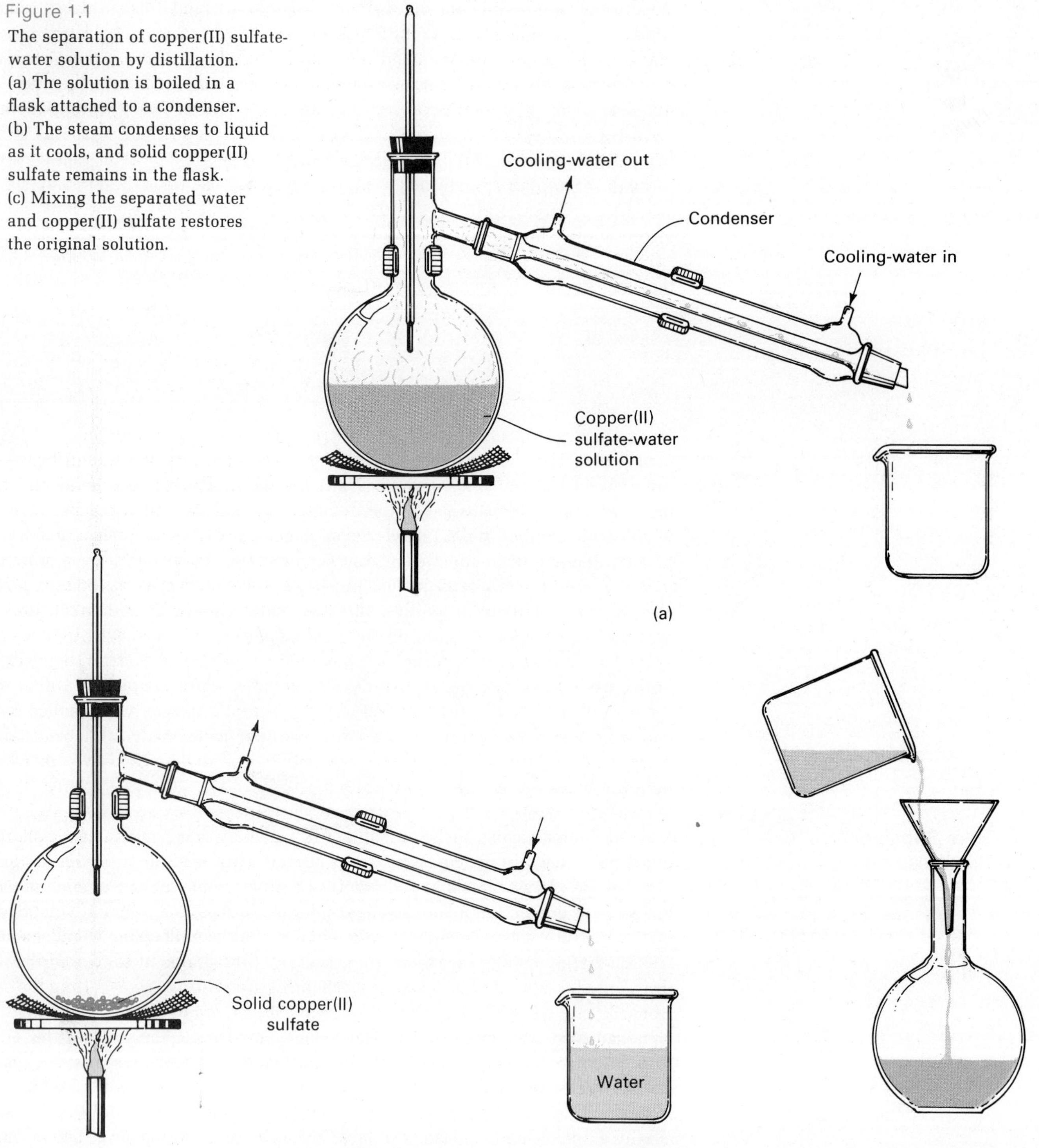

Figure 1.1
The separation of copper(II) sulfate–water solution by distillation.
(a) The solution is boiled in a flask attached to a condenser.
(b) The steam condenses to liquid as it cools, and solid copper(II) sulfate remains in the flask.
(c) Mixing the separated water and copper(II) sulfate restores the original solution.

sugar, and copper. These materials resist separation into distinct components by any and all of the great variety of separation techniques at our disposal. The following experiment (Figure 1.1) illustrates the difference between a pure substance (water) and a mixture (copper(II) sulfate* solution). Distillation of copper(II) sulfate solution (a blue liquid) yields separate samples of water (a colorless liquid)

---

*You will not be held responsible for naming compounds until Chapter 7, which contains the rules for nomenclature. Until then, we will present the names of many compounds, simply because we do not like to discuss anonymous substances.

and blue crystals of copper(II) sulfate. Since the physical process of distillation yields two components, water and copper(II) sulfate, each with its own properties, the original sample was a mixture. A mixture exhibits some of the properties of each of its components; copper(II) sulfate solution is a liquid like water and has the blue color of copper(II) sulfate.

If we now distill half of the water sample, we obtain two identical samples of water. They are both colorless liquids, they have identical densities (about 1 gram/cubic centimeter), and they both freeze at 0°C and boil at 100°C. Neither sample conducts electricity well, is attracted by a magnet, nor dissolves oil. Both samples dissolve alcohol and table salt. The identical properties of the two samples show that distillation fails to resolve water into components and indicate that water is a pure substance. (To be on the safe side, we should subject water to other purification techniques before deciding that it is not a mixture.) The characteristic physical properties of water distinguish it from other pure substances. Each pure substance has its own particular set of physical properties—boiling point, freezing point, density, color, and mechanical, thermal, electric, and magnetic properties.

Chemical Reactions

A **chemical reaction** transforms one or more substances into a different set of substances. The substances that are consumed in the reaction are called **reactants**, and those that are formed by the reaction are called **products**. Chemical reactions include the rusting of iron, the burning of gasoline, the metabolism of food, and the synthesis of plastics. Although a mixture shows properties of each of its components, the products of a chemical reaction usually do not resemble the reactants. A mixture of two colorless gases, hydrogen and oxygen, is also a colorless gas (Figure 1.2, p. 6). But when hydrogen and oxygen react chemically, they produce water, a substance with no obvious physical resemblance to either reactant.

One way chemical reactions often differ from physical processes is the large amount of energy that the chemical reaction either produces or consumes. Hydrogen and oxygen mix with practically no change in energy; the formation of water from hydrogen and oxygen releases large amounts of heat. The reverse reaction, the decomposition of water to hydrogen and oxygen, requires a great deal of electric energy. (Some physical processes, such as distillation, take energy, but the energy needed to distill copper(II) sulfate solution is almost the same as the energy needed to distill pure water. The effect of the copper(II) sulfate on the process is relatively small.)

Elements and Compounds

The idea that the multitude of different substances occurring in nature is composed of a relatively few indestructible *elements* goes back to the ancient Greeks. Medieval scholars accepted Aristotle's view that all substances are composed of four elements—water, earth, air, and fire—but many renaissance chemists, frustrated by the lack of insight provided by this theory, tried to identify the fundamental elements by examining the decomposition products of substances. The primacy of experimental observation over preconceived notions about the nature of matter was firmly established in 1661 when Robert Boyle published *The Sceptical Chymist*. In this book, Boyle suggested a criterion for recognizing an element—try to decompose a substance into simpler components or synthesize it by combining other substances. More concisely, an element is a substance that resists decomposition and cannot be synthesized. There are 105 known elements including such important ones as hydrogen, carbon, nitrogen, oxygen, sodium, aluminum, silicon, sulfur, chlorine, iron, and copper. Pure substances that can be decomposed or synthesized are **compounds** of their constituent elements. Water is a compound of hydrogen and oxygen, table salt a compound of sodium and chlorine, and sugar a compound of carbon, hydrogen, and oxygen.

At the end of the nineteenth century it was discovered that some of the substances classified as elements do decompose spontaneously to form other elements.

Figure 1.2
Mixtures and chemical reactions. (a) Mixtures resemble their components. Mixing two colorless gases, hydrogen and oxygen, produces a colorless, gaseous mixture. (b) Products of chemical reactions do not resemble their components. The reaction of hydrogen gas and oxygen gas under suitable conditions produces water. In an electrochemical reaction, water forms hydrogen gas and oxygen gas.

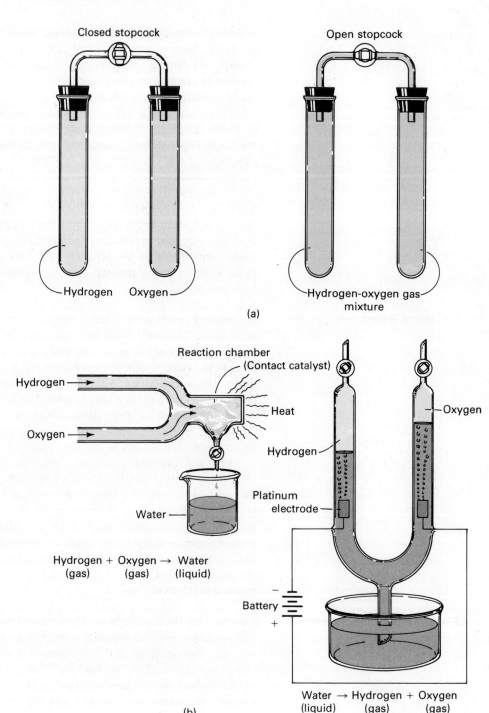

Since then, we have discovered how to induce the decomposition of elements and synthesize others. Of the 105 currently known elements, only hydrogen deserves the name in Boyle's original sense of the word. Rather than deny elemental status to the others, we have changed the definition of an **element** to be a substance that cannot be decomposed, synthesized, or transformed into another element by chemical reactions. The processes that do decompose, synthesize, or transform elements are called **nuclear reactions**.

Merely defining a chemical reaction as a change in substances that does not decompose, synthesize, or transform elements would make any distinction between elements and compounds, or between chemical and nuclear reactions, meaningless. We must have an independent way of differentiating nuclear and chemical reactions. We can classify a change as either chemical or nuclear ac-

Following the style of seventeenth century scientific writing, Boyle wrote *The Sceptical Chymist* in the form of a friendly debate about the number and nature of the elements. Thermistius, an Aristotelian, maintains that all matter is a mixture of fire, air, water, and earth; but Philoponus, a follower of Paracelsus, believes that fire decomposes all matter into sulfur (a water-insoluble and inflammable substance), salt (a water-soluble material with a strong taste), and mercury (a volatile substance). The two opponents agree to submit their arguments to Carneades, a skeptic (and the spokesman for Boyle's views), for judgment, and they also agree to base their debate on experimental evidence rather than "far-fetched and abstracted ratiocination."

The latter condition turns out to be a severe annoyance to Thermistius, who would like to invoke the logical coherence of Aristotle's philosophic system. He believes that experimental evidence is inferior to a logical proof based on first principles—"if men were as perfectly rational as is to be wished they were, this sensible way of probation [experimentation] would be as needless as 'tis wont to be imperfect." Nevertheless, he cites one bit of experimental evidence "to satisfie those that are not capable of a nobler conviction." When green wood burns it produces fire in the form of a flame, air in the form of smoke, water steams and hisses from the burning wood, and earth remains in the form of ashes.

Carneades quickly refutes this argument, but his method also undermines one of the fundamental tenets of Philoponus—he questions the assumption that the substances released during burning must have been present in the wood before it was burned. Although Carneades approves of Philoponus's commitment to experimental studies, he is skeptical about the way he and his fellow Spagyrists (followers of Paracelsus) interpret their observations. They assume fire decomposes all substances into their elements. Carneades wonders if this is always so or if there are not more effective methods for resolving substances into their elements. He points out that classifying an observed decomposition product as either a sulfur, salt, or mercury lumps together substances with very different physical, chemical, and medicinal properties. Faced with the vast variety of different substances present in nature, each with its characteristic properties, Carneades doubts that there is any easily determined number of elements or that all substances contain the same elements.

The questions posed by Boyle (through Carneades) are far more important to modern chemistry than his tentative answers to these questions. (He suggests that water is the only element.) His legacy to chemistry includes his insistence that substances be characterized by their observed properties, the recognition that chemical analysis might alter the components of a substance instead of just separating them from each other, and his rejection of all preconceptions about the number and nature of the elements in a substance.

cording to the current theories about the structure of matter (see Chapters 4 and 5), but a structural theory is not really necessary for this distinction. Nuclear reactions are easy to recognize because they either release or consume tremendous amounts of energy. Put in practical terms, if we can carry out a reaction with the heat of a Bunsen burner or if we can safely watch the reaction of gram quantities of material close up, we are definitely not dealing with nuclear reactions.

1.3 Mass Relationships for Chemical Reactions

Mass

You may have entered this course with an intuitive idea that the term *mass* means a measure of the amount of matter in a sample. You may also have the idea that mass is closely associated with the weight of a sample. Both of these ideas are useful; however, they do not comprise the precise definition of mass that we need to understand many of the properties of molecules. According to its physical definition, **mass** is the tendency of an object to resist a change in motion. The more massive an object, the harder it is to start it moving, slow it down, or change its direction once it is in motion. Unlike weight, the mass of an object does not change with its environment.* A brick in a space craft may be "weightless" but kicking it would be just as painful as on earth—the brick retains its mass and still resists a change in motion.

The **weight** of an object is the force of gravity acting on it. Although weight is proportional to mass, it also depends on the environment where weight is measured. This dependence of weight on environment is not a serious problem for an earth-bound chemist. Since weight at a given location is proportional to mass, the easiest way to estimate the mass of a sample is to compare its weight to that of a known mass. This is the rationale behind the laboratory balance (Figure 1.3).

In chemistry, almost all masses are measured in the metric system of units. Like any other unit system, the metric mass scale is based on an arbitrarily chosen standard that can be reproduced precisely. In this case, the standard unit is the standard **kilogram**, the mass of a piece of platinum alloy that is carefully preserved

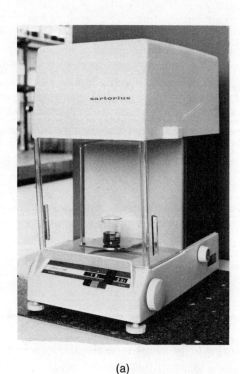

Figure 1.3
Laboratory balances. (a) Analytical balance. Standard masses, kept within the top of the balance, are removed by turning knobs until the amount removed exactly equals the mass of the sample placed on the pan. (b) Electronic top-loading balance. Rapid determination of mass is accomplished by electronic measurement of the capacitance change resulting from a mass load placed on the pan. *Source:* Photographs courtesy of Sartorius Balances, Brinkmann Instruments, Inc., Westbury, New York.

(a)   (b)

*Strictly speaking, this statement applies only to the mass of a motionless object. However, the difference between rest mass and the actual mass of a moving object is immeasurably small at the speeds usually encountered in a chemistry laboratory.

at the International Bureau of Weights and Measures in Paris. The National Bureau of Standards of the United States has a precise copy of the standard kilogram, and this copy is used to prepare other copies, which, in turn, are used to calibrate the standard weights and single-pan balances used in chemical laboratories. The kilogram, weighing about 2.2 lb, is too large to be a convenient unit for many laboratory operations, so we often use the smaller units called grams and milligrams. There are 1000 grams in a kilogram, and each gram contains 1000 milligrams.

Example **1.1**

Perform the following unit conversions: **a.** 0.0274 kilograms to grams **b.** 30.85 milligrams to grams.

Solution

a. Since there are 1000 grams (abbreviated g) per kilogram (kg), we multiply the mass in kg by 1000. You may have more confidence in this and other more complicated solutions to problems if you check them by performing the same mathematical operations you use for the numbers on the units. In this case, we multiply a mass expressed in kilograms by a factor expressed in g/kg to obtain an answer expressed in grams.

$$(0.0274 \text{ kg})\left(1000 \frac{\text{g}}{\text{kg}}\right) = 27.4 \text{ g}$$

b. Since each milligram (mg) is 1/1000 g, we divide the mass in grams by 1000.

$$\frac{30.85 \text{ mg}}{1000 \frac{\text{mg}}{\text{g}}} = 0.03085 \text{ g}$$

SI Units

Scientists perform many other sorts of measurements involving quantities such as length, electric current, and temperature. These measurements resemble mass measurements in being based on arbitrarily defined units. Scientists have recently agreed to an interrelated system of units for a variety of types of measurements. This system, called the Système International d'Unités—the **SI system** for short—is based on the metric system and has as its fundamental units the seven quantities shown in Table 1.1. These fundamental units are multiplied by an appropriate power of ten to form smaller and larger units.* The prefixes and corresponding

Table 1.1 **Fundamental SI Units**

Quantity	Name of unit	Symbol
Length	Meter	m
Mass	Kilogram	kg
Time	Second	sec or s
Temperature	Kelvin	K
Electric current	Ampere	A
Luminous intensity	Candela	cd
Amount of substance	Mole	mol

*Appendix A presents a brief summary of basic mathematics for students who need to review operations with exponents.

Table 1.2 Prefixes for Large and Small SI Units

Prefix	Symbol	Multiplication factor	Example
tera-	T	10^{12}	1 terasecond (Ts) = 10^{12} sec
giga-	G	10^{9}	1 gigameter (Gm) = 10^{9} m
mega-	M	10^{6}	1 megagram (Mg) = 10^{6} g = 10^{3} kg
kilo-	k	10^{3}	1 kilomole (kmol) = 10^{3} mol
hecto-	h	10^{2}	1 hectometer (hm) = 10^{2} m
deka-	da	10	1 dekagram (dag) = 10 g = 10^{-2} kg
deci-	d	10^{-1}	1 decimole (dmol) = 10^{-1} mol
centi-	c	10^{-2}	1 centimeter (cm) = 10^{-2} m
milli-	m	10^{-3}	1 milliampere (mA) = 10^{-3} A
micro-	μ	10^{-6}	1 microgram (μg) = 10^{-6} g = 10^{-9} kg
nano-	n	10^{-9}	1 nanosecond (ns) = 10^{-9} sec
pico-	p	10^{-12}	1 picometer (pm) = 10^{-12} m
femto-	f	10^{-15}	1 femtomole (fmol) = 10^{-15} mol
atto-	a	10^{-18}	1 attometer (am) = 10^{-18} m

multiplication factors are summarized in Table 1.2. The exception to the prefix rule occurs for mass. The standard mass unit, the kilogram (kg), has the prefix *kilo-*, while the simple unit, the gram (g), is a secondary unit in the SI system.

Example 1.2

Convert 500 nm into cm units.

Solution

Since each nm equals 10^{-9} m and since there are 10^{2} cm in 1 m,

$$(500 \text{ nm})\left(10^{-9} \frac{\text{m}}{\text{nm}}\right)\left(10^{2} \frac{\text{cm}}{\text{m}}\right) = 5.00 \times 10^{-5} \text{ cm}$$

Units of many other quantities such as force and energy are built up from these fundamental units in a manner consistent with definitions of the quantities. For example, since velocity is a ratio of distance to time, it has the dimensions of distance divided by time, and its SI unit is m/sec (or m/s). Additional important derived units are listed in Appendix C.

Since the SI system is currently under consideration for adoption and may be changed in the future, units that have been commonly used in the past will most likely continue to be used for some time. Units that may remain are: the liter (L), a unit of volume equal to a cubic decimeter (dm^{3}) and roughly equivalent to a quart; the angstrom (Å), a unit of length equal to 10^{-1} nm (or 100 pm); the torr, a unit of pressure we will deal with in Chapter 3; the atomic mass unit (amu), the unit we use to describe the relative masses of atoms and molecules. Other units, such as the calorie (cal) and the electron volt (eV)—both units of energy— are likely to be replaced by the SI system, but they occur frequently in recent books. Although, in general, we have used SI units in this book, we have retained some non-SI units such as °C, Å, torr, and milliliter because of the convenience and widespread use of these units.

The Law of Conservation of Mass

When a chemical reaction occurs, the masses of the reactants decrease and the product masses increase. The **law of conservation of mass** is a generalization of many careful observations of the changes of masses of substances due to chemical reactions. The data in Table 1.3 illustrate this law for the formation of water by the reaction of hydrogen and oxygen. The masses of the reactants (hydrogen and

1.3 MASS RELATIONSHIPS FOR CHEMICAL REACTIONS

Table 1.3 Mass Changes in the Reaction Hydrogen + Oxygen → Water

	Experiment 1		Experiment 2	
	Initial mass (g)	Final mass (g)	Initial mass (g)	Final mass (g)
Hydrogen	1.000	0.370	1.000	0.000
Oxygen	5.000	0.000	10.000	2.063
Water	0.000	5.630	0.000	8.937
Total mass	6.000	6.000	11.000	11.000
Change in mass		0.000		0.000

oxygen) decrease and the mass of the product (water) increases, but there is no change in the *total* mass of matter. The mass of hydrogen and oxygen consumed exactly equals the mass of water produced. This law applies to all chemical reactions, and there is no known case of a measurable change in total mass during a chemical reaction.

To apply the law of conservation of mass to any particular reaction, it is necessary to keep track of all the reactants and products, including those that are gases. Consider a common and useful chemical reaction, the burning of gasoline in a car motor. It is clear that gasoline is being consumed or else you would never have to stop at a gas station. Since you cannot see any products being formed, it might seem that this consumption of gasoline violates the law of conservation of mass. But if you were to measure the masses of the gaseous products—mainly carbon dioxide and water vapor—coming out of the exhaust, you would find that they exceed the mass of the gasoline burned. That is because there is another reactant besides the gasoline—the oxygen in the air passing through the motor. The chemical reaction is

Oxygen + Gasoline ⟶ Carbon dioxide + Water

and

Mass oxygen and gasoline consumed = Mass carbon dioxide and water produced

Example 1.3

When 2.00 g of iron are heated with chlorine gas, 5.81 g of a compound of iron and chlorine form. How much chlorine does this sample of compound contain?

Solution

According to the law of conservation of mass, the increase in mass of the sample must all be due to the mass of chlorine that has reacted with the iron:

Mass chlorine = Mass compound − Mass iron
  = 5.81 g − 2.00 g = 3.81 g

The Law of Definite Proportions

The data in Table 1.3 also point out another important feature of chemical reactions. Even though there was initially 1.000 g of hydrogen in experiment 1, only 0.630 g was converted to water while all 5.000 g of oxygen were consumed. Adding more hydrogen or increasing the reaction time does not alter this situation. The oxygen and the hydrogen react in a *fixed ratio of masses*, and the only way to consume more hydrogen is to add more oxygen. In other words, oxygen is the **limiting reactant** in experiment 1. The reaction stops when it runs out. If there is enough oxygen present, hydrogen becomes the limiting reactant, and some unreacted oxygen remains after the reaction (experiment 2). But no matter what the limiting reactant and regardless of the initial ratio of the two reactants, hydrogen and oxygen are consumed in the same characteristic ratio.

$$\frac{\text{Mass oxygen}}{\text{Mass hydrogen}} = \frac{5.000 \text{ g}}{0.630 \text{ g}} = 7.937 = \frac{7.937 \text{ g}}{1.000 \text{ g}}$$

We can vary the composition of a mixture by adding more of one of its components, and this causes a gradual change in the properties of the mixture. But we cannot change the composition of most compounds* by adding more of one of the reactants. The water produced in experiment 1 is identical in all its properties with that from experiment 2. Changing the ratio of the reactants just changes the amount of reactant in excess and has no influence on the properties of the product.

Analysis of the water samples from the two experiments would show the same mass composition—88.81% oxygen and 11.19% hydrogen. Generalization of this observation leads to the most common statement of the **law of definite proportions**: A pure compound has a definite and characteristic composition by mass of its constituent elements. Chemists use this law routinely. For example, we can measure the mass of hydrogen in a sample of an organic substance by burning the sample and collecting and weighing the water produced. Relying on the fact that water contains 11.19% hydrogen, we can compute the mass of hydrogen obtained from the sample.

We have, however, learned through additional observations that this simple statement of the law of definite proportions needs modification. The 11.19% hydrogen content of water is an average value for naturally occurring water samples, and it is not true that all water samples have this composition. For example, by using a special form of hydrogen called deuterium we can prepare "heavy water," which contains nearly 20% hydrogen. The 11.19% value is not an absolute or invariant value. Nevertheless, we can use it with a good deal of confidence because there is very little variation in the composition of water derived from natural sources, and the slight fluctuations that do occur are too small to upset most calculations. Variations in composition for most other compounds are also too small to upset the usefulness of the law of definite proportions.

Example 1.4

It is found that 3.21 g of sulfur react with 4.80 g of oxygen to form a particular compound. How many grams of oxygen are required to react with 10.0 g of sulfur to form the same compound?

Solution

The data show that the ratio of the mass of oxygen to the mass of sulfur is

$$\frac{\text{Mass oxygen}}{\text{Mass sulfur}} = \frac{4.80 \text{ g}}{3.21 \text{ g}} = 1.50$$

Therefore,

$$\text{Mass oxygen} = (1.50)(\text{Mass sulfur})$$
$$= (1.50)(10.0 \text{ g}) = 15.0 \text{ g}$$

1.4 Dalton's Atomic Theory

The Quantization of Mass

A **quantum** (plural quanta) is an indivisible unit of definite size. (We will encounter this term often in this book. It is one of the most important concepts in physical science.) The possibility that mass occurs in quanta had interested natural philosophers since ancient times, and one school of thought maintained that

---

*In Chapter 17 we will consider a class of substances, called nonstoichiometric compounds, whose compositions can be varied by changing the ratios of the reactants. We hope by that time you will be familiar enough with chemistry to deal with the ambiguities these substances present.

matter cannot be divided infinitely but occurs as quanta of mass called **atoms**. In 1805 John Dalton proposed an atomic theory that explained the laws of conservation of mass and definite proportions. The success of his theory settled the old controversy over the existence of atoms.

This theory has been modified since 1805 to account for later observations, such as the variations in the composition of a compound. The present form of the atomic theory contains the following hypotheses:

1. All matter consists of small particles called atoms.
2. Atoms cannot be chemically synthesized, decomposed, or transformed into another type of atom.
3. All atoms of a particular element are identical in those properties that influence chemical behavior.
4. Atoms can chemically react to form clusters of atoms called **molecules.** All molecules of a particular compound are identical in terms of the type, number, and geometric arrangement of the constituent atoms.
5. A chemical reaction is a change in molecular structure.

Applications of Dalton's Theory

Dalton explained the difference between elements and compounds in terms of the difference between atoms and molecules. An element such as oxygen contains only one type of atom; chemical reactions cannot synthesize, decompose, or transform oxygen atoms into atoms of any other element. Since chemical reactions cannot alter atoms, there can be no chemical synthesis, decomposition, or transformation of oxygen or of any other element.

Compounds can be decomposed. Water is a compound whose molecules consist of hydrogen and oxygen atoms joined together by chemical bonds. The weakness of these bonds compared to the forces that maintain an atom's identity allows molecules to be destroyed or formed in a chemical reaction, and water, a typical compound, can be both chemically synthesized and decomposed.

The law of conservation of mass is a logical consequence of Dalton's theory. No atoms are created or destroyed during a chemical reaction, and since all matter consists of atoms, a chemical reaction neither creates nor destroys mass.

The data in Table 1.3 show that 7.94 g of oxygen react for every gram of hydrogen, regardless of the ratio of these elements in the reaction mixture. We can explain this according to Dalton's theory by picturing water molecules that consist of 2 hydrogen atoms and 1 oxygen atom. To indicate the presence of 2 hydrogen atoms and 1 oxygen atom in each water molecule, we say that water has the molecular formula H_2O. (We will discuss the rules for writing formulas in Chapter 2.) If the mass of an oxygen atom is 15.87 times as great as the mass of a hydrogen atom, then oxygen and hydrogen must be present in the mass ratio

$$\frac{\text{Mass O}}{\text{Mass H}} = \frac{\text{Mass of 1 O atom}}{\text{Mass of 2 H atoms}} = \frac{15.87}{2} = 7.94$$

If we mix more than 7.94 g of oxygen with each gram of hydrogen, there will be an excess of unreacted oxygen atoms. Likewise, unreacted hydrogen atoms will remain after the reaction if oxygen is the limiting reactant. Figure 1.4 (p. 14) shows a schematic representation of this explanation.

Isotopes

Dalton believed in the complete validity of the law of definite proportions—experimental evidence contradicting it was not discovered until a century later. One hypothesis of his original theory, that all atoms of a particular element have the same mass, has been rejected in view of this more recent evidence. This hypothesis, which would lead to the prediction that a compound's composition could never vary, must be modified to account for the law's failures.

Since Dalton's time, we have discovered that atoms of the same element *can*

Figure 1.4

An atomic explanation of the law of definite proportions for the reaction Hydrogen + Oxygen → Water.

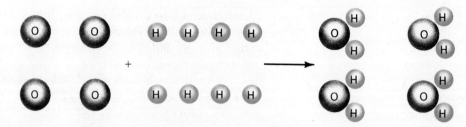

(a) Hydrogen and oxygen in the proper ratio

$$\frac{\text{Mass O reacted}}{\text{Mass H reacted}} = \frac{4(15.87 \times \text{Mass H atom})}{8(\text{Mass H atom})} = 7.94$$

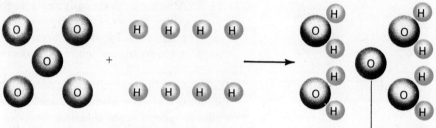

(b) An excess of oxygen mixed with hydrogen

$$\frac{\text{Mass O reacted}}{\text{Mass H reacted}} = \frac{4(15.87 \times \text{Mass H atom})}{8(\text{Mass H atom})} = 7.94$$

One oxygen atom remains unchanged

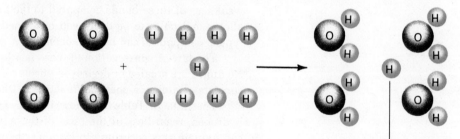

(c) An excess of hydrogen mixed with oxygen

$$\frac{\text{Mass O reacted}}{\text{Mass H reacted}} = \frac{4(15.87 \times \text{Mass H atom})}{8(\text{Mass H atom})} = 7.94$$

One hydrogen atom remains unchanged

differ in mass. Atoms of the same element with different masses are called **isotopes**. The three isotopes of hydrogen are protium,* deuterium, and tritium. Protium is the most common isotope and makes up most of the atoms in a natural sample of hydrogen. Deuterium, a minor component of natural hydrogen, is nearly twice as massive as protium, and the very rare isotope tritium has about three times protium's mass. If the ratio of these three isotopes fluctuated greatly from sample to sample, the mass composition of water would vary accordingly. A sample of water prepared from pure deuterium contains nearly twice the percentage of hydrogen as a sample prepared from pure protium (Figure 1.5). "Heavy water" is the water derived from deuterium.

The law of definite proportions works as well as it does because there is rarely any large variation in the isotopic composition of naturally occurring materials, and we can use the average atomic masses of the elements in composition calculations.

*Although protium is a specific name for the lightest isotope of hydrogen, it is not commonly used. Instead, the name "hydrogen" is used to mean the lightest isotope as well as the naturally occurring mixture of isotopes.

Figure 1.5
The variation of % H in water with the hydrogen isotope.

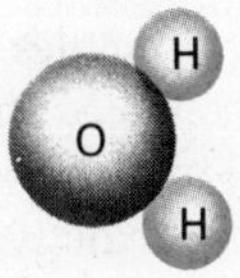

(a) Using pure protium
Mass of 1 protium atom = 1.00
Mass of 1 O atom = 15.87(Mass of 1 protium atom) = (15.87)(1.00) = 15.87
Mass of H$_2$O = 2(1.00) + 15.87 = 17.87

% H in H$_2$O = $\dfrac{2(\text{Mass of 1 protium atom})}{\text{Mass H}_2\text{O}} \times 100 = \dfrac{2(1.00)}{17.87} \times 100 = 11.2\%$

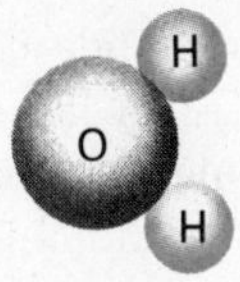

(b) Using pure deuterium (D)
Mass of 1 deuterium atom = 2.00(Mass of 1 protium atom) = 2.00(1.00) = 2.00
Mass of 1 oxygen atom = 15.87(Mass of 1 protium atom) = 15.87(1.00) = 15.87
Mass of D$_2$O = 2(2.00) + 15.87 = 19.87

% D in D$_2$O = $\dfrac{2(\text{Mass of 1 deuterium atom})}{\text{Mass D}_2\text{O}} \times 100 = \dfrac{2(2.00)}{19.87} \times 100 = 20.1\%$

Formulas and Atomic Mass

The explanation we have given for the composition of water works, but it is not the only one possible. Instead of assuming that the formula for water is H$_2$O, we could propose the formula HO with oxygen atoms 7.94 times as massive as hydrogen atoms. This second hypothesis explains the mass data as well as the first explanation. A hypothesis based on the formula HO$_2$ and oxygen atoms with mass 3.97 times that of hydrogen fits the composition data equally well.

This presents a dilemma. There is only one piece of experimental data, the composition, but the theory requires a knowledge of two quantities, the relative atomic mass *and* the molecular formula. Knowing either one of these, we could easily compute the other; however, in the absence of clues concerning either formula or atomic mass, we can find neither of them.

Dalton tried to resolve this dilemma by suggesting that abundant substances should have the simplest possible formulas. Accordingly, water—the most common compound of oxygen and hydrogen—should be HO. But this is an overextension of the criterion of simplicity, which cannot be related to the natural abundance of substances. The true formula for water is H$_2$O, and oxygen atoms are nearly 16 times as massive as hydrogen atoms. The earliest experimental evidence for this formula, which comes from gas volume data, will be presented in Chapter 3.

Atomic Mass Units

Having to talk about one atom being so many times as massive as another is awkward; a unit system for measuring atomic mass is more convenient. The **atomic mass unit** system (abbreviated amu) resembles the kilogram mass system because it is based on an arbitrarily chosen standard mass. Instead of a mass of platinum alloy, the standard for the atomic mass scale is a single atom of the most abundant isotope of carbon, known as ^{12}C. By common agreement, an atom of this isotope has a mass of exactly 12 atomic mass units, and 1 atomic mass unit is 1/12 the mass of the ^{12}C isotope. The mass of any other atom can be measured in amu. On this basis, the lightest isotope of hydrogen has a mass of 1.0078 amu, deuterium 2.0140 amu, and the average atomic mass of natural hydrogen is 1.0080 amu. Average atomic masses are much more useful in chemistry than the individual atomic masses of the isotopes, because chemists rarely work with isotopically pure samples. The table on the inside back cover lists the average atomic masses (also called the atomic weights) of the elements.

The Law of Multiple Proportions

The predictive value of the atomic theory demonstrates its validity. Dalton used logic based on his atomic theory to derive mass relationships between different compounds of the same elements. According to atomic theory, dissimilar compounds of the same elements differ in their molecular structures, and this difference may involve different numbers of atoms. For example, carbon and oxygen form both carbon monoxide, CO, and carbon dioxide, CO_2. Since 1 carbon atom combines with 2 oxygen atoms to form CO_2 but only 1 oxygen atom to form CO, when 1 g of carbon forms CO_2 it combines with exactly twice the mass of oxygen required to form CO. Another example is the mass relationship between two compounds of nitrogen and hydrogen, ammonia (NH_3) and hydrazine (N_2H_4). Three hydrogen atoms are required for every nitrogen atom in the case of ammonia, whereas hydrazine contains 2 hydrogen atoms for each nitrogen atom. The combining mass ratio of hydrogen to nitrogen for hydrazine will be 2/3 the ratio for ammonia.

These logical results of the atomic theory are summarized by the **law of multiple proportions**: A ratio of whole numbers exists between the masses of one element that can combine with a certain mass of another element. This ratio cannot be predicted without knowing the formulas of the compounds, and the formulas must be consistent with the mass composition data. Even though it is impossible to know what the ratio will be until the mass compositions are determined, it is certain that it will be a ratio of whole numbers.

Example 1.5

Use the following data to verify the law of multiple proportions:
a. CO is 42.88% C and 57.12% O; CO_2 is 27.29% C and 72.71% O.
b. NH_3 is 82.24% N and 17.76% H; N_2H_4 is 87.41% N and 12.59% H.

Solution

Since the data give the mass compositions and the formulas, we can find the following ratios.

a. $\dfrac{\text{Mass O}}{\text{Mass C}}$ in CO $= \dfrac{57.12}{42.88} = 1.332$; $\dfrac{\text{Mass O}}{\text{Mass C}}$ in $CO_2 = \dfrac{72.71}{27.29} = 2.664$

$\dfrac{\text{Combining ratio for } CO_2}{\text{Combining ratio for CO}} = \dfrac{2.664}{1.332} = 2$

b. $\dfrac{\text{Mass H}}{\text{Mass N}}$ in $NH_3 = \dfrac{17.76}{82.24} = 0.2160$; $\dfrac{\text{Mass H}}{\text{Mass N}}$ in $N_2H_4 = \dfrac{12.59}{87.41} = 0.1440$

$\dfrac{\text{Combining ratio for } N_2H_4}{\text{Combining ratio for } NH_3} = \dfrac{0.1440}{0.2160} = 0.667 = \dfrac{2}{3}$

In both (a) and (b), the combining ratios of the masses are ratios of whole numbers as stated in the law of multiple proportions.

1.5 Molecular Motion and Heat

Understanding the properties of substances and the nature of chemical reactions requires an understanding of **heat**—the energy of molecular motion. The temperature of a sample is a measure of how vigorously the chemical bonds in its molecules

1.5 MOLECULAR MOTION AND HEAT

are vibrating and how rapidly the molecules are spinning and hurtling through space. These various sorts of molecular motion—vibration, rotation, and movement through space—are major factors governing the physical behavior and chemical properties of a sample.

In describing molecular motion, we use many of the ideas and definitions of Newton's classical physics. Newtonian physics is only partially successful in dealing with molecules. It accounts for the motion of molecules through space adequately and predicts most of the observed properties of gases, but it fails to describe molecular rotations and vibrations accurately. In spite of these limitations of Newtonian physics in describing molecular motions, we retain much of its vocabulary, in particular the terms mass, force, and momentum, when discussing molecules. Two other very important definitions, work and energy, are based on Newtonian terminology.

Force and Momentum

A **force** is any influence that alters the speed or direction of an object. Until it encounters a force, an object at rest remains at rest, and a body in motion continues to move in a straight line at constant speed. (This statement is called Newton's first law of motion.) Force is measured in units called newtons, each newton being equal to 1 kg-m/sec^2. It is possible for an object to be subjected to several forces at the same time and still remain at rest. This is because two forces of equal size pulling in opposite directions cancel each other's effect so that the object experiences no *net* force. In an example we will encounter in Chapter 4, an oil drop is pulled downward by the force of gravity and upward by an electric force (Figure 1.6). When these two opposing forces just balance each other, the oil drop remains suspended.

An important property of moving objects, one that has great significance for the behavior of the constituent parts of atoms (Chapter 5), is momentum. **Momentum** is the mass (m) of an object multiplied by its velocity (v).

$$\text{Momentum} = m \times v \tag{1.1}$$

(Velocity is nearly the same as speed, but it depends on the direction in which the object is moving as well as how fast it is going.) An object can have a large momentum by virtue of its great mass, as in the case of a slowly moving train, or because of its high speed, as in the case of a bullet.

When a moving object, such as a cue ball, collides with an object at rest, such as another billiard ball (Figure 1.7), the moving object slows down and loses momentum while the stationary object picks up speed and gains momentum. The momentum lost by the first object equals exactly the momentum given to the second. A general statement of this phenomenon is called the **law of conservation of**

Figure 1.6
Balanced opposing forces.

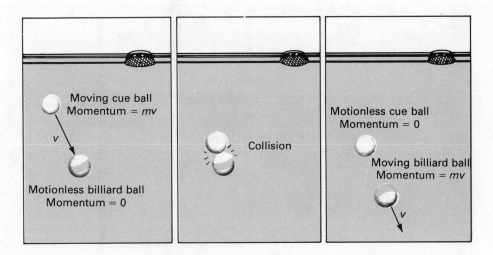

Figure 1.7
The law of conservation of momentum.

Work and Energy

momentum; although the momentum of an individual object may change, the *total momentum* of all interacting objects always remains the same. Momentum can be transferred from one object to another, but it is never created or destroyed.

Whenever a force moves an object through a distance, it performs **work** on the object. The amount of work done on the object is the distance (d) it moves while under the influence of the force multiplied by the force (F).

$$\text{Work} = w = F \times d \tag{1.2}$$

Work is measured in units called joules, each joule (J) being equal to 1 newton-meter or 1 kg-m^2/sec^2. Motion without a force does not involve work; neither does force applied without a resulting motion. Holding a heavy object above your head may be physically exhausting, but it is not work according to the physical definition. Lifting the weight against the gravitational force is work.

Energy is the ability to do work, and it is also expressed in units of joules. During the performance of work, the worker expends energy and the object of the work gains it.

Example 1.6

a. How much work is required to push a piston of a gas cylinder (see Figure 1.8) 0.15 m if the resisting force is 200 newtons?
b. How much energy does the cylinder gain in the process?

Solution

a. To compute the work, we must multiply the distance the piston moves times the resisting force (equation (1.2)).

$$w = (0.15 \text{ m})(200 \text{ newtons})$$
$$= 30 \text{ newton-m} = 30 \text{ J}$$

b. Energy keeps track of the expenditure of work. We must expend 30 J energy to move the piston, and the cylinder is the beneficiary of this energy.

$$\text{Energy gained by cylinder} = \text{Work done on piston} = 30 \text{ J}$$

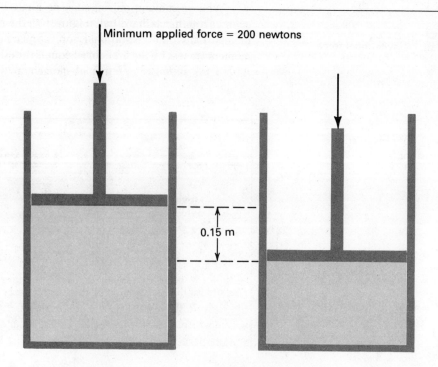

Figure 1.8

1.5 MOLECULAR MOTION AND HEAT

There are two types of mechanical energy an object can possess, kinetic energy and potential energy. **Kinetic energy** is the energy an object possesses due to its own motion. The kinetic energy (KE) of an object depends on its mass and velocity according to the equation

$$KE = \tfrac{1}{2} mv^2 \tag{1.3}$$

Potential energy is energy an object derives from its position and the forces acting on it. A weight suspended on a wire 3 m above the floor has more energy than when it rests on the floor. This extra energy—potential energy of the weight—equals the work used in hoisting the weight 3 m. Should the wire snap, the weight will fall and gain kinetic energy as it speeds up. Its potential energy changes to kinetic energy as it falls.

The example of the hanging weight illustrates gravitational potential energy. Most potential energy changes involving molecules and atoms are due to changes in the positions of particles subject to electrostatic forces (electric potential energy). Nevertheless, the general principle of potential energy is the same as in the gravitational example. A particle occupies a certain position and a force acts on it. If the particle moves in the same direction as the force, some of its potential energy converts to some other energy form—kinetic energy in some cases, light energy in others. Moving the particle in the opposite direction requires work and increases the potential energy of the particle.

The Law of Conservation of Energy

The **law of conservation of energy** says that the total energy in the universe remains constant. One object may yield energy to another, potential energy may change to kinetic energy, light, or some other energy form (the reverse processes also occur), but no creation or destruction of energy occurs. This law has been replaced by a newer law that incorporates Einstein's theory of relativity by allowing for conversions of mass to energy. However, mass—energy conversions are significant only in nuclear reactions, and the older law is satisfactory for most of our purposes.

Heat

An **elastic collision** conserves kinetic energy (Figure 1.9). The kinetic energy lost by one of the colliding particles just equals the kinetic energy gained by the other. However, collisions between visible objects are never completely elastic. Some kinetic energy changes to heat—the kinetic energy of molecules within the objects. Kinetic energy is transferred to particles we cannot see, and our only way of sensing the increased molecular kinetic energy is by the temperature increase associated with it. The heat energy produced in inelastic collisions or by friction equals the loss of kinetic and potential energy of the visible objects involved in the process.

Temperature measures the intensity of the heat energy in an object. The difference between heat and temperature is that the heat of an object depends on (among other things, including temperature) the size of the object. In contrast, temperature does not depend on the size of the object. A kilogram of boiling water contains 1000 times as much heat as a gram of boiling water, but the temperature of the two samples is the same. A dramatic way of illustrating the relationship between heat and temperature is to consider whether you would rather have a drop of boiling water touch you or have a bucket of boiling water dumped on you. The water temperature is the same in either case, so neither experience would be particularly pleasant, but the bucket of water, containing so much more heat than the drop, would do a lot more damage.

We measure temperature by observing physical phenomena that depend on temperature. The common laboratory thermometer uses the fact that the expansion of the volume of a sample of mercury is nearly proportional to the increase in temperature. A calibrated scale on the stem of the thermometer allows us to read the temperature directly without measuring the total mercury volume.

Figure 1.9
Elastic collision.

$KE = \tfrac{1}{2}(1\,kg)(1\,m/sec)^2 = \tfrac{1}{2}\,J \quad KE = 0$

Total KE = $\tfrac{1}{2}$ J

Collision

$KE = 0 \quad KE = \tfrac{1}{2}(1\,kg)(1\,m/sec)^2 = \tfrac{1}{2}\,J$

Total KE = $\tfrac{1}{2}$ J

Figure 1.10

The calibration of a thermometer in the Celsius scale. (a) Mark the 0°C mercury level in ice water. (b) Mark the 100°C mercury level in recondensing steam at atmospheric pressure. (c) Divide the region of the scale between 0°C and 100°C into 100 equal divisions.

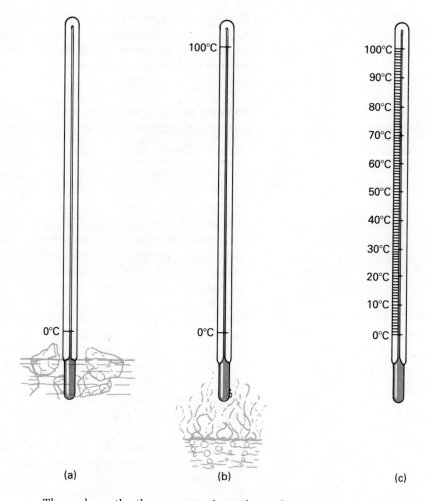

The scale on the thermometer depends on the temperature unit system used. The common laboratory temperature scale is the **Celsius scale** (°C). Like any other temperature scale, it is based on arbitrary, convenient, and reproducible standard temperatures. One of these is the temperature of a mixture of ice and water. The mercury in a given thermometer rises to the same level in any ice-water mixture, whatever the proportions of ice and water. This ice-water temperature is defined as 0°C and marked on the thermometer stem. When the thermometer bulb is placed in the vapor above rapidly boiling water at atmospheric pressure, the mercury rises to a higher reproducible level. This second standard state has a defined temperature of 100°C. We get a reasonably good temperature scale by dividing the distance between the 0° and 100° marks into 100 equal divisions, each division representing a change of 1°C (Figure 1.10).

We still use the **Fahrenheit scale** (°F) for many everyday temperature measurements. On this scale, ice water has a temperature of 32°F, and the temperature of boiling water is 212°F. Since it takes a temperature range of 180°F (212°F − 32°F) to cover the same temperature difference as 100°C, a change of 1.8°F is the same as a change of 1°C. The temperature in °C is the temperature above the ice-water temperature (32°F), so the relationship between temperature in °F and temperature in °C is

$$\text{Temperature in °F} = 1.8(\text{Temperature in °C}) + 32° \tag{1.4}$$

Example 1.7 Convert 40°C into the Fahrenheit scale.

Solution Temperature in °F = 1.8(40°C) + 32° = 104°F

1.5 MOLECULAR MOTION AND HEAT

Table 1.4 Heat Capacities of Common Substances

Substance	Heat capacity (J/g·°C)	Specific heat
Water	4.18	1.00
Ethyl alcohol	2.42	0.579
Calcium carbonate	0.818	0.196
Aluminum	0.903	0.216
Iron	0.452	0.108
Copper	0.385	0.092
Mercury	0.104	0.025
Diamond	0.505	0.121

The quantity of heat in a sample depends on the size and chemical composition of the sample as well as its temperature. The amount of heat needed to raise the temperature of a particular substance by a certain amount is the temperature increase multiplied by the mass and heat capacity of the substance.

$$\text{Heat change} = (\text{Temperature change})(\text{Mass})(\text{Heat capacity}) \tag{1.5}$$

Equation (1.5) can also be used to estimate the heat given off by a substance as it cools. The **heat capacity** of a substance is the energy needed to raise the temperature of 1 g of the substance 1°C. Table 1.4 lists the heat capacities of several common substances. It also contains a column showing **specific heat**—the ratio of the heat capacity of a substance compared to that of water.*

Example 1.8

How much heat does 400 g of water release as it cools from 75°C to 30°C?

Solution

The temperature decreases by 45°C. Applying equation (1.5), we find

$$\text{Heat released} = (45°C)(400\,g)\left(4.18\,\frac{J}{g\cdot°C}\right)$$
$$= 7.5 \times 10^4\,J = 75\,kJ$$

Mechanical energy can be transformed into heat and vice versa. Every joule of mechanical energy lost in an inelastic or frictional process shows up as a joule of heat energy, and a joule of work is done for each joule of heat consumed.

Special Resource Section: Newton's Laws of Motion

Motion is a change in position characterized by its direction and speed. Taken together, the direction and speed of an object define its **velocity.** Velocity is expressed in units of length divided by time (m/sec in SI units). Quantities, such as velocity, for which direction must be specified are vector quantities. As long as we are working in only one dimension, we can indicate the direction of a vector by an algebraic notation, such as + for motion to the right and − for motion to the left.

Any velocity change, whether a change in speed or direction or both, is an **acceleration.** An object beginning to move from rest, slowing to a stop, or following a curved path is accelerating. Quantitatively, acceleration is the rate of change of velocity and it is a vector quantity expressed in units of velocity divided by time (m/sec^2).

*The energy unit called the calorie is defined as the heat needed to raise the temperature of 1 g of water by 1°C. Because of this definition, water has a heat capacity of 1 cal/g-°C, and the heat capacity of any other substance measured in cal/g-°C is nearly identical to its specific heat.

Figure 1.11

Newton's third law. (a) Molecule is moving to the right toward the container wall. (b) On impact, the molecule exerts a force on the container wall and the wall exerts a force on the molecule. (c) After collision, the molecule rebounds from the wall and accelerates to the left.

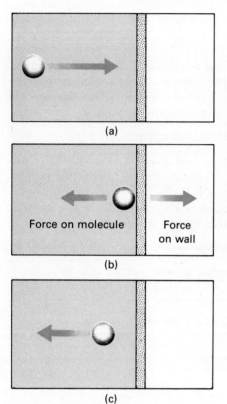

$$\text{Acceleration} = a = \frac{\text{Change in velocity}}{\text{Time}}$$

$$= \frac{v_{\text{final}} - v_{\text{initial}}}{\text{Time}} \qquad (1.6)$$

We have already encountered Newton's first law of motion—a body maintains a constant state of motion unless a force acts on it—earlier in this section. Newton's second law states the quantitative relationship between acceleration, mass, and force: The force acting on an object equals the mass of the object multiplied by the acceleration it receives.

$$\text{Force} = F = ma \qquad (1.7)$$

Force is a vector quantity with the same direction as the acceleration it produces.

According to Newton's third law, when one object exerts a force on another, the second object exerts an equal force acting in the opposite direction of the first. One visible example of this law is the billiard ball collision that we used to illustrate the law of conservation of momentum (Figure 1.7). As the force of the collision accelerates the billiard ball to the right, setting it in motion, there is an equal and opposite force on the cue ball, which makes it lose speed by accelerating it to the left. Another example involves the collision of a molecule with a container wall (Figure 1.11). As the molecule rebounds from the wall, it experiences an acceleration to the left. Therefore the wall must exert a force on the molecule from right to left. According to Newton's third law, the molecule simultaneously exerts an equal force from left to right on the wall.

Summary

1. Physical methods (such as distillation and selective solubility) can separate a mixture into two or more pure substances, each with its own characteristic set of physical properties (melting point, boiling point, density, and so forth).
2. A chemical reaction transforms one set of substances (the reactants) into other substances (the products). Whereas mixtures resemble their components, the products of a reaction do not usually resemble the reactants. Chemical reactions may generate or absorb considerable amounts of energy.
3. Elements (such as carbon) resist chemical decomposition, synthesis, and transformation into other elements. Compounds (such as sodium chloride) are pure substances that can be decomposed into their elements, synthesized from these elements, and transformed into other compounds.
4. Nuclear reactions, which release or consume tremendous amounts of energy, can decompose, synthesize, and transform elements.
5. Mass is the tendency of an object to resist a change in motion. On the earth's surface, weight is proportional to mass and is a useful measure of relative mass.

SUMMARY

6. The current basis for scientific measurement is a form of the metric system called the SI unit system. Decimal multiples of the basic units are obtained by adding suitable prefixes.
7. The law of conservation of mass: There is no change in the total mass of a system during a chemical reaction. The mass of the products formed exactly equals the mass of the reactants consumed.
8. The law of definite proportions: A pure compound has a definite and characteristic mass composition of its constituent elements. Although this law is not strictly correct, it is valid for almost all practical problems in chemistry.
9. A quantum is an indivisible unit of definite size. Matter consists of quanta of mass called atoms.
10. Dalton's atomic theory:
    a. All matter consists of small particles called atoms.
    b. Atoms cannot be chemically synthesized, decomposed, or transformed into another type of atom.
    c. All atoms of a particular element are identical in those properties that influence chemical behavior.
    d. Atoms can react to form clusters of atoms called molecules. All molecules of a particular compound are identical.
    e. A chemical reaction is a change in molecular structure.
11. Isotopes are atoms of the same element with different masses. Variations in the isotopic composition of an element can cause deviations from the law of definite proportions. These variations are rare in natural samples.
12. It is impossible to determine the formula of a compound and the relative atomic masses of its constituent elements from mass composition data alone.
13. One atomic mass unit (amu) is defined as exactly 1/12 the mass of the most common isotope of carbon (^{12}C).
14. The law of multiple proportions: A ratio of whole numbers exists between the masses of one element that can combine with a certain mass of another element.
15. A force is any influence that alters the speed or direction of an object.
16. Momentum is the mass of an object multiplied by its velocity.

    $$\text{Momentum} = mv$$

    The law of conservation of momentum: Momentum can never be created or destroyed.
17. The amount of work done on an object equals the distance it moves while a force acts on it multiplied by the force.

    $$w = Fd$$

18. Energy is the ability to do work. Kinetic energy is energy due to motion of an object.

    $$KE = \tfrac{1}{2}mv^2$$

    Potential energy is energy due to position of an object and the forces acting on it.
19. The law of conservation of energy: The total energy in the universe remains constant.
20. Kinetic energy is conserved in an elastic collision.
21. Heat is the kinetic energy of the molecules in a sample. Temperature is a measure of the intensity of heat in a sample and is independent of the size of the sample.
22. The Celsius temperature scale: 0°C is the temperature of ice water; 100°C is the temperature of boiling water at atmospheric pressure. To convert temperature in °C to the Fahrenheit scale (°F), use the equation

    $$\text{Temperature in °F} = 1.8(\text{Temperature in °C}) + 32°$$

23. The heat needed to raise the temperature of a substance is the temperature change multiplied by the mass and the heat capacity of the substance. The heat capacity is the energy needed to raise the temperature of 1 g of the substance 1°C.

Vocabulary List

scientific law	kilogram	law of conservation of momentum
theory	SI system	work
mixture	law of conservation of mass	energy
heterogeneous	limiting reactant	kinetic energy
homogeneous	law of definite proportions	potential energy
pure substance	quantum	law of conservation of energy
chemical reaction	atom	elastic collision
reactant	molecule	temperature
product	isotope	Celsius scale
element	atomic mass unit	Fahrenheit scale
compound	law of multiple proportions	heat capacity
nuclear reaction	heat	specific heat
mass	force	velocity
weight	momentum	acceleration

Exercises

1. Decide on the basis of the following observations which samples might be pure substances.
   a. Liquid sample A boils at 102°C. After half of the sample has boiled away, the remaining liquid has a boiling point of 106°C, and the recondensed vapor boils at 96°C.
   b. When 10 g of solid sample B are added to 100 milliliters (mL) of water, 2.5 g dissolve, leaving a residue of 7.5 g. When this residue is added to a second 100 mL portion of water, an additional 2.5 g dissolve.
   c. Solid sample C has a melting point of 114°C. The sample dissolves completely in hot water, but part of it precipitates as the solution cools. The melting point of the precipitate is 114°C.
   d. When red liquid D passes through a column of absorbing material, the red color remains on the column, and a colorless liquid drips out the bottom of the column.

2. Which of the following processes are definitely chemical reactions?
   a. Sugar added to water produces a sweet-tasting liquid.
   b. Heating violet iodine crystals produces a violet vapor that, on cooling, reforms violet crystals.
   c. Ethyl alcohol (boiling point 78°C) burns in air, producing a great deal of heat. The liquid remaining in the combustion vessel after cooling boils at 100°C.
   d. Benzoic acid (melting point 121°C) dissolves in benzene. Evaporating the solution leaves a solid residue that melts at 121°C.
   e. As a solution of barium hydroxide absorbs carbon dioxide gas, a white, insoluble solid forms.
   f. Sodium bicarbonate is a white, crystalline powder that foams vigorously when treated with hydrochloric acid. Evaporation of the solution formed in this process leaves a white, crystalline powder that does not foam on contact with hydrochloric acid.
   g. Heating blue crystals of copper(II) nitrate produces a brown gas and a black residue.
   h. A dilute solution of potassium permanganate in water slowly loses its violet color and brown, insoluble material forms.

3. Which of the following processes are chemical reactions and which are nuclear reactions? Refer to the list of elements on the inside back cover. [Note: Compounds of two elements are named by listing the names of the elements and changing the suffix of the second element to -ide as in sodium chloride.]
   a. Potassium and sulfur react to form potassium sulfide.
   b. Potassium changes to calcium.
   c. Aluminum fluoride decomposes to aluminum and fluorine.

EXERCISES

      d. Potassium reacts with water to form hydrogen and potassium oxide.
      e. Uranium decomposes to barium and krypton.
      f. Hydrogen reacts with lithium to produce helium.

4. In *The Sceptical Chymist*, Boyle describes an experiment in which he grew a plant in a carefully weighed amount of soil and reweighed the soil after the plant had grown. Finding no measurable decrease in the mass of the soil, he concluded that the substances in the plant must be compounded exclusively from the water supplied to the plant. What chemical law formed the basis for Boyle's reasoning? Can you think of a flaw in his experimental method?

5. Which of the following processes involve work?
      a. A 20-kg block of Dry Ice slides over a horizontal, frictionless glass surface at a constant speed of 5.00 m/sec.
      b. A 7.5-kg stone falls 15 m.
      c. Brakes applied on a 500-kg car originally moving at 15 m/sec bring it to a complete halt after it moves 20 m.
      d. A spring is held at a steady tension of 40 newtons.

6. An upward electric force of 1.6×10^{-26} newton and a downward magnetic force of 1.6×10^{-26} newton simultaneously act on an electron (a constituent part of an atom). What is the net force on the electron?

7. What is the momentum of a particle with a mass of 9.1×10^{-31} kg if it is moving at 4.5×10^5 m/sec?

8. What is the kinetic energy of a HCl molecule (mass = 6.1×10^{-26} kg) moving at a speed of 500 m/sec?

9. a. How much energy is required to move a particle 2.5 m against a resisting force of 2.5×10^{-19} newton?
      b. What is the change in the potential energy of the particle?

10. How far can a piston be moved against a resisting force of 3.00 newtons if 0.0800 J work is expended?

11. A gas exerts an upward force of 0.200 newton on a piston. How much work must we do to push the piston down 0.350 m?

12. a. Convert into meters: 5.0×10^3 millimeters (mm); 3.5×10^{-3} kilometers (km); 2.0×10^6 angstroms (Å).
      b. Convert into kilograms (kg): 430 milligrams (mg); 5893 hectograms (hg); 85 picograms (pg).
      c. Convert into seconds (sec): 1200 picoseconds (psec); 38.25 nanoseconds (nsec); 53.5×10^{-2} megaseconds (Msec).

13. a. Convert 454 g into kilograms.
      b. Convert 1 year into gigaseconds and into microseconds.
      c. How many picojoules equal 450 megajoules?
      d. How many moles are present in a 5.2 mmol sample?

14. One gram of element X reacts with 11.8 g of element Y (not real chemical symbols) to form the compound XY_4. What is the ratio of the atomic masses of X and Y?

15. One gram of metal M (not a real chemical symbol) reacts with 0.272 g of oxygen to form an oxide with the formula MO. Compute the ratio of the atomic masses of M and O.

16. The reaction of 5.79 g of the element Z and 1.00 g of oxygen yields a pure sample of the compound Z_2O_3. Given that the atomic mass of oxygen is 16.0 amu, compute the atomic mass of Z.

17. The compound X_2O_5 contains 56.0% X and 44.0% O. The atomic mass of oxygen is 16.00 amu. What is the atomic mass of X?

18. One gram of the element X reacts with 1.42 g of element Y to form the compound XY_3. How many grams of Y must react with 1 g of X to form the compound XY_5?

19. A sulfur atom is twice as massive as an oxygen atom. What is the ratio of the mass of sulfur to the mass of oxygen in the compound SO_2?

20. The fermentation of 180 g of glucose,

    Glucose + Oxygen $\longrightarrow$ Carbon dioxide + Water

produces 108 g of water and 264 g of carbon dioxide. How much oxygen does this reaction consume?

21. The reaction between 85.0 g of silver nitrate and 29.2 g of sodium chloride,

    Silver nitrate + Sodium chloride $\longrightarrow$ Silver chloride + Sodium nitrate

produces 71.7 g of silver chloride. How much sodium nitrate is produced?

22. Calcium + Oxygen ⟶ Calcium oxide

Substance	Initial mass	Mass after reaction
Calcium	4.008 g	1.503 g
Oxygen	1.000 g	0.000 g
Calcium oxide	0.000 g	3.505 g

Which reactant, and how much of it, remains after the reaction of 4.008 g of calcium and 2.000 g of oxygen?

23. The following data apply to the reaction between hydrogen and sulfur to form hydrogen sulfide.

	Initial	Final
Mass of hydrogen	1.00 g	0.84 g
Mass of sulfur	2.50 g	0.00 g
Mass of hydrogen sulfide	0.00 g	2.66 g

a. How much hydrogen can react with 20.00 g of sulfur?
b. How much sulfur can react with 1.00 g of hydrogen?
c. What is the percentage of sulfur in hydrogen sulfide?

24. It takes 4.64 g of nitrogen to react completely with 1 g of hydrogen to form ammonia.
a. What is the percentage of hydrogen in ammonia?
b. What mass of hydrogen is needed to react completely with 20.0 g of nitrogen?

25. A sample containing 6.00×10^{23} hydrogen atoms reacts completely with a sample containing 2.00×10^{23} phosphorus atoms to form a pure sample of a phosphorus-hydrogen compound.
a. Suggest a formula for the compound.
b. A sample of 1.5×10^{20} hydrogen atoms and a sample of 1.5×10^{20} phosphorus atoms are mixed and the reaction occurs. Which element is the limiting reactant?
c. How many atoms of the other element remain unreacted?

26. What is normal body temperature (98.6°F) in the Celsius scale?

27. Express 75°C in the Fahrenheit scale.

28. A total of 12.0 kJ of heat flows into 500 g of water originally at 25.0°C. What is the final temperature of the water?

29. a. How much heat is needed to warm 250 g of water from 20°C to 35°C?
b. How much mechanical work is needed to generate this heat?

30. How much will the temperature rise when 3.50 kJ of heat flow into a 2.75-kg block of iron originally at 27°C?

31. A 454-g sample of copper is cooled from 72°C to 23°C. How much heat does the copper give up?

---

32. Equal masses of ethyl alcohol and mercury are heated at the same rate. In which substance will the temperature rise more rapidly? (Refer to Table 1.4.)

33. How much heat is produced in an elastic collision?

34. Two balls, each with a mass of 2.00 kg and moving in opposite directions at speeds of 20.0 m/sec, collide head-on and rebound at speeds of 15.0 m/sec. How much heat is generated in the process?

35. Show that the following data fit the law of multiple proportions. Compound A: 63.6% N, 36.4% O; Compound B: 46.7% N, 53.3% O

36. Naturally occurring chlorine consists of two isotopes, one with a mass of 34.97 amu and the other with a mass of 36.97 amu. The average atomic mass of natural chlorine is 35.45 amu. What is the percentage of the heavier isotope in natural chlorine?

37. A metal M forms two chlorine compounds. The first compound has a ratio of the mass of chlorine to the mass of M equal to 2.22. In the second compound, this ratio is 2.95. Frame a hypothesis for the formulas of the two compounds.

38. Methane is a compound containing 75% C and 25% H. Propose two different hypotheses, each concerning both the formula of methane and the relative atomic masses of carbon and hydrogen.

39. An older atomic mass scale, used until 1961, was based on an atomic mass unit equal to 1/16 the average atomic mass of oxygen. What is the ratio between this atomic mass unit and the current one? The average atomic mass of oxygen on the current scale is 15.9994 amu.

40. a. What is the percentage of oxygen in water (H_2O) prepared from natural hydrogen (average atomic mass = 1.008 amu) and natural oxygen (15.999 amu)?

EXERCISES

    b. What is the percentage of oxygen in water prepared from natural hydrogen and a pure oxygen isotope whose atomic mass is 17.999 amu?

41. What is the percentage of hydrogen in the following?
    a. NH_3 formed from natural sources of nitrogen and hydrogen
    b. NH_3 formed from natural nitrogen and pure deuterium (atomic mass = 2.014 amu)

42. During a period of 2.00×10^{-3} sec, an electron moving through an electric field increases its speed from 2.50×10^5 m/sec to 7.50×10^5 m/sec while maintaining its original direction of motion. What acceleration does the electron undergo?

43. A molecule moving upward at 400 m/sec hits the top of a container and rebounds downward at 400 m/sec. If the molecule is in contact with the container for 3.00×10^{-6} sec, what is the acceleration during the collision?

44. What force is needed on a particle with a mass of 6.4×10^{-27} kg to accelerate it 3.50×10^{10} m/sec^2?

45. What acceleration will a force of 2.5×10^{-16} newton produce on a proton (an atomic fragment with a mass of 1.67×10^{-27} kg)?

46. A force of 5.0×10^{-17} newton accelerates a particle 2.2×10^9 m/sec^2. What is the mass of the particle?

47. A molecule with a mass of 3.0×10^{-26} kg moves downward at 350 m/sec, strikes the bottom of its container, and bounces upward at the same speed. If the molecule is in contact with the container for 2.50×10^{-6} sec, what force does it exert on the container?

48. Two molecules, A and B, have identical kinetic energies. The mass of A is twice that of B. Which molecule is moving faster? What is the ratio of the velocities of the two molecules? Which molecule has the greater momentum?

Note: Throughout this book, exercise numbers that appear in color indicate that the answer to the exercise is given in the back of the book. The rule that appears above Exercise 32 indicates that the exercises below are of greater difficulty than those above. Rules of this nature are used throughout this book.

chapter 2

Stoichiometry

NH$_3$, an important industrial product prepared by the reaction of nitrogen and hydrogen

In the previous chapter, we have presented Dalton's atomic theory and the atomic mass scale for the elements. We can now apply these fundamental concepts to perform a variety of operations that are essential in chemistry: the calculation of the formula of a compound from its composition, the computation of the composition of a compound from its formula, and determination of the relationship between the quantities of reactants consumed and products formed in a reaction. All these quantitative relationships are known by the general term **stoichiometry**. Stoichiometric calculations are the basic operations of chemistry—without them, we would be unable to analyze mixtures, design economical methods for synthesizing the products that are essential to our technology, or to develop the theoretical tools needed to understand the nature of atoms and molecules.

2.1 Formulas

Molecular Formulas

A **molecular formula** is a simple way of representing the number of atoms of each element in a molecule. To do this we write the chemical symbol for each element present in the substance and show the number of atoms of that element per molecule by a subscript number at the lower right of the symbol. We do not write a subscript if there is only one atom of an element; that is, the absence of a subscript means that only one atom is present per molecule. For example, the molecular formula for the vitamin niacin is $C_6H_5NO_2$—each niacin molecule contains six carbon atoms, five hydrogen atoms, a single nitrogen atom, and two oxygen atoms.

The appropriate subscript shows the number of atoms in a molecule of an element. For example, hydrogen, nitrogen, oxygen, and chlorine are all diatomic (two atoms in each molecule) and have the formulas H_2, N_2, O_2, and Cl_2, respectively. Molecules of sulfur containing eight atoms have the formula S_8. Helium, a substance whose molecules are individual, uncombined atoms, has the molecular formula He.

In some cases molecular formulas are written in a way that tells something about the structure of the molecule, that is, how the individual atoms are arranged within the molecule. For example, the formula for acetic acid (the flavor and odor component of vinegar) is often written CH_3CO_2H even though the formula $C_2H_4O_2$ would provide an adequate description of the number and types of atoms in the molecule. The more complicated formula shows that the two carbon atoms occupy different environments in the molecule. One is bound to three hydrogen atoms. The other, in addition to its bond to the first carbon atom, is bound to the oxygen atoms. Three of the hydrogen atoms are bound to carbon, the fourth to oxygen. It takes experience and a familiarity with molecular structures to interpret formulas of this type, and you are not expected to make structural interpretations at this stage of the course. We are presenting these formulas because they are used so often that you are bound to encounter them. If you do, you can use the formula in a stoichiometry problem by counting the total number of atoms of each element present.

Molecular Mass

The **molecular mass** of a substance is the sum of the masses of the atoms that make up one of its molecules. For example, the molecular mass of water is 18.0 amu —twice the atomic mass of hydrogen (2 × 1.0 amu) plus the atomic mass of oxygen (16.0 amu).

Example 2.1

What is the molecular mass of CH_3CO_2H?

Solution

As pointed out earlier, this formula can be rewritten as $C_2H_4O_2$. The mass of the molecule is the total of the masses of two carbon atoms, four hydrogen atoms, and two oxygen atoms.

Molecular mass in amu = 2(12.0 amu) + 4(1.0 amu) + 2(16.0 amu) = 60.0 amu

Determination of Molecular Formulas

To determine the molecular formula of a compound, it is necessary to know both its molecular mass and percentage composition. First multiply the molecular mass by the percentage composition for each element to find the mass of each element in a molecule. Dividing these masses by the atomic masses tells how many atoms of each element are in a molecule.

Example 2.2

Ethylene glycol (common antifreeze) is a compound of C, H, and O. It has a molecular mass of 62 amu and contains 38.7% C and 9.7% H by weight. What is its molecular formula?

Solution

To find the number of carbon atoms in the molecule, first multiply the molecular mass by the percentage of carbon to find the mass of carbon in the molecule.

$$\frac{\text{Mass C}}{\text{Molecule}} = \frac{\left(62 \frac{\text{amu}}{\text{Molecule}}\right)(38.7\% \text{ C})}{100} = 24 \frac{\text{amu C}}{\text{Molecule}}$$

The number of carbon atoms per molecule is this mass divided by the atomic mass of carbon.

$$\frac{\text{C atoms}}{\text{Molecule}} = \frac{24 \frac{\text{amu C}}{\text{Molecule}}}{12 \frac{\text{amu C}}{\text{Atom}}} = 2.0 \frac{\text{Atoms}}{\text{Molecule}}$$

A similar sequence of operations yields the number of hydrogen atoms in the molecule.

$$\frac{\text{Mass H}}{\text{Molecule}} = \frac{\left(62 \frac{\text{amu}}{\text{Molecule}}\right)(9.7\% \text{ H})}{100} = 6.0 \frac{\text{amu H}}{\text{Molecule}}$$

$$\frac{\text{H atoms}}{\text{Molecule}} = \frac{6.0 \frac{\text{amu H}}{\text{Molecule}}}{1.0 \frac{\text{amu H}}{\text{Atom}}} = 6.0 \frac{\text{Atoms}}{\text{Molecule}}$$

We can determine the mass of oxygen in the molecule by subtracting the masses of carbon and hydrogen present in the molecule from the molecular mass.

$$\frac{\text{Mass O}}{\text{Molecule}} = 62 \text{ amu} - 24 \text{ amu} - 6.0 \text{ amu} = 32 \frac{\text{amu O}}{\text{Molecule}}$$

$$\frac{\text{O atoms}}{\text{Molecule}} = \frac{32 \frac{\text{amu O}}{\text{Molecule}}}{16 \frac{\text{amu O}}{\text{Atom}}} = 2.0 \frac{\text{Atoms}}{\text{Molecule}}$$

Therefore, the molecular formula for ethylene glycol is $C_2H_6O_2$.

Simplest Formulas

The **simplest formula** (also called the **empirical formula**) of a compound is the simplest whole-number ratio of atoms that is consistent with its composition. It takes the place of the molecular formula when there is no available information about the molecular mass. Many compounds have well-defined simplest formulas but no definable molecular formula. For instance, SiO_2 is the simplest formula of a compound containing large molecular aggregates in which oxygen and silicon atoms are present in a 2:1 ratio. The size of these molecules varies with the conditions of formation, and every sample of the compound contains a range of molecular sizes. Rather than write an incompletely defined formula such as Si_xO_{2x}, we write SiO_2 with the understanding that this formula does not necessarily represent the molecular formula.

One unfortunate feature of writing formulas is that there is no indication in a formula whether it is a simplest formula or a molecular formula. For example, Na_2O represents a simplest formula—there are two sodium atoms for every oxygen atom but there is no definite molecular size. On the contrary, H_2O is a molecular formula because each molecule contains one oxygen atom and two hydrogen atoms. This ambiguity is something that confronts all chemists. It is an act of courtesy to specify whether a formula is molecular or simplest, but even the most considerate chemists tend to leave this to the reader's knowledge when it is a matter of a common substance such as H_2O, NH_3 (the molecular formula for ammonia), or NaCl (the simplest formula for table salt). You will gradually learn to recognize certain molecular and simplest formulas by experience and by theoretical knowledge about the nature of the chemical bonds in a substance.

Compounds without definite molecular sizes sometimes contain clusters of atoms* with characteristic sizes and formulas. When more than one of these clusters is present in the simplest formula of the compound, the cluster's formula is enclosed in parentheses and a subscript indicates the number of such groups relative to the other atoms. Thus the formula for magnesium nitrate is written $Mg(NO_3)_2$ to show that there are two nitrate groups, each consisting of one nitrogen atom and three oxygen atoms, for every magnesium atom.

Many crystalline substances contain intact water molecules, known as **water of crystallization** or **water of hydration**, combined with other atoms in the crystal. We can display these substances, called **hydrates,** by writing a formula such as

$$Ba(OH)_2 \cdot 8H_2O$$

This formula means that there are 8 water molecules for every barium atom and every two hydroxide groups.

Example 2.3

How many atoms of each element are present in one formula unit of the hydrate $Mg(NO_3)_2 \cdot 6H_2O$?

Solution

Since there is no subscript for Mg, there is only 1 Mg atom in the formula unit. The formula unit contains 2 NO_3 groups, each containing 1 nitrogen atom, so there are 2 nitrogen atoms per formula unit. The formula unit contains a total of 12 oxygen atoms (6 from the 2 NO_3 units and 6 from the 6 molecules of water of hydration), and there are 12 hydrogen atoms from the 6 molecules of water of hydration.

The Mole

If we place a certain number (N_{O_2}) of oxygen molecules (O_2) in one container, N_{NH_3} ammonia (NH_3) molecules in a second container, and N_{H_2O} water molecules in a third, the masses of the three samples are

$$\text{Mass } O_2 = N_{O_2} m_{O_2} \qquad \text{Mass } NH_3 = N_{NH_3} m_{NH_3} \qquad \text{Mass } H_2O = N_{H_2O} m_{H_2O}$$

where m_{O_2}, m_{NH_3}, and m_{H_2O} are the masses of the respective molecules in grams. The mass of each sample is proportional to its molecular mass. This proportionality applies to any collection of N molecules of any substance.

We call any collection of 6.023×10^{23} particles a **mole** (symbol mol). The number of particles in a mole (6.023×10^{23}) is called **Avogadro's number** and is given the symbol N_{Avog}. A mole of H_2 contains 6.023×10^{23} hydrogen molecules, there are 6.023×10^{23} oxygen atoms in a mole of O atoms, and we would have to collect 6.023×10^{23} ball bearings to have a mole of ball bearings. (The mole is comparable to a dozen since both stand for a specific number of things.)

Counting out 6.023×10^{23} molecules would be tedious, if not impossible, but we can use the proportionality between the molecular mass of a substance and the mass of a mole of molecules of that substance to measure out moles fairly easily. *The mass of 1 mole of molecules of any substance is the mass of the substance in grams that numerically equals its molecular mass in atomic mass units.* For example, the molecular mass of water is 18.0 amu, and the mass of a mole of water is 18.0 g. The number of moles, n, in a sample is the mass (in grams) of the sample divided by its molecular mass in amu.

$$n = \frac{m}{\text{Molecular mass}} \tag{2.1}$$

---

*These groups are usually electrically charged clusters of atoms called polyatomic ions.

One important feature of the mole definition is that we can express molecular mass in g/mol as well as in amu.

Example 2.4

What is the mass, in grams, of 0.500 mol C_2H_4?

Solution

The molecular mass of C_2H_4 is the total mass of two carbon atoms and four hydrogen atoms.

$$\text{Molecular mass} = 2(12.0 \text{ amu}) + 4(1.0 \text{ amu}) = 28.0 \text{ amu}$$

The mass of 1 mol C_2H_4 is therefore 28.0 g, and the mass of 0.500 mol is

$$m = (0.500 \text{ mol})\left(28.0 \frac{\text{g}}{\text{mol}}\right) = 14.0 \text{ g}$$

Example 2.5

How many moles are present in 24.0 g O_2?

Solution

The molecular mass of O_2 is twice the atomic mass of oxygen.

$$\text{Molecular mass} = 2(16.0 \text{ amu}) = 32.0 \text{ amu} \left(\text{or } 32.0 \frac{\text{g}}{\text{mol}}\right)$$

According to equation (2.1),

$$n = \frac{24.0 \text{ g}}{32.0 \frac{\text{g}}{\text{mol}}} = 0.750 \text{ mol}$$

Unless we want to calculate properties of individual particles, it is not necessary to use Avogadro's number. For the problems that will concern us most—those involving formulas and compositions of substances—it is sufficient to know that the number of molecules in a sample is proportional to the number of moles of molecules. In fact, the mole concept was used successfully years before there was a precise estimate of Avogadro's number.

In performing stoichiometry problems, it is important to keep the distinction between a mole of atoms and a mole of molecules clearly in mind. A mole of oxygen atoms is 6.023×10^{23} atoms, regardless of the chemical state in which they exist. The mass in grams of a mole of atoms equals the *atomic* mass of the element in atomic mass units. The number of moles of atoms in a certain mass of an element is the mass divided by the atomic mass of the element.

$$\text{Moles of atoms} = \frac{m}{\text{Atomic mass}} \tag{2.2}$$

The mass of a mole of atoms is always the same for a particular element no matter what molecular form it has. For example, oxygen can exist in one of two elemental forms, the common diatomic form (O_2) and a less stable form called ozone (O_3). The molecular mass of O_2 is 32.00 amu, and 1 mol O_2 *molecules* is 32.00 g. Ozone has a molecular mass of 48.00 amu, and 1 mol O_3 molecules is 48.00 g. But whether oxygen is present as O_2, O_3, or in any oxygen-containing compound, a mole of O atoms has a mass of 16.00 g. Avoid confusion between moles of atoms and moles of molecules by taking a cue from the way the formula is written—a mole of O stands for a mole of oxygen atoms, whereas a mole of O_2 stands for a mole of diatomic molecules.

Example 2.6

What is the mass of 1.50 mol Mo?

Solution

Since the atomic mass of Mo is 95.9 amu, there are 95.9 g Mo per mole of atoms.

$$\text{Mass Mo} = (1.50 \text{ mol})\left(95.9 \frac{\text{g}}{\text{mol}}\right) = 144 \text{ g}$$

Example 2.7

How many moles of S are present in 128 g S_8?

Solution

The atomic mass of S is 32.1 amu, so 1 mol S atoms is 32.1 g, regardless of its formula. Applying equation (2.1), we find

$$\text{Moles S} = \frac{128 \text{ g}}{32.1 \text{ g/mol}} = 3.99 \text{ mol}$$

Example 2.8

How many moles of O are present in 2.0 mol CO_2?

Solution

Since there are two oxygen atoms in each CO_2 molecule, there must be twice as many moles of O as moles of CO_2.

$$\text{Moles O} = \frac{2 \text{ mol O}}{\text{mol } CO_2} (2.0 \text{ mol } CO_2) = 4.0 \text{ mol O}$$

The mass of a mole of a compound that does not have a molecular mass is the mass in grams that equals its formula mass in amu. The **formula mass** is the sum of the masses of all the atoms in a formula unit. For example, the formula mass of sodium chloride, NaCl, is the sum of the atomic mass of sodium and the atomic mass of chlorine.

$$\text{Formula mass} = 23.0 \text{ amu} + 35.5 \text{ amu} = 58.5 \text{ amu}$$

One mole of sodium chloride has a mass of 58.5 g.

Determination of Simplest Formulas

The procedure for determining the simplest formula from composition data is to compute the masses of the constituent elements in a conveniently chosen sample (for example, 100 g) and the number of moles of atoms of each element in this sample. The ratio of atoms of the elements in the compound is the same as the ratio of moles of atoms in the sample, and this ratio equals a ratio of small whole numbers.

Example 2.9

Magnetite, a major iron ore, contains 72.4% iron and 27.6% oxygen. What is the simplest formula for magnetite?

Solution

A 100.0-g sample of magnetite contains 72.4 g Fe and 27.6 g O. The atomic masses of Fe and O are 55.8 amu and 16.0 amu, respectively.

$$\text{Moles Fe in sample} = \frac{72.4 \text{ g Fe}}{55.8 \text{ g Fe/mol}} = 1.30 \text{ mol}$$

$$\text{Moles O in sample} = \frac{27.6 \text{ g O}}{16.0 \text{ g O/mol}} = 1.73 \text{ mol}$$

$$\frac{\text{Atoms O}}{\text{Atoms Fe}} = \frac{\text{Moles O}}{\text{Moles Fe}} = \frac{1.73 \text{ mol}}{1.30 \text{ mol}} = 1.33$$

The ratio of 1.33 expressed as a ratio of small whole numbers is approximately 4:3. Therefore, the simplest formula for magnetite is Fe_3O_4.

Once having determined a simplest formula, we can later use the molecular mass—even an approximate estimate—to determine the molecular formula. The molecular formula has the same atomic ratios as the simplest formula and the mass that is closest to the experimental value.

Example 2.10

The simplest formula of ethyl acetate is C_2H_4O. An estimate, based on gas density, for its molecular mass is 91 amu. What is its molecular formula?

Solution

$$\text{Simplest formula mass} = 2(12.0) + 4(1.0) + 16.0 = 44.0 \text{ amu}$$

The estimated molecular mass is approximately twice this value, so the molecular formula is $C_4H_8O_2$ with a mass of 88.0 amu. The 5% error in the experimental molecular mass, possibly due to nonideal experimental conditions, is not enough to upset our conclusion about the formula.

Determination of Compositions from Formulas

The lack of information about molecular mass limits the theoretical usefulness of a simplest formula, but simplest formulas still have important practical applications. They summarize the elemental compositions of compounds. For example, quartz (which is crystalline SiO_2) has a formula mass of 60.09 amu (the atomic mass of Si plus twice the atomic mass of O), and each mole of quartz contains 1 mol Si and 2 mol O. Therefore, the % O in quartz is twice the atomic mass of O divided by the formula mass of quartz.

$$\text{Formula mass } SiO_2 = 28.09 \text{ amu} + 2(16.00 \text{ amu}) = 60.09 \text{ amu}$$

$$\% \text{ O} = \frac{\left[\frac{2 \text{ mol O}}{\text{mol } SiO_2}\right]\left[\frac{16.00 \text{ g O}}{\text{mol O}}\right]}{\left[\frac{60.09 \text{ g}}{\text{mol } SiO_2}\right]} \times 100 = 53.25\%$$

Example 2.11

Find the % Ba, % O, % H, and % water of crystallization in $Ba(OH)_2 \cdot 8H_2O$. Compute the percentages to four significant figures.

Solution

$$\text{Formula mass } Ba(OH)_2 \cdot 8H_2O = 137.3 \text{ amu} + 10(16.00 \text{ amu})$$
$$+ 18(1.008 \text{ amu}) = 315.4 \text{ amu}$$

One mole of the compound (315.4 g) contains 1 mol Ba (137.3 g).

$$\% \text{ Ba} = \frac{\left[\dfrac{1 \text{ mol Ba}}{\text{mol compound}}\right]\left[\dfrac{137.3 \text{ g}}{\text{mol Ba}}\right]}{\left[\dfrac{315.4 \text{ g}}{\text{mol compound}}\right]} \times 100 = 43.53\%$$

One mole of the compound contains 10 mol O—2 from the OH and 8 from the water of crystallization.

$$\% \text{ O} = \frac{\left[\dfrac{10 \text{ mol O}}{\text{mol compound}}\right]\left[\dfrac{16.00 \text{ g}}{\text{mol O}}\right]}{\left[\dfrac{315.4 \text{ g}}{\text{mol compound}}\right]} \times 100 = 50.73\%$$

Each mole of the compound contains 18 mol H—2 from the OH groups and 16 from the 8 molecules of the water of crystallization.

$$\% \text{ H} = \frac{\left[\dfrac{18 \text{ mol H}}{\text{mol compound}}\right]\left[\dfrac{1.008 \text{ g}}{\text{mol H}}\right]}{\left[\dfrac{315.4 \text{ g}}{\text{mol compound}}\right]} \times 100 = 5.753\%$$

There are 8 mol of water of crystallization in each mole of compound. Water has a molecular mass of 18.016 amu.

$$\% \text{ H}_2\text{O} = \frac{\left[\dfrac{8 \text{ mol H}_2\text{O}}{\text{mol compound}}\right]\left[\dfrac{18.016 \text{ g}}{\text{mol H}_2\text{O}}\right]}{\left[\dfrac{315.4 \text{ g}}{\text{mol compound}}\right]} \times 100 = 45.70\%$$

2.2 Chemical Equations

A **chemical equation** summarizes the molecular changes occurring in a chemical reaction. To fulfill its descriptive function, it must meet several conditions.

1. It must show all reactants and products.
2. It must not show any substance not consumed or produced.
3. It must represent all substances by accurate formulas, using molecular formulas when available.
4. It must be consistent with atomic theory—atoms cannot be created, destroyed, or transformed.

To understand how chemical reactions are written, consider an important chemical reaction, the preparation of aluminum metal by decomposing aluminum oxide (Al_2O_3) into its elements. If you are not adept at writing reactions, it may help to start by writing a word equation that shows reactants on the left and products on the right. An arrow indicates that reactants are converted to products. In this particular case, aluminum oxide is the only reactant.

Aluminum oxide $\longrightarrow$ Aluminum + Oxygen

We now replace the names of these substances with their correct formulas. The simplest formula of aluminum oxide is Al_2O_3 (it has no known molecular formula). The molecular formula for oxygen is O_2 and, as is usual for metallic elements, we represent aluminum by its simplest formula, Al.

The equation

$$Al_2O_3 \longrightarrow Al + O_2$$

properly represents the formulas of reactants and products but it is *unbalanced* and therefore incorrect. It is unbalanced because, as it stands, it is inconsistent with atomic theory—it seems to show the disappearance of one aluminum atom and one oxygen atom. We balance the equation by adjusting the number of Al atoms, O_2 molecules, and Al_2O_3 formula units involved in the reaction. These numbers appear in front of the formulas. We can balance aluminum by showing the production of two aluminum atoms.

$$Al_2O_3 \longrightarrow 2\,Al + O_2$$

The three oxygen atoms in Al_2O_3 are enough to form one and one-half O_2 molecules.

$$Al_2O_3 \longrightarrow 2\,Al + \frac{3}{2}\,O_2$$

This equation is balanced, but it might give the impression that half molecules of oxygen exist in the product. (It is correct if taken to mean that 1 mol Al_2O_3 will decompose to 2 mol Al and $\frac{3}{2}$ mol O_2.) We avoid using fractional molecules by multiplying all coefficients by 2.

$$2\,Al_2O_3 \longrightarrow 4\,Al + 3\,O_2$$

There are now four Al atoms and six O atoms on each side of the equation, so the equation is balanced.

The information in an equation may also include the physical states of the substance, indicated by appropriate notations in parentheses: solid (s), liquid (l), and gas (g). Species in a water solution are often designated by the notation (aq) for aqueous. Solid aluminum oxide is added to the reaction involved in preparing aluminum, and molten aluminum is drained out of the reaction vessel. The oxygen that forms is, of course, a gas. Therefore, we could write the equation as

$$2\,Al_2O_3(s) \longrightarrow 4\,Al(l) + 3\,O_2(g)$$

There are strict rules, consistent with the need for descriptive accuracy, about how *not* to balance an equation. We cannot balance an equation by changing formulas or adding reactants or products not really present. The equation

$$AlO_2 \longrightarrow Al + O_2$$

is inaccurate for the decomposition of aluminum oxide because aluminum oxide is Al_2O_3, not AlO_2. In the same way,

$$Al_2O_3 \longrightarrow 2\,Al + O_2 + O$$

and

$$O + Al_2O_3 \longrightarrow 2\,Al + 2\,O_2$$

are wrong because there is no evidence for the presence of monatomic oxygen, either as a reactant or product. The equation

$$Al_2O_3 \longrightarrow 2\,Al + O_3$$

describes the decomposition of aluminum oxide into aluminum and ozone (the less common molecular form of oxygen), but common oxygen is the product of the reaction, and we must use its correct formula in the equation.

Although more sophisticated techniques exist (and will be presented later), a trial-and-error procedure is adequate for balancing many equations—even quite complicated ones. Example 2.12 illustrates this method.

Example 2.12

Balance the equation

$$HCl + KI + KIO_3 \longrightarrow I_2 + KCl + H_2O$$

Solution

It is best to start by balancing elements that appear in only one reactant and one product, and delay balancing elements appearing in several substances on one side of the equation—K and I in this case—until last. In this equation there must be three water molecules for each KIO_3 formula unit to balance oxygen, and six HCl molecules are necessary to balance the six hydrogen atoms in the three water molecules. It takes six KCl formula units to balance the chlorine atoms from the six HCl molecules. Five KI formula units reacting with each KIO_3 unit balance potassium, and three iodine molecules balance the total of six iodine atoms on the left-hand side of the equation.

$$6\,HCl + 5\,KI + KIO_3 \longrightarrow 3\,I_2 + 6\,KCl + 3\,H_2O$$

For reactions in which groups of atoms such as the nitrate group (NO_3) are transferred intact from one compound to another, it is easiest to treat these groups as entities to be balanced rather than trying to keep track of the individual atoms within them.

Example 2.13

Balance the equation

$$(NH_4)_2SO_4(aq) + Ba(NO_3)_2(aq) \longrightarrow BaSO_4(s) + NH_4NO_3(aq)$$

Solution

We need two NH_4 (ammonium) groups and two NO_3 groups on the right-hand side of the equation. This is done by placing the coefficient 2 before NH_4NO_3:

$$(NH_4)_2SO_4(aq) + Ba(NO_3)_2(aq) \longrightarrow BaSO_4(s) + 2\,NH_4NO_3(aq)$$

2.3 Stoichiometric Calculations for Reactions

Mass Calculations

A balanced equation helps us estimate the relationships between masses of reactants and products, and these calculations form the basis for many techniques of quantitative chemical analysis. To compute the mass of product for a given mass of reactant consumed in a particular reaction (1) balance the equation to determine the relative number of moles of reactant and product,

$$\frac{\text{Mol reactant}}{\text{Mol product}} = \frac{\text{Coefficient of reactant in balanced equation}}{\text{Coefficient of product in balanced equation}} \qquad (2.3)$$

(2) compute the number of moles of reactant, (3) use the balanced equation to determine the moles of product, (4) compute the mass of this number of moles. Variations of this procedure give the mass of reactant for a given mass of product, or the mass of second reactant required for the complete consumption of a given mass of first reactant.

Example 2.14

a. When a mixture containing KIO_3 reacts with an excess of KI and HCl according to the equation in Example 2.12, 0.635 g I_2 form. How much KIO_3 was in the mixture?
b. What mass of KI was consumed in the reaction?

Solution

a. First find the number of moles of I_2 present in 0.635 g.

$$\text{Molecular mass } I_2 = 2(127 \text{ amu}) = 254 \text{ amu}$$

$$\text{Moles } I_2 = \frac{0.635 \text{ g}}{254 \text{ g/mol}} = 2.50 \times 10^{-3} \text{ mol}$$

Since each mole of KIO_3 produces 3 mol I_2,

$$\text{Moles } KIO_3 = \frac{1 \text{ mol } KIO_3}{3 \text{ mol } I_2} (2.50 \times 10^{-3} \text{ mol } I_2) = 8.33 \times 10^{-4} \text{ mol}$$

Multiplying the moles of KIO_3 by the molecular mass of KIO_3 gives the mass of KIO_3.

$$\text{Molecular mass } KIO_3 = 39 \text{ amu} + 127 \text{ amu} + 3(16 \text{ amu}) = 214 \text{ amu}$$

$$\text{Mass } KIO_3 = (8.33 \times 10^{-4} \text{ mol})(214 \text{ g/mol}) = 0.178 \text{ g}$$

b. $$\text{Moles KI consumed} = \frac{5 \text{ mol KI}}{3 \text{ mol } I_2}(2.50 \times 10^{-3} \text{ mol } I_2) = 4.17 \times 10^{-3} \text{ mol}$$

$$\text{Molecular mass KI} = 39 \text{ amu} + 127 \text{ amu} = 166 \text{ amu}$$

$$\text{Mass KI consumed} = (4.17 \times 10^{-3} \text{ mol})(166 \text{ g/mol}) = 0.692 \text{ g}$$

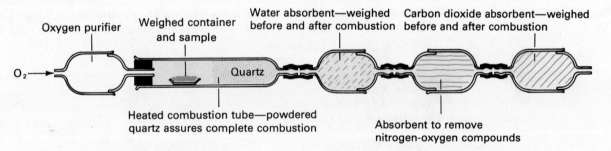

Figure 2.1
An apparatus for analysis of carbon and hydrogen by combustion.

In some cases a balanced equation is not necessary. This sort of situation occurs when a compound whose formula is unknown is analyzed for carbon and hydrogen by burning it in O_2, collecting the CO_2 and H_2O formed on separate absorbents (Figure 2.1), and measuring the weight increase of the absorbent tubes to estimate the masses of H_2O and CO_2 formed. Since the formula of the compound is not known, we cannot write a balanced equation, but we can still use stoichiometry to analyze the compound.

Example 2.15

A 15.35-mg sample of a compound containing only C, H, and O is burned in O_2. The only products are 29.33 mg CO_2 and 18.01 mg H_2O. What is the simplest formula of the compound?

| Solution | Rather than try to write a balanced equation involving an unknown formula, we can use the following stoichiometric relationships:

Moles C in sample = Moles CO_2 produced

Molecular mass CO_2 = 12.01 amu + 2(16.00 amu) = 44.01 amu

$$\text{Moles } CO_2 = \frac{29.33 \times 10^{-3} \text{ g } CO_2}{44.01 \text{ g } CO_2/\text{mol}} = 6.664 \times 10^{-4} \text{ mol}$$

Moles C = 6.664×10^{-4} mol

Moles H in sample = 2(Moles H_2O produced)

Molecular mass H_2O = 2(1.008 amu) + 16.00 amu = 18.02 amu

$$\text{Moles } H_2O = \frac{18.01 \times 10^{-3} \text{ g } H_2O}{18.02 \text{ g } H_2O/\text{mol}} = 9.994 \times 10^{-4} \text{ mol}$$

Moles H = $2(9.994 \times 10^{-4}$ mol$) = 1.999 \times 10^{-3}$ mol

Subtracting the mass of C and the mass of H from the mass of the sample gives the mass of O.

$$\text{Mass C} = \left(12.01 \frac{\text{g}}{\text{mol}}\right)(6.664 \times 10^{-4} \text{ mol}) = 8.00 \times 10^{-3} \text{ g} = 8.00 \text{ mg}$$

$$\text{Mass H} = \left(1.008 \frac{\text{g}}{\text{mol}}\right)(1.999 \times 10^{-3} \text{ mol}) = 2.01 \times 10^{-3} \text{ g} = 2.01 \text{ mg}$$

Mass O = 15.35 mg − 8.00 mg − 2.01 mg = 5.34 mg

$$\text{Moles O} = \frac{5.34 \times 10^{-3} \text{ g O}}{16.0 \text{ g O}/\text{mol}} = 3.34 \times 10^{-4} \text{ mol}$$

(Atoms C):(Atoms H):(Atoms O) = (mol C):(mol H):(mol O)

$\qquad\qquad\qquad\qquad\qquad\qquad = (6.664 \times 10^{-4}):(1.999 \times 10^{-3}):(3.34 \times 10^{-4})$

To reduce this ratio to simple whole numbers, take the smallest of the three numbers (3.34×10^{-4} mol O) and divide it into the others. For example, the ratio of C atoms to O atoms reduces to

$$\frac{6.664 \times 10^{-4}}{3.34 \times 10^{-4}} = 2.00$$

The ratios are

(Atoms C):(Atoms H):(Atoms O) = (2.00):(5.99):(1.00)

The simplest formula is C_2H_6O.

| Limiting Reactants and Theoretical Yields | One of the major contributions of chemistry to technology is that we use it to make the materials we need—iron, aluminum, plastics, fibers, medicines, and a variety of other important products. Because of the economic importance of the reactions used to prepare these substances, it is essential that we be able to predict how much product we can obtain, and stoichiometry provides the answer. One complication in these calculations is that, for several chemical and economic reasons, the reactants are usually not brought together in the ratio in which they react. The product yield depends on the amount of **limiting reactant** present. The limiting reactant is the reactant that is *not* in excess, and we can identify

2.3 STOICHIOMETRIC CALCULATIONS FOR REACTIONS

the limiting reactant by using stoichiometric principles. Example 2.16 is a typical limiting-reactant problem.

Example 2.16

Iron is prepared in a blast furnace by reacting hematite (Fe_3O_4) with carbon monoxide (CO).

$$Fe_3O_4(s) + 4\ CO(g) \longrightarrow 3\ Fe(l) + 4\ CO_2(g)$$

How much iron can be produced by this reaction when 2.00×10^3 kg Fe_3O_4 and 1.50×10^3 kg CO are brought together in the blast furnace?

Solution

The first thing to do in this sort of problem is to compute the number of moles of each reactant. You may find it more convenient to convert the masses in kilograms to grams (1 kg = 10^3 g).

$$\text{Molecular mass } Fe_3O_4 = 3(55.84\text{ amu}) + 4(16.0\text{ amu}) = 231.5\text{ amu}$$

$$\text{Moles } Fe_3O_4 = \frac{2.00 \times 10^6\text{ g }Fe_3O_4}{231.5\text{ g }Fe_3O_4/\text{mol}} = 8.64 \times 10^3\text{ mol}$$

$$\text{Molecular mass CO} = 12.0\text{ amu} + 16.0\text{ amu} = 28.0\text{ amu}$$

$$\text{Moles CO} = \frac{1.5 \times 10^6\text{ g CO}}{28.0\text{ g CO}/\text{mol}} = 5.36 \times 10^4\text{ mol}$$

We can now use the balanced equation to decide whether Fe_3O_4 or CO is the limiting reactant. According to the balanced equation, each mole of Fe_3O_4 requires 4 mol CO.

$$\text{Moles CO required to react with all the } Fe_3O_4 = \frac{4\text{ mol CO}}{\text{mol }Fe_3O_4}(8.64 \times 10^3\text{ mol }Fe_3O_4) = 3.46 \times 10^4\text{ mol}$$

Since the number of moles of CO in the blast furnace is larger than this required amount, CO is in excess and Fe_3O_4 is the limiting reactant. The maximum amount of iron that can be formed depends on the supply of Fe_3O_4. According to the balanced equation,

$$\text{Moles Fe} = \frac{3\text{ mol Fe}}{\text{mol }Fe_3O_4}(8.64 \times 10^3\text{ mol }Fe_3O_4) = 2.59 \times 10^4\text{ mol}$$

$$\text{Mass Fe} = \left(55.8\ \frac{\text{g Fe}}{\text{mol Fe}}\right)(2.59 \times 10^4\text{ mol Fe}) = 1.45 \times 10^6\text{ g} = 1.45 \times 10^3\text{ kg}$$

Another feature of many industrial reactions is that the yield of product obtained in a pure form is less than the yield predicted from the limiting reactant. This can be due to several factors. Some limiting reactant might not be consumed either because it is too inconvenient and costly to wait for complete reaction or because the reacting system reaches an equilibrium state (see Chapter 13) in which a significant amount of reactant remains unchanged. There might also be competing reactions that use up the limiting reactant but convert it to undesired products. Some of the product may be lost due to incomplete recovery during purification. Chemists assess the effectiveness of a chemical process by measuring its **percent**

yield. The percent yield is the ratio, expressed in percent, of the actual yield to the **theoretical yield**. The theoretical yield is the yield predicted by stoichiometry using a limiting reactant calculation.

$$\text{Percent yield} = \frac{\text{Actual yield of product}}{\text{Theoretical yield of product}} \times 100 \tag{2.4}$$

Although the percent yield is not the sole factor deciding the value of a chemical process—the costs of reactants, energy, labor, and equipment, as well as safety and environmental factors are also important—a high percent yield is a desirable feature in an industrial reaction. Example 2.17 illustrates the calculation of percent yield.

Example 2.17

An older method for preparing the explosive trinitrotoluene (TNT) is the reaction of dinitrotoluene ($C_7H_6N_2O_4$) with nitric acid (HNO_3).

$$C_7H_6N_2O_4 + HNO_3 \longrightarrow \underset{\text{TNT}}{C_7H_5N_3O_6} + H_2O$$

After 500 kg of dinitrotoluene and 300 kg of nitric acid are mixed in an industrial reactor and the mixture heated for several hours, 530 kg of pure TNT are obtained. What is the percent yield for this process?

Solution

We identify the limiting reactant and calculate the theoretical yield as in Example 2.16.

$$\text{Molecular mass } C_7H_6N_2O_4 = 7(12.0 \text{ amu}) + 6(1.0 \text{ amu}) + 2(14.0 \text{ amu})$$
$$+ 4(16.0 \text{ amu}) = 182 \text{ amu}$$

$$\text{Moles } C_7H_6N_2O_4 = \frac{5.00 \times 10^5 \text{ g } C_7H_6N_2O_4}{182 \text{ g } C_7H_6N_2O_4/\text{mol}} = 2.75 \times 10^3 \text{ mol}$$

$$\text{Molecular mass } HNO_3 = 1.0 \text{ amu} + 14.0 \text{ amu} + 3(16.0 \text{ amu}) = 63.0 \text{ amu}$$

$$\text{Moles } HNO_3 = \frac{3.00 \times 10^5 \text{ g } HNO_3}{63.0 \text{ g } HNO_3/\text{mol}} = 4.76 \times 10^3 \text{ mol}$$

Since 1 mol $C_7H_6N_2O_4$ reacts with 1 mol HNO_3, the HNO_3 is in excess and $C_7H_6N_2O_4$ is the limiting reactant. Since each mole of dinitrotoluene yields 1 mol TNT,

$$\text{Moles TNT} = \frac{1 \text{ mol TNT}}{1 \text{ mol } C_7H_6N_2O_4} (2.75 \times 10^3 \text{ mol } C_7H_6N_2O_4) = 2.75 \times 10^3 \text{ mol}$$

$$\text{Molecular mass TNT} = 7(12.0 \text{ amu}) + 5(1.0 \text{ amu}) + 3(14.0 \text{ amu}) + 6(16.0 \text{ amu})$$
$$= 227 \text{ amu}$$

$$\text{Theoretical yield of TNT} = \left(227 \frac{\text{g TNT}}{\text{mol TNT}}\right)(2.75 \times 10^3 \text{ mol TNT})$$
$$= 6.24 \times 10^5 \text{ g} = 624 \text{ kg}$$

Since only 530 kg of TNT are obtained in the reaction,

$$\text{Percent yield TNT} = \frac{530 \text{ kg}}{624 \text{ kg}} \times 100 = 84.9\%$$

Summary

1. In a molecular formula, each element present in the molecule is shown by its chemical symbol, and a subscript number to the lower right of the symbol shows the number of atoms of that element in a molecule.
2. The simplest formula (or empirical formula) of a compound is the simplest whole-number ratio of atoms that is consistent with its composition. We use simplest formulas in place of molecular formulas when we lack information about molecular mass.
3. Using simplest formulas, we can determine the mass composition of a compound.
4. A chemical equation must (1) show all reactants and products by correct formulas, (2) be balanced—the number of atoms of each element must be the same on both sides of the equation.
5. Balanced equations display the relationships between the number of moles of each reactant and product in the reaction, and these relationships allow us to compute mass relationships between reactants and products.

Vocabulary List

stoichiometry
molecular formula
molecular mass
simplest formula
empirical formula

water of crystallization
water of hydration
hydrate
mole
Avogadro's number

formula mass
chemical equation
limiting reactant
theoretical yield
percent yield

Exercises

1. How many ammonia molecules are present in 0.030 mol NH_3?

2. How many molecules are present in 0.48 g SO_2?

3. What is the mass, in grams, of 2.00×10^{21} water molecules?

4. What is the mass, in grams, of 3.25 mol Sn?

5. The most common molecular form of phosphorus is P_4. What is the mass, in grams, of 0.58 mol of phosphorus molecules?

6. How many oxygen atoms are present in 0.0300 mol O_2?

7. How many moles of O_3 are present in a sample of pure O_3 containing 0.500 mol of O atoms?

8. What is the ratio of moles of S to moles of N in a sample of sulfanilamide, $C_6H_8N_2SO_2$.

9. What is the ratio of nitrogen atoms to chlorine atoms in a compound that contains 13.8% N and 11.7% Cl?

10. How many atoms of each element are present in the amino acid methionine, $C_5H_{11}NSO_2$?

11. How many atoms of each type are present in $(NH_4)_2Cr_2O_7$?

12. How many atoms of each type are present in $K_4Fe(CN)_6 \cdot 3H_2O$? How many molecules of water of hydration are present in each formula unit?

13. How many atoms of each type are present in $CsAl(SO_4)_2 \cdot 12H_2O$? How many molecules of water of crystallization are present in each formula unit?

14. How many chlorine atoms are present in five CCl_4 molecules?

15. How many oxygen atoms are present in 15 SO_3 molecules?

16. Find the molecular mass of:
    a. S_6 b. N_2O_4 c. C_2H_6O

17. Find the molecular mass of:
    a. P_4 b. XeF_4 c. H_2NOH

18. What is the mass of each?
    a. 1.00 mol P_4 b. 0.350 mol CH_4
    c. 7.0×10^{-4} mol C_2H_5OH

19. Compute the mass of:
    a. 2.50 mol N_2H_4 b. 4.3×10^{-5} mol $(NH_4)_2SO_4$
    c. 0.035 mol $SOCl_2$

20. How many moles are present in each?
    a. 1.00 g HCl   b. 2.70 kg glucose, $C_6H_{12}O_6$

21. Find the number of moles in:
    a. 3.50 g H_2S b. 0.175 mg glycine, $C_2H_5NO_2$

22. How many moles of fluorine atoms are present in 9.5 g F_2?

23. How many moles of nitrogen atoms are there in 21.0 g N_2?

24. How many moles of C are present in 0.200 mol glucose, $C_6H_{12}O_6$?

25. How many moles of hydrogen atoms are present in 5.0×10^{-3} mol $Ba(H_2PO_2)_2 \cdot H_2O$?

26. How many moles of N are present in 20.0 g of pyrazine, $C_4H_4N_2$?

27. How many moles of fluorine atoms are in 15.0 g SF_6?

28. A compound contains 85.8% boron and 14.2% hydrogen and its molecular mass is 63 amu. What is its molecular formula?

29. What is the molecular formula of lactose, a compound of carbon, hydrogen, and oxygen? Its molecular mass is 342 amu and it contains 42.1% C and 6.5% H.

30. The simplest formula of diborane is BH_3 and its molecular mass is 28 amu. What is the molecular formula of diborane?

31. The simplest formula of hydroquinone is C_3H_3O and its molecular mass is 110 amu. What is its molecular formula?

32. Sodium thiosulfate contains 29.1% Na, 40.6% S, and 30.4% O. What is its simplest formula?

33. What is the simplest formula of a compound containing 44.43% Pt, 19.14% N, 4.13% H, and 32.30% Cl?

34. Tartaric acid, a compound of C, H, and O, contains 32.0% C and 4.0% H.
    a. Find the simplest formula of tartaric acid.
    b. The molecular mass of tartaric acid is 150 amu. What is its molecular formula?

35. a. Find the simplest formula of a substance containing 46.2% C and 53.8% N.
    b. The molecular mass is 52 amu. What is its molecular formula?

36. What is the percentage of boron in $C_{10}H_{22}B_2$?

37. Find the percentage of nickel in nickel dimethylglyoxime, $Ni(C_4H_7N_2O_2)_2$.

38. Find the percentage of each element in sodalite, $Na_4Al_3Si_3O_{12}Cl$.

39. a. What is the percentage of each element in Mohr's salt, $(NH_4)_2SO_4 \cdot FeSO_4 \cdot 6H_2O$?
    b. What is the percentage of water of hydration in Mohr's salt?

40. Balance the following equations.
    a. $P_4(g) + O_2(g) \longrightarrow P_4O_{10}(s)$
    b. $C_2H_6(g) + O_2(g) \longrightarrow CO_2(g) + H_2O(g)$
    c. $CaCl_2(aq) + Na_2HPO_4(aq) \longrightarrow Ca_3(PO_4)_2(s) + NaCl(aq) + HCl(aq)$
    d. $H_2O + AgNO_3(aq) + Na_2SO_3(aq) \longrightarrow Ag(s) + Na_2SO_4(aq) + HNO_3(aq)$
    e. $CrO_3(aq) + HCl(aq) \longrightarrow CrCl_3(aq) + H_2O + Cl_2(g)$

41. Balance the following equations.
    a. $NO(g) + O_2(g) \longrightarrow N_2O_5(g)$
    b. $C_6H_{12}(g) + O_2(g) \longrightarrow CO_2(g) + H_2O(g)$
    c. $Ag_3PO_4(s) + HCl(aq) \longrightarrow H_3PO_4(aq) + AgCl(s)$
    d. $Al_2(SO_4)_3(aq) + SrI_2(aq) \longrightarrow AlI_3(aq) + SrSO_4(s)$
    e. $KI(aq) + CuSO_4(aq) \longrightarrow I_2(aq) + CuI(s) + K_2SO_4(aq)$

42. a. Balance the equation
    $$FeCl_3(aq) + KI(aq) \longrightarrow FeCl_2(aq) + I_2(aq) + KCl(aq)$$
    b. This reaction occurs when 1.00 g each of $FeCl_3$ and KI are mixed in the same solution. Which of these substances is the limiting reactant?
    c. What mass of I_2 is formed in the reaction in part b?

43. When a solution containing 5.00 g $Ba(NO_3)_2$ is mixed with one containing 7.00 g KH_2PO_4, the following reaction occurs:
    $$3\,Ba(NO_3)_2(aq) + 2\,KH_2PO_4(aq) \longrightarrow Ba_3(PO_4)_2(s) + 2\,KNO_3(aq) + 4\,HNO_3(aq)$$
    How much $Ba_3(PO_4)_2$ is produced in the reaction?

44. A stream of hydrogen gas reacts with 2.075 g of an oxide of cobalt, forming water and converting the oxide into 1.472 g of cobalt metal.
    a. What is the simplest formula of the oxide?
    b. Write a balanced equation for the reaction.

45. a. Balance the equation
    $$FeCl_2(aq) + KMnO_4(aq) + HCl(aq) \longrightarrow KCl(aq) + MnCl_2(aq) + FeCl_3(aq) + H_2O$$
    b. What mass of $KMnO_4$ reacts with 0.381 g $FeCl_2$?

EXERCISES

46. A chemist mixes 15 g Br_2 and 45 g $CHCl_3$ in order to synthesize $CBrCl_3$ by the reaction

$$Br_2(l) + CHCl_3(l) \longrightarrow CBrCl_3(l) + HBr(g)$$

    a. What is the theoretical yield of $CBrCl_3$ that can be obtained from this reaction?
    b. In practice, 6.0 g $CBrCl_3$ were obtained. What is the percent yield of the reaction?

47. A student reacts 8.00 g of bromine with 50.0 g of benzene (C_6H_6),

$$Br_2(l) + C_6H_6(l) \longrightarrow C_6H_5Br(l) + HBr(g)$$

to obtain 4.35 g of bromobenzene (C_6H_5Br). What is the percent yield of this reaction?

48. Potassium thiocyanate (KSCN) reacts with mercury(II) nitrate $[Hg(NO_3)_2]$ to form mercury(II) thiocyanate $[Hg(SCN)_2]$ and potassium nitrate (KNO_3).
    a. What mass of KSCN is needed to react with 1.000 g $Hg(NO_3)_2$?
    b. How much $Hg(SCN)_2$ can be produced from 1.000 g $Hg(NO_3)_2$?

49. When 100 ml of a solution of barium nitrate $[Ba(NO_3)_2]$ is mixed with an excess of sodium sulfate (Na_2SO_4) solution, 1.34 g of solid barium sulfate $(BaSO_4)$ precipitates.

$$Ba(NO_3)_2(aq) + Na_2SO_4(aq) \longrightarrow BaSO_4(s) + 2\,NaNO_3(aq)$$

    a. What mass of barium nitrate was in the solution?
    b. What mass of sodium sulfate was consumed in the reaction?

50. A compound contains C, H, and I. Burning 18.42 mg of this compound produces 2.06 mg CO_2 and 0.42 mg H_2O. Treatment of 13.27 mg of the compound with excess $AgNO_3$ produces 23.75 mg of AgI.
    a. What is the elemental composition of the compound?
    b. What is its simplest formula?

51. A compound contains C, H, N, and O. Burning 17.35 mg of this compound produces 44.15 mg CO_2 and 9.04 mg H_2O. A second sample of the compound, weighing 14.38 mg, is decomposed to form 1.66 mg of nitrogen. What is the simplest formula of the compound?

52. Write balanced equations for the following reactions. If you are unfamiliar with the formulas of any of these substances and cannot find them in the index of this book, look them up in the *Handbook of Chemistry and Physics* (Chemical Rubber Publishing Co.).
    a. Benzene burns in the presence of oxygen to produce carbon dioxide and water.
    b. Ordinary oxygen is converted to ozone.
    c. Hydrogen chloride reacts with aluminum to produce hydrogen gas and aluminum chloride.
    d. A reaction between barium nitrate, potassium dichromate, and water produces barium chromate, nitric acid, and potassium nitrate.
    e. Ammonia and hypochlorous acid react to form nitrogen gas, chlorine gas, and water.

53. The molecular mass of a compound is 268 amu. Chemical analysis indicates a composition of 36.0% C, 4.6% H, and 59.9% Br. Find the molecular formula of the compound. (*Note:* The total analysis adds up to 100.5%. This indicates some experimental error but not enough to confuse the determination of the formula.)

54. A gaseous compound containing carbon and hydrogen has a molecular mass of 26 amu. Combustion of 25.3 mg of this compound yields 85.8 mg CO_2 and 17.5 mg H_2O.
    a. What is the molecular formula of the compound?
    b. Write the balanced equation for its combustion.

55. When a sample of sodium sulfate containing water of hydration is heated to convert it to anhydrous sodium sulfate (Na_2SO_4), its mass decreases from 10.00 g to 4.40 g. How many molecules of water of hydration are present in the simplest formula unit of the hydrate?

56. By mixing 3.16 g $KMnO_4$ and 2.5 g C_6H_{10}, a chemist synthesizes *cis*-1,2-cyclohexanediol, $C_6H_{12}O_2$, by the reaction

$$4\,H_2O + 3\,C_6H_{10} + 2\,KMnO_4 \longrightarrow 3\,C_6H_{12}O_2 + 2\,MnO_2(s) + 2\,KOH$$

The reaction runs generally as planned. However, 0.20 g $C_6H_{10}O_2$ is formed as a "side product" due to the undesired reaction

$$2\,H_2O + 3\,C_6H_{10} + 4\,KMnO_4 \longrightarrow 3\,C_6H_{10}O_2 + 4\,MnO_2(s) + 4\,KOH$$

What is the maximum amount of $C_6H_{12}O_2$ that can be obtained?

chapter 3

The Behavior of Gases

Molecules of a typical gas, HCl

Depending on the external conditions of temperature and pressure, most substances can exist as a solid, a liquid, or a gas.* Gases differ from solids and liquids in having no well-defined volume. A gas expands until it completely fills its container, unless there is an external constraining force. (The earth's gravity inhibits the expansion of the atmosphere into the rest of the universe.) Another difference between gases and other forms of matter is that a mixture of gases is always homogeneous. In contrast, there are many heterogeneous solid or liquid mixtures such as oil and water or sand and gravel. Compared to liquids and solids, gases are easy to compress. It takes very intense pressures to alter the volume of most liquids and solids, and the volume changes that result are quite small. In the case of a gas, however, even a slight increase in pressure causes an easily observed volume decrease.

3.1 Gas Laws

Pressure

Pressure is the force acting on a surface divided by the surface area and is expressed in units of force over area (newton/m² or lb/in.²).

$$P = \frac{F}{A} \tag{3.1}$$

---

*While most materials are either solids, liquids, or gases at a given temperature and pressure, there are less common forms of matter such as liquid crystals and amorphous solids. We will discuss these less common physical states in later chapters.

48   3 THE BEHAVIOR OF GASES

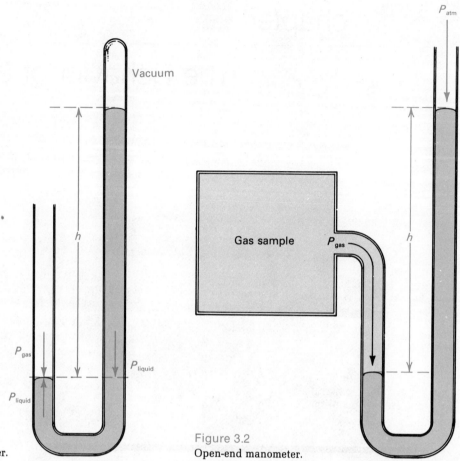

Figure 3.1
Closed-end manometer.

Figure 3.2
Open-end manometer.

A gas exerts a homogeneous, perpendicular pressure on any surface in contact with it. One of the easier ways of measuring gas pressure is to balance it with the pressure of a liquid column. This is the basis of a **closed-end manometer**, a column of liquid (usually mercury) in an uneven U-tube whose long end is evacuated and sealed (Figure 3.1). The gravitational pressure of the liquid at any level depends on the weight of the liquid column at higher levels and is transmitted equally in all directions. Upward liquid pressure in the short, open end of the U-tube balances the gas pressure so there is no net force and no movement of the liquid column. The difference between the liquid levels in the closed and open ends is proportional to the gas pressure and independent of the cross-sectional area of the manometer.

The liquid in a manometer should have a density that gives a convenient and easily measurable column height. A water manometer is inconvenient for measuring atmospheric pressure because the balancing water column is over 30 ft in height. Mercury, with a high density (13.6 g/cm^3), has a much more convenient balancing height—about 0.75 m, or 30 in.—and it is used in manometers to measure atmospheric pressures. This type of device for measuring atmospheric pressure is called a mercury **barometer**.

The use of manometers has led to the use of another system of pressure units: 1 **torr** is enough pressure to support a mercury column 1 mm in height at sea level and 0°C. Because atmospheric pressure varies slightly with time and place, we have also defined a unit called the **standard atmosphere** (atm), equal to 760 torr.

Example **3.1**   Convert a pressure of 650 torr into atmospheres.

3.1 GAS LAWS

Solution

Since there are 760 torr in an atmosphere, we divide the pressure in torr by 760 to find the number of atmospheres.

$$\frac{650 \text{ torr}}{760 \text{ torr/atm}} = 0.855 \text{ atm}$$

An open-end manometer (Figure 3.2) measures the pressure difference between two gases. The gas with the greater pressure lifts a liquid column against the weaker pressure of the other gas until the weight of the column is sufficient to compensate for the pressure difference.

Boyle's Law

An increase in pressure decreases the volume of a gas sample, and the sample expands when the pressure is reduced. This phenomenon has a great deal of practical importance. For example, scuba divers are exposed to increased pressures as they descend under water. At 33 ft (10 m) under water, the pressure on the air in their lungs is about 2 atm. To prevent this increased pressure from collapsing their lungs, divers must take in more air and increase their breathing rates as they go deeper.

Perhaps you have seen pictures of a helium-filled research balloon and noticed that the balloon is only half-filled before launching. This is because the gas volume will increase as the balloon encounters the low pressures of the upper atmosphere. If the balloon were completely inflated at sea level, it would burst due to expansion long before it reached the desired altitude.

The apparatus shown in Figure 3.3 offers a simple way of measuring the

Figure 3.3

The pressure acting on the gas is adjusted by raising or lowering the movable bulb. (a) The level in the bulb is lower than the level in the gas tube, so the pressure of the gas is less than the atmospheric pressure. (b) The bulb level is above the level in the gas tube, so the gas pressure is greater than the atmospheric pressure.

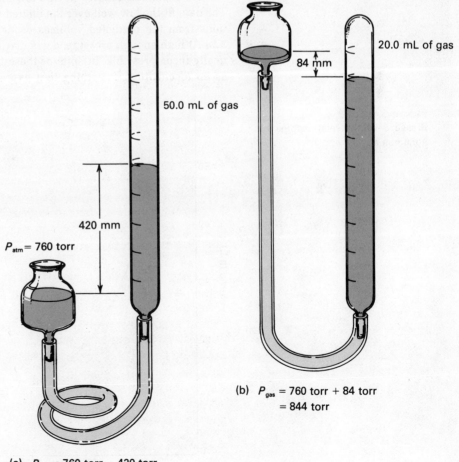

Table 3.1 Pressure—Volume Data for 17 mg of Ammonia at Constant Temperature of 0°C

Volume (mL)	Pressure (torr)	PV (torr-mL)
50.0	340	1.70×10^4
45.0	378	1.70×10^4
40.0	424	1.70×10^4
35.0	484	1.69×10^4
30.0	564	1.69×10^4
25.0	676	1.69×10^4
20.0	844	1.69×10^4

effect of pressure on volume. Generalization of experimental results, such as those in Table 3.1, leads to **Boyle's law:** The volume of a gas sample is inversely proportional to its pressure.

$$PV = \text{Constant} \tag{3.2}$$

The particular value of the constant applies only to a specific gas at a given temperature. The constant changes value if a chemical reaction changes the composition of the gas, if some of the gas leaks out of the apparatus, or if the temperature changes. Figures 3.4 and 3.5 show graphs of the pressure–volume data in Table 3.1. The data form a rectangular hyperbola when V is plotted against P, and a horizontal line when PV is plotted against P.

No substance follows Boyle's law exactly at all temperatures and pressures. The data fit the law well over the limited range of pressure in Table 3.1, but deviations from the predicted volumes occur at high pressure. (See Figures 3.4 and 3.5.) At high enough pressure, a gas may condense into a liquid and become practically incompressible. In spite of these failures, we retain the law and regard it as an *idealization*, something that expresses an important generalization about

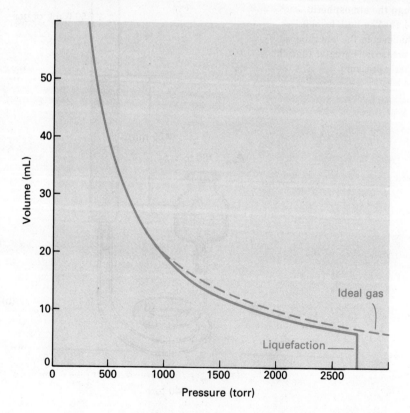

Figure 3.4
Pressure–volume relationships for ammonia at 0°C.

Figure 3.5
A graph of PV versus P for ammonia at 0°C.

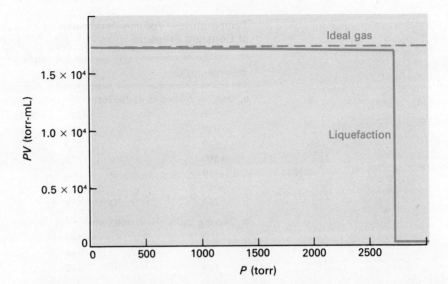

gases without taking into account the secondary effects that cause the observed failures of the law. Boyle's law describes a nonexistent **ideal gas**; real gases approximate ideal behavior but do not exactly match it. A gas at *high temperature* and *low pressure* matches ideal behavior best.

Example 3.2

Predict the volume of ammonia at 920 torr from the data in Table 3.1.

Solution

To find the constant in Boyle's law, we multiply any observed pressure by its corresponding volume. As shown in Table 3.1, the value of PV is nearly independent of pressure, but there is a slight downward trend due to nonideality at high pressure. To minimize the nonideality error, it is best to use a value of PV from an observation at a pressure close to 920 torr.

$$\text{Constant} = PV = (844 \text{ torr})(20.0 \text{ mL}) = 1.69 \times 10^4 \text{ torr-mL}$$

We predict the volume at a particular pressure by dividing the constant by the pressure.

$$V = \frac{\text{Constant}}{P}$$

$$V = \frac{1.69 \times 10^4 \text{ torr-mL}}{920 \text{ torr}} = 18.4 \text{ mL}$$

Example 3.3

A gas occupies 255 mL volume at 3.25 atm and 33°C. What pressure is needed to reduce its volume to 160 mL at 33°C?

Solution

Since the temperature of the gas does not change, we can use Boyle's law.

$$\text{Constant} = PV = (3.25 \text{ atm})(255 \text{ mL}) = 829 \text{ atm-mL}$$

Dividing this constant by the new volume gives us the pressure required to maintain this volume.

$$P = \frac{\text{Constant}}{V} = \frac{829 \text{ atm-mL}}{160 \text{ mL}} = 5.18 \text{ atm}$$

Table 3.2 Temperature—Volume Relationships at Constant Pressure

Temperature (°C)	Volume (mL)	V/T (mL/K)
a. 24.6 mg hydrogen at 760 torr		
0	273	1.00
+10	283	1.00
+50	323	1.00
+100	373	1.00
−10	263	1.00
−50	223	1.00
−100	173	1.00
b. 24.6 mg hydrogen at 380 torr		
+100	746	2.00
0	546	2.00
−100	346	2.00
c. 195 mg oxygen at 760 torr		
+100	187	0.50
+50	162	0.50
0	137	0.50
−100	87	0.50

Absolute Temperature

A gas at constant pressure expands on heating and contracts when cooled. Table 3.2 and Figure 3.6 show data and a graph for the effect of temperature on the volume of various gas samples at constant pressure. The change in volume for any given gas sample at a given pressure is proportional to the change in temperature (**Charles's law**). Graphs of volume against temperature are straight lines whose slopes depend on the composition and mass of the sample and the pressure. If we continue the straight lines downward beyond the range of the data, they all reach $V = 0$ at a temperature of $-273°C$. While this represents ideal behavior—any real gas liquefies at some temperature above $-273°C$—this temperature (determined more precisely to be $-273.15°C$) is the lowest conceivable temperature. We have come close to this absolute minimum temperature but have never reached it.

Recognizing that $-273°C$ is an **absolute zero** for temperature, the **Kelvin** temperature scale (also called the absolute temperature scale) avoids the mathematical complications of negative temperatures. Zero in the Kelvin scale equals $-273.15°C$, and since the temperature change represented by a change of 1 Kelvin (K) is the same as 1°C, the conversion between the two scales is simply

$$\text{Temperature in K} = \text{Temperature in °C} + 273.15 \qquad (3.3)$$

If the temperature is given to the nearest degree, we round 273.15 to three significant digits when converting between the two scales.

Example 3.4

Convert $-152°C$ to the Kelvin scale.

Solution

To obtain the Kelvin temperature, add 273 to the Celsius temperature.

$$\text{Temperature in K} = -152 + 273 = 121 \text{ K}$$

Figure 3.6
Volume–temperature relationships at constant pressure.

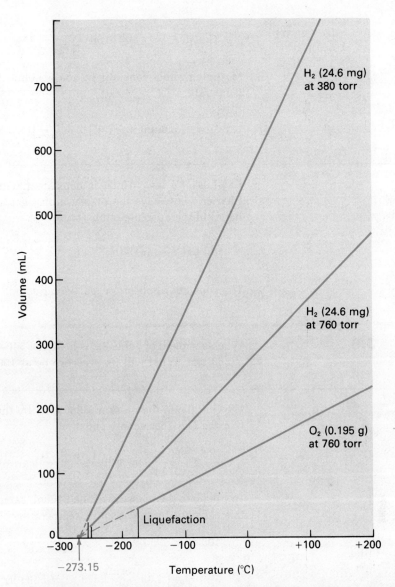

Charles's Law and Gay-Lussac's Law

Using the Kelvin scale, we can express Charles's law as a simple proportionality between volume and temperature: The volume of a sample of gas at constant pressure is directly proportional to its temperature, expressed in K.

$$\frac{V}{T} = \text{Constant} \tag{3.4}$$

The constant applies to a specific mass of sample of a particular composition held at a given temperature.

Example 3.5

What is the volume of sample c in Table 3.2 at 300°C?

Solution

In dealing with temperature effects, we must always use temperatures expressed in K. We use equation (3.3) to convert °C to K.

$$T = 300 + 273 = 573 \text{ K}$$

The Charles's law constant, which can be calculated from any corresponding volume and temperature given in Table 3.2 for sample c, is

$$\text{Constant} = \frac{V}{T} = 0.50 \, \frac{\text{mL}}{\text{K}}$$

Multiplying this constant by the absolute temperature gives the volume at that temperature.

$$V = \text{Constant} \times T = \left(0.50 \, \frac{\text{mL}}{\text{K}}\right)(573 \, \text{K}) = 287 \, \text{mL}$$

Gay-Lussac's law, which is consistent with the predictions of Boyle's law and Charles's law, says the pressure of a gas sample at constant volume is proportional to the absolute temperature.

$$\frac{P}{T} = \text{Constant} \tag{3.5}$$

The constant applies to a particular sample and volume.

Example 3.6

When sample b of Table 3.2 is held in a 546-mL rigid container at 0°C, its pressure is 380 torr. What will its pressure be at 100°C in the same container?

Solution

We can divide the pressure (380 torr) by the absolute temperature (273 K) to obtain the Gay-Lussac law constant.

$$\frac{P}{T} = \frac{380 \, \text{torr}}{273 \, \text{K}} = 1.39 \, \frac{\text{torr}}{\text{K}}$$

To calculate the pressure of the 546-mL sample at any other temperature, multiply this constant by the absolute temperature.

$$100°\text{C} = 373 \, \text{K}$$

$$P = \left(1.39 \, \frac{\text{torr}}{\text{K}}\right)(373 \, \text{K}) = 518 \, \text{torr}$$

A single equation can express all three laws (Boyle's, Charles's, and Gay-Lussac's).

$$\frac{PV}{T} = \text{Constant} \tag{3.6}$$

In this case, the constant depends only on the mass and chemical composition of the sample.

Example 3.7

A gas occupies 500 mL at 127°C and 500 torr. What is its volume at −23°C and 600 torr?

Solution

The initial pressure, volume, and temperature (400 K) give enough information to compute the constant.

$$\text{Constant} = \frac{PV}{T} = \frac{(500 \, \text{torr})(500 \, \text{mL})}{400 \, \text{K}} = 625 \, \frac{\text{torr-mL}}{\text{K}}$$

Multiplying this constant by the new temperature (250 K) and dividing by the new pressure yields the volume under the new conditions.

$$V = \frac{\text{Constant} \times T}{P} = \frac{(625 \text{ torr-mL/K})(250 \text{ K})}{600 \text{ torr}} = 260 \text{ mL}$$

Standard Temperature and Pressure

Because gas volume is so sensitive to temperature and pressure, it is meaningless to report the volume of a gas sample without also specifying the temperature and pressure at which this volume is measured. Any report of a gas sample's volume, temperature, and pressure describes the sample adequately, but it would be very inconvenient if gas volumes were reported at a variety of different conditions. To avoid this inconvenience, it is accepted practice to report gas volumes at **standard temperature and pressure** (abbreviated **STP**)—760 torr (1 atm) and 0°C (273 K). We can perform the volume measurements at any temperature and pressure, but we should use the gas laws to predict the volume the samples would occupy at STP.

Example 3.8

What is the volume at STP of a gas that occupies 37.0 mL at 23°C and 737 torr?

Solution

As in Example 3.7, we apply equation (3.6).

$$\text{Constant} = \frac{PV}{T} = \frac{(737 \text{ torr})(37.0 \text{ mL})}{296 \text{ K}} = 92.1 \frac{\text{torr-mL}}{K}$$

The volume at STP is

$$V_{STP} = \frac{\text{Constant} \times 273 \text{ K}}{760 \text{ torr}}$$

$$= \frac{(92.1 \text{ torr-mL/K})(273 \text{ K})}{760 \text{ torr}} = 33.1 \text{ mL at STP}$$

Dalton's Law of Partial Pressures

The following experiment (Figure 3.7) illustrates a property of gases that was first observed by John Dalton. We first fill a container with a particular sample of oxygen and find that its pressure at 0°C is 750 torr. We then remove the oxygen and replace it with enough nitrogen to produce a pressure of 250 torr. The oxygen sample is then added back into the container so that it intermingles with the nitrogen to form a homogeneous mixture. The pressure exerted by this mixture is 1000 torr—the exact sum of the individual pressures of the two gases.

Figure 3.7
Dalton's law of partial pressures.

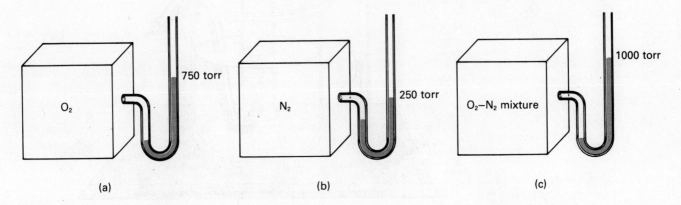

(a) (b) (c)

According to **Dalton's law of partial pressures,** the total pressure of a gas mixture is the sum of the partial pressures of the individual components. The **partial pressure** of a component is the pressure it exerts when it is the sole occupant of the container. In the preceding experiment, oxygen has a partial pressure of 750 torr, nitrogen has a partial pressure of 250 torr, and the total pressure of 1000 torr is the sum of these two partial pressures. In addition to forming a homogeneous mixture, two gaseous substances occupying the same container exert their pressures independently of each other. For a mixture of n components

$$P_{total} = P_1 + P_2 + P_3 + \cdots + P_n \tag{3.7}$$

One of the more common laboratory applications of Dalton's law occurs when we collect a gas sample by trapping it over water (Figure 3.8). Because the water evaporates to some extent, the gas sample is contaminated with water vapor. For reasons we will discuss later (Chapter 9), the partial pressure of the water vapor depends solely on the temperature of the system. Table 3.3 (and Appendix G) lists the vapor pressure of water at various temperatures. To find the pressure that the uncontaminated gas would exert, we subtract the partial pressure of the water vapor from the total pressure of the sample.

$$P_{gas} = P_{total} - P_{water}$$

Figure 3.8
Collection of a gas by trapping it over water.

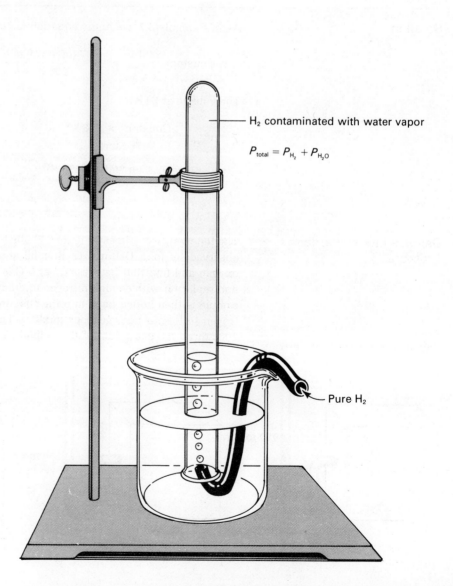

H$_2$ contaminated with water vapor

$P_{total} = P_{H_2} + P_{H_2O}$

Pure H$_2$

Table 3.3   Vapor Pressure of Water at Various Temperatures

T (°C)	Vapor pressure (torr)	T (°C)	Vapor pressure (torr)
15	12.8	26	25.2
16	13.6	27	26.7
17	14.5	28	28.3
18	15.5	29	30.0
19	16.5	30	31.8
20	17.5	31	33.7
21	18.7	32	35.7
22	19.8	33	37.7
23	21.1	34	39.9
24	22.4	35	42.2
25	23.8		

Example 3.9

The total pressure of a sample of nitrogen collected over water is 725 torr at 27°C. What is the pressure exerted by the nitrogen?

Solution

According to Table 3.3, the partial pressure of water vapor at 27°C is 27 torr. Subtract this value from the total pressure to find the pressure of the nitrogen.

$$P_{nitrogen} = 725 \text{ torr} - 27 \text{ torr} = 698 \text{ torr}$$

3.2 Volume Relationships in Gas Reactions

The Experiments of Gay-Lussac

Shortly after Dalton presented his atomic theory, Joseph Gay-Lussac studied the quantitative relationship between the volumes of chemically reacting gases and generalized these results as the **law of combining volumes:** Gases combine chemically in simple ratios of volume (measured at the same temperature and pressure). For example, a sample of oxygen reacts with twice its own volume of hydrogen to form water vapor. The ratio between the volume of a gaseous reactant consumed and the volume of a gaseous product is also simple—the volume of water vapor formed from hydrogen and oxygen equals the volume of hydrogen consumed. In the examples in Table 3.4, the ratios are either 1:1, 1:2, 1:3, or 2:3, but more complicated whole-number ratios occur in other reactions.

Table 3.4   Volume Relationships in Gas-Phase Reactions[a]

Reactants		Product
2 L Hydrogen +	1 L Oxygen	⟶ 2 L Water vapor
1 L Hydrogen +	1 L Chlorine	⟶ 2 L Hydrogen chloride
1 L Nitrogen +	3 L Hydrogen	⟶ 2 L Ammonia
2 L Nitrogen +	1 L Oxygen	⟶ 2 L Nitrous oxide
1 L Nitrogen +	1 L Oxygen	⟶ 2 L Nitric oxide
2 L Carbon monoxide +	1 L Oxygen	⟶ 2 L Carbon dioxide
2 L Sulfur dioxide +	1 L Oxygen	⟶ 2 L Sulfur trioxide

[a] All reactants and products are gases under the conditions of measurement. All volumes are measured at the same temperature and pressure.

Avogadro's Hypothesis

In 1811 Amadeo Avogadro explained Gay-Lussac's combining-volume data by proposing what is known as **Avogadro's hypothesis**: Equal volumes of gases at the same temperature and pressure contain equal numbers of molecules, regardless of their identity. For example, Avogadro interpreted the combining volumes for the formation of water from hydrogen and oxygen in the following way. Since 2 volumes of hydrogen react with each volume of oxygen, there must be twice as many hydrogen molecules reacting as oxygen molecules. Likewise, since each volume of oxygen produces two volumes of water vapor, each oxygen molecule must be capable of forming two water molecules.

$$2 \text{ Hydrogen molecules} + 1 \text{ Oxygen molecule} \longrightarrow 2 \text{ Water molecules}$$

At the time, one of the more controversial aspects of this interpretation was the idea that oxygen molecules contain pairs of oxygen atoms. In other words, oxygen molecules are diatomic and have the formula O_2. The interpretation

$$2 \text{ Hydrogen atoms} + 1 \text{ Oxygen atom} \longrightarrow 2 \text{ Water molecules}$$

requires that an oxygen atom split in half so that each fragment could combine with hydrogen atoms to form a water molecule. Rather than discard Dalton's atomic theory, Avogadro said that identical atoms can be joined by chemical bonds that can be split, and although oxygen contains only oxygen atoms, these atoms cluster as diatomic molecules. Not all gaseous elements are diatomic, but Avogadro's hypothesis allows us to use combining-volume data to determine the molecular formula of an element. If one volume of a gaseous element produces two volumes of a gaseous product in any reaction, it is at least diatomic. Combining-volume data show that many gases, such as hydrogen and chlorine, are diatomic (H_2, Cl_2). We can also use these data to work out the formulas of many compounds. For example,

$$2 H_2 + O_2 \longrightarrow 2 H_2O$$

and

$$H_2 + Cl_2 \longrightarrow 2 HCl$$

Example 3.10

Use data in Table 3.4 to write the molecular formulas and the equations for the formation of **a.** ammonia **b.** nitrous oxide.

Solution

a. According to the combining-volume data, three hydrogen molecules react with one nitrogen molecule to form two ammonia molecules. Since one nitrogen molecule can react to form two ammonia molecules, nitrogen must be diatomic. The six hydrogen atoms from the three H_2 molecules are divided equally between the two ammonia molecules. Therefore, the formula for ammonia is NH_3, and the equation for the reaction is

$$3 H_2(g) + N_2(g) \longrightarrow 2 NH_3(g)$$

b. Two N_2 molecules react with each O_2 molecule to form 2 nitrous oxide molecules, and each nitrous oxide molecule must contain 2 nitrogen atoms and 1 oxygen atom.

$$2 N_2(g) + O_2(g) \longrightarrow 2 N_2O(g)$$

> One of the attractive things about science is its cool, unemotional objectivity. We resolve conflicts by the force of experimental facts, and the power of reason counts for more than the relative prestige of the opponents. Well, that's a pleasant ideal, but scientists don't always live up to it. Consider the case of Avogadro's hypothesis.
>
> This elegant solution to the formula–atomic mass dilemma ran into stiff opposition from Dalton. Dalton had some theories of his own about gases, and they didn't match Avogadro's ideas. Furthermore, Dalton didn't believe Gay-Lussac's law of combining volumes because his own inaccurate data didn't fit it. Because Dalton's atomic theory gave him so much prestige, his opposition was decisive for most of his contemporaries, and Avogadro's hypothesis was neglected for nearly 50 years.
>
> In the following years chemists discovered many new compounds, determined their compositions, and assigned formulas and atomic masses according to Dalton's idea of simplicity. But the choice of the simplest formula became more ambiguous as more compounds were discovered; the assignment of simplest formulas was anything but simple. Several competing atomic mass scales appeared, each based on different formula assignments. In the resulting confusion and exasperation, some chemists were ready to throw out the whole atomic theory.
>
> By 1860 this situation was so untenable that the leading chemists held a conference to straighten out the question of formula assignments. One of the participants, Stanislao Cannizzaro, was an enthusiastic supporter of Avogadro's ideas and had experimentally demonstrated their application in determining formulas. Although the conference reached no official decision, Cannizzaro won important converts, and Avogadro's hypothesis won general acceptance within a few years. Almost immediately, masses of previously confusing data fell into place, and rapid progress occurred in the field of chemical similarity between elements and the theory of molecular structure.
>
> Scientists have all the human failings (perhaps a few extra), but at least the trial-and-error nature of science protects them from their own blunders. Dalton may have led his fellow scientists up a blind alley, but before long they began to suspect something was wrong. They were willing to backtrack and examine their hypotheses to find out where they had taken a wrong turn. Theories are tools for clarifying data, not objects of veneration, and when they don't do their job effectively, we modify or scrap them.

Avogadro's hypothesis solves the problem of assigning molecular formulas for any element that can exist as a gas and any compound formed in a gaseous reaction. Once the formulas of the compounds are assigned, the calculation of the atomic masses of the constituent elements from mass composition data is easy.

According to Avogadro's hypothesis, 1 mol of any gas, regardless of its identity, occupies the same volume at a given temperature and pressure. This **molar volume** is 22.4 L at STP. It does not matter whether we have 1 mol of hydrogen, 1 mol of nitrogen, or 1 mol of HCl. Any one of these gas samples occupies 22.4 L at STP.

Example 3.11

How many moles are present in 56.0 L of oxygen at STP?

Solution

Since each mole of oxygen occupies 22.4 L, we divide the volume by 22.4 L.

$$\text{Moles O}_2 = \frac{56.0 \text{ L}}{22.4 \text{ L/mol}} = 2.50 \text{ mol}$$

Example 3.12

What is the volume at STP of 1.4 g N_2?

Solution

The molecular mass of N_2 is twice the atomic mass.

$$\text{Molecular mass} = 2(14 \text{ amu}) = 28 \text{ amu} \left(\text{or } 28 \frac{\text{g}}{\text{mol}}\right)$$

We use equation (2.1) to find the number of moles in the sample.

$$\text{Moles N}_2 = \frac{1.4 \text{ g}}{28 \text{ g/mol}} = 0.050 \text{ mol}$$

The volume is the number of moles of N_2 multiplied by the molar volume.

$$V = (0.050 \text{ mol})\left(22.4 \frac{\text{L}}{\text{mol}}\right) = 1.12 \text{ L at STP}$$

The Ideal Gas Law

Avogadro's hypothesis, coupled with the mole definition, allows us to write a single equation describing the relationship between temperature, pressure, volume, and the number of moles in a sample of ideal gas,

$$PV = nRT \tag{3.8}$$

where n is the number of moles in the sample and R is the **universal gas constant**. If P is expressed in atmospheres and V in liters, R has the value 0.08205 L-atm/K-mol.

This equation, a mathematical statement of the **ideal gas law**, summarizes all the information contained in Boyle's law, Charles's law, Gay-Lussac's law, and Avogadro's hypothesis. If the number of moles and the temperature are constant, then the ideal gas equation indicates that pressure and volume are inversely proportional (Boyle's law).

$$PV = nRT = \text{Constant}$$

The volume of a gas containing a constant number of moles at constant pressure is directly proportional to temperature (Charles's law).

$$\frac{V}{T} = \frac{nR}{P} = \text{Constant}$$

At constant volume, the pressure exerted by a certain number of moles of gas is proportional to temperature (Gay-Lussac's law).

$$\frac{P}{T} = \frac{nR}{V} = \text{Constant}$$

The volume of a gas at fixed temperature and pressure is proportional to the number of moles of the gas (Avogadro's hypothesis).

$$\frac{V}{n} = \frac{RT}{P} = \text{Constant}$$

The ideal gas law is as good an approximation of real gas behavior as any of the other gas laws we have discussed.

Example 3.13

Calculate the pressure of 64.0 g O_2 in a 5.00-L container at 27°C.

Solution

$T = 27 + 273 = 300$ K

$$n = \frac{64.0 \text{ g}}{32.0 \text{ g/mol}} = 2.00 \text{ mol}$$

$$P = \frac{nRT}{V} = \frac{(2.00 \text{ mol})(0.0821 \text{ L-atm/K-mol})(300 \text{ K})}{5.00 \text{ L}} = 9.85 \text{ atm}$$

Determining Molecular Mass from Gas Density

Using the ideal gas law, we can determine the molecular mass of any gas by measuring its density (the ratio of mass to volume) at a measured temperature and pressure. Rearrangement of the ideal gas law gives

$$n = \frac{PV}{RT}$$

Dividing the number of moles of the gas into the mass of the gas gives the molecular mass.

$$\text{Molecular mass} = \frac{m}{n} = \frac{m}{(PV/RT)} = \frac{mRT}{PV} = \left(\frac{m}{V}\right)\left(\frac{RT}{P}\right)$$

Since density (symbolized by ρ, the Greek letter rho) equals m/V,

$$\text{Molecular mass} = \rho \frac{RT}{P} \tag{3.9}$$

Example 3.14

A gas has a density of 2.0 g/L at 740 torr and 200°C. What is its molecular mass?

Solution

It is important to express pressure in units (atm) consistent with the units of R.

$$P = \frac{740 \text{ torr}}{760 \text{ torr/atm}} = 0.97 \text{ atm}$$

$T = 200 + 273 = 473$ K

$$\text{Molecular mass} = \frac{(2.0 \text{ g/L})(0.0821 \text{ L-atm/K-mol})(473 \text{ K})}{0.97 \text{ atm}}$$

$$= 80 \frac{\text{g}}{\text{mol}} \quad \text{(or 80 amu)}$$

There is an alternative procedure based on the molar volume. First use the combined gas law [equation (3.6)] to compute the molar volume at 740 torr and 200°C (473 K).

$$\text{Constant} = \frac{(760 \text{ torr})(22.4 \text{ L/mol})}{273 \text{ K}} = 62.4 \frac{\text{torr-L}}{\text{K-mol}}$$

$$V = \frac{(62.4 \text{ torr-L/K-mol})(473 \text{ K})}{740 \text{ torr}} = 39.9 \text{ L/mol}$$

Then multiply this molar volume by the gas density to find the molecular mass.

$$\text{Molecular mass} = \left(39.9 \frac{\text{L}}{\text{mol}}\right)\left(2.0 \frac{\text{g}}{\text{L}}\right) = 80 \frac{\text{g}}{\text{mol}} \quad \text{(or 80 amu)}$$

Gas Volume Stoichiometry

Gay-Lussac's law of combining volumes and Avogadro's hypothesis offer a simple way of calculating volume–volume and mass–volume relationships in chemical reactions. The ratio of volumes of gaseous reactants and products is the ratio of molecules in the balanced equation.

Example 3.15

Propane, C_3H_8, an important component of natural gas, burns in oxygen to produce carbon dioxide and water. What volume of oxygen is needed to burn completely 2.50 L propane if all volumes are measured under the same conditions?

Solution

The balanced equation is

$$C_3H_8(g) + 5\, O_2(g) \longrightarrow 3\, CO_2(g) + 4\, H_2O(g)$$

Since five molecules of oxygen react with each propane molecule, five volumes of oxygen react with one volume of propane.

$$\frac{V_{O_2}}{V_{C_3H_8}} = \frac{5 \text{ mol } O_2}{1 \text{ mol } C_3H_8} = 5$$

$$V_{O_2} = \left(5 \frac{\text{L } O_2}{\text{L } C_3H_8}\right)(2.50 \text{ L } C_3H_8) = 12.5 \text{ L } O_2$$

The mole concept is useful in predicting mass–volume relationships. After determining the number of moles of one reactant or product from its mass, we can determine the moles and volumes of other gaseous reactants or products from the balanced equation and the ideal gas equation. Reversing this procedure, we can calculate the mass of a substance from the measured volume of a gaseous reactant or product.

Example 3.16

What volume of Cl_2 at STP reacts with 1.00 g Fe to produce $FeCl_3$?

Solution

$$3\, Cl_2(g) + 2\, Fe(s) \longrightarrow 2\, FeCl_3(s)$$

$$\text{Moles Fe} = \left(\frac{1.00 \text{ g Fe}}{55.8 \text{ g Fe/mol}}\right) = 1.79 \times 10^{-2} \text{ mol Fe}$$

$$\text{Moles Cl}_2 \text{ required} = \left(\frac{3 \text{ mol Cl}_2}{2 \text{ mol Fe}}\right)(1.79 \times 10^{-2} \text{ mol Fe}) = 2.69 \times 10^{-2} \text{ mol}$$

One mole of ideal gas occupies 22.4 L at STP so, assuming ideal behavior for Cl_2,

$$V_{Cl_2} = (2.69 \times 10^{-2} \text{ mol})(22.4 \text{ L/mol}) = 0.603 \text{ L}$$

Example 3.17 illustrates the analytic application of mass–volume relationships and some of the associated experimental problems.

Example 3.17

A solution of the amino acid glycine ($C_2H_5NO_2$) in water reacts with sodium nitrite ($NaNO_2$) and HCl according to the following equation:

$$C_2H_5NO_2(aq) + NaNO_2(aq) + HCl(aq) \longrightarrow$$
$$N_2(g) + C_2H_4O_3(aq) + NaCl(aq) + H_2O$$

The nitrogen produced by treating a glycine solution with an excess of $NaNO_2$ is collected over water. The measured volume of N_2 is 35.4 mL at 740 torr total gas pressure and 27°C. What is the mass of glycine in the solution?

Solution

Since the nitrogen is collected over water, we must use Table 3.3 and Dalton's law to correct for the water vapor pressure.

$$P_{N_2} = P_{total} - P_{H_2O} = 740 \text{ torr} - 27 \text{ torr} = 713 \text{ torr}$$

Expressed in atmospheres, this pressure is

$$P_{N_2} = \frac{713 \text{ torr}}{760 \text{ torr/atm}} = 0.938 \text{ atm}$$

$$T = 27°C = 300 \text{ K}$$

The ideal gas equation yields the number of moles of nitrogen formed.

$$\text{Moles N}_2 = \frac{PV}{RT} = \frac{(0.938 \text{ atm})(0.0354 \text{ L})}{(0.0821 \text{ L·atm/K·mol})(300 \text{ K})} = 1.34 \times 10^{-3} \text{ mol}$$

From the balanced equation, we know that the number of moles of $C_2H_5NO_2$ reacted equals the number of moles of N_2 produced.

$$\text{Moles } C_2H_5NO_2 = \text{Moles N}_2 = 1.34 \times 10^{-3} \text{ mol}$$

$$\text{Molecular mass } C_2H_5NO_2 = 2(12.0 \text{ amu}) + 5(1.0 \text{ amu}) + 14.0 \text{ amu}$$
$$+ 2(16.0 \text{ amu}) = 75.0 \text{ amu}$$

$$\text{Mass glycine} = (1.34 \times 10^{-3} \text{ mol})\left(75.0 \frac{\text{g}}{\text{mol}}\right) = 0.101 \text{ g}$$

3.3 The Kinetic-Molecular Theory of Gases

Brownian Motion

A small but visible dust particle suspended in apparently still air goes through rapid, erratic motion known as **Brownian motion** (Figure 3.9). The dust particle's

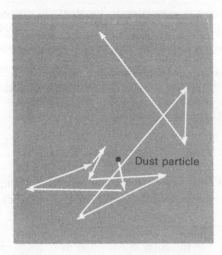

Figure 3.9
Brownian motion is illustrated by the random movement of a visible dust particle suspended in still air.

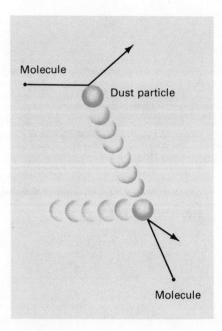

Figure 3.10
Collisions between invisible gas molecules and a visible dust particle result in Brownian motion.

frequent changes in direction, which are random and completely unpredictable, are caused by collisions between invisible gas molecules and the visible dust particle (Figure 3.10). Each collision accelerates the dust particle and changes its velocity, and although we cannot see gas molecules, we can observe their effects on small visible particles. Larger objects resist Brownian motion because their large masses experience very little acceleration from a collision with a molecule. Another factor keeping us from bouncing around due to Brownian motion is that the force of the many molecules hitting us on one side is balanced by the force of the collisions occurring at the same time on the other side. In a strong wind, many more molecules are striking us from one side than from the other, and it is quite easy to feel the effect of these forces.

The Kinetic-Molecular Theory

This theoretical picture of gases—a gas contains many molecules in rapid, random motion through empty space—coincides with the kinetic-molecular theory of heat, which says that heat is the *kinetic* energy of molecules. We cannot see the molecules, but we can experience an increase in their motion as an increase in temperature. The **kinetic-molecular theory of gases** contains the following hypotheses:

1. A sample of gas contains molecules (point masses) in random motion through empty space. The molecules are far apart on the average, they have negligible volume, and the sample is mostly empty space with relatively few molecules speeding around and intermittently colliding with the container walls or each other. This explains why a gas expands to fill its container, why gases have low densities, and why they are so compressible. The molecules in solids and liquids are closely packed, giving dense, incompressible phases with definite volumes.

2. The average kinetic energy of the gas molecules is proportional to the absolute temperature. This is the heart of the theory of heat: The total heat in a sample is the amount of molecular kinetic energy, and temperature is a measure of the average kinetic energy per molecule. Accordingly, 0 K (absolute zero) represents the temperature at which all molecular

motion in the sample ceases. We now believe that molecules still would vibrate weakly at 0 K, but they would lack any motion through space.

3. Collisions between molecules are completely elastic. The assumption of elasticity is necessary because our theoretical gas model would "run down" if molecules lost kinetic energy by collisions. It may seem strange to assign total elasticity, a property never seen in visible objects, to molecules, yet total elasticity for molecules is inherent in the theory of heat. Visible collisions are inelastic because some visible kinetic energy converts to heat—molecular kinetic energy. If heat is molecular kinetic energy, it is meaningless to talk about conversions of molecular kinetic energy to heat.

4. The only forces acting between molecules are the forces of collision. This is an oversimplification that ignores secondary effects. There are weak intermolecular attractions in real gases that cause deviations from ideal behavior and cause real gases to liquefy at low temperatures. If the theory is successful in explaining ideal gases, we can then consider the secondary effects of intermolecular forces.

5. Pressure is due solely to the forces of collisions between molecules and the container walls. A logical consequence of the model is that colliding molecules will exert forces on the walls. We now assume this is the only source of gas pressure.

Qualitative Applications of the Kinetic-Molecular Theory

As the volume of a gas is decreased, the average distance between a molecule and a container wall also decreases, and collisions occur more frequently, causing an increase in pressure. When a rigid container is heated, the molecules of the gas inside move more rapidly, striking the walls more often and more forcefully, and gas pressure increases. The more molecules in a unit volume, the more frequent the collisions, and the greater the pressure.

These qualitative arguments show that gas pressure is inversely related to volume and directly related to temperature and the number of molecules in a sample. Although this is consistent with the ideal gas law, it explains neither the strict proportionality of the law nor Avogadro's hypothesis, that is, pressure is independent of the identity of the molecules. To demonstrate that the kinetic-molecular theory adequately explains the ideal gas law, we need a more involved mathematical analysis of the sort presented in the next section.

Special Resource Section: The Derivation of the Ideal Gas Law

The best way to deal with a complicated theoretical treatment is to start with a simple but unrealistic model. If this model gives the correct answer, we can refine it to conform to our ideas of reality. We begin with a simple model of an ideal gas —a single molecule of mass m that is moving horizontally and to the right with velocity v in a cubic container of edge length l (Figure 3.11). This molecule will bounce off the right-side wall, and the collision will change its velocity to the left ($-v$). A later collision with the left-side wall will change the velocity back to v.

To compute the pressure on the right-side wall, we use the equation

$$P = \frac{F_{ave}}{A} \tag{3.10}$$

where F_{ave} is the average force of collision on the wall, and A is the area of the wall. According to Newton's second and third laws,

$$F_{coll} = -ma \tag{3.11}$$

where F_{coll} is the force of a single collision on the wall, and a is the acceleration of the molecule during the collision. (The algebraic sign keeps track of the direction of the force.)

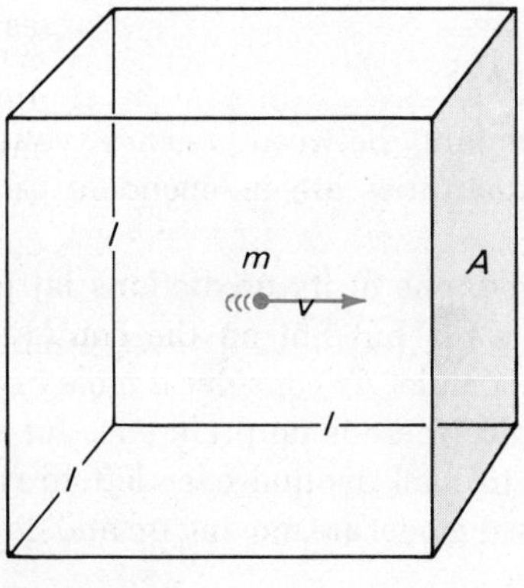

Figure 3.11
Simple model for an ideal gas.

$$a = \frac{(-v) - (v)}{t} = \frac{-2v}{t} \qquad (3.12)$$

and

$$F_{coll} = \frac{2mv}{t} \qquad (3.13)$$

The wall experiences a force of $2mv/t$ (to the right) for each collision. The time, t, during which the molecule interacts with the wall, is difficult to estimate. We get around this difficulty by considering the average force due to many collisions over a period of time, so that t represents the time interval between two successive collisions. This depends on the dimensions of the box and the velocity of the molecule. Since the molecule must travel to the opposite wall and back between collisions,

$$t = \frac{2l}{v} \qquad (3.14)$$

By substituting equation (3.14) into equation (3.13), we obtain an equation for the average force due to many individual collisions.

$$F_{ave} = \frac{2mv}{2\dfrac{l}{v}} = \frac{mv^2}{l} \qquad (3.15)$$

Therefore, the pressure is

$$P = \frac{mv^2}{lA} \qquad (3.16)$$

Since $lA = V$,

$$P = \frac{mv^2}{V}$$

and

$$PV = mv^2 \qquad (3.17)$$

According to the kinetic theory, the molecular kinetic energy is proportional to the absolute temperature,

$$\tfrac{1}{2} mv^2 = CT \qquad (3.18)$$

where C is a universal constant of proportionality between temperature and kinetic energy. By combining equations (3.17) and (3.18), we derive a relationship between P, V, and T for the model gas.

$$PV = 2CT \qquad (3.19)$$

Equation (3.19) shows both the proper proportionality between pressure, volume, and temperature and also shows that these quantities are independent of the molecular mass (Avogadro's hypothesis).

Because this model is grossly oversimplified, one of its predictions is ridiculous—there would be pressure on the side walls but not on the container's other four walls. To develop a more realistic model, let us consider a mole of gas (in other words N_{Avog} molecules) in the box. Since there is no preference for vertical motion over horizontal motion* or front to back motion over left to right motion, a third of the molecules in the hypothetical model are moving up and down,

---

*Gravity does have an effect, creating differences in pressure between the top and bottom of a very tall container. For any container that fits in a laboratory, this effect is negligible because the average molecular kinetic energy is many times greater than the difference in the potential energy of a molecule between the top and bottom of the container.

a third move back and forth, and a third move sideways. Since the pressure on any wall will be proportional to the number of molecules striking it, the pressure will be $N_{\text{Avog}}/3$ times as great as predicted for one molecule [equation (3.19)].

$$PV = \frac{2N_{\text{Avog}}CT}{3} \tag{3.20}$$

This is the ideal gas law. The value of $R(0.0821$ L-atm/K-mol$)$ is the experimentally determined value of the constant quantity $2N_{\text{Avog}}C/3$, and the derived law for 1 mol of gas is $PV = RT$; for n moles it is $PV = nRT$.

This derivation gives us an important insight into the nature of R. It is a universal conversion factor between absolute temperature and molecular kinetic energy and shows up in many problems and applications apparently unrelated to gases.

The model we have developed is still a long way from reality. The molecules in a real gas sample are moving in every possible direction, not just perpendicular to the container walls, and they travel with a variety of many speeds. A single molecule will reach every part of the container in the course of its many collisions, both with the walls and with other molecules. Dealing with a more realistic model is more complicated, but a mathematical analysis confirms the conclusions based on the less realistic model.

Graham's Law of Diffusion

It takes time for a gas to spread from one region to another. For example, if a bottle of ammonia were opened across a large room, it would take several minutes for the pungent ammonia odor to reach us. The ammonia molecules must travel to our noses before we can smell them, and this occurs by a process called **diffusion**. Ammonia molecules leave the bottle because of their random motion. Each molecule incurs many collisions with air molecules in the course of its travels, and each collision alters its motion. Eventually ammonia molecules reach all parts of the room.

Rates of diffusion depend on many factors such as temperature, the concentration of molecules in the surrounding gases, and the dimensions of the space through which the gas diffuses. **Graham's law** describes the influence of molecular mass on diffusion rates. The law says that the relative rates of diffusion of two gases are inversely proportional to the square root of their molecular masses.

$$\frac{\text{Rate}_1}{\text{Rate}_2} = \sqrt{\frac{\text{Molecular mass}_2}{\text{Molecular mass}_1}} \tag{3.21}$$

Figure 3.12 shows a simple experiment illustrating Graham's law. Two cotton

Figure 3.12
A demonstration of Graham's law.

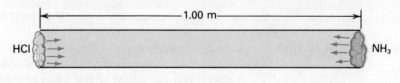

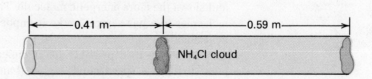

$$\frac{\text{Rate NH}_3}{\text{Rate HCl}} = \frac{0.59}{0.41} = 1.44$$

$$\frac{\text{Molecular mass HCl}}{\text{Molecular mass NH}_3} = \frac{36.5}{17.0} = 2.15$$

$$\sqrt{2.15} = 1.47$$

wads, one soaked with ammonia (NH_3) solution, the other with hydrochloric acid (HCl) solution, are inserted in opposite ends of a long glass tube at the same time. As the NH_3 vapor diffuses into the HCl vapor coming from the other side, a white cloud of ammonium chloride (NH_4Cl) crystals forms from the reaction of the two gases. The cloud appears closer to the HCl end of the tube, showing that the lighter NH_3 molecules have diffused faster and traveled farther. The relative distances traveled by the two gases agree with Graham's law.

The kinetic theory explains Graham's law. The rate of diffusion is proportional to the speed of a molecule with average kinetic energy. Since all gases at the same temperature have the same average kinetic energy ($\frac{1}{2}mv^2 = CT$), we can write the equation

$$\tfrac{1}{2}m_1 v_1^2 = \tfrac{1}{2}m_2 v_2^2 \tag{3.22}$$

in which m_1 and m_2 describe the molecular masses in grams of two different gases, and v_1 and v_2 are their respective average molecular speeds. Since the rates of diffusion are proportional to the average molecular speeds, we can rearrange equation (3.22) to obtain the relationship

$$\frac{\text{Rate}_1}{\text{Rate}_2} = \frac{v_1}{v_2} = \sqrt{\frac{m_2}{m_1}}$$

To obtain equation (3.21), we just rewrite the preceding equation in terms of the ratio of molecular masses in amu. Heavy gases diffuse more slowly because their molecules move more slowly.

Example 3.18

If H_2 diffuses through a porous barrier at a rate of 52 mL/min at 0°C, what is the rate of diffusion of O_2 through the same barrier at this temperature?

Solution

The molecular mass of O_2 is 32.0 amu and that of H_2 is 2.0 amu. Since oxygen molecules are 16 times as massive as hydrogen molecules, they diffuse at 1/4 the rate.

$$\frac{\text{Rate } O_2}{\text{Rate } H_2} = \sqrt{\frac{2.0 \text{ amu}}{32.0 \text{ amu}}} = \sqrt{\frac{1}{16}} = \frac{1}{4}$$

$$\text{Rate } O_2 = \frac{1}{4}\left(52 \, \frac{\text{mL}}{\text{min}}\right) = 13 \, \frac{\text{mL}}{\text{min}}$$

Random Molecular Motion

At any moment, some of the molecules in a gas sample are moving much more rapidly than others. Each molecule collides with other molecules many times per second, and the speed of the molecule changes with each collision. A collision between a slower molecule and a faster molecule speeds up the slower molecule and slows the more energetic molecule. These frequent energy exchanges between molecules in a gas sample make it impossible to be sure of the speed of any one molecule.

We know it is possible for a molecule to receive several successive hard jolts, giving it a speed much higher than the average, but we also know such a sequence of events is not very likely. Once the energy of a molecule rises above the average, this higher-energy molecule has a better chance of colliding with a slower molecule and losing energy than gaining energy from a faster molecule. Even if we cannot be sure what is going to happen to any particular molecule, we can use the mathematics of probability and cite gambler's odds on the chances

Figure 3.13
Statistical distribution of molecular speeds of nitrogen molecules at different temperatures.

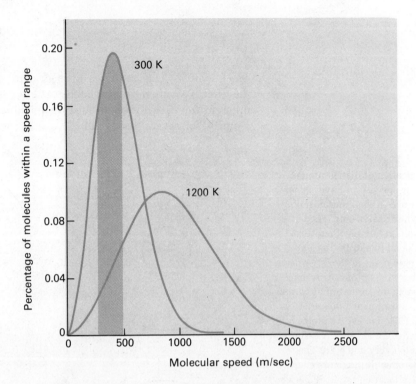

that its speed falls within a certain range. Using these odds, we can estimate the percentage of molecules in the sample whose speeds are within this range.

Figure 3.13 shows the predicted distribution of speeds of nitrogen molecules at two different temperatures. These bell-shaped curves represent the probability that a molecule will have a particular speed. The most likely speed for nitrogen molecules at 300 K is about 420 m/sec. Only relatively few molecules are moving as fast as 1000 m/sec or as slow as 100 m/sec. The shaded area shows the percentage of nitrogen molecules at 300 K moving at speeds between 300 and 500 m/sec (37.6%). Raising the temperature shifts the peak of the statistical distribution to higher speeds and broadens the distribution so that the percentage of high-speed molecules rises. Figure 3.14 (p. 70) outlines an experiment used to show that the distribution of molecular speeds in a gas sample does match the predicted statistical distribution.

Nonideality

The kinetic-molecular theory explains the ideal gas law, but real gases behave differently. The theory as stated neglects the significant secondary factors that cause nonideality. One of these factors is molecular size. Molecules do occupy small volumes, and they collide with each other. The total volume of all the other molecules in a gas decreases the space available to any one molecule, and because it collides with these other molecules, it collides with the container slightly more often than predicted from the ideal model. The small increase in collision frequency causes the gas pressure to be slightly larger than predicted from the ideal gas law.

The other secondary effect, the weak attraction between gas molecules, restricts molecular motion to a slight degree and lowers the pressure. These intermolecular forces are effective only at short distances, so this effect is most important at high compression, when the average distance between the molecules is small. It is also more important at lower temperatures because slower moving molecules are more affected by the attractive forces than the rapidly moving ones.

There are several equations that correct for the nonideal effects and that accurately predict the behavior of real gases. One of these is the **van der Waals equation:**

$$\left(P + \frac{n^2 \mathbf{a}}{V^2}\right)(V - n\mathbf{b}) = nRT \tag{3.23}$$

The quantity **a** is a measure of the strength of the intermolecular forces and **b** is the effective volume occupied by 6.02×10^{23} molecules. These quantities are different for each different substance. Table 3.5 lists van der Waals constants, **a** and **b**, for several gases.

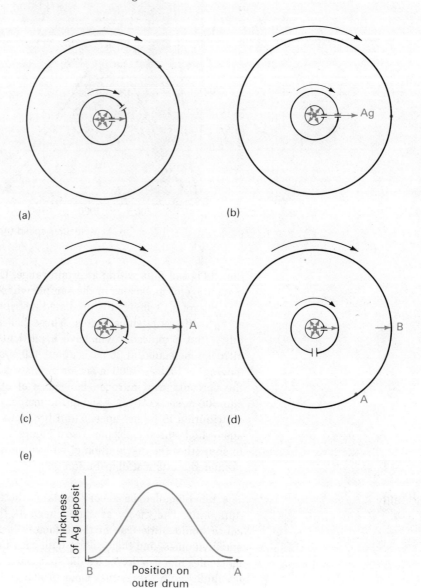

Figure 3.14
Demonstration of distribution of molecular speeds. (a) Silver vapor is contained in the innermost of three cylindrical drums. The outer two drums rotate together. (b) A slit in the middle drum lines up briefly with a slit in the inner drum. A number of Ag atoms, traveling in the same direction but at various speeds, move toward the outer drum. (c) The fastest Ag atoms hit the outer drum at point A before it has rotated very much. (d) The slower atoms hit in a more counterclockwise location (point B). (e) The variation in the thickness of the silver deposit on the outer drum fits the predicted distribution of molecular speeds.

Table 3.5   Van der Waals Constants for Various Gases

Substance	a (L²-atm/mol²)	b (mL/mol)
He	0.0341	23.7
Ar	1.345	32.2
H_2	0.244	26.6
O_2	1.36	31.8
CO	1.49	39.9
CO_2	3.59	42.7
C_2H_4	4.47	57.1
Hg	8.09	17.0

Summary

1. Pressure is the force acting on a surface divided by the surface area. A gas exerts a homogeneous pressure on any surface in contact with it. Gas pressure is measured with a manometer. The height of a liquid column in a manometer is proportional to the gas pressure. One torr is sufficient pressure to support a column of mercury 1 mm in height. One standard atmosphere equals 760 torr.

2. Boyle's law: The volume of a sample of gas at constant temperature is inversely proportional to its pressure.

    $PV =$ Constant

3. The lowest attainable temperature, $-273°C$, equals zero on the Kelvin temperature scale.

    Temperature in K = Temperature in °C + 273

4. Charles's law: The volume of a gas sample at constant pressure is directly proportional to temperature, expressed in K.

    $\dfrac{V}{T} =$ Constant

    Gay-Lussac's law: The pressure of a gas sample at constant volume is directly proportional to the Kelvin temperature.

    $\dfrac{P}{T} =$ Constant

5. Gas volumes are usually reported at standard temperature and pressure (STP)—760 torr (1.0 atm) and 0°C (273 K).

6. Dalton's law of partial pressures: The total pressure of a gas mixture is the sum of the partial pressures of the individual components. The partial pressure of a component is the pressure it exerts when it is the sole occupant of the container.

7. Gay-Lussac's law of combining volumes: Gases react chemically in simple volume ratios.

8. Avogadro's hypothesis: Equal volumes of gases at the same temperature and pressure contain equal numbers of molecules.

9. A mole of any gas occupies 22.4 L at STP.

10. Ideal gas law: $PV = nRT$, where R is the universal gas constant and equals 0.08205 L-atm/K-mol.

11. Gas density (ρ) can be used to estimate molecular mass.

    Molecular mass $= \dfrac{\rho RT}{P}$

12. We can use a balanced equation to determine volume relationships involving gaseous reactants and products.

13. Brownian motion is the erratic motion of small but visible particles due to their collisions with molecules.

14. The kinetic-molecular theory: (a) A sample of gas contains molecules with negligible volume in random motion through empty space. (b) The average kinetic energy of the gas molecules is proportional to the absolute temperature. (c) Molecular collisions are completely elastic. (d) The only forces acting between molecules are the forces of collision. (e) Pressure is due solely to the forces of collision between molecules and the container walls.

15. Graham's law of diffusion: The relative rates of diffusion of two gases are inversely proportional to the square roots of their molecular masses.

16. The molecules of a gas move at a variety of speeds and collisions change their speeds. Although we know nothing about the speed of any particular molecule, we can estimate reliable statistical distributions of molecular speeds.

17. The behavior of real gases differs from behavior of ideal gases because real gas molecules occupy some space in the container and because there are weak attractive forces between the molecules. The van der Waals equation accurately describes the behavior of real gases.

Vocabulary List

pressure	Kelvin scale	molar volume
closed-end manometer	Gay-Lussac's law	universal gas constant
barometer	standard temperature and pressure (STP)	ideal gas law
torr		Brownian motion
standard atmosphere	Dalton's law of partial pressures	kinetic-molecular theory of gases
Boyle's law		
ideal gas	partial pressure	diffusion
Charles's law	law of combining volumes	Graham's law
absolute zero	Avogadro's hypothesis	van der Waals equation

Exercises

1. Briefly define or explain each of the following:
   a. Pressure b. Manometer c. Barometer d. Absolute zero e. STP
   f. Partial pressure g. Gas density h. Kinetic-molecular theory i. Ideal gas
   j. Diffusion

2. The common volume unit is the liter. This is not the official SI system unit; what is the SI unit equivalent? How many cubic centimeters are there in 1 L?

3. A gas sample exerts a pressure of 950 torr. What is its pressure in atmospheres?

4. How many torr are present in 1.35 atm?

5. If water were used in a barometer rather than mercury, what would be the height of the water column in centimeters for 1.0 atm, 273 K?

6. Propose equations and formulas for each of the following gas-phase reactions.
   a. 1 Phosphorus + 6 Hydrogen $\longrightarrow$ 4 Phosphine
   b. 3 mL Oxygen $\longrightarrow$ 2 mL Ozone
   (Ozone is a less common molecular form of oxygen.)

7. Explain each of the following in terms of the expected behavior of gases:
   a. An automobile tire increases its pressure after a long, high-speed trip.
   b. A weather balloon is filled to only about half its capacity before it is launched into the upper atmosphere.
   c. A rock placed too close to a campfire may explode due to trapped moisture.

8. Make the following temperature conversions.
   a. 159°C to K   b. −68°C to K   c. 465 K to °C
   d. 180 K to °C

9. Make the following temperature conversions.
   a. 37°C to K   b. −188°C to K   c. 210 K to °C
   d. 815 K to °C

10. What is the pressure of sample A in Figure 3.15? The liquid in the manometer is mercury, and the atmospheric pressure is 733 torr.

11. The open-end manometer measuring the pressure of sample B in Figure 3.15 contains mercury. The outside pressure is 747 torr. What is the pressure of sample B?

EXERCISES

Figure 3.15

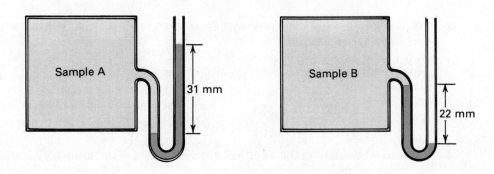

12. A 10.0-mg sample of a gas occupies 45.6 mL at 240 torr and 18°C. What volume will it occupy at 598 torr and 18°C?

13. What volume of oxygen, measured at 1.00 atm, is needed to fill a 450-mL vessel at a pressure of 5.20 atm if both measurements are made at the same temperature?

14. A 25.0-mL sample of a gas exerts a pressure of 585 torr. What pressure will it exert if its volume is decreased to 17.2 mL while its temperature is maintained at a constant value?

15. What pressure results when 3.57 L of nitrogen, originally at 775 torr, is forced into a 700-mL container? The temperature remains constant.

16. A gas occupies 25.0 mL at 25°C and 745 torr. What is its volume at 150°C and 745 torr?

17. A gas occupies 52.0 mL at 33°C. What will its volume be at −62°C at the same pressure?

18. A gas occupies 25.0 mL at 25°C and 745 torr. What pressure would the gas exert if its volume were held at 25.0 mL and its temperature raised to 127°C?

19. A gas in a 15.0-mL rigid container exerts a pressure of 1.20 atm at 18°C. What pressure would it exert if it were cooled to −18°C?

20. When a gas, originally at −10°C, is warmed and allowed to expand at constant pressure, its volume increases from 17.5 mL to 25.4 mL. What is the temperature of the warm gas?

21. A sample of air occupying 253 mL at 23°C is cooled at constant pressure until its volume decreases to 216 mL. What is the temperature of the cooled gas?

22. A sample of O_2 in a 5.00-L box is warmed from 0°C until its pressure rises from 640 torr to 950 torr. What is the temperature of the heated gas?

23. A sample of NO gas in a rigid container is cooled from 50°C until its pressure falls from 745 torr to 458 torr. What is the temperature of the cooled gas?

24. A gas occupies 25.0 mL at 25°C and 745 torr. What volume would it occupy at STP?

25. What volume would a gas occupy at 1.75 atm and 325°C if its volume at STP is 2.35 L?

26. A sample of air occupying 4.50 L at 55°C and 958 torr is cooled to −15°C and its pressure is reduced to 720 torr. What is the new volume of the sample?

27. As the pressure on 3.50 L of helium is increased from 0.95 atm to 6.2 atm, its temperature rises from 30°C to 85°C. What is the new volume of the gas?

28. What is the partial pressure of hydrogen collected over water at 21°C and a total pressure of 690 torr?

29. A sample of H_2 is collected over water at 35°C. The measured pressure of the sample is 642 torr. What would the pressure of the sample be if it were dry?

30. A 3.00-L flask holds helium at 340 torr and 0°C. Enough ammonia (a gas) is added to the flask to bring the pressure of the mixture to 485 torr at 0°C. What is the partial pressure of the ammonia in the flask?

31. A mixture of oxygen and helium in a rigid container exerts a pressure of 1.54 atm at 28°C. The oxygen is removed by reacting it with iron to form solid iron oxides. The pressure after the reaction is 0.86 atm at 28°C. What was the partial pressure of oxygen before the reaction?

32. What is the volume of 0.200 mol of an ideal gas at 840 torr and 150°C?

33. What volume would 1.56 mol O_2 occupy at 720 torr and 50°C?

34. A 2.50-L container holds 0.620 mol N_2 at 85°C. What is its pressure?

35. What pressure would 0.350 mol of an ideal gas exert when placed in a 100-mL container at −15°C?

36. What is the volume at STP of a 48.2-mL sample of oxygen collected over water at 23°C and 685 torr?

37. A 27.5-mL sample of nitrogen is collected over water at 29°C and 715 torr. What is the volume at STP of the nitrogen?

38. **a.** What ratio of reactant volumes is required for each of the following gas-phase reactions?
    b. What is the ratio of the volume of the underlined reactant consumed to the volume of the product formed in each reaction?

    $\underline{S_8}(g) + 8\,H_2(g) \longrightarrow 8\,H_2S(g)$

    $\underline{N_2}(g) + 2\,H_2(g) \longrightarrow N_2H_4(g)$

39. Find the ratio of the volume of the underlined reactant consumed to the volume of the underlined product formed in each of the following reactions.

    $2\,\underline{O_2}(g) + CH_4(g) \longrightarrow CO_2(g) + 2\,\underline{H_2O}(g)$

    $2\,\underline{NO}(g) + O_2(g) \longrightarrow \underline{N_2O_4}(g)$

40. A gas sample occupies 2.62 L at 285°C and 3.42 atm. How many moles are present in this sample?

41. How many moles are present in a sample of gas that occupies 25.7 mL at 26°C and 748 torr?

42. What is the density of Cl_2 measured at 35°C and 735 torr?

43. What is the density of N_2O at −10°C and 500 torr?

44. What volume of oxygen is needed to burn 3.00 L N_2H_4 to form H_2O and N_2O? What volume of N_2O is produced? All substances are gases under the conditions of measurement and all volumes are measured at the same temperature and pressure.

45. What volume of hydrogen is needed to convert 75.0 L CO to CH_3OH vapor? What volume of CH_3OH is produced? All volumes are measured at the same temperature and pressure.

46. What volume of oxygen at 220°C and 745 torr reacts with copper to produce 2.00 g CuO?

47. What volume of oxygen, measured at STP, can be obtained by decomposing 2.592 g $KClO_3$ to KCl and oxygen?

48. Calculate the molecular mass of substance X from the following data: volume of gas sample = 435 mL; mass of gas sample = 0.568 g; T = 35°C; P = 749 torr.

49. A sample of substance Y occupies 552 mL at 58°C and 738 torr. Its mass is 1.95 g. What is its molecular mass?

50. What volume will 0.085 g N_2O occupy at 55°C and 795 torr?

51. What volume will 2.35 g HCl gas occupy at 300 torr and 12°C?

52. What pressure will 0.250 g B_2H_6, a gas, exert if it is held in a 1.65-L container at 18°C?

53. What pressure will 5.00 g SO_2 exert if its volume is 8.00 L at 75°C?

54. A gas has a density of 2.2 g/L at 25°C and 770 torr. What is its molecular mass?

55. A gas with a density of 0.82 g/L exerts a pressure of 0.320 atm at 165°C. What is the molecular mass of this gas?

56. What is the ratio of the rates of diffusion of F_2 and Cl_2 at 25°C?

57. What is the ratio of the rates of diffusion of N_2 and butanone, C_4H_8O, at 200°C?

58. What is the formula of gas X, a compound of C and H?

    $1\,L\,X(g) + 5\,L\,O_2(g) \longrightarrow 3\,L\,CO_2(g) + 4\,L\,H_2O(g)$

59. What volume of N_2 at 300°C and 2.50 atm contains as many molecules as 2.50 L O_2 at STP?

60. A mixture contains Cl_2 at a partial pressure of 485 torr and N_2 at 315 torr. What is the ratio of Cl_2 molecules to N_2 molecules in this sample?

61. What is the volume at STP of a mixture of 2.5 g O_2 and 10.0 g N_2?

62. An 8.00-L box contains 5.0 g O_2 and 15.0 g N_2 at 35°C.
    a. What is the pressure of each component?
    b. What is the total pressure of the mixture?

63. A 750-mL container contains NO at a partial pressure of 350 torr and N_2 at 253 torr at 18°C. How many moles of each component are present?

64. The *contact process* for manufacturing sulfuric acid (H_2SO_4) involves the reaction between sulfur dioxide gas (SO_2) and oxygen to form sulfur trioxide (SO_3), which then reacts with water to form sulfuric acid.

    $2\,SO_2(g) + O_2(g) \longrightarrow 2\,SO_3(g)$

    $SO_3(g) + H_2O(l) \longrightarrow H_2SO_4(aq)$

    What volume of oxygen and what volume of sulfur dioxide, both measured at STP, are required for the production of a metric ton (10^3 kg) of a solution that is 98% H_2SO_4 by weight?

65. Hydrogen peroxide (H_2O_2) decomposes into water and oxygen.

$$2\,H_2O_2(aq) \longrightarrow O_2(g) + 2\,H_2O(l)$$

The decomposition of 15.0 g of a solution of hydrogen peroxide yields 183 mL of oxygen, collected over water at 25°C and 702 torr. What is the percentage (by weight) of hydrogen peroxide in the solution?

66. The following reaction has been used to analyze the oxygen content of air.

$$2\,NO(g) + O_2(g) + H_2O(l) \longrightarrow HNO_3(aq) + HNO_2(aq)$$

a. When 10.0 mL NO and 20.0 mL of an oxygen-containing mixture react, the final gas volume is 18.0 mL. How much oxygen was in the gas?
b. What will the final volume be after 10.0 mL NO and 20.0 mL pure O_2 are mixed above water?

67. Natural hydrogen is a mixture of H_2, HD, and D_2 (D represents the deuterium isotope; atomic mass = 2.0 amu). If H_2 diffuses through a particular membrane at 2.0 mL/min, how fast will the other two forms diffuse? Could this method be used for isotope purification?

68. The following experimental data were measured for a gas at 25°C. In what pressure range does the gas exhibit ideality? (Remember that the last significant digit is subject to fluctuations due to rounding.) Is the major source of nonideality outside this range intermolecular attraction or the volume occupied by the molecules?

P (atm)	V (mL)
0.246	100.0
0.307	80.0
0.410	60.0
0.614	40.0
1.223	20.0
2.432	10.0
4.802	5.00
5.965	4.00
7.870	3.00
11.55	2.00
21.64	1.00

69. A manometer used to measure small pressures contains triethylene glycol, a liquid whose density is 1.46 g/cm³. (The density of mercury is 13.6 g/cm³.) What pressure in torr supports a column of triethylene glycol 15.3 mm in height?

70. A student collects 24.7 mL of hydrogen above a column of water 15.4 cm in height. (See Figure 3.16.) The temperature is 28°C and the atmospheric pressure is 743 torr. What volume would the dry hydrogen occupy at STP? The density of water is 1.00 g/cm³, and the density of mercury is 13.6 g/cm³.

71. a. What volume does a mixture of 7.0 g N_2 and 8.5 g NH_3 occupy at 27°C and 730 torr?
b. What is the partial pressure of N_2 in this system?
c. The ammonia is absorbed from the gas mixture by a solution of sulfuric acid. In the process, the remaining nitrogen becomes contaminated with water vapor at a partial pressure of 26 torr. What volume will this gas occupy at 730 torr and 27°C?

72. a. How much does the average molecular speed of nitrogen increase as its temperature is raised from −73°C to 123°C?
b. How much does the average molecular speed of nitrogen change when its pressure decreases from 2.0 atm to 0.50 atm at constant temperature?

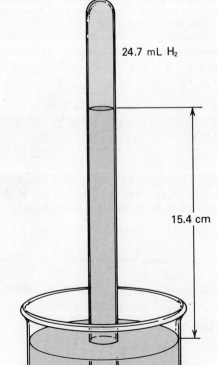

Figure 3.16

Table 3.6 Mass and Composition Data for Molecules of Carbon Compounds

Compound	Molecular mass from gas density (amu)	% C	Mass C / Molecule (amu)	C atoms / Molecule
Carbon monoxide	28.0	42.9	12.0	1
Carbon dioxide	44.0	27.3	12.0	1
Methane	16.0	74.9	12.0	1
Ethane	30.1	79.9	24.0	2
Dimethylamine	45.1	53.3	24.0	2
Propane	44.1	81.7	36.0	3
Butane	58.1	82.7	48.0	4
Ethyl ether	74.1	64.8	48.0	4
Benzene	78.1	92.3	72.1	6

73. A mixture of 1 mol H_2 and 1 mol NO in a 50.0-L container at 800°C reacts completely in the reaction

$$2\,H_2 + 2\,NO \longrightarrow N_2 + 2\,H_2O$$

Both products are gases under the conditions of the reaction. What is the final pressure in the container?

74. Stanislao Cannizzaro extended Avogadro's hypothesis by showing how it could be used to measure the atomic mass of a solid element, provided this element forms several gaseous compounds. The method involves multiplying the percentage of the element in a compound by the molecular mass of the compound (determined from its gas density) to find the mass of the element in a molecule. This value will be a whole-number multiple of the atomic mass, the whole number representing the number of atoms of the element in a molecule of that particular compound. Table 3.6 shows how the method is applied to carbon. Use this method to calculate the atomic mass of element X (not its real symbol). Refer to the following data: Compound A, molecular mass = 32.1 amu, 87.5% X; Compound B, molecular mass = 62.2 amu, 90.4% X; Compound C, molecular mass = 92.3 amu, 91.2% X; Compound D, molecular mass = 170 amu, 16.5% X; Compound E, molecular mass = 88.2 amu, 31.9% X.

75. A cubic container whose edge length is 0.500 m contains two identical molecules. One rebounds from the right-hand wall horizontally at 150 m/sec, the other rebounds horizontally from the left-hand wall at the same speed.
    a. If the two molecules miss each other, how much time will pass before the next collision with the right-hand wall?
    b. How much time will pass before the next collision with the right-hand wall if the two molecules collide head-on and rebound in an elastic collision? (Assume the sizes of the molecules are negligible compared to the dimensions of the box.)

76. a. What pressure is needed to reduce the volume of a mole of an ideal gas at −100°C to 0.300 L?
    b. Use the van der Waals equation to predict the pressure needed to reduce the volume of 1 mol He at −100°C to 0.300 L. (Refer to Table 3.5.)
    c. What pressure is needed to reduce the volume of 1 mol CO to 0.300 L at −100°C?

77. A rigid container holds O_2 at 400 torr and CO at 600 torr and 500°C. The components of the mixture react according to the equation

$$2\,CO(g) + O_2(g) \longrightarrow 2\,CO_2(g)$$

What is the pressure at 500°C after the reaction is complete?

EXERCISES

78. A mixture of NO and H_2 in a rigid 5.00-L container exerts a pressure of 1.00 atm at 700°C. The partial pressure of NO is 500 torr. Hydrogen and NO undergo the following reaction:

$$2 H_2(g) + 2 NO(g) \longrightarrow N_2(g) + 2 H_2O(g)$$

   a. What will be the pressure of the gas mixture after completion of the reaction and restoration of the temperature to 700°C?
   b. What will be the partial pressure of each component of the final mixture?

79. Aluminum reacts with HCl solution to form $AlCl_3$ and hydrogen gas. The reaction between HCl solution and magnesium yields $MgCl_2$ and hydrogen. The reaction of 14.6 mg of an Al-Mg alloy produces 20.7 mL of hydrogen at 29°C and 617 torr. What is the percentage of aluminum in this alloy?

80. Derive the following relationship from the kinetic-molecular theory:

$$\frac{\text{Rate}_1}{\text{Rate}_2} = \sqrt{\frac{T_1}{T_2}}$$

where rate_1 and rate_2 are the diffusion rates of a particular gas at temperatures T_1 and T_2 respectively.

chapter 4

Atomic Structure: The Constituent Parts

Representation of a lithium nucleus, ^7_3Li, containing 3 protons and 4 neutrons

4.1 Properties of Elements

The Periodic Table

Certain groups of elements have similar chemical and physical properties, and we classify these elements as members of the same chemical "family." For example, lithium, sodium, potassium, rubidium, and cesium (the **alkali metal** family) are all soft, shiny metals with low melting points and high reactivity (Table 4.1). They form compounds with similar formulas and properties. For example, the chlorine compound of each (general formula MCl, where M denotes any one of the alkali metals) is a high-melting, water-soluble, crystalline solid.

The **halogens**—fluorine, chlorine, bromine, and iodine—are another example of a chemical family. These diatomic, volatile, chemically reactive nonmetals all react violently with sodium to form saltlike compounds with the general formula NaX, where X denotes a halogen atom. With carbon they form volatile, water-insoluble compounds of the type CX_4 (carbon tetrahalides).

The atomic masses of family members are spaced at fairly regular intervals, a regularity that led several nineteenth-century chemists to look for relationships between chemical behavior and atomic mass. In 1871 Dmitri Ivanovitch Mendeleev* presented the periodic regularity of the properties of elements as a scientific law: If we arrange all elements in sequence according to increasing atomic mass, members of a particular family appear at predictable spots in the sequence. Although Mendeleev's law works reasonably well (there are a few exceptions),

---

*The German chemist Lothar Meyer developed the same law at approximately the same time as Mendeleev, but he was less aggressive in presenting his law and not as daring in making predictions from it.

Table 4.1 Chemical Families

Element	Atomic mass (amu)	Melting point (°C)	Chloride compound Formula	Chloride compound Melting point (°C)	Formula of oxide
Alkali metals					
Li	6.939	186	LiCl	613	Li_2O
Na	22.990	98	NaCl	801	Na_2O
K	39.098	62	KCl	776	K_2O
Rb	85.47	39	RbCl	715	Rb_2O
Cs	132.905	29	CsCl	646	Cs_2O

Element	Atomic mass (amu)	Molecular formula	Boiling point (°C)	Formula of sodium compound	Formula of carbon compound
Halogens					
F	18.998	F_2	−188	NaF	CF_4
Cl	35.453	Cl_2	−34.6	NaCl	CCl_4
Br	79.909	Br_2	58.5	NaBr	CBr_4
I	126.904	I_2	184	NaI	CI_4

Figure 4.1
Periodic table of the elements.

IA	IIA											IIIA	IVA	VA	VIA	VIIA	0
					1 H 1.0												2 He 4.0
3 Li 6.9	4 Be 9.0											5 B 10.8	6 C 12.0	7 N 14.0	8 O 16.0	9 F 19.0	10 Ne 20.2
11 Na 23.0	12 Mg 24.3	IIIB	IVB	VB	VIB	VIIB	←—VIII—→			IB	IIB	13 Al 27.0	14 Si 28.1	15 P 31.0	16 S 32.1	17 Cl 35.5	18 Ar 39.9
19 K 39.1	20 Ca 40.1	21 Sc 45.0	22 Ti 47.9	23 V 50.9	24 Cr 52.0	25 Mn 54.9	26 Fe 55.8	27 Co 58.9	28 Ni 58.7	29 Cu 63.5	30 Zn 65.4	31 Ga 69.7	32 Ge 72.6	33 As 74.9	34 Se 79.0	35 Br 79.9	36 Kr 83.8
37 Rb 85.5	38 Sr 87.6	39 Y 88.9	40 Zr 91.2	41 Nb 92.9	42 Mo 95.9	43 Tc (97)	44 Ru 101.0	45 Rh 102.9	46 Pd 106.4	47 Ag 107.9	48 Cd 112.4	49 In 114.8	50 Sn 118.7	51 Sb 121.8	52 Te 127.6	53 I 126.9	54 Xe 131.3
55 Cs 132.9	56 Ba 137.3	*57 La 138.9	72 Hf 178.5	73 Ta 180.9	74 W 183.9	75 Re 186.2	76 Os 190.2	77 Ir 192.2	78 Pt 195.1	79 Au 197.0	80 Hg 200.6	81 Tl 204.4	82 Pb 207.2	83 Bi 209.0	84 Po (210)	85 At (210)	86 Rn (222)
87 Fr (223)	88 Ra (226)	†89 Ac (227)															

*Lanthanide (rare earth) series

58 Ce 140.1	59 Pr 141.0	60 Nd 144.2	61 Pm (147)	62 Sm 150.4	63 Eu 152.0	64 Gd 157.3	65 Tb 158.9	66 Dy 162.5	67 Ho 164.9	68 Er 167.3	69 Tm 168.9	70 Yb 173.0	71 Lu 175.0

†Actinide series

90 Th 232.0	91 Pa (231)	92 U 238.0	93 Np (237)	94 Pu (242)	95 Am (243)	96 Cm (247)	97 Bk (247)	98 Cf (249)	99 Es (254)	100 Fm (253)	101 Md (256)	102 No —	103 Lr —

Atomic mass appears below the symbol, atomic number above.
Numbers in parentheses are the mass numbers of the most stable, known isotope.

4.1 PROPERTIES OF ELEMENTS

we now know that electrical properties, not the mass, of an atom control its chemical properties. But in spite of his erroneous emphasis on atomic mass, Mendeleev gave chemists a valuable tool for investigating the nature of atoms.

Mendeleev backed up his proposal with daring predictions of the properties of undiscovered elements. For example, carbon, silicon, tin, and lead, all known in 1871, have marked similarities (such as the oxide formulas CO_2, SiO_2, SnO_2, and PbO_2). Nitrogen, phosphorus, arsenic, antimony, and bismuth also form a family. Except for As, each nitrogen-family member had a known carbon-family element of slightly smaller atomic mass (C for N, Si for P, Sn for Sb, and Pb for Bi). Only As seemed to lack a carbon-family element preceding it in the atomic mass sequence. Periodic law says that As must also have a counterpart in the carbon family, and Mendeleev was so sure of his law that he predicted the properties of an undiscovered carbon-family element, which he called eka-silicon, with atomic mass about 72 amu. The discovery of germanium confirmed his predictions (except for the name). This and other successful predictions for elements corresponding to the gaps in the periodic relationship (Table 4.2) demonstrated the usefulness of the periodic law. These gaps were quickly filled, giving the modern **periodic table** its present form (Figure 4.1).

Atomic Number

There are several cases where we must reverse the sequence of atomic mass to preserve the periodic relationship. For example, Ar has a higher average atomic mass (39.9 amu) than K (39.1 amu). If we were to follow the atomic mass sequence exactly, the reactive metal K would appear in the same family as He, Ne, Kr, and

Table 4.2 **Comparison of Mendeleev's Predicted Properties with Observed Properties of Gallium, Scandium, and Germanium**

Property	Predicted	Observed
	Eka-aluminum (Ea)	Gallium[a]
Atomic mass (amu)	about 68	69.72
Density (g/cm^3)	5.9	5.93
Melting point (°C)	low	30
Formula of oxide	Ea_2O_3	Ga_2O_3
Formula of chloride	$EaCl_3$	$GaCl_3$
	Eka-boron (Eb)	Scandium[b]
Atomic mass (amu)	about 44	45.0
Formula of oxide	Eb_2O_3	Sc_2O_3
Density of oxide (g/cm^3)	3.5	3.8
Formula of chloride	$EbCl_3$	$ScCl_3$
	Eka-silicon (Es)	Germanium[c]
Atomic mass (amu)	about 72	72.59
Density (g/cm^3)	5.5	5.47
Formula of oxide	EsO_2	GeO_2
Density of oxide (g/cm^3)	4.7	4.707
Formula of chloride	$EsCl_4$	$GeCl_4$
Boiling point of chloride (°C)	less than 100	86

[a]Gallium was discovered in 1875, four years after Mendeleev's predictions about its properties. Mendeleev also correctly predicted that the element would be volatile and would be slowly attacked by air, water, acids, and bases. It was later observed that the hydroxide, $Ga(OH)_3$, is soluble in acids and bases and that $GaCl_3$ is water soluble.

[b]Scandium was discovered in 1879. Mendeleev also correctly predicted that Sc_2O_3 would be a colorless solid and that $ScCl_3$ would be less volatile than $AlCl_3$.

[c]Discovered in 1886.

Xe—all unreactive gases—and the unreactive gas Ar would lie in the alkali metal family. Mendeleev believed that such inversions indicated experimental errors in determining atomic masses. Equipped with hindsight about isotopes (Chapter 1), we know the mass of an atom is not the decisive factor in its chemical behavior. The inversions of the mass sequence are real.

The chemical behavior of an atom depends on its electrical characteristics, signified by its **atomic number**. The sequence of atomic numbers, which is also the sequence of elements in the periodic table, is almost but not exactly the same as the sequence of atomic masses. As we look at the evidence for atomic structure and develop a picture of how mass and electric charge are distributed within the atom, we will be able to assign a more specific meaning to atomic number. For now, it stands for the position of the atom in the periodic table. The modern form of the periodic law states that members of a particular family appear at predictable spots in the sequence of the elements arranged in order of their atomic numbers.

The Structure of the Periodic Table

The chemical symbols of the members of a particular family of elements lie in the same vertical column of the periodic table, and there are eight long columns representing eight major families. The alkali metals occupy the first column, and a member of the **noble gas** family—the monatomic gases helium, neon, argon, krypton, xenon, and radon—terminates each horizontal row, also called a **period**. The first period consists of two elements, hydrogen and helium. Hydrogen shows similarities both to alkali metals (Na displaces H from HCl to form H_2 and NaCl) and the halogens (hydrogen is a diatomic gas, and Cl displaces H atoms in CH_4 to form CCl_4); consequently, H stands alone with connecting lines to both families.

The second and third periods each contain eight elements. The fourth row consists of 18 elements. The first two elements of this row (potassium and calcium) fit easily into groups IA and IIA, respectively. The next element, scandium, is similar to aluminum, but so is the thirteenth element of the row, gallium. Rather than squeeze both elements into group IIIA, we start a new column (IIIB) for Sc, the notation IIIB emphasizing the similarities between Sc and the elements of the major group IIIA. The next four elements—titanium (group IVB), vanadium (VB), chromium (VIB), and manganese (VIIB)—head short columns but show progressively less similarity to their major families. The only similarities between the metal Mn and the halogens are the similarities in formulas such as Mn_2O_7 and Cl_2O_7 or $KMnO_4$, $KClO_4$, and KIO_4. The next three elements—iron, cobalt, and nickel—do not correspond to any major family, and they belong to a new group (VIII). Copper and zinc show some similarities to groups IA and IIA, and they head groups IB and IIB. Gallium and the remaining five elements of the row all fit easily into the major families. The 30 elements of groups IIIB through IIB show many similarities. Because of these similarities, we often discuss them as a single group of elements called the **transition metals**.

The fifth row repeats the sequence of groups IA and IIA, ten transition metals, and six elements of groups IIIA through 0. The sixth row is 32 elements long. The 14 additional elements (cerium through lutetium) are all very similar to lanthanum, a member of group IIIB and, along with lanthanum, comprise the **lanthanide**, or **rare earth**, series. The seventh row is still incomplete, and we may never find all the elements of this row because of the nuclear instability of heavy atoms.

Example 4.1

What elements belong to the same family as indium?

Solution

As can be seen from Figure 4.1, indium lies in the same column as boron, aluminum, gallium, and thallium (group IIIA) and therefore these elements are part of the same family.

4.2 Chemistry and Electricity

In order to understand the behavior of elements, the reasons why the periodic law works, and the significance of atomic numbers, we must investigate the internal structures of atoms. To do this, we must know something about the electrostatic forces that maintain the structures of atoms and molecules. By observing electrostatic forces acting on visible objects, scientists have characterized the laws describing electrical phenomena and applied these laws to atomic and molecular systems. This section introduces the definitions and laws of electricity and the related phenomenon of magnetism.

Charge

Electric charges can be produced by rubbing two different nonmetallic substances, such as hair and a rubber comb, together. The rubbing creates attractive electrostatic forces between the hair and the comb. In a somewhat more elaborate experiment, we can rub two pieces of glass with silk and then observe that the two pieces of glass repel each other. The existence of both attractive and repulsive electrostatic forces implies two types of charge, designated **positive** and **negative**. Objects with the same type of charge repel each other; opposite charges attract each other. A glass rod rubbed with silk carries a positive charge and repels any other object that is positively charged (including another piece of charged glass), but the rod attracts a negatively charged object.

All visible objects contain both positive and negative charges, and these opposite charges are equal in a **neutral** object. There are various ways of giving objects charges by forcing negative charge* to flow from one object to another, giving the recipient a net negative charge and the donor object a positive charge. A negative and a positive object can neutralize each other as excess negative charge flows from the negative body to the positive one. There is no creation or destruction of charge during either of these processes (charging or neutralization), just a redistribution of existing charges. This **conservation of charge** applies to chemical processes. There are many chemical reactions that involve transfer of charge between atoms, but the total charge of a system remains constant during any chemical reaction.

Electrostatic Forces

The magnitude of the electrostatic force acting between two charges is directly proportional to the magnitude of each charge and inversely proportional to the square of the distance between them. This statement, known as **Coulomb's law**, can be expressed as

$$F = K \frac{Q_1 Q_2}{r^2} \tag{4.1}$$

where F is the electrostatic force, Q_1 and Q_2 are the two charges, r is the distance separating them, and K is a constant of proportionality equal to 9.0×10^9 newton-m^2/coulomb^2. This fundamental law of electrostatics has great significance for chemistry—the force acting between two charged constituent parts of an atom is strong in proportion to the magnitudes of the two charges, and the force is much stronger if the two particles are close to each other.

---

*Until the discovery of the electron (1897), physicists had no way of telling whether positive charge flows from one object to another or whether current is a flow of negative charge. Lacking any reliable information, they adopted the arbitrary convention of drawing electric currents as movements of positive charge. We now know that this is wrong, but the old convention has persisted, particularly in practical electronics.

The unit of charge in practical electronics is the **coulomb** (abbreviated coul). However, we rarely encounter static charges as large as a coulomb because they exert tremendous electrostatic forces—two charges of 1 coul each 1 m apart exert a force of 9.0×10^9 newtons. Although charges of a coulomb or more may exist in thunderclouds, charges of such magnitude never accumulate in a beaker during a chemical reaction. (Considering the dangerous forces they would cause, it is a good thing.) A static charge of 1 coul is very dangerous, however, it is common to have several coulombs per second flowing through a wire. It is possible to work with large amounts of charge provided the charge does not accumulate in any one place.

Moving a charged object in the vicinity of another charge may involve work because there is a force acting on the object as it moves (work = force × distance). The amount of work required to move the object against a resisting electrostatic force represents the increase in **electric potential energy** of the object. If the object moves with the electrostatic force, its potential energy decreases (Figure 4.2).

In discussing atomic properties, we will be concerned with the amount of work needed to remove a negative charge from the vicinity of a positive charge. This discussion will be a lot clearer to you if you keep in mind that it is hard to remove a negative particle from a place near an intensely charged positive particle, particularly when these two charges are close together.

Magnetism

A **permanent magnet**, such as a compass needle, tends to orient itself in a north–south direction because there is a magnetic field running nearly parallel to the earth's axis of rotation (Figure 4.3). When placed near each other, two permanent

Figure 4.3
Magnetic dipoles.

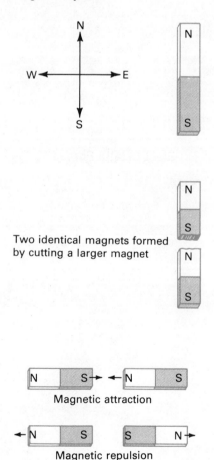

Two identical magnets formed by cutting a larger magnet

Magnetic attraction

Magnetic repulsion

Figure 4.2
Electric potential energy. (a) There are attractive forces between opposite charges. (b) Separating these charges requires work and increases potential energy. (c) If the charges move toward each other, potential energy decreases.

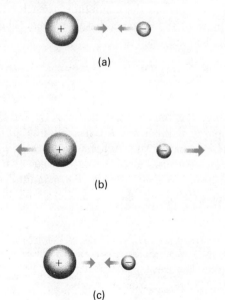

Figure 4.4
The orientation of the magnetic regions in iron. (a) The magnetic regions are in random arrangements in unmagnetized iron. (b) A magnetic field causes partial reorientation of the magnetic regions and temporary magnetism. (c) Long exposure to a magnetic field produces permanent reorientation.

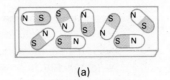

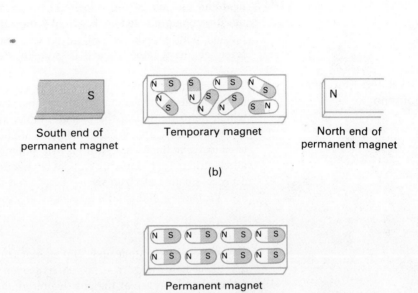

magnets align themselves so their respective magnetic fields oppose each other. The north-seeking end of one magnet attracts the south-seeking end of the other. Repulsions occur between the north-seeking ends or between the south-seeking ends of any two magnets. Magnets occur only as **magnetic dipoles** consisting of both a north-seeking pole and a south-seeking pole. No one has ever found an object with only the north-seeking force and not the south-seeking force. Cutting a bar magnet in half produces two identical smaller magnets, each with a north-seeking and a south-seeking end.

An ordinary iron bar acts as a **temporary magnet**: in the presence of a magnetic field, it acts as a magnet with the opposite orientation, but does not retain its magnetism after removal from the field. This is why a permanent magnet attracts an ordinary iron bar, and why an iron bar attracts other pieces of iron when it is near a permanent magnet. Both the permanent magnet and the ordinary iron bar contain many tiny microscopic magnetic regions, but those in the unmagnetized iron are randomly oriented so their magnetic fields cancel each other, and there is no observable net magnetism (Figure 4.4). An external magnetic field causes a small degree of alignment of the microscopic magnets within the iron bar, resulting in temporary net magnetism. In permanent magnets, the magnetic regions are aligned to produce a significant net magnetic field.

Not all substances (not even all metals) show temporary magnetism. In order to exhibit temporary magnetism (or **paramagnetism**), the atoms of a substance must be magnetic. **Diamagnetic** substances (zinc, for example) show no temporary magnetism, are weakly repelled by a magnetic field, and their atoms are not magnetic. The formation of permanent magnets (**ferromagnetism**) is a phenomenon restricted to a few paramagnetic metals and requires a metal crystal structure that can maintain an orderly arrangement of the magnetic regions.

The clue to the cause of paramagnetism is the phenomenon of **electromagnetism** (Figure 4.5). An electric current creates a magnetic field whose lines of force follow concentric circles perpendicular to the wire. This magnetic field ceases as soon as the current is switched off. A magnetic field always occurs when an electric charge moves. Because of its electromagnetic field, a charged particle moving perpendicular to an external magnetic field experiences a force that is perpendicular to both the direction of its motion and the direction of the magnetic field. The strength of this force depends on the strength of the external magnetic field and the charge and velocity of the particle. A positive charge will be forced in a direction predicted by the "right-hand rule" (Figure 4.6). The force on a negative particle points in the opposite direction, and the direction of an electromagnetic force tells us whether a particle is positive or negative.

Chemical and Electrical Phenomena

The idea that the electrical properties of atoms control their chemical behavior is as old as Dalton's theory. Many observations support this view: (1) metals, which lie to the left of the periodic table, are good electric conductors, whereas most nonmetals, appearing to the right, are poor conductors, (2) many spontaneous chemical reactions can yield electric potential energy directly (as in a battery), (3) electric potential energy can cause reactions that do not occur spontaneously, a process called **electrolysis**.

Water is a stable substance, and the reaction

$$2\ H_2O(l) \longrightarrow 2\ H_2(g)\ +\ O_2(g)$$

does not occur spontaneously. But a moderate electric potential decomposes water into its elements (Figure 4.7, p. 88), hydrogen forming at the negative terminal (the **cathode**) and oxygen at the positive terminal (the **anode**) of the electrolysis apparatus. This and other electrolysis reactions occur by the forced transfer of charge between atoms. Metals form at the cathode and their atoms become more negative as their compounds decompose into elements, while the nonmetallic atoms become more positive at the anode. A great deal of electric charge passes through the system as electrolysis occurs, but there is no destruction or creation of charge.

Figure 4.5
Schematic illustration of the electromagnetic field created by an electric current.

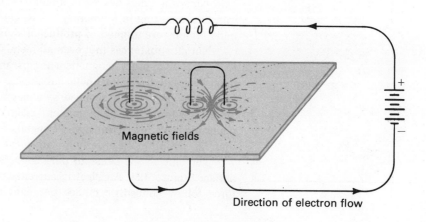

Figure 4.6
Electromagnetic forces.

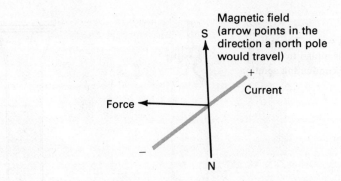

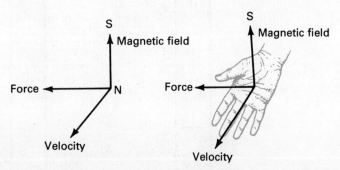

Right-hand rule for a positive charge moving through a magnetic field

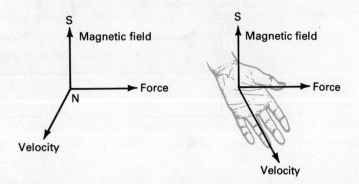

Left-hand rule for a negative charge moving through a magnetic field

Current flows out of the apparatus at one electrode as fast as it flows in at the other.

In the middle of the nineteenth century, Michael Faraday showed experimentally that the charge passing through an electrolysis has a stoichiometric relationship to the amount of reaction, and he formulated what is now known as **Faraday's law**: The charge needed to produce 1 mol of atoms of any element is a whole-number multiple of 9.65×10^4 coul [9.65×10^4 coul = 1 **faraday** ($\mathscr{F}$)]. The whole-number multiplier (n) depends on the formula of the compound being electrolyzed (Table 4.3, p. 88).

$$\frac{\text{Charge}}{\text{Mole of atoms}} = n\mathscr{F} = n(9.65 \times 10^4 \text{ coul}) \tag{4.2}$$

Example 4.2

How much charge is needed to produce 5.0 g of lead from $PbCl_2$?

Solution

Atomic mass Pb = 207 amu

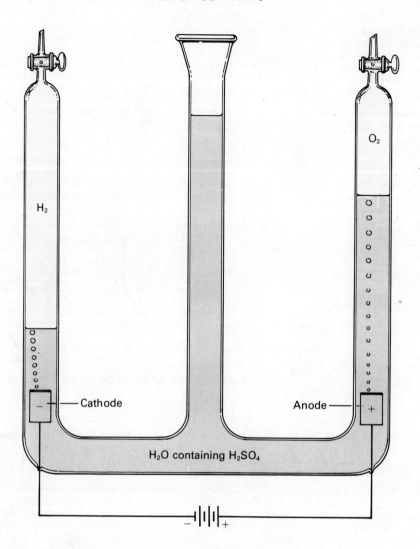

Figure 4.7

Electrolysis of water. The sulfuric acid (H_2SO_4) is added to produce an electrically conducting solution.

Table 4.3 Faraday's Law

Element	Source	Charge/mole of atoms (coulomb)	n
Formed at cathode			
H	HCl	9.65×10^4	1
Na	NaCl	9.65×10^4	1
K	KCl	9.65×10^4	1
Mg	$MgCl_2$	1.93×10^5	2
Ca	$CaCl_2$	1.93×10^5	2
Al	$AlCl_3$	2.90×10^5	3
Fe	$FeCl_2$	1.93×10^5	2
Fe	$FeCl_3$	2.90×10^5	3
Pb	$PbCl_2$	1.93×10^5	2
Pb	PbO	1.93×10^5	2
Pb	PbO_2	3.86×10^5	4
Formed at anode			
Cl	NaCl	9.65×10^4	1
O	H_2O	1.93×10^5	2
O	H_2O_2	9.65×10^4	1
H	NaH	9.65×10^4	1

$$\text{Moles Pb} = \frac{5.0 \text{ g}}{207 \text{ g/mol}} = 2.4 \times 10^{-2} \text{ mol}$$

According to Table 4.3, each mole of Pb requires $2\,\mathscr{F}$ when $PbCl_2$ is electrolyzed.

$$\text{Faradays required} = \left(2\,\frac{\mathscr{F}}{\text{mol}}\right)(2.4 \times 10^{-2} \text{ mol}) = 4.8 \times 10^{-2}\,\mathscr{F}$$

$$\text{Charge required} = (4.8 \times 10^{-2}\,\mathscr{F})\left(9.65 \times 10^4\,\frac{\text{coul}}{\mathscr{F}}\right) = 4.7 \times 10^3 \text{ coul}$$

Faraday's law lends itself to a quantized view of electrochemical charge—n is the number of charge quanta (called **electrons**) transferred per atom in an electrolysis. One mole of electrons has a charge of 9.65×10^4 coul. The electrolysis of water involves the transfer of one electron for each H atom and two electrons for each O atom. The number of electrons involved depends on the reaction, not just the product. For example, the electrolysis of $FeCl_3$ requires three electrons per iron atom, but only two electrons per iron atom are required in the electrolysis of $FeCl_2$. The electric charge of an atom is fundamentally related to the state of chemical combination of the element.

4.3
The Electron

A **cathode-ray tube** (Figure 4.8) consists of an evacuated tube with a large electric potential across it. A tube containing air is a poor conductor, but current flows steadily through a vacuum under the influence of a potential. Invisible rays ema-

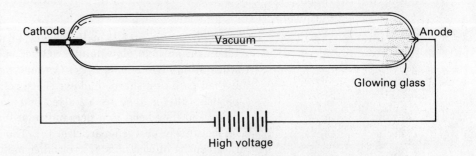

Figure 4.8
Cathode-ray tubes. Cathode rays travel in straight lines from the cathode.

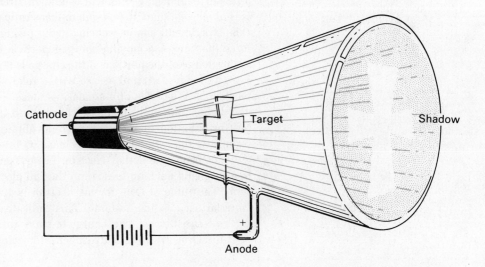

Figure 4.9
Cathode-ray tube for measuring charge-to-mass ratio of the electron.

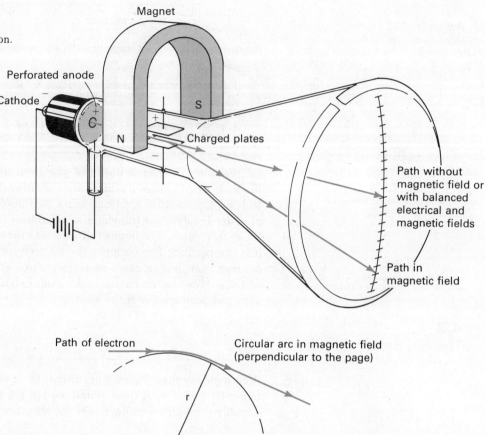

nating from the cathode (**cathode rays**) strike the glass at the opposite end of the tube causing it to glow faintly. (Other substances, such as ZnS, glow brightly when struck by cathode rays.) Cathode rays travel in straight lines and objects in their path cast sharp shadows. They behave like streams of negatively charged particles; that is, they are attracted by a positive plate and repelled by a negative plate, and they bend in a perpendicular magnetic field in the direction expected for negatively charged particles. J. J. Thomson, in 1897, proposed that cathode rays consist of electrons, free of other matter.

By measuring the degree of curvature of a beam of electrons moving through a perpendicular magnetic field of known strength at a particular speed, Thomson was able to estimate the **charge-to-mass ratio** of the electron. The principle behind this experiment, shown schematically in Figure 4.9, is that the electromagnetic force that makes a moving charged particle follow a circular arc is proportional to the charge of the particle. If the charge is large, the particle is subject to a strong force and moves around an arc with a relatively small radius. But the path of the particle also depends on its mass—a massive particle is not accelerated as much as a lighter one with the same charge, and its path will not curve as much. The best information that Thomson could obtain from the curvature of the electron beam was the *ratio* of charge to mass (**Q**/m) of the electron.

Thomson found that electron beams keep their sharp focus when bending in a magnetic field, an indication that all electrons have the same charge-to-mass ratio. The value of **Q**/m for an electron is 1.76×10^{11} coul/kg, regardless of the material used in the cathode. This indicates that electrons are the same in all sorts of substances. The charge-to-mass ratio is much larger than the highest possible **Q**/m ratio for a hydrogen atom since it takes only 9.58×10^7 coul/kg

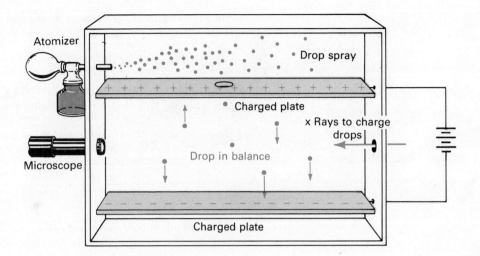

Figure 4.10
Apparatus used in Millikan's charged-drop experiment.

to form H_2 by electrolysis. The intense charge of the electron convinced Thomson that electrons are small fragments of atoms.

The work of Robert A. Millikan (around 1910) confirmed this controversial idea. Millikan used an apparatus shown schematically in Figure 4.10 to measure the charges produced when x-rays struck small drops of liquid that had been sprayed between two charged plates. He adjusted the electric charges on the two plates until a particular drop hung suspended, neither rising nor falling. The drop remained suspended because an upward electric force, which depended on the charge on the drop, just equaled the weight of the drop. Millikan could calculate the charge on the drop from a knowledge of the charges on the plates and the weight of the drop. He found that the charge of a drop is always a whole-number multiple of -1.60×10^{-19} coul (Table 4.4). His explanation was that each drop carries a particular number of extra electrons that give the observed charge,

$$Q = ne \qquad (4.3)$$

where n is the number of extra electrons and e is the charge of 1 electron $(-1.60 \times 10^{-9}$ coul). Experiments with drops of oil, water, glycerol, and mercury give the same result indicating that all electrons have the same charge independent of the matter with which they are associated.

If the electrons studied by Thomson and Millikan are the same as the charge

Table 4.4 Data from Millikan's Experiment

Drop number	Charge (coulomb)	Estimated number of electrons on the drop	Estimated e (coulomb)
1	4.62×10^{-19}	3	1.54×10^{-19}
2	6.08×10^{-19}	4	1.52×10^{-19}
3	3.09×10^{-19}	2	1.55×10^{-19}
4	8.05×10^{-19}	5	1.60×10^{-19}
5	3.25×10^{-19}	2	1.63×10^{-19}
6	9.38×10^{-19}	6	1.56×10^{-19}
		Mean value =	1.57×10^{-19} [a]

[a] Later experimental refinements led to the current estimate of 1.602×10^{-19} coul.
Source: Calculated from data in R. A. Millikan, *Philosophical Magazine*, 6th Ser. **19**, 223 (1910).

quanta involved in electrolysis, $1\,\mathscr{F}$ should contain Avogadro's number of electrons (N_{Avog} electrons react with water to produce N_{Avog} H atoms, or 1 mol H).

$$N_{Avog} \text{ electrons} = 1\,\mathscr{F} \tag{4.4}$$

Since each electron has a charge of 1.60×10^{-19} coul and $1\,\mathscr{F} = 9.65 \times 10^4$ coul, we can write an equation for N_{Avog} using the common unit of coulombs.

$$(N_{Avog})(1.60 \times 10^{-19} \text{ coul/electron}) = 9.65 \times 10^4 \quad \text{coul/faraday}$$

$$N_{Avog} = \frac{9.65 \times 10^4 \text{ coul/faraday}}{1.60 \times 10^{-19} \text{ coul/electron}}$$

$$= 6.02 \times 10^{23} \text{ electrons/faraday}$$

This result agrees with an earlier estimate of Avogadro's number (by J. Perrin) based on the statistical distribution of particles in a colloidal suspension. The agreement confirms that the electrons measured by Millikan are identical with electrons transferred in an electrolysis.

Knowing the charge of an electron, its charge-to-mass ratio, and Avogadro's number, we can compare the mass of an electron with that of a hydrogen atom.

$$\text{Mass of electron} = \frac{1.60 \times 10^{-19} \text{ coul}}{1.76 \times 10^{11} \text{ coul/kg}} = 9.1 \times 10^{-31} \text{ kg} \tag{4.5}$$

$$\text{Mass of H atom} = \frac{1.008 \text{ g/mol}}{6.023 \times 10^{23} \text{ atoms/mol}}$$

$$= 1.67 \times 10^{-24} \text{ g/atom} = 1.67 \times 10^{-27} \text{ kg/atom} \tag{4.6}$$

The electron, much less massive than the smallest atom, is definitely a subatomic fragment.

4.4 The Nucleus

Rutherford's Experiments

The positive charge in a neutral atom must equal the total negative charge of the electrons in the atom. To determine how these charges and the mass of an atom are distributed, Ernest Rutherford, in 1910, aimed narrow beams of alpha particles at thin metal foils and observed the degree of deflection of the alpha particles.

An alpha particle is a positively charged product of certain nuclear reactions.

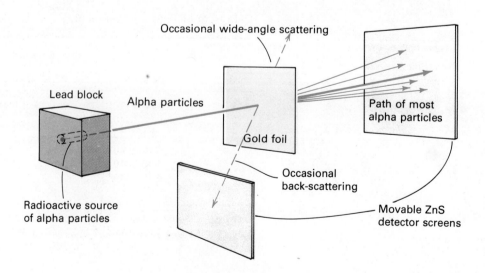

Figure 4.11
Rutherford's experiment on the scattering of alpha particles.

Figure 4.12
Early models of atomic structure.

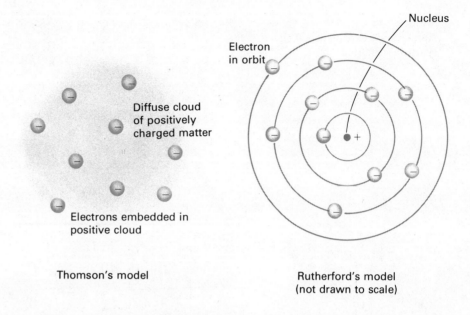

Thomson's model

Rutherford's model
(not drawn to scale)

Its mass is 4 amu, and the magnitude of its charge is twice that of the electron (+2). When an alpha particle strikes a ZnS screen, it causes a visible flash on the screen. By observing ZnS screens at different positions around the metal foil (Figure 4.11), Rutherford determined the paths of individual alpha particles that had struck the foil. His early experiments, which showed that most alpha particles pass through the screen with little or no deflection, seemed to support Thomson's model for atomic structure (Figure 4.12). In Thomson's model the mass and positive charge of the atom are spread evenly in a thin cloud throughout the atom, with electrons embedded in this diffuse cloud of positive matter. Such a model would explain the apparently poor stopping power of thin foils against alpha particles. An electron, being so much less massive than an alpha particle, would not offer much resistance to it.

A meticulous experimenter, Rutherford had an associate look for scattering through wide angles, although the slight degree of scattering for most alpha particles seemed to make such an event highly unlikely. Wide-angle scattering and even cases where alpha particles bounce back from the foil did occur for a very small percentage of alpha particles. Since alpha particles are much more massive than electrons, Thomson's diffuse atoms could not be capable of ever bouncing an alpha particle backward. (Rutherford described the event of back-scattering as "almost as incredible as if you fired a 15-inch shell at a piece of tissue paper and it came back and hit you.")

Rutherford's model of the atom (Figure 4.12) explains why wide-angle scattering occurs and why it is so rare. In this model, all of the positive charge and nearly all of the mass are concentrated in a small, dense, central region of the atom called the **nucleus**. The radius of an atom is more than 10^4 times as large as the nuclear radius. The rest of the space in the model is empty except for electrons moving in well-defined orbits around the nucleus.

A sheet of atoms seen from the viewpoint of an approaching alpha particle looks like nearly empty space studded with widely separated nuclei. Most alpha particles pass through a thin foil without coming near a nucleus, and they experience little or no net force from the nuclei. Occasionally an alpha particle happens to pass near a nucleus that, because of its positive charge, strongly repels the alpha particle. Near misses cause wide-angle scattering. The even rarer occurrence of a head-on collision causes back-scattering (Figure 4.13). (Because of the strong repulsive force between a nucleus and an alpha particle, the alpha particle bounces away before it contacts the nucleus.)

Figure 4.13
Rutherford's explanation of alpha-particle scattering.

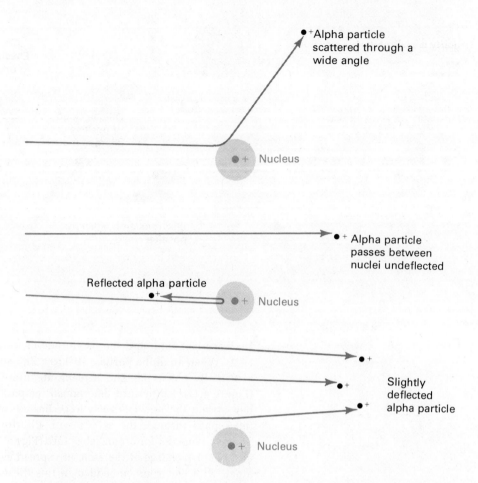

According to Rutherford's explanation, nuclei with greater charge should scatter alpha particles more effectively. The scattering capabilities of elements increase with their atomic numbers, indicating a correlation between nuclear charge and atomic number. H.G.J. Moseley, a co-worker of Rutherford's, proposed in 1914 that the atomic number of an element (Z) is its nuclear charge, expressed in terms of positive electronic charges.

$$\text{Nuclear charge} = Z(+1.60 \times 10^{-19} \text{ coulomb}) \tag{4.7}$$

All atoms of a particular element have the same nuclear charge, and a neutral atom contains Z electrons. Nuclear charge is quantized.

Nuclear charge determines the chemical identity of an atom. Chemical reactions involve changes in the arrangements of electrons and may alter the total charge of the atom, but they do not change the nucleus. Nuclear reactions, which change nuclear charge and transform one element into another, involve much greater energy changes than do distortions in electronic arrangements.

Isotopes and Nuclear Mass

An electron beam passing through a gas at low pressure produces positively charged atoms (positive ions) by colliding with electrons in the atoms and knocking them free of the electrostatic attraction of the nucleus. These positive ions accelerate toward a cathode and their paths bend in a magnetic field. We can calculate the charge-to-mass ratio of an ion by its deflection in a magnetic field. When He^+ ions (He atoms lacking an electron each) formed from a natural sample of He pass through a magnetic field, most bend along a path consistent with the charge-to-mass ratio we would expect for an atom with a mass of 4.0 amu, but a few atoms bend to a greater degree, indicating an atomic mass of 3.0 amu.

Figure 4.14
Schematic diagram of a mass spectrometer.

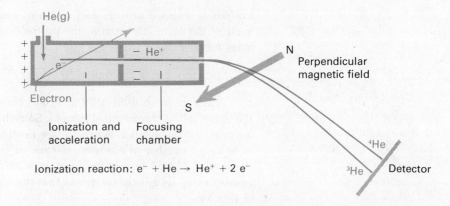

This experiment shows that natural helium consists of two different isotopes, ^{4}He (the most abundant) and the minor isotope, ^{3}He.

Isotopes have the same nuclear charge but differ in mass. The **mass spectrometer** (Figure 4.14) measures the magnetic deflection of ions, giving information about the masses and abundances of isotopes.

Protons and Neutrons

The exact atomic masses, in atomic mass units, of individual isotopes are very close to whole numbers (Table 4.5). This discovery revived **Prout's hypothesis** (1816): Atoms are constructed of simpler units. (Prout believed these units were hydrogen atoms.) The average atomic mass of Cl (35.5 amu) seemed to contradict Prout's hypothesis, but mass spectrometry shows that natural Cl is 75.4% ^{35}Cl and 24.6% ^{37}Cl.

The particles that comprise nuclei cannot all be alike in both mass and charge, or the charge-to-mass ratios of all nuclei would be the same, and isotopes would not exist. We believe that two different particles, the **proton** and the **neutron,** make up the nucleus. The proton is identical to the ^{1}H$^+$ ion; it has a mass of 1.0073 amu and a +1 charge. The neutron has nearly the same mass (1.0087 amu) but it is neutral.

The number of protons in a nucleus equals the atomic number and is the same for all atoms of an element. Isotopes differ in the number of neutrons present in their nuclei. For example, the ^{1}H nucleus is a proton, and the ^{2}H (deuterium)

Table 4.5   Nuclear Composition of Isotopes

Isotope	Protons	Neutrons	Atomic mass (amu)
^{1}H	1	0	1.007825
^{2}H	1	1	2.01410
^{3}He	2	1	3.01603
^{4}He	2	2	4.00260
^{12}C	6	6	12.00000
^{13}C	6	7	13.00335
^{14}N	7	7	14.00307
^{15}N	7	8	15.00011
^{16}O	8	8	15.99491
^{17}O	8	9	16.99914
^{18}O	8	10	17.99916
^{35}Cl	17	18	34.96885
^{37}Cl	17	20	36.96590
^{36}Ar	18	18	35.96755
^{40}Ar	18	22	39.96238
^{39}K	19	20	38.96371

nucleus contains a neutron and a proton, each of which contribute 1 amu to the mass of the atom. The sum of the protons and neutrons in a nucleus equals the **mass number** of an atom.

$$\text{Mass number} = \text{Protons} + \text{Neutrons} \tag{4.8}$$

When we wish to identify a certain isotope, we display its mass number at the upper left of the chemical symbol. Some books print the atomic number at the lower left of the symbol. When we are dealing with ions, we indicate the charge, in terms of number of electron charges, at the upper right of the symbol. Thus, the symbol $^{94}_{40}Zr^{2+}$ stands for an atom with an atomic number of 40 and a mass number of 94. Its 2+ charge means that there are two fewer electrons than protons in this ion.

Example 4.3 How many protons, neutrons, and electrons are present in ^{101}Rh?

Solution The atomic number of Rh is 45, so it has 45 protons. Since it is a neutral atom, the number of electrons equals the number of protons—45 in this case. To find the number of neutrons, we subtract the atomic number from the mass number.

$$\text{Neutrons} = 101 - 45 = 56$$

Example 4.4 How many protons, neutrons, and electrons does $^{37}K^+$ contain?

Solution The atomic number of K is 19, so there are 19 protons.

$$\text{Neutrons} = 37 - 19 = 18$$

Since the atom loses one electron to attain the 1+ charge,

$$\text{Electrons} = \text{Protons} - 1 = 19 - 1 = 18$$

Example 4.5 How many protons, neutrons, and electrons are in $^{127}I^-$?

Solution
$$\text{Protons} = Z = 53$$
$$\text{Neutrons} = 127 - 53 = 74$$

Since the atom gains an extra electron to attain its 1− charge,

$$\text{Electrons} = 53 + 1 = 54$$

Summary

1. Elements belonging to the same chemical family exhibit similar chemical properties. Periodic law: Elements of a particular family appear at predictable places in a sequence of all the elements arranged according to their atomic numbers. The members of a family occupy the same column of the periodic table.

2. There are two types of electric charge, positive and negative. Opposite charges attract each other; like charges repel. Positive and negative charges are balanced in a visible neutral object. A chemical reaction cannot create or destroy charge.

3. Two magnets align themselves so that their magnetic fields point in opposite directions. Paramagnetic substances contain microscopic magnetic dipoles and are attracted by magnetic fields. Diamagnetic substances lack magnetic dipoles and are weakly repelled by magnetic fields. Ferromagnetic metals can form permanent magnets.

4. A moving charged particle creates a magnetic field. A charged particle moving through a perpendicular magnetic field is subject to an electromagnetic force whose strength is proportional to the charge and velocity of the particle and the strength of the magnetic field. The direction of this force indicates the sign of the charge on the particle.

5. Faraday's law: The charge needed to produce a mole of atoms of an element by electrolysis is a whole-number multiple of 9.65×10^4 coul. The whole number represents the number of electrons transferred per atom, and 9.65×10^4 coul is the charge of a mole of electrons.

6. An electron is a subatomic particle whose charge is -1.60×10^{-19} coul and whose mass is 9.1×10^{-31} kg.

7. Most of the mass and all of the positive charge of an atom is concentrated in a very small, dense nucleus. Most of the atom is empty space in which the electrons move. The atomic number of an element equals its nuclear charge and the number of electrons in the neutral atom. Nuclear charge determines the chemical identity of an atom.

8. A mass spectrometer measures the masses and abundance of isotopes of an element by measuring the degree of deflection of positively charged ions in a magnetic field.

9. There are two types of constituent particles in a nucleus, protons and neutrons. A proton has a +1 charge and a mass of 1 amu. A neutron is a neutral particle with a mass of 1 amu. The atomic number of an atom equals the number of protons in its nucleus. The mass number of an atom is the sum of the protons and neutrons in its nucleus. Isotopes have the same number of protons but different numbers of neutrons.

Vocabulary List

alkali metal	coulomb	Faraday's law
halogen	electric potential	faraday
periodic table	permanent magnet	electron
atomic number	magnetic dipoles	cathode-ray tube
noble gas	temporary magnet	cathode ray
period	paramagnetism	charge-to-mass ratio
transition metals	diamagnetic	nucleus
lanthanide (rare earth) series	ferromagnetism	mass spectrometer
positive charge	electromagnetism	proton
negative charge	electrolysis	neutron
conservation of charge	cathode	mass number
Coulomb's law	anode	

Exercises

1. Which elements are family members of selenium?

2. Which of the following pairs of elements are members of the same family?
   a. Ar and Cs **b.** Si and P **c.** N and As **d.** B and Tl **e.** K and Sr
   f. V and Ta

3. Which of the following elements are transition metals?
   a. Sn **b.** Ru **c.** Mg **d.** Cr **e.** U **f.** Pt

4. Which of the following elements is a lanthanide?
   a. Pm **b.** Ca **c.** W **d.** Ho **e.** Th **f.** Kr

5. Which of the following is a noble gas?
   a. Cl **b.** He **c.** O **d.** Sm **e.** N **f.** CO **g.** Rn

6. We often see electrostatic attraction between two objects that have been rubbed together, but we never see repulsion between two objects rubbed together. Explain this in terms of the nature of neutral objects and the conservation of charge.

7. Is the statement, A neutral piece of metal contains no charges, true? If not, correct this statement.

8. Explain why it takes twice as much charge to produce 1 mol of copper from a solution of $Cu(NO_3)_2$ as is needed to produce 1 mol of silver from $AgNO_3$ solution.

9. How many faradays are present in 500 coul?

10. How many coulombs make up $0.0350\,\mathscr{F}$?

11. Why does the existence of isotopes show that mass is not a major factor in determining the chemical behavior of an atom?

12. Explain why the path of a heavy isotope bends less than the path of a light one in a mass spectrometer.

13. Why is the following process impossible?

    $Ag^+ + 2\,Cl^- \longrightarrow AgCl_2$

14. What is wrong with the following equation?

    $(NH_4)_2SO_4(s) \longrightarrow 2\,NH_4^+(aq) + SO_4^-(aq)$

15. An atom contains 59 protons, 82 neutrons, and 56 electrons. What is its atomic number?

16. An atom contains 51 protons, 72 neutrons, and 54 electrons. What is its mass number?

17. An atom contains 58 protons, 82 neutrons, and 54 electrons. What is the charge on this atom?

18. An alpha particle consists of two protons, two neutrons, and no electrons. Write a chemical symbol for an alpha particle showing the atomic symbol for the element, the mass number, and the charge.

19. How many protons, neutrons, and electrons are present in each?
    a. ^{98}Tc **b.** $^{167}Lu^{3+}$ **c.** ^{208}At

20. How many protons, neutrons, and electrons are present in each?
    a. $^{45}Sc^{3+}$ **b.** $^{106}Pd^{2+}$ **c.** ^{174}Hf

21. How many protons, neutrons, and electrons are present in each?
    a. $^{32}S^{2-}$ **b.** $^{15}N^{3-}$ **c.** $^{81}Br^-$

22. How many protons, neutrons, and electrons are present in each?
    a. $^{37}Cl^-$ **b.** ^{238}U **c.** $^{40}Ca^{2+}$

23. From the data that follow, determine which elements (W, X, Y, or Z) belong to the same family.

    W: atomic mass = 24.31 amu; melting point = 651°C; boiling point = 1107°C. This element corrodes rapidly in water. It forms the following compounds: WBr_2 (melting point = 700°C; very soluble in water); WCO_3 (slightly soluble in water; very soluble in acid); WO (melting point = 2800°C; insoluble in water; very soluble in acid).

    X: atomic mass = 127.6 amu; melting point = 450°C; boiling point = 990°C. This element does not corrode in water but reacts with H_2SO_4 or KOH solution. It forms the following compounds: XBr_2 (melting point = 210°C; boiling point = 339°C; decomposes in water); XBr_4 (melting point = 380°C; slightly soluble in water); XO decomposes below 400°C without melting; insoluble in water; slightly soluble in acids or in KOH solution).

    Y: atomic mass = 137.4 amu; melting point = 725°C; boiling point = 1140°C. Element Y corrodes rapidly in water. It forms the following compounds: YBr_2 (melting point = 847°C; very soluble in water); YCO_3 (insoluble in water; very soluble in acid); YO (melting point = 1923°C; slightly soluble in water; very soluble in acid).

    Z: atomic mass = 26.98 amu; melting point = 660°C; boiling point = 2467°C. Element Z corrodes slowly in water, rapidly in acids or in KOH solution. It forms the following compounds: ZBr_3 (melting point = 93°C; boiling point = 263°C; reacts violently with water; soluble in CS_2); Z_2O_3 (melting point = 2045°C; insoluble in water; soluble in acids or KOH solution).

24. The periods of the periodic table are 2, 8, 8, 18, 18, and 32 elements long. Without looking at a periodic table, place the following elements in their proper groups.
    a. Element A, atomic mass = 32.06 amu, atomic number = 16
    b. Element B, atomic mass = 87.62 amu, atomic number = 38
    c. Element C, atomic mass = 121.75 amu, atomic number = 51
    d. Element D, atomic mass = 52.0 amu, atomic number = 24

25. Without looking at a periodic table, write the atomic numbers of the elements of group IVA.

26. An atom contains 25 protons, 30 neutrons, and 23 electrons. Without looking at a periodic table, locate the position of this atom in the periodic table.

27. When Pierre and Marie Curie were isolating radium from pitchblende, a uranium ore, they noted that radium salts tend to precipitate along with barium salts. This observation led them to postulate, even before the element's final purification, that radium is a member of group IIA and has an atomic mass of about 225 amu. Reconstruct the logic behind their postulate.

28. At the end of the nineteenth century, Lord Rayleigh and Sir William Ramsay discovered the element argon, an unreactive monatomic gas. Rayleigh and Ramsay immediately set about searching for other unreactive monatomic gases. Why should they have any confidence in the belief that these other gases exist?

29. How much charge, in coulombs, must pass through an electrolysis apparatus to form 5.0 g of barium from $BaCl_2$?

30. Electrolysis of a solution of $Cr_2(SO_4)_3$ by a total charge of 900 coul produces 0.162 g of chromium at the cathode. How many electrons are transferred per chromium atom?

31. How would electron beams in J. J. Thomson's experiment behave if all electrons in a particular substance have identical charges but the substance contains two types of electrons, heavy and light?

32. How would electron beams behave in J. J. Thomson's experiment if all electrons have the same mass but there are two types of electrons, one with a strong negative charge and one with a weak negative charge?

33. Suppose that electrons differ in both mass and charge but all have the same charge-to-mass ratio. How would electron beams behave in J. J. Thomson's experiment?

34. Suppose each metallic element possesses its own type of electron with characteristic mass and charge properties. What would J. J. Thomson have observed in this case?

35. Using Millikan's apparatus, we measure the charge on an oil drop to be 7.95×10^{-19} coul. How many electrons does the drop have? (Allow for some experimental error.)

36. What experimental result would Millikan have obtained if the electrons of mercury had twice the electric charge of those of oil?

37. a. What is the mass in grams of an average aluminum atom?
    b. What is the charge in coulombs of an Al^{3+} ion?
    c. What is the charge-to-mass ratio of an average Al^{3+} ion?
    d. Compare your answer for part c with the charge-to-mass ratio for the formation of aluminum by electrolysis of $AlCl_3$. (See Table 4.3.)

38. Why should atoms with higher atomic numbers scatter alpha particles more effectively?

39. a. How many grams are there in an atomic mass unit?
    b. What is the mass in grams of an ^{16}O atom? (Answer to two significant figures.)
    c. What is the nuclear charge in coulombs of an ^{16}O atom?
    d. What is the charge-to-mass ratio of an $^{16}O^+$ ion?

40. One case of the sequence of atomic numbers not matching the sequence of atomic masses involves the elements Te and I. Look up the properties of these elements and their compounds and compare them to other elements of periodic groups VIA and VIIA to show that following the sequence of atomic masses would put these elements in the wrong groups. (The *Handbook of Chemistry and Physics*, published by the Chemical Rubber Co. contains a table of properties of elements and their compounds.)

41. The force on two charged particles, one positive and one negative, 2×10^{-10} m apart is 1.2×10^{-8} newton.
    a. Is this force attractive or repulsive?
    b. What will the force be if the particles are 4×10^{-10} m apart?
    c. What will the force be if the charge of one of the particles is doubled?
    d. Will the potential energy increase, decrease, or stay the same as the distance between the particles is increased?

chapter 5

Quantum Mechanics

Cross-section of a 2s orbital, one of the wave functions of the hydrogen atom

The early twentieth century witnessed a fundamental revolution in the way physical scientists look at natural phenomena. This revolution allowed scientists to reexamine their fundamental hypotheses and, by doing so, they were able to see previously unsuspected relationships between such diverse phenomena as light and chemical behavior. Chemists used the ideas of the new theory called **quantum mechanics** to develop our current ideas about atomic and molecular structures. In this chapter we will present the basic ideas of the quantum mechanical description of physical systems and how they apply to atomic structure.

5.1 The Nature of Light

Light is a unique form of energy that can travel through empty space. (Sound, in contrast, cannot be transmitted in the absence of matter.) There are many different types of light—visible light of various colors, the infrared light emitted by hot objects, and the highly penetrating x rays, just to name a few. All varieties of light move through space with the same speed (3.00×10^8 m/sec) and can change into other energy forms upon contact with matter. Under the right circumstances, an object can emit some of its energy in the form of light.

Wave Phenomena

At the beginning of the eighteenth century there was an apparent conflict between two theories describing the nature of light. Newton's followers believed that a light beam is a stream of small particles, while the disciples of Huygens (a Dutch physicist) held that light was a periodic wave disturbance. The wave theory

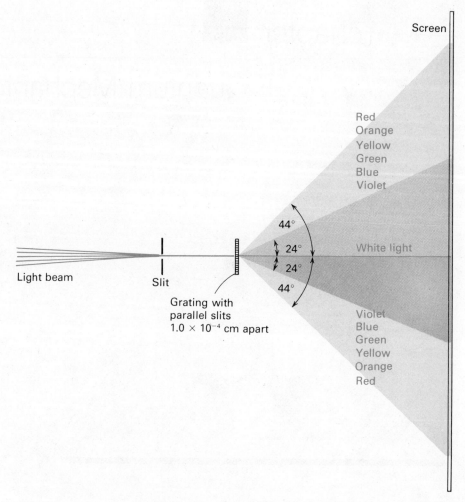

Figure 5.1
Separation of light into a spectrum using a diffraction grating.

gained a temporary dominance over the particle theory because it could account for the **interference phenomena** of light.

An example of interference occurs when a beam of monochromatic light (light of one color) passes through two parallel slits and then strikes a screen. If the slits are far enough apart, images of the two slits appear on the screen, but if the slits are closer (about 10^{-4} cm apart) the passage of light through them produces a **diffraction pattern** on the screen (Figure 5.1). This pattern consists of a band of light in line with the original beam and successively fainter parallel bands to either side of the central band. The positions of the bands depend on the color of the light and the distance between the slits. Both violet and red light produce diffraction patterns consisting of parallel bands, but the spacings of the violet bands are narrower than those between the red bands. White light, a mixture of all the visible colors, can be separated into a spectrum of its colors by this process.

Diffraction phenomena are a consequence of the properties of waves. For example, water waves exhibit similar interference patterns when passed through slits. When a regularly spaced series of water waves passes through two narrow openings whose separation approximates the distance between two adjacent wave peaks (called the **wavelength**), some regions beyond the openings are in violent wave motion, and other regions are calm (Figure 5.2). The largest waves form in the same direction as the original waves, and in bands at specific angles to the right and left. The angles of wave reinforcement depend on the distance between the openings and the wavelength.

Waves are periodically repeating disturbances. Water waves are periodic disturbances in the water level, the level at any point rising and falling in a well-

Figure 5.2
Diffraction of water waves.
Source: Reproduced courtesy Education Development Center, Newton, MA 02160.

defined time cycle (Figure 5.3). The wave crests appear to move in a definite direction at a certain velocity and the number of wave cycles occurring at any point in a unit of time, called the **frequency**, v (nu), is related to the wavelength, λ (lambda), and velocity, v, of the wave.

$$v = \lambda v \tag{5.1}$$

Interference between water waves after they emanate from the narrow openings results in the diffraction phenomenon. Two waves undergo **constructive interference** when their crests appear at the same point at the same time, and the result of these two *in-phase* waves is a wave disturbance twice as large. When two waves are *out of phase*, the crest of one occurs at the same place and time as the trough of another, and the two waves cancel each other. This **destructive interference** produces a calm region (Figure 5.4).

Constructive interference in water-wave diffraction occurs at points whose distances from the two openings differ by exactly one wavelength (Figure 5.5), or a whole-number multiple of the wavelength ($n\lambda$). These points form straight lines at angles that can be predicted by trigonometry from the distance between the openings and the wavelength. Knowing the distance between the openings and the angles of diffraction, we can estimate the wavelength. The calm regions

Figure 5.3
Waves rise and fall in well-defined time cycles and the distance between peaks is the wavelength.

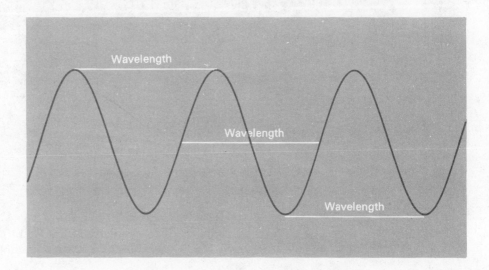

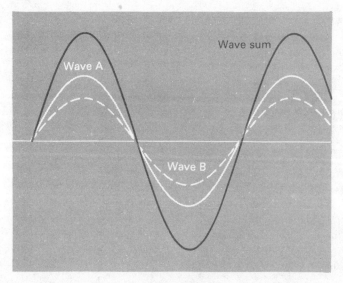
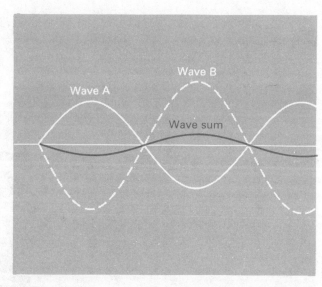

(a) Constructive interference—waves are in phase

(b) Destructive interference—waves are out of phase

Figure 5.4
Wave interference.

Figure 5.5
Interference in diffraction.
(a) Constructive interference occurs at 1λ or $n\lambda$. (b) Destructive interference occurs at odd-numbered multiples of $\lambda/2$.

in the diffraction pattern are composed of points whose distances from the two openings differ by half a wavelength, or an odd-number multiple of $\lambda/2$. Destructive interference between the two waves occurs at these points.

Water-wave diffraction serves as a model for the diffraction of light. By measuring the reinforcement angles for a known slit separation, we can assign a wavelength to each type of light. Although water waves can travel at various speeds, light always travels through a vacuum at a velocity (c) of 3.00×10^8 m/sec. (The speed of light through any medium, such as air, depends on the medium and the wavelength.) The relationship between the frequency of light and its wavelength is given by the equation

$$\nu = \frac{c}{\lambda} = \frac{3.00 \times 10^8 \text{ m/sec}}{\lambda} \tag{5.2}$$

The wavelength, λ, is measured in meters and ν (frequency) in sec^{-1} (also called hertz, abbreviated Hz). The various forms of light differ in their wavelengths and frequencies. Table 5.1 shows the approximate ranges of wavelengths in nano-

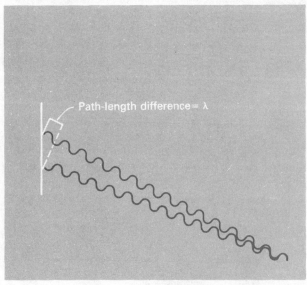
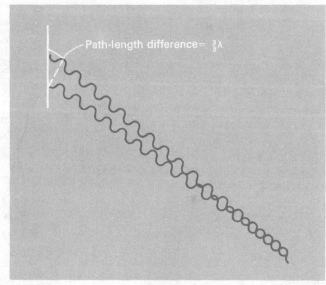

(a) Wave reinforcement

(b) Wave cancellation

5.1 THE NATURE OF LIGHT

Table 5.1 Forms of Light

Form of light	Wavelength (nm)	Frequency (sec^{-1})	Energy per photon (J)
Gamma rays	< 0.01	> 3 × 10^{19}	> 2 × 10^{-14}
x Rays	0.01–1	3 × 10^{19}–3 × 10^{17}	2 × 10^{-14}–2 × 10^{-16}
Ultraviolet	1–400	3 × 10^{17}–7.5 × 10^{14}	2 × 10^{-16}–5 × 10^{-19}
Visible (violet)	400	7.5 × 10^{14}	5 × 10^{-19}
Visible (red)	700	4.3 × 10^{14}	2.8 × 10^{-19}
Infrared	700–5 × 10^6	4.3 × 10^{14}–6 × 10^{10}	2.8 × 10^{-19}–4 × 10^{-23}
Radio waves	> 5 × 10^6	< 6 × 10^{10}	< 4 × 10^{-23}

meters (1 nm = 10^{-9} m) and the frequencies in sec^{-1} for the various forms of light.

Example 5.1

What is the frequency of light whose wavelength is 590 nm?

Solution

Our answer for frequency will not be very useful unless we use consistent units for the speed of light and the wavelength. Consequently, we convert the wavelength from nm to m.

$$\lambda = 590 \text{ nm} \left(10^{-9} \frac{\text{m}}{\text{nm}}\right) = 5.90 \times 10^{-7} \text{ m}$$

Frequency is the speed of light divided by its wavelength.

$$\nu = \frac{c}{\lambda} = \frac{3.00 \times 10^8 \text{ m/sec}}{5.90 \times 10^{-7} \text{ m}} = 5.08 \times 10^{14} \text{ sec}^{-1}$$

Quantum Theory

The success of the wave model in explaining the diffraction of light seemed complete, but toward the end of the nineteenth century and in the beginning of the twentieth century experimental observations that could not be explained by the wave model began to accumulate. These bits of experimental evidence, involving a variety of specific phenomena, had one thing in common—they were associated either with the way light is generated from other forms of energy or with what happens when light is absorbed by matter and converted into another energy form. The inability of the wave model to explain these observations led to the revival of a particle model for light. This theory, developed by Max Planck and Albert Einstein, states that a light beam is a stream of quanta of energy, called **photons**. The energy of each photon is proportional to the frequency of the light (**Planck's law**) and is given by the equation

$$E_{\text{photon}} = h\nu \tag{5.3}$$

where h, a universal constant called **Planck's constant**, equals 6.626×10^{-34} J-sec.

Example 5.2

What is the energy of a photon of wavelength 590 nm?

Solution

$\nu = 5.08 \times 10^{14}$ sec^{-1} (See Example 5.1.)

$E = h\nu = (6.63 \times 10^{-34} \text{ J-sec})(5.08 \times 10^{14} \text{ sec}^{-1}) = 3.37 \times 10^{-19}$ J

Figure 5.6
The photoelectric effect.

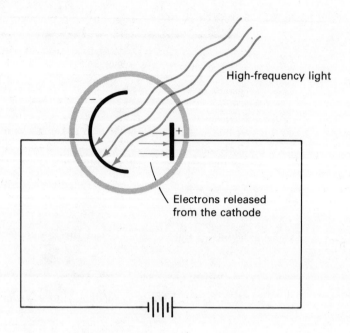

One of the phenomena that stumps the wave model but that can be treated by the photon model is the photoelectric effect (Figure 5.6). When light with a high enough frequency strikes a metal surface, it sets electrons in the metal into motion. The energy gained by an electron depends on the frequency rather than the intensity of the light—the higher the frequency of the light, the greater the energy gained by the electron. This conflicts with the wave model, which says that the energy in a light beam depends on the intensity of the light but has no connection with its frequency. The photon model pictures each absorbed photon giving all its energy, which is proportional to its frequency, to a single electron and provides a straightforward explanation of the photoelectric effect.

The photon model describes some important aspects of light such as the production and transformation of light into other energy forms, but it cannot explain diffraction and other interference phenomena. Because of its limited application, the photon model supplements rather than displaces the wave model. Indeed, in relating the energy of a photon to the frequency of light, the photon model invokes a wave property of light. This ambiguity as to whether light is a stream of particles or a wave is a frustrating affair for students as well as scientists, but no one has resolved it, and it appears to be a fundamental feature of light. In fact, as we will see in Section 5.3, particles such as electrons also show wave properties.

5.2 The Quantized Atom

Atomic Emission Spectra

Hot gases emit light at characteristic and precisely defined wavelengths (Figure 5.7). These **emission spectra** depend on the elements present. A particular pattern of emission wavelengths is so characteristic of a particular element that the nineteenth-century chemist R. W. Bunsen used these patterns to discover new elements. **Atomic emission spectroscopy** is a current method for routine chemical analysis, and it is used to learn what elements are present in stars that are light-years away.

One of the major failures of a pure wave model for light is its inability to explain why the light emitted by a particular type of atom should be restricted to a relatively small set of characteristic wavelengths rather than covering all possible wavelengths. Niels Bohr, working in Rutherford's laboratory in 1913, explained the emission spectrum of hydrogen in a way that was not only a further blow to the wave model but also changed the foundations of physical science. Bohr based his explanation on Rutherford's nuclear model for the atom and Coulomb's law, but he also made the radical assumption that the motion of an electron in an atom is *quantized*.

From the time of Newton, physics had been based on the idea of continuous change. According to the continuous view, a planet can occupy any orbit in which its centripetal acceleration matches the gravitational acceleration from the sun. No observation on visible systems had ever contradicted the continuous view. Bohr said it does not apply at the atomic level, and he replaced it with the idea that an electron in an atom is restricted to discrete orbits, or **energy levels**. Heat or electrical energy, or absorbed light can raise the energy of an electron, but the electron must receive the exact amount of energy needed to jump from a low-energy level to a higher one. An electron in a high-energy level (an **excited state**) can fall to a lower energy state, emitting a photon of light whose frequency is related to the energy change according to Planck's law.

$$E_{high} - E_{low} = h\nu \tag{5.4}$$

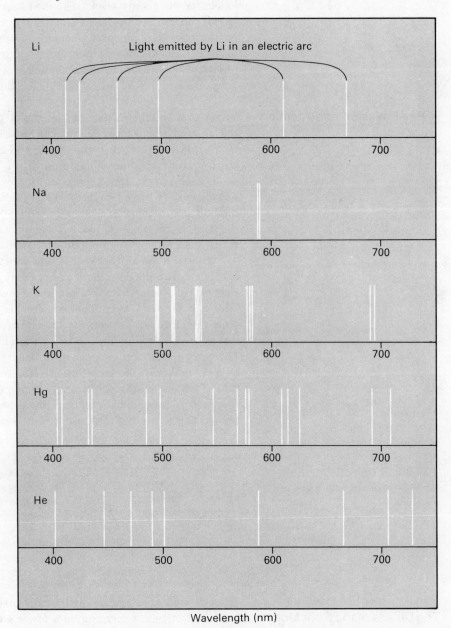

Figure 5.7
Atomic emission spectra.

Each energy level is associated with a whole number called a **quantum number**. The quantum number, **n**, can equal any positive whole number, but it cannot have any other values.

$$\mathbf{n} = 1, 2, 3, \ldots$$

Bohr derived equations that allowed him to calculate the energy of a particular energy level from its quantum number. Using these energies, he calculated the emission frequencies from equation (5.4), and the calculated frequencies were in excellent agreement with the experimental values.

Later theoretical advances, some guided by Bohr himself, have changed many features of the quantized model for the atom. We no longer believe in well-defined electron orbits, but the most striking feature of the Bohr theory—the quantization of energy—remains. We have encountered other examples of quantization (mass and charge), but we could explain these with comfortably solid models involving particles (protons, neutrons, and electrons). It was hard for Bohr and his contemporaries to imagine what quantum numbers represent, other than identifying labels for the energy levels. The idea that numbers could influence physical behavior ran counter to the views of nineteenth-century physics, which were based on the idea that physical phenomena were explicable in terms of mechanical models. Because no one has successfully explained emission spectra without numerical quantization, we accept Bohr's fundamental views and no longer attempt descriptions of atoms and molecules based on classical mechanics.

x-Ray Emission and Atomic Number

When high-speed electrons bombard a target, high-intensity x rays emanate from the material in the target (Figure 5.8). Moseley found that x-ray frequencies emitted depend on the element in the target, and there is a precise relationship between the frequency and the atomic number of the element. This relationship (Figure 5.9) is useful in determining atomic numbers of elements and placing the elements in their proper places in the periodic table. Because of their similarity, it is very difficult to determine the periodic sequence of the lanthanide elements solely on the basis of chemical evidence. The x-ray emission frequencies of the lanthanides reveal their atomic numbers unambiguously.

The relationship between x-ray emission frequency and atomic number is consistent with the Bohr theory. Suppose that, by chance, a high-speed bombarding electron collides with an electron in the lowest energy level of the target atom and knocks it out of the atom. Then an electron in the second energy level of the target atom falls into the first energy level to replace the missing electron. The energy lost by the electron in moving to a lower energy level will be released as a photon. The energy loss, and consequently the frequency of the photon, depend on the atomic number of the target atom because an intensely charged nucleus will attract electrons much more strongly than one with a small charge. For atoms with reasonably large atomic numbers, the energy change is so large that the emitted photon has a very high frequency that is in the x-ray range.

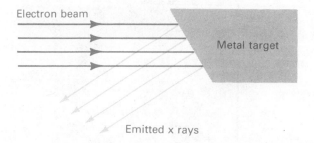

Figure 5.8
Emission of x rays from a metal target bombarded with high-speed electrons.

Special Resource Section:
The Details of the Bohr Model

The emission spectrum of hydrogen consists of several series of lines (Figure 5.10), each series converging on a lower wavelength limit. The wavelengths fit a precise mathematical relationship called Balmer's law:

5.2 THE QUANTIZED ATOM

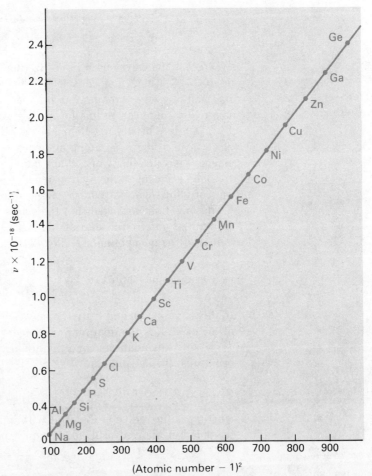

Figure 5.9
The x-ray emission frequencies of some elements.

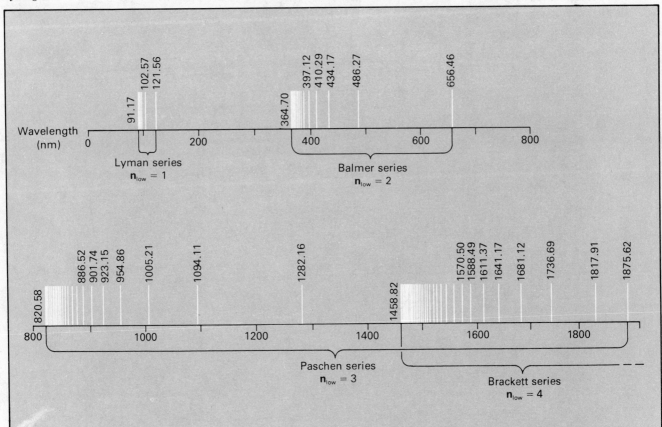

Figure 5.10
Atomic emission spectrum of hydrogen.

$$\frac{1}{\lambda} = R_H\left[\frac{1}{\mathbf{n}_{\text{low}}^2} - \frac{1}{\mathbf{n}_{\text{high}}^2}\right] \tag{5.5}$$

where λ is the wavelength in m, R_H (the Rydberg constant, not the gas constant) is 10,973,731 m^{-1} (one of science's most precisely determined constants), and $\mathbf{n}_{\text{low}}$ and $\mathbf{n}_{\text{high}}$ are two quantum numbers, $\mathbf{n}_{\text{high}}$ being the larger of the two. There is a hydrogen emission line for every possible pair of values for $\mathbf{n}_{\text{low}}$ and $\mathbf{n}_{\text{high}}$.

To apply his theory, Bohr had to calculate the allowed energy levels for an electron in a hydrogen atom, and these energy levels had to be consistent with Balmer's law. In other words, there had to be a possible energy jump corresponding to each frequency in the hydrogen emission spectrum, and there could not be any pair of allowed orbits whose energy difference did not match an emission line. To do this, Bohr assumed that the orbits were circular and that the angular momentum of an electron in a stable orbit must be quantized; that is, only certain values of the momentum are allowed.

$$\text{Angular momentum} = mvr = \frac{nh}{2\pi} \tag{5.6}$$

In this equation, m and v are the mass and velocity of the electron, r is the radius of the orbit, $\mathbf{n}$ is the quantum number for that orbit, and h is Planck's constant.

Using this quantum condition, Bohr calculated that the radius of an allowed orbit must fit the equation

$$r_{\mathbf{n}} = \mathbf{n}^2 r_1 = \mathbf{n}^2(0.053 \text{ nm}) \tag{5.7}$$

where $r_{\mathbf{n}}$ is the radius of the orbit with quantum number $\mathbf{n}$ and r_1 is the radius of the smallest allowed orbit (0.053 nm). An electron can occupy any orbit whose

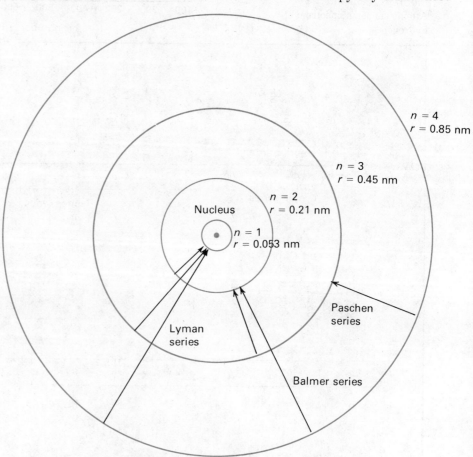

Figure 5.11
Bohr's model for the hydrogen atom.

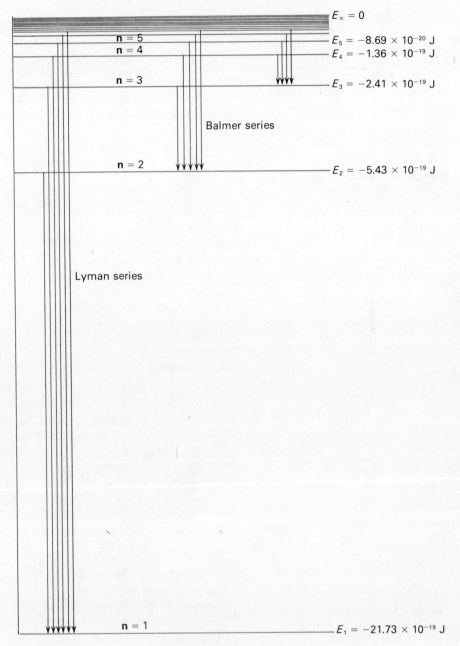

Figure 5.12
Energy levels of the hydrogen atom.

radius is $4r_1$, $9r_1$, $16r_1$, and so forth (Figure 5.11). The energies associated with each of these orbits fit the equation

$$E_n = \frac{E_1}{n^2} = \frac{-21.73 \times 10^{-19} \text{ J}}{n^2} \tag{5.8}$$

where E_1 (-21.73×10^{-19} J) is the energy of the most stable orbit of the hydrogen atom (Figure 5.12). (The negative value of the energy means that an electron is more stable in a hydrogen atom than when it is free in space.) The energies of higher orbits are $E_1/4$, $E_1/9$, $E_1/16$, ...

A hydrogen atom emits a photon when the electron falls from a higher energy orbit (n_{high}) to a lower energy orbit (n_{low}). Heat or light absorption can raise the electron to a higher energy state, and the subsequent fall of the electron releases light according to Planck's law.

$$E = E_{high} - E_{low} = h\nu = \frac{hc}{\lambda}$$

$$\frac{hc}{\lambda} = \frac{-21.73 \times 10^{-19}\text{ J}}{n_{high}^2} - \frac{-21.73 \times 10^{-19}\text{ J}}{n_{low}^2}$$

$$\frac{1}{\lambda} = \frac{21.73 \times 10^{-19}\text{ J}}{hc}\left[\frac{1}{n_{low}^2} - \frac{1}{n_{high}^2}\right] \tag{5.9}$$

Not only does this equation have the same algebraic form as Balmer's law, but the value of $(21.73 \times 10^{-19}\text{ J})/hc$ is in excellent agreement with the experimental value of R_H.

5.3 Wave–Particle Duality

Particles and Waves

Although Newtonian physics adequately describes visible objects moving at moderate velocities, it cannot deal with the nature of light, with objects moving at nearly the speed of light, or with very small particles. In order to explain atomic behavior, scientists had to replace Newton's classical system with a more generally applicable one. This system was presented in 1924 by Louis de Broglie who proposed that any particle (including a photon) has wave properties, the wavelength being inversely proportional to the momentum according to the equation

> Sit Jessica. Look how the floor of heaven
> Is thick inlaid with patines of bright gold;
> There's not the smallest orb which thou beholds't
> But in his motion like angel sings,
> Still quiring to the young-eyed cherubins;
> Such harmony is in immortal souls,
> But whilst this muddy vesture of decay
> Doth grossly close it in, we cannot hear it.
>   Shakespeare, *Merchant of Venice*, Act V, Scene I

This romantic speech contains a cosmological theory—the Pythagorean view of the universe. We don't have much firm information about the Pythagoreans, a group of ancient Greeks (sixth century B.C.) involved in religion, politics, and philosophy, but we credit them with discovering the mathematical relationships between the musical intervals. The discovery of the connection between musical harmony and numerical harmony led to the belief that "all things are numbers"—the universe is a harmonious one, following numerical laws, and the heavenly bodies move according to "the music of the spheres." Examine the geometric regularity of crystals, seashells, and plants and you may feel the attraction of this idea. It played an important part in classical and medieval thought and was still taken seriously in Shakespeare's day. The rise of Newtonian physics displaced the Pythagorean view—planets move according to mechanical, not musical, principles. By the late nineteenth century, the belief in the physical influence of numbers was reduced to a few superstitions about lucky and unlucky numbers or treated as something quaint. (*Music of the Spheres* is the title of a Strauss waltz.) But that was before de Broglie, considering orbs much smaller than the ones Shakespeare refers to, turned us into young-eyed cherubins. The atomic world is a Pythagorean one.

5.3 WAVE–PARTICLE DUALITY

$$\lambda = \frac{h}{\text{Momentum}} = \frac{h}{mv} \tag{5.10}$$

Wave phenomena are not part of our daily experience with visible particles, which have such large momenta that their wavelengths are too small to measure. Wave properties are apparent only for very small particles such as electrons.

Example 5.3

What is the wavelength of an electron traveling at 2.7×10^6 m/sec?

Solution

Since the mass of an electron is 9.1×10^{-31} kg, the momentum is

$$mv = (9.1 \times 10^{-31} \text{ kg})\left(2.7 \times 10^6 \frac{\text{m}}{\text{sec}}\right) = 2.5 \times 10^{-24} \frac{\text{kg-m}}{\text{sec}}$$

According to equation (5.10),

$$\lambda = \frac{h}{mv} = \frac{6.6 \times 10^{-34} \text{ J-sec}}{2.5 \times 10^{-24} \text{ kg-m/sec}}$$

$$= 2.6 \times 10^{-10} \text{ m} = 0.26 \text{ nm}$$

Since 1 J = 1 kg-m^2/sec^2, a J-sec equals a kg-m^2/sec. Dividing this unit by a kg-m/sec leaves m as the unit.

Example 5.4

What is the momentum of a 0.25 nm x ray?

Solution

$$0.25 \text{ nm} = 2.5 \times 10^{-10} \text{ m}$$

$$\text{Momentum} = \frac{6.6 \times 10^{-34} \text{ J-sec}}{2.5 \times 10^{-10} \text{ m}} = 2.6 \times 10^{-24} \frac{\text{kg-m}}{\text{sec}}$$

De Broglie's theory was quickly verified experimentally. The regular geometric patterns of nuclei in crystals act as gratings suitable for diffracting waves in the 0.1–1 nm range. When x rays are reflected from crystals, they produce interference patterns (Figure 5.13). A beam of electrons reflected from the same crystals shows similar interference patterns, and the measured wavelength is in exact agreement with de Broglie's prediction. Since interference phenomena are the sole evidence

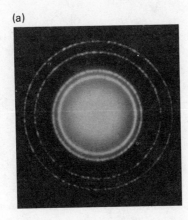
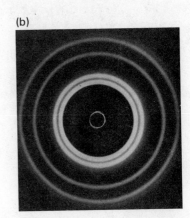

Figure 5.13
Electron diffraction. (a) The diffraction pattern produced by an electron beam on polycrystalline aluminum. (b) The diffraction pattern produced by an x-ray beam on polycrystalline aluminum. Note the similarity of the two patterns.
Source: Reproduced courtesy Education Development Center, Newton, MA 02160.

Standing Waves and Quantization

for the wave nature of light, and since similar interference patterns occur for electrons, an electron has as much claim to wave properties as a photon.

Bohr's model of the atom depends on whole numbers; so do phenomena involving waves. One example of these integer-dependent wave phenomena is familiar to any musician. A string fixed firmly at both ends can vibrate in many different ways (**standing waves**) to produce many different frequencies, but *not* all frequencies. There are precise restrictions on these frequencies; that is, they are quantized (Figure 5.14). In its simplest vibration (principal tone, $n = 1$), the center of the string vibrates most noticeably; the string forms a half-wave and the wavelength is $2l$, where l is the length of the string. Plucked differently, the string vibrates in its first harmonic ($n = 2$) and the wavelength is l. The center of the string is a stationary point called a *node*. The points at $\frac{1}{4}l$ and $\frac{3}{4}l$ vibrate most strongly, one moving up as the other moves down (this is the *phase* of the wave). The frequency of the first harmonic is twice that of the principal tone. The second harmonic ($n = 3$) has a frequency three times the principal tone. The string can produce many different vibrations, but all must fulfill the quantization condition

$$\nu = n\nu_{\text{principal}}$$

where n is a whole number.

De Broglie applied the principle of standing waves to electrons in atoms and explained Bohr's energy quantization in the following way. There must be a whole number of electron waves on the circumference of a stable orbit, or else the waves interfere destructively with themselves (Figure 5.15). (A metal ring vibrates in quantized tones because of this effect.) The only energy levels that can exist in Bohr's model of the hydrogen atom are those whose circumferences equal a whole-number multiple of the wavelength of the electron. The quantum number, according to this interpretation, would be the number of waves per orbit circumference.

Figure 5.14

Standing waves of a vibrating string.

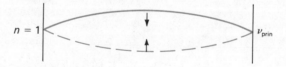

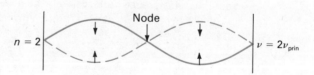

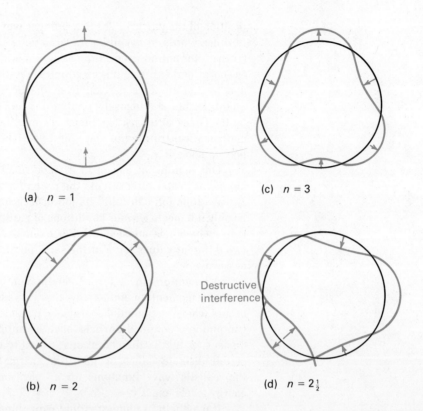

Figure 5.15
Standing electron waves on circular orbits. In (a), (b), and (c), the number of electron waves per circumference is a whole-number for a stable, circular orbit. In (d), where n is not a whole-number, the wave undergoes destructive interference and cancels itself.

The Heisenberg Uncertainty Principle

Scientists must plan carefully their methods of measurement, analyze the potential sources of uncertainty in these techniques, and consider the degree to which the act of observation might disturb the system under study. All visual observations involve reflections of photons from the observed object. Although disturbance of a large object by photon collisions is insignificant, this disturbance is very important for small particles such as electrons.

The problem of observing small particles and the effects of wave–particle duality limit the information we can obtain about a particle. Because of its wave nature, a light beam cannot resolve two points less than a wavelength apart, no matter how powerful a microscope we use. If we want to locate a particle within a certain range of uncertainty, we must use light with a wavelength smaller than the uncertainty range. But the shorter the wavelength, the greater the momentum, and the photon disturbs the particle more severely. In other words, the more precisely we determine a particle's position, the more we disturb its motion. This is the basis for the **Heisenberg uncertainty principle**, which states that it is impossible to determine both the position and momentum of an object with absolute precision. The more precisely we determine the position, the less we know about the motion and vice versa.

$$\text{(Position uncertainty)}(\text{Momentum uncertainty}) \approx h \qquad (5.11)$$

The recognition of the uncertainty principle changed the goals of atomic science. If it is impossible to determine the paths of electrons in atoms, it is impossible to check experimentally the accuracy of the Bohr model and its well-defined orbits. The best we can do in describing atoms is to estimate the probability of finding an electron in any region of an atom.

The Schrödinger Wave Equation

Erwin Schrödinger generalized de Broglie's idea of standing particle waves by developing a harmonic equation (similar to the ones describing the quantized vibrations of a string) that allows us to calculate the possible energy values a particle can have, provided we know the relationship between the energy and position of the particle. The **Schrödinger wave equation** is a differential equation.* In other words, it is unlike the equations we have used so far in two respects: (1) the information used in the equation is not a set of numerical quantities such as energy or distance but is an *equation* relating the particle's energy to its position, and (2) the answers derived from the Schrödinger equation are not numerical quantities but are themselves equations called **wave functions** and are represented by the Greek letter psi (ψ). Since the solution of the wave equation is beyond the mathematical training of many first-year chemistry students, we will not discuss how to solve it.

One significant aspect of a wave function is that its value squared (ψ^2) at any point in space represents the *probability* of finding the electron at that point. A wave function can fulfill its job of locating where an electron is most likely to be only if it meets various conditions of reality, and we must reject a wave function if its behavior is nonsensical. For example, no realistic wave function can have two different values for ψ at the same point; this is equivalent to self-interference of a wave.†

All solutions for ψ for a freely moving particle are real, and the particle can have any momentum and energy; that is, its motion is not quantized. Quantization occurs when the potential energy of a particle varies with its position. For a system containing a confined particle, such as an electron held by the electric field of a nucleus, ψ fulfills the conditions of reality only if the energy of the system has certain discrete values, or quantized energy levels. The energies corresponding to the realistic wave functions for an electron in a hydrogen atom match Bohr's energy levels exactly.

It may help in understanding wave functions to think of the vibrating string in a musical instrument as an analogy for an electron whose motion is restricted by an electrostatic force. The string's motion is limited because its ends are tied down. It can go through a variety of wave motions, but the only possible ones are those that avoid wave interference. The analogy to a quantized electron is purely mathematical—electrons do not vibrate like strings. Nevertheless, electrons do have wave properties (although no one has presented a satisfactory mechanical analogy for these waves), and these waves are related to the motions (momenta) of the electron. If an electrostatic attraction limits this motion, the possibility of self-interference of waves arises. The only real waves (ψ functions) are those that avoid this interference. The waves of an electron are also more complicated than those of a string because the electron wave is in three dimensions, whereas the string vibrates in only one dimension.

Atomic Orbitals of Hydrogen

Unlike the Bohr model, Schrödinger's picture of the atom does not have any precise paths for the electrons. Instead, each energy level is represented by a three-dimensional graph of its ψ^2 function. Each of these **orbitals** resembles a cloud whose density in a certain region represents the chance of finding the electron in that part of the atom. An electron spends most of its time where the orbital cloud is thickest, but there is still some chance of finding it in regions where the orbital is very diffuse.

The lowest energy orbital of the hydrogen atom, called the 1s orbital (Figure 5.16), is a spherical cloud whose density is greatest near the nucleus, and the chance of finding an electron in a certain volume is greatest near the nucleus. This orbital

---

*The Schrödinger wave equation in three dimensions is

$$\nabla^2 \psi + \frac{8\pi^2 m}{h^2}(E - V)\psi = 0$$

E is the energy associated with the wave function ψ, and V is an equation relating the potential energy to the position of a particle.

†Other unreal conditions occur if ψ^2 is imaginary (involves $\sqrt{-1}$) at any point, if ψ is discontinuous (changes abruptly from one position to the next), or if ψ is infinite.

5.3 WAVE–PARTICLE DUALITY

Figure 5.16
The 1s orbital of the hydrogen atom.

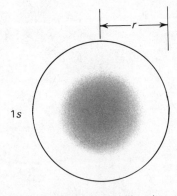

There is a 90% probability that the 1s electron is within the sphere with radius r

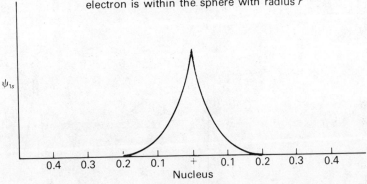

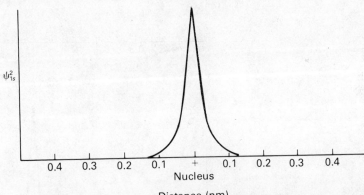

differs from a sphere in having no finite dimension—it just continues to become more and more diffuse as the distance from the nucleus increases. The greater the distance from the nucleus, the less probability of finding the electron, but the chance of finding the electron very far from the nucleus never quite becomes zero. This orbital picture is very different from Bohr's planetary model. Instead of moving in the circular orbitals described by Bohr, the electron might be found all over the atom and it spends a considerable portion of time near the nucleus.

Higher-Energy Hydrogen Orbitals

Four different solutions of the wave equation are associated with the second energy level of hydrogen. Orbitals having identical energies are termed **degenerate**. One of these degenerate orbitals (the 2s orbital) is a sphere whose density fluctuates with the distance from the nucleus (Figure 5.17). Near the nucleus is a spherical region of high electron density surrounded by a spherical **nodal surface** on which the electron is never found. Another spherical region of electron density exists outside the nodal surface. The other three orbitals of the second energy level (the 2p orbitals) have identical shapes but point at 90° angles to each other. Each has

Figure 5.17
The 2s and 2p orbitals of hydrogen.

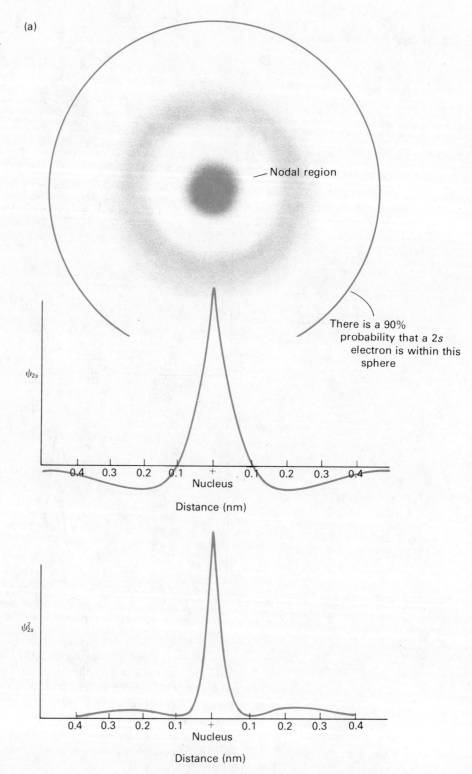

its electron density concentrated in two lobes (one on either side of the nucleus) lying on a single axis.

The third energy level consists of nine degenerate orbitals. In addition to a 3s orbital and three 3p orbitals, there are five 3d orbitals (Figure 5.18, p. 120). The number of degenerate orbitals increases with each energy level.

Quantum Numbers

In the same sense that one quantum number is required for a one-dimensional vibrating string, three quantum numbers are needed to describe an electron orbital

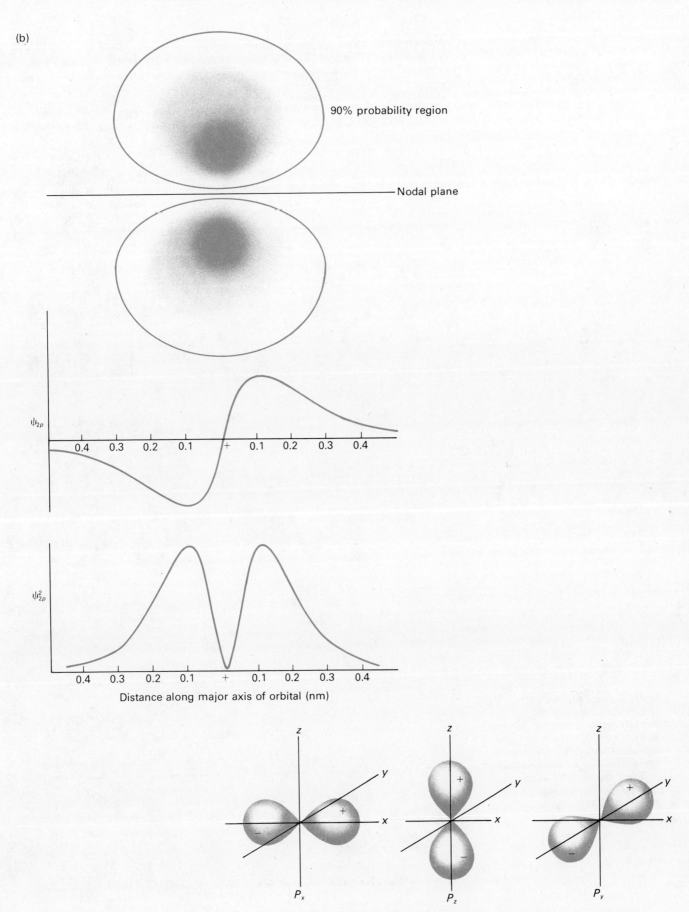

The orientations of 2p orbitals

in three dimensions, and each orbital has a characteristic set of these three quantum numbers. The first is the **principal quantum number (n)**. Hydrogen orbitals with the same value for **n** have the same energy and the same approximate average distance to the nucleus, and we say that they belong to the same **shell** of orbitals. The only member of the first shell is the 1s orbital, while the 2s and three 2p orbitals form the second shell and the 3s, three 3p, and five 3d orbitals constitute the third shell.

Orbitals of the same shell may differ in shape, and electrons in different orbitals of a shell may have different quantized angular momenta. A second quantum number, called the **azimuthal quantum number (l)**, denotes the angular momentum. All orbitals with the same **n** and **l** values form a **subshell** of orbitals.

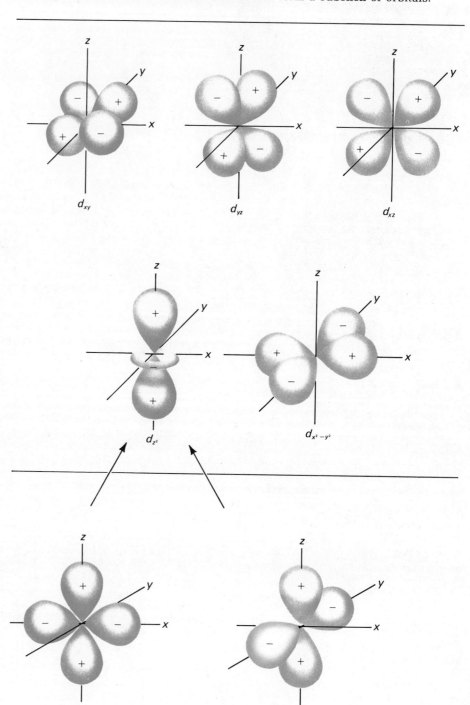

Figure 5.18
The five 3d orbitals.

There are actually two sets of orbitals (three orbitals in each set, four lobes in each orbital) that make up a d orbital subshell; however, not all the orbitals are independent and two are usually merged to give the five-orbital set commonly seen.

5.3 WAVE–PARTICLE DUALITY

Table 5.2 **Quantum Numbers**

Shell	n	Subshell	l	m_l	m_s
First	1	1s	0	0	$+\frac{1}{2}, -\frac{1}{2}$
Second	2	2s	0	0	$+\frac{1}{2}, -\frac{1}{2}$
		2p	1	−1, 0, +1	$+\frac{1}{2}, -\frac{1}{2}$
Third	3	3s	0	0	$+\frac{1}{2}, -\frac{1}{2}$
		3p	1	−1, 0, +1	$+\frac{1}{2}, -\frac{1}{2}$
		3d	2	−2, −1, 0, +1, +2	$+\frac{1}{2}, -\frac{1}{2}$
Fourth	4	4s	0	0	$+\frac{1}{2}, -\frac{1}{2}$
		4p	1	−1, 0, +1	$+\frac{1}{2}, -\frac{1}{2}$
		4d	2	−2, −1, 0, +1, +2	$+\frac{1}{2}, -\frac{1}{2}$
		4f	3	−3, −2, −1, 0, +1, +2, +3	$+\frac{1}{2}, -\frac{1}{2}$

The **l** values depend on the value of the principal quantum number. The first shell can have only one **l** value (**l** = 0), indicating a spherical orbital. The orbitals of the second shell can be either spherical (**l** = 0) or the double-lobed *p* orbitals (**l** = 1). The value of **l** can be zero or any whole number up to and including **n** − 1.

$$l = 0, 1, 2, \ldots, n - 1$$

For example, the third shell contains three subshells: an *s* subshell (**l** = 0), a *p* subshell (**l** = 1), and a *d* subshell (**l** = 2). The number of subshells in any particular shell is equal to the principal quantum number **n** of that shell.

A third quantum number, called the **magnetic quantum number** (m_l), describes the direction the orbital points in space, each value of m_l representing one possible orientation. (The name "magnetic quantum number" comes from the observation that an external magnetic field influences the energies of the orbitals of a particular subshell differently, and the difference arises from the different orientations of the orbitals.) The values of m_l are whole numbers that can range from the negative value of **l**, through zero, and up to the positive value of **l**.

$$m_l = -l, -(l - 1), \ldots, -1, 0, +1, \ldots, l - 1, l$$

For example, a *d* subshell (**l** = 2) has five orbitals with the m_l values −2, −1, 0, +1, +2. The number of different orbitals in a subshell equals 2**l** + 1, and addition of all the orbitals in all the subshells shows that there are n^2 orbitals in each shell (Table 5.2).

Electron Spin

If we pass a beam of evaporated silver atoms through a strong, uneven magnetic field, the beam splits in two, one beam moving toward the north pole and the other toward the south (Stern-Gerlach experiment). This shows that silver atoms are weak quantized magnets; half of them have magnetic fields in one direction and half in the other.

Electrons in atomic orbitals act like quantized magnets and can have a magnetic field pointing in one of two directions—either in the same general direction as the external magnetic field or in the opposite direction. The difference in the magnetic behavior of the two types of silver atoms is due to the difference in the magnetic properties of a single electron in each atom.

We are not sure of the exact source of the magnetic fields since we do not know anything about the internal structure of the electron, but we draw a pictorial analogy with a charged sphere spinning on its own axis (Figure 5.19). The direction of the magnetic field would depend on the direction of rotation of the electron.

Figure 5.19
Electron spin. (a) The Stern-Gerlach experiment illustrates how silver atoms behave as weak quantized magnets. (b) An electron behaves like a charged sphere spinning on its own axis. The direction of the magnetic field depends on the direction of rotation.

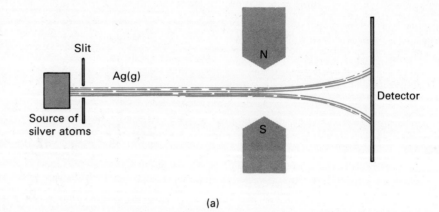

(a)

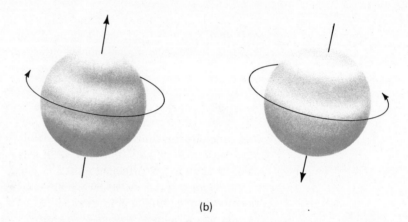

(b)

We say the **electron spin** is quantized and describe the spin by a **spin quantum number**, m_s, which has one of two possible values, either $+\frac{1}{2}$ or $-\frac{1}{2}$.*

5.4 Electron Configuration and the Periodic Table

Polyelectron Atoms

The major problem with quantum mechanics is the mathematical complexity of the Schrödinger wave equation. A student who knows differential equations can solve the Schrödinger equation for the wave functions of the hydrogen atom, but exact solutions for systems with three charged particles (for example, the one nucleus and two electrons in the helium atom) bewilder the best mathematicians.

The problem with systems containing several electrons is that we must take into account the energy effect of repulsions between electrons as well as the attractions between the nucleus and the electrons. We cannot write an equation for the electron repulsion energy until we know the average distance between the electrons, but we will not know this average distance until we solve the wave equation. In other words, we cannot get started until we know the answer.

One way around this problem is to use a series of mathematical approximations. These approximations require the use of computers, and the solutions to the Schrödinger wave equation for most atoms are still beyond the capabilities of the best computers. Another way of obtaining wave functions is to use intuitive approximations. These approximations are based on the form of the hydrogen wave functions and spectroscopic data for the different elements.

*If relativistic considerations are included in the solution of the Schrödinger equation, the fourth quantum number, m_s, follows naturally.

5.4 ELECTRON CONFIGURATION AND THE PERIODIC TABLE

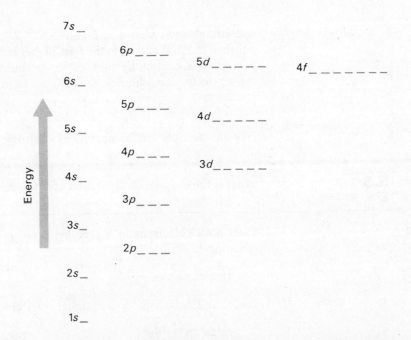

Figure 5.20
Relative subshell energies. These energy spacings are not drawn to scale. The difference in energy between two subshells is smaller for the larger principal quantum numbers.

Hydrogenlike Orbitals

The first intuitive approximation is that any atom has a set of wave functions similar in most respects to the hydrogen orbitals. The same rules govern the n, l, and m_l quantum numbers, and the orbitals have nodal surfaces and orientations similar to corresponding hydrogen orbitals. Every atom has a spherical 1s orbital, a spherical 2s orbital, three mutually perpendicular 2p orbitals, and all the corresponding higher energy orbitals.

Because of the effects of the greater nuclear charge and repulsions between electrons, orbitals of other atoms differ from hydrogen orbitals in their dimensions and energies. One important effect is the difference in energies between the subshells of a shell. The s subshell has the lowest energy, followed by the p subshell, the d subshell, and higher subshells. Electron repulsion affects the subshells of a particular shell differently.

Figure 5.20 is a graph of the increasing order of energies of subshells. Note that the 3d subshell has higher energy than the 4s subshell, and 4d electrons have higher energy than 5s. Similarly, 6s electrons are more stable than 4f electrons.

The Pauli Exclusion Principle

The **Pauli exclusion principle** is the key to explaining the periodic table: Two electrons in the same atom cannot be in identical quantum states. In other words, electrons in an atom must differ from each other in at least one of the four quantum numbers, n, l, m_l, or m_s. This means that (1) a particular orbital cannot hold more than two electrons, and an electron pair in an orbital must have opposite spins, (2) since a subshell contains $2l + 1$ orbitals, it can hold up to $4l + 2$ electrons, and (3) since a shell contains n^2 orbitals, it can hold up to $2n^2$ electrons.

A complete first shell consists of a single orbital with 2 electrons. Two of the 8 electrons in the second shell are in the 2s subshell, and the remaining 6 are in the 2p subshell. There are 18 electrons in a filled third shell—2 in the 3s subshell, 6 in the 3p, and 10 in the 3d. The fourth shell can hold 32 electrons—2 4s, 6 4p, 10 4d, and 14 4f ($l = 3$).

Electron Configuration

An atom can exist in any one of a large variety of quantized electron arrangements, but most atoms are in their lowest energy state called the **ground state**. We can write out the electron configuration of the ground state by a "building" procedure (called the Aufbau principle), using the capacities of the subshells and the relative subshell energies shown in Figure 5.20.* We add electrons to the lowest energy subshell (1s) first and go to the next higher subshell when we have filled the lower energy levels, continuing until we have accounted for all the electrons. For example, the 17 electrons of chlorine are distributed in the ground state as follows:

$$1s^2 2s^2 2p^6 3s^2 3p^5$$

(The superscripts are the number of electrons in each subshell.)

Example 5.5

What is the electron configuration of V (atomic number 23)?

Solution

There are 23 electrons in V. The first two shells fill up in order and contain ten electrons.

$$1s^2 2s^2 2p^6 \ldots$$

The 3s and 3p subshells, containing eight electrons, also fill in order.

$$1s^2 2s^2 2p^6 3s^2 3p^6 \ldots$$

According to the relative subshell energies shown in Figure 5.20, the next most stable subshell is the 4s, rather than the 3d.

$$1s^2 2s^2 2p^6 3s^2 3p^6 4s^2 \ldots$$

The remaining three electrons of V fit into the 3d subshell to complete the electron configuration.

$$1s^2 2s^2 2p^6 3s^2 3p^6 4s^2 3d^3$$

If we follow the style of grouping the subshells according to their shells rather than in the sequence of increasing energy, the electron configuration would be written

$$1s^2 2s^2 2p^6 3s^2 3p^6 3d^3 4s^2$$

Example 5.6

What is the electron configuration of europium?

Solution

The atomic number of Eu is 63, so this atom contains 63 electrons. To write the electron configuration, we continue along the lines of Example 5.5, remembering that the 5s subshell fills before the 4d. Summing the number of electrons through the 5p subshell, we can account for 54 electrons.

---

*It is useful to memorize the sequence of subshell energies: A convenient memory device is to list the subshells, placing all s subshells in the first column, all p subshells in the second, and so on. A series of arrows running from the upper right to the lower left gives the subshell energies in ascending sequence.

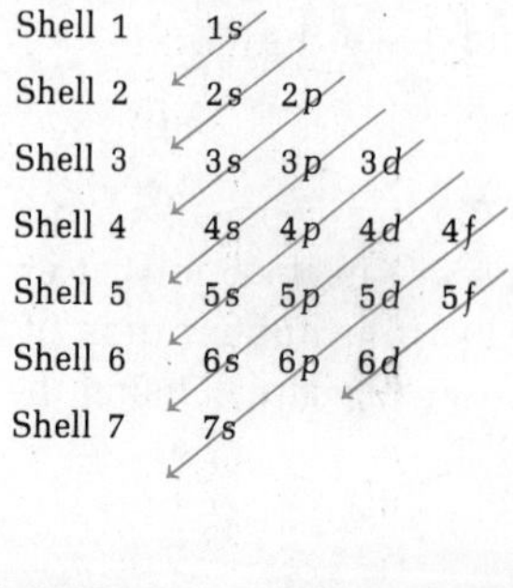

Shell 1	1s			
Shell 2	2s	2p		
Shell 3	3s	3p	3d	
Shell 4	4s	4p	4d	4f
Shell 5	5s	5p	5d	5f
Shell 6	6s	6p	6d	
Shell 7	7s			

$$1s^22s^22p^63s^23p^64s^23d^{10}4p^65s^24d^{10}5p^6\ldots$$

Next comes the 6s subshell followed by the 4f (not the 5d). Two of the remaining nine electrons fit in the 6s subshell, and the last seven are 4f electrons.

$$1s^22s^22p^63s^23p^64s^23d^{10}4p^65s^24d^{10}5p^66s^24f^7$$

or

$$1s^22s^22p^63s^23p^63d^{10}4s^24p^64d^{10}4f^75s^25p^66s^2$$

Table 5.3 shows the spectroscopically determined ground-state configuration of the elements. There are occasional discrepancies involving the positions of one or two electrons between the observed configuration and that predicted by the building method (for example, Cr with configuration $3d^54s^1$). The building method is an oversimplification because it neglects the effects of repulsion between electrons in the same orbital. Neglecting this electron-repulsion effect produces the errors in the predicted ground-state configurations.

Hund's Rules

In addition to revealing the arrangements of electrons among subshells, spectroscopic and magnetic evidence show how electrons tend to be distributed within the orbitals of a partially filled subshell and how the electron spins are oriented. **Hund's rules** summarize these tendencies. These rules state that (1) an orbital holds an electron pair only if every other degenerate (same energy) orbital holds at least one electron and (2) electrons in degenerate orbitals have as many parallel spins as possible. In other words, in the ground state, electrons of a partially filled subshell occupy as many orbitals as possible and maintain the maximum number of parallel spins. An example of this is the ground-state arrangement for a $3d^5$ configuration:

$$\underline{\uparrow}\;\underline{\uparrow}\;\underline{\uparrow}\;\underline{\uparrow}\;\underline{\uparrow}$$

not

$$\underline{\uparrow\downarrow}\;\underline{\uparrow}\;\underline{\uparrow}\;\underline{\uparrow}\;\underline{} \quad\text{or}\quad \underline{\uparrow}\;\underline{\uparrow}\;\underline{\uparrow}\;\underline{\uparrow}\;\underline{\downarrow}$$

In this notation, each horizontal line represents a 3d orbital. Each half-arrow pointing up (↑) represents an electron spinning in one direction, and a half-arrow pointing down (↓) represents an electron spinning in the opposite direction.

Hund's rules explain why the elements in the middle of the transition metal rows, such as Mn, Fe, and Co, are the most paramagnetic. The magnetism of an atom is due to the spin of its unpaired electrons. The high degree of magnetism of manganese is due to the $3d^5$ arrangement with its five parallel spins. The other highly magnetic elements have a large number of parallel electron spins.

Example 5.7

Show the arrangement of electrons and their spins in a $4f^9$ configuration.

Solution

According to Hund's rules, the first seven electrons must all be in separate orbitals and have parallel spins. The remaining two electrons will each pair up with one of these electrons, giving a configuration with two electron pairs and five parallel, unpaired electrons.

$$\underline{\uparrow\downarrow}\;\underline{\uparrow\downarrow}\;\underline{\uparrow}\;\underline{\uparrow}\;\underline{\uparrow}\;\underline{\uparrow}\;\underline{\uparrow}$$

Table 5.3   Electron Configurations of the Elements

Atomic number (Z)	Element	Shell 1	Shell 2		Shell 3			Shell 4				Shell 5				Shell 6				Shell 7
		s	s	p	s	p	d	s	p	d	f	s	p	d	f	s	p	d	f	s
1	H	1																		
2	He	2																		
3	Li	2	1																	
4	Be	2	2																	
5	B	2	2	1																
6	C	2	2	2																
7	N	2	2	3																
8	O	2	2	4																
9	F	2	2	5																
10	Ne	2	2	6																
11	Na	2	2	6	1															
12	Mg	2	2	6	2															
13	Al	2	2	6	2	1														
14	Si	2	2	6	2	2														
15	P	2	2	6	2	3														
16	S	2	2	6	2	4														
17	Cl	2	2	6	2	5														
18	Ar	2	2	6	2	6														
19	K	2	2	6	2	6		1												
20	Ca	2	2	6	2	6		2												
21	Sc	2	2	6	2	6	1	2												
22	Ti	2	2	6	2	6	2	2												
23	V	2	2	6	2	6	3	2												
24	Cr	2	2	6	2	6	5	1												
25	Mn	2	2	6	2	6	5	2												
26	Fe	2	2	6	2	6	6	2												
27	Co	2	2	6	2	6	7	2												
28	Ni	2	2	6	2	6	8	2												
29	Cu	2	2	6	2	6	10	1												
30	Zn	2	2	6	2	6	10	2												
31	Ga	2	2	6	2	6	10	2	1											
32	Ge	2	2	6	2	6	10	2	2											
33	As	2	2	6	2	6	10	2	3											
34	Se	2	2	6	2	6	10	2	4											
35	Br	2	2	6	2	6	10	2	5											
36	Kr	2	2	6	2	6	10	2	6											
37	Rb	2	2	6	2	6	10	2	6			1								
38	Sr	2	2	6	2	6	10	2	6			2								
39	Y	2	2	6	2	6	10	2	6	1		2								
40	Zr	2	2	6	2	6	10	2	6	2		2								
41	Nb	2	2	6	2	6	10	2	6	4		1								
42	Mo	2	2	6	2	6	10	2	6	5		1								
43	Tc	2	2	6	2	6	10	2	6	5		2								
44	Ru	2	2	6	2	6	10	2	6	7		1								
45	Rh	2	2	6	2	6	10	2	6	8		1								
46	Pd	2	2	6	2	6	10	2	6	10										
47	Ag	2	2	6	2	6	10	2	6	10		1								
48	Cd	2	2	6	2	6	10	2	6	10		2								
49	In	2	2	6	2	6	10	2	6	10		2	1							
50	Sn	2	2	6	2	6	10	2	6	10		2	2							
51	Sb	2	2	6	2	6	10	2	6	10		2	3							
52	Te	2	2	6	2	6	10	2	6	10		2	4							

Atomic number (Z)	Element	Shells																			
		1	2		3			4				5				6				7
								Subshells												
		s	s	p	s	p	d	s	p	d	f	s	p	d	f	s	p	d	f	s
53	I	2	2	6	2	6	10	2	6	10		2	5							
54	Xe	2	2	6	2	6	10	2	6	10		2	6							
55	Cs	2	2	6	2	6	10	2	6	10		2	6			1				
56	Ba	2	2	6	2	6	10	2	6	10		2	6			2				
57	La	2	2	6	2	6	10	2	6	10		2	6	1		2				
58	Ce	2	2	6	2	6	10	2	6	10	2	2	6			2				
59	Pr	2	2	6	2	6	10	2	6	10	3	2	6			2				
60	Nd	2	2	6	2	6	10	2	6	10	4	2	6			2				
61	Pm	2	2	6	2	6	10	2	6	10	5	2	6			2				
62	Sm	2	2	6	2	6	10	2	6	10	6	2	6			2				
63	Eu	2	2	6	2	6	10	2	6	10	7	2	6			2				
64	Gd	2	2	6	2	6	10	2	6	10	7	2	6	1		2				
65	Tb	2	2	6	2	6	10	2	6	10	9	2	6			2				
66	Dy	2	2	6	2	6	10	2	6	10	10	2	6			2				
67	Ho	2	2	6	2	6	10	2	6	10	11	2	6			2				
68	Er	2	2	6	2	6	10	2	6	10	12	2	6			2				
69	Tm	2	2	6	2	6	10	2	6	10	13	2	6			2				
70	Yb	2	2	6	2	6	10	2	6	10	14	2	6			2				
71	Lu	2	2	6	2	6	10	2	6	10	14	2	6	1		2				
72	Hf	2	2	6	2	6	10	2	6	10	14	2	6	2		2				
73	Ta	2	2	6	2	6	10	2	6	10	14	2	6	3		2				
74	W	2	2	6	2	6	10	2	6	10	14	2	6	4		2				
75	Re	2	2	6	2	6	10	2	6	10	14	2	6	5		2				
76	Os	2	2	6	2	6	10	2	6	10	14	2	6	6		2				
77	Ir	2	2	6	2	6	10	2	6	10	14	2	6	7		2				
78	Pt	2	2	6	2	6	10	2	6	10	14	2	6	9		1				
79	Au	2	2	6	2	6	10	2	6	10	14	2	6	10		1				
80	Hg	2	2	6	2	6	10	2	6	10	14	2	6	10		2				
81	Tl	2	2	6	2	6	10	2	6	10	14	2	6	10		2	1			
82	Pb	2	2	6	2	6	10	2	6	10	14	2	6	10		2	2			
83	Bi	2	2	6	2	6	10	2	6	10	14	2	6	10		2	3			
84	Po	2	2	6	2	6	10	2	6	10	14	2	6	10		2	4			
85	At	2	2	6	2	6	10	2	6	10	14	2	6	10		2	5			
86	Rn	2	2	6	2	6	10	2	6	10	14	2	6	10		2	6			
87	Fr	2	2	6	2	6	10	2	6	10	14	2	6	10		2	6			1
88	Ra	2	2	6	2	6	10	2	6	10	14	2	6	10		2	6			2
89	Ac	2	2	6	2	6	10	2	6	10	14	2	6	10		2	6	1		2
90	Th	2	2	6	2	6	10	2	6	10	14	2	6	10		2	6	2		2
91	Pa	2	2	6	2	6	10	2	6	10	14	2	6	10	2	2	6	1		2
92	U	2	2	6	2	6	10	2	6	10	14	2	6	10	3	2	6	1		2
93	Np	2	2	6	2	6	10	2	6	10	14	2	6	10	4	2	6	1		2
94	Pu	2	2	6	2	6	10	2	6	10	14	2	6	10	5	2	6	1		2
95	Am	2	2	6	2	6	10	2	6	10	14	2	6	10	7	2	6			2
96	Cm	2	2	6	2	6	10	2	6	10	14	2	6	10	7	2	6	1		2
97	Bk	2	2	6	2	6	10	2	6	10	14	2	6	10	8	2	6	1		2
98	Cf	2	2	6	2	6	10	2	6	10	14	2	6	10	9	2	6	1		2
99	Es	2	2	6	2	6	10	2	6	10	14	2	6	10	10	2	6	1		2
100	Fm	2	2	6	2	6	10	2	6	10	14	2	6	10	11	2	6	1		2
101	Md	2	2	6	2	6	10	2	6	10	14	2	6	10	12	2	6	1		2
102	No	2	2	6	2	6	10	2	6	10	14	2	6	10	14	2	6			2
103	Lr	2	2	6	2	6	10	2	6	10	14	2	6	10	14	2	6	1		2
104																				
105																				
106																				

Electron Configuration and Chemical Similarity

This analysis applies only to the magnetic properties of uncombined atoms, not to atoms in molecules or in crystals. The forces that hold molecules and crystals together alter the electron configuration of an atom and may change the number of unpaired electrons. A single hydrogen atom (a very unstable and short-lived species) is paramagnetic, but H_2 is diamagnetic. The magnetic properties of solid metals depend on the details of their crystal structures. Nevertheless, it is apparent from the magnetism of manganese, iron, and cobalt that many of the unpaired electron spins remain in the crystal.

Chemical reactions involve changes in the arrangement of the electrons of an atom. Only the outer-shell electrons, also called the **valence electrons,** are affected to any large extent, and atoms with similar outer-shell configurations react similarly to form similar compounds.

Members of a chemical family usually have the same number of valence electrons equal to the group number (Table 5.4). An exception is the noble gas family in which helium has two valence electrons and the other elements eight. The electron configurations in this family have a characteristic feature—the most stable unfilled orbital in each element is in a higher shell than the outer occupied shell.

The first row of the periodic table contains hydrogen and helium, the only elements with one occupied shell. Hydrogen resembles group IA elements because it has only one outer-shell electron, but it also resembles the halogens because it has one less outer-shell electron than a noble gas (in other words, room for one more low-energy electron).

Second-row elements differ from each other in their outer-shell arrangements. They have anywhere from one (Li) to eight (Ne) valence electrons and belong to eight different families. The elements of the third period have corresponding configurations in the third shell.

Potassium, an alkali metal, has one outer-shell electron ($4s^1$), and calcium (group IIA) has a complete $4s$ subshell. The next ten elements have different numbers of $3d$ electrons. The similarity in outer-shell configuration ($4s^2$) of the transition metals agrees with the similarity found for these elements, but their $3d$ electrons are also accessible for chemical reactions. As a result, scandium can act as if it has three outer-shell electrons, titanium four, vanadium five, chromium six, and manganese seven. This explains the similarities between subgroups IIIB through VIIB and their corresponding major groups. Copper (group IB) resembles the alkali metals in having one outer-shell electron, and Zn is similar to group IIA elements in having two outer-shell electrons and filled inner subshells.

The lanthanide elements have two outer-shell electrons ($6s^2$) and similar fifth shells (some lanthanides have a single $5d$ electron), but differ in the number of $4f$ electrons. The lanthanides are very similar chemically because of their similarities in the two outermost shells. Differences in $4f$ configurations do not exert much influence on chemical properties.

Table 5.4 Outer-Shell Electron Configurations

Group	IA	IIA	IIIA	IVA	VA	VIA	VIIA	0
	H $1s^1$						H $1s^1$	He $1s^2$
	Li $2s^1$	Be $2s^2$	B $2s^22p^1$	C $2s^22p^2$	N $2s^22p^3$	O $2s^22p^4$	F $2s^22p^5$	Ne $2s^22p^6$
	Na $3s^1$	Mg $3s^2$	Al $3s^23p^1$	Si $3s^23p^2$	P $3s^23p^3$	S $3s^23p^4$	Cl $3s^23p^5$	Ar $3s^23p^6$

Summary

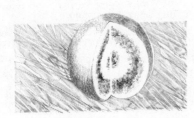

1. Light is a unique form of energy that travels through space at a constant velocity (c) of 3.00×10^8 m/sec.

2. Because light exhibits interference phenomena such as diffraction, it can be regarded as a wave phenomenon. Each form of light has a characteristic wavelength (λ) and frequency (v).

$$v = \frac{c}{\lambda}$$

3. Light energy occurs as quanta called photons. Planck's law: The energy of a photon is proportional to its frequency.

$$E = hv$$

4. When heated, each element emits a characteristic spectrum of light of discrete wavelengths.

5. The Bohr theory explained the emission spectrum of hydrogen by assuming that the electron in the atom was restricted to one of a set of discrete energy levels. A hydrogen atom emits a photon corresponding to a particular emission line when its electron falls from a higher energy level to a lower one.

6. The frequencies of the x rays emitted by an element are related to the atomic number of the element.

7. Any particle has a wavelength that is inversely proportional to its momentum.

$$\lambda = \frac{h}{\text{Momentum}}$$

8. The quantization of energy of a particle restricted in space (such as an electron in an atom) arises from the requirement that its wave not interfere with itself.

9. Heisenberg's uncertainty principle: It is impossible to measure both the position and momentum of an object with absolute precision.

$$(\text{Position uncertainty})(\text{Momentum uncertainty}) \approx h$$

10. Provided we know the relationship between the potential energy and position of a particle, we can describe its wave properties and quantized energy levels by solving the Schrödinger wave equation to obtain the wave function of the particle.

11. The square of the value of the wave function of a particle at any point is the probability of finding the particle at that point.

12. The allowed energy levels for a particle are those corresponding to realistic wave functions.

13. An orbital is a three-dimensional graph of a wave function of an electron. This graph resembles a cloud whose density in a particular region is proportional to the probability of finding the electron in that region.

14. The lowest energy level for an electron in a hydrogen atom is the 1s orbital. The second energy level of hydrogen contains four degenerate orbitals, that is, orbitals with identical energies. These four orbitals are a single 2s orbital and three 2p orbitals. The 2p orbitals have identical shapes but point in different directions. The third energy level consists of one 3s orbital, three 3p orbitals, and five 3d orbitals.

15. Three quantum numbers are needed to describe an orbital. The principal quantum number **n** can be any positive whole number. The azimuthal quantum number **l** can have any whole-number value ranging from zero to **n** − 1. The magnetic quantum number $\mathbf{m_l}$ can have whole-number values (including zero) ranging from −**l** to +**l**.

16. All orbitals with the same principal quantum number constitute a shell of orbitals. A subshell is a set of orbitals with the same **n** and **l** values.

17. An electron spin, a quantum number that determines the magnetic moment of an electron, can have one of two possible quantized values.

18. The order of subshell energies in an atom containing several electrons is

    $1s\, 2s\, 2p\, 3s\, 3p\, 4s\, 3d\, 4p\, 5s\, 4d\, 5p\, 6s\, 4f\, 5d\, 6p\, 7s\, 5f$

19. The Pauli exclusion principle: No two electrons in the same atom can be in identical quantum states; that is, they must differ in at least one quantum number. Only two electrons can occupy the same orbital, and they must have opposite spins. An s subshell can hold a maximum of 2 electrons, a p subshell 6 electrons, a d subshell 10, and an f subshell 14.

20. We can make a reasonable prediction of the ground-state electron configuration of an atom by filling the lowest energy subshell (the 1s subshell) first, going on to the next higher subshell, and continuing until all the electrons have been assigned to their energy levels.

21. Hund's rules: In a ground-state electron configuration, no orbital holds a pair of electrons unless every other orbital in the subshell holds at least one electron. The electrons in a subshell will have the maximum number of parallel spins.

22. Atoms with similar outer-shell, or valence, electron configurations are chemically similar. Members of the same chemical family usually have the same number of outer-shell electrons.

Vocabulary List

quantum mechanics	energy level	principal quantum number
light	excited state	shell
interference phenomena	quantum number	azimuthal quantum number
diffraction pattern	standing wave	subshell
wavelength	Heisenberg uncertainty principle	magnetic quantum number
photon	Schrödinger wave equation	electron spin
Planck's law	wave function	Pauli exclusion principle
Planck's theory	orbital	ground state
emission spectrum	degenerate orbital	Hund's rules
	nodal surface	valence electrons

Exercises

1. Define or describe the following terms.
   a. Wave function   b. Subshell   c. Shell   d. Stern-Gerlach experiment
   e. Quantum number

2. Which has the longer wavelength, ultraviolet light or infrared light?

3. Which has the higher frequency, a radio wave or an x ray?

4. Which has the greater photon energy, violet light or red light?

5. Which of the following statements are false? Provide correct statements in their place.
   a. A 1s electron in a hydrogen atom is always 0.053 nm from the nucleus.
   b. The Schrödinger equation is a general mathematical statement of the ideas of de Broglie.
   c. The 2s and 2p states of the hydrogen atom have the same energy.
   d. Degenerate orbitals are inherently unstable.

    e. There are no restrictions on the number of electrons an orbital can hold.
    f. A photon of red light has more momentum than a photon of an x ray.
    g. The wavelengths of visible objects are so much smaller than the dimensions of these objects that we cannot detect their wave properties.
    h. Although the wave nature of light can be demonstrated by experiment, the wave nature of particles is a matter of pure speculation.
    i. If we know the exact position of a particle at a given time, we know nothing about how it is moving.

6. The crests of a series of water waves are 2.5 m apart and appear to move at a speed of 8.0 m/sec. What is the frequency of these waves?

7. A calm spot occurs at a point 50.0 cm from one slit in a water wave barrier and 47.0 cm from the other. What is the largest possible wavelength for the water waves?

8. What is the frequency of light whose wavelength is 405 nm?

9. A light beam has a wavelength of 258 nm. What is its frequency?

10. What is the wavelength of a radio wave with a frequency of 1300 kHz (1.300×10^6 sec^{-1})?

11. An x ray has a frequency of 1.2×10^{18} sec^{-1}. What is its wavelength?

12. What is the energy of a photon of light whose frequency is 2.0×10^{14} sec^{-1}?

13. A warm object emits a photon having an energy of 4.0×10^{-20} J. What is the frequency of the emitted light?

14. What is the energy of a photon of light whose wavelength is 305 nm?

15. An x ray photon has an energy of 3.5×10^{-15} J. What is the wavelength of the x ray?

16. What is the wavelength of a proton (mass = 1.67×10^{-27} kg) moving at 4.0×10^3 m/sec?

17. What is the wavelength of a neutron (mass = 1.68×10^{-27} kg) moving at 500 m/sec?

18. An electron whose total energy (kinetic and potential) is -2.42×10^{-19} J gives up a photon as its energy falls to -21.76×10^{-19} J. What is the wavelength of the emitted light?

19. A major visible line in an atomic emission spectrum occurs at 405 nm. How much does the energy of an electron decrease as this photon is emitted?

20. How much uncertainty will there be in the velocity of an electron whose position is known within an uncertainty range of 0.02 nm. (An approximate answer is sufficient.)

21. If we determine the velocity of an electron to be $1.5 \times 10^6 \pm 2 \times 10^5$ m/sec, what is the approximate uncertainty in its position?

22. a. What is the maximum number of electrons that the fifth shell of an atom can hold?
    b. What subshells would the fifth shell contain? How many electrons would each subshell hold?
    c. How many orbitals would be present in a subshell with $l = 4$? How many electrons could each orbital hold?

23. Draw diagrams showing the arrangement of electrons and their spins in the ground-state configurations for the following subshells.
    a. A 2p subshell with four electrons
    b. A 3d subshell with four electrons
    c. A 4f subshell with ten electrons

24. What is the principal quantum number of the lowest shell that contains each of the following?
    a. A subshell with $l = 5$ b. An orbital with $m_l = -3$
    c. An electron with an upward spin

25. a. What are the possible values of l for a fourth shell electron?
    b. What is the value of l for a 4d electron?
    c. What possible values of m_l could a 4d electron have?

26. Which of the following sets of quantum numbers are impossible? State why in each case.

	n	l	m_l	m_s
a.	1	0	1	$+\frac{1}{2}$
b.	3	0	0	$-\frac{1}{2}$
c.	1	2	2	$+\frac{1}{2}$
d.	4	3	−3	$+\frac{1}{2}$
e.	5	2	1	$-\frac{1}{2}$
f.	3	2	1	0

27. A student lists the following sets of quantum numbers for the three electrons of lithium. What is wrong with these assignments?

	n	l	m_l	m_s
Electron 1	1	0	0	$+\frac{1}{2}$
Electron 2	1	0	0	$-\frac{1}{2}$
Electron 3	1	0	0	$+\frac{1}{2}$

28. Which of the following electron configurations are impossible? State why in each case. Which represent ground states? Which represent excited states?
    a. $1s^2 2s^2 2p^6 3s^2 3p^6 4s^1$
    b. $1s^2 2s^2 2p^6 3s^2 3p^6 3d^1$
    c. $1s^2 2s^2 2p^6 3s^2 3p^7$
    d. $1s^2 2s^2 2p^6 3s^2 3p^6 3d^{10} 4s^2 4p^5$
    e. $1s^2 2s^2 2p^6 3s^2 3p^6 3d^{11} 4s^2 4p^5$
    f. $1s^2 2s^2 2p^6 3s^2 3p^6 3d^{10} 4s^2 4p^6 4d^{10} 4f^1 5s^2 5p^6$

29. Which of the following electron configurations represent elements of the same periodic group? Which represents a noble gas? Which represents a member of group IVA? Which represents a transition metal? Which represents a lanthanide rare earth?
    a. $1s^2 2s^2 2p^6 3s^2 3p^6 3d^{10} 4s^2 4p^6 5s^2$
    b. $1s^2 2s^2 2p^6 3s^2 3p^6 3d^{10} 4s^2 4p^6 4d^{10} 4f^{10} 5s^2 5p^6 6s^2$
    c. $1s^2 2s^2 2p^6 3s^2 3p^6 3d^{10} 4s^2 4p^6 4d^{10} 4f^{14} 5s^2 5p^6 5d^5 6s^2$
    d. $1s^2 2s^2 2p^6 3s^2 3p^6 3d^{10} 4s^2 4p^6 4d^{10} 4f^{14} 5s^2 5p^6 5d^{10} 6s^2$
    e. $1s^2 2s^2 2p^6 3s^2 3p^6 3d^{10} 4s^2 4p^6 4d^{10} 4f^{14} 5s^2 5p^6 5d^{10} 6s^2 6p^2$
    f. $1s^2 2s^2 2p^6 3s^2 3p^6 3d^{10} 4s^2 4p^6 4d^{10} 4f^{14} 5s^2 5p^6 5d^{10} 6s^2 6p^6$
    g. $1s^2 2s^2 2p^6 3s^2 3p^6 3d^{10} 4s^2 4p^6 4d^{10} 5s^2 5p^6 6s^2$

30. How many unpaired electron spins are present in an atom of the following elements? Which of these atoms are paramagnetic?
    a. Ar   b. K   c. Ca   d. Cr

31. How many unpaired electron spins does each of the following atoms contain? Which of these atoms are paramagnetic?
    a. Mn   b. Fe   c. Co   d. N

32. Use the building principle to predict the electron configurations for each.
    a. Ca   b. Fe   c. Ba

33. Using the building principle, work out the electron configurations for each of the following atoms.
    a. Al   b. Pm   c. Br

34. What is the difference between the electron configuration of Zr (group IVB, fifth row) and Mo (group VIB, fifth row)?

35. Write out the outer-shell electron configuration (showing the number of electrons in each outer subshell) of an element belonging to periodic group VA.

36. Would a beam of calcium atoms be split in the Stern-Gerlach experiment? Explain your answer.

37. What result would have been obtained in the Stern-Gerlach experiment if each silver atom could have a magnetic dipole with any possible value over a wide range?

38. What is the momentum of a photon whose wavelength is 12.0 nm?

39. An electron has a wavelength of 0.85 nm. How fast is it moving?

40. Atomic absorption spectroscopy is an analytic technique based on the observation that each element absorbs characteristic wavelengths of light. The absorption wavelengths correspond exactly to the wavelengths of the emission lines. Use the Bohr model to explain atomic absorption spectroscopy.

41. Suppose the order of increasing energies of the subshells were 1s 2s 2p 3s 3p 3d 4s 4p 4d 4f.... (In other words, a shell must fill completely before electrons can go into the next higher shell.)
    a. Draw the periodic table that would result from these electron configurations. Make it large enough to hold 92 elements.
    b. What are the lengths of the periods of this table?
    c. What are the atomic numbers of the elements that belong to the same family as element number 5? How many outer-shell electrons do these elements have?

42. Why is it necessary to know something about the relative positions of the electrons in an atom in order to estimate the potential energies of the electrons?

43. Without referring to a table of electron configurations or using the building method, locate at least one position in the periodic table representing an element that has the following characteristics.
    a. A partially filled 4f subshell   b. A partially filled 4d subshell
    c. Four valence electrons   d. Two vacant positions in its outer p subshell
    e. No vacancies in its outer shell
    f. An element, other than your answer for part e, that has no partially filled subshells

44. a. What is the change in energy as an electron falls from the fifth energy level of a hydrogen atom to its first energy level?
    b. What is the frequency of the photon emitted in this process?
    c. What is the wavelength of the emitted light?

EXERCISES

45. What energy level transition in the hydrogen atom produces light with a wavelength of 1005.2 nm?

46. Copper, an element of group IB, resembles the group IA elements in forming a monochloride, but it also resembles the IIA elements (such as Ca) in forming a dichloride. Can you explain this in terms of the electron configuration of copper? [Hint: Unlike group IA elements, the difference between the ground state and the first excited state of copper is quite small.]

47. **a.** Use the building method to predict the electron configurations of the lanthanide elements (atomic numbers 57 through 71).
    b. Assuming that the chemical behavior of these elements depends only on the number of 6s and 5d electrons, predict the formulas of their chlorides.
    c. Look up the actual ground-state configuration of each lanthanide element and predict the formula of its chloride.
    d. Check the validity of your predictions in parts b and c by referring to the *Handbook of Chemistry and Physics* (Chemical Rubber Publishing Co.).

48. A light beam with a wavelength of 5.0×10^{-5} cm passes through two narrow slits 2.0×10^{-4} cm apart.
    a. Make a scale-drawing of these slits and show either by a drawing or by trigonometry that wave reinforcement occurs at angles of 0°, 14.5°, and 30°. [Hint: Measure the angles from a point midway between the two slits. Look at points far from the slits so that the two light beams are nearly parallel.]
    b. Show that destructive interference occurs at 7° and 22°.

49. One of the interesting theoretical problems in quantum mechanics involves the quantized energy levels of an "electron in a box." The electron is kept within a narrow region by very high potential-energy barriers on all sides, but there is no force acting on the electron while it is within this region, so its potential energy is zero.
    a. Use de Broglie's ideas to predict the wavelength, number of nodes, momentum, total energy, and kinetic energy for the lowest-three energy levels of an electron within a one-dimensional box 0.10 nm long. (See Figure 5.14.) [Hint: Consider the standing waves for a guitar string 0.10 nm long.]
    b. The de Broglie waves for the electron in the box are identical to the solutions for the Schrödinger equation for this system. At what positions in the box are you most likely to find an electron in the third energy level? Where is this electron least likely to be?

50. The solutions for the Schrödinger equation for a particle in a one-dimensional box have the general form

$$\psi = \sin\left(\frac{2\pi}{h}\sqrt{2mE} \cdot x\right)$$

where m is the mass of the particle, E its total energy, and x the distance from one wall of the box. (The quantity $(2\pi/h)\sqrt{2mE} \cdot x$ is expressed in radians; there are 2π radians in a complete circle.) The condition of physical reality for the wave functions is that the chances of the electron being at either wall are zero. In other words,

$\psi = 0$ when $x = 0$

$\psi = 0$ when $x = a$

where a is the length of the box.
a. Show that a wave function meets this condition of reality if, and only if, $E = \mathbf{n}^2 h^2/8ma^2$.
b. Does the energy of an electron in its ground state increase, decrease, or stay the same if the size of the box is increased?

chapter 6

The Electronic Structures of Ions and Molecules

Methane, CH$_4$, a molecule of a typical, covalent substance

We used quantum mechanics in the last chapter to predict the electron configurations of isolated atoms. These configurations agree with the observed atomic emission spectra and with the chemical behavior of the elements. But a knowledge of atomic electron configurations gives only a partial explanation of chemical behavior because it tells us nothing about the chemical bonds that hold molecules of an element together and stabilize its compounds. These chemical bonds can be described by applying the principles of quantum mechanics presented in Chapter 5 to molecules. No new forces are necessary to explain chemical bonding; they are electrostatic. The criterion used to decide whether or not a molecule is stable will be its energy—if the energy of a molecule (for example, Cl$_2$) is less than the energy of its constituent atoms (two Cl atoms), we regard it as stable.*

There are a variety of bonds, and the differences between various bonds produce profound differences in the chemical and physical properties of substances. To help understand these differences, we classify a particular bond as one of four types: (1) ionic, (2) covalent, (3) polar covalent, or (4) metallic. We will consider the first three types in this chapter. A discussion of metallic bonding appears in Chapter 9.

---

*This criterion is an approximation that neglects several potentially important factors. Energy is not the only factor governing stability—probability (entropy) also plays a part. (This will be discussed in Chapter 8.) Also, we have not considered the possibility that a compound might decompose into other compounds rather than its elements (for example, 2 NO$_2$ $\longrightarrow$ 2 NO + O$_2$). Finally, some compounds are less stable than their elements but decompose so slowly that we can keep them around indefinitely. However, none of these effects upsets our conclusions about the specific cases discussed in this chapter.

Ionic bonding occurs when one type of atom gives one or more electrons to another type of atom. This produces positively and negatively charged atoms called **ions**, and the electrostatic attraction between the ions is responsible for the bonding.

$$\text{Na}\cdot \; + \; \cdot\ddot{\underset{..}{\text{Cl}}}\!: \; \longrightarrow \; \text{Na}^+ \; + \; :\!\ddot{\underset{..}{\text{Cl}}}\!:^-$$

(The dots in the equation represent valence electrons.) The covalent bond is the result of two atoms sharing electrons between them.

$$:\!\ddot{\underset{..}{\text{Cl}}}\cdot \; + \; \cdot\ddot{\underset{..}{\text{Cl}}}\!: \; \longrightarrow \; :\!\ddot{\underset{..}{\text{Cl}}}\!:\!\ddot{\underset{..}{\text{Cl}}}\!:$$

The polar covalent bond also involves electron sharing between atoms but the sharing is unequal—the shared electrons spend more time around one bond partner than the other. We will examine each of these bond types in detail and discuss the relationship between the physical properties of a substance and the nature of its chemical bonds.

6.1 Ions

Ionization Potential and Electron Affinity

Monatomic ions are single atoms that have net electric charges (for example, Ca^{2+}, I^-, or S^{2-}). A **polyatomic ion** is a group of atoms (such as NH_4^+ or SO_4^{2-}) bearing an electric charge. An atom takes on a charge either by losing one or more of its electrons and becoming a positive ion such as Na^+, or by gaining one or more extra electrons to form a negative ion such as Cl^-. Positive ions are called **cations**, and negative ions are **anions**.

Removing electrons from the vicinity of a positive nucleus requires work against the attractive force of the nucleus. The energy required to remove an electron from a gaseous atom is called the **ionization potential**. The **first ionization potential** of an element can be measured by determining the minimum kinetic energy a colliding electron must have to remove an electron (e^-) from an isolated gaseous atom (M), or we can estimate it from spectroscopic data. The first ionization potential can be represented as

$$M(g) \longrightarrow M^+(g) + e^-$$

The first ionization potential depends on the electron configuration of an atom, reflected in its position in the periodic table. Figure 6.1 shows how first ionization potential varies with atomic number. Elements to the left of the periodic table (metals) have low ionization potentials, while ionization potentials of the elements to the right (nonmetals) are higher. Generally, ionization potential increases across a period from left to right because nuclear charge increases and atomic radius decreases slightly. Both of these factors increase the attraction between the nucleus and the outer-shell electron, making it harder to remove the electron. The ionization potential of a group IA element is always much lower than that of its preceding noble gas because the valence electron of group IA elements occupies a new shell that is far from the nucleus and shielded against nuclear attractive force by the repulsions from the inner-shell electrons.

There is a modest tendency for the ionization potentials of members of a periodic family to decrease going down the column. The increase in the average distance of the valence electrons from the nucleus, coupled with the increased shielding of the positive nuclear charge by the inner-shell electrons, more than compensates for the increased nuclear charge.

Example 6.1

Which of the following pairs of atoms has the higher first ionization potential? Do not refer to a table of ionization potentials, but make your judgment on the position of the atoms in the periodic table.   **a.** Mg or S   **b.** N or As

Solution

a. Magnesium and sulfur both occupy the third period, but sulfur is a member of group VIA and magnesium a member of group IIA. Since sulfur lies to the right of magnesium in the periodic table, it has the higher ionization potential.

b. Nitrogen and arsenic are both group VA elements. Nitrogen is a member of the second period and arsenic lies in the fourth, therefore nitrogen has the higher ionization potential.

The **second ionization potential** of an element is the energy needed to remove a second electron from the gaseous M^+ ion to form an M^{2+} ion. Higher ionization potentials have corresponding definitions.

$$M^+(g) \longrightarrow M^{2+}(g) + e^- \quad \text{(second ionization potential)}$$

$$M^{2+}(g) \longrightarrow M^{3+}(g) + e^- \quad \text{(third ionization potential)}$$

The second ionization potential of an element is always larger than the first ionization potential, and the third ionization potential is still larger. As long as valence electrons are involved, the difference between successive ionization potentials is fairly small (500–1000 kJ/mol). However, the increase in ionization potential when an inner-shell electron is removed is particularly large—several tens of thousands of kilojoules per mole. The inner-shell electron is much closer

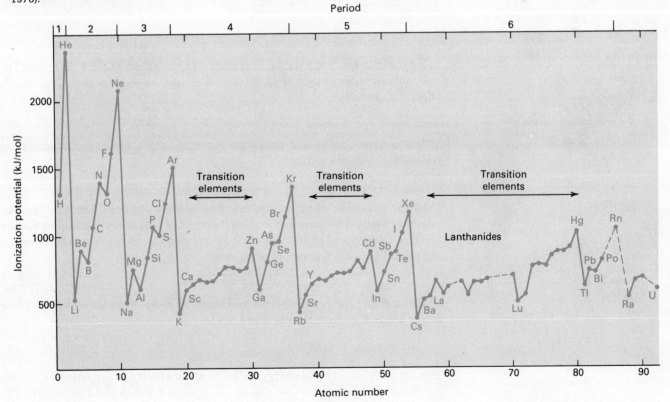

Figure 6.1
First ionization potentials of the elements. *Source:* Document NSRDS-NBS 34 (Washington, D.C.: National Bureau of Standards, 1970).

Figure 6.2
First, second, and third ionization potentials of some elements of groups I, II, and III.

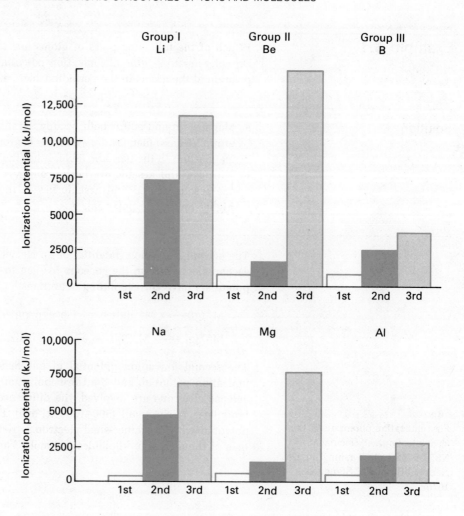

to the nucleus and much harder to remove. A comparison of the first, second, and third ionization potentials of Li, Be, and B as well as Na, Mg, and Al shows this effect (Figure 6.2).

Example 6.2

Which is larger, the second ionization potential of potassium, or the second ionization potential of calcium?

Solution

Calcium, a member of group IIA, has two valence electrons and can lose both relatively easily. The second ionization potential of potassium, a group IA element, involves the removal of an inner-shell electron and is therefore much larger than the second ionization potential of calcium.

When an atom captures an electron and forms an anion, potential energy decreases as the nucleus attracts the electron.

$$e^- + M(g) \longrightarrow M^-(g)$$

This decrease in potential energy, called **electron affinity**, is estimated by measuring the ionization potential of the anion. Table 6.1 shows the electron affinities of several elements. Electron affinities are small compared to ionization potentials. They appear to be largest for the halogens. A halogen anion, such as Cl^-, has the

6.1 IONS

Table 6.1 **Electron Affinities**

Element	Electron affinity (kJ/mol)
H	72
Li	59
C	121
O	142
S	200
F	331
Cl	348
Br	324
I	296

smallest radius of any anion in its period and, with the exception of the adjacent noble gas, the largest nuclear charge. This combination of large nuclear charge and small size gives the halogens the greatest electrostatic attraction for electrons. The electron affinity of a noble gas is low because the electron must enter a higher energy level far from the nucleus.

Ionic Compounds

The most characteristic physical property of an **ionic compound**, such as sodium chloride (table salt), is the ability to conduct an electric current when it is in a liquid state or dissolved in water. In addition, ionic compounds are crystalline and tend to have high melting points and low volatilities, and they are usually more soluble in water than in nonpolar solvents such as benzene.

The model of an ionic crystal is a **lattice** array—a regular three-dimensional arrangement of cations surrounded by anions and each anion by cations (Figure 6.3). There are no recognizable molecules in the ionic crystal, just a continuous array of ions extending through the entire crystal. This arrangement gives the maximum attractive force between oppositely charged ions while minimizing repulsion between like ions. The net attractive force in the lattice is a major source of stability of an ionic compound.

Molten ionic compounds are good electric conductors, but the solid crystals

Figure 6.3
Ionic crystal lattice.

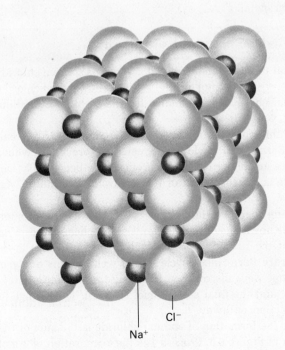
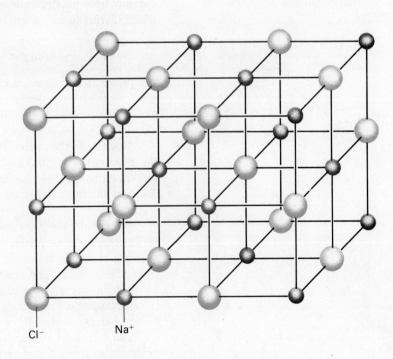

Figure 6.4
The conduction of electric current by ions in the molten or dissolved states.

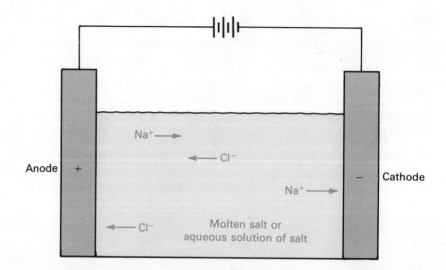

do not conduct well. Electrical conduction requires freely moving charged particles; when ions are held in their lattice positions, they cannot respond to an electric field. Melting frees the ions from the lattice, and their movement in the appropriate direction in an electric field produces a current (Figure 6.4). Dissolving an ionic substance also destroys the crystal lattice, freeing the ion and producing the same conducting effect.

The high melting point and low volatility of an ionic compound result from the strong attractive forces that hold the ions in their lattice positions and maintain the crystal's shape even when the ions are vibrating energetically at high temperature. Ionic solids like KCl and LiBr tend to dissolve in water because the water molecules interact with ions in a manner that insulates their electric charges. This greatly reduces the energy needed to separate the ions from the crystal lattice, and the ions can break away into the solution relatively easily. Other solvent molecules, particularly poor electric insulators such as benzene, cannot interact effectively with ions, and these solvents do not dissolve ionic substances readily.

The Stability of Ionic Compounds

The release of large quantities of energy accompanies the formation of sodium chloride from its elements—a fact that is obvious to anyone foolhardy enough to pass chlorine gas over sodium metal.

$$\text{Na(s)} + \frac{1}{2}\text{Cl}_2\text{(g)} \longrightarrow \text{NaCl(s)} + 410 \text{ kJ heat released}$$

The production of so much heat indicates that 1 mol of crystalline sodium chloride has a much lower potential energy than 1 mol of solid sodium and $\frac{1}{2}$ mol of chlorine gas. This decrease in potential energy is responsible for the stability of sodium chloride.

To analyze the source of this potential energy decrease, we arbitrarily break the reaction down into the **Born-Haber cycle**—a sequence of processes whose individual heats either are measurable or can be theoretically estimated. The net result of the Born-Haber cycle is the formation of NaCl(s) from Na(s) and $\text{Cl}_2\text{(g)}$ (Figure 6.5).

We should avoid confusing the Born-Haber cycle with the way the reaction really occurs (its *mechanism*). Knowledge of the mechanism of a chemical reaction requires studying the rates of the reaction and the factors that influence these rates (Chapter 12). The arbitrary Born-Haber cycle works well for this analysis because the overall heat change depends only on the nature of the initial state (1 mol Na(s) and $\frac{1}{2}$ mol $\text{Cl}_2\text{(g)}$) and the final state (1 mol NaCl(s)). The change in heat is independent of the reaction pathway.

The total heat change for the formation of sodium chloride is the sum of the heats of processes 1–4 shown in Figure 6.5. Process 1 is the evaporation of 1 mol

Figure 6.5
The Born-Haber cycle for the formation of NaCl(s).

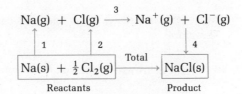

1. Heat of sublimation of Na(s) = 109 kJ/mol required
2. Heat of dissociation of Cl_2(g) = 121 kJ/mol Cl required
3. Heat of electron transfer from Na(g) to Cl(g)
   = Ionization potential of Na − Electron affinity of Cl
   − 493 kJ/mol − 364 kJ/mol = 129 kJ/mol
4. Theoretical estimate of the crystal-lattice energy of NaCl(s) = 769 kJ/mol released

Total heat change = 109 + 121 + 129 − 769 = −410 kJ/mol

Measured heat of formation = −411 kJ/mol

(The negative value indicates that heat is released.)

of sodium metal to form the vapor. Since sodium atoms are held in the metallic crystal lattice by attractive forces, this process both increases potential energy and requires external heat energy. Process 2, the dissociation of ½ mol of molecular chlorine to monatomic chlorine, requires heat to break the chemical bonds of the chlorine molecules. We can break the energy of process 3 down into the energy needed for ionization of sodium atoms and the energy released by the capture of electrons by chlorine atoms, and the heat for process 3 can be computed by subtracting the electron affinity of chlorine from the ionization potential of sodium.

The bookkeeping completed so far shows that gaseous Na^+ and Cl^- have a much higher potential energy than sodium metal and chlorine gas. But to compute the potential energy of solid sodium chloride relative to the starting reactants we must also consider the strong attractive forces between the positive Na^+ ions and the negative Cl^- ions within the crystal lattice. This **crystal-lattice energy**, which is the energy that would be released in process 4, is the major source of stability of an ionic compound. The crystal-lattice energy, estimated by theoretical calculations in which the electrostatic forces in the crystal are computed on the basis of the ionic sizes and charges, agrees well with the experimental value obtained from the Born-Haber calculation. Crystal-lattice energy is greatest for crystals containing small, highly charged ions because electrostatic attraction is greatest when opposite charges are close together and when these charges are large.

Example **6.3**

Calculate the heat of formation of CsF from the following data:

Heat of sublimation of Cs(s) = 79 kJ/mol

Heat of dissociation of F_2(g) = 77 kJ/mol required for each mole F(g) produced

Ionization potential of Cs = 376 kJ/mol

Electron affinity of F = 333 kJ/mol

Crystal-lattice energy of CsF(s) = 730 kJ/mol released

Solution

The total heat change is the sum of the three energy requiring processes (sublimation of Cs, bond dissociation of F_2, and ionization of Cs) less the energies of the two energy releasing steps (electron capture by F and crystal-lattice formation). Therefore,

$$\text{Heat of formation} = 79\,\frac{\text{kJ}}{\text{mol}} + 77\,\frac{\text{kJ}}{\text{mol}} + 376\,\frac{\text{kJ}}{\text{mol}} - 333\,\frac{\text{kJ}}{\text{mol}} - 730\,\frac{\text{kJ}}{\text{mol}}$$

$$= -531\,\frac{\text{kJ}}{\text{mol}}$$

The formation of 1 mol CsF releases 531 kJ/mol of heat.

Stable Ions and Electron Configuration

The charges of the cation and anion control the formula of an ionic compound because the ionic crystal is electrically neutral; there must be an equal amount of positive and negative charge. Sodium chloride, composed of Na^+ and Cl^- ions, has the formula NaCl, whereas magnesium chloride, containing Mg^{2+} and Cl^-, is $MgCl_2$.

Example 6.4

What is the formula of the compound formed from K^+ and S^{2-} ions?

Solution

Since the charge on each S^{2-} ion is twice the charge on a K^+ ion, there must be twice as many K^+ ions to balance the charge. The simplest formula of the compound is therefore K_2S.

The charge of an ion in a stable ionic compound depends on the electron configuration and thus on the location of the element in the periodic table. Although the correlation of ionic charge with the position in the periodic table is not always obvious (the Born-Haber cycle shows that the stability of an ionic compound depends on many factors), there are many cases in which ionic charge is related in a simple way to the periodic group. The alkali metals (group IA) form 1+ ions, alkaline earth elements (group IIA) 2+ ions, and halogens (group VIIA) 1− ions. These relationships can be explained by the dominant factors in the Born-Haber cycle. The second ionization potentials of sodium and the other alkali metals are very large, and these energy-demanding second ionization potentials prevent the formation of Na^{2+} and $NaCl_2$. Magnesium and the other alkaline earths have small second and large third ionization potentials. The loss of both valence electrons by magnesium produces a small, intensely charged ion that forms stable crystals, such as $MgCl_2$, with large lattice energies. In this case, the increase in lattice energy dictates the loss of two electrons, while the large third ionization potential prevents the loss of a third electron.

6.2 Covalent Bonds

The Properties of Covalent Substances

A prime example of a **covalent substance** is methane (CH_4), the major component of natural gas. The properties of methane, like those of other covalent substances, show clearly that there are no ions present in the substance. The most obvious indication of the absence of ions in methane is that its liquid state does not conduct electricity—there are no charged particles free to respond to an electric force. Another important difference between methane and ionic substances is that we can prove experimentally that methane molecules, each consisting of one carbon atom and four hydrogen atoms, do exist. In contrast, there are no definable molecules in an ionic crystal lattice. Furthermore, the physical properties of methane (it is a gas at room temperature and does not dissolve in water) show that no strong interionic attractions hold the crystal lattice of solid CH_4 together.

All covalent substances, including many polyatomic elements such as N_2, H_2, O_2, and the halogens, show the characteristic covalent property of not conducting electricity in their liquid states. Other physical properties of these substances, which depend on their molecular structures and masses, vary greatly. Methane is a gas, benzene (C_6H_6) a liquid, naphthalene ($C_{10}H_8$) a solid with a low melting point, and diamond (a form of carbon) has a very high melting point. Some covalent substances—quartz (SiO_2) and diamond are two examples—do not have definite molecular masses but instead consist of collections of molecules of various sizes. Covalent substances with low molecular masses have low melting points and high volatilities. Table 6.2 gives comparative data for a covalent substance, butane (C_4H_{10}, molecular mass = 58 amu), and an ionic substance, NaCl (formula mass = 58.5 amu).

In 1916, G. N. Lewis proposed the existence of a chemical bond to explain the nature of covalent substances. He suggested that two atoms could share pairs of electrons, creating an attractive force called a **covalent bond** between the atoms. In the simplest example, two hydrogen atoms each contribute an electron to a covalent bond that holds the H_2 molecule together.

$$H\cdot \; + \; \cdot H \longrightarrow H:H$$

The covalent bond is also commonly represented by a line joining the two atoms, each line representing a shared electron pair.

$$H—H$$

A molecule of methane is held together by four covalent bonds, one from each of the four hydrogen atoms to the carbon atom.

$$\begin{array}{c} H \\ | \\ H—C—H \\ | \\ H \end{array}$$

The covalently bonded structure for methane explains its properties. There is no conduction of electricity through methane because its molecules are neutral and will not respond to an electric force. Methane maintains its molecular structure through various physical changes of state such as liquefaction and evaporation because the strong covalent bonds within the molecule maintain its molecular identity. The absence of strong electrostatic forces between methane molecules means that solid methane has a weak crystal lattice and explains the low melting point and low boiling point of methane.

Table 6.2   Comparison of Properties of Butane and Sodium Chloride

Substance	Melting point (°C)	Boiling point (°C)	Solubility in water	Electrical conductance
Butane (C_4H_{10})	−135	−0.5	Insoluble	Nonconducting liquid
Sodium chloride (NaCl)	801	1413	Soluble	Conducting liquid and conducting aqueous solution

Figure 6.6
The vector addition of electrostatic forces in a model H_2 molecule.

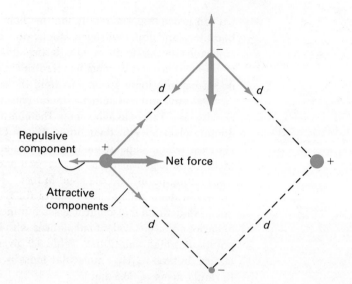

Molecular Quantum Mechanics

Lewis proposed the existence of covalent bonds but did not give a physical explanation of their stability; the concepts of electrostatic attraction and quantization of energy provide an explanation. A situation in which two electrons remain between two nuclei most of the time will lead to attraction rather than repulsion. A highly stylized model of this situation shows a square with two hydrogen nuclei at opposite corners and two electrons at the other corners (Figure 6.6). Calculation of the electrostatic forces yields a net force for each particle that points toward the center of the square.

A realistic model must take into account the motion of the electrons but, according to the Heisenberg uncertainty principle, we cannot determine the exact motion of electrons. By assuming the same quantum mechanical principles used in describing atoms, we can characterize the wave functions associated with the various electron energy levels in a molecule. The most stable wave function (lowest energy state) will give a high probability of finding electrons between the nuclei. Lewis's idea of a bond consisting of a shared pair of electrons translates well into the language of quantum mechanics, and this description does not require any new concepts or hypotheses.

It is not enough to show that there might be a favorable wave function for the model hydrogen molecule. We also should be able to compute accurately the physical properties of the real molecule on the basis of our model. Two important measurable properties of a molecule are **internuclear distance** (the distance between the nuclei of the two bond partners, also called the **bond length**) and **bond-dissociation energy** (the energy needed to split the bond to form two neutral fragments). Table 6.3 lists these two properties for a number of diatomic molecules.

Table 6.3 Physical Properties of Some Diatomic Molecules

Molecule	Bond-dissociation energy (kJ/mol)	Internuclear distance (nm)
H_2	435	0.074
N_2	941	0.110
O_2	493	0.121
F_2	150	0.144
Cl_2	238	0.200
Br_2	192	0.229
I_2	150	0.267
CO	1070	0.113
NO	677	0.115

The internuclear distance in H_2 is 0.074 nm, and it takes 434 kJ/mol to split a mole of hydrogen molecules into atoms.

By solving the Schrödinger wave equation for the wave functions and energy levels of our model H_2 molecule, we can calculate internuclear distance and bond energy. In practice, this is very difficult for even the simplest molecule and so we resort to approximate methods. In dealing with diatomic molecules, we will use an approximate method called the molecular-orbital approach (abbreviated MO). It is very useful for describing diatomic molecules, but there are other useful approximate systems, such as the valence-bond (VB) method, that you will encounter later in this course and in more advanced courses.

A simplification of the MO method uses the approximation that the wave functions of the H_2 molecule will be a combination of the wave functions of two hydrogen atoms. When two wave functions interact they, like water waves, can interact with constructive or destructive interference. The two ways in which the 1s wave functions of two hydrogens can be combined are (1) in-phase combination giving constructive interference and (2) out-of-phase combination giving destructive interference (Figure 6.7). There are always as many different interference combinations as there are atomic orbitals interacting, and each interference combination is associated with a **molecular orbital**.

In-phase combination produces the low-energy **bonding molecular orbital** (designated σ) of H_2 that, according to the Pauli exclusion principle, can hold both electrons of H_2, provided they have opposite spins. Electrons in the bonding orbital are more stable than they are in the atomic orbitals, partly because they spend a great deal of time *between* the nuclei and partly because they are less confined in space. One of the general results of the Schrödinger equation is that particles are more stable when allowed to move in larger regions.

An **antibonding molecular orbital** (designated σ^*) is associated with out-of-phase combination of the atomic orbitals. Electrons in the antibonding orbital are less stable than in the atomic orbital because they tend to stay *outside* of the internuclear region, causing a repulsion of the nuclei. In the ground state of H_2, both electrons are in the bonding orbital, forming a stable bond, and the antibonding orbital is empty. At high temperature, or after the absorption of the proper photon of light, an electron could jump from the bonding to the antibonding orbital to produce an "excited molecule" that would be reactive and have a strong tendency to dissociate into atoms.

Figure 6.7
The combination patterns and molecular orbitals resulting from two 1s atomic orbitals of hydrogen. (a) In-phase combination leading to a bonding molecular orbital. (b) Out-of-phase combination leading to an antibonding molecular orbital.

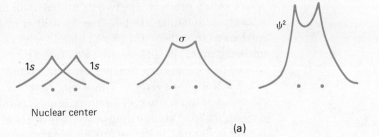
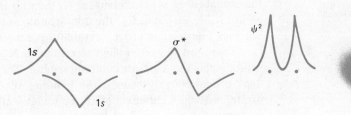

146   6 THE ELECTRONIC STRUCTURES OF IONS AND MOLECULES

Figure 6.8
Molecular-orbital energy levels and electron distributions of diatomic molecules of H and He.

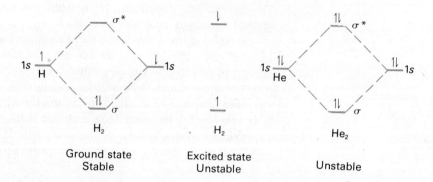

Covalent Bonding and Electron Configuration; He$_2$

Hydrogen exists as a covalent molecule, whereas He is monatomic; He atoms have little tendency to form covalent bonds. An MO analysis of a hypothetical He$_2$ molecule explains why. The joining of two helium atoms would involve the interaction of two atomic 1s orbitals to produce, as in the case of H$_2$, a bonding σ orbital and an antibonding σ^* orbital. The He$_2$ molecule would contain four electrons, and since each orbital can hold only two electrons, the bonding orbital would contain one electron pair and the antibonding orbital the other. The repulsive effect of the electrons in the antibonding orbital would cancel the attractive effect of the bonding electrons, and there would be no marked stability for He$_2$. Figure 6.8, which shows the energy levels and electron distributions for models of ground state H$_2$, excited H$_2$, and He$_2$, illustrates the effect of the relative number of bonding and antibonding electrons.

Diatomic Molecules of Second-Row Elements

Of the eight elements in the second row of the periodic table, one (Ne) exists as a monatomic gas, three (N$_2$, O$_2$, F$_2$) are diatomic gases, and four (Li, Be, B, C) are solids in which the sharing of electrons between adjacent atoms plays an important part in maintaining their crystal structures. These observations, as well as more detailed ones, are consistent with the model of the covalent bond.

As in the case of the first-row elements, we can work out bonding and antibonding orbitals for models of the diatomic molecules. As a first approximation, only valence orbitals play a significant part in bonding because the inner-shell 1s orbitals do not approach each other closely enough to produce any significant interaction. Another approximation is that only interaction between orbitals of similar energy is important; s orbitals interact with s orbitals, and p orbitals interact with p orbitals. The two 2s orbitals will interact to form a 2sσ bonding and a 2sσ^* antibonding orbital. One 2p orbital of each atom can point toward the nucleus of the other; this pair of atomic orbitals forms a 2pσ bonding and a 2pσ^* antibonding orbital. There are two other 2p orbitals on each atom, perpendicular to the line connecting the nuclei, that cannot overlap by the efficient head-on manner (called σ overlap). These can, however, overlap in a parallel way, called π overlap, forming two degenerate (same energy) 2pπ bonding orbitals and two degenerate 2pπ^* antibonding orbitals. Pictures of the types of 2p molecular orbitals appear in Figure 6.9. Figure 6.10 shows the energy levels for the molecular orbitals and the ground-state electron configurations of N$_2$, O$_2$, F$_2$, and the hypothetical Ne$_2$. Because of electrostatic effects, which we will not go into here, the order of the orbital energies is different for N$_2$ than for the other three cases. In N$_2$ (as well as in Li$_2$, Be$_2$, B$_2$, and C$_2$) the 2pπ orbitals are more stable than the 2pσ orbital, whereas this relative order of stability is reversed in O$_2$, F$_2$, and Ne$_2$.

There are ten outer-shell electrons in N$_2$ distributed among four bonding orbitals and the 2sσ^* antibonding orbital. The effect of the electron pair in the 2sσ^* orbital roughly cancels the bonding effect of the electron pair in the 2sσ orbital, leaving a net effect of six bonding electrons; N$_2$ is a stable molecule. Put in terms of Lewis covalent bonds, the six electrons form three covalent bonds, or a **triple covalent bond.**

In O$_2$ there are twelve valence electrons distributed among the four bonding orbitals, the 2sσ^* and the 2pπ^* antibonding orbitals. Four antibonding electrons roughly cancel the effect of four bonding electrons and leave a net bonding due to four electrons. This is equivalent to a **double covalent bond**. The model for the oxygen molecule, following Hund's rules (Chapter 5), has two electrons in the $2p\pi^*$ antibonding orbitals with parallel and unpaired electron spin. The model is paramagnetic, and the real O_2 molecule is paramagnetic. This agreement between the properties of the real molecule and the molecular model gives us confidence in the relevance of quantum mechanics to chemical bonding.

Figure 6.9
Molecular orbitals derived from $2p$ atomic orbitals.

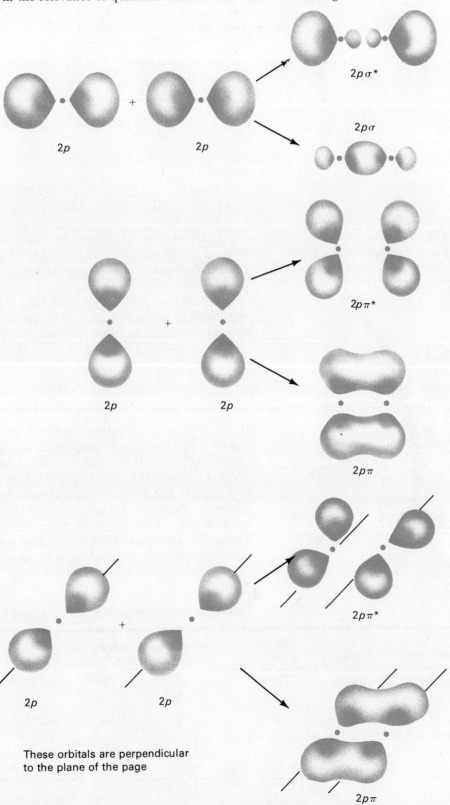

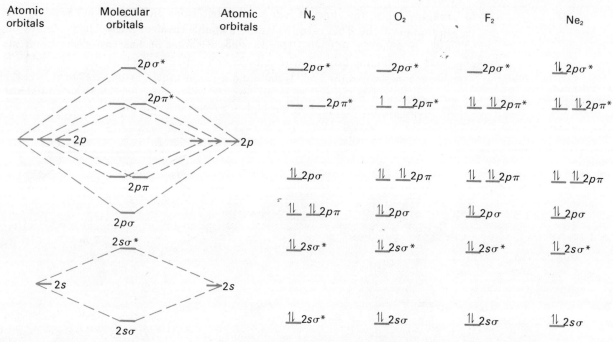

Figure 6.10
Electron configurations of several second-row diatomic molecules.

There are six antibonding and eight bonding electrons in F_2, and the net effect of one bonding pair is equivalent to a **single covalent bond**. In the hypothetical Ne_2 molecule there would be equal numbers of bonding and antibonding electrons and no significant covalent bonding.

Although we have not computed any bond distances or energies, we can predict the trend in these properties: the covalent bond will be longer and weaker if there are fewer net bonding electrons. The prediction that bonds get longer and weaker in the sequence N_2, O_2, F_2, and Ne_2 is borne out by the data in Table 6.3.

6.3 Polar Covalent Compounds

Properties of Polar Compounds

Hydrogen chloride (HCl) shows many of the typical properties of a covalent substance. It is a gas at room temperature and a poor conductor in the liquid state, and it has a definite molecular formula. However, several of its properties are atypical for a normal covalent substance. It has a higher melting point, a higher boiling point, and a higher degree of water solubility than a normal covalent substance (see Table 6.4). Although pure liquid hydrogen chloride is a better electrical insulator than normal covalent substances, its aqueous solutions are very effective conductors. Hydrogen chloride is a **polar covalent substance**. The differences in properties between covalent and polar covalent substances indicate some difference in the nature of their chemical bonds.

Table 6.4 The Physical Properties of Some Polar and Nonpolar Substances

Substance	Nature	Molecular mass (amu)	Melting point (°C)	Boiling point (°C)	Solubility in water	Electrical conductance
HCl	Polar	36.5	−112	−84	Soluble	Conducting aqueous solution
O_2	Nonpolar	32.0	−218	−183	Slightly soluble	Nonconducting aqueous solution
F_2	Nonpolar	38.0	−223	−188	Reacts with water to form O_2 and HF	

Polar Covalent Bonds

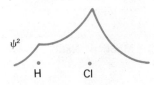

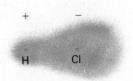

Figure 6.11
The electron distribution of bonding electrons in HCl.

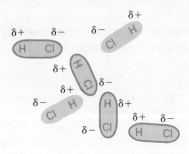

Figure 6.12
The intermolecular attraction of polar molecules.

Figure 6.13
The orientation of polar molecules in an electric field.

The chemical bond in a hydrogen chloride molecule is essentially covalent; it involves the sharing of an electron pair between two nuclei. The difference between the bond in HCl and the one in H_2 is that the electrons in HCl are shared between two dissimilar nuclei, whereas those in H_2 are shared between two similar nuclei. Because hydrogen and chlorine atoms differ in their nuclear charges and the number of inner-shell electrons, the tendency of a chlorine nucleus to attract electrons is greater than the attracting force of a hydrogen nucleus. As a result, there is a much greater probability of finding a shared electron near the chlorine nucleus than near the hydrogen nucleus (Figure 6.11).

Such an unevenly shared electron pair is a **polar covalent bond**. In HCl and other diatomic polar covalent molecules, there is a separation of charge between the two ends of the **dipolar** molecule. The chlorine atom, with an excess of electrons, is negatively charged, and the hydrogen atom, deficient in electrons, is positively charged. The total charge of the molecule is zero, but the two atoms have opposite charges that are smaller than the charge of one electron. We use the Greek letter δ (delta) and the appropriate sign to designate these partial positive or partial negative charges.

$$^{\delta+}H\!\!-\!\!Cl^{\delta-}$$

The physical properties of HCl reflect the charge separation in the molecule. The higher melting and boiling points are due to stronger intermolecular attractions, which result from the electrostatic attraction of the positive end of one dipole molecule for the negative end of another (Figure 6.12). Neutral molecules that are dipolar tend to orient themselves with their positive ends pointing toward the negative pole of an external electric field and their negative ends toward the positive pole. Although, due to thermal motion, only a small fraction of the molecules line up in this way, the number is definitely more than predicted by probability. The oriented molecules set up an opposing electric field resulting in the observed insulating ability of polar substances (Figure 6.13).

Solutions of hydrogen chloride in water are good conductors because a chemical reaction between water and hydrogen chloride produces the ions needed for conduction. The polar bond is split unevenly; the chlorine atom retains both electrons and forms Cl^-, while the hydrogen nucleus combines with a water molecule to form H_3O^+.

$$H_2O + HCl \longrightarrow H_3O^+ + Cl^-$$

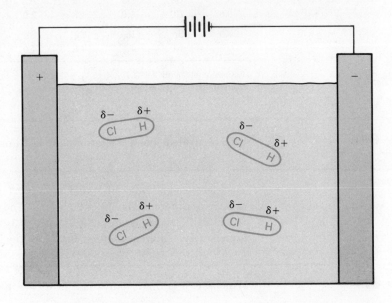

Figure 6.14
The sequence of polarities of third-row chlorides.

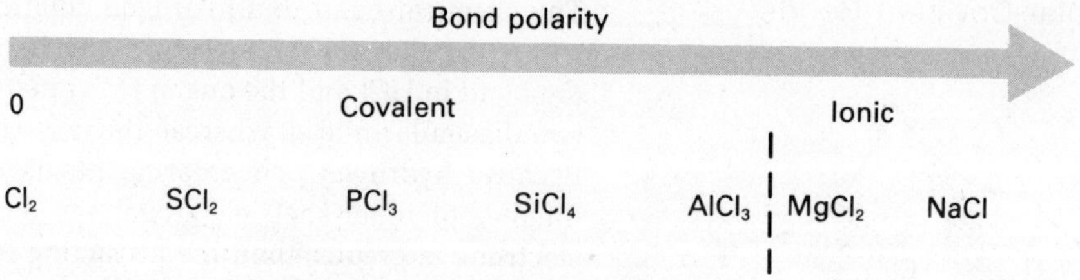

Bond polarity →

0 — Covalent — Ionic

Cl_2 SCl_2 PCl_3 $SiCl_4$ $AlCl_3$ | $MgCl_2$ $NaCl$

Electronegativity

The concept of the polar covalent bond raises the possibility of a unified view of chemical bonding. Nonpolar covalent and ionic substances, although quite different in properties, may be regarded as extreme cases in a series of substances of different polarities. Ionic substances, such as sodium chloride, have "bonds" that are so polar that we have difficulty detecting independent molecules. Nonpolar substances have no measurable charge separation. In between, we have a sequence of covalent molecules of differing polarities (Figure 6.14).

Bond polarity depends on the two bond partners and their relative abilities to attract a covalent bond electron pair. **Electronegativity** is a unitless number that expresses, on a *relative scale*, a bonded atom's electron-attracting ability. Electronegativities were calculated by Linus Pauling from covalent bond energies, which increase with polarity, and from a comparison of the ionization potential and electron affinity of an atom. Electronegativity increases across a row of the periodic table from left to right, and generally decreases from top to bottom down a column (Table 6.5).

The *difference* in electronegativity between two atoms is a measure of the polarity of a covalent bond between them. The more electronegative atom is the negative end of the polar bond.

Table 6.5 Electronegativities (Pauling Scale)

			H			
			2.1			
Li	Be	B	C	N	O	F
1.0	1.5	2.0	2.5	3.0	3.5	4.0
Na	Mg	Al	Si	P	S	Cl
0.9	1.2	1.5	1.8	2.1	2.5	3.0
K	Ca	Ga	Ge	As	Se	Br
0.8	1.0	1.6	1.8	2.0	2.4	2.8
Rb	Sr	In	Sn	Sb	Te	I
0.8	1.0	1.7	1.8	1.9	2.1	2.5

Transition elements 1.3–2.4
Lanthanides and actinides ≈ 1.2

Example 6.5

Which covalent bond is more polar, P–Cl or S–Br?

Solution

The electronegativities of P and Cl are 2.1 and 3.0, respectively, so the electronegativity difference is $3.0 - 2.1 = 0.9$. The electronegativity difference between S (2.5) and Br (2.8) is only $2.8 - 2.5 = 0.3$. The P–Cl bond has the greater electronegativity difference and is therefore the more polar.

Example 6.6

Which atom is the negative end of a N–S bond?

6.3 POLAR COVALENT COMPOUNDS

Solution

Nitrogen (electronegativity 3.0) is more electronegative than sulfur (2.5), so the nitrogen atom has a partial negative charge and the sulfur atom a partial positive one.

$$^{\delta+}S-N^{\delta-}$$

When the electronegativity difference is very large, the substance shows typical ionic behavior. It is difficult to give an exact borderline value for the electronegativity difference of an ionic substance. In general, a bond with an electronegativity difference of more than 2.0 is usually ionic, and a bond with an electronegativity difference less than 1.5 is usually covalent. (The chemical bonds of hydrogen are the major exceptions to these rules of thumb.)

Molecular Shape and Molecular Polarity

When a molecule contains three or more atoms, there is a complicating effect in polarity. For example, carbon dioxide should, according to the electronegativity difference between C and O, have polar bonds, and the chemical reactions of CO_2 bear this out. However, CO_2 is a poor electrical insulator and has a dipole moment of zero, the **dipole moment** being a measure of the charge separation between the two ends of a molecule. How do we resolve this contradiction of bond polarity without molecular polarity?

The problem is the existence of two polar $C=O$ bonds in CO_2 (the double line stands for a double covalent bond) and the orientation of these two bonds. The CO_2 molecule is linear—the molecule is straight with the oxygen atoms at either end and the carbon atom in the middle—rather than bent. The dipole moments of the two $C=O$ bonds, with identical polarities but pointing in opposite directions, cancel each other producing a nonpolar molecule.

Water is also a molecule with very polar bonds, and its chemical reactions reflect its polarity. It is, unlike CO_2, a good insulator and has a high dipole moment. This can happen only if water has a bent structure (Figure 6.15). Independent evidence confirms that water molecules are bent.

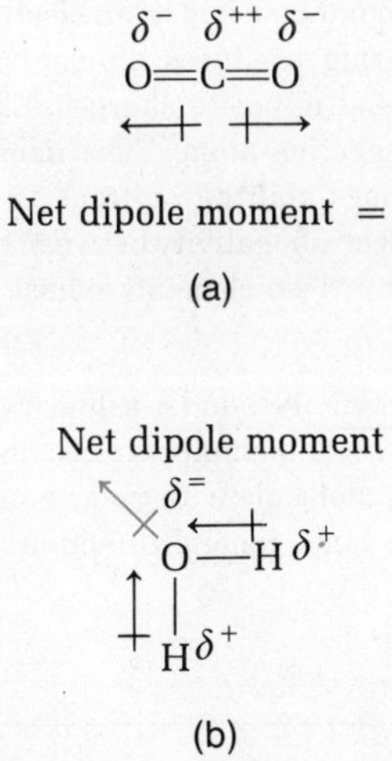

Figure 6.15
Relationships between molecular shape and dipole moment.
(a) Linear triatomic molecule of CO_2. (b) Bent triatomic molecule of H_2O.

$$\overset{\delta^-}{O}=\overset{\delta^{++}}{C}=\overset{\delta^-}{O}$$
$$\longleftarrow \quad \longrightarrow$$

Net dipole moment = 0

(a)

Net dipole moment

$$\overset{\delta^-}{O}-H^{\delta^+}$$
$$\quad | \quad$$
$$\quad H^{\delta^+}$$

(b)

Summary

1. Ions are atoms or groups of atoms that have net electric charges. Cations are positive ions formed by removal of one or more electrons from the atom. Anions are negative ions formed when an atom combines with one or more extra electrons.

2. Ionization potential is the energy needed to remove an electron from an atom in the gas phase to form a monatomic cation in the gas phase. Elements to the left of the periodic table have low ionization potentials, those to the right high ionization potentials. The ionization potentials for the removal of inner-shell electrons are particularly high.

3. Electron affinity is the energy released when an atom in the gas phase attracts an electron to form an anion in the gas phase. Electron affinities are small compared to ionization potentials. The halogens have the largest electron affinities.

4. An ionic compound conducts electricity when molten or dissolved in water. It is a crystalline substance with a high melting point and low volatility. A crystal of an ionic substance consists of a three-dimensional lattice of cations and anions in which each cation is surrounded by anions and each anion by cations. The major source of stability of an ionic substance is its crystal-lattice energy—the net energy of attraction between cations and anions in the crystal lattice. The crystal-lattice energy is greatest for ionic crystals containing small intensely charged ions.

5. Covalent substances consist of neutral molecules. They do not conduct electricity in their liquid states. The atoms in a covalent molecule are held together by covalent bonds—the attractive forces due to the sharing of pairs of electrons between atoms.

6. Outer-shell electrons in a molecule are present in molecular orbitals—orbitals that encompass at least two nuclei. We can compute the wave functions of molecular orbitals by combining the wave functions of the orbitals of the interacting atoms. An in-phase combination of two atomic orbitals produces a bonding molecular orbital that is more stable than the atomic orbitals because of its high electron density between the nuclei. Out-of-phase combination produces a less stable antibonding orbital whose electron density between the nuclei is low. The stability of a molecule depends on the number of net bonding electrons, that is, the number of electrons in bonding orbitals less the number of electrons in antibonding orbitals.

7. A polar covalent bond consists of an electron pair shared unequally between two atoms. This unequal sharing produces dipolar molecules—neutral molecules whose opposite ends bear equal and opposite charges. Electrons in a polar bond tend to remain near the more electronegative atom. Electronegativity is a number that is an approximate measure of an atom's ability to attract an electron pair in a covalent bond. The greater the difference in electronegativity between two atoms, the greater the polarity of a covalent bond between them. Two elements whose electronegativities are very different form an ionic compound.

8. No matter how polar its bonds, a linear triatomic molecule with two identical bonds (for example, CO_2) will be nonpolar because the dipole moments of its bonds will cancel each other. A bent molecule with polar bonds, such as H_2O, is polar because its bond dipoles point in the same general direction.

Vocabulary List

ionic bond	ionic compound	antibonding molecular orbital
ion	Born-Haber cycle	triple covalent bond
monatomic ion	crystal-lattice energy	double covalent bond
polyatomic ion	covalent substance	single covalent bond
cation	internuclear distance	polar covalent substance
anion	bond length	polar covalent bond
ionization potential	bond-dissociation energy	dipolar
first ionization potential	molecular orbital	electronegativity
second ionization potential	bonding molecular orbital	dipole moment
electron affinity		

Exercises

1. By referring to a periodic table (but not Figure 6.1), predict which element from each of the following pairs has the higher ionization potential.
   a. Mg or Cl   b. He or Ar   c. Ar or K

2. Predict the charges of the monatomic ions formed in the ionic compounds of the following.
   a. Ba   b. Cs   c. O   d. Br

3. Place the following molecules in a sequence of increasing bond polarity: Cl_2O, I_2, HCl, H_2O, CH_4.

4. Which of the following reactions is involved in the measurement of a first ionization potential?
   a. $e^- + F(g) \longrightarrow F^-(g)$
   b. $Na(s) \longrightarrow Na^+(aq) + e^-$
   c. $Rb(g) \longrightarrow Rb^+(g) + e^-$
   d. $Sr^+(g) \longrightarrow Sr^{2+}(g) + e^-$
   e. $e^- + K^+(g) \longrightarrow K(g)$

5. Which of the following processes require relatively large amounts of energy? Explain the reason for each case.
   a. $Rb^+(g) \longrightarrow Rb^{2+}(g) + e^-$
   b. $Sr^+(g) \longrightarrow Sr^{2+}(g) + e^-$
   c. $Cs(g) \longrightarrow Cs^+(g) + e^-$
   d. $Ba^{2+}(g) \longrightarrow Ba^{3+}(g) + e^-$

6. Why does calcium form compounds containing Ca^{2+} ions but none containing Ca^+ even though its second ionization potential is larger than its first ionization potential?

7. Compare the bonding and antibonding orbitals formed from the same pair of atomic orbitals with respect to:
   a. Energy   b. Electron density between the nuclei
   c. Contribution to the stability of the molecule when occupied.

8. Why is there no such thing as a $2s\pi$ molecular orbital?

9. An electric current through a wire is due to the flow of electrons through the metal. How does an electric current through a solution of an ionic substance differ from this process?

10. Even though a crystal of KBr contains both positive and negative ions, it does not conduct electricity. Why not?

11. Draw a picture that explains the conducting properties of molten KF. Show the charge on each type of particle present in the liquid and the direction of its motion relative to the two electrodes.

12. Draw a picture to explain the insulating properties of HF.

13. Why is the first ionization potential of Rb so much lower than that of Kr?

14. Which has the higher second ionization potential, Sr or Cs? Why?

15. How do the following factors affect the stabilities of ionic halides?
    a. An increase in the sublimation energy of the metal
    b. An increase in the bond dissociation energy of the halogen molecule
    c. An increase in the ionization potential of the metal
    d. An increase in the electron affinity of the halogen
    e. An increase in the crystal-lattice energy

16. How do the following factors affect crystal-lattice energy?
    a. An increase in the charge of the cation
    b. An increase in the size of the cation

17. What is the direction of polarity (if any) for each of the following bonds? N—H; P—H; As—H

18. Which is the negative end of each of the following bonds? C—N; N—O; C—Al

19. Propose a molecular shape consistent with the dipole moment for each.
    a. SO_3; dipole moment = 0
    b. PCl_3; dipole moment greater than zero

20. Nitrous oxide is a linear molecule with the structure N=N=O. Does this molecule have a dipole moment?

21. The cyanogen molecule, N≡C—C≡N, is linear. Does this molecule have a dipole moment? Explain.

22. The sulfur dichloride molecule is bent.

    S—Cl
    |
    Cl

    Does this molecule have a dipole moment? Explain.

23. Strontium sulfate, $SrSO_4$, has a melting point greater than 1580°C (the temperature at which it decomposes chemically). It dissolves in water to form a conducting solution. Is this substance ionic, covalent, or polar covalent?

24. Silicon tetrachloride ($SiCl_4$) is a nonconducting liquid with a freezing point of −70°C and a boiling point of 58°C. It is not an effective electrical insulator. Is this compound ionic, covalent, or polar covalent?

25. Predict the formulas of the ionic compounds of:
    a. Ba and Cl   b. Sr and S   c. K and O
    d. Li and F

26. Predict the formulas of the following ionic compounds.
    a. A compound containing Na^+ and O^{2-} ions
    b. A compound of K^+ and I^- ions
    c. A compound of Ba^{2+} and Br^- ions
    d. A compound of Ca^{2+} and S^{2-} ions

27. Although ionization potentials generally increase across a row of the periodic table, the first ionization potential of boron is less than that of beryllium. Can you explain this inversion in the general trend in terms of the electron configurations of the two elements?

28. Can you explain why the first ionization potential of oxygen is lower than that of nitrogen? [Hint: Consider repulsions between electrons.]

29. Make rough sketches (similar to Figure 6.7) of the bonding and antibonding wave functions formed by overlap of a 1s and a 2p orbital. Draw the ψ^2 functions corresponding to these molecular orbitals.

30. A $2p\sigma$ orbital has two nodes, and a $2p\sigma^*$ three. Use rough sketches of the in-phase and out-of-phase overlap of the atomic orbitals to explain this.

31. Use the molecular orbital method (and the energy level pattern for O_2) to predict the number of net bonding electrons and the magnetic properties of each.
    a. NO   b. NF

32. According to a molecular orbital picture, how many net bonding electrons and unpaired electron spins are there in
    a. NO^+?   b. O_2^{2-}?
    Use the electron energy level pattern for O_2.

33. By referring to the energy levels of N_2, verify that gaseous lithium should be diatomic and gaseous beryllium monatomic.

34. Use the MO method and the energy level pattern for N_2 to show that B_2 should have one covalent bond and C_2 two covalent bonds.

35. Would you expect He_2^+ to be more stable than a helium atom and a He^+ ion? Justify your answer by the MO method.

36. Use the MO method to decide whether or not the HHe molecule should be more stable than isolated atoms of H and He.

37. A diatomic boron molecule (B_2) has two unpaired electron spins, whereas C_2 is diamagnetic.
    a. Would you predict this using the building method and the energy level diagram for N_2?
    b. Predict the magnetic properties of C_2^{2+}.

38. Compare the bond-dissociation energies and bond distances of CO and N_2 (Table 6.3). What does this indicate about the number of net bonding electrons in CO? Is this consistent with a molecular orbital picture for CO? [Hint: Assume the energy level pattern for CO is similar to that of N_2.]

39. Use the Born-Haber cycle and the following data to compute the heat of formation of KCl(s).

    Heat of sublimation of K = 88 kJ/mol

    Heat of dissociation of Cl_2 = 121 kJ/mol Cl

    Ionization potential of K = 414 kJ/mol

    Electron affinity of Cl = 364 kJ/mol

    Crystal lattice energy of KCl = 681 kJ/mol

40. The heat of sublimation of sodium is 109 kJ/mol and its ionization potential is 493 kJ/mol. The bond-dissociation energy of $I_2(g)$ is 71 kJ/mol I and its electron affinity is 297 kJ/mol. The crystal-lattice energy of NaI is 694 kJ/mol. What is the amount of energy released in the process

    $$Na(s) + \frac{1}{2}I_2(g) \longrightarrow NaI(s)$$

EXERCISES

41. Boron trifluoride (BF_3) is shaped like a triangle with the boron atom at the center and a fluorine atom at each corner. Does this molecule have a dipole moment?

42. Both valence electrons of magnesium occupy the same energy level (the 3s). Why then, is the second ionization potential of magnesium greater than its first?

43. The two main types of electrostatic forces in an ionic crystal are the attraction between oppositely charged ions and the repulsion between ions of the same charge. Why do the attractive forces outweigh the repulsive ones?

44. Use the MO method to decide which should have the shortest internuclear distance, NF, OF, or F_2.

45. Perchloric acid ($HClO_4$) is a high-boiling liquid that does not conduct electricity, but it dissolves in water to form a conducting solution. Its melting point is $-112°C$. Is this substance ionic, covalent, or polar covalent?

46. By using a periodic table but not a table of electronegativities, decide which element of each of the following pairs is the more electronegative.
    a. Si and S   b. S and Te   c. Ba and C

47. Draw an energy level diagram showing the arrangement of electrons in the first (lowest energy) excited state of N_2. How many covalent bonds would this excited molecule have?

48. Which of the following statements is false? State why in each case.
    a. Crystal-lattice energy is the only important factor in influencing the formulas of ionic compounds.
    b. A polar substance is an electric insulator regardless of the direction of its dipole moment.
    c. The first ionization potentials of the elements rise steadily with increasing atomic number.
    d. Every electron transfer between neutral atoms in the gas phase requires some energy because the largest electron affinity is smaller than the lowest ionization potential.
    e. One indication of the ionic nature of a compound is its low melting point.
    f. Ionic substances always form by a sequence of reactions beginning with the sublimation of the metal and ending with the crystallization of the ions from the gas phase.
    g. Compounds of carbon and sulfur are ionic.
    h. The application of quantum mechanics to problems of molecular structure, while more complicated than for atomic structures, does not involve any new fundamental hypotheses.
    i. A double bond is stronger than a single covalent bond.
    j. The bond energy holding a diatomic molecule together depends only on the number of occupied bonding orbitals.
    k. The fact that CF_4 has no molecular dipole moment shows that C–F bonds are not polar.

49. Using electronegativities, predict which of the following compounds are ionic and which covalent. (You might not be able to make a clear decision in all cases.) Look up the physical properties of these substances to check on the accuracy of your predictions.
    a. $SiCl_4$ b. MgF_2 c. PH_3 d. $AlCl_3$ e. BF_3

50. The heat of formation of CsCl(s) is -433 kJ/mol. Compute the crystal-lattice energy of CsCl from the following data:

    Heat of sublimation of Cs = 79 kJ/mol

    Ionization potential of Cs = 376 kJ/mol

    Heat of dissociation of Cl_2 = 121 kJ/mol Cl

    Electron affinity of Cl = 364 kJ/mol

chapter 7

The Symbolic Representation of Ions and Molecules

The trigonal bipyramidal structure of PCl$_5$ is one of the common molecular geometries encountered in this chapter

7.1 Structural Formulas

The previous few chapters have presented modern atomic theory, the periodic behavior of atoms, and the arrangement of electrons in atoms, ions, and molecules. Although the quantum mechanical theory developed in these chapters does effectively describe molecular structures, the descriptions, which require maps of the relative positions of the atomic nuclei, diagrams of the electron energy levels, and pictures of the wave functions for the valence electrons, are cumbersome. A simpler, although less precise, device for routinely representing ions and molecules is a **structural formula** showing the types of chemical bonds and the arrangement of valence electrons in a substance.

The symbol for an element in a structural formula stands for the nucleus and inner-shell electrons of the atom, and dots around the symbol show the unshared valence electrons. A straight line between two symbols represents a single covalent bond (a shared electron pair) between atoms. Figure 7.1 (p. 158) shows examples of structural formulas (also called **electron-dot structures** or **Lewis structures**) for some typical atoms, ions, and molecules.

Example **7.1** Write Lewis structures for **a.** an uncombined oxygen atom **b.** N^{3-} **c.** Cs$^+$.

Solution **a.** Oxygen, a group VIA atom, has six valence electrons.

$:\overset{..}{\underset{..}{O}}:$

Figure 7.1
Structural formulas.

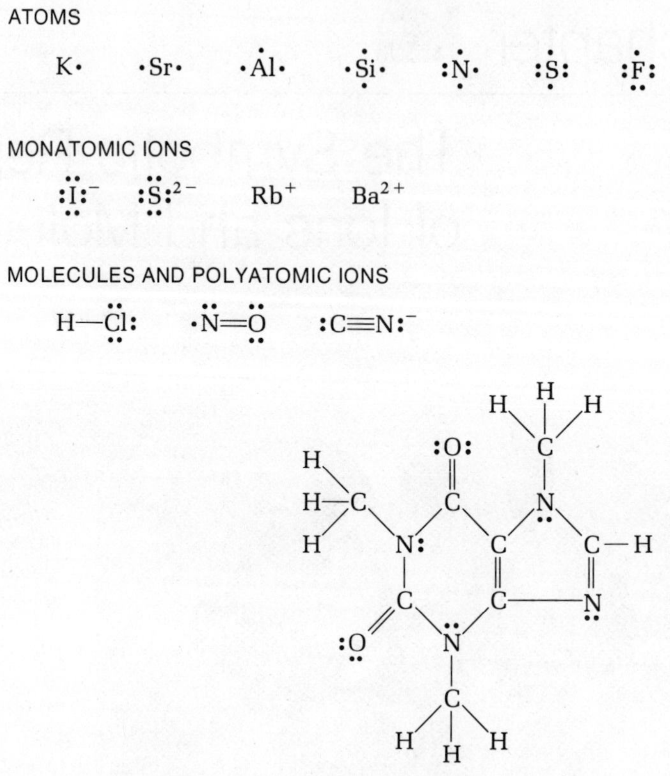

ATOMS

K· ·Sr· ·Äl· ·S̈i· :N̈· :S̈: :F̈:

MONATOMIC IONS

:Ï:⁻   :S̈:²⁻   Rb⁺   Ba²⁺

MOLECULES AND POLYATOMIC IONS

H—C̈l: ·N̈=Ö   :C≡N:⁻

[Caffeine structure]

Caffeine

b. Nitrogen (group VA) has five valence electrons when neutral but gains an additional three to produce the 3− charge.

:N̈:³⁻

c. Cesium (group IA) loses its only valence electron when it forms the positive ion, so the Lewis structure for the ion does not have any dots. We simply write Cs⁺.

Example 7.2

The water molecule is held together by two single covalent bonds, each holding a hydrogen atom to the oxygen atom. Write a structural formula for H_2O.

Solution

Each hydrogen atom uses its only electron in a covalent bond to oxygen. Two of oxygen's six valence electrons participate in the two covalent bonds, leaving the oxygen atom with four unshared electrons.

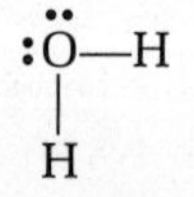

:Ö—H
|
H

A complete description of an ionic compound requires a three-dimensional drawing of the ionic lattice (see Chapter 9). For convenience, we use a simpler notation showing the types of ions present and the ratio of the cations and anions. (This ratio must give a neutral crystal.) The electron-dot structures of the cation and anion each appear in brackets with subscript numbers showing the ratio of these ions. Two examples, the structural formulas of NaCl and MgI_2, appear below.

[Na⁺][:C̈l:⁻]   [Mg²⁺][:Ï:⁻]₂

Example **7.3**   Write the Lewis structure for Cs_2S, an ionic compound consisting of Cs^+ and S^{2-} ions.

Solution   Each cesium atom loses its only valence electron to form Cs^+. Each sulfur atom, originally containing six valence electrons, gains two additional electrons in forming S^{2-}.

$$[Cs^+]_2[:\!\ddot{\underset{..}{S}}\!:^{2-}]$$

Although electron-dot structures describe many substances well, lines and dots are neither subtle nor flexible enough to adequately represent some substances, and we must develop more elaborate symbols to display their electronic structures. Even when a line-and-dot structure gives a reasonable picture of the number of covalent bonds and unshared electrons, it fails to convey the energies, lengths, or polarities of the bonds. The covalent bonds in H_2 and HCl are both represented by lines, but these two bonds differ greatly in length, energy, and polarity.

7.2 The Lewis Octet Rule

The **Lewis octet rule** is a simple method for predicting the chemical behavior of an atom, ion, or molecule from its structural formula: Atoms, ions, and molecules tend to react so that each atom attains the outer-shell electron configuration of a noble gas. (The total number of shared and unshared electrons in the outer shell of an atom equals the number of valence electrons in a noble gas.) With the exception of helium (two valence electrons), all noble gases have eight valence electrons—an "octet." Those elements near helium in the periodic table (hydrogen and lithium) tend to gain a configuration of two electrons, whereas other elements tend to an octet of electrons.

Many (but not all) stable substances have octet structures (noble-gas configurations around all the atoms). We can use the octet rule to predict the types of ions and how many covalent bonds an atom might form. For example, in most of its compounds as well as its elemental form (H_2), hydrogen forms one covalent bond to attain the helium configuration.

$$H-H \qquad H-\ddot{\underset{|}{O}}: \qquad H-\underset{|}{\overset{|}{C}}-H$$
$$\qquadH\qquad H$$

A hydrogen atom can also pick up an extra electron and form the hydride ion ($H\!:^-$).

In most of its compounds, carbon (four valence electrons) gains its octet by forming four covalent bonds.

$$H-\underset{H}{\overset{H}{\underset{|}{\overset{|}{C}}}}-\underset{H}{\overset{H}{\underset{|}{\overset{|}{C}}}}-H \qquad H-\underset{:\ddot{O}-H}{\overset{\ddot{O}:}{\underset{\|}{C}}} \qquad \ddot{O}=C=\ddot{O} \qquad H-C\equiv N:$$

Oxygen (six valence electrons) picks up two electrons to form the oxide ion,

$$:\!\ddot{\underset{..}{O}}\!:^{2-}$$

or it forms one covalent bond and picks up one extra electron as in the hydroxide ion,

$$H-\ddot{\underset{..}{O}}:^-$$

or it forms two covalent bonds as in H_2O or CO_2. In the case of the hydronium ion (H_3O^+), oxygen forms a third covalent bond but maintains its octet because it provides both electrons to the bond.*

$$:\overset{..}{\underset{..}{O}}:^{2-} \qquad H-\overset{..}{\underset{..}{O}}:^{-} \qquad H-\overset{..}{\underset{|}{O}}-H^{+}$$
$$H$$

Oxide ion Hydroxide ion Hydronium ion

The Lewis octet rule describes a tendency in chemical behavior, not an absolute requirement for molecular stability. There are many stable structures that do not fit the octet rule, and it is wrong to deny the existence of a molecule or ion because it lacks an octet arrangement.

Example 7.4

Write an equation, using structural formulas, that describes the reaction of Cl_2 gas with calcium metal to form $CaCl_2$ (an ionic substance).

Solution

The chlorine atoms in Cl_2 have octet arrangements because of a single covalent bond between them. This bond is broken during the reaction, and each chlorine atom takes an electron from calcium to form Cl^- and thus regain its octet. The loss of its two valence electrons to form Ca^{2+} gives calcium a noble-gas configuration. The cations and anions cluster together in a crystal lattice.

$$:\overset{..}{\underset{..}{Cl}}-\overset{..}{\underset{..}{Cl}}: \; + \; Ca: \longrightarrow [Ca^{2+}][:\overset{..}{\underset{..}{Cl}}:^{-}]_2$$

Applications of the Octet Rule

Even though there are many stable substances that do not fit the octet rule, chemists have enough confidence in it to write octet structures unless there is chemical or physical evidence to the contrary. The first step in writing an octet structure is to determine the number of covalent bonds in the molecule or ion. We arrive at this number by a simple arithmetic procedure: (1) find the total number of valence electrons in the structure; (2) compute the total number of valence electrons that would be needed to give each atom a noble-gas configuration in the absence of covalent bonds; (3) the number of covalent bonds equals half the difference between the value for (2) and that of (1).

Example 7.5

How many covalent bonds are present in PH_3?

Solution

Since phosphorus has five valence electrons and each of the three hydrogen atoms has one electron, the number of valence electrons in PH_3 is eight. Phosphorus needs eight valence electrons, and each H atom needs two.

$$8 + 3(2) = 14$$

$$\text{Covalent bonds} = \frac{1}{2}(14 - 8) = 3$$

Example 7.6

Find the number of covalent bonds in $C_2H_4O_2$.

---

*Covalent bonds in which one bond partner provides both electrons in the bond are often called *coordinate covalent bonds*. We will not use this term because it might obscure the fact that there are no identifiable differences in the properties of the three covalent bonds of H_3O^+.

Solution

Each carbon atom has four valence electrons, each hydrogen atom one, and each oxygen atom six.

Total valence electrons = 2(4) + 4(1) + 2(6) = 24

Each carbon atom and each oxygen atom need eight electrons, and each hydrogen atom needs two.

$$2(8) + 2(8) + 4(2) = 40$$

$$\text{Covalent bonds} = \frac{1}{2}(40 - 24) = 8$$

Example **7.7**

How many covalent bonds are present in the polyatomic ion SO_4^{2-}?

Solution

Two extra electrons account for the double negative charge. Since both the sulfur atom and the oxygen atoms carry six valence electrons each,

Total valence electrons = 6 + 4(6) + 2 = 32

Each of the five atoms in the ion needs an octet, so a total of 40 electrons would be needed in the absence of covalent bonding.

$$\text{Covalent bonds} = \frac{1}{2}(40 - 32) = 4$$

Ionic octet substances composed of monatomic ions have formulas that contain whole-number multiples of eight valence electrons (for example, KCl has 8; BaF_2 has 16; Cs_2S has 8). Transfer of all the valence electrons of the metal atoms to the nonmetal atoms gives every atom an octet without requiring covalent bonding.

The decision whether to write an ionic or covalent structure depends on the observed physical properties of the substance. For example, it is easy to write an ionic octet formula for $BeCl_2$,

$$[Be^{2+}][:\overset{..}{\underset{..}{Cl}}:^-]_2$$

but this structure is wrong because beryllium chloride refuses to act like an ionic substance—it melts at a rather low temperature (about 400°C), and its liquid is a poor electrical conductor. Beryllium chloride has a large molecular mass, and its molecules have long, chainlike structures (Figure 7.2). Note that this molecule fits the octet rule.

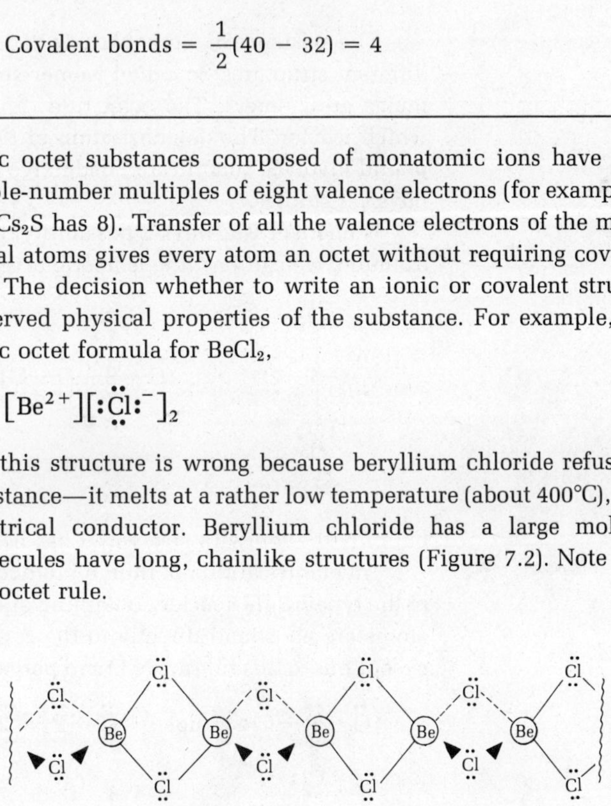

Figure 7.2
A portion of the beryllium chloride molecule. The increasingly thick line indicates a bond projecting forward; dotted lines indicate bonds that project backward.

Isomerism

Once we decide on the number of covalent bonds in a molecule, we can try to write its structural formula. In some cases this is unambiguous. The octet structure of PH_3 is easy to write because hydrogen forms only one covalent bond. Each of the three covalent bonds must connect a hydrogen atom to the phosphorus atom. An unshared pair of electrons completes the octet of phosphorus.

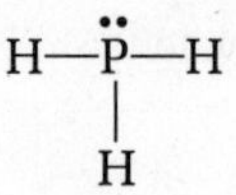

H—P̈—H
|
H

Several structural formulas corresponding to $C_2H_4O_2$ fit the octet rule. Two of the possibilities represent the structures of two common compounds—acetic acid and methyl formate. These two substances have different physical and chemical properties (Table 7.1) and can be separated by physical methods.

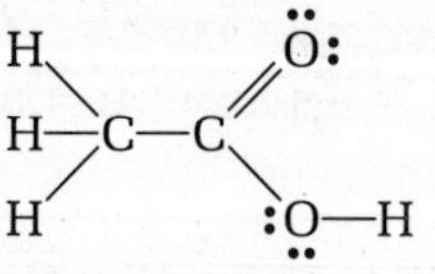

Acetic acid

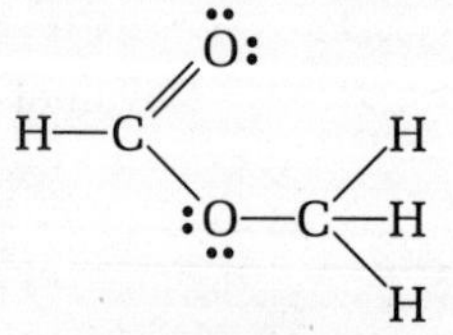

Methyl formate

Table 7.1 Physical Differences between Isomers

Isomer	Melting point (°C)	Boiling point (°C)	Density (g/cm³)
Acetic acid	16.6	118.5	1.05
Methyl formate	−99	31.5	0.97

The existence of more than one compound with the same molecular formula (but different structures) is called **isomerism**; compounds with the same molecular formulas are **isomers**. The octet rule cannot tell us which structure corresponds to which isomer. The demonstration of the molecular structure for a particular compound involves complicated deductive arguments based on extensive chemical and physical studies.

Isomerism is always a possibility, but often only one of the isomers exists. For example, we can imagine isomeric octet structures for SO_4^{2-},

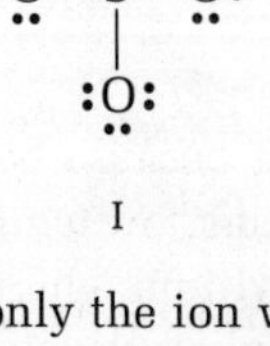

I 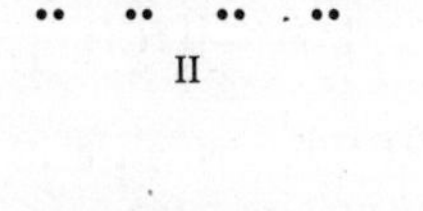 II

but only the ion with structure I has been found.

An empirical rule of thumb predicts the structures of stable ions and molecules of the type AB_n (a species containing one atom of element A and n atoms of B): The B atoms are all bound directly to the A atom, not to each other. (The only common exceptions to this rule are N_2O and peroxy compounds such as H_2SO_5.) For example,

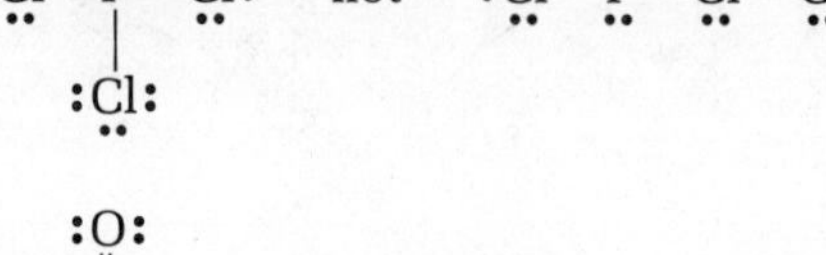

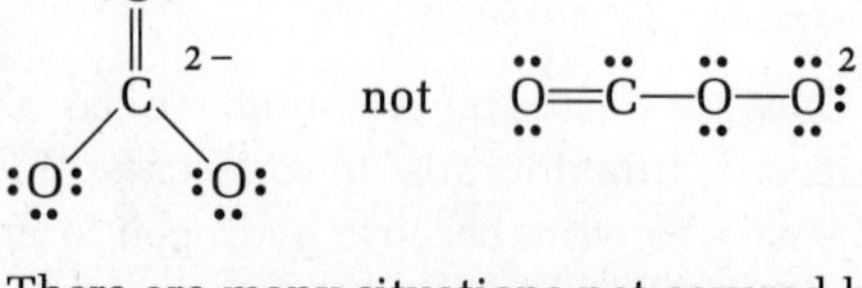

:C̈l—P̈—C̈l: not   :C̈l—P̈—C̈l—C̈l:
|
:C̈l:

There are many situations not covered by this rule, and there is no general rule for choosing the stable structure. We can see from the examples below that hydrogen atoms sometimes are bound to the more electronegative atom, sometimes to the less electronegative one.

$$H\text{—}\ddot{\underset{..}{O}}\text{—}\ddot{\underset{..}{Cl}}: \quad \text{not} \quad H\text{—}\ddot{\underset{..}{Cl}}\text{—}\ddot{\underset{..}{O}}:$$

But

$$\begin{array}{c} :\ddot{O}: \\ | \\ H\text{—}\ddot{\underset{..}{O}}\text{—}P\text{—}\ddot{\underset{..}{O}}\text{—}H \\ | \\ H \end{array} \quad \text{not} \quad \begin{array}{c} :\ddot{O}\text{—}H \\ | \\ H\text{—}\ddot{\underset{..}{O}}\text{—}P\text{—}\ddot{\underset{..}{O}}\text{—}H \end{array}$$

Instead of trying to generalize about isomeric stability, we will discuss the structures of individual molecules and ions as part of the survey of the chemistry of the elements (Chapters 17 through 21).

Nonoctet Structures

The many molecules and ions whose structures do not fit the octet rule fall into four classes: (1) **transition metal ions**, (2) molecules and ions with more than an octet of electrons around an atom (**superoctet species**), (3) molecules and ions with less than an octet around an atom (**electron-deficient species**), and (4) species containing at least one unpaired electron spin (**free radicals**).

Transition metal ions. It is hard to apply the octet rule to the transition metals because it is difficult to decide how many valence electrons they have. Most transition metals have two outer-shell electrons and, as expected from the octet rule, they form doubly-charged positive ions. But because the inner-shell d electrons and the empty d orbitals can also participate in chemical bonds, transition metals can also form ions with larger charges (Sc^{3+}, Cr^{3+}, Fe^{3+}) and polyatomic ions (TiO^{2+}, VO_3^-, CrO_4^{2-}, MnO_4^-) that do not readily fit the octet rule.

Superoctet species. Nonmetallic elements of the third and higher periods can react with electronegative elements to form structures in which the central atom has more than eight valence electrons. Some examples are SiF_6^{2-}, PCl_5, and SF_6.

Elements such as silicon, phosphorus, and sulfur can form more than four covalent bonds because their empty $3d$ orbitals have low enough energies to participate in stable bond orbitals. Since they lack low energy d orbitals, the elements of the second row (C, N, and O) do not form stable superoctet structures.

Electron-deficient species. Electron-deficient compounds such as boron trifluoride

have empty outer-shell orbitals and fewer than eight electrons around one of their atoms. These compounds readily form covalent bonds with molecules or ions containing unshared electron pairs, and the products of these reactions have octet structures. For example, BF_3 reacts with ammonia, NH_3, to form BF_3NH_3.

$$\text{:F:} \quad \text{H} \quad \text{:F:} \quad \text{H}$$
$$\text{:F—B} \quad + \quad \text{:N—H} \longrightarrow \text{:F—B—N—H}$$
$$\text{:F:} \quad \text{H} \quad \text{:F:} \quad \text{H}$$

In this reaction, the boron atom has acquired a partial negative charge and the N atom has acquired a partial positive charge. Another example of an electron-deficient molecule is $AlCl_3$, which reacts with chloride ion to form $AlCl_4^-$.

$$\text{:Cl:} \quad \quad \quad \text{:Cl:}$$
$$\text{:Cl—Al} \quad + \quad \text{:Cl:}^- \longrightarrow \text{:Cl—Al—Cl:}$$
$$\text{:Cl:} \quad \quad \quad \text{:Cl:}$$

Free radicals. A free radical contains an unpaired electron spin and is paramagnetic. Most free radicals are very short-lived, rapidly combining with another radical to form an octet structure. For example, whenever two methyl radicals (CH_3) collide, they form a covalent bond to satisfy the octet of each carbon atom.

$$\text{H} \quad \quad \text{H} \quad \text{H}$$
$$2\ \text{H—C·} \longrightarrow \text{H—C—C—H}$$
$$\text{H} \quad \quad \text{H} \quad \text{H}$$

Several free radicals are quite stable. Nitrogen oxide (NO) contains an odd number of electrons and has an unpaired spin,

$$·\text{N}=\text{Ö}$$

but has no apparent tendency to combine to form N_2O_2 in the gas phase. As stated in Chapter 6, O_2 molecules have two unpaired electron spins each and are called **diradicals**.*

The existence of nonoctet structures leaves us with the problem of recognizing them. In most cases it is impossible to write any octet structure for a nonoctet formula. Any structure with an odd number of electrons must be a radical because it is impossible to divide an odd number into pairs. We can recognize many electron-deficient molecules because there are not enough electrons to fill the noble-gas configurations of all the atoms—the number of shared electrons needed exceeds the number of valence electrons. Superoctet structures stand out because the number of bonds required for the octet structure is not enough to hold all the atoms together. An octet structure for PCl_5 would contain four covalent bonds, but the phosphorus atom cannot hold five chlorine atoms with only four bonds.

7.3
The Shapes of Covalent Molecules

Spectroscopic and other evidence show that covalent molecules of three or more atoms have definite geometric shapes: water is bent, CO_2 is linear, and ammonia

---

*There is no adequate line and dot structure for O_2. The octet structure

$$\text{Ö}=\text{Ö}$$

shows the double bond but misses the diradical character. A nonoctet structure showing the unpaired spins (with arrows)

$$\uparrow \quad \uparrow$$
$$\text{:Ö—Ö:}$$

makes the bond appear weaker than it really is.

Figure 7.3
Molecular shapes of (a) ammonia and (b) methane.

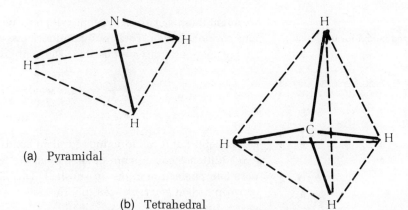

(a) Pyramidal

(b) Tetrahedral

(NH$_3$) has a pyramidal structure with the nitrogen atom at the apex and the hydrogen atoms at the base. Methane (CH$_4$) is in the form of a tetrahedron with carbon at the center and four hydrogen atoms at the corners, equidistant from the carbon atom (Figure 7.3). These shapes reflect the nature of the covalent bonds in the molecules.

The simplest structural formulas written on paper do not show molecular shapes. Although CCl$_4$ has a tetrahedral shape, its formula is often written

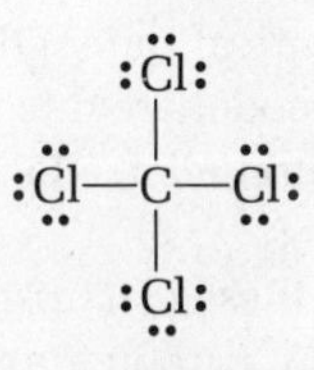

$$\overset{\cdot\cdot}{:}\overset{\cdot\cdot}{\underset{\cdot\cdot}{Cl}}:$$
$$:\overset{\cdot\cdot}{\underset{\cdot\cdot}{Cl}}-C-\overset{\cdot\cdot}{\underset{\cdot\cdot}{Cl}}:$$
$$:\overset{\cdot\cdot}{\underset{\cdot\cdot}{Cl}}:$$

More elaborate (and more expensive) perspective drawings (Figure 7.4) depict the true shape.

The two general methods used to explain molecular shapes are the hybrid orbital method and the electron-pair repulsion method. The hybrid orbital method, an attempt to describe the bonding orbitals within a molecule, is currently widely used by organic chemists. The electron-pair repulsion method is simpler and can predict the shape of a molecule from its line-and-dot structure. There are no major conflicts between the shapes predicted by the two methods; the differences that do exist are subtle, merely a matter of a few degrees in a bond angle.

Hybrid Orbital Method

Ideally, we should be able to solve the Schrödinger wave equation for a system containing one carbon nucleus, four hydrogen nuclei, and ten electrons to find that the lowest energy results from the same tetrahedral arrangement that is observed for methane. Since this solution is beyond our current mathematical abilities, we rely on an approximate method in which we mathematically combine several orbitals similar to those in a hydrogen atom to generate new orbitals called **hybrid orbitals**. These hybrid orbitals point in directions that are consistent with the observed shape of the molecule and are used to construct bond orbitals.

The hybridized model for a CH$_4$ molecule illustrates this method. Given the configuration of the valence electrons of carbon,

2p^2 ↑ ↑ __

2s^2 ↑↓

Figure 7.4
Three-dimensional representations of a carbon tetrachloride molecule.

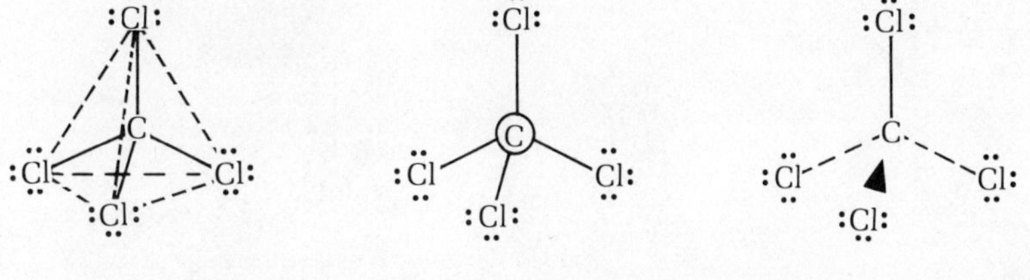

Figure 7.5
Unhybridized model for methane.

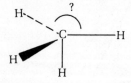

we might think that a carbon atom, with only two unpaired electrons, could not form four covalent bonds. This configuration, however, is the ground state, and higher-energy configurations are possible. The configuration

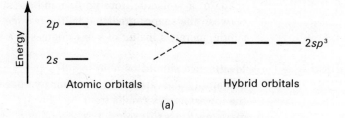

is less stable but it can form four covalent bonds, and the energy dividends from the two additional bonds more than compensate for the energy investment needed to raise one 2s electron to the 2p subshell. This is *not* the mechanism of the reaction between carbon and hydrogen to form methane. We are merely keeping track of the energy involved.

If this excited-electron configuration is used to form four C—H bonds, we would expect two types of bonds. Three bonds would involve 2p orbitals, and these bonds would be mutually perpendicular, reflecting the geometric arrangement of the atomic orbitals. Because the fourth bond would involve the spherical 2s orbital, it is not easy to predict its direction in relation to the other bonds. Furthermore, it would be shorter and stronger than the bonds constructed from the more elongated, less stable 2p orbitals. This model (Figure 7.5) does not accurately represent methane, whose four bonds are identical in length and energy and are arranged tetrahedrally. The inaccuracies of this model are due to our neglect of the electrostatic repulsions between the four hydrogen nuclei, a factor that increases the average angle between the C—H bonds.

In order to adjust for these electrostatic repulsions, the wave functions of the 2s and the three 2p orbitals of carbon are averaged to produce a set of four sp^3 hybrid orbitals (Figure 7.6). The number of hybrid orbitals (in this case, four) is always the same as the number of orbitals averaged (one s and three p orbitals), a principle called the *conservation of orbitals*. These hybrid orbitals are identical to each other in shape and energy, point toward the corners of a tetrahedron, and a model of methane based

Figure 7.6
Formation of sp^3 hybrid orbitals. (a) Energy level of atomic and hybrid orbitals. (b) Shape of orbital. (c) Geometrical arrangement of four sp^3 hybrid orbitals. For simplicity, only the large lobe of each orbital is shown.

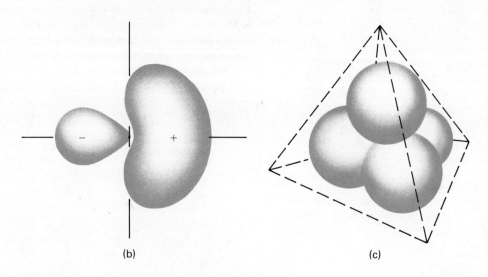

Figure 7.7
The sp^3 hybrid models for water and ammonia.

on the bonds formed from these hybrid orbitals is a good approximation of the real molecule. In the sp^3 hybridization model for ammonia (Figure 7.7), three of the hybrid orbitals form N—H bonds, and the fourth orbital contains the unshared electron pair. Since the relative positions of the nuclei (not its electrons) define the shape of the molecule, the hybridized model of NH_3 has a pyramidal shape. In the hybridized model of water, two sp^3 orbitals form O—H bonds to give the molecule a bent shape while the other two orbitals hold unshared electron pairs. The shape of any octet molecule that contains only single bonds can be explained by sp^3 hybridization.

Additional types of hybrid orbitals are used to describe other molecular shapes. One such hybrid set involves combination of an s orbital with two p orbitals to form three sp^2 hybrid orbitals. These orbitals are planar and point toward the corners of an equilateral triangle, 120° apart. This sort of hybridization is consistent with the shapes of electron-deficient, single-bonded molecules with six electrons around the central atom (for example, BCl_3, Figure 7.8) and octet molecules containing one double bond (formaldehyde). In the case of BCl_3, the boron atom uses each of its sp^2 hybrid orbitals to form a B—Cl bond. In the formaldehyde molecule, the three hybrid orbitals form three σ bonds—two to hydrogen and one to oxygen—while the unhybridized p orbital, which is perpendicular to the plane of the hybrid orbitals, forms a π bond with one of the oxygen orbitals.

In electron-deficient, singly bonded species with only four electrons around the central atom (for example, gaseous $BeCl_2$), sp hybridization (Figure 7.9) is consistent with the linear shape of the molecule. The two hybrid orbitals, constructed from one s and one p orbital, point in opposite directions and form the Be—Cl bonds. This type of hybridization also occurs in octet species with two double bonds to the same atom (CO_2) and in triply bonded octet species (HCN). In these cases, π bonding by the unhybridized p orbitals completes the multiple bonds.

Although hybridization is widely used for explaining molecular shapes, many chemists question whether it is the best way to handle this problem. They feel that it is a cumbersome revision of a faulty initial approximation (that bond orbitals can be constructed from hydrogenlike atomic orbitals), and they prefer methods that yield more direct predictions of molecular shapes. The next section presents an alternative, the electron-pair repulsion method.

Figure 7.8
The sp^2 hybrid orbitals.

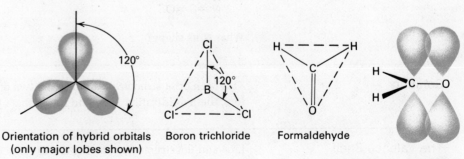

Orientation of hybrid orbitals (only major lobes shown)   Boron trichloride   Formaldehyde

$p\pi$ bonding in formaldehyde

Figure 7.9
The *sp* hybrid orbitals.

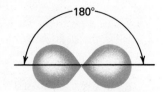
Orientation of hybrid orbitals
(only major lobes shown)

H—C≡N: :C̈l—Be—C̈l:
Hydrogen cyanide   Beryllium chloride

Ö=C=Ö
Carbon dioxide

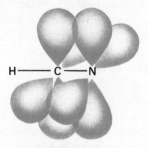
pπ bonding in HCN

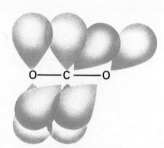
pπ bonding in CO₂

Example **7.8**	Use the hybridization method to predict the shape of PF$_3$. The line-and-dot structure of PF$_3$ is
	
	:F̈—P̈—F̈:
	           |
	           :F̈:

Solution	Since PF$_3$ has an octet structure and no multiple bonds, its hybridization is sp^3. Three of the hybrid orbitals form the three P—F bonds, and the fourth contains the unshared electron pair. The nuclei form a pyramid with the phosphorus atom at the peak and the fluorine atoms at the triangular base.
Example **7.9**	The cyanate ion has the structure N̈=C=Ö⁻ What is its shape?
Solution	This is an octet structure, and there are two double bonds to the carbon atom. Therefore, the hybridization is sp and the cyanate ion has a linear structure.

The Valence-Shell Electron-Pair Repulsion Method (VSEPR)

Line-and-dot structures can offer an easy way to predict the shapes of molecules if we consider repulsions between electrons. The valence electrons of an atom are either covalently bound to an adjacent atom or exist as unshared pairs. Molecules have shapes that minimize repulsion between electrons by maintaining the maximum average angles between orbitals. The orbitals in a multiple bond must all point toward the same atom, but the different bound atoms and the unshared electron pairs point as far away from each other as possible. The molecule takes on an orbital arrangement that gives the largest average angle between its orbitals, and this arrangement depends on the total number of bonded atoms and unshared electron pairs (Table 7.2, p. 170). This approach to determining molecular geometry is called the **valence-shell electron-pair repulsion** method.

For example, carbon dioxide has two oxygens and no unshared electron pairs around the carbon atom, so it takes on a linear shape to maintain the maximum separation between the bonding electrons. In any molecule whose bound atoms and unshared electron pairs add up to three, the orbitals point toward the corners of an equilateral triangle, a shape that has the largest possible average angle (120°) between the orbitals. Both BF_3 and NO_2^- are examples of triangular geometry. Since molecular shape is defined by the positions of the nuclei, NO_2^- has a bent shape with a O—N—O bond angle near 120°. When the number of bonded atoms and unshared electron pairs equals four (as in SO_4^{2-}, NH_3, and H_2O), the molecule has a shape and bond angles related to a tetrahedral arrangement of orbitals.

The trigonal bipyramidal arrangement formed by an atom with a total of five bound atoms and electron pairs has two distinct types of orbitals—those in the horizontal triangular plane (the equatorial positions, sometimes called *basal*) and the two orbitals projecting perpendicular to this triangle (the *axial* positions). In order to predict the shapes of I_3^-, ClF_3, and $TeCl_4$, we must know whether an unshared electron pair tends to occupy an axial or an equatorial position. It appears that repulsions due to an unshared electron pair are more severe than those due to an electron pair in a covalent bond. An unshared electron pair in an equatorial position causes less repulsion than one in an axial position because the axial position lies at 90° angles to its three adjacent pairs, whereas the equatorial position has only two adjacent pairs with 90° angles and two more distant pairs with 120° angles. Consequently, unshared pairs occupy the equatorial positions and the bound atoms the axial positions. This explains the linear shape of I_3^- and the shapes of ClF_3 and $TeCl_4$.

Similar considerations apply to the arrangements of the six bonded atoms and unshared electron pairs in an octahedron. Since all the angles between adjacent orbitals in the octahedron are the same (90°), it makes no difference which orbital contains the unshared pair in ICl_5. But in XeF_4, the unshared electron pairs occupy orbitals on the opposite sides of the octahedron in order to be as far apart as possible. This accounts for the square planar shape of XeF_4.

Example 7.10

Use the VSEPR method to predict the shape of SO_3. The line-and-dot structure of SO_3 is

$$:\ddot{O}-S-\ddot{O}:$$
$$\hspace{1.2cm}\|$$
$$\hspace{1.2cm}:\ddot{O}:$$

Solution

In this example, the central atom (S) has three bonded atoms (the O atoms) and no unshared pairs. Therefore, it forms a triangular planar arrangement with each of the three oxygen atoms occupying a vertex of the triangle. The molecule is flat and the O—S—O bond angle is 120°.

7.4 Resonance

The detailed properties of the carbonate ion are inadequately represented by a single line-and-dot structure. All three carbon–oxygen bonds are alike in length, energy,

7 THE SYMBOLIC REPRESENTATION OF IONS AND MOLECULES

Table 7.2 Molecular Shapes

HYBRIDIZATION

Total number of bonded atoms and unshared electron pairs	Arrangement with minimum repulsion	Number of bonded atoms	Number of unshared electron pairs	Shape	Example
SP — 2	Linear 180°	2	0	Linear	O=C=O Cl—Hg—Cl
SP² — 3	Trigonal planar 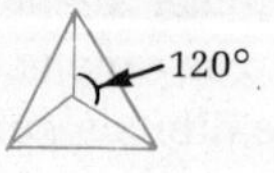 120°	3	0	Trigonal planar	F—B—F ; H₂C=O
		2	1	Bent	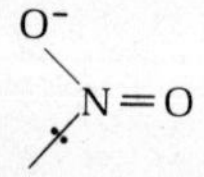 O⁻—N=O
SP³ — 4	Tetrahedral 109° 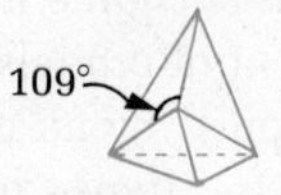	4	0	Tetrahedral	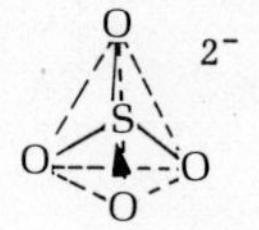 SO₄²⁻
		3	1	Trigonal pyramidal	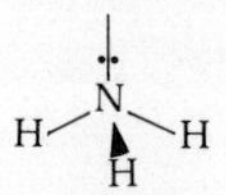 NH₃
		2	2	Bent	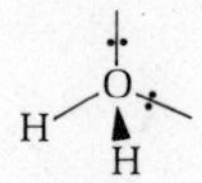 H₂O
5	Trigonal bipyramidal 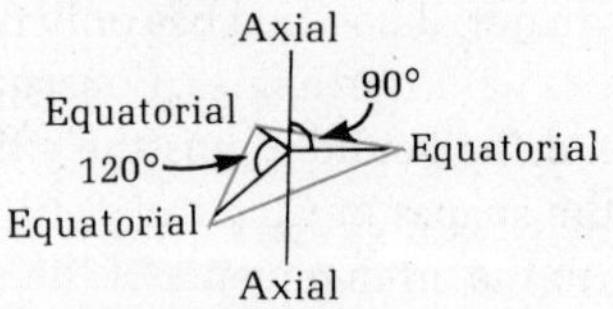 (Axial, Equatorial 90°, 120°)	5	0	Trigonal bipyramidal	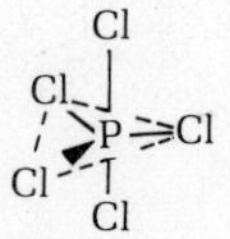 PCl₅
		4	1		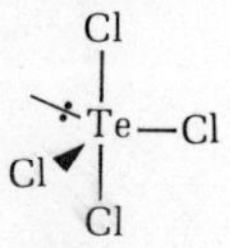 TeCl₄
		3	2	T-shaped	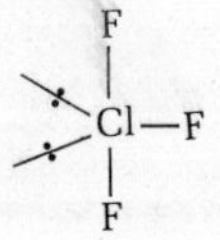 ClF₃
		2	3	Linear	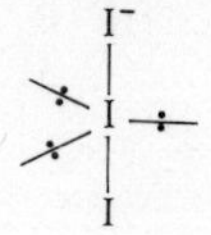 I₃⁻
6	Octahedral 90° 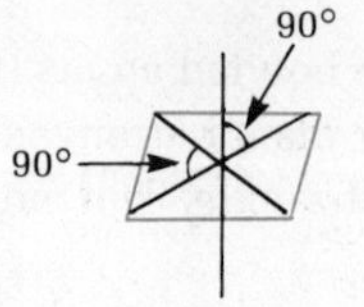	6	0	Octahedral	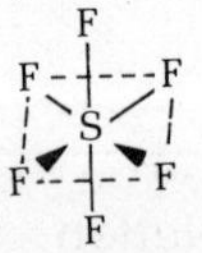 SF₆
		5	1	Square pyramidal	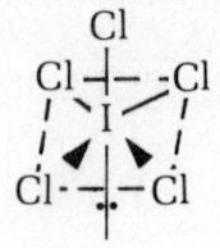 ICl₅
		4	2	Square planar	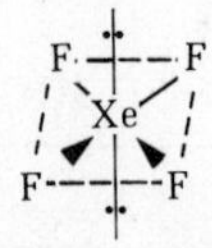 XeF₄

> Imagine a fourteenth-century European who has just returned from a voyage to Africa and wants to describe a rhinoceros. He has observed a rhinoceros carefully and can go into precise details, but he is afraid of boring his audience. Looking for a quick way to convey the strange beast's essential features, he says "It's part dragon, part unicorn." This doesn't mean that a rhinoceros fluctuates between a dragon and a unicorn, but that it has some dragon characteristics (a thick skin resembling armor plate) and some features of a unicorn (a horn in the middle of its head).
>
> The images of a dragon and a unicorn are used because the listeners are familiar with both—it doesn't matter that they are images of mythical beasts. This is a resonance description. A chemist who uses resonance describes a real molecule in terms of the structures of mythical molecules, relying on the audience's knowledge of the myths. The chemist avoids one problem facing our medieval friend; there's no chance that the chemist's audience will think of a rhinoceros as a fire-breathing horse.

and polarity. They are shorter and stronger than a carbon–oxygen single bond (as in CH_3OH) but weaker and longer than a double bond (as in CH_2O). A single octet structure cannot express this because it shows one of the three oxygen atoms being held by a double bond while the other two are singly bonded.

$$\begin{array}{c} :\ddot{O}: \\ \| \\ C \\ / \quad \backslash \\ :\ddot{O}: \quad :\ddot{O}: \end{array} \quad 2-$$

This is not a failure of our understanding of the bonding in CO_3^{2-}, that is, the molecular orbital model contains three carbon–oxygen bonds with identical properties, but there is no way to symbolize this bonding with one octet structure.

One method of representing this is to describe the structure of CO_3^{2-} as an average of three octet structures (Figure 7.10, p. 172). None of these structures by itself can describe a carbonate ion, which acts as though it were a **resonance hybrid** of all three. (The double-headed arrow is a specific symbol for an average between different line-and-dot structures.) The individual structures, called **resonance forms**, are not isomers of each other because the nuclei in all three structures are in exactly the same geometric relationship. The only differences involve the positions of the electrons,* and since we cannot exactly locate electrons in any molecule (Heisenberg's uncertainty principle), we cannot detect these forms as different species. There is only one structure for the carbonate ion and it is an average of the three resonance forms.

7.5 Oxidation–Reduction

Oxidation State

The **oxidation state** of an atom in a particular chemical environment (a molecule or an ion) is the number of valence electrons in a free, neutral atom of that element minus the number of valence electrons assigned to the atom in the particular environment. This whole number (which is some times positive and in other cases nega-

---

*The three resonance forms of carbonate ion are not merely different perspective views of the same ion. They represent three distinct model structures whose wave functions are averaged together to show that an electron pair is equally distributed among the three bonds.

Figure 7.10
Resonance hybrid descriptions.

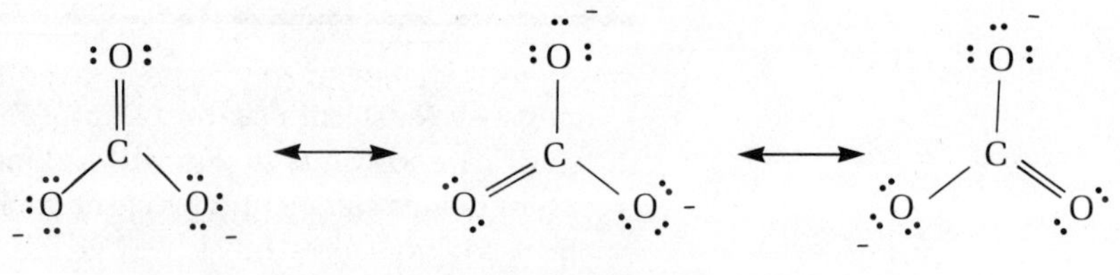

CO_3^{2-} C—O bond order = $1\tfrac{1}{3}$

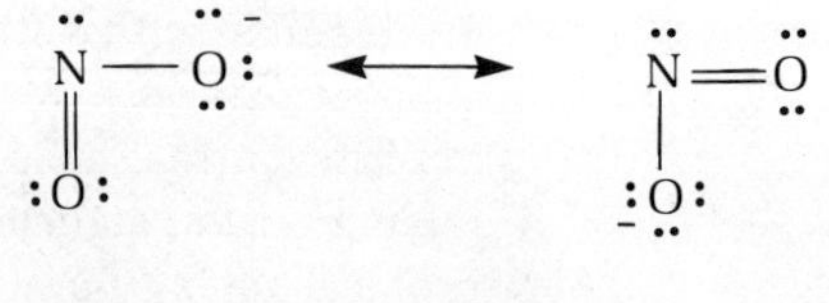

NO_2^- N—O bond order = $1\tfrac{1}{2}$

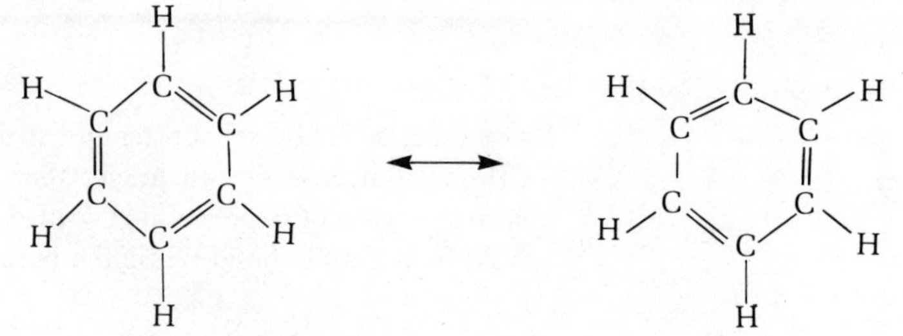

Benzene (C_6H_6)   C—C bond order = $1\tfrac{1}{2}$

tive) is useful for predicting the formulas of compounds, classifying them, comparing the chemical properties of elements, and describing chemical reactions.

The calculation of the oxidation state of an atom is simplest for monatomic ions (an ion containing only one atom). For example, calcium ion (Ca^{2+}) has no valence electrons, whereas a free calcium atom has two; therefore the oxidation state of calcium in Ca^{2+} is +2 (2 − 0). Iodide ion (I^-) has a −1 oxidation state because there are eight valence electrons in the ion compared to only seven in the neutral atom. These examples can be generalized as the first of six rules for assigning oxidation states to atoms in molecules or ions.

Rule 1. The oxidation state of a monatomic ion always equals its charge.

When an atom participates in one or more covalent bonds, we assign the shared electrons of each bond to its more electronegative bond partner. If two bond partners have the same electronegativity, the electron pair is divided equally, one electron to each atom. For example, in HCl the electronegativity of the chlorine atom is 3.0 and the hydrogen atom is 2.1, so the chlorine atom is assigned both the shared electrons and the hydrogen atom none. The chlorine atom has eight electrons and is in the −1 oxidation state, whereas the hydrogen atom, with no electrons assigned to it, is in the +1 state.

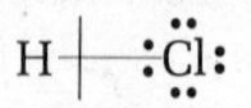

In the case of the Cl_2 molecule, both bond partners have the same electronegativity, and each is assigned one of the two shared electrons to give each atom seven electrons and a zero oxidation state.

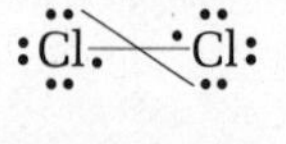

Example 7.11

What is the oxidation state of each atom in CCl_4?

Solution

The octet structure of CCl_4 consists of four chlorine atoms, each singly bonded to a central carbon atom.

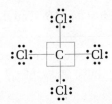

Since chlorine is more electronegative than carbon (3.0 versus 2.5), each chlorine atom gets both electrons from its C—Cl bond and is assigned eight electrons, while the carbon atom has none.

Oxidation state of $Cl = 7 - 8 = -1$

Oxidation state of $C = 4 - 0 = +4$

Using some useful approximate rules, we can assign oxidation states without the difficulty of drawing the line-and-dot structures.

Rule 2. The oxidation state of an atom in its elemental state is zero. This assignment applies no matter what the molecular form of the element; H_2, S_8, P_4, and O_2 are all examples of zero oxidation states.

Rule 3. Hydrogen present in compounds is in the $+1$ oxidation state (never greater) except when combined with a less electronegative atom in which case it is -1 (for example, in hydride salts such as LiH).

Rule 4. Oxygen in its compounds has a -2 oxidation state except when it is covalently bonded to a fluorine atom or to another oxygen atom. The most common exceptions to the -2 oxidation state are hydrogen peroxide (H_2O_2) and the peroxide ion (O_2^{2-}). As can be seen from their line-and-dot structures, these two species represent the -1 oxidation state.

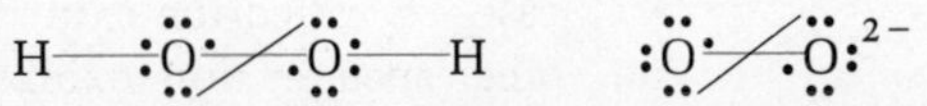

$$H—\ddot{\underset{..}{O}}—\ddot{\underset{..}{O}}—H \qquad [\ddot{\underset{..}{O}}—\ddot{\underset{..}{O}}]^{2-}$$

Rule 5. When present in compounds, the elements of periodic groups IA and IIA have oxidation states equal to their group numbers.

Rule 6. The sum of the oxidation states of all the constituent atoms of a polyatomic ion must equal the charge of the ion. The sum of the oxidation states in a molecule equals zero.

Example 7.12

Find the oxidation state of N in NO_3^-.

Solution

The sum of the oxidation states of the atoms in NO_3^- equals -1, the ion's charge (rule 6). According to rule 4, each of the three oxygen atoms is in the -2 state. Therefore,

(Oxidation state of N) $+ 3(-2) = -1$

Oxidation state of N $= +5$

Example 7.13

What is the oxidation state of carbon in C_2H_6O?

Solution	Since we are dealing with a neutral molecule, the sum of the oxidation states is zero. Assigning the -2 state to oxygen and $+1$ to hydrogen, we find

$$2(\text{Oxidation state of C}) + 6(+1) + (-2) = 0$$

$$\text{Oxidation state of C} = -2$$ |
| Example 7.14 | What is the oxidation state of oxygen in Na_2O_2? |
| Solution | Since sodium has a $+1$ oxidation state (rule 5) and since the Na_2O_2 formula unit is neutral, each oxygen must be in the -1 state. This compound is ionic and contains the peroxide ion. |
| Oxidation States and Chemical Similarity | The possible oxidation states of an element depend on its periodic family. With the exception of the noble gases, the highest possible oxidation state of an element is its group number, and the lowest possible oxidation state is its group number minus eight. Chlorine (group VIIA) has oxidation states ranging from -1 (Cl^-) to $+7$ (ClO_4^-), sulfur (VIA) from -2 (H_2S) to $+6$ (SO_4^{2-}), and nitrogen (VA) from -3 (NH_3) to $+5$ (NO_3^-).

There is no guarantee that an element will have compounds with these extreme oxidation states. Fluorine has no positive oxidation states, and the oxidation states of oxygen range from -2 (H_2O) to $+2$ (F_2O) but none higher. Metals do not exhibit negative oxidation states. No reliable method exists for telling which of the oxidation states between the group limits occur for a particular element. Table 7.3 lists some common oxidation states of common elements. |
| Redox Reactions | An atom undergoes **oxidation** when it increases its oxidation state. **Reduction** is a decrease in oxidation state. Since the number of electrons in a chemical reaction remains the same, reduction of one atom must accompany the oxidation of another. Any reaction involving changes in oxidation states is an **oxidation–reduction reaction**, or **redox reaction**. An example of a redox reaction is

$$Zn(s) + Cu^{2+}(aq) \longrightarrow Zn^{2+}(aq) + Cu(s)$$

In this reaction the oxidation state of zinc increases from zero to $+2$, and copper is reduced from the $+2$ state to the zero state. The species that is reduced—Cu^{2+} in this case—is called the **oxidizing agent** (or **oxidant**) because it takes electrons from another species, thus oxidizing it. Using the same logic, we call the species that is oxidized—zinc metal—the **reducing agent** (or **reductant**) because it gives electrons to the oxidizing agent. |
| Example 7.15 | Which of the following is a redox reaction? Identify the species being oxidized and the species being reduced in each redox reaction.
a. $H_3O^+(aq) + OH^-(aq) \longrightarrow 2\ H_2O$
b. $Ca(s) + 2\ H_2O \longrightarrow Ca(OH)_2(s) + H_2(g)$
c. $2\ OH^-(aq) + Ba^{2+}(aq) + CO_2(g) \longrightarrow BaCO_3(s) + H_2O$
d. $2\ H^+(aq) + NO_3^-(aq) + 2\ Fe^{2+}(aq) \longrightarrow 2\ Fe^{3+}(aq) + NO_2^-(aq) + 3\ H_2O$ |
| Solution | a. This is not a redox reaction. Hydrogen remains in the $+1$ state and oxygen in the -2.
b. A redox reaction. Calcium (the reductant) is oxidized from 0 to the $+2$ state, and the hydrogen ions (oxidant) are reduced from $+1$ to 0.
c. There is no change in oxidation state, thus it is not a redox reaction.
d. It is a redox reaction because NO_3^- is reduced and Fe^{2+} is oxidized. |

7.5 OXIDATION–REDUCTION

Table 7.3 Common Oxidation States of Common Elements

Periodic group number	Common members	Oxidation states	Examples
IA	Li, Na, K, Rb, Cs	+1	LiCl, Na^+, K_2S
IIA	Be, Mg, Ca, Sr, Ba	+2	$BeCl_2$, Mg^{2+}, $BaSO_4$
IIIA	B, Al	+3	B_2H_6, Al^{3+}, Al_2O_3
IVA	C, Si, Sn, Pb	Carbon shows a variety of oxidation states ranging from -4 (CH_4) to $+4$ (CO_2).	
		+2	$SnCl_2$, Pb^{2+}
		+4	$SnCl_4$, PbO_2
VA	N, P, As, Sb, Bi	-3	NH_3, NH_4^+
		+2 (N only)	NO
		+3	NO_2^-, As_2O_3, BiO^+
		+4 (N only)	NO_2, N_2O_4
		+5	NO_3^-, H_3PO_4, As_2O_5
VIA	O, S, Se	-2	H_2O, oxides, Na_2S
		-1	H_2O_2 and peroxides
		+4 (except O)	SO_2, SO_3^{2-}, H_2SeO_3
		+6 (except O)	Na_2SO_4
VIIA	F, Cl, Br, I	-1	HCl, F^-, MgI_2
		+1 (except F)	HOCl, OI^-
		+3 (except F)	$HClO_2$
		+4 (Cl only)	ClO_2
		+5 (except F)	Cl_2O_5, IO_3^-
		+7 (Cl, I)	ClO_4^-, KIO_4
Transition metals	Cr	+2	Cr^{2+} (chromous)
		+3	Cr^{3+} (chromic)
		+6	CrO_4^{2-}, $Cr_2O_7^{2-}$
	Mn	+2	Mn^{2+} (manganous)
		+4	MnO_2
		+7	MnO_4^-
	Fe	+2	Fe^{2+} (ferrous)
		+3	Fe^{3+} (ferric)
	Co	+2	Co^{2+} (cobaltous)
		+3	Co_2O_3 (cobaltic)
	Ni	+2	Ni^{2+}, $NiCO_3$
	Cu	+1	Cu^+, Cu_2Cl_2 (cuprous)
		+2	Cu^{2+} (cupric)
	Zn	+2	Zn^{2+}
	Ag	+1	Ag^+, AgCl
	Hg	+1	Hg_2^{2+}, Hg_2Cl_2 (mercurous)
		+2	Hg^{2+} (mercuric)

Net Ionic Equations

When an aqueous solution of sodium sulfate is mixed with a solution of barium chloride, barium sulfate precipitates. Evaporation of the supernatant solution produces sodium chloride. A balanced equation based on the formulas of the reactants used and the products isolated is

$$BaCl_2(aq) + Na_2SO_4(aq) \longrightarrow BaSO_4(s) + 2\ NaCl(aq)$$

Our knowledge of the nature of ionic compounds gives us a better insight into this reaction. Ionic substances dissociate into their constituent ions in solution; a sodium sulfate solution contains Na^+ and SO_4^{2-}, and a barium chloride solution Ba^{2+} and Cl^-. The precipitation reaction is the combination of barium and sulfate ions to form the barium sulfate lattice. If we were to write the formulas of all the species dissolved in the two reacting solutions, the equation would be

$$Ba^{2+}(aq) + 2\ Cl^-(aq) + 2\ Na^+(aq) + SO_4^{2-}(aq) \longrightarrow$$
$$BaSO_4(s) + 2\ Cl^-(aq) + 2\ Na^+(aq)$$

but this form of the equation is needlessly complicated. Since Na^+ and Cl^- appear on both sides of the equation, we can eliminate them to obtain the **net ionic equation**.

$$Ba^{2+}(aq) + SO_4^{2-}(aq) \longrightarrow BaSO_4(s)$$

A net ionic equation displays only those species involved in the chemical change. Sodium and chloride ions are not involved in this reaction and do not appear in its net ionic equation; they are often called **spectator ions**.

Balancing a net ionic equation carries the additional requirement that charge must be conserved. Since chemical reactions cannot create or destroy charge, the total charge of the reactants must equal the total charge of the products. The equation

$$Fe^{2+}(aq) + I_2(aq) \longrightarrow Fe^{3+}(aq) + 2\ I^-(aq)$$
Total charge = +2   Total charge = +1

is out of balance with respect to charge. We cannot balance the equation by adding an electron to the left side of the equation; we must change the ratio of the ions. The correct equation is

$$2\ Fe^{2+}(aq) + I_2(aq) \longrightarrow 2\ Fe^{3+}(aq) + 2\ I^-(aq)$$
Total charge = +4   Total charge = +4

Balancing Redox Equations

There are two popular methods for balancing redox equations: the oxidation number method and the half-reaction method.

Oxidation number method. We will illustrate the oxidation number method by balancing the following equation.

$$Fe(s) + O_2(g) + H^+(aq) \longrightarrow Fe^{3+}(aq) + H_2O$$

The procedure is outlined below.

1. Identify the elements being oxidized and reduced by assigning oxidation states.

    $$Fe(s) + O_2(g) + H^+(aq) \longrightarrow Fe^{3+}(aq) + H_2O$$
    $\quad\ 0 \quad\quad\ \ 0 \quad\quad\ +1 \quad\quad\quad\ +3 \quad\quad +1\ -2$

    Iron is oxidized from the 0 to the +3 state and oxygen is reduced from 0 to −2.
2. Determine the number of electrons gained or lost per atom. Each oxygen atom gains two electrons, and each iron atom loses three.

    2 e⁻ gained per atom
    $$Fe(s) + O_2(g) + H^+(aq) \longrightarrow Fe^{3+}(aq) + H_2O$$
    3 e⁻ lost per atom

3. Determine the number of electrons gained or lost per formula unit.

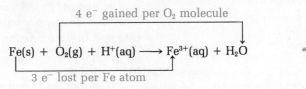

Since each oxygen atom gains two electrons, each O_2 molecule gains four.

4. Balance the electrons gained versus the electrons lost by using appropriate coefficients in front of the oxidizing and reducing agents. To do this, we look for a number that is a whole-number multiple of both the number of electrons gained per formula unit by the oxidizing agent and the number of electrons lost per formula unit by the reducing agent. In this example, the number is 12 and represents the number of electrons transferred in the balanced equation. It takes 4 iron atoms to give up these 12 electrons and 3 oxygen atoms to receive them, so the coefficients of Fe and O_2 in the balanced equation are 4 and 3, respectively.

$$3 \times 4 = 12 \text{ e}^- \text{ gained}$$
$$4\ Fe(s) + 3\ O_2(g) + H^+(aq) \longrightarrow 4\ Fe^{3+}(aq) + 6\ H_2O$$
$$4 \times 3 = 12 \text{ e}^- \text{ lost}$$

Note that we have maintained balance in iron atoms by producing 4 Fe^{3+} ions and kept oxygen in balance by forming 6 water molecules.

5. Balance charge and matter by adjusting the coefficients of the other species in the equation. As it now stands, the equation is unbalanced with respect to hydrogen and to charge. Both can be balanced by placing 12 H^+ ions on the left-hand side of the equation.

$$4\ Fe(s) + 3\ O_2(g) + 12\ H^+(aq) \longrightarrow 4\ Fe^{3+}(aq) + 6\ H_2O$$

As a final step, check to make sure that all atoms and charges balance. If not, look for an error in the earlier steps.

Half-reaction method. The half-reaction method involves the balancing of two separate **half-reactions**. The oxidation half-reaction describes the oxidation of the reducing agent and shows one or more electrons as a product on the right-hand side of the equation. The reduction half-reaction shows the chemical change of the oxidizing agent and contains one or more electrons on the left-hand side. After balancing these two half-reactions, we combine them in such a way that the electrons produced in the oxidation half-reaction are consumed in the reduction half-reaction. While it is permissible to show electrons as products or reactants in the half-reactions, electrons can never appear in the final balanced equation.

We will illustrate this method by applying it to the same reaction used in the oxidation number method. The oxidation half-reaction involves the conversion of iron to Fe^{3+}.

$$Fe(s) \longrightarrow Fe^{3+} + e^-$$

The reduction half-reaction involves the conversion of O_2 to water.

$$e^- + O_2(g) \longrightarrow H_2O$$

The method is as follows.

1. Identify the half-reactions by the changes in oxidation states. This has already been done in this case.

2. Balance each half-reaction individually. First, change the coefficients of all species except H^+, OH^-, and H_2O to balance the atoms oxidized or reduced. In this example, neither half-reaction requires any change. The iron atoms are balanced in the oxidation half-reaction. Since balancing the oxygen atoms in the reduction half-reaction involves water molecules and H^+ ions, we treat it in the next step.

   Then add H_2O molecules to the side deficient in oxygen atoms. The oxidation half-reaction does not involve oxygen, so no change is necessary. The reduction half-reaction becomes

   $$e^- + O_2(g) \longrightarrow 2\,H_2O$$

   Add H^+ to the side deficient in hydrogen atoms.

   $$e^- + 4\,H^+(aq) + O_2(g) \longrightarrow 2\,H_2O$$

   Increase the number of electrons, if necessary, to balance the charge. This step yields the two balanced half-reactions

   $$Fe(s) \longrightarrow Fe^{3+} + 3\,e^-$$

   $$4\,e^- + 4\,H^+(aq) + O_2(g) \longrightarrow 2\,H_2O$$

3. If necessary multiply the coefficients of each half-reaction by an appropriate constant factor and add the two half-reactions so that the electrons lost equal the electrons gained. If we multiply the coefficients in the oxidation half-reaction by 4, there will be 12 electrons lost.

   $$4\,Fe(s) \longrightarrow 4\,Fe^{3+}(aq) + 12\,e^-$$

   Multiplication of the reduction half-reaction by 3 will show a gain of 12 electrons.

   $$12\,e^- + 12\,H^+(aq) + 3\,O_2(g) \longrightarrow 6\,H_2O$$

   Adding these two equations yields

   $$4\,Fe(s) + 12\,e^- + 12\,H^+(aq) + 3\,O_2(g) \longrightarrow 4\,Fe^{3+}(aq) + 6\,H_2O + 12\,e^-$$

4. Cancel species occurring on both sides of the equation. If the equation has been properly balanced, the electrons should cancel completely. In some cases (but not this example), there may be some cancellation between water molecules or between H^+ ions.

   $$4\,Fe(s) + 12\,H^+(aq) + 3\,O_2(g) \longrightarrow 4\,Fe^{3+}(aq) + 6\,H_2O$$

5. If one of the reactants or products is hydroxide ion (OH^-), change each H^+ ion to a water molecule and add an equal number of OH^- ions to the opposite side of the equation. This can be done either in the half-reaction or in the final reaction. For example, the half-reaction for the reduction of CrO_4^{2-} to $Cr(OH)_3$ in the presence of OH^- can be balanced according to the steps shown above:

   $$3\,e^- + 5\,H^+(aq) + CrO_4^{2-}(aq) \longrightarrow Cr(OH)_3(s) + H_2O$$

   Since the reaction occurs in the presence of OH^-, we write instead

   $$3\,e^- + 5\,H_2O + CrO_4^{2-}(aq) \longrightarrow Cr(OH)_3(s) + H_2O + 5\,OH^-(aq)$$

   Cancellation of the water molecules on opposite sides of the equation yields the proper form of the balanced half-reaction.

   $$3\,e^- + 4\,H_2O + CrO_4^{2-}(aq) \longrightarrow Cr(OH)_3(s) + 5\,OH^-(aq)$$

7.5 OXIDATION–REDUCTION

The half-reaction method is so reliable that we can balance reactions even if H_2O, H^+, or OH^- are left out of the unbalanced equation. All we need know is that the reaction occurs in an aqueous solution and whether this solution is acidic (contains excess H^+) or basic (contains excess OH^-). Because H^+ and OH^- react with each other to form water, only one of the two ions can be present in large amounts in a given solution. Reactions in acid solutions can involve H^+ as either a reactant or a product, and OH^- may appear on either side of the equation for a reaction in basic solution. Water is, of course, present in all aqueous solutions and can be either a reactant or a product.

Example 7.16

Balance the equation for the following reaction in acid solution.

$$IO_3^-(aq) + Cl^-(aq) \longrightarrow Cl_2(g) + I_2(aq)$$

Solution

We shall use the half-reaction method for this example.
Identify the half-reactions.

$e^- + IO_3^-(aq) \longrightarrow I_2(aq)$ $Cl^-(aq) \longrightarrow Cl_2(g) + e^-$

Balance all atoms except H and O.

$e^- + 2\ IO_3^-(aq) \longrightarrow I_2(aq)$ $2\ Cl^-(aq) \longrightarrow Cl_2(g) + e^-$

Balance the oxygen atoms using H_2O.

$e^- + 2\ IO_3^-(aq) \longrightarrow I_2(aq) + 6\ H_2O$ Nothing needed here

Balance the hydrogen atoms using H^+.

$e^- + 2\ IO_3^-(aq) + 12\ H^+(aq) \longrightarrow I_2(aq) + 6\ H_2O$ Nothing needed here

Balance the charge using e^-.

$10\ e^- + 2\ IO_3^-(aq) + 12\ H^+(aq) \longrightarrow I_2(aq) + 6\ H_2O$ $2\ Cl^-(aq) \longrightarrow Cl_2(g) + 2\ e^-$

Now balance half-reactions so that the number of electrons balances.

$1\ [10\ e^- + 2\ IO_3^-(aq) + 12\ H^+(aq) \longrightarrow I_2(aq) + 6\ H_2O]$

$\underline{5\ [2\ Cl^-(aq) \longrightarrow Cl_2(g) + 2\ e^-]}$

$10\ e^- + 2\ IO_3^-(aq) + 10\ Cl^-(aq) + 12\ H^+(aq) \longrightarrow I_2(aq) + 5\ Cl_2(g) + 6\ H_2O + 10\ e^-$

Cancel species duplicated on each side of the reaction.

$$2\ IO_3^-(aq) + 10\ Cl^-(aq) + 12\ H^+(aq) \longrightarrow I_2(aq) + 5\ Cl_2(g) + 6\ H_2O$$

Example 7.17

Balance the equation for the following reaction.

$$MnO_4^{2-}(aq) + SO_3^{2-}(aq) \longrightarrow SO_4^{2-}(aq) + MnO_2(s) \quad \text{(basic solution)}$$

Solution

Identify the half-reactions.

$e^- + MnO_4^{2-}(aq) \longrightarrow MnO_2(s)$ $SO_3^{2-}(aq) \longrightarrow SO_4^{2-}(aq) + e^-$

Balance all atoms except H and O.
No change in either half-reaction.

Balance O atoms with H_2O.

$e^- + MnO_4^{2-}(aq) \longrightarrow MnO_2(s) + 2\ H_2O$ $SO_3^{2-}(aq) + H_2O \longrightarrow$
$\qquad\qquad\qquad\qquad\qquad\qquad\qquad\qquad\qquad\qquad SO_4^{2-}(aq) + e^-$

Balance H atoms with H^+.

$e^- + 4\ H^+(aq) + MnO_4^{2-}(aq) \longrightarrow MnO_2(s) + 2\ H_2O$ $SO_3^{2-}(aq) + H_2O \longrightarrow$
$\qquad\qquad\qquad\qquad\qquad\qquad\qquad\qquad\qquad\qquad SO_4^{2-}(aq) + 2\ H^+(aq) + e^-$

Balance charge using e^-.

$2\ e^- + 4\ H^+(aq) + MnO_4^{2-}(aq) \longrightarrow$ $SO_3^{2-}(aq) + H_2O \longrightarrow$
$\qquad\qquad\qquad MnO_2(s) + 2\ H_2O$ $SO_4^{2-}(aq) + 2\ H^+(aq) + 2\ e^-$

Balance the two half-reactions against each other so that the electrons gained and lost are equal. Since both half-reactions involve two electrons, they can be added without changing the coefficients.

$$2\ e^- + 4\ H^+(aq) + MnO_4^{2-}(aq) \longrightarrow MnO_2(s) + 2\ H_2O$$
$$SO_3^{2-}(aq) + H_2O \longrightarrow SO_4^{2-}(aq) + 2\ H^+ + 2\ e^-$$
$$\overline{2\ e^- + 4\ H^+(aq) + MnO_4^{2-}(aq) + SO_3^{2-}(aq) + H_2O \longrightarrow}$$
$$MnO_2(s) + 2\ H_2O + SO_4^{2-}(aq) + 2\ H^+ + 2\ e^-$$

Cancel species duplicated on each side of the reaction.

$$2\ H^+(aq) + MnO_4^{2-}(aq) + SO_3^{2-}(aq) \longrightarrow MnO_2(s) + H_2O + SO_4^{2-}(aq)$$

Since the reaction occurs in basic solution, H^+ should not show in the final equation. Add $2\ OH^-$ to the right side and replace the $2\ H^+$ on the left with $2\ H_2O$. Since there is also $1\ H_2O$ on the right, $1\ H_2O$ can be eliminated from each side.

$$H_2O + MnO_4^{2-}(aq) + SO_3^{2-}(aq) \longrightarrow MnO_2(s) + SO_4^{2-}(aq) + 2\ OH^-(aq)$$

Check to see that the equation is completely balanced.

7.6 Chemical Nomenclature

Substances were once named in arbitrary and unsystematic ways. Some of these traditional names such as water, ammonia (NH_3), lime (CaO), lye (NaOH), potash (K_2CO_3), baking soda ($NaHCO_3$), and the names of many minerals are still used. Others such as quicksilver (mercury), hartshorn [$(NH_4)_2CO_3$], and oil of vitriol (H_2SO_4) are quaint relics of the past. The major problem with these names is that each must be memorized, and this is an unacceptable burden. Instead, we have developed naming systems in which most names are related to their formulas by a few easily mastered rules.

Chemical nomenclature is still evolving toward a system that conveys as much information as possible about the formula of a substance. Since this evolution is not complete, you will probably encounter examples of two naming systems—the IUPAC (International Union of Pure and Applied Chemistry) system and an older system. There are also a few exceptions to the systematic rules because chemists persist in using a few unsystematic names. The rules that follow are in current use.

Binary compounds of nonmetals. The name of a compound of two nonmetals consists of the name of the less electronegative element followed by the number of atoms

7.6 CHEMICAL NOMENCLATURE

Table 7.4 Prefixes Used in Naming Compounds

Prefix	Number	Prefix	Number
Mono-	1	Hexa-	6
Di-	2	Hepta-	7
Tri-	3	Octa-	8
Tetra-	4	Nona-	9
Penta-	5	Deca-	10

and name of the other element. The name of the second element is usually preceded by a Greek prefix (Table 7.4) indicating the number of atoms present in the molecule or formula unit, and the ending of this element's name is replaced by *-ide*. For example,

CO	carbon monoxide		CS_2	carbon disulfide
SO_3	sulfur trioxide		$SiCl_4$	silicon tetrachloride
PCl_5	phosphorus pentachloride		SF_6	sulfur hexafluoride
Cl_2O_7	dichlorine heptoxide*			

Some of the oxides of nitrogen are given common names: nitrous oxide for N_2O and nitric oxide for NO.

Acids. Hydrogen-containing compounds with a tendency to dissociate into H^+ and an anion are called acids. For acids that do not contain oxygen, the pure compound is named as a binary compound of two nonmetals—hydrogen followed by the name of the anion. The aqueous solutions of these acids are named with a *hydro-* prefix and an *-ic* suffix.

Acid	Pure substance	Aqueous solution
HI	Hydrogen iodide	Hydroiodic acid
H_2S	Hydrogen sulfide	Hydrosulfuric acid
HCN	Hydrogen cyanide	Hydrocyanic acid

An oxyacid is one containing hydrogen, oxygen, and a third element. They are named using a word derived from the name of the third element followed by *acid*. If there are two oxyacids of the same element, each representing a different oxidation state, the name of the one with the lower oxidation state ends in *-ous acid*, and the one with the higher oxidation state in *-ic acid*.

HNO_2	nitrous acid	HNO_3	nitric acid
H_2SO_3	sulfurous acid	H_2SO_4	sulfuric acid

If an element forms more than two oxyacids, the prefix *per-* designates the highest oxidation state and *hypo-* the lowest.

HClO	hypochlorous acid	$HClO_2$	chlorous acid
$HClO_3$	chloric acid	$HClO_4$	perchloric acid

*Until fairly recently, it was common practice to leave out the prefix telling how many atoms of the first element are present. According to this style, Cl_2O_7 would be called chlorine heptoxide. A recent practice, which we are not using, is to name the elements in a way that describes their molecular formula. Thus O_2, H_2, and N_2 can be called dioxygen, dihydrogen, and dinitrogen, respectively.

If two oxyacids each contain one atom of the same element in the same oxidation state, the name of the one containing more oxygen (because it has an additional water molecule) begins with *ortho-*, the other with *meta-*.

HPO_3 metaphosphoric acid H_3PO_4 orthophosphoric acid

HBO_2 metaboric acid H_3BO_3 orthoboric acid

It is common practice to omit the prefix for the more common acid. A chemist seeing the name phosphoric acid assumes it refers to H_3PO_4.

Some oxyacids are polymeric. In other words, they are formed from several molecules of a simpler acid. The degree of polymerization is indicated by an appropriate prefix.

$H_4P_2O_7$ diphosphoric acid (also called pyrophosphoric acid)

$H_2Cr_2O_7$ dichromic acid

Organic acids, which consist of carbon, hydrogen, and oxygen, often have specific names. The most common examples are formic acid (HCO_2H), acetic acid (CH_3CO_2H), and oxalic acid ($H_2C_2O_4$).

Cations. If a metal forms only one cation, the cation takes its name directly from the metal.

K^+ potassium ion

Ca^{2+} calcium ion

Al^{3+} aluminum ion

There are two ways of distinguishing cations of the same metal. The method recommended by the IUPAC, also called the Stock system, is to list the oxidation state of the metal as a Roman numeral in parentheses after its name. The older method is to give the ion with the lower oxidation state the *-ous* suffix, and the one with the higher oxidation state *-ic* (see Table 7.5).

There are relatively few common polyatomic cations, the most important being ammonium ion, NH_4^+. Oxygen-containing cations are named by replacing the ending of the name of the element with the suffix *-yl*.

TiO^{2+} titanyl ion VO_2^+ vanadyl ion

Anions. Monatomic anions are named by replacing the ending of the name of the element with the suffix *-ide*.

Cl^- chloride ion O^{2-} oxide ion

S^{2-} sulfide ion N^{3-} nitride ion

Table 7.5 **Nomenclature of Cations**

Cation	IUPAC name	Traditional name
Cr^{2+}	Chromium(II) ion	Chromous ion
Cr^{3+}	Chromium(III) ion	Chromic ion
Fe^{2+}	Iron(II) ion	Ferrous ion
Fe^{3+}	Iron(III) ion	Ferric ion
Cu^+	Copper(I) ion	Cuprous ion
Cu^{2+}	Copper(II) ion	Cupric ion
Hg_2^{2+}	Mercury(I) ion	Mercurous ion
Hg^{2+}	Mercury(II) ion	Mercuric ion

7.6 CHEMICAL NOMENCLATURE

Several important polyatomic anions are

OH^-	hydroxide ion	O_2^{2-}	peroxide ion
CN^-	cyanide ion	I_3^-	triodide ion

Anions of oxyacids are named by replacing -ous endings with -ite and -ic endings with -ate.

NO_2^-	nitrite ion	IO_4^-	periodate ion
NO_3^-	nitrate ion	CO_3^{2-}	carbonate ion
BrO^-	hypobromite ion	SO_3^{2-}	sulfite ion

When naming anions formed by removing one H^+ from acid capable of yielding two H^+, insert either the word hydrogen or the prefix bi- in front of the name of the ion.

HCO_3^-	hydrogen carbonate ion or bicarbonate ion
HSO_3^-	hydrogen sulfite ion or bisulfite ion

The following notation indicates the number of protons remaining on an anion of an acid that yields three H^+.

$H_2PO_4^-$	dihydrogen phosphate ion
HPO_4^{2-}	monohydrogen phosphate ion
PO_4^{3-}	(ortho)phosphate ion

Salts. A salt is named by listing the name of the cation followed by the name of the anion. This rule applies to many substances that have very little tendency to dissociate into ions.

CaH_2	calcium hydride
Fe_2O_3	ferric oxide or iron(III) oxide
K_2SO_4	potassium sulfate
NH_4HS	ammonium bisulfide or ammonium hydrogen sulfide
$LiClO_4$	lithium perchlorate
Hg_2I_2	mercurous iodide or mercury(I) iodide
$PbCrO_4$	lead chromate
$Na_2Cr_2O_7$	sodium dichromate
$SnCl_2$	stannous chloride or tin(II) chloride
$SnCl_4$	stannic chloride or tin(IV) chloride

An appropriate prefix followed by -hydrate indicates the number of molecules of water of hydration per formula unit of salt.

$Ba(OH)_2 \cdot 8H_2O$ barium hydroxide octahydrate

Additional rules and examples for naming more complex substances are given in Appendix F.

Summary

1. In a structural formula, the symbol for an element represents the nucleus and inner-shell electrons of the atom, a straight line represents a shared pair of electrons, and dots stand for unshared valence electrons. These are called Lewis structures.

2. According to the Lewis octet rule, atoms, ions, and molecules tend to react so that each atom attains the electron configuration of a noble gas. An atom can attain its noble-gas configuration either by losing all its valence electrons and forming a positive ion, by gaining electrons and forming a negative ion, by forming covalent bonds, or by some combination of electron transfer and covalent-bond formation.

3. We can find the number of covalent bonds in an octet structure by subtracting the number of valence electrons in the species from the number of electrons needed to give each atom a noble-gas configuration and then dividing this number by two.

4. Isomers are two different compounds with the same molecular formula but different arrangements of the atoms in the molecules.

5. Many stable substances do not fit the octet rule. These include transition metal ions; superoctet species such as SF_6; electron-deficient species such as BF_3; and free radicals, which are species such as NO containing at least one unpaired electron spin.

6. One way of developing quantum mechanical models that are consistent with the shapes and relative bond energies of covalent molecules is to construct bond orbitals from hybrid atomic orbitals. Hybrid orbitals are orbitals formed by averaging the wave functions of a set of atomic orbitals derived from hydrogen.

7. Molecules have shapes that produce the maximum average angle between covalently bound atoms and orbitals containing unshared electron pairs. These shapes minimize valence-shell electron-pair repulsions between electrons.

8. Many species whose structures cannot be described by a single octet structure can be described as a resonance hybrid of several octet structures.

9. The oxidation state of an atom equals the number of valence electrons in the free atom minus the number of valence electrons assigned to the atom in the combined state. Both electrons of a covalent bond are assigned to the more electronegative atom.

10. The following approximate rules are useful for estimating oxidation states:
    Rule 1. The oxidation state of a monatomic ion always equals its charge.
    Rule 2. The oxidation state of an atom in its elemental state is zero.
    Rule 3. Hydrogen present in compounds is in the +1 oxidation state (never greater) except when combined with a less electronegative atom in which case it becomes −1 (hydride ion; for example, CaH_2, LiH).
    Rule 4. Oxygen has a −2 oxidation state except when it is combined with fluorine (it becomes positive), or when covalently bound to another oxygen.
    Rule 5. When present in compounds, the elements of periodic groups IA and IIA have oxidation states equal to their group numbers.
    Rule 6. The sum of the oxidation states of all the constituent atoms of a polyatomic ion must equal the charge on the ion and in a molecule this sum must be zero.

11. The maximum possible oxidation state of an element is its group number, and its minimum possible oxidation state is its group number minus eight.

12. Oxidation is an increase in the oxidation state of an atom. Reduction is a decrease in oxidation state. Any reaction involving changes in oxidation states is a redox reaction.

13. A net ionic equation displays only those species directly involved in the chemical reaction. Ionic reactions must be balanced with respect to charge.

14. A half-reaction is an equation in which electrons appear either as reactants (a reduction half-reaction) or as products (an oxidation half-reaction). We can balance a redox reaction by first balancing the two separate half-reactions and then combining them so that electrons are neither consumed nor produced in the total reaction.

15. Compounds are named in systematic ways that describe their formulas.

EXERCISES

Vocabulary List

structural formula	diradical	oxidizing agent
electron-dot structure	hybrid orbital	oxidant
Lewis octet rule	valence-shell electron-pair repulsion	reducing agent
isomerism	resonance hybrid	reductant
isomer	resonance forms	net ionic equation
transition metal ion	oxidation state	spectator ion
superoctet species	oxidation	half-reaction
electron-deficient species	reduction	
free radical	oxidation-reduction (redox) reaction	

Exercises

1. **a.** What are the possible ways (formation of ions, covalent bonding, or any combination of these) by which sulfur can attain the same electron configuration as argon?
   b. Find examples (in this textbook or any other source) of compounds in which sulfur atoms attain their octets by each of the ways cited in part a.
   c. Sulfur forms octet species in which it forms more than the minimum number of covalent bonds required for its octet. Find at least two species in which this occurs.
   d. What is the maximum number of covalent bonds sulfur can form and still follow the octet rule?

2. Draw Lewis structures for each of the compounds described in Exercise 49, Chapter 6. Make sure that each structure is consistent with the type of bonding indicated by the physical properties of each compound.

3. What reaction would you expect between BF_3 and F^- ion? Show line and dot structures for the reactants and products. [*Hint:* The reaction satisfies boron's need for an octet.]

4. Which of the following reactions are redox reactions? Identify the oxidizing agent and the reducing agent in each case.
   a. $H_2(g) + Br_2(g) \longrightarrow 2\ HBr(g)$
   b. $Cr_2O_7^{2-}(aq) + 2\ OH^-(aq) \longrightarrow 2\ CrO_4^{2-}(aq) + H_2O$
   c. $CaCO_3(s) + 2\ H^+(aq) \longrightarrow Ca^{2+}(aq) + H_2O + CO_2(g)$
   d. $2\ MnO_4^-(aq) + 5\ H_2S(aq) + 6\ H^+(aq) \longrightarrow 2\ Mn^{2+}(aq) + 5\ S(s) + 8\ H_2O$
   e. $2\ H_2O_2(aq) \longrightarrow 2\ H_2O + O_2(g)$

5. How many shared and unshared valence electrons surround each of the following?
   a. Rb·   **b.** Se in H—S̈e—H   **c.** Sr^{2+}
   d. N in Ö=N̈—Ö: **e.** ·S̈e·

6. How many shared and unshared valence electrons surround each of the following?
   a. :S̈e:$^{2-}$ **b.** O in :C̈l—Ö:$^-$
   c. Al in :B̈r—Al—B̈r: with :B̈r: below Al   **d.** C in $:C\equiv C:^{2-}$

7. Write electron dot structures for each.
   a. A phosphorus atom   **b.** Ca^{2+} **c.** O^{2-}
   d. $SrCl_2$ (an ionic compound composed of Sr^{2+} and Cl^- ions)
   e. A hydroxide ion

8. Write electron dot structures for each.
   a. A tin atom   **b.** Ga^{3+} **c.** Br^-
   d. CaH_2 (an ionic compound composed of Ca^{2+} and H^- ions)
   e. NH_2^-

9. Write an equation, using line and dot structures, that describes the reaction between solid potassium and solid I_2 to form solid KI.

10. Write an equation, using line and dot structures, to describe the formation of NH_3 from N_2 and H_2.

11. **a.** How many covalent bonds are present in a molecule with the formula C_2H_6O?
    b. Draw structures for two isomers of C_2H_6O.

12. **a.** How many covalent bonds are present in a molecule with the formula C_2H_3N?
    b. Draw structures for two isomers of C_2H_3N.

13. Draw octet structures for each.
    a. ClO_3^- **b.** C_2H_6 **c.** H_3S^+ **d.** SeO_3

14. Draw octet structures for each.
    a. C_2H_4 **b.** $N_2H_5^+$ **c.** BH_4^- **d.** CS_2

7 THE SYMBOLIC REPRESENTATION OF IONS AND MOLECULES

15. Which of the structures below are the following?
    a. Octet structures b. Free radicals
    c. Superoctet structures d. Electron-deficient species

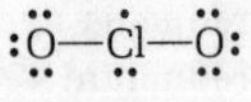

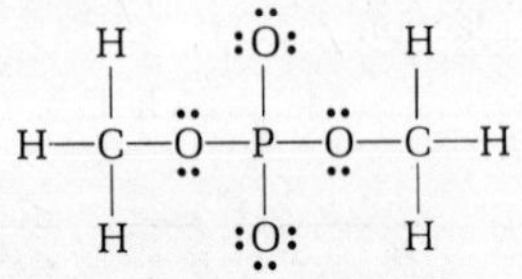

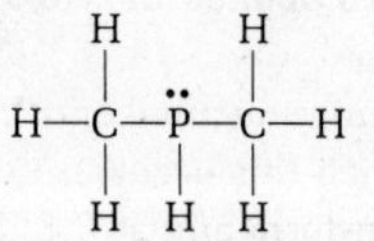

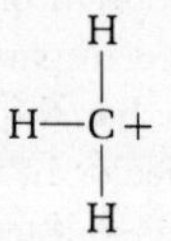

    I: :Ö—Ċl—Ö:

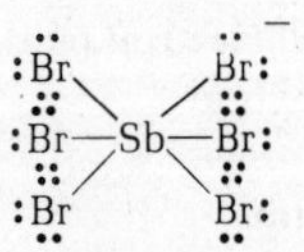

    II: SbBr₆⁻ (Br around Sb, six Br)

    III: (CH₃O)₂P(O)(OCH₃) — dimethyl phosphate-like structure with H—C—O—P(=O)(—O:)—O—C—H

    IV: H—C(H₂)—P̈—C(H₂)—H (with H's)

    V: H—C⁺(H₃) (methyl cation, H—C+ with three H)

    VI: H—C≡C—H

    VII: Na⁺

16. Which of the structures below:
    a. Must be free radicals?
    b. Must be superoctet structures?
    c. Must be electron-deficient structures?

$C_2H_5^+$	NO_2	N_2O_4	XeF_2	$GeCl_4$	IF_7
I	II	III	IV	V	VI

17. Without drawing line and dot structures, find the oxidation state of
    a. O in O_2 b. O in O_3 c. P in P_2O_5
    d. Ti in TiO^{2+} e. Fe in Fe^{3+} f. S in HS^-

18. Find the oxidation state of each of the following elements. Structures are not necessary to answer this question.
    a. I in HI   b. I in HIO_4 c. P in $P_2O_7^{4-}$
    d. C in HCO_2^- e. Se in Se^{2-} f. Cr in $Cr(OH)_4^-$

19. Find the oxidation state of each atom in the following molecules and ions.
    a. 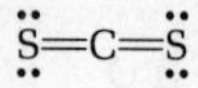 S=C=S   b. 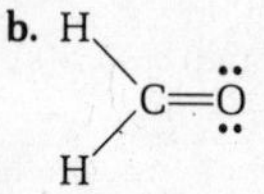 H₂C=O (formaldehyde)   c. La^{3+}
    d. 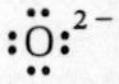 $:\!\ddot{O}\!:^{2-}$ e. 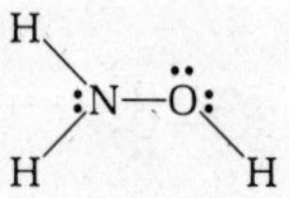 H₂N—OH   f. 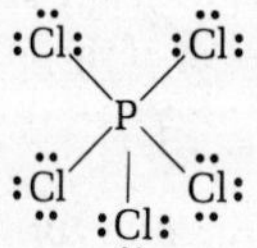 PCl_5

20. Find the oxidation state of each atom in the following molecules and ions.
    a. 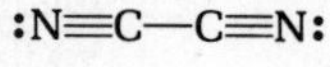 :N≡C—C≡N: b. $SOCl_2$ structure   c. :F—S̈—F:
    d. H—Ö—Ö:⁻   e. :O—S(=O)—O—H⁻ (with extra O)   f. :Ï—Cl:

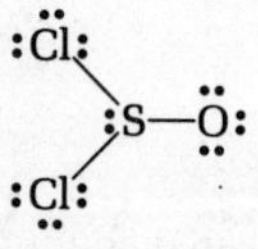

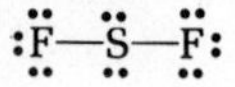

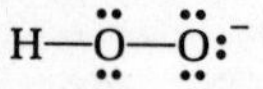

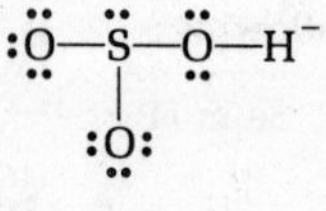

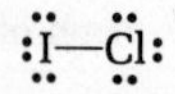

21. Use the periodic table to predict the highest possible oxidation state for each.
    a. Si   b. Te   c. Al   d. I

22. According to the periodic table, what is the highest possible oxidation state for each?
    a. P   b. Ba   c. Rb   d. Bi

23. Which is most likely to be the stable structure for PO_4^{3-}?

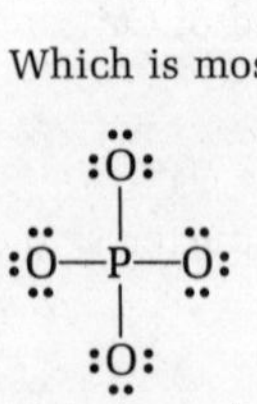

    I: O=P(O⁻)₃ type tetrahedral
    II: :O—P(=O)(O⁻)—O—O:
    III: :O—O—P(=O)(O⁻)—O—O:
    IV: :Ö—P—Ö—Ö—Ö:
    V: :P—O—O—O—O:

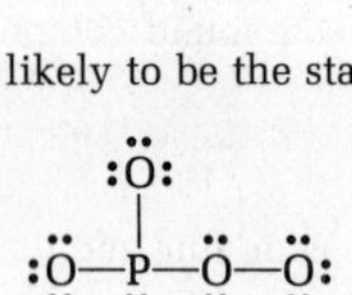

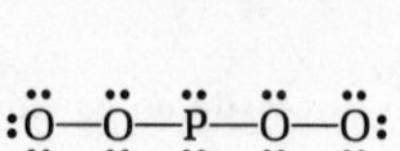

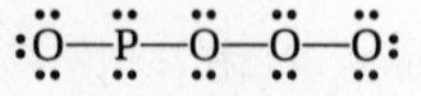

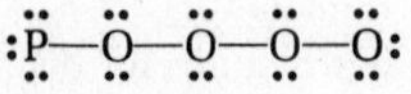

24. Three possible octet structures for silicon dioxide are

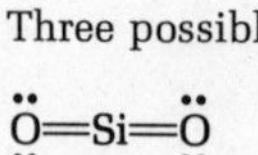

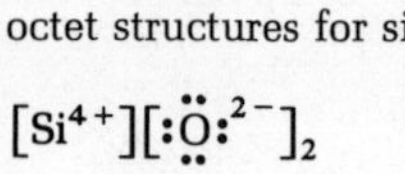

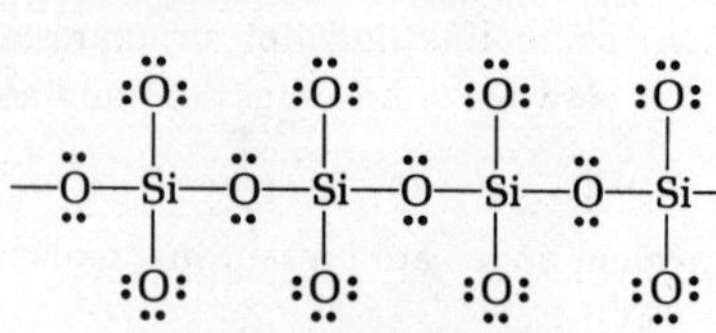

    O=Si=O $[Si^{4+}][:\!\ddot{O}\!:^{2-}]_2$

    —O—Si(O)(O)—O—Si(O)(O)—O—Si(O)(O)—O—Si— (Polymeric structure)

    Silicon dioxide melts at 1700°C, but molten SiO_2 does not conduct electricity. Which of these structures fits the properties of SiO_2 best? Explain your answer.

EXERCISES

25. What shapes would minimize repulsions between electrons in each of the following species?
    a. CH_3^+ b. CH_3^- c. CS_2 d. ClO_3^- e. PBr_5

26. Use the VSEPR method to predict the shapes of the following species.
    a. SiF_6^{2-} b. PO_4^{3-} c. SO_3 d. $TeCl_4$
    e. $POCl_3$ (P is the central atom)

27. Which of the following are resonance forms for the same species?

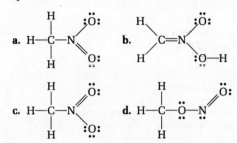

28. All of the nitrogen–oxygen bonds in NO_3^- have the same length, strength, and polarity. Present a resonance description for the structure of the nitrate ion.

29. What shape would you expect for XeF_2? (Remember that electron pairs tend to occupy the equatorial positions in a trigonal bipyramid.)

30. What shape would you predict for the molecule BrF_5?

31. What is the state of hybridization of the central atom and the shape of each of the following species?
    a. $:\ddot{O}-\ddot{S}-\ddot{O}:^{2-}$ with $:O:$ below (double bond)
    b. $:\ddot{O}-N-\ddot{O}:^-$ with $:O:$ below (double bond)
    c. $:\ddot{Cl}-Mg-\ddot{Cl}:$

32. The azide ion, N_3^-, can be described by averaging the following three resonance forms.

    $\ddot{N}=N=\ddot{N}^-$ $:N\equiv N-\ddot{N}:^-$

    $:\ddot{N}-N\equiv N:^-$

    a. What is the state of hybridization of the central atom in each of these resonance forms?
    b. What is the shape of each of these three resonance forms?

33. When solutions of sodium hydroxide and $MgSO_4$ are mixed, a precipitate of $Mg(OH)_2$ forms. Write a net ionic equation for this reaction.

34. Magnesium metal reacts with hydrochloric acid solution to produce hydrogen gas and a solution of magnesium chloride. Write a net ionic equation for this reaction.

35. Balance the half-reactions for the following transformations.
    a. $Li^+(aq)$ to $Li(s)$
    b. $Ca(s)$ to $Ca(OH)_2(s)$ in the presence of a basic solution
    c. $Mg(s)$ to $Mg^{2+}(aq)$
    d. $Cd(s) + CN^-(aq)$ to $Cd(CN)_4^{2-}(aq)$
    e. $Zn(NH_3)_4^{2+}(aq)$ to $Zn(s) + NH_3(aq)$

36. Balance the half-reactions for the following transformations.
    a. $Sn(OH)_6^{2-}(aq)$ to $HSnO_2^-(aq)$ in a basic solution
    b. $BrO_3^-(aq)$ to $Br_2(aq)$ in an acid solution
    c. $Br_2(l)$ to $BrO^-(aq)$ in a basic solution
    d. $NO_3^-(aq)$ to $NO(g)$ in an acid solution
    e. $H_2C_2O_4(aq)$ to $CO_2(g)$ in an acid solution

37. Balance the following equations. All reactants and products are shown.
    a. $I^-(aq) + Cu^{2+}(aq) \longrightarrow I_2(aq) + CuI(s)$
    b. $Ag^+(aq) + Fe^{2+}(aq) \longrightarrow Ag(s) + Fe^{3+}(aq)$
    c. $HNO_2(aq) + H^+(aq) + I^-(aq) \longrightarrow I_3^-(aq) + NO(g) + H_2O$
    d. $Hg^{2+}(aq) + CuCl(s) \longrightarrow Hg_2Cl_2(s) + Cu^{2+}(aq)$
    e. $Cr(OH)_4^-(aq) + HO_2^-(aq) \longrightarrow CrO_4^{2-}(aq) + OH^-(aq) + H_2O$

38. Balance the following equations. All reactants and products are shown.
    a. $H^+(aq) + MnO_4^-(aq) + Br^-(aq) \longrightarrow Br_2(aq) + MnO_2(s) + H_2O$
    b. $Cr_2O_7^{2-}(aq) + H^+(aq) + Fe^{2+}(aq) \longrightarrow Cr^{3+}(aq) + Fe^{3+}(aq) + H_2O$
    c. $Cu(s) + H^+(aq) + NO_3^-(aq) \longrightarrow Cu^{2+}(aq) + NO_2(g) + H_2O$
    d. $Al(s) + OH^-(aq) + H_2O \longrightarrow Al(OH)_4^-(aq) + H_2(g)$
    e. $MnO_4^{2-}(aq) + H^+(aq) \longrightarrow MnO_2(s) + MnO_4^-(aq) + H_2O$

39. Balance the equations for the following reactions. All of these reactions take place in acid solutions.
    a. $MnO_4^-(aq) + H_2C_2O_4(aq) \longrightarrow CO_2(g) + Mn^{2+}(aq)$
    b. $Cu(s) + Ag^+(aq) \longrightarrow Cu^{2+}(aq) + Ag(s)$
    c. $Cr_2O_7^{2-}(aq) + H_2CO(aq) \longrightarrow HCO_2H(aq) + Cr^{3+}(aq)$
    d. $H_2SO_3(aq) + I_3^-(aq) \longrightarrow HSO_4^-(aq) + I^-(aq)$
    e. $MnO_2(s) + H_2O_2(aq) \longrightarrow Mn^{2+}(aq) + O_2(g)$

40. Balance the equations for the following reactions, all of which occur in acid solutions.
    a. $Au(s) + Cl^-(aq) + NO_3^-(aq) \longrightarrow AuCl_4^-(aq) + NO_2(g)$
    b. $IO_3^-(aq) + I^-(aq) \longrightarrow I_2(aq)$
    c. $O_2(g) + Ti^{3+}(aq) \longrightarrow TiO^{2+}(aq)$
    d. $Ce^{4+}(aq) + HNO_2(aq) \longrightarrow NO_3^-(aq) + Ce^{3+}(aq)$
    e. $HNO_2(aq) \longrightarrow NO_2(g) + NO(g)$

41. All of the following reactions occur in basic solutions. Balance their equations.
    a. $BrO_3^-(aq) + I^-(aq) \longrightarrow Br^-(aq) + OI^-(aq)$
    b. $H_2CO(aq) + Ag(NH_3)_2^+(aq) \longrightarrow$
       $HCO_2^-(aq) + Ag(s) + NH_3(aq)$
    c. $Mg(s) \longrightarrow Mg(OH)_2(s) + H_2(g)$
    d. $O_2(g) + Cr(s) \longrightarrow Cr(OH)_4^-(aq)$
    e. $S_2O_3^{2-}(aq) + MnO_2(s) \longrightarrow SO_4^{2-}(aq) + Mn(OH)_2(s)$

42. Balance the following equations. All of these reactions occur in basic solutions.
    a. $CrO_4^{2-}(aq) + Fe(OH)_2(s) \longrightarrow Fe(OH)_3(s) + Cr(OH)_4^-(aq)$
    b. $OCl^-(aq) + NH_3(aq) \longrightarrow N_2(g) + Cl_2(g)$
    c. $MnO_4^-(aq) + CN^-(aq) \longrightarrow MnO_2(s) + CNO^-(aq)$
    d. $Ag_2O(s) + AsH_3(aq) \longrightarrow Ag(s) + H_2AsO_3^-(aq)$
    e. $Br_2(l) \longrightarrow BrO^-(aq) + Br^-(aq)$

43. Name the following compounds.
    a. H_2Se b. NI_3 c. $(NH_4)_2Cr_2O_7$ d. SrI_2
    e. H_3AsO_4 f. $HAsO_3$ g. $Cu(IO_3)_2$ h. $Zn(OH)_2$
    i. $Sr(NO_3)_2$ j. Sb_2O_5

44. Name the following compounds.
    a. SiO_2 b. $Ba(CN)_2$ c. Ag_3PO_4 d. $AlBr_3 \cdot 6H_2O$
    e. KH_2PO_4 f. $Cd(NO_3)_2$ g. Hg_2F_2 h. CaC_2O_4
    i. KH j. $NaOCl$

45. Write formulas for the following compounds.
    a. Barium arsenide   b. Mercuric chlorate
    c. Nickel(II) sulfate   d. Ammonium acetate
    e. Magnesium perchlorate   f. Sodium oxalate
    g. Hydrobromic acid
    h. Zinc sulfide   i. Barium chromate
    j. Periodic acid

46. Write formulas for the following compounds.
    a. Diiodine pentoxide   b. Cesium oxide
    c. Ferrous bromide
    d. Potassium hydrogen sulfide   e. Chromium(II) oxide
    f. Stannous sulfate   g. Arsenic pentafluoride
    h. Boron oxide   i. Sodium diphosphate
    j. Copper(II) sulfate hexahydrate

47. What is the lowest possible oxidation state for the following elements?
    a. Br   b. C   c. P   d. Se

48. a. What is the average oxidation state of carbon in acetic acid ($C_2H_3O_2H$)?
    b. What is the oxidation state of each carbon atom in the acetic acid molecule? (See the structure on page 162.)
    c. Are the answers in parts a and b consistent with each other?

49. Mercury(II) chloride molecules are linear. What is the state of hybridization of mercury in Hg_2Cl_2?

50. Draw energy level diagrams showing the valence electrons in the following atoms:
    a. An unhybridized nitrogen atom in its ground state
    b. The ground state arrangement of a nitrogen atom in the sp^3 state of hybridization
    c. The ground state arrangement of a nitrogen atom in the sp^2 state of hybridization
    d. A nitrogen atom in the sp^2 state of hybridization with one electron occupying a p orbital

chapter 8

Chemical Thermodynamics

Thermodynamics is the study of the energies of transformation of atoms and molecules, such as the dimerization of NO_2 to form N_2O_4

The discovery of fire was one of the most important technological advances of early man, a benefit so essential to human welfare that the ancient Greeks regarded it as a divine gift stolen from miserly and hostile gods. Modern society has, until very recently, tended to take it for granted and forgotten how dependent we are on the heat released by many chemical reactions. Chemically generated heat from the burning of coal, natural gas, and petroleum not only provides us with warmth and mechanical energy, but it also furnishes the energy necessary for the numerous important chemical reactions that absorb heat.

There are a number of examples of the chemical applications of heat. For example, in the laboratory we can convert $MgCO_3$ to MgO by heating it with a Bunsen burner. The burning of natural gas in the Bunsen flame involves heat-producing reactions such as

$$CH_4(g) + 2\ O_2(g) \longrightarrow CO_2(g) + 2\ H_2O(g)$$

Some of the released heat is absorbed by the decomposition of magnesium carbonate

$$MgCO_3(s) \longrightarrow MgO(s) + CO_2(g)$$

A reaction that releases heat is called an **exothermic reaction**; one that consumes heat is an **endothermic reaction**.

The theoretical significance of the heat transferred in chemical reactions matches its practical importance. **Thermodynamics**—the study of energy relationships in chemical reactions—gives us key insights about why certain chemical and physical processes occur while others do not under the same conditions. By analyzing the thermodynamic properties of a given system, chemists can predict whether or not conditions are favorable for a particular reaction. A careful consideration of thermodynamic factors can also tell us whether or not a particular compound will be stable under a given set of conditions. (However, it does not tell us how long it will take for an unstable compound to decompose, a question to be discussed in Chapter 12.)

The significance of the relationship between heat and chemical reactions became apparent after the realization that heat is molecular kinetic energy. Since the law of conservation of energy (also called the **first law of thermodynamics**) applies to chemical reactions, a decrease in molecular potential energy must balance the increase in molecular kinetic energy during an exothermic process. An endothermic reaction converts molecular kinetic energy into molecular potential energy.

Molecular potential energy is a measure of the strength of the chemical bonds in the system. The heat of a reaction depends on the energies of the chemical bonds in the reactant molecules relative to those of the product molecules. For example, the burning of CH_4 is an exothermic reaction because the bonds of one CO_2 and two H_2O molecules are much stronger than those of one CH_4 and two O_2 molecules. When the atoms break away from the weaker bonds to form the stronger ones, molecular potential energy changes to molecular kinetic energy, a process analogous to the conversion of gravitational potential energy to kinetic energy as an object falls. We experience the more rapid motion and vibrations of the product molecules as an increase in heat.

Because the bonds and lattice forces in $MgCO_3$ are stronger than those in MgO and CO_2, the decomposition of magnesium carbonate to magnesium oxide and carbon dioxide is endothermic. Stronger forces are replaced by weaker forces, and we must supply heat to raise the system's molecular potential energy.

The potential energy of a suspended weight depends only on its position and is independent of its past—it does not make any difference whether we hoisted it slowly or rapidly, or whether we first raised it higher and then lowered it. In the same way, the chemical potential energy of a molecule is independent of the history of its formation, and the change in potential energy during a reaction depends only on the reactants and products, not the details of how the reaction occurs. Because of this, we say that energy is a **state function**. State functions depend only on the system's chemical composition, physical state, temperature, and pressure but are independent of the system's past. We will encounter several other state functions in this chapter.

8.1
Energy and Enthalpy

Thermodynamic Systems

We begin a thermodynamic analysis by directing our attention to a system that interests us. The system might be a flask in which a reaction occurs, a car's engine, a live organism, or the earth and its atmosphere. Whatever the details of the system, we try to keep track of the ways in which it exchanges matter and energy with its surroundings.

Systems that exchange matter with their surroundings, such as car engines, are open; those that do not, such as a sample of gas in a sealed container, are closed (Figure 8.1). Since the flow of material in and out of a system complicates the interpretation of thermodynamic data, all of the theory presented in this chapter applies to closed systems. In the absence of a well-developed theory for open systems,* we apply the theory developed for closed systems to open ones and hope the errors inherent in this approximation are small.

A closed system, although it does not exchange matter with its surroundings,

---

*Chemists are working hard on the development of the theory of nonequilibrium thermodynamics, a theory they hope will be able to explain the open systems.

Figure 8.1

(a) A closed system. (b) An open system.

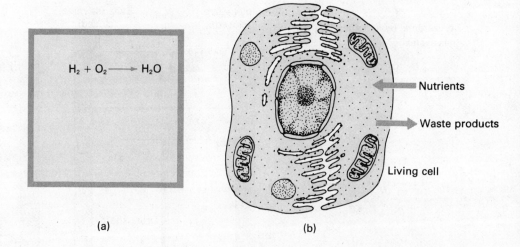

can exchange energy. Depending on the nature of the system, this exchange may involve the flow of heat, the performance of mechanical work, the flow of an electric current, or the emission or absorption of light. (We are not going to consider exchanges in electric or light energy in this chapter.) When performing thermodynamic experiments, we can design a system so as to either inhibit or favor these exchanges. One procedure is to insulate the system (Figure 8.2) so no heat flows in or out of it. A system that does not exchange heat energy with its surroundings is called **adiabatic**, and the processes occurring within it are adiabatic processes.

An alternate procedure is to design an **isothermal** system that is in effective thermal contact with its surroundings so that heat flows quickly in or out of the system to maintain a constant temperature within it. Isothermal processes occur without temperature changes. Many systems are neither strictly isothermal or adiabatic—some heat exchange occurs, but it is not enough to keep the temperature constant.

Work and Energy

Let us examine a system consisting of a gas in a cylinder with a movable piston. A cylinder in a car motor during its ignition cycle (Figure 8.3) is an example. Ignition initiates a chemical reaction that makes the system expand by heating and by produc-

Figure 8.2

(a) An adiabatic system. (b) An isothermal system.

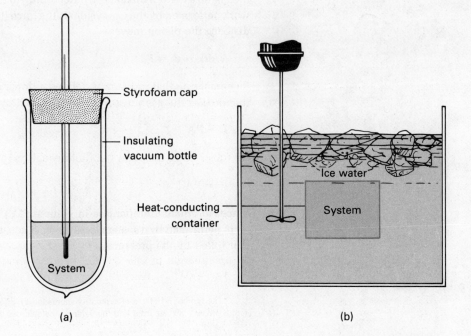

Figure 8.3
Work done by an expanding gas system.

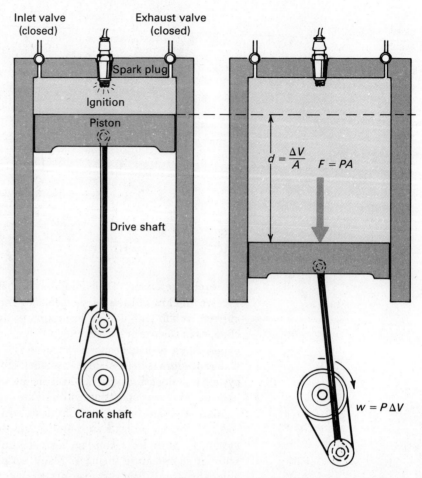

ing more moles of gas than were initially present. One of the typical reactions occurring during the burning of gasoline is

$$2\ C_8H_{18}(g) + 25\ O_2(g) \longrightarrow 16\ CO_2(g) + 18\ H_2O(g)$$

As the gases in the system expand against the piston, they apply a force (F) to move the piston a certain distance (d). According to the definition of work (Section 1.5), the work performed by the gas system is the force it exerts on the piston multiplied by the distance the piston moves.

$$\text{Work} = w = Fd \tag{8.1}$$

Remembering that pressure is defined as force per area (Section 3.1), we can describe the force as the gas pressure (P) multiplied by the surface area (A) of the piston.

$$F = PA \tag{8.2}$$

By substituting equation (8.2) into equation (8.1), we obtain equation (8.3).

$$w = PAd \tag{8.3}$$

Since Ad equals the increase in volume (ΔV)* of the system due to expansion, the work performed by a system expanding at constant pressure is the increase in volume multiplied by the pressure.

$$w = P\Delta V \tag{8.4}$$

*The symbol Δ in front of a quantity stands for the difference between the value of that quantity after a certain process has occurred and the value of the quantity before the process. For example,

$$\Delta E = E_{\text{final}} - E_{\text{initial}}$$

Table 8.1 Thermodynamic Sign Conventions

$+q$	An endothermic process in which the system absorbs heat
$-q$	An exothermic process in which the system produces heat
$+w$	The system does work on the surroundings
$-w$	The system receives work from the surroundings

Equation (8.4) also applies to a system in which a gas is contracting. In such cases, the surroundings compress the gas, producing a negative volume change and performing work on the system. Work done by the system is a positive quantity; work done on the system by the surroundings is negative (Table 8.1).

Although we derived equation (8.4) for a gaseous system, it applies equally well to solid and liquid systems. However, the volume changes in these nearly incompressible and nonexpanding systems are so small that it takes extreme pressures to produce significant amounts of work.

Example 8.1

A sample of gas in a cylinder expands 22.4 L against an average pressure of 2.50 atm. How much work is done in the process? Is it done by the gas sample or its surroundings?

Solution

Multiplying the volume change by the average resisting pressure gives us the work done.

$$w = P \Delta V = (22.4 \text{ L})(2.50 \text{ atm}) = 56.0 \text{ L-atm}$$

Although the L-atm is a legitimate energy unit, it is more convenient to express this answer in joules. One L-atm = 101.3 J.*

$$(56.0 \text{ L-atm}) \left(101.3 \ \frac{\text{J}}{\text{L-atm}} \right) = 5.67 \times 10^3 \text{ J, or } 5.67 \text{ kJ}$$

Since the work is positive, the gas sample does work on its surroundings.

Many systems, including the car cylinder, exchange both heat and mechanical work with their surroundings. Both of these factors—heat and work—contribute to the total energy, and we must keep track of them in order to describe the energy changes occurring in the system. If heat (q) flows into a system, the system gains energy, and it loses energy as heat flows out to the surroundings. When an expanding system performs work, it transfers that much of its own energy to the surroundings. To measure the energy change (ΔE) of the system, we must subtract the work performed by the system from the heat flowing into the system.

$$\Delta E = q - w \tag{8.5}$$

In this equation, heat flowing into the system is positive, and heat flowing out is negative. The car cylinder undergoes a decrease in energy during ignition because heat from the exothermic reaction flows out of the cylinder (negative q) as the cylinder performs work (positive w).

Neither heat nor work are state functions. Although the energy change is independent of how the change in the system is carried out, there are an infinite number of combinations of heat and work that will produce this energy change. The distribution of the energy change between heat and work depends on the details of how the

---

*To check this conversion factor, use the definition of an atmosphere, the density of mercury (13.6 g/cm³), and the acceleration due to gravity (9.79 m/sec²).

Example 8.2

A sample of gas expands 500 mL against a constant pressure of 0.500 atm while 60.0 J heat flow into it. What is the change in energy of the system?

Solution

The work done by the expanding gas is

$$w = P\Delta V = (0.500 \cancel{L})(0.500 \cancel{atm})\left(101.3 \frac{J}{\cancel{L\text{-atm}}}\right)$$

$$= 25.3 \text{ J}$$

According to equation (8.5), the energy change in the system is

$$\Delta E = 60.0 \text{ J} - 25.3 \text{ J} = 34.7 \text{ J}$$

Since the system has performed less work than it gains in heat, its energy increases.

Calorimetry

One of the fundamental measurements of chemical thermodynamics is the measurement of how much heat is gained or released by a system as a chemical reaction occurs within it. The measurement of these heat changes is called **calorimetry** and the devices for performing these measurements are called calorimeters. In the bomb calorimeter (Figure 8.4), an instrument useful for measuring heats of combustion, the

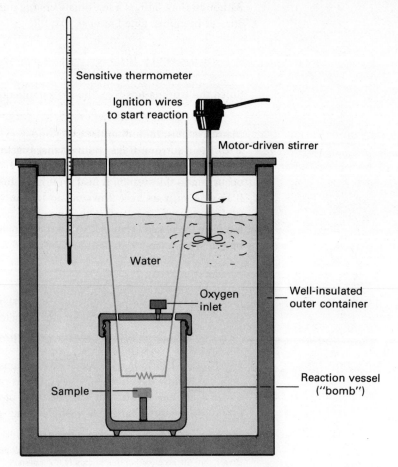

Figure 8.4
Laboratory precision calorimeter.

chemical system is enclosed in a strong steel container called the bomb. The bomb sits in a measured amount of water held in a well-insulated outer container. After the combustion reaction is initiated by an electric spark, the heat generated by the reaction flows from the bomb into the surrounding water, raising its temperature slightly. A sensitive thermometer measures the temperature increase of the water. The heat change in the system is related to the temperature increase of the calorimeter by equation (8.6).

$$q = -C\Delta T \tag{8.6}$$

The quantity C is the total heat capacity of the calorimeter and the water within it. In Section 1.5 we introduced heat capacities of substances, expressed in J/g-°C. We can also talk about the total heat capacity of the calorimeter system in J/°C; that is, the amount of heat needed to raise its temperature one degree Celsius.

The right-hand side of equation (8.6) has a negative sign because if the system within the bomb gives up heat, q will be negative and the thermometer will record a positive temperature change. If an endothermic process occurs within the calorimeter, q will increase in the system and the temperature of the surroundings will fall. Therefore, q and ΔT always have opposite signs.

Example 8.3

When a chemical reaction occurs in a bomb calorimeter whose total heat capacity is 2.40 kJ/°C, the temperature of the calorimeter rises 0.117°. What is the heat change for the chemical system?

Solution

Using equation (8.6), we find

$$q = -\left(2.40\frac{\text{kJ}}{°\text{C}}\right)(0.117°\text{C}) = -0.281 \text{ kJ, or } -281 \text{ J}$$

A simpler version of a calorimeter, useful for studying reactions in solution, can be fashioned from two styrofoam cups as in Figure 8.5 (p. 196). (You will probably use this type of calorimeter in your first-year laboratory.) Since styrofoam is a remarkably good insulator, the system inside the cups is essentially adiabatic. After initiating the reaction by mixing the solutions in the calorimeter, we observe the temperature change due to the reaction. For example, a solution of NaOH at 25.0°C could be mixed with a 25.0°C solution of HCl. The heat released by the reaction is the temperature increase multiplied by the heat capacity of the system (the reaction solution and the styrofoam cups).

Example 8.4

When 50 g of a solution containing 0.050 mol HCl reacts with 50 g of a solution containing 0.050 mol NaOH,

$$\text{HCl(aq)} + \text{NaOH(aq)} \longrightarrow \text{NaCl(aq)} + \text{H}_2\text{O}$$

the temperature of the styrofoam-cup calorimeter rises from 23.0°C to 29.1°C. The heat capacity of the mixed solution is 4.1 J/g-°C, and the heat capacity of the styrofoam cups is 40 J/°C. What is the heat of the reaction?

Solution

We first calculate the heat capacity of the entire system. The heat capacity of the solution is its mass multiplied by the heat capacity expressed in J/g-°C. Since we mixed 50 g of each solution, the total mass is 100 g.

Figure 8.5
Student calorimeter.

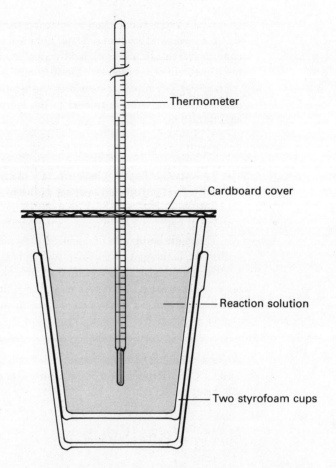

$$\text{Heat capacity of solution} = (100\,g)\left(4.1\ \frac{J}{g\cdot°C}\right)$$
$$= 4.1 \times 10^2\ J/°C$$

Adding the heat capacity of the cups to this value gives us the total heat capacity of the system.

$$\text{Heat capacity of the system} = 4.1 \times 10^2\ J/°C + 40\ J/°C$$
$$= 4.5 \times 10^2\ J/°C$$

The temperature increase is

$$\Delta T = 29.1°C - 23.0°C = 6.1°C$$

and the heat of reaction is

$$q = -(6.1°C)\left(4.5 \times 10^2\ \frac{J}{°C}\right) = -2.7 \times 10^3\ J,\ \text{or}\ -2.7\ kJ$$

Enthalpies of Reaction

Because an expanding system does work, an exothermic reaction in a system that is expanding produces less heat than the same reaction in a rigid container. In a rigid system,

$$\Delta V = 0\ \text{and}\ w = P\Delta V = 0$$

so that

$$\Delta E = q - w = q$$

An expanding system converts some energy to work, and the heat change is less negative than it would be for a rigid system. (An exothermic reaction generates less heat, an endothermic one consumes more heat.) More heat will be generated (or less consumed) in a contracting system than in a rigid system because the work performed by the surroundings heats the system.

Since the conditions under which a reaction occurs—whether it occurs in a rigid container, or whether the system is free to expand or contract—can influence the amount of heat involved, we must specify whether the heat of a reaction is computed for constant volume or constant pressure. For a rigid system, the measured heat represents the energy change for the reaction. To describe heat changes for systems maintained at a constant pressure, we define a thermodynamic quantity called enthalpy. The **enthalpy** (H) of a system is its energy plus the product of its pressure and volume.

$$H = E + PV \tag{8.7}$$

Enthalpy is a state function—it is not influenced by the history of the system. The usefulness of enthalpy is that, as can be seen from equation (8.8), the heat of a reaction in a system at constant pressure equals its change in enthalpy.

$$\Delta H = \Delta E + P\Delta V = q \tag{8.8}$$

Example 8.5

The energy change due to the reaction

$$NH_4Cl(aq) + NaNO_2(aq) \longrightarrow N_2(g) + 2\,H_2O$$

is -333 kJ for each mole of N_2 produced. The formation of the mole of nitrogen expands the system 22.4 L at 1.00 atm pressure. What is the enthalpy change due to this reaction?

Solution

As in Examples 8.1 and 8.2, we calculate $P\Delta V$ by multiplying the volume increase by the pressure and converting to joules.

$$P\Delta V = (22.4\,\cancel{L})(1.00\,\cancel{atm})\left(101.3\,\frac{J}{\cancel{L\text{-atm}}}\right)$$
$$= 2.27 \times 10^3\,J = 2.27\,kJ$$

We add this quantity to the energy change to compute the enthalpy change.

$$\Delta H = \Delta E + P\Delta V$$
$$= -333\,kJ + 2.27\,kJ = -331\,kJ$$

Since most of the practical systems we will encounter occur at constant pressure or, as in the case of a reaction in an open flask, something approximating constant pressure, we will develop thermodynamics based on enthalpy changes rather than energy changes. However, most calorimetric measurements on reactions involving gases are performed in rigid containers such as a bomb calorimeter. To use this sort of calorimetric data, we must predict the volume change that would occur at constant pressure and use this volume change to calculate the enthalpy change. If we are dealing with a gas phase reaction, we can use the ideal gas law to perform these calculations.

Consider the following example of calculating enthalpy changes from data obtained from bomb calorimeters. The reaction between 1.00×10^{-3} mol of hydrogen and excess oxygen in a bomb calorimeter

$$2\ H_2(g) + O_2(g) \longrightarrow 2\ H_2O(l)$$

releases 282 J heat. Since the system is rigid, this heat represents ΔE.

$$\Delta E = -282\ J$$

Now if the volume could have changed at constant pressure, the system would have contracted because hydrogen and oxygen (both gases) are converted to water (a liquid). (The volume of water produced—about 0.018 mL—is so small compared to the volume of gases consumed that we neglect it.) The ideal gas law tells us what the value of $P\Delta V$ would be if the reaction were run at constant pressure.

$$P\Delta V = \Delta n_{gas}\ RT \tag{8.9}$$

The quantity Δn_{gas} is the change in the number of moles of gas due to the reaction.

$$\Delta n_{gas} = n_{gas,\ product} - n_{gas,\ reactant} \tag{8.10}$$

Since 1.00×10^{-3} mol H_2 reacts with 0.50×10^{-3} mol O_2, the change in the number of moles of gas is

$$\Delta n_{gas} = 0\ \text{mol product gas} - 1.00 \times 10^{-3}\ \text{mol}\ H_2 - 0.50 \times 10^{-3}\ \text{mol}\ O_2$$
$$= -1.50 \times 10^{-3}\ \text{mol gas}$$

In using equation (8.9), it is more convenient to use R expressed in units of J/K·mol. This value is 8.31 J/K·mol.

If the calorimetric measurement was performed at 25°C (298 K), the value of $P\Delta V$ is

$$P\Delta V = (-1.50 \times 10^{-3}\ \text{mol})\left(8.31\ \frac{J}{K\cdot mol}\right)(298\ K)$$
$$= -3.72\ J$$

The enthalpy of the reaction is

$$\Delta H = \Delta E + P\Delta V$$
$$= -282\ J - 3.72\ J = -286\ J$$

The enthalpy of a reaction is proportional to the amount of material reacting. Therefore, ΔH for the production of 1 mol of water from its elements is

$$\Delta H = \frac{-286\ J}{1.00 \times 10^{-3}\ \text{mol}} = -2.86 \times 10^5\ J/\text{mol, or } -286\ kJ/\text{mol}$$

Because melting a solid or evaporating a liquid requires energy, the physical states of the reactants and products have a significant effect on the enthalpy change. The molar enthalpy for the reaction of hydrogen and oxygen to form water vapor (as opposed to liquid water) is -242 kJ/mol.

Example 8.6

Burning 0.1676 g of iron in a bomb calorimeter at 25°C produces 1.228 kJ heat.

$$4\ Fe(s) + 3\ O_2(g) \longrightarrow 2\ Fe_2O_3(s)$$

What is the enthalpy per mole of Fe_2O_3 for this reaction?

Solution

The heat released in the bomb calorimeter is ΔE for the reaction so, in addition to determining the heat that would be released when 1 mol Fe_2O_3 forms, we must correct for the work that would be done at constant pressure. We will first compute ΔE for 1 mol of Fe_2O_3.

$$\text{Moles of Fe} = \frac{0.1676 \text{ g}}{55.85 \text{ g/mol}} = 3.00 \times 10^{-3} \text{ mol}$$

$$\text{Moles of Fe}_2\text{O}_3 \text{ produced} = \frac{1}{2}(\text{moles of Fe consumed})$$

$$= \frac{1}{2}(3.00 \times 10^{-3} \text{ mol}) = 1.50 \times 10^{-3} \text{ mol}$$

Since ΔE is proportional to the amount of Fe_2O_3 produced,

$$\Delta E = \frac{-1.228 \text{ kJ}}{1.50 \times 10^{-3} \text{ mol}} = -819 \text{ kJ/mol}$$

Each mole of Fe_2O_3 produced consumes 1.50 mol O_2 gas, and

$$\Delta n_{gas} = 0 \text{ mol product gas} - 1.50 \text{ mol } O_2 = -1.50 \text{ mol}$$

$$P\Delta V = \left(-1.50 \frac{\text{mol gas}}{\text{mol Fe}_2\text{O}_3}\right)\left(8.31 \frac{\text{J}}{\text{K-mol}}\right)(298 \text{ K})$$

$$= -3.72 \text{ kJ/mol}$$

$$\Delta H = \Delta E + P\Delta V$$

$$= -819 \text{ kJ/mol} - 3.72 \text{ kJ/mol} = -823 \text{ kJ/mol}$$

Enthalpies of Formation

The **enthalpy of formation** (ΔH_f) of any substance is the enthalpy of the isothermal reaction needed to produce 1 mol of this substance from its elements. Because enthalpies of reactions depend on temperature and (particularly when gases are involved) pressure, we define **standard enthalpies of formation** ($\Delta H_f°$). These standard enthalpies (Table 8.2) refer to reactions in which the reactants are initially present at the standard state of 1 atm pressure and 25°C (298 K), and the products are also at this standard condition of temperature and pressure. (The superscript zero in the symbol indicates that the measurements are related to the standard states.) The most common elemental form of an element—molecular states such as gaseous H_2 or liquid Br_2—have, by definition, enthalpies of formation equal to zero.

Hess's Law

Hess's law says that we can add the enthalpies of two or more reactions to find the enthalpy of the net reaction that would result from these reactions occurring consecutively. For example, the reaction between hydrogen gas and solid copper(II) oxide to produce solid copper and water vapor

$$CuO(s) + H_2(g) \longrightarrow Cu(s) + H_2O(g)$$

produces the same change as two reactions occurring in sequence—the decomposition of CuO to its elements followed by the reaction between oxygen and hydrogen to form water vapor.

$$CuO(s) \longrightarrow Cu(s) + \frac{1}{2} O_2(g) \quad \Delta H° = \quad 155 \text{ kJ}$$

$$\frac{1}{2} O_2(g) + H_2(g) \longrightarrow H_2O(g) \quad \underline{\Delta H° = -241.83 \text{ kJ}}$$

$$- 87 \text{ kJ}$$

The enthalpy of decomposition (an endothermic reaction) of 1 mol CuO is 155 kJ and the enthalpy of formation of 1 mol H_2O is −241.83 kJ. Adding these two heats

Table 8.2 Standard Thermodynamic Quantities at 25°C

Substance	ΔH_f° (kJ/mol)	S° (J/K-mol)	ΔG_f° (kJ/mol)
Ag(s)	0.00	42.72	0.00
AgCl(s)	−127.0	96.11	−109.7
$AgNO_3$(s)	−123.1	140.9	−32.2
Al(s)	0.00	28.3	0.00
Al_2O_3(s)	−1669.8	51.00	−1576.4
Ba(s)	0.00	67	0.00
$BaCO_3$(s)	−1219	112	−1139
$BaCl_2$(s)	−860.06	126	−810.9
BaO(s)	−558.1	70	−528.4
$BaSO_4$(s)	−1465	132	−1353
Br_2(l)	0.00	152	0.00
Br_2(g)	+30.7	245.35	+3.13
C(graphite)	0.00	5.69	0.00
C(diamond)	+1.9	2.4	+2.9
CH_4(g)	−74.85	186.2	−50.79
C_2H_4(g)	+52.30	219.5	+68.12
C_2H_6(g)	−84.68	229.5	−32.9
CCl_4(l)	−139.5	214.4	−68.74
CO(g)	−110.523	197.91	−137.268
CO_2(g)	−393.51	213.64	−394.38
Ca(s)	0.00	41.6	0.00
$CaCO_3$(s)(calcite)	−1206.9	92.9	−1128.8
$CaCl_2$(s)	−795.0	114	−750.2
$CaSO_4$(s)	−1432.7	107	−1320.3
Cl(g)	+121.39	165.084	+105.40
Cl_2(g)	0.00	222.95	0.00
Cr(s)	0.00	23.8	0.00
Cr_2O_3(s)	−1128	81.2	−1047
Cu(s)	0.00	33.3	0.00
CuO(s)	−155	43.5	−127
Cu_2O(s)	−166.7	101	−146.4
F_2(g)	0.00	203	0.00
Fe(s)	0.00	27.2	0.00
FeO(s)	−267	54.0	−244
Fe_2O_3(s)	−822.2	90.0	−741.0
Fe_3O_4(s)	−1117	146	−1014
FeS(s)	−95.06	67.4	−97.57
H_2(g)	0.00	130.59	0.00
HBr(g)	−36.2	198.5	−53.22
HCl(g)	−92.30	186.7	−95.26
HF(g)	−269	173.5	−271
HI(g)	+25.9	206.3	+1.3
H_2O(l)	−285.84	69.940	−237.19
H_2O(g)	−241.83	188.72	−228.60
Hg(l)	0.00	77.4	0.00
Hg_2Cl_2(s)	−264.9	196	−210.7
HgO(s)	−90.71	72.0	−58.53
I_2(s)	0.00	117	0.00
I_2(g)	+62.26	260.6	+19.4
K(s)	0.00	63.6	0.00
KBr(s)	−392.2	96.44	−379.2
KCl(s)	−435.89	82.68	−408.32
KF(s)	−562.58	66.57	−533.13
KI(s)	−327.6	104.35	−322.3
Mg(s)	0.00	32.5	0.00
$MgCl_2$(s)	−641.83	89.5	−592.3
$Mg(NO_3)_2$(s)	−789.60	164	−588.40

8.1 ENERGY AND ENTHALPY

Substance	ΔH_f° (kJ/mol)	S° (J/K-mol)	ΔG_f° (kJ/mol)
MgO(s)	−601.83	27	569.57
MgSO$_4$(s)	−1278	91.6	−1174
Mn(s)	0.00	31.8	0.00
MnO(s)	−385	60.2	−363
MnO$_2$(s)	−520.9	53.1	−466.1
N$_2$(g)	0.00	191.5	0.00
NH$_3$(g)	−46.19	192.5	−16.7
NH$_4$Cl(s)	−315.3	94.6	−203.9
NO(g)	+90.37	210.6	+86.69
NO$_2$(g)	+33.05	240.5	+51.840
N$_2$O(g)	+81.55	220.0	+103.6
N$_2$O$_4$(g)	+9.66	304.3	+98.29
Na(s)	0.00	51.0	0.00
NaF(s)	−569.0	58.6	−541.0
NaCl(s)	−410.90	72.4	−384.0
Na$_2$CO$_3$(s)	−1131	136	−1048
NaHCO$_3$(s)	−947.7	102	−851.9
Na$_2$SO$_4$(s)	−1384.5	149.5	−1266.8
O$_2$(g)	0.00	205.03	0.00
O$_3$(g)	+142	238	+163
Pb(s)	0.00	64.89	0.00
PbO(s)	−219.2	67.8	−189.3
PbO$_2$(s)	−276.6	76.6	−219.0
PbSO$_4$(s)	−918.39	147	−811.24
S(s)	0.00	31.9	0.00
SO$_2$(g)	−296.9	248.5	−300.4
SO$_3$(g)	−395.2	256.2	−370.4
Sn(s)	0.00	51.5	0.00
SnO(s)	−286	56.5	−257
SnO$_2$(s)	−580.7	52.3	−519.7
Zn(s)	0.00	41.6	0.00
ZnCl$_2$(s)	−415.9	108	−369.3
ZnO	−348.0	43.9	−318.2

together, we estimate the enthalpy of the reaction between hydrogen and copper(II) oxide to be −87 kJ per mole H$_2$O produced, an exact agreement with the direct measurement of the heat of this reaction.

The reaction between CuO and H$_2$ does not occur by a sequence of decomposition of CuO and formation of H$_2$O, but that does not upset the validity of Hess's law. The mechanism by which a mole of CuO and a mole of hydrogen change to a mole of water vapor and a mole of copper makes no difference in the enthalpy change of the system. (We used a similar argument in Chapter 6 when we presented the Born-Haber cycle—a more complicated example of Hess's law.)

Example 8.7

The enthalpy of decomposition of HCl gas into its elements is +92.30 kJ/mol.

$$HCl(g) \longrightarrow \frac{1}{2} H_2(g) + \frac{1}{2} Cl_2(g)$$

The enthalpy of formation of solid BaCl$_2$ is −860.06 kJ/mol.

$$Ba(s) + Cl_2(g) \longrightarrow BaCl_2(s)$$

What is the enthalpy of the reaction between HCl and barium to produce barium chloride and hydrogen gas?

$$2 \, HCl(g) + Ba(s) \longrightarrow BaCl_2(s) + H_2(g)$$

Solution

Since the reaction consumes 2 mol HCl, we must multiply the enthalpy of decomposition of HCl by two.

$$2\text{ HCl(g)} \longrightarrow \text{H}_2\text{(g)} + \text{Cl}_2\text{(g)} \qquad \Delta H° = (2\text{ mol})(+92.30\text{ kJ/mol})$$
$$= +184.6\text{ kJ}$$

$$\text{Ba(s)} + \text{Cl}_2\text{(g)} \longrightarrow \text{BaCl}_2\text{(s)} \qquad \Delta H° = (1\text{ mol})(-860.06\text{ kJ/mol})$$
$$= -860.06\text{ kJ}$$

$$2\text{ HCl(g)} + \text{Ba(s)} \longrightarrow \text{BaCl}_2\text{(s)} + \text{H}_2\text{(g)}$$

$$\Delta H° = +184.6\text{ kJ} - 860.06\text{ kJ} = -675.5\text{ kJ}$$

Using enthalpies of formation, we can predict the standard enthalpies of a large number of reactions. By applying Hess's law, we can show that the enthalpy of a reaction equals the sum of the enthalpies of formation of the products less the sum of the enthalpies of formation of the reactants.*

$$\Delta H°_{\text{reaction}} = \Sigma \Delta H°_{f,\text{ products}} - \Sigma \Delta H°_{f,\text{ reactants}} \tag{8.11}$$

In carrying out these calculations, we must be careful to multiply the enthalpies of formation of each substance by the number of moles of that substance involved in the reaction.

Example 8.8

Predict the standard enthalpy of the reaction

$$2\text{ Fe(s)} + 3\text{ H}_2\text{O(g)} \longrightarrow \text{Fe}_2\text{O}_3\text{(s)} + 3\text{ H}_2\text{(g)}$$

Solution

According to Hess's law, we subtract the standard enthalpies of formation of 2 mol Fe and 3 mol of water vapor from the enthalpies of formation of 1 mol Fe_2O_3 and 3 mol H_2.

$$\Delta H° = (\Delta H°_{f,\text{ Fe}_2\text{O}_3}) + 3(\Delta H°_{f,\text{ H}_2}) - 2(\Delta H°_{f,\text{ Fe}}) - 3(\Delta H°_{f,\text{ H}_2\text{O}})$$

The heats of formation of the elements Fe and H_2 are zero so the equation reduces to

$$\Delta H° = (\Delta H°_{f,\text{ Fe}_2\text{O}_3}) - 3(\Delta H°_{f,\text{ H}_2\text{O}})$$

Referring to the data in Table 8.2 (and being careful to use the heat of formation of water *vapor* rather than the liquid), we obtain

$$\Delta H° = (1\text{ mol})\left(-822.2\frac{\text{kJ}}{\text{mol}}\right) - (3\text{ mol})\left(-241.83\frac{\text{kJ}}{\text{mol}}\right)$$
$$= -96.7\text{ kJ}$$

8.2 Entropy

Spontaneous Change

Enthalpy is an important factor in determining the stability of a system, and a decrease in enthalpy (a negative ΔH) for a process favors that process. There was a period in the early development of thermodynamics when it was believed that enthalpy is the only thing that dictates the stabilities of systems and the direction of spontaneous change. According to this view, all spontaneous processes should be

---

*The symbol Σ stands for the sum of a group of quantities.

exothermic. However, this is simply not true; many spontaneous processes (such as ice melting at room temperature) are endothermic and increase the enthalpy of a system. Another thermodynamic factor, called entropy, also influences the direction of spontaneous change in a system.

It is easiest to observe the effect of entropy on the direction of spontaneous change by studying adiabatic processes. Since these processes do not involve enthalpy changes, their directions of spontaneous change depend solely on the changes of entropy due to these processes.

The first system (Figure 8.6, part a) consists of two pieces of metal, one hot and the other cold, in contact with each other. Heat flows spontaneously from the hot region to the cold one, never from cold to hot. Once the temperature has become uniform throughout the system, it takes work to transfer heat from one piece of metal to the other and create a temperature difference between them. This is why it is necessary to plug a refrigerator into an electric outlet. The refrigerator does pump heat from a colder region (its interior) to a warmer one (the room), but it must perform work to do this and requires electrical energy.

We can see this "one-way" aspect of spontaneous change in the behavior of gases. A gas flows into a vacuum (Figure 8.6, part b), but once it spreads throughout the container it will never congregate on one side, leaving a vacuum in the rest of the vessel. Two gases mix spontaneously (Figure 8.6, part c); gas mixtures do not sort themselves out so that pure gases concentrate at opposite ends of the container. To understand what entropy is, we must ask ourselves what these three examples have

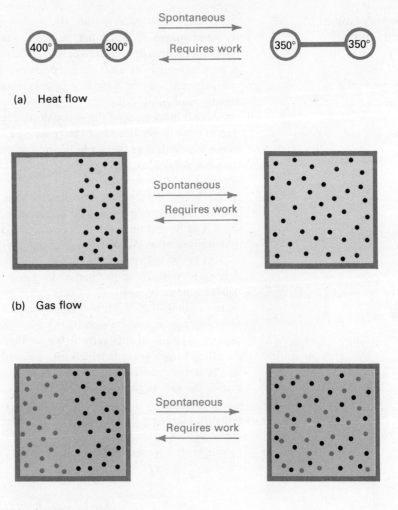

Figure 8.6
Spontaneous changes.

(a) Heat flow

(b) Gas flow

(c) Gas mixing

Figure 8.7
Maxwell's demon.

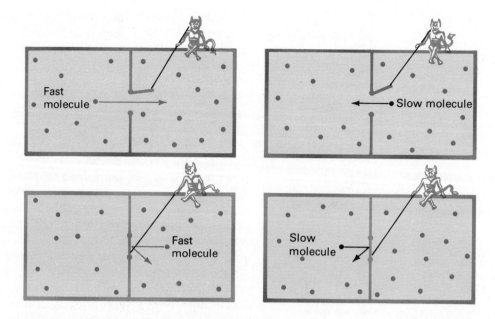

in common. The answer, which we shall develop in detail in the next section, is that the preferred direction of change is the one that produces a more random system, and **entropy** is a measure of that randomness.

The Second Law of Thermodynamics

James Clerk Maxwell explained the one-way nature of heat flow by the hypothesis of the randomness of molecular motion. According to the kinetic-molecular theory, there is a random distribution of molecular energies at any temperature (Figure 3.13). If we place a gas at uniform temperature in an insulated vessel composed of two compartments A and B, with a hole between them (Figure 8.7), the distribution of rapidly and slowly moving molecules is the same in both compartments. For chamber B to warm up at the expense of A, fast molecules must leave A and collect in B, and slow molecules must congregate in A. But a fast molecule is just as likely to move from B to A as from A to B, and the same lack of preferred direction applies to the slow molecules. Maxwell said that to violate the one-way limitation of heat flow, we must have the services of "a being whose facilities are so sharpened that he can follow every molecule in its course,"* and this being (later nicknamed Maxwell's demon) "opens and closes this hole, so as to allow only the swifter molecules to pass from A to B, and only the slower ones to pass from B to A. He will then, without expenditure of work, raise the temperature of B and lower that of A."

As far as we know, Maxwell's demon—the personification of a conscious and deliberate control of individual molecular events—does not exist. Because molecules behave randomly, ordered systems (such as hot gas in A and cold gas in B) tend to more random, less organized arrangements unless some physical force (such as a closed stopcock, heat insulation, or a positive enthalpy change) opposes this. A more prosaic example of this **second law of thermodynamics** deals with playing cards. Shuffling a deck of cards which is in ordered suits gives random sequences of cards, and it is very unlikely that further shuffling (unaided by Maxwell's demon) will restore the ordered suits.

The second law is based on probabilities; violations can, in theory, happen, but violations large enough to significantly affect average properties for large numbers of molecules are so improbable that we need not worry about them. The reliability of the second law increases with the number of molecules in the system and is very high for the usual laboratory samples.

*Quotations are from James Clerk Maxwell, *Theory of Heat* (New York: D. Appleton and Company, 1872), p. 308.

Entropy and Spontaneous Change

Entropy (S), a thermodynamic state quantity, is a measure of the molecular disorder in a system. Rephrased, the second law of thermodynamics says that the entropy of the universe always *increases*. There can be entropy decreases in particular processes (for example, freezing water makes the molecules more ordered), but they must be compensated by larger entropy increases elsewhere in the universe. An orderly universe is continuously losing its order, although some small sections of it may become more organized.

When heat (q) flows into a system that is not undergoing chemical or physical changes (including mixing or compression), the entropy increases by the quantity q/T. Heat flowing out of a system produces a corresponding entropy loss.

$$\Delta S_{system} = \frac{q}{T_{system}} \tag{8.12}$$

When heat flows into a system, the heat change of the surroundings is $-q$, and the entropy change for the surroundings is

$$\Delta S_{surroundings} = -\frac{q}{T_{surroundings}}$$

The total entropy change of the universe is the sum of these two entropies.

$$\Delta S_{total} = \frac{q}{T_{system}} - \frac{q}{T_{surroundings}}$$

If T_{system} is less than $T_{surroundings}$—in other words, if the heat goes from a warm region to a cold one—the total entropy change is positive and the heat flow occurs spontaneously. The opposite process, heat flow from a colder region to a warmer one, requires work. The performance of this work leads to an increase in entropy that is greater than the entropy decrease due to the heat flow.

8.3 Chemical Stability

Entropy and Chemical Reactions

Since entropy measures disorder, its change is inherent in any change in molecular structure. For example, the reaction

$$N_2O_4(g) \longrightarrow 2\ NO_2(g)$$

has a positive entropy change because it produces twice as many molecules, which can be arranged in many more ways. Although the change in entropy of a system equals q/T if no chemical or physical change occurs, the entropy change for a reaction does not bear such a simple relationship to the heat of reaction. We compute the standard entropy of a reaction by subtracting the standard entropies of the reactants from the standard entropies of the products. The **standard entropy** ($S°$) of a substance (Table 8.2) is an estimate of the entropy increase that occurs when a mole of the substance is heated from a perfectly crystalline state at 0 K (a completely orderly state) and undergoes the various changes of state (such as melting and evaporation) needed to reach its standard state at 25°C and 1 atm pressure. Pressure has very little effect on the entropy of a solid or liquid but is very important in determining the entropy of a gas. The entropy of a gas increases as it expands isothermally, and an increase in pressure lowers its entropy.

Example 8.9

What is the standard entropy for the decomposition of N_2O_4 into NO_2?

$$N_2O_4(g) \longrightarrow 2\ NO_2(g)$$

Solution

According to Table 8.2, the standard entropy of NO_2 is 240.5 J/K-mol and N_2O_4 is 304.3 J/K-mol. Since 2 mol NO_2 are formed for each mole of N_2O_4 that decomposes,

$$S° = 2 S°_{NO_2} - S°_{N_2O_4}$$

$$= (2 \text{ mol}) \left(240.5 \frac{J}{K\text{-mol}}\right) - (1 \text{ mol})\left(304.3 \frac{J}{K\text{-mol}}\right)$$

$$= 176.7 \text{ J/K per mol of } N_2O_4 \text{ decomposed}$$

Since the reaction has a positive entropy, the decomposition of N_2O_4 would always occur spontaneously if it were not for the fact that it is an endothermic reaction. Breaking the chemical bond that holds N_2O_4 together requires 58.04 kJ per mole of N_2O_4 and raises the molecular potential energy of the system. The tendency of the system to maintain a low potential energy favors the stability of N_2O_4, while the trend toward increased molecular disorder favors its decomposition. The result is that N_2O_4 is stable at low temperature but forms NO_2 at higher temperatures.

Gibbs Free Energy

The thermodynamic quantity that determines whether or not conditions are favorable for a particular chemical or physical process is the **Gibbs free energy** (G), which is the enthalpy of a system less the product of its temperature and entropy.

$$G = H - TS \tag{8.13}$$

If the change in free energy for an isothermal process

$$\Delta G = \Delta H - T\Delta S \tag{8.14}$$

is negative, the reaction occurs spontaneously. We must perform work to bring about a reaction whose free energy change is positive.

The decomposition of N_2O_4 into two molecules of NO_2 (see Example 8.9) illustrates the relationship between the free energy of a reaction and its tendency to occur. Since ΔH for this reaction is positive, its occurrence at constant pressure requires that heat flow from the surroundings into the reacting system.

Heat change in system $= q_{reaction} = \Delta H_{reaction}$

Heat change in surroundings $= q_{surroundings} = -\Delta H_{reaction}$

Since the surroundings give up heat, their entropy decreases.

$$\Delta S_{surroundings} = \frac{-\Delta H_{reaction}}{T}$$

But the total entropy change of the universe must be positive.

$$\Delta S_{universe} = \Delta S_{reaction} + \Delta S_{surroundings} > 0$$

$$= \Delta S_{reaction} - \frac{\Delta H_{reaction}}{T} > 0$$

Therefore, the entropy change due to the spontaneous reaction must be greater than $\Delta H_{reaction}/T$.

$$\Delta S_{reaction} > \frac{\Delta H_{reaction}}{T}$$

or

$$T\Delta S_{reaction} > \Delta H_{reaction}$$

8.3 CHEMICAL STABILITY

Consequently, the free energy change must be negative for the reaction to occur spontaneously.

$$\Delta G_{\text{reaction}} = \Delta H_{\text{reaction}} - T\Delta S_{\text{reaction}} < 0$$

As in the case of enthalpy and entropy, we define the standard free energy of a substance (G°) as the free energy at the standard conditions of 25°C and 1 atm pressure. The standard free energy change for a reaction ($\Delta G°$) is the free energy change measured under the standard conditions. Table 8.2 contains the standard free energies of formation for some important substances.

Example 8.10

The standard enthalpy of decomposition of N_2O_4 to NO_2 is 58.04 kJ, and the standard entropy for this reaction is 176.7 J/K. What is the standard free energy change for this reaction at 25°C?

Solution

$$\Delta G° = \Delta H° - T\Delta S°$$

$$= 58.04 \text{ kJ} - (298 \text{ K})\left(176.7 \frac{\text{J}}{\text{K}}\right)$$

$$= 58.04 \text{ kJ} - 5.265 \times 10^4 \text{ J}$$

$$= 58.04 \text{ kJ} - 52.65 \text{ kJ} = +5.39 \text{ kJ}$$

The positive free energy change for this reaction indicates that it is unfavorable at 25°C, and N_2O_4 at 1 atm and 25°C is more stable than NO_2 at the same conditions.

Another way of estimating the standard free energy change of a reaction is from the standard free energies of formation of the substances involved in the reaction. The free energy change of the reaction is the sum of the free energies of the products minus the sum of the free energies of the reactants.

$$\Delta G°_{\text{reaction}} = \Sigma \Delta G°_{f,\text{ products}} - \Sigma \Delta G°_{f,\text{ reactants}} \tag{8.15}$$

Example 8.11

Use the standard free energies of formation in Table 8.2 to compute the standard free energy of the reaction

$$2\text{ Fe(s)} + 3\text{ H}_2\text{O(g)} \longrightarrow \text{Fe}_2\text{O}_3\text{(s)} + 3\text{ H}_2\text{(g)}$$

Solution

The solution to this problem is analogous to Example 8.8 except that we use the free energies of formation rather than the enthalpies.

$$\Delta G° = (\Delta G°_{f,\text{ Fe}_2\text{O}_3}) + 3(\Delta G°_{f,\text{ H}_2}) - 2(\Delta G°_{f,\text{ Fe}}) - 3(\Delta G°_{f,\text{ H}_2\text{O}})$$

Since the free energies of formation of elements are zero, this equation reduces to

$$\Delta G° = (\Delta G°_{f,\text{ Fe}_2\text{O}_3}) - 3(\Delta G°_{f,\text{ H}_2\text{O}})$$

$$= (1 \text{ mol})\left(-741.0 \frac{\text{kJ}}{\text{mol}}\right) - (3 \text{ mol})\left(-228.60 \frac{\text{kJ}}{\text{mol}}\right)$$

$$= -55.2 \text{ kJ}$$

Figure 8.8
Temperature effect on equilibrium.

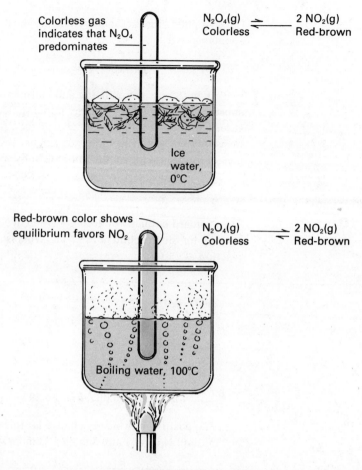

It is clear from equation (8.14) that temperature plays an important part in the free energy of a reaction and helps to determine whether or not a reaction occurs. For example, the standard free energy for the decomposition of N_2O_4 is positive at 25°C because the unfavorable enthalpy of the reaction dominates the favorable (positive) entropy. However, at higher temperatures $T\Delta S$ becomes larger than ΔH, and the free energy of the reaction is negative. The temperature increase reverses the relative stability of N_2O_4 and NO_2 (Figure 8.8).

Another factor governing free energy change is the relative concentrations of the reactants and products. As the reaction proceeds, the concentrations of reactants decrease and those of the products increase. These changes in concentration make the free energies of the reactants approach the free energies of the products, and the reaction free energy eventually becomes zero. At this point the reaction reaches a state known as chemical equilibrium and all net chemical change ceases. A system at equilibrium does not change its composition even though in many cases considerable amounts of reactants may still be present. We will not discuss the details of concentration effects and equilibrium until we have the background to discuss concentrations (Chapter 10) and the measurement of the extents of reactions (Chapter 13) more precisely.

Because of concentration effects and chemical equilibrium, the stability of a compound is not an "either-or" question but rather a question of what portion of the molecules of a compound have decomposed at chemical equilibrium. For example, even though the standard free energy for the decomposition of N_2O_4 to NO_2 at 25°C is positive, indicating an unfavorable reaction, some N_2O_4 molecules in a sample will decompose to NO_2 to form an equilibrium mixture.

Free Energy and Work

When entropy increases, energy becomes less useful. Any engine based on heat flow (Figure 8.9) takes heat from a hot region (such as a boiler), converts some of it into

Figure 8.9
Heat–work conversion.

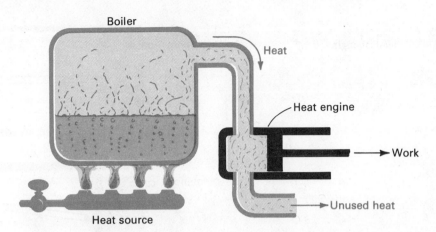

work, and deposits the rest in a cooler region. Although the energy passed on to the cooler region is conserved, it can no longer be used to run this engine or any other engine operating in the same temperature range. If a particular amount of heat should flow directly from the hot to the cold region without passing through the engine, it would do no work and the opportunity to operate the engine with that heat would be lost forever. This loss of work equals the entropy change due to the heat flow times the cooler temperature.

$$\text{Loss of useful work} = T\Delta S \tag{8.16}$$

Equation (8.16) applies to chemical reactions as well as to heat flow. Since $T\Delta S$ represents the minimum amount of energy rendered useless to the system, the free

---

In the last few years we have all been urged to conserve energy, and yet an important scientific law says that energy is always conserved—it never decreases but just changes from one form to another. Is this a contradiction? Not really, the energy shortage is a matter of the *quality* rather than the *quantity* of energy. Energy is conserved, but as it changes from one form to another it often becomes less useful. When we burn petroleum to operate an engine some energy is dissipated as heat, and although this energy is still around, the entropy increase caused by the combustion of the fuel and the friction of the motor renders the dissipated energy useless at room temperature.

One of the scenarios cosmologists have imagined for the end of the universe is its "heat death." The heat death would occur if the universe's entropy increases so much that its free energy reaches a minimum value, and there's no usable energy left in the world. Obviously this event isn't about to happen on a cosmic scale, and not many people are losing sleep over it. But now we are faced with the heat death of our technological society on earth, the conversion of the earth's usable energy resources to dissipated heat. That's why the idea of using solar energy (energy that comes from beyond the earth) is so attractive.

Another way of conserving usable energy is to trap the heat before it is completely dissipated. One possible method of doing this is to take advantage of the temperature difference between the ocean's surface and its lower depths. Engines placed between the ocean's surface and its floor could use the surface waters as a heat source and the depths as a heat sink.

Figure 8.10
An ideal chemical engine.

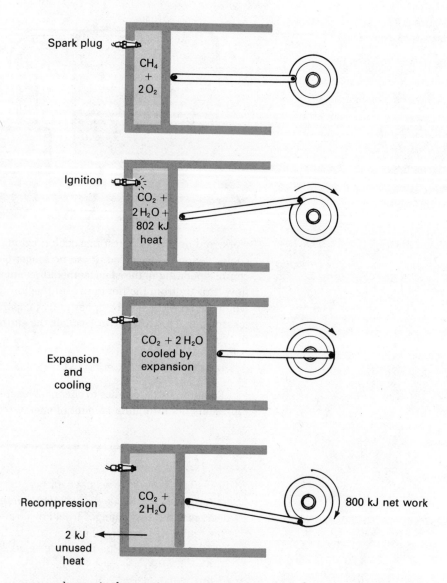

energy change is the maximum amount of work the system can gain from its surroundings. If ΔG is negative, the system can do work on its surroundings, and we can extract a maximum of $-\Delta G$ work from the reaction.

Figure 8.10 shows a model system for extracting the maximum amount of work from the reaction

$$CH_4(g) + 2\ O_2(g) \longrightarrow CO_2(g) + 2\ H_2O(g)$$

$\Delta G° = -800$ kJ $\Delta H° = -802$ kJ

The engine consists of a cylinder containing 1 mol CH_4 and 2 mol O_2, each at a partial pressure of 1 atm and 25°C. Ignition with a spark causes the reaction to occur rapidly, generating 802 kJ heat which warms the mixture of product gases (CO_2 and H_2O). The hot gases expand against the piston and drive the flywheel as they cool. Because of its momentum, the flywheel continues to turn, compressing the gas mixture to its original volume. In this stage of the engine's operation, the flywheel does some work on the gas system and generates some heat that flows out of the engine. By the end of the cycle, the flywheel will have gained 800 kJ in net work, and the engine will have given up 2 kJ heat. A car's engine works on the same general principles but is considerably more complicated. An important difference between the car engine (or any other practical power source) and the theoretical model engine is that the car engine is less efficient. The heat expelled from the car engine is considerably greater than $T\Delta S$ for the reaction, and the work performed is only a fraction of $-\Delta G$.

Summary

1. An exothermic reaction releases heat; an endothermic reaction absorbs heat.
2. A state function, such as the energy of a system, depends only on its chemical composition and physical conditions, not on the history of its formation.
3. An open system exchanges matter with its surroundings, a closed system does not.
4. A process that occurs without heat exchange is adiabatic. An isothermal process occurs at constant temperature.
5. The energy change experienced by a system equals the heat (q) it receives from its surroundings minus the work (w) it performs on its surroundings.

    $$\Delta E = q - w$$

    For an expanding system

    $$\Delta E = q - P\Delta V$$

6. We can measure the heat generated by a reaction by measuring the change in temperature of a calorimeter in which the reaction occurs. The heat produced by the reaction ($-q$) equals the temperature increase (ΔT) of the calorimeter times its heat capacity (C).

    $$-q = C\Delta T$$

7. Enthalpy (H) is the energy of the system plus the product of its pressure and volume.

    $$H = E + PV$$

    The heat change due to a reaction at constant pressure is the enthalpy change of the system.

    $$\Delta H = \Delta E + P\Delta V$$

8. Hess's law: The sum of the enthalpies of two or more reactions equals the enthalpy of the net reaction that would result from these reactions occurring consecutively.
9. Entropy is a measure of the disorder of a system. The second law of thermodynamics: A system tends to change to a state of higher entropy unless this change is opposed by some physical force.
10. When heat (q) flows into a system that is not undergoing a chemical or physical change, the entropy (S) increases by q/T_{system}.

    $$\Delta S = \frac{q}{T_{system}}$$

    The total entropy of the universe never decreases.

11. The standard entropy change due to a chemical reaction equals the standard entropies (at 1 atm and 25°C) of the products minus the standard entropies of the reactants.
12. The Gibbs free energy (G) of a system is the enthalpy of the system minus the product of its temperature and entropy.

    $$G = H - TS$$

    The change in Gibbs free energy for an isothermal process is

    $$\Delta G = \Delta H - T\Delta S$$

    Reactions with negative free energy changes are favorable processes. Those with positive ΔG require work. A system in which the products have the same free energy as the reactants is in equilibrium.

13. An increase in entropy renders energy less useful. The free energy of a reaction is the maximum amount of useful work that can be obtained from it.

Vocabulary List

exothermic reaction
endothermic reaction
thermodynamics
first law of thermodynamics
state function
adiabatic

isothermal
calorimetry
enthalpy
enthalpy of formation
standard enthalpy of formation
Hess's law

entropy
second law of thermodynamics
standard entropy
Gibbs free energy

Exercises

1. The following reaction is endothermic.

    $$C_2H_6(g) \longrightarrow C_2H_4(g) + H_2(g)$$

    a. Will heat flow into or out of a system when this reaction occurs within it?
    b. Will q be positive or negative for the system?

2. The reaction

    $$3 H_2(g) + N_2(g) \longrightarrow 2 NH_3(g)$$

    occurs at constant pressure so that the volume of the system contracts.
    a. Does the system do work on its surroundings or vice versa?
    b. Is w positive or negative for the system?

3. Which of the following are state quantities?
    a. Enthalpy   b. Heat   c. Work   d. Entropy

4. The following reaction is exothermic.

    $$2 NO(g) + 2 H_2(g) \longrightarrow N_2(g) + 2 H_2O(g)$$

    Which has the stronger chemical bonds, one N_2 and two water molecules or two NO molecules and two H_2 molecules?

5. What is the net energy change in a system in which an ideal gas is first warmed from 0°C to 100°C, allowed to expand to twice its volume, then cooled to 0°C and compressed back to its original volume while maintaining a constant temperature? (Calculations are not required. This question can be answered simply on the basis of general principles.)

6. Can an adiabatic process ever cause a negative entropy change within a system? Explain.

7. Which of the following systems are open and which are closed?
    a. A jet engine   b. A pressure cooker   c. The earth   d. A tree
    e. A piece of rusting iron   f. A rocket during propulsion   g. A battery
    h. A gas range

8. In which direction will the temperature of an adiabatic system change if an endothermic reaction takes place within it?

9. Compute the total change in work and energy for each of the following processes.
    a. 5.00 L of air is expanded to 20.0 L by heating it from 300 K to 1200 K at a constant pressure of 3.00 atm. The heat required for this process is 17.8 kJ.
    b. 5.00 L of air, initially at 3.00 atm is heated from 300 K to 1200 K while its volume is held constant. The heat required is 12.8 kJ.
    c. The hot gas from part b is expanded isothermally until its pressure drops to 3.00 atm and its volume reaches 20.0 L. The heat required in this step is 9.7 kJ. [Hint: Energy is a state function.]

10. How much work (in J) does a sample of an ideal gas do if it initially occupies 3.00 L at 25°C and 2.00 atm and then expands to 9.00 L at a constant pressure of 2.00 atm?

11. How many joules work are involved in the contraction of a system by 25.0 L due to a constant pressure of 45.0 atm? Which does the work, the system or its surroundings?

EXERCISES

12. The heat capacity of 1 mol He at constant volume is 12.5 J/°C. How much heat does 2.00 g He lose when it cools from 150°C to −40°C at constant volume?

13. The heat capacity of S_2Cl_2, a liquid, is 124 J/°C-mol. How much will the introduction of 2.00 kJ heat raise the temperature of 100 g S_2Cl_2, originally at 12°C?

14. The energy change when 1 mol of argon gas is heated from 273 K to 546 K as it expands at a constant pressure of 1 atm from 22.4 L to 44.8 L is 3.42 kJ. What will the energy change be if 1 mol of argon at 273 K and 1 atm is first heated to 546 K at constant volume and then allowed to expand to 44.8 L by reducing the pressure? [Hint: Energy is a state function.]

15. What is the energy change in a sample of gas as it expands adiabatically from 750 mL to 800 mL at an average pressure of 700 torr?

16. When a chemical reaction occurs within a bomb calorimeter whose heat capacity is 3.00 kJ/°C, the temperature decreases 0.230°C. What is the change in heat for the chemical system?

17. When 25.0 g of an aqueous solution containing 0.0100 mol KBr reacts with 25.0 g of an aqueous solution containing 0.1000 mol $AgNO_3$ to produce the reaction,

$$Ag^+(aq) + Br^-(aq) \longrightarrow AgBr(s)$$

the temperature of the styrofoam-cup calorimeter rises from 21.8°C to 22.6°C. The heat capacity of the mixed solution is 4.1 J/g-°C, and the heat capacity of the styrofoam cups is 32 J/°C. How much heat is generated by this reaction? (Neglect the heat capacity of the solid AgBr.)

18. The enthalpy of the reaction

$$2\ Al(s) + 3\ O_2(g) \longrightarrow 2\ Al_2O_3(s)$$

is −1204 kJ at 25°C. What is the energy of this reaction?

19. The enthalpy of the reaction

$$2\ H_2(g) + C_2H_2(g) \longrightarrow C_2H_6(g)$$

is −311.4 kJ. What is the energy of this reaction at 25°C?

20. Compute the enthalpy of the reaction

$$Zn(s) + CuO(s) \longrightarrow ZnO(s) + Cu(s)$$

from heats of formation in Table 8.2.

21. Use heats of formation in Table 8.2 to compute the enthalpy of the following reaction.

$$Fe_2O_3(s) + 2\ Al(s) \longrightarrow Al_2O_3(s) + 2\ Fe(s)$$

22. The heat of the reaction

$$MnO_2(s) \longrightarrow MnO(s) + \frac{1}{2} O_2(g)$$

is +136.0 kJ. The heat of the reaction

$$SO_2(g) + \frac{1}{2} O_2(g) \longrightarrow SO_3(g)$$

is −125.3 kJ. What is the heat of the following reaction?

$$MnO_2(s) + SO_2(g) \longrightarrow MnO(s) + SO_3(g)$$

23. Compute the heat of the reaction

$$Zn(s) + 2\ AgCl(s) \longrightarrow 2\ Ag(s) + ZnCl_2(s)$$

24. How much does the entropy of an ideal gas sample change when 75 J are drawn from it at 28°C while its volume remains constant? The sample is so large that this small heat loss does not change its temperature significantly.

25. a. What is the entropy change in 500 kg of water at 55°C when 85 J heat flow out of it into the surroundings at 25°C?
    b. What is the entropy change in the surroundings?
    c. What is the total entropy change in the universe due to this process?

26. A large block of steel at −15°C receives 37.0 J heat from its surroundings. The temperature of the surroundings is +15°C. What is the total entropy change for this process?

27. The heat capacity of xenon is 12.5 J/°C-mol when heated at constant volume. How much will the entropy of 0.500 mol Xe change if its temperature is raised from 25.00°C to 26.00°C while its volume remains constant? [Hint: Use the average temperature of the sample in equation (8.12).]

28. When a mixture of carbon monoxide and oxygen reacts isothermally at 0°C and a constant pressure of 2.00 atm to form carbon dioxide, 14.6 kJ heat flow out of the system, and the volume contracts from 3.00 L to 2.00 L.
    a. What energy change does the system experience?
    b. What is the enthalpy change in the system?

29. The reaction of 1 mol SO_3 gas with solid NaOH

$$SO_3(g) + NaOH(s) \longrightarrow NaHSO_4(s)$$

at 0°C and a constant pressure of 1.00 atm releases 305 kJ heat as the volume of the system decreases by 22.4 L. What is the energy change in this system?

30. Use the data in Table 8.2 to calculate the standard entropies of each of the following reactions.
    a. $2\ SO_2(g) + O_2(g) \longrightarrow 2\ SO_3(g)$
    b. $CH_4(g) + 4\ Cl_2(g) \longrightarrow CCl_4(l) + 4\ HCl(g)$
    c. $2\ FeS(s) + 3\ O_2(g) \longrightarrow 2\ FeO(s) + 2\ SO_2(g)$
    d. $2\ Zn(s) + O_2(g) \longrightarrow 2\ ZnO(s)$
    e. $CO(g) + NO_2(g) \longrightarrow CO_2(g) + NO(g)$

31. What is the change in the standard entropy at 25°C for each of the following reactions? (Refer to Table 8.2.)
    a. $Cl_2(g) \longrightarrow 2\ Cl(g)$
    b. $Br_2(l) \longrightarrow Br_2(g)$
    c. $2\ H_2(g) + O_2(g) \longrightarrow 2\ H_2O(l)$
    d. $Na(s) + \frac{1}{2} Cl_2(g) \longrightarrow NaCl(s)$
    e. $CO_2(g) + H_2(g) \longrightarrow CO(g) + H_2O(g)$

32. What is the standard Gibbs free energy at 25°C for each of the reactions in Exercise 30?

33. Compute the change in standard Gibbs free energy at 25°C for each reaction in Exercise 31.

34. Which of the following processes perform significant amounts of work when carried out at constant temperature and pressure? Which make the system receive work from the surroundings?
    a. $2\ Cr(s) + 3\ Cl_2(g) \longrightarrow 2\ CrCl_3(s)$
    b. $2\ Cr(s) + 6\ HCl(aq) \longrightarrow 2\ CrCl_3(aq) + 3\ H_2(g)$
    c. $2\ Al(s) + Fe_2O_3(s) \longrightarrow Al_2O_3(s) + 2\ Fe(s)$
    d. $AgNO_3(aq) + NaCl(aq) \longrightarrow AgCl(s) + NaNO_3(aq)$
    e. $H_2O(l) \longrightarrow H_2O(g)$
    f. $2\ H_2(g) + CO(g) \longrightarrow CH_3OH(g)$
    g. $N_2(g) + O_2(g) \longrightarrow 2\ NO(g)$

35. For which of the following reactions is the enthalpy of reaction the same as the energy of reaction except at very high pressure?
    a. $3\ H_2(g) + N_2(g) \longrightarrow 2\ NH_3(g)$
    b. $2\ C_6H_6(l) + 9\ O_2(g) \longrightarrow 12\ CO_2(g) + 6\ H_2O(l)$
    c. $F_2(g) + NaCl(s) \longrightarrow 2\ NaF(s) + Cl_2(g)$
    d. $Ag^+(aq) + Cl^-(aq) \longrightarrow AgCl(s)$
    e. $2\ H^+(aq) + Mg(s) \longrightarrow Mg^{2+}(aq) + H_2(g)$
    f. $Cl_2(g) + H_2(g) \longrightarrow 2\ HCl(g)$

36. a. Why is the amount of heat needed to raise the temperature of an ideal gas by a certain amount (ΔT) larger if the gas is kept at constant pressure than if its volume is constant?
    b. Show that the difference between the heat needed at constant pressure and the heat needed at constant volume equals $nR\ \Delta T$.

37. The heat of neutralization between a strong acid and a strong base

    $$H^+(aq) + OH^-(aq) \longrightarrow H_2O$$

    is -56.3 kJ per mole of H^+. When 0.0250 mol H^+ neutralizes 0.0250 mol OH^- in a calorimeter, the temperature rises from 25.000°C to 25.528°C. What is the heat capacity of the calorimeter?

38. Use data in Table 8.2 to compute the heat of vaporization of water.

    $$H_2O(l) \longrightarrow H_2O(g)$$

39. The heat capacity of neon is 20.8 J/°C-mol at constant pressure.
    a. How much heat is needed to warm 0.500 mol Ne from -50°C to $+27$°C while maintaining a constant pressure of 2.50 atm?
    b. How much will the volume of this sample change during heating?
    c. How much will the energy of the sample change during heating?
    d. How much will the enthalpy of the sample change during this process?

40. The reaction of 0.1053 g Fe in the presence of excess acid

    $$3\ H^+(aq) + Fe(s) \longrightarrow Fe^{3+}(aq) + \frac{3}{2} H_2(g)$$

    generates 97.1 J heat when run in a rigid system at 25°C.
    a. What is the energy of the reaction expressed in kJ per mole of Fe?
    b. What is the enthalpy of this reaction?

EXERCISES

41. Imagine Maxwell's demon in control of the apparatus shown in Figure 8.7 and that it is filled uniformly with O_2 at 25°C and 1 atm. What would the demon do to compress all the oxygen into chamber B without performing any work on the gas?

42. What sort of entropy change in the system occurs with each of the following processes? Your answer should be either (i) definitely positive, (ii) definitely negative, or (iii) approximately zero.
    a. Mixing of helium and neon   b. The solidification of molten iron
    c. The evaporation of methyl alcohol
    d. Each step in the Born-Haber cycle for the formation of solid NaCl from its elements. (See Section 6.1 and Figure 6.5.)

43. Calculate the standard enthalpy values for each of the reactions in Exercise 30.

---

44. The heat of the reaction

    $$CO(g) + Cl_2(g) \longrightarrow COCl_2(g)$$

    is −112.5 kJ at 25°C, and the standard entropy is −132 J/K-mol.
    a. Which reaction, the formation of $COCl_2$ or its decomposition into CO and Cl_2, will be favorable in a system containing CO, Cl_2, and $COCl_2$ each at a partial pressure of 1 atm?
    b. Which thermodynamic factor, the tendency toward lower energy or the tendency toward greater molecular disorder, is dominant in determining the direction of chemical change in this system?

45. It is possible to add heat to a system in which a chemical reaction is in equilibrium so slowly that the equilibrium is maintained at all times during the course of the heating. Show that in such a case, equation (8.12) correctly describes the increase in the entropy of the system provided the reaction is run at constant pressure.

46. The heat of the reaction

    $$2 HCN(g) \longrightarrow C_2N_2(g) + H_2(g)$$

    is +46.9 kJ/mol and its standard entropy change is −30.9 J/K-mol.
    a. Compute the standard free energy change for this reaction at 25°C.
    b. Assuming that $\Delta H°$ and $\Delta S°$ do not change with temperature, compute the standard free energy change at 500°C.
    c. Is there any temperature at which this reaction is favorable?

47. The heat of decomposition of $MgCO_3$ to MgO and CO_2

    $$MgCO_3(s) \longrightarrow MgO(s) + CO_2(g)$$

    is +117 kJ, and the standard entropy change is +174 J/K. Assume that the standard entropy and the heat of the reaction do not change significantly with temperature.
    a. Find the standard free energy change for this reaction at 0°C.
    b. Find the standard free energy change at 800°C.
    c. At what temperature will a system containing solid $MgCO_3$, solid MgO, and CO_2 at 1 atm be at equilibrium?

48. Because crystalline monatomic iodine, I(s), is not stable, we cannot estimate the standard entropy of I by computing the entropy increase as we heat I(s) from absolute zero. Instead, we use the following data to compute the standard entropy of I(g) at 25°C. The standard free energy change for the reaction

    $$I_2(g) \longrightarrow 2 I(g)$$

    at 25°C is +120.9 kJ. (This free energy can be calculated from the concentrations of I_2 and I in systems at equilibrium.) The heat of the reaction is +151.0 kJ. The standard entropy of I_2 gas at 25°C is given in Table 8.2.
    a. What is the standard entropy change for the given reaction?
    b. What is the standard entropy of monatomic iodine gas at 25°C?

49. A sample of helium occupies 8.00 L at STP. Adiabatic compression of this sample to 7.00 L raises its temperature to 25.6°C. The heat capacity of a mole of helium held at constant volume is 12.5 J/°C.
   a. Isothermal compression of the sample from 8.00 L to 7.00 L at 0.0°C requires 1.067 L-atm work, and the system gives up 108 J in heat. What is the energy change in the system due to isothermal compression?
   b. How much energy does the system gain if it is heated from 0.0°C to 25.6°C while its volume is constant?
   c. What is the energy change in the helium sample due to the adiabatic compression from 8.00 L to 7.00 L?
   d. How much work does the adiabatic compression require?

50. a. In 1843, Joule tried to measure the temperature change in an adiabatic system in which a gas expands freely into a vacuum, but he could not detect any such temperature changes. Explain this observation in terms of the kinetic-molecular theory for ideal gases.
   b. During the 1850s, Joule and Thomson performed more precise experiments and found that small temperature changes do occur during the unrestrained expansion of gases. At room temperature, most gases cool slightly as they expand, but the amount of cooling differs for each substance. Explain these observations in terms of the molecular properties of real gases.
   c. At high temperatures, many gases warm up on free expansion. The magnitude of this phenomenon and the temperature at which it appears differ for each substance. The warming is related to the distortions in the structures of the gas molecules when they collide with each other. Try to develop a detailed explanation of this effect.

chapter 9

Solids, Liquids, and Changes of State

The orderly arrangement of atoms in solids such as quartz (SiO_2) is a theme of this chapter.

All substances can exist in any of the three states of matter—solid, liquid, or gas—depending upon external and internal conditions. We have already examined the gaseous state in some detail; now we shall study the solid and liquid states in which *intermolecular forces* play a significant role.

9.1 The Regularity of Crystal Structures

When a pure substance forms a solid phase, either by freezing of the liquid or precipitation from solution, it often develops in the form of **crystals**—*particles with well-defined planar surfaces*. While no scientific training is needed to appreciate the elegant regularity of a crystal's surfaces, this regularity has special meaning to chemists. The external shape of a crystal is the visible manifestation of the orderly, and often symmetric, way molecules* are packed into its interior. It is this orderly internal structure that distinguishes crystals from other solid states such as alloys and amorphous solids or glasses.

A substance crystallizes in characteristic shapes. These shapes are not always obvious from casual observation since crystals of a particular substance may vary considerably in appearance. For example, the characteristic shape of alum, $KAl(SO_4)_2 \cdot 12H_2O$, is an octahedron as shown in Figure 9.1(b). We can, with care, grow octahedral alum crystals, but if no special efforts are made the crystals that form usually have quite different shapes. For instance, the crystals may fall from the

---

*The term "molecule" as used in this context can also refer to atoms or ions.

Figure 9.1

The symmetry of a crystal is one of its most revealing properties. Symmetry is present all around us, in plant leaves, pine cones, flower petals, etc. (a) Maurits Escher presented symmetry in many unusual ways, this plate is but one. How many symmetry elements can you identify? *Source:* M. C. Escher's "Study of regular division of the plane with angels and devils," Escher Foundation, Haags Gemeentemuseum, The Hague. (b) An octahedral alum crystal. (c) A distorted alum crystal with the relationship to the octahedron shown in color.

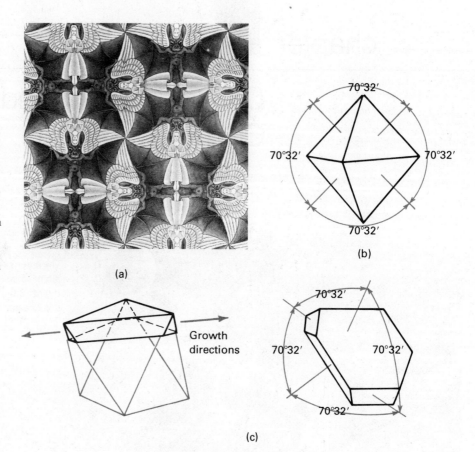

Crystal Symmetry

solution to rest on the bottom of the container and grow faster sideways than vertically as represented in Figure 9.1(c). Nevertheless, all of these alum crystals have one feature in common—the angle between adjacent faces of an octahedron is 70° 32′, and corresponding faces of the distorted crystals also have this angle.

Ions, molecules, and atoms coalesce into geometrically regular crystal arrays, the nature of the particles dictating the geometric pattern of the lattice. A **lattice** is a periodic, repeating array of points with identical environments about each similar point in the lattice. We use the properties of a lattice to describe a regular array of objects, whether it is the two-dimensional array of patterns in wallpaper or the three-dimensional array of molecules in a crystal. The lattice geometry controls the angles between the faces of a crystal even when variations in the conditions of crystallization alter the crystal's appearance.

Well-formed sodium chloride crystals are cubes, highly symmetric structures with many symmetry elements. A **symmetry element** is an imagined operation that relates one part of an object or group of objects to another, or others. Symmetry operations include rotating a crystal to change the perspective with which we view it, reflecting one part of the crystal onto another, or even turning the crystal inside out. Two common types of symmetry elements are rotation axes and mirror planes. A mirror plane relates one part of an object to the other by a reflection. For example, if we cut a cube along any one of its mirror planes and hold either half against a mirror, we see the original cube (Figure 9.2). To understand the nature of a rotation axis, consider the following operation. Look directly at one of the faces of a cube, then close your eyes and have someone rotate the cube face by 90°. When you open your eyes, the cube will look exactly as it did before, and you would be unable to tell whether the cube had been rotated or not. This particular symmetry element is called a fourfold rotation axis because *four* symmetry operations return the cube to its original posi-

9.1 THE REGULARITY OF CRYSTAL STRUCTURES   219

Figure 9.2
A most unusual mirror plane and some symmetry elements of a cube. *Source:* The mirror plane is from M. C. Escher's "Toverspiegel—Magic Mirror," Escher Foundation, Haags Gemeentemuseum, The Hague.

Figure 9.3
Mirror planes and rotation axes.

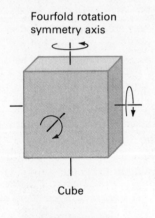

Cube

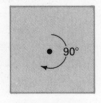

Cube face

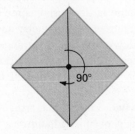

Octahedron (looking down on a vertex)

Mirror planes

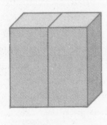
Mirror plane

Half cube

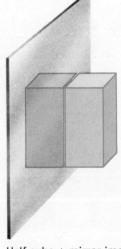
Half cube + mirror image

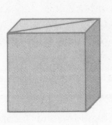
Mirror plane

Half cube

Half cube + mirror image

tion (Figure 9.3). There are other types of rotation axes requiring different rotation angles, such as 120° or 180°, to restore the same apparent perspective.

It is interesting to note that a change in crystallization conditions can produce octahedral rather than cubic sodium chloride crystals. Even though these two shapes—an octahedron and a cube—are apparently quite different, examination shows that they possess exactly the same type and number of symmetry elements. For example, there is a fourfold rotation axis through opposite vertices (corners) of an octahedron. Looking down the axis at one of the vertices and rotating the octahedron 90° re-creates the original figure (see Figure 9.3).

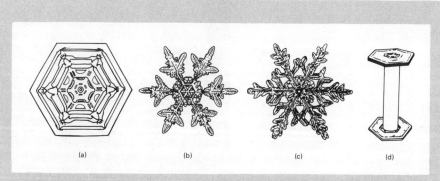

Snow crystals. (a) Hexagonal plate. (b) Stellar. (c) Dendritic stellar. (d) Capped hexagonal prism. *Source:* W. A. Bently and W. J. Humphrey, *Snow Crystals* (New York: Dover, 1962).

The earliest known report (second century B.C.) of the six-pointed shape of snowflakes came from China. Johannes Kepler was the first European to recognize the hexagonal symmetry of snow crystals when, according to his humorous account, he happened to look at some snowflakes that had fallen on his coat. In the essay, *A New Year's Gift, or On the Six-Cornered Snowflake* (published in 1611), he developed and rejected several explanations, concluding, "In sulphates of metals the rhomboid cubic shape is common, and saltpetre has its own shape. So let the chemist tell us whether there is any salt in a snowflake and what kind of salt, and what shape it assumes otherwise."

How well have we met Kepler's challenge (and followed his shrewd guess that this symmetry had something to do with crystal structure)? Snowflakes are aggregates of ice crystals; x-ray diffraction shows the hexagonal symmetry of the ice-crystal lattice. Ice crystal aggregates grow along axes that run through the corners of a hexagon and branch from these main axes to produce myriad snowflake shapes: hexagonal plates, six-pointed stars, and hexagonal prisms as shown in the figure.

Crystal aggregates of other substances grow along symmetry axes and branch from these axes, but the way in which the different arms branch are independent of each other. Only snowflakes preserve their symmetry when branching; whatever happens in one of the arms happens precisely the same way in the other five. How does information concerning crystal development in one arm of the snowflake reach the other arms? This is still a mystery to us.

Crystal Lattices

Although they lacked experimental techniques for observing the positions of molecules in a crystal, nineteenth-century scientists and mathematicians applied mathematical principles to characterize the three-dimensional lattices used as models for crystal structures. It is important to realize that each lattice has symmetry and geometrical properties of its own. The structure of a crystal, with its regular arrangement of molecules, is patterned after one particular lattice type. While the term lattice refers to a mathematical, geometrical array of points, the regular arrangement of molecules in a crystal is usually called a **lattice array**.

To understand the properties of a lattice, consider the two-dimensional array in Figure 9.4, a regular pattern of sevens. All the lattice points chosen in this figure are at the centers of triangular arrangements of sevens and the environment about each point is identical. Because the lattice is repetitive, a small portion of the lattice called the **unit cell** characterizes the entire lattice. The unit cell is the smallest part of a

Figure 9.4
A two-dimensional lattice array. One unit cell is shown with solid lines.

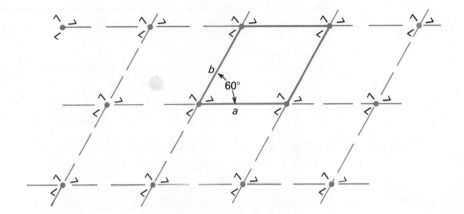

lattice that contains all the properties of the lattice. Once we know the shape, dimensions, and arrangement of the objects in one unit cell, we can construct the rest of the array by joining identical unit cells together, or alternatively we could translate the unit cell. **Translation** means moving the unit cell exactly one edge length in a direction along the edge. (It is not necessary to place lattice points at the centers of groups of sevens; they could be anywhere with convenience directing our choice. The unit cell should, however, indicate the symmetry of the lattice.)

Crystal Systems and Bravais Lattices

Any crystal fits into one of seven classes (**crystal systems**) based on its symmetry elements. For example, any crystal with four threefold symmetry axes, a symmetry operation involving rotation around the axis by 120°, belongs to the cubic system even though it might not look like a cube. Two examples of cubic crystal shapes that do not resemble cubes are the octahedron and the tetrahedron.

In addition, the internal structure of any crystal belongs to one of the 14 three-dimensional lattices, called **Bravais lattices** (Figure 9.5, p. 222), which include the seven crystal systems. The outlines of all the unit cells of the Bravais lattices are parallelepipeds (figures whose opposite faces are parallel) with lattice points at their corners. Some unit cells have additional lattice points at their centers, I or **body-centered lattices**, or on their faces, F or **face-centered lattices** and C or **side-centered lattices**.

9.2 x-Ray Crystallography

Experimental verification of the existence of periodically repeating crystal structures came in 1912 when Max von Laue suggested that if crystals have regular arrays, they should act as diffraction gratings toward x rays (light with wavelengths comparable to the interatomic distances in crystals). Since visible light can be diffracted by passing through a narrow pair of open slits to produce a characteristic pattern (Figure 9.6, p. 223), crystals should also give characteristic diffraction patterns. Von Laue's co-workers focused x-ray beams on crystals and photographed the transmitted light. The crystals did indeed diffract the x-ray beams, producing regular patterns of spots on the photographic plates (Figure 9.7, p. 224). Not only did the success of the experiment prove that crystals have repetitive structures, but the symmetry of the lattice was apparent from the symmetry of the spot pattern on the photographic plate.

Each crystalline substance has a characteristic x-ray diffraction pattern that depends on the positions of atoms in the lattice array. These diffraction patterns can reveal the symmetry and dimensions of the unit cell and even the structures of individual molecules. Because the mathematical analysis involved in x-ray crystallography is very complicated, we will consider only the most general aspects of how to interpret diffraction patterns.

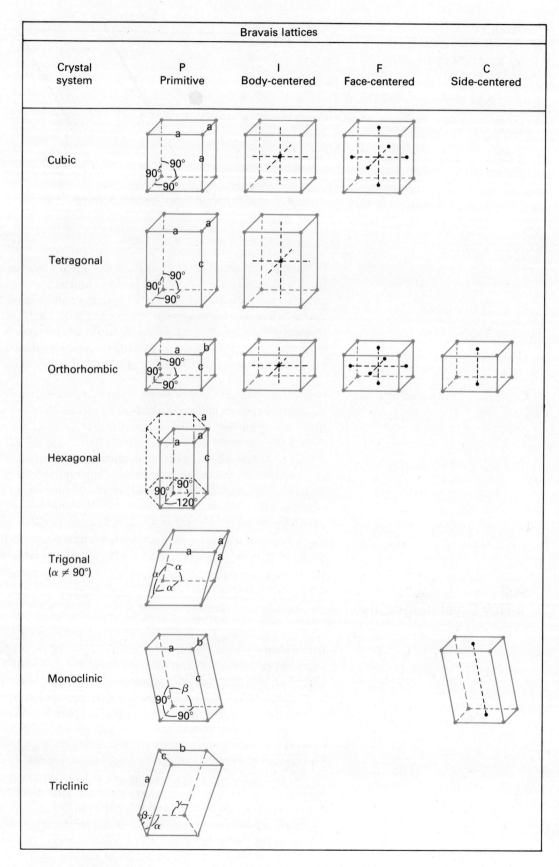

Figure 9.5
The seven crystal systems and the fourteen Bravais lattices.

Figure 9.6
Diffraction of light passing through two slits.

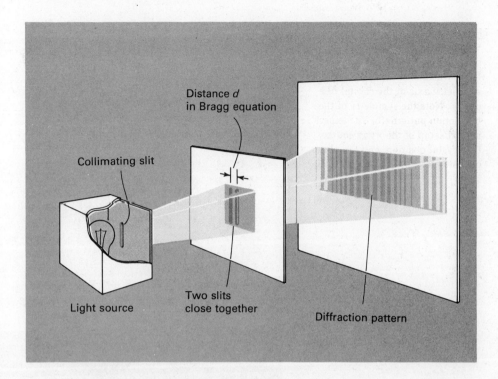

Bragg's Law

A simple way to visualize how spots occur in a diffraction pattern is to imagine a beam of parallel x rays (wavelength λ) penetrating a crystal and interacting with the regularly spaced parallel planes of atoms (Figure 9.8, p. 225). Some of the x rays reflect from the first plane, but others penetrate to interior planes before bouncing back. Compared to an x ray reflected from the first plane, one bouncing off the second plane will travel an extra distance equal to

$$2d \sin \theta$$

where d is the perpendicular distance between the planes and θ is the angle at which the x rays approach the planes. The two reflected x rays will be in phase—that is reinforce each other—only if the two path lengths differ by a whole-number multiple, n, of the wavelength.

$$n\lambda = 2d \sin \theta \tag{9.1}$$

Whether or not destructive interference occurs depends on the values of λ, d, and θ. The value of n can be 1, 2, 3, . . . (called the order of the reflection), which indicates that there are several different, but related values of θ for any one d and λ.

The usual experimental procedure in x-ray crystallography is to rotate the crystal slowly in a monochromatic x-ray beam (only one value of λ). The degree of interference in the x rays reflected from a given set of atomic planes changes as θ changes. Measuring the values of θ for the spots on the photograph reveals the distances between parallel planes of atoms.

Example 9.1

A spot in the diffraction pattern of a crystal obtained using copper x radiation, $\lambda = 0.15418$ nm, occurs at an angle of 30°. What is the interplanar distance associated with this spot? (Assume $n = 1$.)

Solution

We have all the information needed to apply Bragg's law given in equation (9.1). Using a table of trigonometric functions, we find that sin (30°) = 0.500. Therefore,

Figure 9.7
The Laue photograph for a sapphire crystal. (a) The x-ray beam was aimed along a threefold symmetry axis of the crystal (the c-axis). Note the symmetry of the diffraction pattern. (b) The experimental setup of the x-ray source, the crystal position, and the photographic plate.

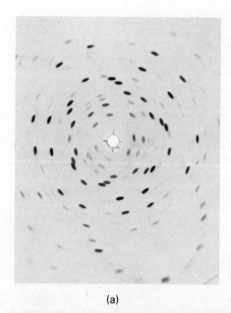

(a)

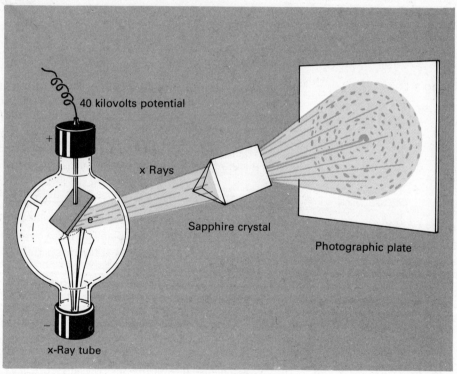

(b)

$$d = \frac{n\lambda}{2 \sin \theta}$$

$$= \frac{0.15418 \text{ nm}}{2(0.500)} = 0.154 \text{ nm}$$

Example **9.2**   What diffraction angle would be observed from the planes whose d spacing was evaluated in Example 9.1 if molybdenum radiation, $\lambda = 0.07107$ nm, were used instead?

Solution   To calculate an angle knowing d, we can rearrange Bragg's law.

Figure 9.8
Bragg's law.

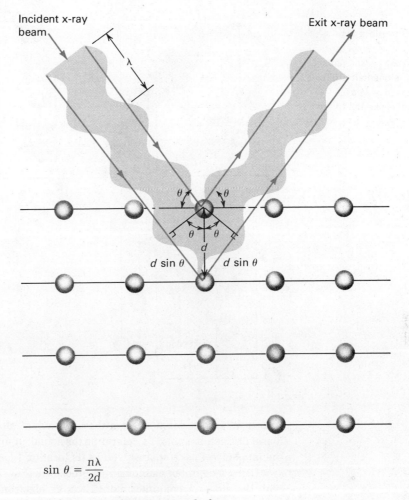

$$\sin\theta = \frac{n\lambda}{2d}$$

If molybdenum radiation is used, then

$$\sin\theta = \frac{0.07107 \text{ nm}}{2(0.154 \text{ nm})} = 0.231$$

According to a trigonometric table, the angle whose sine is 0.231 is 13.3°. Therefore, $\theta = 13.3°$.

Determination of Crystal Structures

Every crystalline structure has characteristic sets of parallel planes formed by the regular arrangements of atoms, like the rows we can see at different angles when driving past an orchard. These planes are responsible for Bragg reflections. They either run through lattice points or are parallel to planes through lattice points. For example, the major planes running through the lattice of a face-centered cube (F-cubic lattice) include the face and mirror planes of the cube, and the planes running diagonally across three adjacent faces (Figure 9.9, p. 226).

The symmetry of the lattice produces a corresponding symmetry in the diffraction patterns. By observing the symmetry of the diffraction spots, an x-ray crystallographer assigns the crystal lattice to the proper symmetry system. The presence of body-centering or face-centering in the lattice alters the symmetry in a way that cancels out Bragg reflections for certain planes, and the absence or presence of spots corresponding to these reflections identifies the Bravais lattice of the crystal. The measured values of θ are used to calculate the dimensions of the unit cell.

The specific mechanism of x-ray scattering involves interaction with electrons. The more electrons in an atom, the more effectively it scatters x rays, and reflections

Figure 9.9

Some atomic planes in a face-centered cubic lattice structure.
(a) Planes of closest-packed atoms.
(b) Planes that are one-half unit cell edge length apart ($d = \frac{1}{2} a$).
(c) Diagonal planes parallel to the z axis.

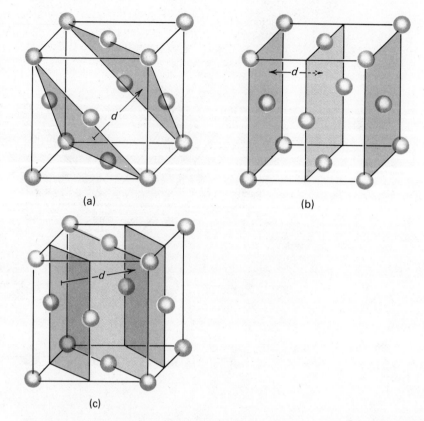

(constructive interference) from these "heavier" atoms produce brighter (more intense) diffraction spots. To determine the actual atomic positions within the unit cell, the intensity of each spot as well as its location (θ angle) must be measured.

Using a complex mathematical procedure (Fourier analysis), a crystallographer, with the aid of a computer, constructs an **electron-density function**, which is an equation describing the variations of electron density within the unit cell. A contour map of the electron-density function (Figure 9.10) shows the regions of intense concentrations of electrons—the atomic positions—and a cluster of atoms close together represents a molecule. In this manner the actual molecular structure can be displayed.

Because rapid advances in computer technology have extended the capabilities of the method, x-ray crystallographers can determine the structures of very large, complicated molecules. Using this technique Dorothy Hodgkins proved the structure of penicillin even before the traditional methods, based on indirect chemical evidence, could solve the problem. In conjunction with chemical evidence, x-ray crystallographic techniques have been used to determine the structures of a number of proteins—molecules with over 1000 atoms (Figure 9.11, p. 228). With the more recent development of automated x-ray diffractometers utilizing minicomputers, x-ray crystallography is rapidly becoming an almost routine analytical technique as a method of structure proof.

9.3 Some Typical Unit Cells

Close-Packed Arrangements

Crystal lattice arrays are often pictured as regular stacks of solid spheres, the spheres representing atoms or ions. It is impossible to completely fill a volume in space with a collection of spheres, and there will always be some free space in a stacked sphere model of a crystal. Two arrangements, *cubic closest-packed* and *hexagonal closest-packed*, have the smallest free space possible (26%) for a regular array of spheres of one size.

A sphere in a close-packed plane (Figure 9.12, p. 229) touches six other spheres, these nearest neighbors forming a regular hexagon. We can build a three-dimensional array of spheres by stacking two close-packed planes together so the spheres of one plane rest in the spaces (or **interstices**) formed by the spheres of the lower plane. Each atom touches three nearest neighbors in the adjacent plane.

Placing a third close-packed plane on the stack surrounds the spheres in the middle plane with twelve nearest neighbors (three below, six in the middle plane, and three above), but there are two ways to place this third plane. If the spheres in the third plane lie above the interstices of the first, the arrangement is **cubic closest-packed** (ccp) as shown in Figure 9.13. Continued stacking in this fashion produces a lattice in which the atoms of a particular plane lie directly above the atoms in the plane *three* below it. In other words, similar planes repeat in the sequence ABCABC .... The ccp arrangement has a face-centered (F) cubic lattice. The close-packed planes run diagonally through the cubic unit cell (see Figure 9.14). The twelve nearest neighbors of an atom do not all belong to the same unit cell. The nearest neighbors of a face-centered atom are the four corner atoms of its own face and the eight adjacent face-centered atoms, four in each of the two unit cells that share the central atom. The number of nearest neighbors of an atom in a crystal lattice is called the **coordination number**. The coordination number for an atom in a ccp lattice is therefore twelve.

The **hexagonal closest-packed** (hcp) arrangement occurs when the atoms of alternate planes lie directly above each other (Figure 9.15, p.230). The planes are stacked in the repeating sequence ABAB .... The hcp lattice has a hexagonal unit cell with atoms of alternate (A) close-packed planes at the lattice points and an atom

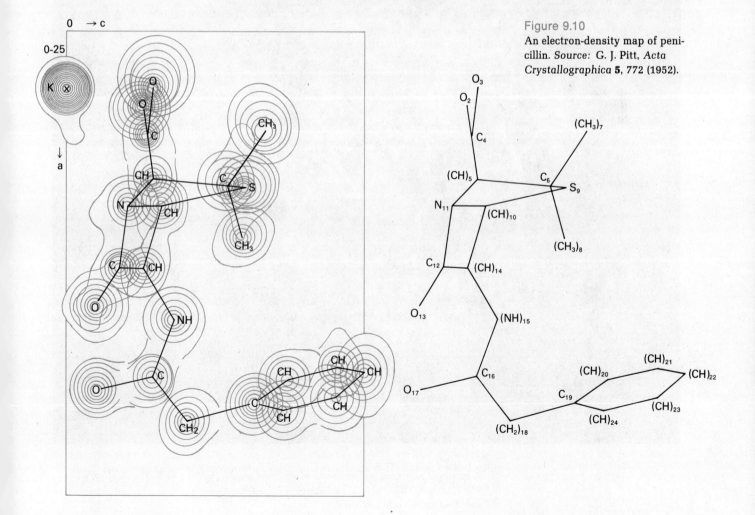

Figure 9.10
An electron-density map of penicillin. *Source:* G. J. Pitt, *Acta Crystallographica* **5**, 772 (1952).

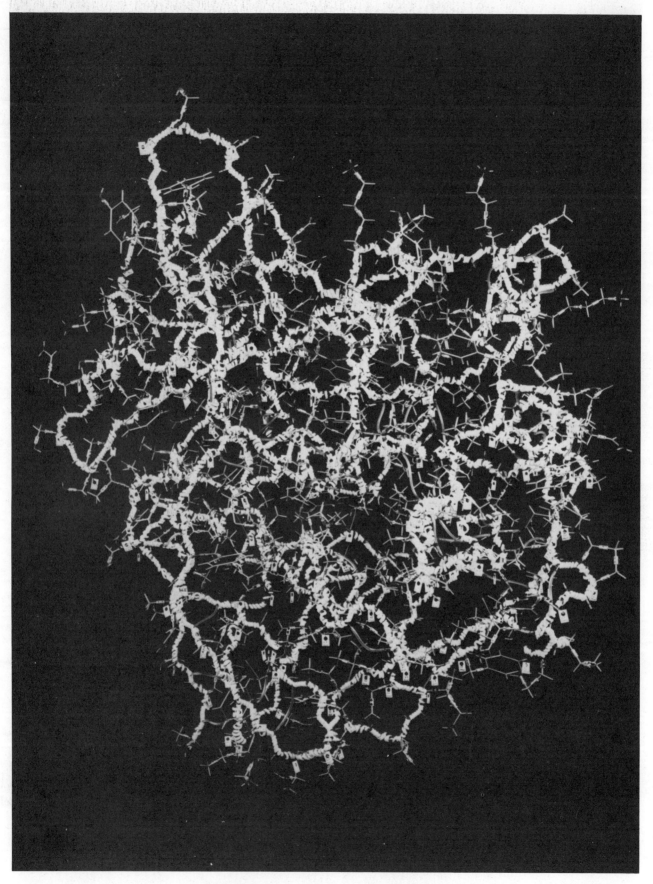

Figure 9.11
A model of a protein molecule (subtilisin). *Source:* H. Holter and K. Max Møller, eds., *The Carlsberg Laboratory* (The Carlsberg Foundation, 1976), p. 159. The photograph was supplied by Novo Industries A/S.

Figure 9.12
Closest packing of spheres.

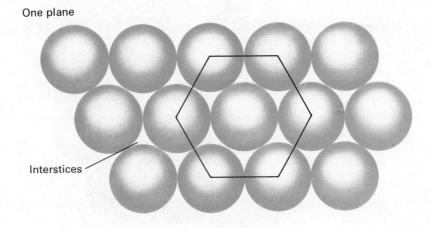

Figure 9.13
Cubic closest-packed arrangement of three planes of spheres.

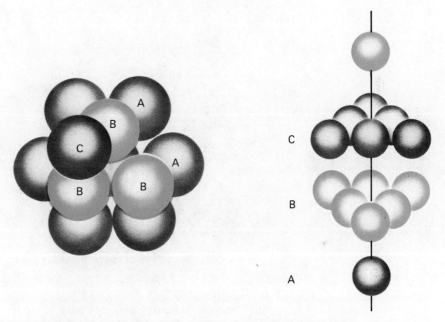

Figure 9.14
Nearest neighbors of a face-centered atom.

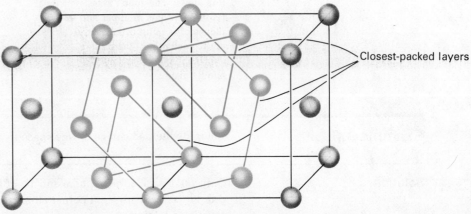

of the middle (B) plane within the cell (Figure 9.16). As in the case of a ccp lattice, each atom has a coordination number of twelve (including nearest neighbors in adjacent unit cells) and the minimum possible free space (26%).

Free Space and Cubic Closest Packing

It is easiest to calculate the percentage of free space for a close-packed lattice for the ccp case. To compute this value, compare the volume of the F-cubic unit cell with the total volume of those parts of the spheres that lie within it (Figure 9.17).

230   9 SOLIDS, LIQUIDS, AND CHANGES OF STATE

Figure 9.15
Hexagonal closest-packed arrangement of three planes of spheres.

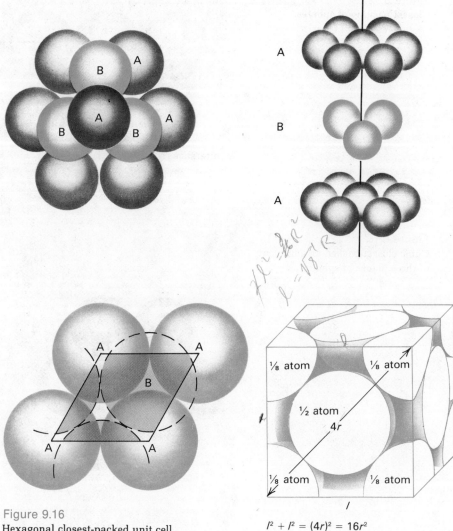

Figure 9.16
Hexagonal closest-packed unit cell (top view). The solid circles represent atoms in upper and lower planes of the unit cell directly above each other. The broken circles indicate atoms in the middle plane.

$l^2 + l^2 = (4r)^2 = 16r^2$
$l = \sqrt{8}\,r$ (cube edge length)

Figure 9.17
Calculating free space in a face-centered cubic cell.

Example 9.3

Compute the percentage of free space in a F-cubic unit cell.

Solution

Only 1/8 of each corner sphere and 1/2 of each face sphere lie within any single unit cell. Since there are eight corner atoms and six face atoms in an F-cubic unit cell, it contains a total of

$$8\left(\frac{1}{8}\right) + 6\left(\frac{1}{2}\right) = 4 \text{ spheres}$$

Since the volume of each sphere is $(4/3)\pi r^3$, the total volume occupied by the four spheres is

$$4\left(\frac{4}{3}\pi r^3\right) = \frac{16}{3}\pi r^3$$

where r is the sphere radius (the atomic radius). A corner sphere touches a face-centered sphere, so the distance between their centers is 2r, and the length of the face diagonal is 4r. The Pythagorean theorem shows that the cube edge length is $\sqrt{8}\,r$ (see Figure 9.17). The volume of the cube is therefore

$$V = l^3 = (\sqrt{8}\,r)^3 = 8\sqrt{8}\,r^3$$

The percentage of *filled* space is therefore

$$\frac{\text{Volume of spheres}}{\text{Volume of cube}} \times 100 = \frac{(16/3)\,\pi r^3}{8\sqrt{8}\,r^3} \times 100 = \frac{2\pi}{3\sqrt{8}} \times 100 = 74\%$$

The remaining 26% of the volume is free space.

A modification of this type of calculation can be used to calculate the theoretical density of a substance. Agreement between the theoretical and observed densities confirms the validity of an assignment of a particular unit cell and its dimensions.

Example 9.4

Calcium forms a cubic closest-packed crystal whose unit cell is 5.576 Å on a side. What is its theoretical density?

Solution

To compute the density in g/cm³, we should first convert the cell dimensions to cm.

$$l = (5.576\,\text{Å})\left(1.000 \times 10^{-8}\,\frac{\text{cm}}{\text{Å}}\right) = 5.576 \times 10^{-8}\,\text{cm}$$

The total volume of the cell is

$$\text{Volume} = l^3 = (5.576 \times 10^{-8}\,\text{cm})^3 = 1.734 \times 10^{-22}\,\text{cm}^3$$

The cell contains the equivalent of four calcium atoms, and since the atomic mass of calcium is 40.08 amu, the mass of calcium in the cell is 160.32 amu. Expressed in grams, this mass is

$$\text{Mass} = (160.32\,\text{amu})\left(\frac{1\,\text{g}}{6.023 \times 10^{23}\,\text{amu}}\right) = 2.662 \times 10^{-22}\,\text{g}$$

The theoretical density is the mass divided by the volume.

$$\text{Density} = \frac{2.662 \times 10^{-22}\,\text{g}}{1.734 \times 10^{-22}\,\text{cm}^3} = 1.535\,\frac{\text{g}}{\text{cm}^3}$$

This value agrees well with the measured density of 1.55 g/cm³ for calcium.

The general formula for computing the theoretical density of a substance from its unit cell is

$$\text{Density} = \frac{\text{Formula mass} \times Z}{\text{Unit cell volume} \times N_{\text{Avog}}} \tag{9.2}$$

where Z is the number of formula units of the substance in the unit cell and N_{Avog} is Avogadro's number.

Body-Centered Cubic Lattice

Some metals, particularly the alkali metals, crystallize in body-centered cubic lattices (Figure 9.18). The coordination number of each atom is eight, the edge length of the cell is $(4/\sqrt{3})\,r$, and the cell contains the equivalent of two atoms.

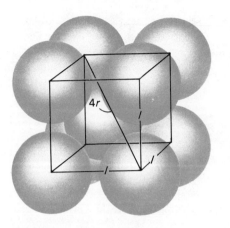

Figure 9.18
A body-centered cubic unit cell.

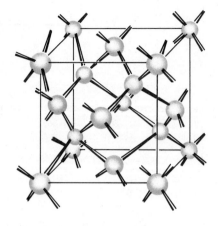

Figure 9.19
Diamond unit cell.

Diamond Structure

One of the crystal forms of carbon is diamond, a crystal with a face-centered cubic lattice. Carbon atoms occupy the lattice points and four of the eight positions halfway between the cube corners and the cube center. If an atom lies in one of these interior positions, the adjacent interior positions are empty (Figure 9.19). Each atom has a coordination number of four, the nearest neighbors forming a regular tetrahedron. (To verify this, observe one of the interior atoms. It is equidistant from a corner atom and three face atoms.)

Ionic Lattices

The unit cell of an ionic compound must be electrically neutral and have the same simplest formula as the compound. We will examine the ways in which the lattices of sodium chloride, cesium chloride, and calcium fluoride meet these requirements.

Sodium chloride has a face-centered lattice with chloride ions at the lattice points and sodium ions at the cube center and the midpoints of the cube edges, Figure 9.20. (We can just as well describe the crystal structure as an F-cubic lattice of Na$^+$ with Cl$^-$ ions at the center and edges.) The face-centered cubic arrangement of Cl$^-$ contributes four ions to the unit cell (see Example 9.3). Since 1/4 of each of the 12 edge-centered Na$^+$ ions belongs to the unit cell, the number of Na$^+$ ions in the cell is

$$1 \text{ center ion} + \frac{1}{4}(12 \text{ edge ions}) = 4 \text{ ions}$$

Figure 9.20
NaCl unit cell.

Figure 9.21
CsCl unit cell.

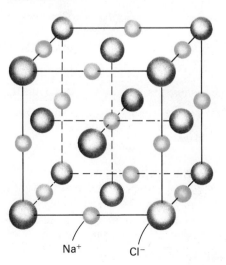

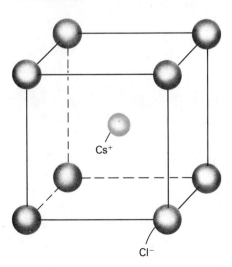

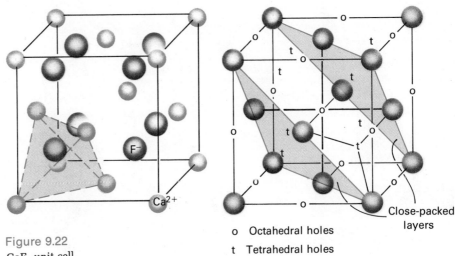

Figure 9.22
CaF₂ unit cell.

o Octahedral holes
t Tetrahedral holes

Figure 9.23
Octahedral and tetrahedral holes in a close-packed arrangement of spheres.

Because there are as many Na^+ ions as Cl^- ions, the unit cell of sodium chloride is electrically neutral and has the simplest formula NaCl. Each sodium ion has a coordination number of six. The surrounding chloride ions form an octahedron, and each chloride ion is at the center of an octahedron of sodium ions.

The cesium chloride unit cell is simple cubic* with chloride ions at the corners and a cesium ion at the center (Figure 9.21), and it contains one Cs^+ ion and the equivalent of one Cl^- ion. The coordination number of each ion is eight.

A unit cell of calcium fluoride, CaF_2, has an F-cubic lattice of calcium ions (equivalent to four Ca^{2+} ions) and eight F^- ions at the midpoints between the corners and the center (Figure 9.22). Each fluoride ion is surrounded by a tetrahedron of calcium ions, and eight fluoride ions surround each calcium ion.

Close-packed structures such as NaCl and CaF_2 contain two sorts of "holes"— tetrahedral holes and octahedral holes—lying between the close-packed planes (Figure 9.23). In NaCl the Na^+ ions occupy the octahedral holes in a face-centered cubic lattice of Cl^- ions, and the fluoride ions in CaF_2 occupy the tetrahedral holes in a face-centered cubic calcium ion lattice.

Polymorphism

Many compounds form several different crystal lattices called **polymorphic** structures. For example, zinc sulfide may crystallize as either zinc blende or wurzite (Figure 9.24, p. 234). Zinc blende has a lattice that resembles the diamond structure, with sulfide ions occupying the face-centered cubic lattice and zinc ions at four of the eight interior tetrahedral positions. The wurzite lattice is hexagonal, with Zn^{2+} ions lying between the planes of a hcp-like lattice of S^{2-} ions. In both crystal forms, each ion is surrounded by a tetrahedron of the oppositely charged ions. The difference between the two structures is in the orientation of these tetrahedra.

9.4 Crystal Properties

Lattice Forces

The nature of the lattice forces holding a crystal together depends on the substance and the nature of its chemical bonds.

---

*Cesium chloride is *not* body centered because the atom at the center, cesium, is different from the atoms at the corners, chloride ions.

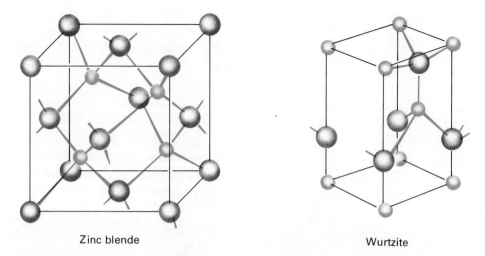

Figure 9.24
Polymorphism of zinc sulfide.

Ionic crystals. Electrostatic attractions between oppositely charged ions are the lattice forces in an **ionic crystal**. Since cations are surrounded by anions and anions by cations, the attractive forces outweigh repulsions between ions with the same charge. The strength of these electrostatic attractions gives ionic crystals their characteristic hardness and high melting points.

Covalent crystals. Diamond, quartz (SiO_2), and other hard, high melting, nonionic crystals are composed of extended networks of covalently bound atoms to form **covalent crystals**. An ideal picture of the diamond crystal is a single giant molecule in which each carbon atom is covalently bound to its four nearest neighbors. The tetrahedral coordination around each carbon atom matches the tetrahedral arrangement of the single covalent bonds of carbon.

Molecular crystals. Many crystals are arrangements of discrete molecules. In these **molecular crystals**, no ionic attractions or covalent bonds maintain the lattice structure. The only forces maintaining the crystal arrangement are the same weak **van der Waals forces** causing nonideality in gases. The motion of its electrons can momentarily turn a neutral, nonpolar molecule into a dipole. Attractions between these temporary dipoles slightly outweigh repulsions and produce van der Waals forces.

Because of the weakness of their lattice forces, crystals held together by van der Waals attractions are soft and have low melting points. The strength of van der Waals forces increases with the number of electrons in a molecule, and the intermolecular lattice forces of nonpolar substances tend to rise with molecular weight.

Dipole–dipole attractions maintain the crystal lattices of highly polar substances such as HCl. These lattice forces are weaker than ionic attractions but considerably stronger than van der Waals forces.

Metallic crystals. Electron sharing holds the atoms of a **metallic crystal** together, but the lattice forces are not typical covalent bonds. A metallic crystal can be described as a giant molecule whose outer-shell molecular orbitals extend through the whole crystal. These outer-shell molecular orbitals have very similar energies and form a cluster of energy levels known as the **conduction band.**

This model for metallic crystals can explain the most useful properties of metals; electrical conductivity, luster, ductility, and malleability. A metal is a good conductor because only the lower-energy orbitals of the conduction band are occupied. Electrons can easily jump into the empty higher-energy orbitals and flow through the crystal from a negative pole to a positive pole. Metals are malleable (it is fairly easy to deform their shapes) and ductile (can be drawn into wire) because the interatomic attractions have no strong preference for a particular direction. Planes of atoms can slide over each other under the influence of a deforming force (Figure 9.25).

9.4 CRYSTAL PROPERTIES

Crystal Imperfections

A completely regular crystal is an idealization that is nearly impossible to achieve in practice. A variety of errors occur during the growth of a lattice. Imagine 10^{18} ions forming an orderly lattice arrangement in a matter of a few minutes! Impurities may be trapped in the lattice, particularly if the size of the impurity is approximately the same as that of the available lattice sites in the crystal. An outer plane of molecules may cover an incomplete inner plane, producing "holes" in the lattice. A molecule may become trapped in the spaces between the planes instead of occupying its proper position in the plane. Several planes of a ccp lattice sometimes stack together in a hcp arrangement. (The opposite can happen in a hcp lattice.)

Slow growth of a crystal minimizes the chance of crystal imperfections, and a rapidly formed crystal incorporates the most errors. Crystal growth is a dynamic process—some molecules break away from the crystal surface as others deposit on it from the surrounding liquid. Given enough time, this process will repair errors on the surface of the crystal. But rapid crystal growth buries errors deep within the crystal before there is a chance of rectifying them.

Crystal defects have significant effects on the physical properties of a substance. They lower the tensile strength of a metal and increase its brittleness and corrosion rate, all undesirable effects. Some crystal imperfections can be useful; they can make metals more effective catalysts and are responsible for the desirable properties of semiconductors and phosphors.

Amorphous Solids and Crystallinity

A piece of glass (a mixture of silica and silicates), with a definite shape and volume, appears to be a solid yet differs from crystalline solids in important respects. Crystalline solids have sharp melting points, tend to break along specific fracture planes, and give well-defined x-ray diffraction patterns. Glass softens gradually over a wide temperature range, shatters in unpredictable ways (as you may have observed in the laboratory), and has very little ability to diffract x rays.

The lack of a clear x-ray diffraction pattern shows that silica glass has a *disordered* internal structure. (Quartz is the crystalline form of silica.) There is no definable unit cell for glass; the arrangement of one portion of the solid gives no reliable information about the arrangement of molecules in the rest of the solid. Because of this lack of periodic repetition in its molecular arrangement, glass is classified as an **amorphous solid**, although it can also be described as a supercooled liquid.

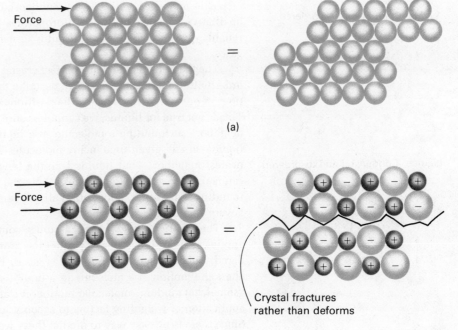

Figure 9.25
Ease of deformation of metals versus ionic crystals. (a) In metals, electrons "glue" the atoms together so there are no strong directive bonds or forces, and the planes are easily moved. (b) In ionic crystals, there are localized electrostatic forces. Deformation results in strong repulsion interactions, and the crystal fractures.

We can make substances with intermediate degrees of internal order. Although most synthetic polymers are amorphous, recent advances in polymer chemistry have produced synthetic fibers with increased internal periodicity or **crystallinity**, as shown by the greater resolution in their x-ray diffraction patterns. This increase in crystallinity greatly increases the tensile strength of the fibers.

Since crystallinity is a variable quantity expressing the degree of internal order of a solid, a perfect crystal has the highest crystallinity and an amorphous solid the lowest. Between these extremes are several intermediate crystalline states, such as imperfect crystals and semicrystalline substances. "Graphitized" carbon is amorphous carbon that has been made partially crystalline by heat treatment. It is very strong and reasonably inert and has been used for the nose cones of reentry rockets.

9.5
The Disorder of Liquids

As in the case with solids, the molecules in a liquid are also close to each other, but molecules in the liquid state possess considerably more kinetic energy and are relatively free to move through the liquid. The *inter*molecular forces in liquids are strong enough to hold the molecules near to each other but too weak to keep them in fixed positions. The motion of molecules through liquids gives them their *disordered* character and deprives them of any definite shape. Because of the closeness of the molecules, liquids—like solids—are **condensed phases**; they have definite volumes and are difficult to compress.

The phenomenon of diffusion offers the clearest evidence for molecular migration in a liquid. When a drop of ink is placed at the top of a beaker of water, its color spreads steadily throughout the entire beaker without any apparent mechanical agitation of the liquid. The speed of liquid diffusion contrasts with the extremely slow diffusion in solids and illustrates the freedom of molecules to move in liquids.

Liquid diffusion is considerably slower than diffusion in the gas phase. Gas molecules diffuse rapidly because they are far apart, and collisions between them are relatively rare. A molecule in a liquid is always in contact with other molecules. Frequent collisions and the attractive forces of the surrounding molecules retard the progress of the diffusing molecules (Figure 9.26).

This model of a liquid as a condensed and disordered phase gives a simple qualitative explanation of liquid properties. Unfortunately, no one has developed a reliable quantitative model for liquids to match those for crystalline solids and for gases.

The theoretical model of a perfect crystal is based on the assumption of complete order within the crystal. Once we describe the shape, dimensions, and atomic positions of one unit cell, we know the positions of every other atom in the ideal crystal. This is not true for liquids; we cannot be sure of the exact number and location of the neighbors surrounding a molecule in a liquid. There is some degree of order in a liquid—at any given time, many molecules do have a close-packed arrangement of nearest neighbors—but liquids lack the long-range order of crystals. Although we can make a reasonably good guess about the number of neighbors immediately surrounding a molecule, information about the arrangement of more distant molecules is very unreliable.

The kinetic theory of gases assumes completely random molecular motion and applies statistical methods to predict the properties of gas samples. Although disordered, liquids are not as random as gases. Because the attraction of its neighbors alters the motion of a molecule in a liquid, we must deal with the vague notion of "somewhat random" molecular motion. A statistical treatment of liquids requires the application of weighting factors to account for the effects of intermolecular motion. Since there is no easy way to predict these weighting factors because they vary from substance to substance, we have no general quantitative theory for liquids.

Figure 9.26
Molecular models for solids, liquids, and gases.

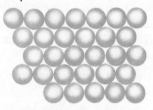

Solids: Condensed and ordered

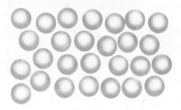

Liquids: Condensed and disordered

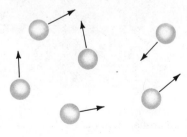

Gases: Dispersed and disordered

9.6 Changes of State

Because many important reactions take place in the liquid phase, the lack of detailed quantitative liquid models is a serious handicap for chemists. Not understanding the medium limits our knowledge of these liquid-phase reactions.

Heating Curves

Figure 9.27 shows the changes in temperature as we heat 1 mol (18 g) of ice at a rate of 10 J/sec. The temperature rises steadily until it reaches 0°C, the melting point of ice, and then remains constant at 0°C as the ice melts and liquid forms. It takes 6.02 kJ to melt a mole of ice. That is why ice at 0°C is much better for cooling something than liquid water at 0°C. As it melts, ice draws heat from its surroundings even though its own temperature does not increase. The 6.02 kJ heat absorbed as 1 mol of ice melts is called the **molar enthalpy of fusion** (or the molar heat of fusion) of water. The molar enthalpy of fusion of a substance is the heat needed to melt 1 mol of the substance. When a liquid freezes, it releases its heat of fusion.

After all the ice has melted, continued heating steadily raises the temperature of the water until it reaches 100°C. At this temperature the water begins to boil, forming steam. The temperature again stays constant until all the liquid evaporates, and 40.7 kJ heat are absorbed in the process. This heat is known as the **molar enthalpy** (or heat) **of vaporization** of water. The molar enthalpy of vaporization of a substance is the heat needed to evaporate 1 mol of liquid. Once all the water is in the gas phase, further heating will raise the temperature above 100°C. The reverse occurs when steam at 100°C condenses to the liquid—40.7 kJ heat are released. Because of this, steam is a much more effective heating agent than hot water.

The behavior described for water is typical for the changes of state of most substances. A typical substance has a characteristic melting point, boiling point, molar enthalpy of fusion, and molar enthalpy of vaporization (Table 9.1, p. 238).

Melting

As the temperature of a solid rises, its molecules vibrate more violently within the crystal lattice. At the melting point, a sufficient number of molecules vibrate so

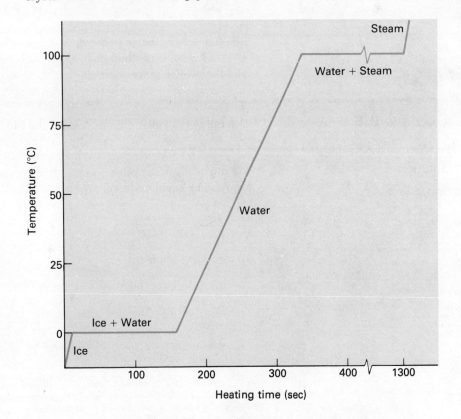

Figure 9.27
Heating curve for water.

Table 9.1 Thermodynamic Quantities for Changes of State

Substance	Melting point (°C)	Boiling point (°C)	ΔH_{fusion} (kJ/mol)	$\Delta H_{vaporization}$ (kJ/mole)
H_2O	0.0	100	6.02	40.7
Na	97.5	880	2.6	94.1
CCl_4	−23.0	76.8	2.7	33.0
$C_2H_6O_2$ (ethylene glycol)	−13.2	198	11.3	49.8
C_6H_6 (benzene)	5.5	79.6	9.92	31
$SiCl_4$	−70.3	57	5.77	25.5

strongly that they tear free of their lattice positions, and the ordered crystal disintegrates into the disordered liquid. Since the molecules must move against opposing intermolecular attractions to break away from the lattice, the molecular potential energy of the sample increases, and melting requires heat.*

The Thermodynamics of Melting

The **melting point** is the temperature at which the solid and liquid phases are in equilibrium. The solid does not melt nor does the liquid solidify in an adiabatic (one insulated against heat flow) system at the melting point.

Stated in terms of Gibbs free energy, melting equilibrium occurs when the free energy change due to melting (ΔG_{melt}) equals zero.

$$\Delta G_{melt} = \Delta H_{melt} - T\Delta S_{melt} = 0 \tag{9.3}$$

(The symbol ΔS_{melt} stands for the entropy increase when the solid melts to liquid, and ΔH_{melt} is the heat of fusion.) The tendency of systems to form the more disordered state favors melting, whereas the tendency toward lower potential energy favors solidification. Temperature determines which of these two effects dominates. They are balanced at the melting point (T_{melt}).

$$T_{melt} = \frac{\Delta H_{melt}}{\Delta S_{melt}} \tag{9.4}$$

The free enery change is negative at higher temperatures and melting occurs. Substances freeze below their melting points because the melting free energy is positive, and the reverse process occurs.

Example 9.5

Determine the molar entropy for melting of ice.

Solution

We can estimate the value of ΔS_{melt} from the heat of fusion and the melting point by rearranging equation (9.4).

$$\Delta S_{melt} = \frac{\Delta H_{melt}}{T_{melt}}$$

According to Table 9.1,

$\Delta H_{melt} = 6.02$ kJ/mol

$T_{melt} = 0°C = 273$ K

$$\Delta S_{melt} = \frac{6.0 \text{ kJ/mol}}{273 \text{ K}} = 22.1 \text{ J/K-mol}$$

---

*Because of the quantized nature of bond vibrations and rotations in a molecule, kinetic energy can also change during melting. The hypothesis in the kinetic-molecular theory that the average molecular kinetic energy of a sample is proportional to its temperature applies to translational kinetic energy, but not to vibrational and rotational energy.

9.6 CHANGES OF STATE

Example 9.6

Calculate the molar free energy for melting of ice at **a.** −5°C   **b.** +10°C.

Solution

a. We can use the value of ΔS_{melt} at 0°C to estimate ΔG_{melt} at −5°C (268 K).

$$\Delta G_{melt} = \Delta H_{melt} - T\Delta S_{melt}$$

$$= \frac{6.02 \text{ kJ}}{\text{mol}} - \left(\frac{22.1 \text{ J}}{\text{K-mol}}\right)(268 \text{ K})$$

$$= \frac{6.02 \text{ kJ}}{\text{mol}} - \frac{5.91 \text{ kJ}}{\text{mol}} = \frac{+0.11 \text{ kJ}}{\text{mol}}$$

Since the free energy change for melting is positive, at −5°C water freezes.

b. At +10°C,

$$\Delta G_{melt} = \frac{6.02 \text{ kJ}}{\text{mol}} - \left(\frac{22.1 \text{ J}}{\text{K-mol}}\right)(283 \text{ K}) = \frac{-0.22 \text{ kJ}}{\text{mol}}$$

Ice melts at +10°C according to our calculations.

Supercooling

It is sometimes possible to cool a liquid below its melting point without freezing it. This phenomenon is known as **supercooling**. Supercooled liquids are not stable and may suddenly and unpredictably begin to freeze. Once freezing starts, the released heat of fusion raises the temperature to the melting point, and it remains there until the last of the liquid solidifies (Figure 9.28).

Obtaining a supercooled liquid is often a matter of luck (bad luck, if you want the solid). To supercool a liquid, you should work with clean, smooth glassware, avoid jarring or stirring the liquid, and keep the system free of dust. Agitation of the sample, solid impurities, or rough surfaces can all initiate freezing.

Before a crystal can develop, a few molecules that have low enough energy to coalesce and are situated in the right arrangement to form the lattice must collide. A supercooled liquid, although less stable than the solid, can exist during the time it takes this chance event to occur. Without assistance, this improbable event could take a long time to occur (particularly if the liquid is viscous). Agitation increases the number of orientations in which molecules collide and raises the probability of initiating crystal growth. Rough surfaces and solid contaminants may adsorb

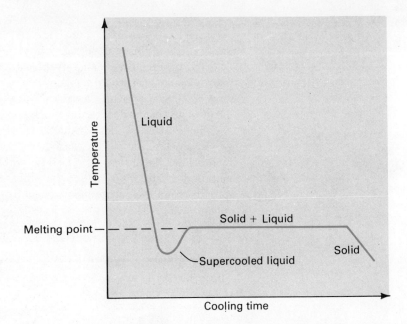

Figure 9.28
Supercooling.

molecules in an arrangement resembling the lattice. Once the crystal lattice begins to form, it acts like a template for the further growth of the crystal and freezing occurs rapidly. The surest way to stop supercooling is to add a "seed crystal" of the substance.

Pressure Effects on Melting

The effect of pressure on the melting point depends on how the volume of a substance changes as it melts. Most substances, including CO_2, expand as they melt and contract as they freeze. Increased pressure increases the melting points of these substances. Water is a notable exception because liquid water occupies less volume

---

"Now suppose," chortled Dr. Breed, enjoying himself, "that there were many possible ways in which water could crystallize, could freeze. Suppose the sort of ice we skate upon and put into highballs—what we call *ice-one*—is only one of several types of ice. Suppose water always froze as *ice-one* on Earth because it never had a seed to teach it how to form *ice-two*, *ice-three*, *ice-four*...? And suppose," he rapped on his desk with his old hand again, "that there were one form, which we will call *ice-nine*—a crystal as hard as this desk—with a melting point of, let us say, one-hundred degrees Fahrenheit, or better still, a melting point of one-hundred-and-thirty degrees.". . .

"If the streams flowing through the swamp froze as *ice-nine*, what about the rivers and lakes the streams fed?"

"They'd freeze. But there is no such thing as *ice-nine*."

"And the oceans the frozen rivers fed?"

"They'd freeze, of course," he snapped.

Kurt Vonnegut, Jr., *Cat's Cradle**

How close is this bit of science fiction to reality? Ice has at least six polymorphic forms. One of them (ice VI) melts at 100°C—provided the pressure is 20,000 atm! Although water molecules stack into various crystal forms, pressure dictates the form, and old-fashioned, everyday ice-one is the only form known to exist below 1,000 atm.

But that's only what we know from past experience. Could a seed crystal of ice-nine, as yet unformed, freeze the oceans? It's not too likely. Water would freeze to ice-nine only if ice-nine were more stable than ice-one. It's improbable that ice-nine, if it's that stable, wouldn't have formed spontaneously by now.

Although ice-nine, or anything like it, isn't likely to exist at normal temperatures and pressures, Vonnegut certainly deserves credit for exploring the implications of this stuff so imaginatively. In *Cat's Cradle*, a piece of ice-nine does fall into the ocean, and all the earth's oceans, rivers, and lakes quickly freeze. That's what would happen if a crystal of a more stable, higher melting form of ice ever does touch water. If there's a flaw in Vonnegut's treatment of the idea, it's that he doesn't push it far enough. In the book, the survivors of the ice-nine disaster can choose suicide by touching ice-nine and freezing themselves to death. But such a conscious act wouldn't really be necessary. Being more stable than ordinary ice, ice-nine would have a lower vapor pressure, and most of the water vapor in the air would condense to ice-nine. Everyone would soon die from the dehydration caused by the low water content of the atmosphere.

---

*Excerpted from the book *Cat's Cradle* by Kurt Vonnegut, Jr. Copyright © 1963 by Kurt Vonnegut, Jr. Reprinted by permission of Delacorte Press/Seymour Lawrence.

than ice and water expands on freezing. Consequently, the melting point of ice decreases as pressure is applied to it.

One application of the effect of pressure on the melting point is ice skating. The intense pressure under the skate blade melts a small amount of the surface ice, and the water that forms acts as a lubricant, helping the blade to glide over the surface. This would not work on solid CO_2 because a pressure increase would solidify liquid CO_2 rather than melt the solid.

Le Chatelier's Principle

Le Chatelier's principle is a simple rule for predicting the qualitative effect of a change in external conditions on an equilibrium system: A system at equilibrium responds to an external stress so as to minimize the effects of that stress. For example, substances melt when heated at their melting points because the solid-liquid equilibrium shifts to absorb the added heat (the external stress). Cooling shifts this equilibrium in the exothermic direction and freezing occurs. A pressure increase forces an equilibrium system toward the state with the smaller volume, thus relieving some of the applied pressure.

Evaporation

Evaporation occurs when a molecule at the surface of a liquid (or solid) gains enough kinetic energy to break away from its neighbors and into the gas phase. Because some molecules have much higher kinetic energies than the average for the sample, evaporation occurs below the boiling point of a liquid. Since evaporation selectively removes "hot" molecules from the liquid, the average molecular kinetic energy and temperature of the liquid drop. Evaporation in an adiabatic system slows down as the liquid cools by evaporation. That is why liquid nitrogen (boiling point $-196°C$) stays around for a long time in an open but well-insulated flask (Figure 9.29). Evaporation continues at a steady rate in an isothermal system because of the flow of heat from the surroundings. Consequently, a liquid can evaporate completely if left long enough in the open.

Equilibrium Vapor Pressure

Evaporation in a closed container reaches equilibrium when the partial pressure of the substance reaches its **equilibrium vapor pressure**. This pressure depends only on the substance and the temperature, not the size of the sample or the volume of the container. For example, the equilibrium vapor pressure of water at 20°C is 17.5 torr (see Appendix G). If a container holds enough pure water at 20°C, the vapor pressure will reach 17.5 torr (Figure 9.30). If there is not enough water to fill the container with vapor at 17.5 torr, all the liquid evaporates before equilibrium is attained. If

Figure 9.29
Evaporation. (a) In an insulated flask, slow-moving molecules remain in the liquid, the liquid cools, and evaporation slows. (b) In a beaker, heat flows into the liquid to restore distribution of fast and slow molecules and evaporation continues.

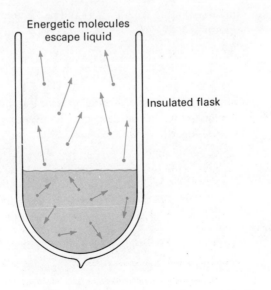
(a)

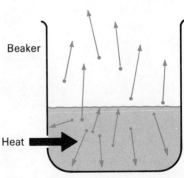
(b)

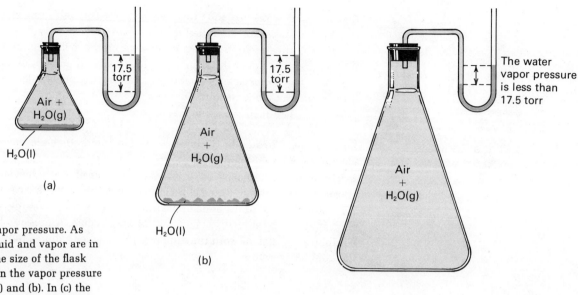

Figure 9.30
Equilibrium vapor pressure. As long as the liquid and vapor are in equilibrium, the size of the flask has no effect on the vapor pressure as shown in (a) and (b). In (c) the liquid has completely evaporated and the system is not in equilibrium.

water vapor at 20°C is compressed to more than 17.5 torr, it condenses until the pressure drops to the equilibrium value.

Evaporation equilibrium is dynamic, with molecules condensing from the vapor phase as rapidly as they evaporate from the liquid. Condensation occurs when a molecule, moving too slowly to break away from intermolecular attractions, collides with the liquid surface and is captured. At any given temperature, the percentage of these low-energy molecules is constant, and the recondensation rate depends on the concentration of molecules in the vapor. As the partial pressure increases, so does the rate of recondensation until it reaches the rate of evaporation, and equilibrium is established.

Special Resource Section:
Temperature Effects on Vapor Pressure

Equilibrium vapor pressure rises with temperature. The temperature increase shifts the kinetic energy distribution of the molecules (Figure 9.31) so a higher percentage have sufficient energy to escape into the vapor. The percentage of gas-phase molecules that can recondense drops, and equilibrium occurs at a higher partial pressure.

The **Clausius-Clapeyron equation** describes the logarithmic form of the pressure increase (Figure 9.32, p. 244) and its dependence on the heat of vaporization, ΔH_{vap} (Figure 9.33),

$$\ln\left(\frac{P_2}{P_1}\right) = -\frac{\Delta H_{vap}}{R}\left(\frac{1}{T_2} - \frac{1}{T_1}\right)$$

where P_2 is the equilibrium pressure at T_2, and P_1 the pressure at T_1. It is more convenient to express this equation in terms of base-ten logarithms (log) rather than natural logarithms (ln).

$$\ln\left(\frac{P_2}{P_1}\right) = (\ln 10)\log\left(\frac{P_2}{P_1}\right) = 2.303 \log\left(\frac{P_2}{P_1}\right)$$

$$\log\left(\frac{P_2}{P_1}\right) = -\frac{\Delta H_{vap}}{2.303\,R}\left(\frac{1}{T_2} - \frac{1}{T_1}\right)$$

A graph of $\log P$ versus $1/T$ is a straight line whose slope equals $[-(\Delta H_{vap}/2.303\,R)]$.*

---

*The heat of vaporization may vary slightly with temperature, producing a slight curvature in the graph. The estimates of ΔH_{vap} in Figure 9.33 are average values for the entire temperature range.

Figure 9.31
The effect of temperature on the distribution of molecular kinetic energies.

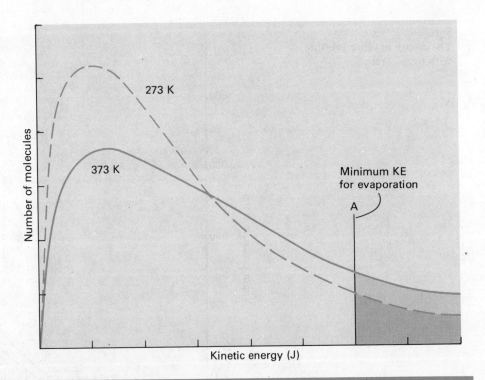

Example 9.7

The vapor pressure of toluene is 36.7 torr at 30°C and 139.5 torr at 60°C.
a. Calculate the heat of vaporization of toluene.
b. What is the vapor pressure of toluene at 100°C?

Solution

a. First estimate the log of the ratio of the two pressures.

$$\log\left(\frac{139.5 \text{ torr}}{36.7 \text{ torr}}\right) = \log(3.80) = 0.5799$$

This value equals

$$-\left(\frac{\Delta H_{vap}}{2.30R}\right)\left(\frac{1}{333 \text{ K}} - \frac{1}{303 \text{ K}}\right)$$

Since we want an answer in J/mol, R must be expressed in J/K-mol.

$$R = 8.31 \frac{\text{J}}{\text{K-mol}}; \quad T_2 = 60°C = 333 \text{ K}; \quad T_1 = 30°C = 303 \text{ K}$$

$$0.5799 = -\left[\frac{\Delta H_{vap}}{2.30 \times 8.31 \text{ J/K-mol}}\right]\left(\frac{1}{333 \text{ K}} - \frac{1}{303 \text{ K}}\right)$$

$$= -\left[\frac{\Delta H_{vap}}{19.1 \text{ J/mol}}\right](3.00 \times 10^{-3} - 3.30 \times 10^{-3})$$

$$= -\left[\frac{\Delta H_{vap}}{19.1 \text{ J/mol}}\right](-3.0 \times 10^{-4})$$

$$\Delta H_{vap} = \frac{(0.5799)(19.1 \text{ J/mol})}{3.0 \times 10^{-4}} = 3.7 \times 10^4 \text{ J/mol} = 37 \text{ kJ/mol}$$

b. You can use either of the known vapor pressures and the corresponding temperature to find the vapor pressure at 100°C (373 K). We will compare the predicted pressure at 100°C to the known pressure at 30°C (36.7 torr at 303 K). Assuming that ΔH_{vap} does not vary with temperature, we can write the equation

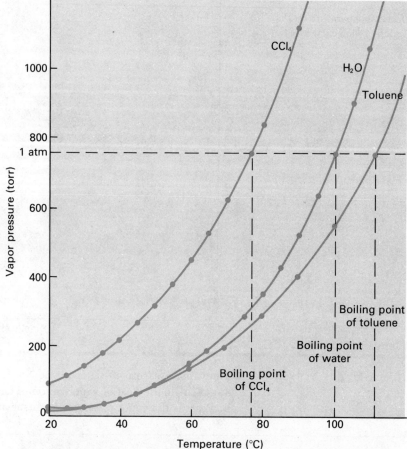

Figure 9.32
The change of vapor pressure with temperature.

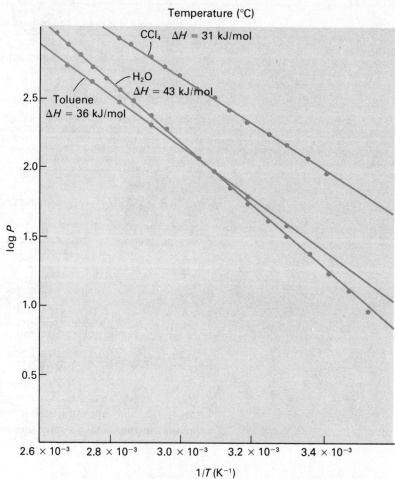

Figure 9.33
Plot of log P versus 1/T.

9.6 CHANGES OF STATE

$$\log\left(\frac{P_{100}}{36.7 \text{ torr}}\right) = -\left[\frac{3.7 \times 10^4 \text{ J/mol}}{19.1 \text{ J/K-mol}}\right]\left(\frac{1}{373 \text{ K}} - \frac{1}{303 \text{ K}}\right) = 1.20$$

The antilog of 1.20 is approximately 16, so

$$\frac{P_{100}}{36.7 \text{ torr}} = 16$$

$$P_{100} = 16(36.7 \text{ torr}) = 5.9 \times 10^2 \text{ torr}$$

The Thermodynamics of Evaporation

The thermodynamic principles governing evaporation are the same as those that apply to melting or any other reversible process. If the change in free energy that accompanies evaporation is negative, the process is favorable, and further evaporation occurs. Condensation, the reverse of evaporation, takes place when the evaporation free energy is positive. A liquid-vapor system is at equilibrium if the free energy for evaporation is zero. Like melting, evaporation has a positive entropy because it leads to a more disordered state, and its enthalpy is positive because it takes work to overcome the attractive intermolecular forces.

The evaporation equilibrium illustrates Le Chatelier's principle. Since evaporation is endothermic, a temperature increase shifts the equilibrium in that direction and raises the vapor pressure. As we increase the volume of a liquid-gas system, the system responds by increased evaporation to produce more gas to fill the void. When compressed, the vapor in the system condenses because the liquid phase occupies less volume.

Boiling

A liquid cannot exist at equilibrium if its equilibrium vapor pressure exceeds the external pressure. Figure 9.34 illustrates the case of water and air in a cylinder under a constant pressure of 1 atm. Up to 100°C, the cylinder can maintain pressure equilibrium by expanding, but no amount of expansion can maintain the liquid-vapor equilibrium above 100°C, since the vapor pressure of water is more than 1 atm. The liquid will continue to evaporate as the cylinder expands until all the liquid boils away.

The **boiling point** is the temperature at which equilibrium vapor pressure of a liquid equals the external pressure. The boiling point at 1 atm is called the **normal boiling point**. Boiling points are very sensitive to the external pressure. For example, water has a normal boiling point of 100°C, but it boils at 20°C under an external pressure of 17.5 torr. This dependence of boiling point on external pressure enables us to distill a high-boiling substance at a reduced pressure and temperature, avoiding the possibility of thermal decomposition of the material. It is also possible to heat a liquid at high pressure above its normal boiling point. Water boils at 121°C at 2 atm pressure.

Figure 9.34
Boiling.

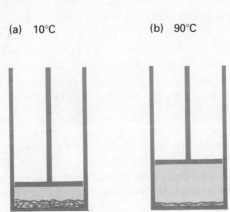

Figure 9.35
The phase diagram for H₂O.

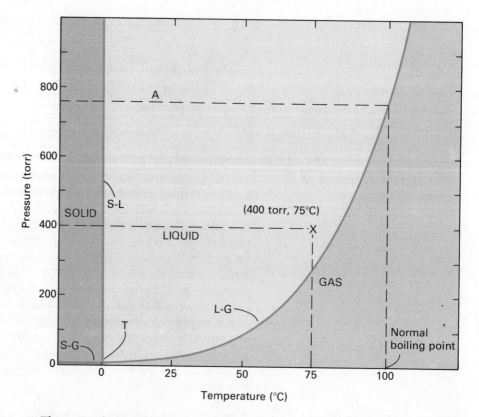

The most obvious features of a boiling liquid are its violent bubbling and the rapidity of its evaporation. Vapor bubbles form in the interior of a boiling liquid and expand against the slightly weaker external pressure. Because these bubbles offer additional surfaces for evaporation, the liquid vaporizes more rapidly. Evaporation below the boiling point is slow because it occurs only at the outer surface of the liquid.

Phase Diagrams

A **phase diagram** (Figure 9.35) of a substance is a graph showing the conditions of temperature and pressure necessary for the existence of its various phases. The vertical axis shows pressure and the horizontal axis temperature. The solid phase occurs at high pressure and low temperature (the upper left of the diagram), and the gas phase exists at high temperature and low pressure (lower right) with the liquid requiring intermediate conditions of temperature and pressure. We can use a phase diagram to tell what the physical state of a substance will be at a given temperature and pressure. To do this, we locate the point on the graph with the proper coordinates of temperature and pressure and see in which region it is located. For example, the point X in the phase diagram in Figure 9.35 representing 75°C and 400 torr is in the liquid region, so we know that water is liquid under these conditions.

If the point is located on one of the lines separating two phase regions, it means that the two phases are in *equilibrium* at this temperature and pressure. There are three such lines in Figure 9.35: the temperature dependence of the vapor pressure of the solid (line S–G), the vapor pressure curve of the liquid (L–G), and the line representing the variation of the melting point with pressure (S–L). Because the effect of pressure on the melting point is so slight, the S–L line appears vertical for moderate pressure ranges. The decrease in the melting point of water with increased pressure is apparent only if we extend the diagram to very high pressures.

The solid, liquid, and gas phases of a pure substance can exist together at equilibrium under only one set of conditions, known as the **triple point**. This is a point at which three phase equilibrium lines meet. Water has a triple point (point T in Figure 9.35) at 4.58 torr and 0.0075°C. Heating a piece of ice in an otherwise empty box raises the temperature and vapor pressure along the S–G line until the triple point is reached. As long as ice, water, and vapor are present, the temperature stays at

Figure 9.36
Triple-point region shown on an expanded scale.

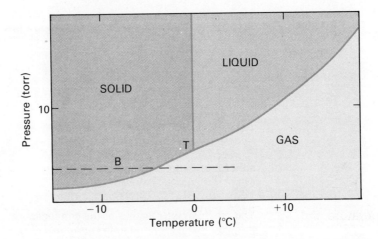

0.0075°C and the vapor pressure remains at 4.58 torr. After the last of the ice melts, further heating raises the temperature and vapor pressure along the L–G line.

Dotted line A shows the changes that occur on heating a cylinder containing ice and no empty space at an external pressure of 1 atm. Melting occurs when the temperature reaches the intersection of A and the S–L line. After all the solid melts, the cylinder contains only liquid until the temperature reaches the intersection of A and the L–G line (the normal boiling point). Water then evaporates, leaving only vapor.

Heating ice in a cylinder at 3 torr (dotted line B, Figure 9.36) produces different behavior. Because the external pressure is below the triple point pressure, ice **sublimes** (evaporates without melting) when the temperature reaches the intersection of B and the S–G line.

Each substance has its own characteristic phase diagram. Carbon dioxide (Figure 9.37) has a triple-point pressure above 1 atm. Solid CO_2 (Dry Ice) sublimes at atmo-

Figure 9.37
The phase diagram for CO_2.

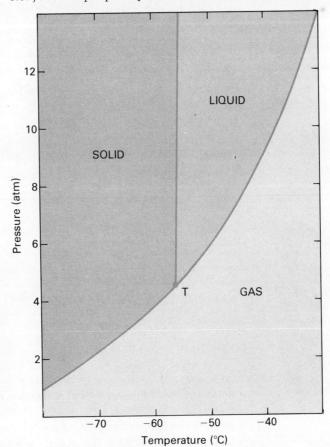

Table 9.2 Critical Temperatures and Pressures

Gas	Normal boiling point (°C)	Critical temperature (°C)	Critical pressure (atm)
Hydrogen	−252.8	−239.9	12.8
Oxygen	−183.0	−118.8	49.7
Carbon dioxide	−78.5	31.3	73.0
Water	100.0	374.0	217.7

spheric pressure, and we must raise the external pressure in order to obtain liquid CO_2.

Critical Temperature and Pressure

The most obvious feature of the condensation of a gas is the abrupt change in properties as the gas, completely filling its container, is replaced by a small liquid phase with a much greater density. An increase in pressure, however, cannot cause this sudden liquefaction if the gas is maintained at a temperature above its **critical temperature**. Each gas has its own characteristic critical temperature above which it cannot be made liquid, no matter how great the pressure. The pressure that must be applied to cause liquefaction *at* the critical temperature is called the **critical pressure**. Table 9.2 lists some critical temperatures and pressures.

9.7 Liquid Crystals

Some substances melt to form phases with both the fluidity of liquids and a considerable degree of the internal order we associate with crystals. These **liquid crystals** (Figure 9.38) show the mechanical properties of liquids and the optical properties of crystals. (These optical properties are too complicated to explain briefly. They involve the influence of liquid crystals on plane polarized light—light whose wave disturbance is in a single plane.)

The molecules that form liquid crystals are long and asymmetric (see Figure 9.38). Molecules in **smectic liquid crystals** group together in well-defined planes and lie perpendicular to these planes. These planes slide past each other easily, giving the phase its fluid character. **Nematic liquid crystals** contain parallel molecules that are not restricted to any particular plane. In **cholesteric liquid crystals**, the molecules lie lengthwise and parallel in planes.

Although liquid crystals were first described nearly a century ago, they were generally treated as insignificant curiosities. We are now just beginning to recognize their technological potential. Because its color is very sensitive to temperature

Figure 9.38
Liquid crystals.

(a) Smectic

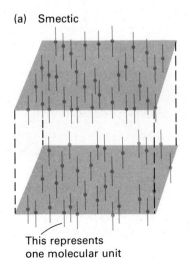

This represents one molecular unit

(b) Nematic

(c) Cholesteric

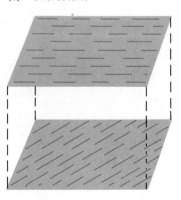

changes, a layer of a liquid crystal can, by showing variations in the heat conductivity in a piece of metal, locate structural flaws in the metal. Low concentrations of impurities also can change the color of a liquid crystal, and this response could provide a sensitive test for air pollutants. Low voltages alter the color and opacity of some liquid crystals. We use this phenomenon to display numbers in portable electronic calculators and digital watches. Liquid crystals may one day be used to produce a compact television set with obvious advantages over one based on the cumbersome, high-voltage cathode-ray tube.

The greatest technological applications of liquid crystals may be occurring right now in living systems. The fluids in living cells, especially cell membranes, have a much greater degree of molecular order, which is necessary for the functioning of the cell, than simple liquids. Studies on liquid crystals prepared in the laboratory may give us insights into the subtleties of cell structure.

Summary

1. A crystal is a solid particle with well-defined planar surfaces. The shape of a crystal reflects its orderly internal structure.

2. A lattice is a periodic repeating array of points with identical environments about each similar point in the lattice.

3. Symmetry elements: If we cut a crystal along a mirror plane and hold either half against a mirror, we will see the original crystal. Rotation of a crystal around one of its axes of symmetry, by a certain amount, produces identical perspectives.

4. There are seven different crystal systems based on symmetry elements. A unit cell is the smallest portion of a lattice that can characterize it by translation. We can describe any crystal by one of 14 three-dimensional Bravais lattices.

5. When x rays pass through crystals, they are diffracted. This shows that crystals have repeating lattices. The positions and intensities of the spots in a diffraction pattern can be used to discover the lattice arrangement of a crystal and the structure of the molecules within it.

6. Bragg's law: Spots in an x-ray diffraction pattern occur whenever the reflection angle θ fits the condition

   $$n\lambda = 2d \sin \theta$$

   where n is a whole number, λ the wavelength of the x ray, and d the distance between two parallel planes of atoms.

7. There are two lattices, cubic closest-packed (ccp) and hexagonal closest-packed (hcp), that have the smallest free space possible (26%) for a regular array of identical atoms. The unit cell of a ccp lattice is the face-centered cubic cell. An atom in either closest-packed lattice type has a coordination number (the number of nearest neighbors) of 12.

8. Unit cells of ionic compounds are electrically neutral and have the same simplest formula as the compound. Many ionic lattices consist of closest-packed arrangements of one type of ion in which the oppositely charged ions occupy some of the holes in the lattice. A closest-packed lattice has two types of holes, tetrahedral and octahedral.

9. Polymorphic structures are different crystal forms of the same compound.

10. Ionic crystals (for example, NaCl) are held together by electrostatic forces, covalent crystals (for example, diamond) by a network of covalent bonds, and molecular crystals by van der Waals forces.

11. The outer-shell electrons of metallic crystals occupy some of a group of orbitals with nearly identical energies called the conduction band. The higher-energy levels of the conduction band are empty, and these orbitals extend through the entire crystal. Electrons can easily jump into these empty orbitals and migrate through the metal.

12. Crystals contain many imperfections that have important effects on their properties.

13. An amorphous solid has no periodic order in its internal structure. Crystallinity is a measure of the degree of internal order in a substance.
14. A liquid is a condensed and disordered phase; its molecules are close together but do not have the regular arrangements characteristic of crystalline solids.
15. Every substance has a characteristic melting point, boiling point, molar enthalpy (heat) of fusion and molar enthalpy (heat) of vaporization. The molar enthalpy of fusion is the enthalpy change occurring when 1 mol of solid melts. Molar enthalpy of vaporization is the enthalpy change associated with the evaporation of 1 mol of liquid.
16. The melting point (T_{melt}) is the temperature at which the solid and liquid phases are in equilibrium. At the melting point, the free energy for melting is zero, and the entropy for melting is

$$\Delta S_{melt} = \frac{\Delta H_{melt}}{T_{melt}}$$

17. A supercooled liquid is a liquid that exists below its melting point. Introduction of a seed crystal into a supercooled liquid initiates freezing.
18. Le Chatelier's principle: A system at equilibrium responds to an external stress so as to minimize the effect of that stress.
19. Evaporation occurs when a molecule has enough energy to break away from its neighbors in the liquid phase and enter the gas phase. Evaporation in a closed container reaches equilibrium when the partial pressure of the substance equals its equilibrium vapor pressure. At equilibrium, the rate of evaporation equals the rate of condensation.
20. The Clausius-Clapeyron equation describes the influence of temperature on the equilibrium vapor pressure of a substance.

$$\log\left(\frac{P_2}{P_1}\right) = -\frac{\Delta H_{vap}}{2.30R}\left(\frac{1}{T_2} - \frac{1}{T_1}\right)$$

21. The boiling point of a substance is the temperature at which its equilibrium vapor pressure equals the external pressure. The normal boiling point is the boiling point at 1 atm.
22. A phase diagram is a graph showing the conditions of temperature and pressure necessary for the existence of the various phases of a substance. The triple point is the unique set of conditions (temperature and pressure) at which three phases of a pure substance can coexist.
23. The critical temperature is the temperature above which it is impossible to liquefy a substance no matter how great the pressure. The critical pressure is the pressure required to liquefy a gas at the critical temperature.
24. Liquid crystals show the mechanical properties of liquids and the optical properties of crystals.

Vocabulary List

crystal	hexagonal closest-packed	supercooling
lattice	coordination number	Le Chatelier's principle
symmetry element	polymorphic	equilibrium vapor pressure
lattice array	ionic crystal	Clausius-Clapeyron equation
unit cell	covalent crystal	boiling point
translation	molecular crystal	normal boiling point
crystal systems	van der Waals forces	phase diagram
Bravais lattice	metallic crystal	triple point
body-centered lattice	conduction band	sublime
face-centered lattice	amorphous solid	critical temperature
side-centered lattice	crystallinity	critical pressure
Bragg's law	condensed phase	liquid crystal
electron-density function	molar enthalpy of fusion	smectic liquid crystal
interstice	molar enthalpy of vaporization	nematic liquid crystal
cubic closest-packed	melting point	cholesteric liquid crystal

Exercises

1. Briefly define the following terms.
   a. Lattice   b. Diffraction   c. Cubic closest packing
   d. Coordination number   e. van der Waals forces
   f. Crystalline

2. Define each of the following terms.
   a. Enthalpy of fusion   b. Enthalpy of vaporization   c. Supercooling
   d. Equilibrium vapor pressure   e. Normal boiling point
   f. Triple point   g. Critical temperature

3. Which of the following statements is true?
   a. Crystals of a particular substance have the same shape no matter what the conditions of crystallization.
   b. There is no regularity in the shapes of the crystals of a particular substance.
   c. While differing in appearance, different crystals of the same substance have identical angles between corresponding faces.
   d. Any difference in the appearance of two crystals of the same substance is known as polymorphism.

4. When a small amount of iodine vapor is introduced into a flask of nitrogen, the entire flask soon takes on the characteristic violet color of I_2. The spreading of the violet color of I_2 through liquid CCl_4 is considerably slower. Explain these observations in terms of the molecular nature of liquids and gases.

5. Why are there no laws, similar to the ideal gas law, describing the physical properties of liquids solely in terms of universal constants rather than parameters that are different for each substance?

6. What relationship is there between the arrangement of carbon atoms in the diamond lattice and the nature of the covalent bonds in methane (Chapter 7)? What is the state of hybridization of a carbon atom in diamond?

7. a. Would you expect solid I_2 to be an ionic crystal, a covalent crystal, a molecular crystal, or a metallic crystal?
   b. What type of forces holds the I_2 crystal together?

8. a. What sort of x-ray diffraction patterns, if any, would liquids produce if they were completely disordered states?
   b. Liquids do produce x-ray diffraction patterns, but these patterns consist of hazy concentric rings rather than the patterns of well-defined spots that make up the diffraction patterns of crystals. Is this consistent with the description of a liquid as a somewhat random state?

9. What are the similarities and the differences between the molecular nature of a liquid and an amorphous solid?

10. Glass is sometimes described as a very viscous liquid. Is this description consistent with the molecular picture of a liquid? Would you expect glass to have a definite melting point?

11. The melting point of benzene increases slightly as pressure is applied to the sample. Which is denser, liquid benzene or solid benzene?

12. Use Le Chatelier's principle to predict how the following physical stresses change the phase equilibria.
    a. The pressure of a mixture of solid and liquid benzene is decreased. The density of solid benzene is 1.01 g/cm³ and the density of liquid benzene is 0.89 g/cm³.
    b. Heat is withdrawn from a mixture of solid and liquid benzene.
    c. The pressure on a flask containing both liquid benzene and its vapor is increased.
    d. The benzene vapor above a sample of the liquid is blown away in a stream of air.

13. Sodium and oxygen form a cubic crystalline oxide in which the coordination number of sodium is four, and the coordination number of oxygen is eight. Draw a picture of this unit cell. Clearly indicate the positions of the sodium ions and the oxide ions.

Figure 9.39

Rutile structure, TiO_2 (tetragonal).

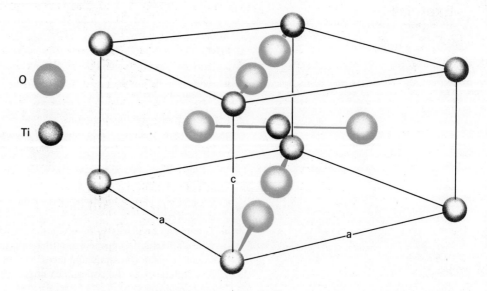

14. Another common crystal lattice is the rutile (TiO_2) structure shown in Figure 9.39. How many atoms of each type lie within the unit cell?

15. Substances such as nylon, plastic, and rubber consist of long, flexible molecules, and these materials do not give x-ray diffraction patterns. However, a severely stretched piece of rubber does produce an x-ray diffraction pattern. Give a plausible explanation for this phenomenon.

16. It often happens that freshly deposited metals or thin films do not produce x-ray diffraction patterns, but if the metal or thin film is heated for a period of time and then cooled (a process called annealing), they then give sharp diffraction patterns. Explain.

17. Is the following statement true or false? The density of two metals will always be directly proportional to their atomic masses and inversely proportional to the unit cell volume.

18. Suppose a particular metal crystallizes in three different polymorphs, body-centered cubic, hcp, and simple cubic.
    a. Which polymorph would be the densest?
    b. Which would be the least dense?

19. Make a heating graph (similar to Figure 9.27) for 1 g of benzene. Start at 0°C and heat the sample to 100°C at a rate of 10 J/sec. In addition to the data in Table 9.1, you need to know: the heat capacity of solid benzene = 6.3 J/°C-g; the heat capacity of liquid benzene = 1.75 J/°C-g; the heat capacity of benzene vapor = 1.25 J/°C-g.

20. Draw a phase diagram for I_2. Show temperatures ranging from 80°C to 200°C and pressures up to 800 torr. Iodine melts at 114°C, and its normal boiling point is 184°C. Use the following vapor pressures.

Temperature (°C)	Vapor pressure (torr)	Temperature (°C)	Vapor pressure (torr)
Solid I_2		Liquid I_2	
80	15	120	111
90	27	130	157
100	46	140	217
114	90	150	294
		160	394
		170	521
		180	679

EXERCISES

21. Draw an appropriate unit cell for each of the following two-dimensional lattices.

    a. # # # # # #
        # # # # # #
        # # # # # #
        # # # # # #

    b. # # # # # #
         # # # # # #
        # # # # # #
         # # # # # #

    c. 8 8 8 8 8
         ∞ ∞ ∞ ∞
        8 8 8 8 8
         ∞ ∞ ∞ ∞
        8 8 8 8 8
         ∞ ∞ ∞ ∞
        8 8 8 8 8

22. Draw a unit cell for each of the following lattices.

    a. (lattice pattern of 8s)

    b. (lattice pattern of 8s and ∞s)

    c. (lattice pattern of 8s)

23. A common error is to describe the CsCl lattice as body-centered cubic. What is wrong with this statement? What is the Bravais lattice of CsCl?

24. What is wrong with the following statement? The lattice points of CaF_2 consist of the eight corners and the six face-centered positions (the site of the calcium ions) as well as the eight tetrahedral holes (the location of the fluoride ions).

25. Describe the changes in temperature, pressure, and phase (if any) in each of the steps shown in the phase diagram in Figure 9.40.
    a. A to B   b. B to C   c. C to D   d. D to E

26. Describe the changes in temperature and pressure and any phase change occurring in each step shown in the phase diagram in Figure 9.41.

Figure 9.40
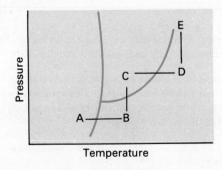

Figure 9.41
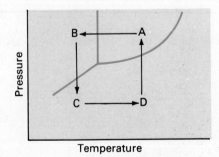

27. The normal boiling point of ethylene glycol ($C_2H_6O_2$) is 198°C. What is the free energy difference between liquid ethylene glycol and ethylene glycol vapor at 1 atm and 198°C?

28. The equilibrium vapor pressure of ethanol is 78.8 torr at 30°C, and its heat of vaporization is 33.2 kJ/mol. What is the difference in free energy between ethanol vapor at 78.8 torr and 30°C and liquid ethanol at this temperature?

29. The melting point of sodium metal is 97.5°C. Which has the higher Gibbs free energy at 100°C, solid or liquid sodium?

30. Which has the higher Gibbs free energy at 0°C, solid or liquid benzene? Refer to Table 9.1.

31. A spot in an x-ray diffraction pattern occurs at a reflection angle of 15.9°. The wavelength of the x ray is 0.1542 nm.
    a. Assuming that $n = 1$, calculate the distance between the planes responsible for this reflection.
    b. What other reflection angle would you expect from these planes?

32. Some observed θ values for platinum using molybdenum x radiation ($\lambda = 0.07107$ nm) are 9.00°, 10.41°, and 14.81°. All of these are first-order reflections. Calculate the interplanar distance for each reflection.

33. The distance between parallel planes of atoms in calcium is 2.283 Å. Calculate the first-order Bragg reflection angles using copper x radiation (0.1542 nm).

34. A set of parallel planes in calcium are 1.971 Å apart. What will the first-order reflection angles be if these planes reflect 0.07107 nm x rays?

35. Aluminum forms a face-centered cubic lattice whose unit-cell edge length is 4.04 Å. Calculate the density of aluminum.

36. Vanadium metal crystallizes in a body-centered cubic lattice, and the length of an edge of a unit cell is 3.028 Å. What is the density of vanadium?

37. Palladium forms a face-centered cubic lattice, and its atomic radius is 1.37 Å. What is the theoretical density of Pd?

38. Platinum has an atomic radius of 1.39 Å. What density should the ccp form of platinum have?

39. If the radius of a rubidium atom is 2.48 Å and the lattice type is body-centered cubic, what is the theoretical density?

40. One polymorph of polonium metal has the simple cubic lattice. The atomic radius of polonium is 1.67 Å. Calculate the density of this polymorph of polonium.

41. The measured density of crystalline copper (F-cubic lattice) is 8.96 g/cm³. What is the unit-cell edge length?

42. The density of crystalline barium is 3.5 g/cm³, and the crystal lattice is body-centered cubic. Calculate the length of an edge of the unit cell.

43. How much heat is required to melt 25.0 g CCl_4 at −23°C? Refer to Table 9.1.

44. Use data in Table 9.1 to predict the amount of heat released when 30.0 g of $SiCl_4$ vapor condense to the liquid at 57°C.

45. a. Use data in Table 9.1 to find the entropy of melting of CCl_4.
    b. The entropy of liquid CCl_4 at −23.0°C is 214.4 J/K-mol. What is the entropy of solid CCl_4 at −23.0°C?

46. What is the entropy of melting of sodium? Refer to Table 9.1.

47. What is the Gibbs free energy for melting of CCl_4 at 0°C? Refer to Table 9.1 for the relevant data.

48. Using data in Table 9.1, compute the Gibbs' free energy for melting of benzene at 25°C.

49. Refer to Figure 9.37. What phase changes, if any, occur when:
    a. CO_2 is cooled from −50°C to −70°C while the pressure is held constant at 4 atm?
    b. The pressure on CO_2 at −70°C is lowered from 4 atm to 0.20 atm?

50. Describe the phase changes, if any, that occur in each of the following steps. Refer to the phase diagram in Figure 9.37.
    a. A sample of CO_2 is warmed from −70°C to room temperature at a constant pressure of 8 atm.
    b. A sample of CO_2 is compressed isothermally at −40°C from 1.0 atm to 12 atm.

51. Why is a face-centered tetragonal lattice *not* one of the unique Bravais lattices? [*Hint:* Draw two adjacent F-centered tetragonal cells and look for a simpler cell.]

52. Explain why the vapor pressure of a substance with a large enthalpy of vaporization increases more for a given temperature increase than a substance with a small enthalpy of vaporization. Base your argument on the behavior of molecules; do not paraphrase the Clausius-Clapeyron equation. [*Hint:* See Figure 9.31.]

53. A 0.100 g drop of water is placed in a 3.00-L container maintained at 40°C.
    a. Will the drop evaporate completely?
    b. What will the water vapor pressure be after evaporation has ceased?

54. A 2.0-L sample of water vapor at 60°C and 100 torr is compressed isothermally to 0.50 L.
    a. What is the pressure of the system at 0.50 L?
    b. What phases are present in the system? [*Hint:* See Appendix G.]

EXERCISES

55. The following table shows vapor pressure data for $CHCl_3$ at various temperatures.

T(°C)	P(torr)	T(°C)	P(torr)	T(°C)	P(torr)
−10	34.8	25	199.1	45	439.0
0	61.0	30	246.0	50	526.0
10	100.5	35	301.3	55	625.2
20	159.6	40	366.4	60	739.6

    a. Prepare a graph of log P versus 1/T for $CHCl_3$.
    b. Calculate the heat of vaporization of $CHCl_3$.
    c. Calculate the boiling point of $CHCl_3$ at 2.00 atm.

56. Use the Clausius-Clapeyron equation to predict the boiling point of water at 15.0 atm.

57. Use the Clausius-Clapeyron equation to compute the boiling point of water at 15.0 torr.

58. When p-azoxyanisole is heated to 84°C, it melts to form a liquid crystal. The liquid crystal phase is stable up to 150°C and then changes to a normal liquid.
    a. On the basis of the structures of liquid crystals, would you expect the phase change from the liquid crystal to the normal liquid to have a positive or negative entropy?
    b. A positive or a negative enthalpy?
    c. Are your answers to parts a and b consistent with the fact that the liquid crystal is more stable than the normal liquid at lower temperatures, but the reverse is true at higher temperatures?

59. Why is the boiling point of a substance with a large heat of evaporation less sensitive to pressure than the boiling point of a substance with a small heat of evaporation? [*Hint:* Consider the Clausius-Clapeyron equation.]

60. Trouton's rule says that the ratio of the heat of vaporization to the boiling point (in K) of a liquid is approximately 100 J/K-mol.
    a. How well does Trouton's rule work for the liquids in Table 9.1?
    b. Paraphrase Trouton's rule with a statement about entropies of vaporization at 1 atm.

61. The carbon atoms at the interior positions of the diamond unit cell are not at lattice points. How do the environments of these carbon atoms differ from those of the carbon atoms at the lattice points?

62. A cube has three fourfold rotation axes, one running through each pair of opposite faces.
    a. How many fourfold rotation axes are there in the tetragonal unit cell?
    b. In the orthorhombic unit cell?

63. What sorts of rotation axes does Figure 9.1, part (a), possess?

64. The entropy of an ideal gas at any pressure is given by the equation

$$S = S° - R(\ln P)$$

where $S°$ is the gas entropy at 1 atm and P is the pressure in atmospheres. Use this equation and the definition of free energy of vaporization to derive the Clausius-Clapeyron equation. Assume that pressure has no effect on the entropy of the liquid or on the heat of vaporization.

65. Although a benzene molecule is about as large and as polar as a toluene molecule (benzene has no dipole moment; toluene is very slightly polar), the melting point of benzene is about 100° higher than that of toluene. A student tries to explain this large difference in melting points by the following plausible argument:

    Since the forces between benzene molecules should be about as strong as those between toluene molecules, the difference in the melting points must reflect a difference in the entropies of fusion of the two substances, not a difference in the heats of fusion. Since benzene molecules are more symmetric than toluene molecules (see the structures below), liquid benzene is less disordered than liquid toluene—if we could look at the molecules moving around in liquid benzene, we would see fewer different perspective views than for toluene. Since the entropy of liquid benzene is lower than that of liquid toluene, the entropy of fusion of benzene is lower and its melting point higher.

Find thermodynamic data in the library to either support or disprove this argument.

Benzene

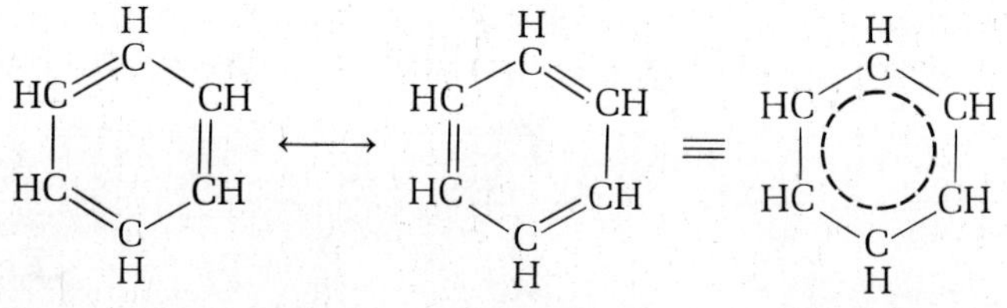

Toluene

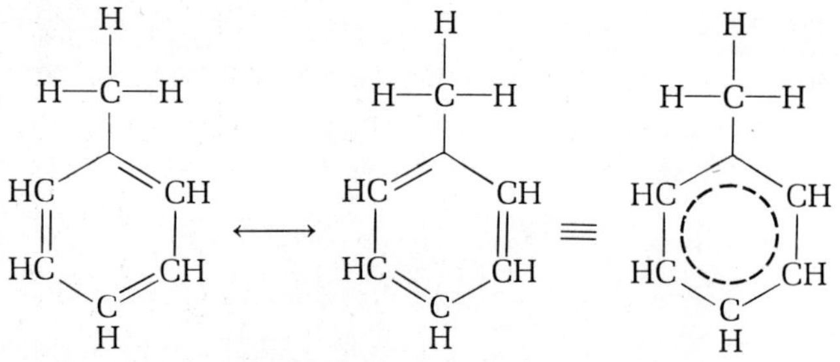

66. The heat of vaporization of $SiCl_4$ is 25.5 kJ/mol, and its vapor pressure at 57°C is 760 torr. What is the vapor pressure of $SiCl_4$ at 0°C?

67. What is the vapor pressure of CCl_4 at 100°C? The heat of vaporization of CCl_4 is 33.0 kJ/mol, and its vapor pressure is 760 torr at 77°C.

68. The vapor pressure of acetone at −10°C is 38.7 torr; at 25°C it is 229.2 torr.
   a. Calculate the heat of vaporization of acetone.
   b. Calculate the normal boiling point of acetone.

69. Liquid I_2 has a vapor pressure of 111 torr at 120°C and 294 torr at 150°C. What is the heat of vaporization of liquid I_2?

chapter 10

Physical Properties of Solutions

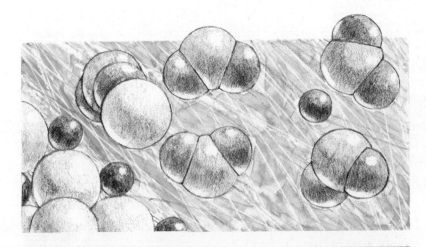

Ionic solutions result when ionic substances such as Na⁺Cl⁻ dissolve and are solvated by H₂O molecules.

10.1 Definitions

A **solution** is a homogeneous mixture. Although any mixture of gases is a solution (air is the most common example), solid or liquid mixtures are not necessarily homogeneous. Ethyl alcohol and water mix to form a solution, water and butterfats do not. Even though a dispersion of butterfats in water (as in homogenized milk) may appear homogeneous to the naked eye, magnification reveals its heterogeneity by showing the small droplets of butterfat. Magnification also shows that a mixture of finely divided powdered solids consists of two types of crystals. Some alloys are solid solutions—magnification reveals only one phase.

The constituents of a solution are randomly dispersed. Molecules in a gas mixture move randomly through the container, making it impossible to predict what sort of molecule will be in a particular place at a particular time. Similar random motion takes place in liquid solutions. In a solid solution every molecule occupies a fixed lattice position, but there is no way to be sure which type of molecule occupies a particular position (Figure 10.1).

The liquid component of a liquid solution is called the **solvent**. If there are several liquid constituents in a solution, the solvent is the one present in the largest quantity. All the other substances in the solution are **solutes**. As commonly used, the term solvent denotes the active agent that "dissolves" solutes. However, no single component of a solution can be given the sole credit for forming the solution. Whether or not a solution forms depends on the mutual interactions of the molecules of all its components.

Concentration

Since a solution is homogeneous, the composition of any portion of it is identical to the composition of any other portion. The **concentrations** of the components describe the composition of the solution. Concentrations either are ratios of the quantities of the components or are ratios of the amount of one component to the total amount of the sample. Being ratios, concentrations are independent of the size of the sample; the concentrations in a microliter of a solution are the same as in a liter of that solution. Chemists use several concentration systems, each with its particular applications.

Figure 10.1 Molecular arrangements in solutions.

Gas solution

Liquid solution

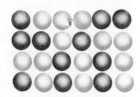
Solid solution

Percent by weight. Labels on commercially available reagents list concentrations in **percent by weight**.

$$\text{Percent by weight} = \frac{\text{Mass of component}}{\text{Mass of solution}} \times 100$$

For example, a grade of aqueous hydrogen peroxide is 30% by weight; that is, every 100 g of this solution contain 30 g of hydrogen peroxide. While useful for commercial transactions, weight percentage does not tell us the relative number of solute and solvent molecules. The following concentrations, based on mole quantities, are much more useful in the laboratory.

Mole fraction. The **mole fraction** of a component in a solution is the ratio of the number of moles of that particular component to the total number of moles in the solution.

$$\text{Mole fraction of A} = X_A = \frac{\text{Moles of component A}}{\text{Total moles in the sample}}$$

For a solution with only two components (A and B), this unitless quantity can be calculated from the equation

$$X_A = \frac{n_A}{n_A + n_B} \tag{10.1}$$

The sum of the mole fractions of all the components of a solution equals one. For a two-component solution,

$$X_B = 1 - X_A \tag{10.2}$$

Example 10.1

Find the mole fractions of H_2O_2 and H_2O in 30% aqueous hydrogen peroxide.

Solution

Each 100 g of the solution contains 30 g H_2O_2 (molecular mass = 34 amu) and 70 g H_2O (molecular mass = 18 amu). From this information, we can compute the number of moles of each component of the solution.

$$\text{Moles } H_2O_2 = \frac{30 \text{ g}}{34 \text{ g/mol}} = 0.88 \text{ mol } H_2O_2$$

$$\text{Moles } H_2O = \frac{70 \text{ g}}{18 \text{ g/mol}} = 3.9 \text{ mol } H_2O$$

According to equation (10.1),

$$X_{H_2O_2} = \frac{n_{H_2O_2}}{n_{H_2O_2} + n_{H_2O}} = \frac{0.88 \text{ mol}}{0.88 \text{ mol} + 3.9 \text{ mol}} = 0.18$$

Since the rest of the solution is water,

10.1 DEFINITIONS

$$X_{H_2O} = 1 - X_{H_2O_2} = 1 - 0.18 = 0.82$$

These mole fractions can also be stated as mole percents (mol %). The composition of the mixture is 18 mol % H_2O_2 and 82 mol % H_2O.

Molality. The **molality** of a solution is the number of moles of solute per 1 kg of solvent.

$$\text{Molality of A} = m_A = \frac{\text{Moles A}}{\text{Mass of solvent in kg}} \tag{10.3}$$

Example 10.2

What is the molality of H_2O_2 in a 30% aqueous solution?

Solution

As we saw in Example 10.1, each 100-g sample of the solution contains 30 g H_2O_2 and 70 g water. The 30 g H_2O_2 represents 0.88 mol of H_2O_2. We divide this figure by the mass of water, expressed in kg.

$$\text{Mass solvent} = \frac{70 \text{ g}}{1000 \text{ g/kg}} = 0.070 \text{ kg solvent}$$

$$m_{H_2O_2} = \frac{\text{Moles solute}}{\text{Mass solvent in kg}} = \frac{0.88 \text{ mol}}{0.070 \text{ kg}} = 12.6 \text{ mol/kg, or } 12.6 \text{ } m$$

Molarity. The concentration expressed as the number of moles of solute in a liter of solution is the **molarity**.

$$\text{Molarity of A} = M_A = \frac{\text{Moles A}}{\text{Liters of solution}} \tag{10.4}$$

To prepare a liter of 0.50 M aqueous solution of glucose ($C_6H_{12}O_6$), weigh out 0.50 mol of glucose (90 g) and add water to it until the volume of the well-mixed solution is 1.00 liter. (For 250 mL of this solution, add water to 0.125 mol of glucose until the volume of the solution is 250 mL.) Because the volume of a solution often differs slightly from the sum of the individual volumes of its components, it is difficult to predict how much solvent is needed for a particular molarity. Instead we prepare the solution in a **volumetric flask** (Figure 10.2, p. 260) by adding solvent to the required mass of solute in the flask until the level of the solution reaches the calibrated mark.

Example 10.3

How would you prepare 50.0 mL of 0.035 M naphthalene dissolved in benzene? The formula of naphthalene is $C_{10}H_8$.

Solution

The moles of naphthalene needed equals the molarity of the solution multiplied by its volume, expressed in liters.

$$V = \frac{50.0 \text{ mL}}{1000 \frac{\text{mL}}{\text{L}}} = 0.0500 \text{ L}$$

$$\text{Moles naphthalene} = MV = \left(0.035 \frac{\text{mol}}{\text{L}}\right)(0.0500 \text{ L}) = 1.75 \times 10^{-3} \text{ mol}$$

The molecular mass of naphthalene is

Figure 10.2
Preparation of 0.50 M glucose solution. Add water to the flask containing 90 g of glucose until the solution level reaches the calibration mark.

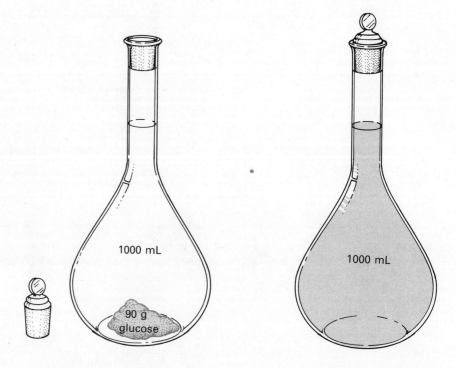

Molecular mass = 10(12.0 amu) + 8(1.0 amu) = 128 amu, or 128 g/mol

Mass naphthalene required = (Molecular mass)(Moles required)

$$= \left(128 \frac{\text{g}}{\text{mol}}\right)(1.75 \times 10^{-3} \text{ mol}) = 0.224 \text{ g}$$

We prepare the solution by weighing 0.224 g of naphthalene into a 50.0 mL volumetric flask and then adding benzene to the calibrated mark.

Formality. *Formality* (a unit not used in this book) is almost the same as molarity but describes the solutes that are mixed with the solvent rather than the solutes present after all the chemical reactions involved in the solution process have reached equilibrium.

$$\text{Formality of A} = F_A = \frac{\text{Moles A added to the solution}}{\text{Liters of solution}}$$

For example, we can prepare 1.00 F HCl by mixing water with 1 mol HCl until the volume is 1 L. All of the HCl dissociates into ions on contact with the water,

$$\text{HCl(g)} \longrightarrow \text{H}^+(\text{aq}) + \text{Cl}^-(\text{aq})$$

so the solution is in fact 1.00 M in H$^+$, 1.00 M in Cl$^-$, and zero molar in HCl. Formality tells us what goes into the solution but makes no statement about what subsequently happens. However most chemists, although aware that HCl dissociates, will label the solution 1.00 M HCl with the understanding that this represents the initial rather than the final concentration.

10.2 Solubility

Saturated Solutions

Some pairs of substances are completely miscible with each other. In other words, we can form solutions of these two compounds with any desired concentration. Other substances form solution only within a limited concentration range. If the ratios of

Figure 10.3
Saturated solution of iodine in ethanol.

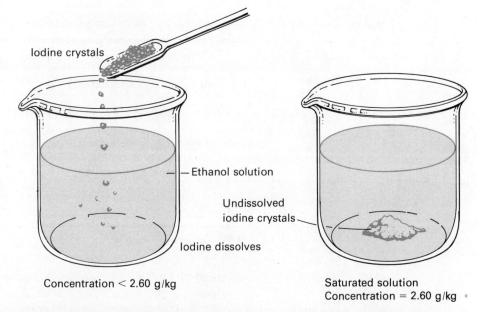

the components exceed the limits of miscibility, a two-phase system forms.

If we add iodine to ethanol at 15°C, the amount we add initially dissolves, and a brown solution forms. The concentration of I_2 in solution increases as more iodine is added until 2.60 g/kg ethanol have dissolved. If we add any more iodine, it sinks to the bottom of the beaker and remains as undissolved solid, and the I_2 concentration remains constant at 2.60 g/kg (Figure 10.3). The solution is now saturated. A **saturated** solution is one which is in equilibrium with the pure solute. (The solute phase does not have to be present; filtering the mixture to remove the solid iodine does not change the solution.) The **solubility** of a substance in a given solvent (Table 10.1) is the concentration of its saturated solution.

The Thermodynamics of Solubility

A solution, being a random dispersion, almost always has a higher entropy than its pure components. In a case in which this entropy increase is the dominant thermodynamic factor involved in the solubility equilibrium, the components will mix to form a solution. That is why benzene and toluene are completely miscible—mixing these two substances does not change their enthalpies significantly, so the solution forms because of the entropy increase.

The **heat of solution**—the enthalpy change occurring when a solution forms—also influences solubility (Table 10.2). If it is large enough, a positive heat of solution

Table 10.1 Solubilities[a]

Solute	Solvent			
	H_2O	C_2H_5OH	Ethyl ether	Other
S	Insoluble	0.78 at 19°	9.72 at 25°	531 in CS_2 at 25°
I_2	0.30 at 25°	2.60 at 15°	2.61 at 17°	
NaCl	357 at 0°	0.65 at 25°	Insoluble	
AgCl	8.9×10^{-4} at 10°	Insoluble	Insoluble	
BaO	43 at 25°	208 at 9°	Insoluble	
$BaSO_4$	2.46×10^{-3} at 25°	Insoluble	Insoluble	
CH_3OH (methanol)	Completely miscible	Completely miscible	Completely miscible	
C_6H_6 (benzene)	1.81 at 20°	Completely miscible	Completely miscible	
CH_4N_2O (urea)	1094 at 20°	53.2 at 20°	Insoluble	

[a] All solubilities are given in grams of solute per kilogram solvent at the indicated temperature in °C.

Table 10.2 Thermodynamic Factors and Solubility

Attractive forces	$\Delta H°$	$\Delta S°$	$\Delta G°$	Solubility effect	Example
Pure components > Solute–solvent	+	+	+ or −	Insoluble / Soluble	PbI_2 in H_2O / NaCl in H_2O
	+	−	+	Insoluble	$PbSO_4$ in H_2O
Pure component = Solute–solvent	0	+	−	Soluble	Benzene and toluene[a]
	0	−	+	Insoluble	$BaSO_4$ in H_2O[a]
Pure components < Solute–solvent	−	+	−	Soluble	H_2SO_4 in H_2O
	−	−	+ or −	Insoluble / Soluble	Benzene in H_2O / HCl in H_2O

[a] $\Delta H°$ is approximately equal to 0.

can counteract a favorable entropy change and limit solubility. For example, silver chloride has a very low solubility in water because its large positive heat of solution (+65.5 kJ/mol) overwhelms the entropy increase. A substance can have a moderately endothermic heat of solution and still be quite soluble; the heat of solution of sodium chloride is slightly positive (+3.9 kJ/mol), but it is very soluble in water because of the entropy effect.

When a solution forms, solvent and solute molecules break away from the intermolecular attractions in their pure phases and come into contact with other types of molecules. The heat of solution depends on the relative strengths of the intermolecular attractions in the solution compared with the attractive forces in the pure components. If solvent and solute molecules attract each other more effectively than they attract molecules of their own type, the heat of solution is negative. Positive heats of solution occur when solvent–solvent and solute–solute attractions dominate over the attractions between solvent and solute molecules.*

The heat of solution will be nearly zero if the attraction of solute molecules for solvent molecules nearly balances the solvent–solvent and solute–solute forces. For example, benzene and toluene, both having low polarities, are held by weak intermolecular forces. No strong attractions are broken or formed when benzene and toluene mix, and the heat of solution is almost zero.

Ionic and polar substances (such as CH_3OH) dissolve in water because the strong attraction of polar water molecules for ions and polar solutes compensates for the energy needed to break up the ionic crystal lattice or to separate polar molecules from each other. Nonpolar solvents, unable to interact strongly with polar solutes, cannot dissolve them. We can summarize the importance of polarity in controlling solubility with the rule of thumb, *like dissolves like*. Polar solvents dissolve polar solutes, and nonpolar substances are soluble in nonpolar solvents, but a polar substance does not mix well with a nonpolar substance.

Temperature influences solubility, but the effect varies from substance to substance. Although the solubilities of most substances increase as temperature rises, many solutes become less soluble at higher temperatures. The temperature effect depends on the heat of solution of the solute in a way that is consistent with Le Chatelier's principle. The solubilities of substances with positive heats of solution increase with temperature; the system responds to the stress of increased temperature by shifting its equilibrium in the heat absorbing direction (solubility). If dissolving a

---

*Rearrangements in the orientations of solvent molecules can increase the attractions between solvent molecules, producing negative heats of solution even though solvent–solute attractions are weak. This occurs in solutions of benzene in water—the introduction of benzene molecules into a sample of water changes the orientations of the water molecules in a way that increases both the attractions between the water molecules and the order of the liquid. The resulting decrease in enthalpy is offset by a decrease in entropy, and benzene has a very low solubility in water.

particular solute is an exothermic process, a temperature increase lowers its solubility because the system absorbs heat as the solute precipitates from the solution. The sensitivity of solubility to temperature is also related to the heat of solution; the larger the heat of solution, the more temperature-sensitive the solubility.

Supersaturated Solutions

It is possible to prepare solutions, called **supersaturated** solutions, whose solute concentrations exceed their solubility. For example, the solubility of sodium acetate in water decreases with cooling (139 g/L at 50°C, 76 g/L at 0°C). As a warm solution of 100 g/L of sodium acetate cools to 0°C, some solute should precipitate as its solubility falls below its concentration. No precipitation occurs if the solution is kept free from dust and not shaken; the solution is supersaturated. Addition of a seed crystal of sodium acetate breaks down the supersaturated state, and rapid precipitation occurs.

Like a supercooled liquid, a supersaturated solution is **metastable**; a solute is less stable in the supersaturated solution than in its crystalline phase. In spite of the higher free energy of the solution, precipitation does not occur until a few molecules (or ions) form the beginning of a crystal lattice, and the solution remains supersaturated during the period that elapses before this occurs. Because **nucleation**—the start of crystal lattice growth—is a random event, it is impossible to predict how long a supersaturated solution will persist.

10.3 Phase Changes and Solutions

Vapor Pressure

The equilibrium vapor pressure of a substance in solution depends on its concentration and is lower than the vapor pressure of the pure substance. **Raoult's law** expresses an ideal relationship between vapor pressure and concentration,

$$P_A = X_A P_A^\circ \tag{10.5}$$

where P_A is the vapor pressure of component A present at a mole fraction X_A in solution, and P_A° is the vapor pressure of pure A at the same temperature. A graph of P_A versus X_A for an **ideal solution** (one obeying Raoult's law) is a straight line passing through zero pressure at zero concentration and through P_A° at $X_A = 1.00$. Figure 10.4 (p. 264) compares the vapor pressures of solutions of carbon tetrachloride and benzene (C_6H_6) at 50°C with the predictions of Raoult's law.

Example 10.4

The vapor pressure of pure ethylene bromide ($C_2H_4Br_2$) at 85°C is 173 torr, and propylene bromide ($C_3H_6Br_2$) has a vapor pressure of 127 torr at this temperature. Predict the vapor pressure of ethylene bromide in a 60.0% (by weight) solution of ethylene bromide in propylene bromide at 85°C. What is the vapor pressure of the propylene bromide above this solution and the total vapor pressure of the solution?

Solution

To apply Raoult's law, we must first find the mole fraction of ethylene bromide in the solution. We use the composition of the solution and the molecular masses of the two components to perform this calculation. Let us consider 100 g of solution. Since the solution is 60.0% ethylene bromide, it contains 60.0 g $C_2H_4Br_2$ and 40.0 g $C_3H_6Br_2$.

Molecular mass $C_2H_4Br_2 = 188$ amu

$$\text{Moles } C_2H_4Br_2 = \frac{60.0 \text{ g}}{188 \text{ g/mol}} = 0.319 \text{ mol}$$

Molecular mass $C_3H_6Br_2 = 202$ amu

$$\text{Moles } C_3H_6Br_2 = \frac{40.0 \text{ g}}{202 \text{ g/mol}} = 0.198 \text{ mol}$$

Figure 10.4
Raoult's law.

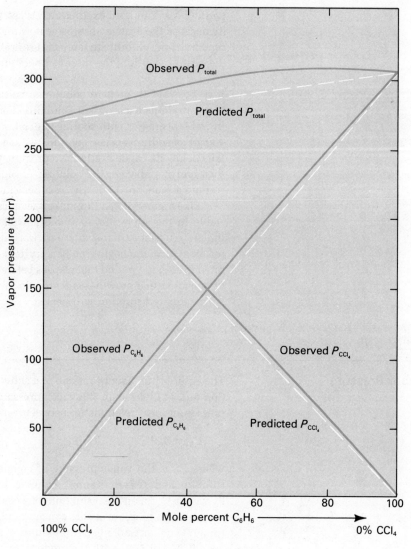

$$X_{C_2H_4Br_2} = \frac{0.319 \text{ mol}}{0.319 \text{ mol} + 0.198 \text{ mol}} = 0.617$$

Applying Raoult's law, we find

$$P_{C_2H_4Br_2} = (0.617)(173 \text{ torr}) = 107 \text{ torr}$$

The mole fraction of propylene bromide is

$$X_{C_3H_6Br_2} = 1 - X_{C_2H_4Br_2} = 1 - 0.617 = 0.383$$

$$P_{C_3H_6Br_2} = X_{C_3H_6Br_2} P°_{C_3H_6Br_2} = (0.383)(127 \text{ torr}) = 48.6 \text{ torr}$$

According to Dalton's law of partial pressures,

$$P_{total} = P_{C_2H_4Br_2} + P_{C_3H_6Br_2} = 107 \text{ torr} + 48.6 \text{ torr} = 156 \text{ torr}$$

Distillation

Raoult's law provides the theory behind **distillation**—the separation of a mixture by selective evaporation. The solution in Example 10.4 can be separated by distillation, but the process is not as simple as it might seem at first glance. Even though $C_2H_4Br_2$ is the more volatile of the two components, the solution will not boil at the boiling point of this substance. As shown in Example 10.4, the vapor pressure of the solution is less than the vapor pressure of pure $C_2H_4Br_2$ so the solution has a higher boiling point. When the solution does boil, the vapor formed still contains considerable amounts of $C_3H_6Br_2$.

Distillation works because the vapor above the solution has a higher mole fraction of the more volatile component than the liquid. Because the partial pressure of a component is proportional to the number of moles of this substance in the gas phase (ideal gas law), we can rewrite the mole fraction equation for gases as

$$X_A = \frac{P_A}{P_A + P_B} \tag{10.6}$$

For the vapor above the solution in Example 10.4

$$X_{C_2H_4Br_2} = \frac{107 \text{ torr}}{107 \text{ torr} + 48.6 \text{ torr}} = 0.688$$

compared to the 0.617 value in the liquid. By making use of this modest enrichment, we can obtain pure $C_2H_4Br_2$ after repeated evaporations and condensations.

Figure 10.5 is a liquid-vapor phase diagram for solutions of benzene (boiling point 79.7°C) and toluene (boiling point 110.7°C) at a constant pressure of 1 atm. The lower curve shows the variation in the boiling point as the composition of the solution changes. The upper curve gives the composition of the vapor boiling away at a given temperature. To purify a 50.0 mol % solution of benzene and toluene, we must recondense the distilling vapor and then redistill it. Several of these repetitions of evaporation and condensation will eventually yield pure samples of benzene and toluene. We achieve this repeated distillation by using a fractional distillation apparatus (Figure 10.6) with a long, vertical column containing an inert packing material (such as glass beads) to provide a large surface area. The vapor condenses when it comes into contact with the cold surface of the column, the recondensed vapor drips down the column until it reaches a region warmed by hot vapors rising from the boiling flask, and redistills. Each successive redistillation enriches the vapor in benzene, and the first vapor to reach the top of the column is pure benzene.

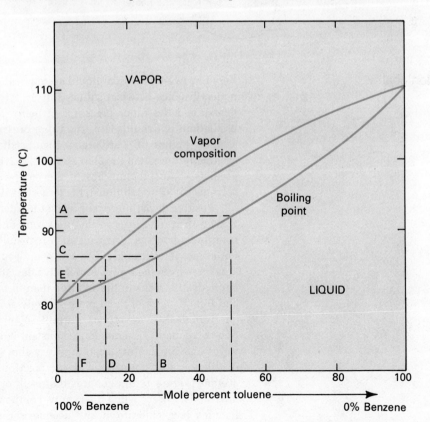

Figure 10.5

Liquid-vapor phase diagram for benzene-toluene mixtures at 1 atm. The following data are obtained from the intersection of the labeled lines with the curves:

A: Boiling point of a 50 mol % solution of benzene and toluene is 92.4°C.
B: Composition of first distillate from 50 mol % benzene and toluene is 71.5 mol % benzene.
C: Boiling point of first distillate is 86.2°C.
D: Composition of second distillate is 86 mol % benzene.
E: Boiling point of second distillate is 83.0°C.
F: Composition of third distillate is 94 mol % benzene.

Figure 10.6
A fractional distillation apparatus.

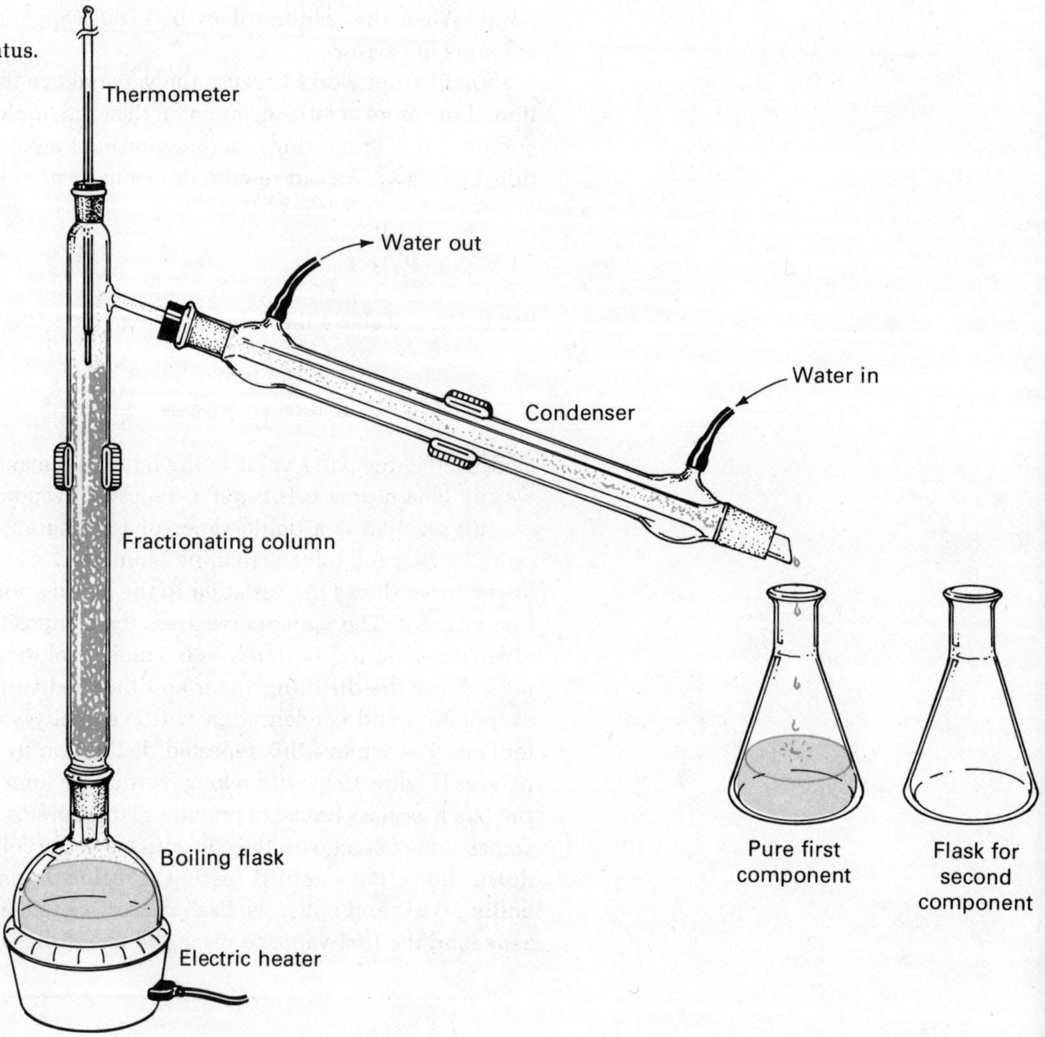

Nonideality

Very few two-component liquid systems fulfill the conditions for an ideal solution (a perfect balance between solute–solvent forces and those between the pure components), and the vapor pressures of many solutions deviate considerably from the predictions of Raoult's law. The vapor pressures of solutions of carbon disulfide and acetone (Figure 10.7) are higher than predicted, while solutions of chloroform and acetone (Figure 10.8, p. 268) show negative deviations from ideality. Raoult's law predicts vapor pressure best if the mole fraction is near one (in other words, for solvents of dilute solutions), the largest deviations occurring at low concentrations.

Positive deviations occur when the intermolecular attractions in the solution are weaker than the forces in the pure components—in other words, when the heat of solution is positive. Bringing an A molecule into contact with B molecules that do not attract it as strongly as other A molecules makes it easier for this molecule to break free from the liquid. A negative deviation from Raoult's law is associated with stronger attractions in the solution than in the pure components. Contact between A and B binds each of them more strongly and inhibits evaporation.

Nonideality sometimes prevents complete purification by distillation. Many two-component systems form constant-boiling mixtures called **azeotropes**. The vapor above a boiling azeotrope has the same mole fractions as the liquid, and distillation does not change its composition. The azeotrope of carbon disulfide and acetone (Figure 10.9), which is the result of large positive deviations from Raoult's law, boils lower than either pure component. Due to negative deviations from Raoult's law, chloroform and acetone form an azeotrope whose boiling point is higher than either component (Figure 10.10, p. 269).

Figure 10.7
Positive deviations from Raoult's law for solutions of carbon disulfide (CS$_2$) and acetone (C$_3$H$_6$O) at 35°C.

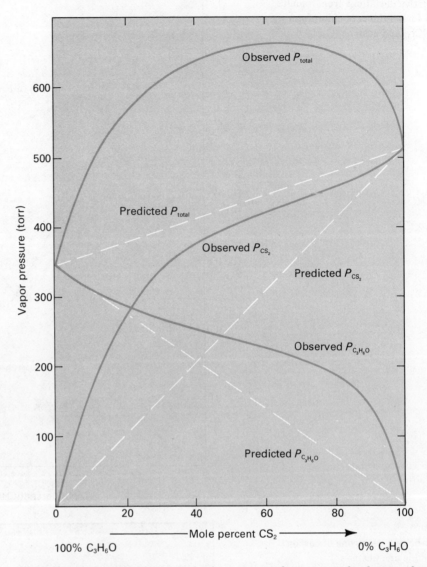

Vapor Pressures of Saturated Solutions

Consider a bell jar holding two beakers, one with a saturated solution of iodine in water, the other with pure iodine (Figure 10.11). The pure iodine saturates the bell jar with I$_2$ vapor at a partial pressure of $P°_{I_2}$. If the solution had a lower equilibrium vapor pressure of iodine, iodine vapor would dissolve in it. But this is impossible in a solution that is already saturated. Therefore, the vapor pressure of the solute of a saturated solution must equal the vapor pressure of the pure solute. This is a general condition: Two states in equilibrium have the same vapor pressure.

Colligative Properties

Vapor-pressure lowering. According to Raoult's law, the relative decrease in solvent vapor pressure depends on the total solute concentration and is independent of the identity of the solute particles.

$$\frac{P_{\text{solvent}}}{P°_{\text{solvent}}} = X_{\text{solvent}} = 1 - X_{\text{solute}} \tag{10.7}$$

By comparing the solvent vapor pressure of a solution and the vapor pressure of the pure solvent, we can estimate the solute concentration. Solvent vapor pressure and other physical properties depending on solute concentration but not solute identity are called **colligative properties**. Several techniques for measuring total solute concentration are based on colligative properties. Knowing the mass composition of a solution, we can use its colligative properties to estimate the molecular mass of the solute.

Vapor-pressure techniques are easiest to use when the solute is not volatile. In such cases, the solvent vapor pressure is the vapor pressure of the solution and can be measured with a manometer.

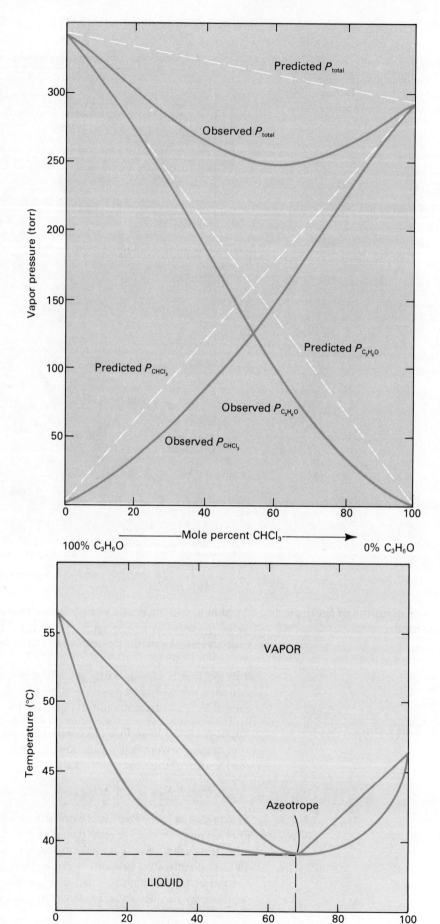

Figure 10.8
Negative deviations from Raoult's law for solutions of chloroform ($CHCl_3$) and acetone (C_3H_6O) at 35°C.

Figure 10.9
Low-boiling azeotrope of carbon disulfide and acetone.

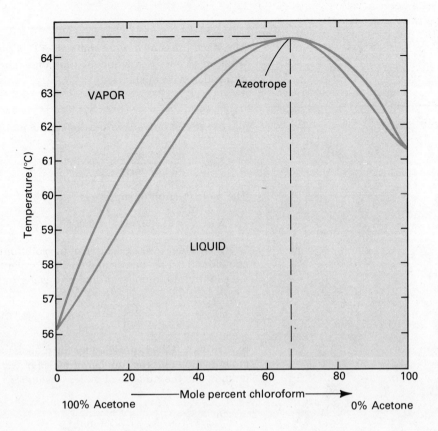

Figure 10.10
High-boiling azeotrope of chloroform and acetone.

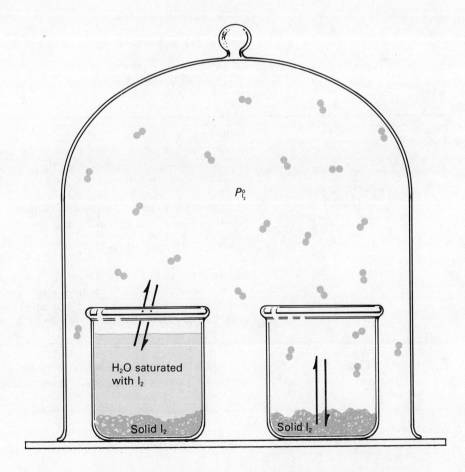

Figure 10.11
Vapor pressure of a saturated solution.

Example 10.5

The vapor pressure of pure water at 25°C is 24.47 torr. The vapor pressure of a 13.8% solution of sucrose, a nonvolatile compound, in water is 24.27 torr. What is the molecular mass of sucrose?

Solution

We can use Raoult's law to estimate the mole fraction of water.

$$X_{H_2O} = \frac{24.27 \text{ torr}}{24.47 \text{ torr}} = 0.9914$$

The mole fraction of sucrose is

$$X_{sucrose} = 1 - 0.9914 = 0.0086$$

Let us consider a 100-g sample of the solution. Since it is 13.8% sucrose, this sample contains 13.8 g of sucrose and 86.2 g of water. Knowing the molecular mass of water, we can find the number of moles of water in the sample.

$$n_{H_2O} = \frac{86.2 \text{ g}}{18.0 \text{ g/mol}} = 4.79 \text{ mol}$$

By writing out the expression for the mole fraction and substituting the values for the mole fraction and the number of moles of water, we obtain an equation whose only unknown is the number of moles of sucrose.

$$0.0086 = \frac{n_{sucrose}}{n_{sucrose} + 4.79 \text{ mol}}$$

Use simple algebra to solve this equation.

$$n_{sucrose} = 0.042 \text{ mol}$$

Since 0.042 mol of sucrose has a mass of 13.8 g,

$$\text{Molecular mass of sucrose} = \frac{13.8 \text{ g}}{0.042 \text{ mol}} = 3.3 \times 10^2 \text{ g/mol, or } 3.3 \times 10^2 \text{ amu}$$

Boiling-point elevation. The decrease in solvent vapor pressure in a solution of a nonvolatile substance raises the boiling point of the solution. (Cooks use the boiling-point-elevation effect when they add salt to boiling water. The dissolved salt raises the liquid's boiling point, and the higher-boiling liquid is a hotter cooking medium.) It is necessary to heat the solution to a higher temperature before the solvent vapor pressure reaches the external pressure (Figure 10.12). The boiling-point increase of a dilute solution is proportional to the solute molality.

$$\Delta T_b = T_b - T_b^\circ = K_b m \qquad (10.8)$$

In equation (10.8) T_b and T_b° are the boiling points of the solution and the pure solvent, respectively, m is the total solute molality, and K_b is the **molal boiling-point elevation** constant of the solvent. We can obtain values of K_b by measuring the boiling points of solutions of known molalities (Table 10.3, p. 272).

Example 10.6

A solution of 14.6 g of benzophenone in 100 g of benzene boils 1.94° higher than pure benzene. Calculate the molecular mass of benzophenone.

Solution

As usual, we can find the molecular mass once we know the number of moles in a given mass of substance. Using the boiling-point elevation, we can find the molality

Figure 10.12
Boiling-point elevation of a non-volatile solute in an aqueous solution.

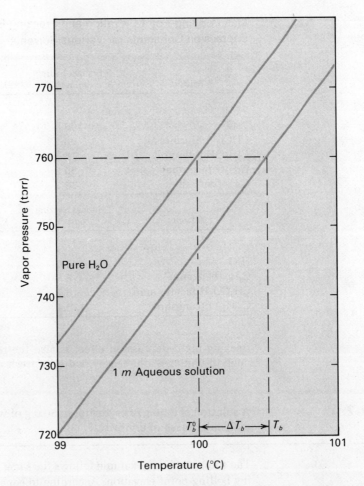

of benzophenone, that is, the number of moles of benzophenone per kg of benzene. According to Table 10.3, the value of K_b is 2.53°C/m. According to equation (10.8),

$$\text{Molality} = \frac{\Delta T_b}{K_b} = \frac{1.94 \text{ °C}}{2.53 \text{ °C/m}} = 0.767 \text{ m}$$

Since the solution contains 100 g (0.100 kg) of solvent,

$$\text{Moles of benzophenone} = \left(0.767 \frac{\text{mol}}{\text{kg solvent}}\right)(0.100 \text{ kg solvent}) = 0.0767 \text{ mol}$$

$$\text{Molecular mass of benzophenone} = \frac{14.6 \text{ g}}{0.0767 \text{ mol}} = 190 \text{ g/mol} = 190 \text{ amu}$$

Freezing-point depression. Any solute, volatile or not, decreases the freezing point of a solution. The solvent freezes when its vapor pressure above the liquid solution equals the vapor pressure of the pure solid solvent. The depression of the vapor pressure of the liquid solvent lowers the temperature of the solid–liquid equilibrium (Figure 10.13, p. 272). For a given solvent, the freezing-point depression of a dilute solution is proportional to the solute molality,

$$\Delta T_f = T_f^\circ - T_f = K_f m \tag{10.9}$$

where K_f is the **molal freezing-point-depression** constant of the solvent (Table 10.3).

There are many practical applications of freezing-point depression. For example, salt is sprinkled on icy roads to melt the ice. The salt depresses the freezing point so much that the salt-ice mixture melts considerably below 0°C. The

10 PHYSICAL PROPERTIES OF SOLUTIONS

Table 10.3 **Molal Boiling-Point-Elevation and Freezing-Point-Depression Constants for Various Solvents**

Solvent	Normal boiling point (°C)	K_b (°C/m)
Molal Boiling-Point Elevation		
H_2O	100.0	0.51
$CHCl_3$ (chloroform)	60.2	3.63
C_2H_5OH (ethanol)	78.3	1.22
C_3H_6O (acetone)	56.0	1.71
C_6H_6 (benzene)	80.2	2.53

Solvent	Normal freezing point (°C)	K_f (°C/m)
Molal Freezing-Point Depression		
H_2O	0.0	1.86
C_6H_6 (benzene)	5.5	5.12
CH_3CO_2H (acetic acid)	16.7	3.9
$C_{10}H_{16}O$ (camphor)	178.4	37.7

freezing-point-depression effect is also the reason why saltwater harbors remain open in winter while nearby bodies of fresh water freeze.

Example 10.7 A solution of 2.50 g of erythritol in 50.0 g of water freezes at $-0.773°C$. What is the molecular mass of erythritol?

Solution The solution of this example follows the same general lines as Example 10.6 involving boiling-point elevation. According to equation (10.9),

$$\text{Molality} = \frac{\Delta T_f}{K_f} = \frac{0.773°C}{1.86 \frac{°C}{m}} = 0.416 \ m$$

Figure 10.13
Freezing-point depression of a solute in an aqueous solution.

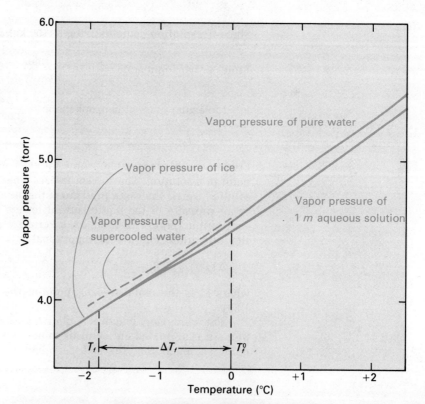

$$\text{Moles of erythritol} = \left(0.416 \; \frac{\text{mol}}{\text{kg H}_2\text{O}}\right)(0.0500 \; \text{kg H}_2\text{O}) = 0.0208 \; \text{mol}$$

$$\text{Molecular mass of erythritol} = \frac{2.509 \; \text{g}}{0.0208 \; \text{mol}} = 120 \; \text{g/mol, or 120 amu}$$

Osmotic Pressure. The most sensitive colligative property is osmotic pressure. When a membrane that is permeable to solvent but retards the passage of solute molecules is placed between a solution and its pure solvent, solvent flows through the membrane into the solution. This solvent flow dilutes the solution, reducing the free energy difference between solvent and solution. The **osmotic pressure** (π) is the pressure that must be applied to prevent solvent flow. Osmotic pressure is proportional to solute concentration,

$$\pi = RTM \tag{10.10}$$

where R is the gas constant, T is the temperature (K), and M is the solute molarity. The osmotic pressure for a 1 M solution at 0°C is 22.4 atm. The size of the osmotic pressure effect makes it very useful for measuring low concentrations, and it is particularly helpful for estimating the mass of a large molecule. Figure 10.14 (p. 274) shows an apparatus for measuring osmotic pressures.

Example 10.8

The osmotic pressure of a solution containing 0.100 g of hemoglobin in 10.0 mL of solution is 2.67 torr at 1°C. Estimate the molecular mass of hemoglobin.

Osmosis and osmotic pressure are extremely important factors in physiology and medicine. Our bodies contain a vast number of membranes, and if any one of these membranes separates two solutions of different concentrations, water will flow from the more dilute solution to the more concentrated one. This flow can create severe problems—when placed in pure water, red blood cells swell until the osmotic pressure bursts them. Edema, which is a severe swelling of tissue that is due to osmotic flow from the blood stream into the tissue, indicates a dangerous concentration difference between blood and the fluid in the tissue. This concentration imbalance could be due to kidney malfunction, producing high salt concentrations in the tissue fluid, or to low protein concentrations in the blood due to a nutritional disorder.

There are ways of reversing osmotic flow. One method, not available within our bodies, is to apply a reverse mechanical pressure, forcing solvent from the more concentrated solution to the less concentrated one. This procedure, known as reverse osmosis, has been used to purify seawater. The biggest problem in the future development of reverse osmosis is finding membranes that are strong enough to stand the intense pressures involved and yet porous enough to allow rapid liquid flow.

Instead of reversing the osmotic flow of water, our bodies use sophisticated biochemical mechanisms to pass solute across a membrane from a dilute solution to a concentrated one. The membrane, playing an active role in this process, performs *osmotic work* by transporting a particular type of ion to the concentrated solution. For example, membranes in the stomach lining carry H^+ from the blood stream into the stomach even though the H^+ concentration of the stomach fluid is about a million times greater than in blood.

Figure 10.14
Measurement of osmotic pressure. Pressure on the piston is increased until the flow meter indicates that osmotic flow has ceased. The measured pressure then equals the osmotic pressure.

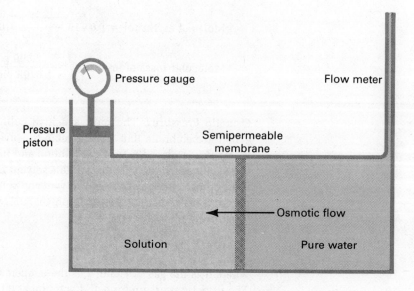

Solution

Use equation (10.10), which gives the relationship between osmotic pressure and concentration, to find the molarity of hemoglobin.

$$\pi = \frac{2.67 \text{ torr}}{760 \frac{\text{torr}}{\text{atm}}} = 3.51 \times 10^{-3} \text{ atm} \qquad T = 274 \text{ K}$$

$$M = \frac{\pi}{RT} = \frac{3.51 \times 10^{-3} \text{ atm}}{\left(0.0821 \frac{\text{L-atm}}{\text{K-mol}}\right)(274 \text{ K})} = 1.56 \times 10^{-4} \text{ M}$$

The number of moles of hemoglobin in the solution equals its molarity times the volume of the solution in liters.

$$\text{Moles of hemoglobin} = \left(1.56 \times 10^{-4} \frac{\text{mol}}{\text{L}}\right)(0.0100 \text{ L})$$

$$= 1.56 \times 10^{-6} \text{ mol}$$

$$\text{Molecular mass of hemoglobin} = \frac{0.100 \text{ g}}{1.56 \times 10^{-6} \text{ mol}}$$

$$= 6.41 \times 10^4 \text{ g/mol, or } 6.41 \times 10^4 \text{ amu}$$

Henry's Law

We have seen that Raoult's law is an ideal law that in many cases does not accurately predict the vapor pressure of a solute in a dilute solution. There is a law, called Henry's law, that is useful for dealing with the vapor pressures of these solutes. According to **Henry's law,** the vapor pressure of a solute in a dilute solution is proportional to its concentration. Since weight concentration, molality, molarity, and mole fraction are approximately proportional to each other in a dilute solution, we can express Henry's law in *any* of these concentration units. For example, the law based on molarity would be:

$$P_{\text{solute}} = kM_{\text{solute}} \tag{10.11}$$

The value of the proportionality constant, k, depends on the identities of both the solute and the solvent as well as on the temperature. It must be determined for each solute–solvent pair by measuring the solute vapor pressure above a solution of known concentration.

10.3 PHASE CHANGES AND SOLUTIONS

Henry's law also describes the effect of external pressure on the solubilities of slightly soluble gases. If the gas pressure above a solution is higher than the equilibrium vapor pressure, more gas will dissolve until the concentration of the gas in the solution equals

$$M_{solute} = \frac{P_{solute}}{k} \qquad (10.12)$$

If the external pressure is less than kM_{solute}, gas escapes from the solution until the equilibrium vapor pressure drops to the external pressure. The solubility of a slightly soluble gas is proportional to its pressure.

Example 10.9

The partial pressure of O_2 in air is 0.21 atm. The solubility of O_2 in water at 25°C in the presence of air is 3.2×10^{-4} M. What would the solubility of O_2 be in water at 25°C in the presence of pure O_2 at 1.00 atm pressure?

Solution

We find the value of k from the O_2 solubility in the presence of air.

$$k = \frac{P_{solute}}{M_{solute}} = \frac{0.21 \text{ atm}}{3.2 \times 10^{-4} \text{ M}} = 6.6 \times 10^2 \text{ atm/M}$$

The solubility of O_2 in the presence of pure O_2 is

$$M_{solute} = \frac{P_{solute}}{k} = \frac{1.00 \text{ atm}}{6.6 \times 10^2 \frac{\text{atm}}{\text{M}}} = 1.5 \times 10^{-3} \text{ M}$$

One of the most common illustrations of the effect of pressure on solubility occurs when you open a bottle of soda water (a solution of CO_2 in water). The pressure in the sealed bottle is above 1 atm. Removing the cap suddenly lowers the pressure, and CO_2 bubbles from the solution (Figure 10.15). Scuba divers are faced with a much

Figure 10.15
The effect of pressure on gas solubility.

Sealed soda bottle
Pressure > 1 atm

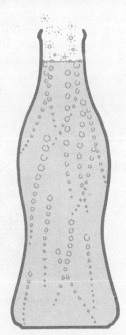

When bottle is opened,
P_{CO_2} decreases and CO_2
bubbles come out of solution

Figure 10.16
An illustration of dialysis.

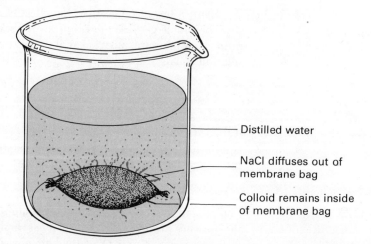

more serious example of the effect of pressure on gas solubility. They breathe compressed air while in the depths of the sea, and the increased gas pressure raises the concentration of nitrogen in their blood streams. A diver who comes to the surface too suddenly suffers an illness, commonly called the bends, from the sudden decompression. As the external pressure drops suddenly, the nitrogen bubbles from the blood stream and causes tissue damage.

10.4 Colloids

Colloidal suspensions (also called **sols**) are dispersions of relatively large particles called **colloids** in a homogeneous medium (a liquid or a gas). The colloidal particles, which can be either very large molecules (such as a protein molecule) or large aggregates of smaller molecules or ions, range in mass from several thousand to several million atomic mass units. These particles are too small to be filtered out of their medium by conventional filter paper but are large enough to give sols some special properties.

Ultrafiltration and dialysis. Although colloidal particles are not held back by ordinary filters, they can be retained by membranes called ultrafilters that do allow the passage of small molecules and ions in a process called **ultrafiltration**. In **dialysis** (Figure 10.16), a colloidal suspension is placed in a membrane bag, and the bag is

Figure 10.17
The Faraday-Tyndall effect. Light scattering occurs in the colloidal sol but not in sodium chloride solution.

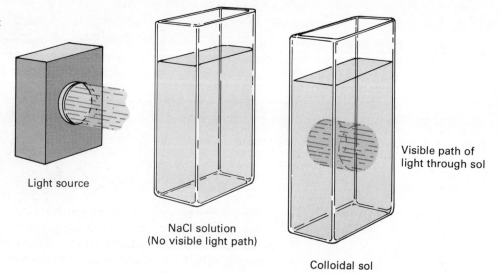

Figure 10.18
Electrophoresis apparatus.

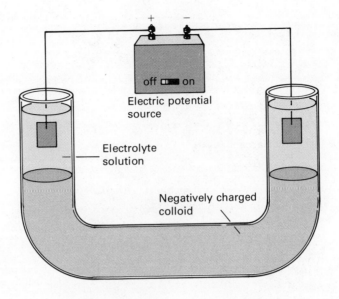

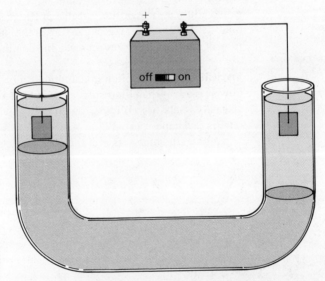

Colloid migrates toward the positive pole

submerged in pure solvent. Any small ions or molecules in the colloidal suspension diffuse out of the bag, but the membrane retains the colloids. This technique is used in the artificial kidney—dialysis removes poisonous ionic impurities from the blood stream but not the proteins essential for the patient's survival.

Light scattering. Light passes through a true solution without producing a visible path, but we can see the path of a light beam passing through a sol (the **Faraday-Tyndall effect**, Figure 10.17). This is why we can see the beam from a spotlight on a foggy night but not on a clear one. The fog, a dispersion of water droplets in air, scatters the light.

Electrophoresis. Many colloids are electrically charged and migrate in an electric field. This is the basis for **electrophoresis** (Figure 10.18), a technique for purifying colloids. The electrophoresis apparatus is a U-tube with the sol in the bottom and electrolyte solution in each arm. When a high-voltage direct current is applied across terminals in each arm, the colloid migrates toward the pole with the opposite charge.

Figure 10.19
Ultracentrifugation.

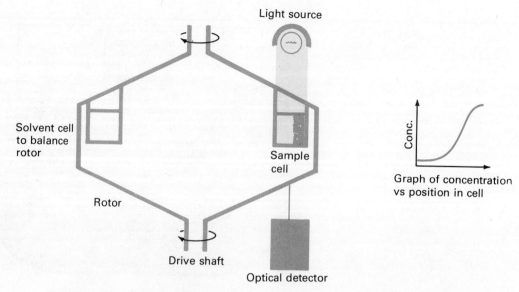

Ultracentrifugation. Rapid rotation of a sample of a sol (Figure 10.19) creates large forces (often a million times the force of gravity) that tend to drive colloids, if denser than the solvent, toward the outer edge of the ultracentrifuge cell. This technique, called **ultracentrifugation**, creates an increase in colloid concentration toward the outer part of the cell. We can estimate the particle mass of a colloid from the steepness of the concentration increase in the outer portion of the cell.

Viscosity. Sols are often considerably more viscous—in other words, they do not flow as easily—than their solvents (Figure 10.20). This resistance to flow is called the **viscosity**. Gels are colloidal suspensions whose viscosity is so great that the sol retains its shape.

Although some sols are nuisances for chemists, others are useful in the labora-

Figure 10.20
Viscosity.

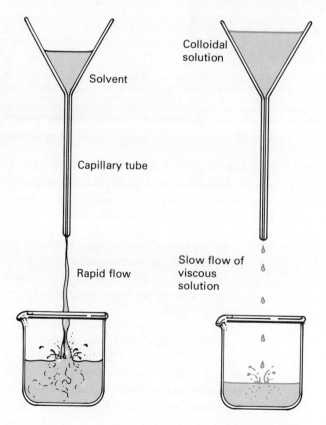

tory or essential in our daily lives. Many reactions between immiscible reactants would be inconveniently slow if we were not able to form colloidal dispersions between them. Some paints are dispersions of insoluble pigments in a homogeneous medium such as oil. Aerosol sprays consist of colloidal particles of liquids or solids dispersed in the gas phase. Sols also play an important part in biochemistry. For example, before we can digest fats, the mechanical and chemical action of our intestines must form an **emulsion** (a dispersion of a liquid in a liquid) of these insoluble fats in water.

Summary

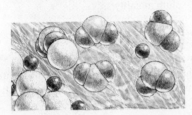

1. A solution is a homogeneous mixture. The solvent of a liquid solution is the liquid component or, if the solution contains several liquids, the major component. All other components of the solution are solutes.

2. Some important concentration definitions are

$$\text{Percent by weight} = \frac{\text{Mass of component}}{\text{Mass of solution}} \times 100$$

$$\text{Mole fraction of A} = X_A = \frac{\text{Moles of component A}}{\text{Total moles in solution}}$$

For a solution containing two components, A and B,

$$X_A = \frac{n_A}{n_A + n_B} \quad \text{and} \quad X_A + X_B = 1$$

$$\text{Molality of A} = \frac{\text{Moles A}}{\text{Mass of solvent in kg}} = m_A$$

$$\text{Molarity of A} = \frac{\text{Moles A}}{\text{Liters of solution}} = M_A$$

$$\text{Formality of A} = \frac{\text{Moles of A added to the solution}}{\text{Liters of solution}} = F_A$$

3. A saturated solution is a solution that is in equilibrium with pure solute. The solubility of a substance in a given solvent is its concentration in a saturated solution.

4. Entropy usually favors solubility because the solution is usually a more random state than the separate pure components. Most cases of limited solubility are due to the fact that the particular solution process is endothermic. The heat of solution depends on the relative strengths of the solvent–solute forces compared to solvent–solvent and solute–solute forces.

5. A supersaturated solution is a metastable solution whose solute concentration exceeds the solubility.

6. Raoult's law: The vapor pressure of a component of an ideal solution equals its pure vapor pressure multiplied by its mole fraction.

$$P_A = X_A P_A^\circ$$

7. Fractional distillation is the separation of the components of a mixture by selective evaporation.

8. Many solutions are nonideal—they do not obey Raoult's law. Large deviations from ideality can produce azeotropes, which are constant-boiling mixtures that cannot be separated by distillation.

9. The vapor pressure of a solute in a saturated solution equals the vapor pressure of the pure solute. Two states in equilibrium have the same vapor pressure.

10. Colligative properties are physical properties of solutions that depend on solute concentration but not solute identity. Colligative properties are useful for estimating the concentrations of solutions and the molecular masses of solutes.

    Solvent vapor pressure in the solution compared to the pure solvent is a function of the total solute concentration.

    $$\frac{P_{solvent}}{P°_{solvent}} = 1 - X_{solute}$$

    Boiling point elevation is proportional to solute molality,

    $$\Delta T_b = T_b - T°_b = K_b m$$

    where K_b is the molal boiling-point-elevation constant of the solvent.
    Freezing point depression is proportional to solute molality,

    $$\Delta T_f = T°_f - T_f = K_f m$$

    where K_f is the molal freezing-point-depression constant of the solvent.
    Osmotic pressure (π), which is the pressure needed to prevent solvent flow from pure solvent through a membrane into a solution, is proportional to the molarity of the solute.

    $$\pi = RTM$$

11. Henry's law: The vapor pressure of a solute in a dilute solution is proportional to its concentration.

    $$P_{solute} = kM_{solute}$$

    The constant, k, depends on the identities of both the solvent and solute.

12. A colloidal suspension or sol is a dispersion of colloidal particles in a homogeneous medium. Colloidal particles are much larger than the ions and molecules found in true solutions.

Vocabulary List

solution	heat of solution	osmotic pressure
solvent	supersaturated	Henry's law
solute	metastable	colloidal suspensions
concentration	nucleation	sols
percent by weight	Raoult's law	colloids
mole fraction	ideal solution	ultrafiltration
molality	distillation	dialysis
molarity	azeotrope	Faraday-Tyndall effect
volumetric flask	colligative properties	electrophoresis
formality	molal boiling-point elevation	viscosity
saturated	molal freezing-point depression	emulsion
solubility		

Exercises

1. A solution is prepared by mixing 10 g of water, 15 g of methyl alcohol (CH_3OH), and 75 g of acetic acid ($HC_2H_3O_2$). Identify the solvent and the solutes.

2. How many moles of CH_3OH are present in each of the following solutions?
   a. 350 mL of a 0.075 M solution in water
   b. A 1.25 m solution containing 50.0 g of ethanol (C_2H_5OH) as the solvent
   c. A 14.5 mol % solution containing 5.0 mol of ethanol, the only other component of the solution

EXERCISES

3. How would you prepare each of the following solutions?
   a. 200 g of 15% NaCl in water
   b. 500 g of 0.250 m naphthalene ($C_{10}H_8$) in benzene (C_6H_6)
   c. 25.0 mL of 0.150 M ethylene glycol ($C_2H_6O_2$) in water
   d. 500 mL of 0.05824 M $CuSO_4 \cdot 6H_2O$ in water

4. Find the mass of the designated component.
   a. KNO_3 in 100 g of an 8.5% solution in water
   b. Water in 300 g of a solution that is 8.50 mol % acetone (C_3H_6O) and the rest water
   c. $LiClO_4$ in a 0.0375 m solution containing 1000 g of water
   d. O_2 in 75.0 L of a 3.6×10^{-4} M solution in water

5. What is the mole fraction of a 17.0% solution of ethylene (C_2H_4) in pentane (C_5H_{12})?

6. What is the molality of a 5.3% solution of NH_3 in water?

7. If 26.7 mL of a 0.187 M solution of NaOH are added to a reaction mixture, how many moles of NaOH are added?

8. a. According to Le Chatelier's principle, will the solubility of a substance whose heat of solution is positive (an endothermic process) increase, decrease, or stay the same if the temperature is raised?
   b. The solubility of Li_2CO_3 is 0.200 m at 0°C and 0.090 m at 100°C. Is dissolving Li_2CO_3 an exothermic or an endothermic process?
   c. Which has the higher entropy, saturated Li_2CO_3 solution or solid Li_2CO_3?

9. Exactly 15.0 mL of 0.1093 M $KMnO_4$ is diluted to 50.0 mL. What is the molarity of the new solution?

10. What is the molarity of HCl in a solution prepared by mixing 5.0 mL of 0.50 M HCl, 15.0 mL of 0.20 M HCl, and enough water to bring the volume to 25.0 mL?

11. A chemist dissolves 38.3 mg $Ba(OH)_2 \cdot 8H_2O$ in enough water to form 25.0 mL of solution.
    a. What is the molarity of the solution?
    b. When $Ba(OH)_2 \cdot 8H_2O$ dissolves, its crystal lattice dissociates completely to Ba^{2+}, OH^-, and water molecules. What is the molarity of hydroxide ion in this solution?

12. A solution is prepared by mixing 7.0 mL of 0.0200 M $BaCl_2$, 3.0 mL of 0.0400 M HCl, and enough water to bring the volume to 50.0 mL. The solute dissociates completely into Ba^{2+} and Cl^- ions.
    a. What is the molarity of $BaCl_2$?
    b. What is the actual molarity of intact $BaCl_2$ molecules?
    c. What is the actual molarity of chloride ion?
    d. What is the molarity of HCl in 15.0 mL of this solution?

13. A solution of pentane (C_5H_{12}), cyclohexane (C_6H_{12}), and methylene chloride (CH_2Cl_2) is 24.7 mol % pentane and 16.2 mol % methylene chloride. What is the mole fraction of cyclohexane in this solution?

14. A solution is made by mixing together 15.0 g of water, 75.0 g of ethanol (C_2H_5OH), and 45.0 g of dimethylformamide (C_3H_7NO). What is the mole fraction of each component of this solution?

15. a. Is the Gibbs free energy of a supersaturated solution of $Na_2S_2O_3$ greater than, less than, or the same as pure solid $Na_2S_2O_3$?
    b. In which direction will the free energy of the dissolved $Na_2S_2O_3$ change as the solute precipitates?

16. The solubility of sucrose ($C_{12}H_{22}O_{11}$) in water is 5.96 m at 20°C.
    a. What is the difference between the Gibbs free energy of solid sucrose and the Gibbs free energy of its saturated solution?
    b. Which has the higher free energy, solid sucrose or a supersaturated solution of sucrose?
    c. Which has the higher free energy, solid sucrose or a solution that is 50% sucrose by weight at 20°C?

17. The vapor pressure of pure ethyl acetate at 25°C is 104 torr and the vapor pressure of pure ethyl propionate is 38 torr at this temperature. What is the partial pressure of ethyl propionate at 25°C above a solution of these two compounds that is 18 mol % ethyl propionate?

18. The vapor pressure of pure methanol (CH_3OH) is 129 torr and that of pure ethanol (C_2H_5OH) is 63 torr at 25°C.
    a. What is the vapor pressure at 25°C of ethanol above a 32.0% solution of methanol in ethanol?
    b. What is the total vapor pressure above the solution?

19. What is the vapor pressure at 25°C of cyclopentane over a solution of cyclopentane and cyclohexane that is 34.0 mol % cyclohexane? Pure vapor pressures at 25°C: cyclopentane—311 torr; cyclohexane—95 torr.

20. The vapor pressures of pure heptane (C_7H_{16}) and pure toluene (C_7H_8) at 50°C are 92.1 torr and 59.1 torr, respectively. What is the partial pressure of heptane at 50°C above a solution that is 15.0% toluene by weight? (Assume ideal behavior.)

21. The vapor pressure of pure methanol (CH_3OH) at 20°C is 96.3 torr, and that of 1-propanol (C_3H_7OH) is 15.1 torr. Assume that solutions of methanol and 1-propanol are ideal and prepare a graph showing the partial pressures of each component and the total vapor pressure of the solution at 20°C versus the mole fraction of methanol.

22. The vapor pressure of pure benzene (C_6H_6) at 40°C is 372 torr, and that of pure nitrobenzene ($C_6H_5NO_2$) is 185 torr. The total vapor pressure of a 50.0 mol % solution of these two compounds is 266 torr at 40°C.
  a. Is this solution ideal? If not, is the deviation positive or negative?
  b. Which effect (if either), attractions between similar molecules or attractions between dissimilar molecules, is dominant in this solution?

23. The vapor pressure of pure ethyl ether at 20°C is 444 torr. The partial pressure of ethyl ether above a solution of ethyl ether and acetone is 350 torr at 20°C. What is the mole fraction of ethyl ether in this solution? (Assume ideal behavior.)

24. The vapor pressure of pure Br_2 at 25°C is 300 torr. The vapor pressure of Br_2 above a solution of Br_2 in 50.0 g CCl_4 is 72.0 torr. What mass of Br_2 is dissolved in the solution? (Assume ideality.)

25. The vapor pressure of a dilute aqueous solution of glucose at 30°C is 30.95 torr compared with 31.82 torr for pure water. What is the mole fraction of glucose in this solution?

26. The vapor pressure of pure isopentane at 50°C is 1536 torr, and the vapor pressure of pure n-octane is 49.4 torr. What is the mole fraction of isopentane in the vapor in equilibrium with a 47.8 mol % solution of isopentane in n-octane at 50°C?

27. The vapor pressure of a 2.05% solution of benzoic acid (a nonvolatile substance) in benzene (C_6H_6) is 93.90 torr at 25°C. The vapor pressure of pure benzene is 95.18 torr at 25°C. What is the molecular mass of benzoic acid?

28. The vapor pressure of water at 85.0°C is 433.6 torr. The vapor pressure above a 6.00% solution of glycerol (a nonvolatile substance) in water is 428.3 torr. What is the molecular mass of glycerol?

29. A solution of 1.00 g of α-naphthalenesulfonic acid in 10.0 g of benzene boils 1.22°C higher than pure benzene. What is the molecular mass of α-naphthalenesulfonic acid?

30. The boiling point of a solution of 1.25 g of ribose in 50.0 g of water is 0.085°C higher than pure water.
  a. What is the molecular mass of ribose?
  b. What is the freezing point of this solution?

31. What is the total molality of solutes in an aqueous solution that freezes at −0.279°C?

32. What is the total molality of nonvolatile solutes in a chloroform solution that boils at 62.0°C?

33. What is the osmotic pressure of 0.050 M sucrose ($C_{12}H_{22}O_{11}$) at 25°C?

34. The osmotic pressure of a solution containing 50 mg of myoglobin in 10 mL of solution is 4.8 torr at 5°C. Estimate the molecular mass of myoglobin.

35. A solution of 0.140 g of pyrogallol in 2.00 g of camphor ($C_{10}H_{16}O$) melts at 155.5°C. What is the molecular mass of pyrogallol?

36. The freezing point of a 2.00% solution of an unknown compound in benzene is 0.84°C lower than the freezing point of pure benzene. What is the molecular mass of the unknown compound?

37. Pure carbon tetrachloride (CCl_4) boils at 76.8°C. A solution of 4.45 g of phenanthrene ($C_{14}H_{10}$), a nonvolatile compound, in 100 g CCl_4 boils at 78.1°C. What is the molal boiling point elevation of CCl_4?

38. The density of 5.00 m methanol (CH_3OH) in water is 0.973 g/mL.
  a. What is the weight percent of methanol in this solution?
  b. What is the mole fraction of methanol?
  c. What is the molarity of methanol?

39. What is the composition, expressed in mole fractions, of the vapor in contact with the solution in Exercise 17?

40. The vapor pressure of pure camphor ($C_{10}H_{16}O$) at 19°C is 0.46 torr. The solubility of camphor in water at 19°C is 0.15% by weight. What is the partial pressure of camphor above a saturated solution in water?

41. What is the percent by weight of ethylene glycol ($C_2H_6O_2$) needed to reduce the freezing point of an aqueous medium to −5.00°C?

42. What is the freezing point of an aqueous solution containing 2.00% CH_3OH and 5.00% glucose ($C_6H_{12}O_6$)?

chapter 11

Ionic Solutions, Acids, and Bases

Ammonia gas (NH$_3$) dissolves in water to form molecules (NH$_3$·H$_2$O), which dissociate to give basic solutions.

Almost all the solutions that you will deal with in your first-year chemistry laboratory work contain ions. A large proportion of the substances that dissolve readily in water are either ionic or molecular substances that dissociate (break apart into ions) as they dissolve. Since several of the subsequent chapters—in particular Chapters 14 (Aqueous Equilibria) and 15 (Electrochemical Reactions)—deal with ionic solutions in detail, it is important to understand the fundamentals of the behavior of ions in solution. In addition to describing the general properties of ionic solutions, this chapter also introduces two particularly important types of ionic solutions: acidic solutions (those containing H$_3$O$^+$) and basic solutions (those containing OH$^-$).

11.1 Electrolytes

The Arrhenius Theory

One of the most useful ways of classifying aqueous solutions is according to their electrical conducting properties (Figure 11.1). Aqueous solutions of **electrolytes** are good conductors of electricity; aqueous solutions of **nonelectrolytes** do not conduct electricity. The Swedish physical chemist, Svante Arrhenius, explained the conductivity of electrolytes by their dissociation into ions in solution. The freely moving ions migrate in the presence of an electric field and produce a current.

Strong electrolytes are either water-soluble ionic substances such as NaCl or covalent substances that react completely with water to form ions. For example,

$$\text{HCl(g)} + \text{H}_2\text{O} \longrightarrow \text{H}_3\text{O}^+(\text{aq}) + \text{Cl}^-(\text{aq})$$

Figure 11.1
Electrolytes and nonelectrolytes.

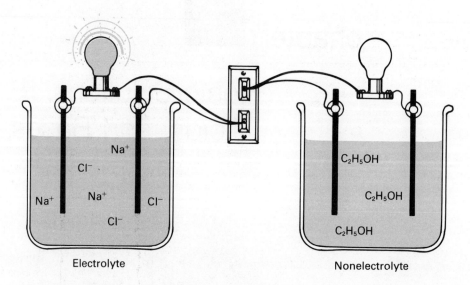

Covalent electrolytes dissolve in nonpolar solvents to produce nonconducting solutions without ions, such as a solution of HCl in benzene.

Arrhenius used colligative properties to show that strong electrolytes dissociate almost completely when dissolved in water. The freezing-point depression (ΔT_f) of a strong electrolyte solution reveals that the total moles of solute in the solution is close to a whole-number multiple of the moles of electrolyte dissolved in the solution (Table 11.1).

$$\Delta T_f = K_f m(i) \tag{11.1}$$

Table 11.1 Observed and Theoretical Values of i from Freezing-Point Depressions

Solute	Molality	$\Delta T(°C)$	$\Delta T/K_f(m)$	$i_{observed}$	$i_{theoretical}$
Strong electrolytes					
NaCl	0.00100	0.00366	0.00197	1.97	2
	0.0100	0.0360	0.0194	1.94	2
	0.100	0.348	0.187	1.87	2
	1.00	3.37	1.81	1.81	2
HCl	0.00100	0.00369	0.00198	1.98	2
	0.0100	0.0360	0.0194	1.94	2
	0.100	0.352	0.189	1.89	2
BaCl$_2$	0.00100	0.00530	0.00285	2.85	3
	0.0100	0.0503	0.0270	2.70	3
	0.100	0.470	0.253	2.53	3
K$_2$SO$_4$	0.00100	0.00528	0.00283	2.83	3
	0.0100	0.0501	0.0269	2.69	3
	0.100	0.432	0.232	2.32	3
LaCl$_3$	0.030	0.22	0.12	4.0	4
	0.10	0.65	0.35	3.5	4
Weak electrolyte					
Acetic acid	0.0100	0.0194	0.0104	1.04	>1
	0.100	0.190	0.102	1.02	>1
Nonelectrolyte					
Ethanol	0.0200	0.0366	0.0196	0.98	1
	0.100	0.183	0.098	0.98	1
	1.00	1.83	0.98	0.98	1

Source: Data are from *International Critical Tables* (1928), Vol. IV, pp. 254–263. Reproduced with the permission of the National Academy of Sciences, Washington, D.C.

Figure 11.2
Electrostatic attractions in an ionic solution. The insulating effect of water (the solvent) diminishes, but does not totally prevent, these attractive forces.

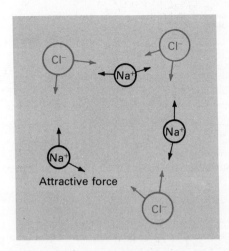

In equation (11.1), m is the molality of the electrolyte, K_f the molal freezing-point-depression constant of water (1.86 °C/m), and i is a whole number equal to the number of moles of ions produced when a mole of the electrolyte dissociates. For example,

$$NaCl(s) \longrightarrow Na^+(aq) + Cl^-(aq) \quad i = 2$$

$$K_2SO_4(s) \longrightarrow 2\,K^+(aq) + SO_4^{2-}(aq) \quad i = 3$$

$$LaCl_3(s) \longrightarrow La^{3+}(aq) + 3\,Cl^-(aq) \quad i = 4$$

Estimates of i based on colligative properties of strong electrolytes are, however, slightly *lower* than the appropriate whole number, and the discrepancy is larger for more concentrated solutions. Arrhenius ascribed this effect to incomplete dissociation. According to this view, a small percentage of the electrolyte remains as neutral, undissociated molecules in solution.

We now believe these deviations from the whole-number value of i are due mainly to nonideality. Even when they are separated by insulating water molecules, oppositely charged ions still attract each other weakly (Figure 11.2). These attractions slightly restrict the motion of the ions so that they do not behave as if they were completely free ions. This retardation of ionic motion lowers the conductivity of the solution, and makes it appear that that total solute-particle concentration is lower than it actually is. The effect is greater at higher concentrations because the ions are, on the average, closer together and attract each other more strongly.

Ions in Solution

In addition to electrical conductivity and colligative properties, there is considerable chemical evidence for Arrhenius's idea of dissociation of strong electrolytes. This evidence can be summarized as follows: The chemistry of a strong electrolyte solution is the chemistry of the cation and the chemistry of the anion, each acting independently. To see this, consider the following four aqueous solutions: 1 M $Cu(NO_3)_2$, 1 M $CuSO_4$, 1 M $Mg(NO_3)_2$, and 1 M $MgSO_4$. The first two solutions show all the typical reactions and properties of Cu^{2+} in solution. For example, both solutions have the characteristic blue color of hydrated Cu^{2+}, and addition of NaOH to either solution will precipitate $Cu(OH)_2$.

$$Cu(NO_3)_2(aq) + 2\,NaOH(aq) \longrightarrow Cu(OH)_2(s) + 2\,NaNO_3(aq)$$

$$CuSO_4(aq) + 2\,NaOH(aq) \longrightarrow Cu(OH)_2(s) + Na_2SO_4(aq)$$

Detailed studies of these precipitation reactions, including such features as the heats of reaction, reveal no difference between them. This is why we can represent either reaction by the same *net ionic equation*.

$$Cu^{2+}(aq) + 2\,OH^-(aq) \longrightarrow Cu(OH)_2(s)$$

Figure 11.3
Identical ionic solutions from two sources.

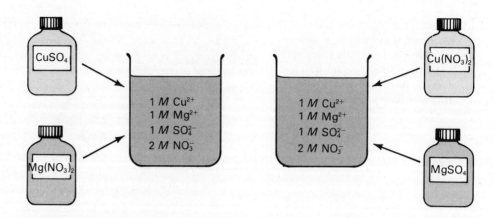

In a similar way, $Mg(NO_3)_2$ and $MgSO_4$ solution show typical Mg^{2+} ion chemistry, $Cu(NO_3)_2$ and $Mg(NO_3)_2$ have the reactions of NO_3^- in common, and $CuSO_4$ and $MgSO_4$ both undergo all the reactions of SO_4^{2-}.

One of the best illustrations of the independence of the ions in a strong electrolyte solution is the behavior of the two solutions shown in Figure 11.3. Solution A is $1\,M$ in $CuSO_4$ and $1\,M$ in $Mg(NO_3)_2$. Solution B is $1\,M$ in $Cu(NO_3)_2$ and $1\,M$ in $MgSO_4$. These two solutions are, in fact, identical in all their properties. Each contains $1\,M$ Cu^{2+}, $1\,M$ Mg^{2+}, $2\,M$ NO_3^-, and $1\,M$ SO_4^{2-}. Since these ions retain no memories of their origins and carry no labels describing their source, it is impossible to devise any analytic test to distinguish between the two solutions or tell us how either solution was prepared.

Weak Electrolytes

A solution of a **weak electrolyte** conducts electricity but, except for very dilute solutions, it does not conduct as effectively as a strong electrolyte at the same concentration. The equilibrium for dissociation of a weak electrolyte is much less favorable than for strong electrolyte dissociation. For example, acetic acid (a weak electrolyte) dissociates when dissolved in water but only to a slight extent.

$$H_2O + CH_3CO_2H\,(aq) \rightleftharpoons H_3O^+(aq) + CH_3CO_2^-(aq)$$

In other words, CH_3CO_2H molecules have a tendency to react with water to produce H_3O^+ and $CH_3CO_2^-$ ions, but there is also a strong tendency for these ions, once formed, to react and regenerate water and undissociated acetic acid. Net dissociation ceases when the rate of the backward reaction equals the rate of the forward reaction, and since the tendency for the backward reaction to occur is considerably stronger than the production of the ions, equilibrium occurs when only a small proportion of the acetic acid molecules have dissociated. Because of its relatively low concentration of ions, $1\,M$ CH_3CO_2H is a much less effective conductor than $1\,M$ HCl (Figure 11.4).

Figure 11.4
Strong and weak electrolytes.

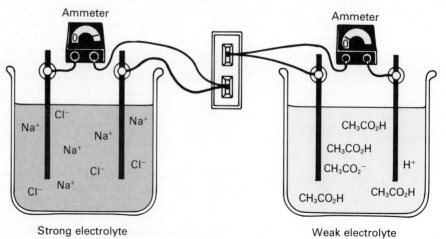

The best way to measure the **degree of dissociation**—the proportion of solute molecules that are dissociated at equilibrium—of a weak electrolyte is to measure the conductivity of its solution and compare it with the conductivity we would expect if the weak electrolyte dissociated completely.* These measurements reveal that the degree of dissociation of a weak electrolyte is greater in a dilute solution than in a concentrated one. For example, 35% of the CH_3CO_2H molecules in 10^{-4} M acetic acid are dissociated, whereas only 0.4% are dissociated in a 1 M solution. This observation is consistent with Le Chatelier's principle. Dilution is a stress that decreases the total concentration of solute particles; the increased degree of dissociation produces more solute particles to partially counteract the stress.

Colligative properties of weak electrolytes are consistent with their small degree of dissociation. Since dissociation raises the total solute concentration by producing two solute particles from one, it also increases the i factor for freezing-point depression above one. Because the degree of dissociation of a weak electrolyte is quite low, this effect is small and the i factor is only slightly greater than one. (See the data for acetic acid in Table 11.1.)

11.2 Acids and Bases

Arrhenius Definitions

An **acid** is an electrolyte whose solutions corrode some metals, producing hydrogen gas in the process. Acid solutions taste sour and turn blue litmus red. A **base** is an electrolyte whose solutions taste bitter, etch glass, and turn red litmus blue. An acid reacts with a base to form a salt. This **neutralization reaction** destroys the characteristic properties of both the acid and the base. Table 11.2 lists typical acids and bases.

Table 11.2 **Acids and Bases**

	Dissociation reactions
Strong acids	
Nitric acid	$HNO_3 \rightleftharpoons H^+(aq) + NO_3^-(aq)$
Hydrochloric acid	$HCl \rightleftharpoons H^+(aq) + Cl^-(aq)$
Weak acids	
Acetic acid	$CH_3CO_2H(aq) \rightleftharpoons H^+(aq) + CH_3CO_2^-(aq)$
Hypochlorous acid	$HOCl(aq) \rightleftharpoons H^+(aq) + OCl^-(aq)$
Hydrocyanic acid	$HCN(aq) \rightleftharpoons H^+(aq) + CN^-(aq)$
Polyprotic acids	
Sulfuric acid	$H_2SO_4 \rightleftharpoons H^+(aq) + HSO_4^-(aq)$
(diprotic, strong)	$HSO_4^-(aq) \rightleftharpoons H^+(aq) + SO_4^{2-}(aq)$
Carbon dioxide	$CO_2(g) + H_2O \rightleftharpoons H^+(aq) + HCO_3^-(aq)$
(diprotic, weak)	$HCO_3^-(aq) \rightleftharpoons H^+(aq) + CO_3^{2-}(aq)$
Hydrosulfuric acid	$H_2S(g) \rightleftharpoons H^+(aq) + HS^-(aq)$
(diprotic, weak)	$HS^-(aq) \rightleftharpoons H^+(aq) + S^{2-}(aq)$
Phosphoric acid	$H_3PO_4(aq) \rightleftharpoons H^+(aq) + H_2PO_4^-(aq)$
(triprotic, moderate)	$H_2PO_4^-(aq) \rightleftharpoons H^+(aq) + HPO_4^{2-}(aq)$
	$HPO_4^{2-}(aq) \rightleftharpoons H^+(aq) + PO_4^{3-}(aq)$
Strong base	
Sodium hydroxide	$NaOH(s) \longrightarrow Na^+(aq) + OH^-(aq)$
Weak base	
Ammonia	$NH_3(aq) + H_2O \rightleftharpoons NH_4^+(aq) + OH^-(aq)$
Diprotic base	
Barium hydroxide	$Ba(OH)_2(s) \rightleftharpoons Ba^{2+}(aq) + 2\,OH^-(aq)$

*The interpretation of these conductivity data is rather complicated since it requires an estimate of the intrinsic ability of the free ions to carry a current, and we must also consider the effects of nonideality.

Arrhenius proposed that acids dissociate in water to produce $H^+(aq)$, called the **hydronium ion**, and an anion. (In fact, the reaction forms hydrated species of the type $H(H_2O)_n^+$, but we still use the symbol $H^+(aq)$ or just H^+ to describe these ions. Many chemists prefer to use $H_3O^+(aq)$ as a simplified formula for a hydrated proton.) All typical acid properties are due to H^+. For example, the corrosion of metals by acids is their oxidation by hydronium ion.

$$Fe(s) + 2\,H^+(aq) \longrightarrow Fe^{2+}(aq) + H_2(g)$$

The sour taste of acids is the physiological response of our tongues to H^+, and the color change of blue litmus in acid is due to a chemical reaction between the dye and hydronium ion.

According to Arrhenius, bases dissociate to form **hydroxide ion** (OH^-) and a cation. The presence of OH^- in basic solutions gives them their bitter taste (the physiological response of our taste buds to OH^-), their ability to etch glass (a reaction of OH^- with the SiO_2 in the glass), and their effect on red litmus (a reaction between OH^- and the red dye). Bases precipitate many metal ions because these metals have insoluble hydroxides.

$$Fe^{3+}(aq) + 3\,OH^-(aq) \rightleftharpoons Fe(OH)_3(s)$$

The acid–base neutralization reaction is the reaction between hydronium ion and hydroxide ion to produce water.

$$H^+(aq) + OH^-(aq) \rightleftharpoons H_2O$$

Acids and bases are classified as strong or weak according to the completeness of their dissociations. Strong acids, such as HCl, dissociate completely in aqueous solutions, and the hydronium-ion concentration of these solutions is relatively high. A common example of a weak acid is acetic acid, a weak electrolyte that dissociates to only a slight extent. Since the degree of dissociation of acetic acid is so low, the concentration of H^+ in 1 M acetic acid is much less than in 1 M HCl. Because of this lower H^+ concentration, acetic acid corrodes metals more slowly than a strong acid. A weak base produces OH^- in aqueous solution, but the extent of this reaction is limited. The unfavorable dissociation equilibrium keeps the concentration of OH^- low compared to the hydroxide-ion concentration of a solution of a strong base.

Although these definitions are adequate for aqueous solutions, there is some difficulty in applying them to aqueous solutions of CO_2, a weak acid containing no hydrogen, and NH_3, a weak base without an OH group. One suggestion is that CO_2 reacts with water to form carbonic acid (H_2CO_3), and that NH_3 forms NH_4OH (ammonium hydroxide) with water.

$$CO_2(aq) + H_2O \rightleftharpoons H_2CO_3(aq) \rightleftharpoons H^+(aq) + HCO_3^-(aq)$$

$$NH_3(aq) + H_2O \rightleftharpoons NH_4OH \rightleftharpoons OH^-(aq) + NH_4^+(aq)$$

Although the existence of NH_4OH remains hypothetical, manufacturers often label ammonia solutions as ammonium hydroxide and write NH_4OH as the formula.

Neutralization

Since strong acids and strong bases are completely dissociated in aqueous solutions, the strong acid–strong base neutralization reaction is

$$H^+(aq) + OH^-(aq) \rightleftharpoons H_2O$$

As evidence for this, the heat of neutralization (an exothermic reaction) is the same for any reaction between a strong acid and strong base. The acid's anion and the base's cation are not involved in the reaction and do not influence the enthalpy change. The neutralized solution, containing mainly water, a cation, and an anion, is identical to a solution of a salt.

11.2 ACIDS AND BASES

The reaction between H^+ and OH^- to form water is nearly complete, but low concentrations (10^{-7} M each) of these ions persist at equilibrium. This solution is defined as **neutral** because the molar concentration of H^+ equals the molar concentration of OH^-. Furthermore, the H^+ and OH^- concentrations are the same as in pure water because water has a very slight tendency to dissociate, producing equal concentrations of H^+ and OH^-.

Because of the nature of equilibrium systems, we cannot increase the H^+ concentration of a solution without also decreasing the OH^- concentration. According to Le Chatelier's principle, the stress of adding H^+ to a solution causes a shift in the neutralization equilibrium to decrease H^+. This reaction, consuming both H^+ and OH^-, counteracts the effect of the added H^+ to some extent but also lowers the OH^- concentration. For similar reasons, the H^+ concentration of a solution decreases as OH^- is added to the solution. Because of this relationship between the concentration of H^+ and OH^-, we define an acidic aqueous solution as one with a higher H^+ concentration and a lower OH^- concentration than pure water. A basic aqueous solution is richer in OH^- and poorer in H^+ than pure water.

Even though a solution of a weak acid contains relatively few H^+ ions compared to the OH^- in a solution of a strong base, it reacts nearly completely with an equal number of moles of the base. This is because the affinity of OH^- for H^+ is so strong that it can remove a proton from an undissociated weak acid molecule. For example,

$$CH_3CO_2H(aq) + OH^-(aq) \rightleftharpoons CH_3CO_2^-(aq) + H_2O$$

However, the neutralization of the strong base by the weak acid is not quite complete, and the neutralized solution is still slightly basic. In other words, there is a small concentration of unreacted OH^- and weak acid at equilibrium. In a similar way, when a mole of a weak base (such as NH_3) reacts with a mole of a strong acid the neutralized solution is slightly acidic.

$$NH_3(aq) + H^+(aq) \rightleftharpoons NH_4^+(aq)$$

Brønsted–Lowry Definitions

The equations we have used to describe the dissociation of an acid such as HCl

$$HCl(aq) \longrightarrow H^+(aq) + Cl^-(aq)$$

are really shorthand notations for reactions in which the acid transfers a hydrogen ion (in other words, a proton) to water.

$$HCl(aq) + H_2O \longrightarrow H_3O^+(aq) + Cl^-(aq)$$

The **Brønsted–Lowry** definitions of **acids and bases** generalize this idea of proton transfers so that it applies to acid–base reactions in nonaqueous systems. Substances are classified as acids or bases in the context of a particular proton transfer reaction. *The acid is the proton donor, and the base is the proton receiver.* A particular species can be an acid in one reaction and a base in another. Water is a base when it reacts with nitric acid, an acid when it reacts with amide ion (NH_2^-).

$$\underset{\text{Acid}}{HNO_3(aq)} + \underset{\text{Base}}{H_2O} \rightleftharpoons \underset{\substack{\text{Conjugate} \\ \text{acid}}}{H_3O^+(aq)} + \underset{\substack{\text{Conjugate} \\ \text{base}}}{NO_3^-(aq)}$$

(Conjugate pairs: HNO_3/NO_3^- and H_2O/H_3O^+)

$$\underset{\text{Acid}}{H_2O} + \underset{\text{Base}}{NH_2^-(aq)} \rightleftharpoons \underset{\substack{\text{Conjugate} \\ \text{acid}}}{NH_3(aq)} + \underset{\substack{\text{Conjugate} \\ \text{base}}}{OH^-(aq)}$$

(Conjugate pairs: H_2O/OH^- and NH_2^-/NH_3)

Example 11.1

Identify the Brønsted–Lowry acid and the Brønsted–Lowry base on the left-hand side of the following equation.

$$HPO_4^{2-}(aq) + HSO_4^-(aq) \rightleftharpoons H_2PO_4^-(aq) + SO_4^{2-}(aq)$$

Solution

Bisulfate ion (HSO_4^-) donates its proton to HPO_4^{2-}, so HSO_4^- is the acid and HPO_4^{2-} is the base.

The species produced by removal of a proton from an acid is its **conjugate base**; proton addition to a base produces its **conjugate acid** (Table 11.3). Strong acids have weak conjugate bases, and weak acids have strong conjugate bases. Any reaction between an acid and a base forms a conjugate acid–conjugate base pair. The direction of favorable equilibrium is toward the weaker acid–base pair. In the preceding reaction, NO_3^- is a weaker base than H_2O and H_3O^+ is a weaker acid than HNO_3, so the equilibrium favors the formation of the weak acid–base pair, NO_3^- and H_3O^+. Because H_2O is more acidic than NH_3 and NH_2^- is more basic than OH^-, the reaction between NH_2^- and H_2O to yield NH_3 and OH^- is favorable.

Example 11.2

What is the conjugate base of CH_4?

Solution

Removal of a proton from CH_4 produces its conjugate base, CH_3^-.

Example 11.3

What is the conjugate acid of H_3AsO_3?

Solution

The addition of a proton to H_3AsO_3 forms $H_4AsO_3^+$, its conjugate acid.

Example 11.4

Does the following reaction have a favorable equilibrium as written from left to right?

$$HCN(aq) + HS^-(aq) \rightleftharpoons H_2S(aq) + CN^-(aq)$$

Solution

According to Table 11.3, H_2S is a stronger acid than HCN and CN^- a stronger base than HS^-. Since the reaction would produce a stronger acid-base pair from a weaker one, its equilibrium is unfavorable.

The Brønsted–Lowry definitions account for the weak basicity of ammonia without invoking any hypothetical intermediate such as NH_4OH.

$$\underset{\text{Acid}}{H_2O} + \underset{\text{Base}}{NH_3(aq)} \rightleftharpoons \underset{\text{Conjugate acid}}{NH_4^+(aq)} + \underset{\text{Conjugate base}}{OH^-(aq)}$$

These definitions apply also to nonaqueous systems such as a solution of HCl in methyl alcohol (CH_3OH),

$$\underset{\text{Acid}}{HCl} + \underset{\text{Base}}{CH_3OH} \rightleftharpoons \underset{\text{Conjugate acid}}{CH_3OH_2^+} + \underset{\text{Conjugate base}}{Cl^-}$$

11.2 ACIDS AND BASES

Table 11.3 Brønsted–Lowry Acids and Bases

	Acid	Conjugate base	
Strong ↑	$HClO_4$	ClO_4^-	Weak
	H_2SO_4	HSO_4^-	
	HNO_3	NO_3^-	
	HCl	Cl^-	
	H_3O^+	H_2O	
	HSO_4^-	SO_4^{2-}	
	H_3PO_4	$H_2PO_4^-$	
	CH_3CO_2H	$CH_3CO_2^-$	
	H_2S	HS^-	
	$H_2PO_4^-$	HPO_4^{2-}	
	NH_4^+	NH_3	
	HCN	CN^-	
	HCO_3^-	CO_3^{2-}	
	HS^-	S^{2-}	
	H_2O	OH^-	
	CH_3OH	CH_3O^-	
	H_2C_2 (acetylene)	HC_2^-	↓
Weak	NH_3	NH_2^-	Strong
	Conjugate acid	Base	

and to gas-phase reactions such as the reaction between ammonia and hydrogen chloride.

$$HCl(g) + NH_3(g) \longrightarrow NH_4Cl(s)$$

Hydrogen chloride, by donating H^+ to NH_3, is the acid, and ammonia (the proton acceptor) is the base.

The strongest acid that can exist in a solution is the conjugate acid of the solvent. Any stronger acid will transfer a proton to the solvent and be converted to its conjugate base. There is no apparent difference between the strengths of strong acids such as HCl, HNO_3, and H_2SO_4 in aqueous solutions because they all react completely with water to produce hydronium ion, the conjugate acid of water.

$$HCl(aq) + H_2O \longrightarrow H_3O^+(aq) + Cl^-(aq)$$

$$HNO_3(aq) + H_2O \longrightarrow H_3O^+(aq) + NO_3^-(aq)$$

$$H_2SO_4(aq) + H_2O \longrightarrow H_3O^+(aq) + HSO_4^-(aq)$$

We can avoid this **leveling effect** and rank these strong acids by comparing their reactions in more acidic solvents. Since HCl acts as a Brønsted–Lowry base in concentrated H_2SO_4,

$$H_2SO_4 + HCl \rightleftharpoons H_2Cl^+ + HSO_4^-$$

it is a weaker acid than H_2SO_4.*

Bases show a similar leveling effect; the conjugate base of the solvent is the strongest possible in solution. The bases CH_3O^-, NH_2^-, and HC_2^- are all stronger than OH^-, react completely with water to produce OH^-,

$$CH_3O^-(aq) + H_2O \longrightarrow OH^-(aq) + CH_3OH(aq)$$

---

*The relative acidity of two acids depends on the nature of the solvent, so the stronger of two acids in one solvent may be the weaker in a second solvent.

$$NH_2^-(aq) + H_2O \longrightarrow OH^-(aq) + NH_3(aq)$$

$$HC_2^-(aq) + H_2O \longrightarrow OH^-(aq) + H_2C_2(aq)$$

and appear equally strong in aqueous solution. In liquid ammonia (a more basic solvent than water), acetylene (H_2C_2) reacts with amide ion (NH_2^-) to form ammonia and acetylide ion, showing that amide ion is a stronger base than HC_2^-.

$$NH_2^- + H_2C_2 \rightleftharpoons NH_3 + HC_2^-$$

Lewis Definitions

G. N. Lewis defined an acid–base reaction as one in which a covalent bond forms between an electron-deficient species (a **Lewis acid**) and a species with an unshared electron pair (a **Lewis base**). A Brønsted–Lowry acid–base reaction involves two Lewis reactions.

$$H\!-\!\ddot{\underset{..}{Cl}}\!: \;\rightleftharpoons\; \underset{\text{Lewis acid}}{H^+} \;+\; \underset{\text{Lewis base}}{:\!\ddot{\underset{..}{Cl}}\!:^-}$$

$$\underset{\text{Lewis acid}}{H^+} \;+\; \underset{\text{Lewis base}}{:\!\ddot{O}\!-\!H \atop |\;\;H} \;\rightleftharpoons\; H\!-\!\ddot{O}\!-\!H^+ \atop |\;\;H$$

Lewis's definitions also apply to reactions that do not involve proton transfer.

$$\underset{\text{Lewis base}}{:\!\ddot{\underset{..}{F}}\!:^-} \;+\; \underset{\text{Lewis acid}}{\underset{:\ddot{F}:}{\overset{:\ddot{F}:}{B\!-\!\ddot{F}\!:}}} \;\longrightarrow\; \underset{:\ddot{F}:}{\overset{:\ddot{F}:}{:\ddot{F}\!-\!B\!-\!\ddot{F}\!:^-}}$$

$$\underset{\text{Lewis acid}}{C_6H_5CH_2^+} \;+\; \underset{\text{Lewis base}}{:\!\ddot{\underset{..}{I}}\!:^-} \;\longrightarrow\; C_6H_5CH_2\!-\!\ddot{\underset{..}{I}}\!:$$

There is no contradiction between the Arrhenius, Brønsted–Lowry, and Lewis definitions. The Brønsted–Lowry definitions are more general than the Arrhenius, and the Lewis definitions are the most general of all. Arrhenius acids engage in proton transfer reactions with water; Brønsted-Lowry definitions apply to any proton transfer. A Brønsted–Lowry base uses an unshared electron pair to bind a proton covalently. Lewis generalized his definitions to cover any reaction in which one substance supplies an electron pair to a covalent bond. Since we will be dealing mainly with reactions in aqueous systems, we will use the Arrhenius definitions most frequently. The Brønsted–Lowry and Lewis definitions describe many important reactions in nonaqueous systems.

Polyprotic Acids

A molecule of a **polyprotic acid** contains more than one dissociable proton. Sulfuric acid, a diprotic acid, can lose two protons, neutralizing two moles of base and forming SO_4^{2-}.

The difficulty of removing successive protons from a polyprotic acid increases with each proton. The dissociation of H_2SO_4 to H^+ and HSO_4^- is complete in water; only a fraction of the HSO_4^- dissociates further unless base is added (Table 11.2). Phosphoric acid (H_3PO_4) dissociates in water to a moderate degree to form $H_2PO_4^-$. Significant further dissociation to HPO_4^{2-} does not occur until the solution is basic, and PO_4^{3-} forms only in very basic solutions. It is harder to pull a positive proton from a negative anion than from a neutral molecule, and proton removal from a doubly charged anion is more difficult than from a monoanion.

An example of a diprotic base is hydrazine (NH_2NH_2), a species that can combine with two protons.

11.3 STOICHIOMETRY IN SOLUTIONS

$$2\,H^+(aq) + NH_2NH_2(aq) \rightleftharpoons NH_3NH_3^{2+}(aq)$$

A mole of $Al(OH)_3$, a triprotic base, neutralizes three moles of acid.

$$3\,H^+(aq) + Al(OH)_3(s) \rightleftharpoons Al^{3+}(aq) + 3\,H_2O$$

Amphoterism

An **amphoteric** species acts as a base under sufficiently acidic conditions and is an acid if the solution is basic enough. Bicarbonate ion (HCO_3^-) both neutralizes acid, forming water and carbon dioxide,

$$H^+(aq) + HCO_3^-(aq) \rightleftharpoons H_2O + CO_2(g)$$

and reacts with hydroxide ion to form carbonate ion.

$$OH^-(aq) + HCO_3^-(aq) \rightleftharpoons H_2O + CO_3^{2-}(aq)$$

The water-insoluble compound, aluminum hydroxide, is also amphoteric, neutralizing and dissolving in both acid and base.

$$3\,H^+(aq) + Al(OH)_3(s) \rightleftharpoons 3\,H_2O + Al^{3+}(aq)$$
$$OH^-(aq) + Al(OH)_3(s) \rightleftharpoons Al(OH)_4^-(aq)$$

11.3 Stoichiometry in Solutions

Titration

We can determine the molarity of hydrogen chloride in a solution by adding sodium hydroxide to a sample of the solution until it is neutralized and calculating the number of moles of acid from the moles of base used. To avoid the clumsy procedure of adding solid NaOH and estimating the mass used, we **titrate** the acid sample with a sodium hydroxide solution of known molarity (a **standard solution**). We add the standard base solution from a buret (Figure 11.5, p. 294) until the color change of an indicator dye (for example, phenolphthalein) in the titration flask signals the consumption of the last acid and the accumulation of the first excess base (the **end point**). Knowing the volume and concentration of the standard base, we calculate the moles of OH^- added.

$$\text{Moles OH}^- = M_{NaOH} V_{NaOH} \qquad (11.2)$$

Because 1 mol of H^+ neutralizes 1 mol of OH^-, the moles of HCl in the acid sample equals the moles of base. The molarity of the HCl solution is

$$M_{HCl} = \frac{\text{Moles HCl}}{V_{HCl}} = \frac{M_{NaOH} V_{NaOH}}{V_{HCl}} \qquad (11.3)$$

Example 11.5

A 25.00-mL sample of dilute H_2SO_4 is titrated with 0.0965 M NaOH so that both protons of H_2SO_4 are neutralized. It takes 17.83 mL of base to reach the end point. What is the molarity of the H_2SO_4?

Solution

When working with the volumes usually encountered in titrations (0 to 50 mL, the capacity of a buret), it is convenient to express amounts of reactants in millimoles (1 mmol = 10^{-3} mol). A 1 M solution contains 1 mmol/mL.

$$\text{Millimoles OH}^- \text{ used} = (17.83\text{ mL})(0.0965\text{ mmol/mL}) = 1.721\text{ mmol}$$

Since two protons are titrated for each H_2SO_4 molecule,

Figure 11.5
Titration of dilute sulfuric acid with a standard solution of sodium hydroxide.

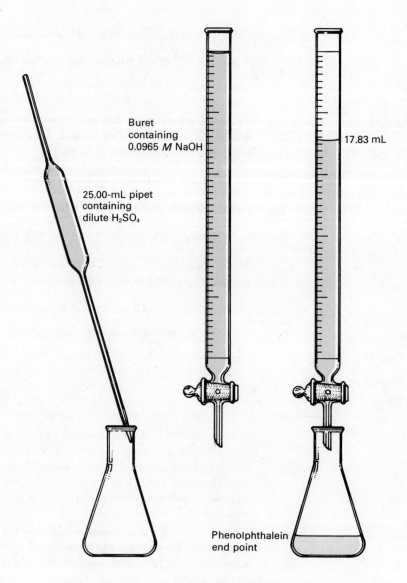

$$\text{Millimoles } H_2SO_4 = \frac{1}{2}(\text{mmol } OH^-) = \frac{1}{2}(1.721 \text{ mmol}) = 0.861 \text{ mmol}$$

The molarity of a solution is the number of mmoles per milliliter.

$$M_{H_2SO_4} = \frac{0.861 \text{ mmol}}{25.00 \text{ mL}} = 0.0344 \frac{\text{mmol}}{\text{mL}}, \text{ or } 0.0344 \, M$$

Titrations may involve precipitation or redox reactions. By titrating a solution containing Ag^+ with standard KSCN solution,

$$Ag^+(aq) + SCN^-(aq) \longrightarrow AgSCN(s)$$

we can determine the Ag^+ concentration present in the solution. Oxalate ion ($C_2O_4^{2-}$) can be titrated by oxidizing it with standard permanganate solution.

$$16 \, H^+(aq) + 2 \, MnO_4^-(aq) + 5 \, C_2O_4^{2-} \longrightarrow 2 \, Mn^{2+}(aq) + 10 \, CO_2(g) + 8 \, H_2O$$

Standardization

We often can prepare standard solutions by weighing the desired reagent and measuring the solution's volume carefully. But there is no source of NaOH pure enough to allow an accurate estimate of the concentration of a NaOH solution from

11.3 STOICHIOMETRY IN SOLUTIONS

the way it is prepared. We must **standardize** NaOH solutions by titrating them against carefully weighed amounts of pure acids with known molecular masses.

Example 11.6

Potassium hydrogen phthalate ($KHC_8H_4O_4$, molecular mass = 204.1 amu) is a monoprotic acid. It takes 28.35 mL of NaOH solution to neutralize 0.6191 g $KHC_8H_4O_4$. What is the concentration of the NaOH solution?

Solution

Since $KHC_8H_4O_4$ is monoprotic, the titration reaction is

$$NaOH + KHC_8H_4O_4 \longrightarrow NaKC_8H_4O_4 + H_2O$$

$$\text{Moles } KHC_8H_4O_4 = \frac{0.6191 \text{ g}}{204.1 \text{ g/mol}} = 3.033 \times 10^{-3} \text{ mol, or } 3.033 \text{ mmol}$$

$$\text{Moles NaOH} = \text{Moles } KHC_8H_4O_4 = 3.033 \text{ mmol}$$

$$M_{NaOH} = \frac{3.033 \text{ mmol}}{28.35 \text{ mL}} = 0.1070 \text{ mmol/mL, or } 0.1070 \, M$$

Normality

We can avoid balancing equations and using stoichiometric ratios in titration calculations by using **normality** (N) as the concentration system. The normality of a solution is the number of **equivalents** of reagent in a liter of solution. In any titration each equivalent of the standard substance combines with one equivalent of unknown, and the number of equivalents of standard reagent used equals the number of equivalents of unknown. This is expressed in the following equation.

$$N_{unk} \cdot V_{unk} = N_{std} \cdot V_{std} \tag{11.4}$$

The **equivalent mass** of a substance is its molecular mass divided by a whole number, n,

$$\text{Equivalent mass} = \frac{\text{Molecular mass}}{n} \tag{11.5}$$

and the number of equivalents in a sample equals the mass of the substance divided by its equivalent mass.

$$\text{Equivalents} = \frac{\text{Sample mass}}{\text{Equivalent mass}} \tag{11.6}$$

It is fairly easy to see that the number of equivalents in a sample is a whole-number multiple of the number of moles, and the normality of a solution equals its molarity multiplied by the same whole number.

$$N = nM \tag{11.7}$$

In some cases n equals one, in others, larger whole numbers. The definition of an equivalent of a particular substance and the related value of n depends on whether the titration involves an acid–base, redox, or precipitation reaction.

Acid-base titrations. An equivalent of acid yields 1 mol of H^+ in a titration. The equivalent mass of an acid equals its molecular mass divided by the number of protons titrated per molecule. In other words, the value of n is the number of H^+ ions titrated per molecule. The equivalent mass of H_2SO_4 is half its molecular mass, 1 mol H_2SO_4 contains 2 equivalents, and the normality of a H_2SO_4 solution is twice its molarity.

11 IONIC SOLUTIONS, ACIDS, AND BASES

Equivalent mass H_2SO_4 = ½ Molecular mass H_2SO_4

$N_{H_2SO_4} = 2\, M_{H_2SO_4}$

An equivalent of base yields 1 mol of OH^- and neutralizes 1 equivalent of acid in a titration. For hydroxides, n is the number of OH^- ions titrated per formula unit. Thus the normality of a solution of a monoprotic base such as NaOH equals its molarity.

Example 11.7

What are the normalities of a. 0.150 M HI as an acid
b. 0.033 M $Ba(OH)_2$ as a base?

Solution

a. Since there is only one proton per molecule, N = M = 0.150 N.
b. Since $Ba(OH)_2$ contains two OH^- ions per formula unit, its normality is twice its molarity.

$$N = 2M = \left(2\, \frac{\text{equiv}}{\text{mol}}\right)\left(0.033\, \frac{\text{mol}}{\text{L}}\right) = 0.066\ \text{equiv/L, or 0.066 N}$$

Redox titrations. An equivalent of an oxidizing agent absorbs 1 mol of electrons; its equivalent mass equals its molecular mass divided by the number of electrons taken up per formula unit. When titrated against oxalate, one permanganate ion absorbs five electrons so its n value is five.

$$5\, e^- + 8\, H^+(aq) + MnO_4^-(aq) \longrightarrow Mn^{2+}(aq) + 4\, H_2O$$

The equivalent mass of $KMnO_4$ is 1/5 its molecular mass, and the normality of a $KMnO_4$ solution is five times its molarity.

$$\text{Equivalent mass } KMnO_4 = \frac{1}{5}\, \text{Molecular mass } KMnO_4$$

$$N_{KMnO_4} = 5\, M_{KMnO_4}$$

One equivalent of a reducing agent gives up a mole of electrons and reduces one equivalent of oxidizing agent. The value of n for a reducing agent is the number of electrons released by a formula unit. For example,

$$Sn^{2+}(aq) \longrightarrow Sn^{4+}(aq) + 2\, e^-$$

$$\text{Equivalent mass } SnCl_2 = \frac{1}{2}\, \text{Molecular mass } SnCl_2$$

$$N_{SnCl_2} = 2\, M\ SnCl_2$$

Example 11.8

What is the normality of 0.0768 M $Na_2C_2O_4$ in a titration with permanganate?

Solution

$$C_2O_4^{2-}(aq) \longrightarrow 2\, CO_2(g) + 2\, e^-$$

Since each oxalate ion releases two electrons, the normality is twice the molarity.

$$N = \left(2\, \frac{\text{equiv}}{\text{mol}}\right)\left(0.0768\, \frac{\text{mol}}{\text{L}}\right) = 0.1536\ \text{equiv/L, or 0.1536 N}$$

11.3 STOICHIOMETRY IN SOLUTIONS

Precipitation titrations. The normality of a solution in a precipitation titration is the molarity of the precipitating ion times its charge. (In this case we ignore the sign of the charge.) For example, solutions of CrO_4^{2-} are used to precipitate silver ion in a titration. In this titration, a CrO_4^{2-} solution has a normality that is twice its molarity.

Example 11.9

What is the normality of 0.1040 M KSCN in the following titration?

$$Hg^{2+}(aq) + 2\ SCN^-(aq) \longrightarrow Hg(SCN)_2(s)$$

Solution

Since the ionic charge of SCN^- is -1, the normality equals the molarity. The solution is 0.1040 N.

Example 11.10

It takes 18.3 mL of 0.115 N H_2SO_4 to titrate 10.0 mL of an unknown ammonia solution. What is the normality of NH_3?

Solution

Since we are dealing in normalities, we can ignore the stoichiometric ratios of the balanced titration equation. The equivalents of ammonia equal the equivalents of sulfuric acid.

$$\text{Equivalents } H_2SO_4 = \left(0.115\ \frac{\text{mequiv}}{\text{mL}}\right)(18.3\ \text{mL}) = 2.10\ \text{mequiv}$$

(A mequiv is 10^{-3} equiv, and normality can be expressed in mequiv/mL.)

$$\text{Equivalents } NH_3 = 2.10\ \text{mequiv}$$

$$N_{NH_3} = \frac{2.10\ \text{mequiv}}{10.0\ \text{mL}} = 0.210\ \text{mequiv/mL} = 0.210\ N$$

The normality of a particular solution may be different for different titrations. Titrating phosphoric acid (H_3PO_4) to the color change of methyl red indicator neutralizes only one proton per molecule, whereas two of the three protons react if titrated to a phenolphthalein end point. The normality of phosphoric acid is equal to its molarity in the first titration, but twice its molarity in the second titration.

The concept of equivalent mass is useful for titrating an acid of unknown structure. Not knowing how many acidic protons the molecule contains, we cannot estimate its molecular mass by titration. Instead we measure the equivalent mass. If we later obtain an estimate of the molecular mass (even an approximate one), we will know how many acidic protons there are in the molecule.

Example 11.11

The titration of 0.2142 g of an unknown acid requires 32.07 mL of 0.1028 N NaOH. What is the equivalent mass of the acid?

Solution

$$\text{Equivalents base} = V_{NaOH} \times N_{NaOH}$$

$$= \left(0.1028\ \frac{\text{mequiv}}{\text{mL}}\right)(32.07\ \text{mL})$$

$$= 3.194\ \text{mequiv, or } 3.194 \times 10^{-3}\ \text{equiv}$$

$$\text{Equivalents acid} = \text{Equivalents base} = 3.194 \times 10^{-3}\ \text{equiv}$$

$$\text{Equivalent mass} = \frac{0.2142\ \text{g}}{3.194 \times 10^{-3}\ \text{equiv}} = 67.06\ \text{g/equiv, or } 67.06\ \text{amu}$$

Summary

1. Electrolytes dissolve in water to form conducting solutions by dissociating into freely moving ions. Colligative properties of electrolyte solutions show that each formula unit of strong electrolyte dissociates into a whole number (i) of ions. The equilibrium for dissociation of a weak electrolyte is much less favorable than for strong electrolytes, and only a small percentage of the dissolved molecules dissociate in solution.

2. Arrhenius definitions: An acid is an electrolyte that reacts with water to produce a hydrated H^+ ion. A base dissociates in water to form OH^- ion. The neutralization reaction between a strong acid and a strong base is

$$H^+(aq) + OH^-(aq) \rightleftharpoons H_2O$$

3. Brønsted–Lowry definitions: An acid–base reaction is a proton transfer from an acid to a base. An acid is a proton donor and a base is a proton receptor. The species formed by removing a proton from an acid is its conjugate base. A conjugate acid of a base is the species formed when the base receives a proton. The direction of favorable equilibrium for a proton transfer reaction is toward the weaker acid–base pair.

4. Lewis definitions: A base is a species that donates an electron pair to form a covalent bond. An acid is an electron-pair receptor.

5. A molecule of a polyprotic acid contains more than one dissociable proton. A polyprotic base can react with several H^+ ions.

6. An amphoteric species acts as a base under sufficiently acidic conditions and is an acid in a sufficiently basic solution.

7. We titrate an acid solution by adding measured volumes of base of known concentration (a standard solution) until the end point is reached. Knowing the concentration of the standard base solution, the relative volumes of the two solutions needed to reach the end point, and the stoichiometry of the neutralization reaction, we can compute the concentration of the acid. Other titrations are based on redox and precipitation reactions.

8. $\text{Normality} = N = \dfrac{\text{Equivalents}}{\text{Liter of solution}}$

   At the end point of any titration the number of equivalents of one reagent equals the number of equivalents of the other, and

   $$N_1 V_1 = N_2 V_2$$

   An equivalent of acid is enough acid to release 1 mol of H^+ in a titration. The equivalent mass of an acid equals its molecular mass divided by the number of protons titrated per molecule. An equivalent of base is enough base to react with 1 mol of H^+. An equivalent of oxidant absorbs 1 mol of electrons. An equivalent of reductant releases 1 mol of electrons. The normality of a solution in a precipitation titration is the molarity of the precipitating ion times its charge.

Vocabulary List

electrolyte	neutral solution	amphoteric
nonelectrolyte	Brønsted–Lowry acid	titration
strong electrolyte	Brønsted–Lowry base	standard solution
weak electrolyte	conjugate base	end point
degree of dissociation	conjugate acid	standardization
acid	leveling effect	normality
base	Lewis acid	equivalent
neutralization reaction	Lewis base	equivalent mass
hydronium ion	polyprotic acid	
hydroxide ion	polyprotic base	

Exercises

1. What is the *i* factor for each of the following?
   a. $Ba(NO_3)_2$ b. $CaSO_4$ c. $MgCl_2$ d. NH_4ClO_4

2. The *i* factor for Na_2CO_3 is approximately three. Write an equation for the dissociation of solid Na_2CO_3 as it dissolves.

3. a. What ions are present in a solution that is 0.200 M in KCl, 0.300 M in $CdSO_4$ and 0.500 M in $FeCl_3$? All the reagents are strong electrolytes.
   b. What is the concentration of each ion?

4. Describe two ways of preparing a solution that is 0.50 M in Ca^{2+}, 0.50 M in Zn^{2+}, 1.00 M in Cl^-, and 1.00 M in NO_3^-.

5. Which has the lower freezing point, 0.500 M HI (a strong acid) or 0.500 M HF (a weak acid)? Why?

6. Phosphorous acid, H_3PO_3, is a diprotic acid in water. Write the equations for its dissociation reactions.

7. A 16.35-mL sample of HCl solution contains 3.725 mmol of HCl. What is the molarity of HCl in this solution?

8. Mercuric ion, Hg^{2+}, can be titrated by precipitating it with thiocyanate ion.

   $$Hg^{2+}(aq) + 2\ SCN^-(aq) \longrightarrow Hg(SCN)_2(s)$$

   A solution containing Hg^{2+} reacts with 17.83 mL of 0.0987 M KSCN. How many millimoles of Hg^{2+} were present in the solution?

9. Describe the neutralization of an aqueous solution of ammonia by an aqueous solution of HI (a strong acid):
   a. As a reaction between an Arrhenius acid and an Arrhenius base
   b. As a Brønsted–Lowry acid–base reaction
   c. As a Lewis acid–base reaction

10. Identify the Lewis acid and the Lewis base in each of the following reactions.

    a. $:\!\ddot{\underset{\cdot\cdot}{Cl}}\!-\!Al\!-\!\ddot{\underset{\cdot\cdot}{Cl}}\!: + :\!\ddot{\underset{\cdot\cdot}{Cl}}\!:^{-} \longrightarrow :\!\ddot{\underset{\cdot\cdot}{Cl}}\!-\!Al\!-\!\ddot{\underset{\cdot\cdot}{Cl}}\!:^{-}$ (with Cl atoms above and below Al)

    b. $H\!:^{-} + H^{+} \longrightarrow H_2$

    c. $H\!-\!\overset{H}{\underset{H}{N\!:}} + H\!-\!\overset{H}{\underset{H}{C^{+}}} \longrightarrow H\!-\!\overset{H\ \ H}{\underset{H\ \ H}{C\!-\!N}}\!-\!H$ (product has + charge)

11. a. How many millimoles of NaOH are present in 37.2 mL of 0.09865 M NaOH?
    b. How many moles of NaOH are present in this sample?

12. a. How many millimoles of H_2SO_4 are present in 20.0 mL of a 0.03765 M solution?
    b. How many moles of NaOH can this solution neutralize?

13. What is the conjugate acid of each?
    a. HSO_3^- b. C_6H_6 c. HNO_3

14. What is the conjugate acid of each?
    a. I^- b. HIO_3 c. $HAsO_4^{2-}$

15. What is the conjugate base of each?
    a. AsH_3 b. $H_2AsO_3^-$ c. H_2ZnO_2

16. What is the conjugate base of each?
    a. HSO_3^- b. C_6H_6 c. OH^-

17. What is the normality of 0.0783 M H_3AsO_4 in the following titration?

    $$H_3AsO_4(aq) + 2\ OH^-(aq) \longrightarrow HAsO_4^{2-}(aq) + 2\ H_2O$$

18. What is the normality of 0.0376 M NaOH when titrated against H_3PO_4?

19. What is the normality of 0.1725 M $K_2Cr_2O_7$ when it acts as an oxidizing agent in the following half-reaction?

$$6\ e^- + 14\ H^+(aq) + Cr_2O_7^{2-}(aq) \longrightarrow 2\ Cr^{3+}(aq) + 7\ H_2O$$

21. A 0.0567 M solution of $Ba(NO_3)_2$ is used to precipitate sulfate ion according to the equation

$$Ba^{2+}(aq) + SO_4^{2-}(aq) \longrightarrow BaSO_4(s)$$

What is the normality of the solution?

23. Monohydrogen arsenate ion, $HAsO_4^{2-}$, is amphoteric. Write equations for its reaction with a strong acid and its reaction with a strong base.

25. Identify the Brønsted–Lowry acid and the Brønsted–Lowry base on the left-hand side of each of the following equations.
    a. $H_2PO_4^-(aq) + HSO_4^-(aq) \rightleftharpoons SO_4^{2-}(aq) + H_3PO_4(aq)$
    b. $CH_3CO_2H(aq) + HCl(aq) \rightleftharpoons Cl^-(aq) + CH_3CO_2H_2^+(aq)$
    c. $H_2O + S^{2-}(aq) \rightleftharpoons HS^-(aq) + OH^-(aq)$

27. Which of the following equations have favorable equilibria? (Refer to Table 11.3.)
    a. $HCO_3^-(aq) + CN^-(aq) \rightleftharpoons HCN(aq) + CO_3^{2-}(aq)$
    b. $NH_4^+(aq) + CO_3^{2-}(aq) \rightleftharpoons NH_3(aq) + HCO_3^-(aq)$
    c. $H_2PO_4^-(aq) + H_3O^+(aq) \rightleftharpoons H_3PO_4(aq) + H_2O$

29. What is the equivalent mass of $Cr(OH)_3$ in the following reaction?

$$3\ H^+(aq) + Cr(OH)_3(s) \longrightarrow Cr^{3+}(aq) + 3\ H_2O$$

31. How many equivalents are present in 45.0 g $Na_2SO_4 \cdot 10H_2O$ if the solution made from it is used to precipitate strontium sulfate?

$$Sr^{2+}(aq) + SO_4^{2-}(aq) \longrightarrow SrSO_4(s)$$

33. What is the equivalent mass of H_2O_2 in a reaction in which it is oxidized to oxygen gas?

$$H_2O_2(aq) \longrightarrow O_2(g) + 2\ H^+(aq) + 2\ e^-$$

35. It takes 22.37 mL of $Ba(OH)_2$ solution to neutralize 10.00 mL of 0.01075 M HCl.
    a. What is the molarity of the $Ba(OH)_2$ solution?
    b. What is its normality?

37. A 15.0-mL sample of an unknown HCl solution is neutralized by 26.3 mL of 0.0767 N KOH. What is the normality of the HCl solution?

20. A 0.00374 M solution of $SnCl_2$ is used in a titration in which it acts as a reducing agent. What is its normality?

$$Sn^{2+}(aq) \longrightarrow Sn^{4+}(aq) + 2\ e^-$$

22. What is the normality of 0.0529 M K_2CrO_4 in the following precipitation reaction?

$$Pb^{2+}(aq) + CrO_4^{2-}(aq) \longrightarrow PbCrO_4(s)$$

24. Phosphate ion, PO_4^{3-}, is a triprotic base. Write equations for the three basic reactions in which water acts as the Brønsted–Lowry acid.

26. Identify the Brønsted–Lowry acid and the Brønsted–Lowry base on the left-hand side of each of the following equations.
    a. $HCO_2H(aq) + NH_3(aq) \rightleftharpoons HCO_2^-(aq) + NH_4^+(aq)$
    b. $CH_3O^- + NH_3 \rightleftharpoons CH_3OH + NH_2^-$
    c. $CH_3^- + CHCl_3 \rightleftharpoons CH_4 + CCl_3^-$

28. Use the information in Table 11.3 to decide which of the following reactions have favorable equilibria.
    a. $H_3O^+(aq) + Cl^-(aq) \rightleftharpoons HCl(aq) + H_2O$
    b. $CH_3CO_2H(aq) + HS^-(aq) \rightleftharpoons H_2S(aq) + CH_3CO_2^-(aq)$
    c. $NH_4^+(aq) + HS^-(aq) \rightleftharpoons NH_3(aq) + H_2S(aq)$

30. What is the equivalent mass of H_3BO_3 in the following titration?

$$OH^-(aq) + H_3BO_3(aq) \longrightarrow H_2O + H_2BO_3^-(aq)$$

32. Succinic acid, $H_2C_4H_4O_4$, is a diprotic acid. How many equivalents of succinic acid are present in a 25.0-g sample if both acidic protons on the molecule are titrated?

34. What is the equivalent mass of $Na_2S_2O_3 \cdot 5H_2O$ when used as a reducing agent?

$$2\ S_2O_3^{2-}(aq) \longrightarrow S_4O_6^{2-}(aq) + 2\ e^-$$

36. In a titration in which two of the three protons of H_3PO_4 are neutralized, 35.0 mL of a phosphoric acid solution are titrated by 17.7 mL of 0.0876 N NaOH.
    a. What is the normality of the phosphoric acid?
    b. What is its molarity?

38. A solution of NaOH is standardized by titrating 0.4870 g of potassium hydrogen phthalate ($KHC_8H_4O_4$), a monoprotic acid, with 22.73 mL of the solution. What is the normality of the solution?

39. Predict the freezing point of 0.150 m $Mg(NO_3)_2$. The molal freezing-point depression of water is 1.86 °C/m. (Assume ideality.)

40. What is the normality of 0.00468 M $HgCl_2$ when it acts as an oxidant?

$$2\ e^- + 2\ Hg^{2+}(aq) + 2\ Cl^-(aq) \longrightarrow Hg_2Cl_2(s)$$

41. A solution of Na_2S is oxidized according to the following half-reaction.

$$8\ OH^-(aq) + S^{2-}(aq) \longrightarrow SO_4^{2-}(aq) + 4\ H_2O + 8\ e^-$$

How many equivalents of Na_2S are present in a solution containing 15.0 g of this substance?

EXERCISES

42. What is the strongest base that can exist in liquid HF?

43. a. What is the strongest acid that can exist in liquid ammonia?
b. Will you be able to detect any differences in the acid strengths of H_3PO_4, CH_3CO_2H, and H_2S in liquid ammonia?

---

44. The concentration of a permanganate solution is determined (standardized) against solid $Na_2C_2O_4 \cdot 2H_2O$. This solid dissolves in water according to

$$Na_2C_2O_4 \cdot 2H_2O(s) \longrightarrow 2\,Na^+(aq) + C_2O_4^{2-}(aq) + 2\,H_2O$$

and the solution containing $C_2O_4^{2-}$ ions is titrated as

$$16\,H^+(aq) + 2\,MnO_4^-(aq) + 5\,C_2O_4^{2-}(aq) \longrightarrow 2\,Mn^{2+}(aq) + 10\,CO_2(g) + 8\,H_2O$$

Exactly 35.27 mL of permanganate solution react with 0.7521 g of $Na_2C_2O_4 \cdot 2H_2O$.
a. Calculate the normality of the permanganate solution. **b.** Calculate its molarity.

45. What volume of 0.0387 N H_2SO_4 is needed to react with 25.84 mL of 0.0589 N NH_3? Both protons on H_2SO_4 are neutralized.

46. It takes 28.73 mL of 0.1403 N NaOH to neutralize 0.7951 g of an unknown acid.
a. What is the equivalent mass of the acid?
b. An approximate estimate of the molecular mass of the acid by freezing-point depression gives the value 385 amu. How many acidic protons does a molecule of the acid contain?

47. A 1.578-g sample of an unknown reducing agent is titrated with 38.47 mL of 0.1046 N $KMnO_4$. In the titration, the permanganate ion is reduced to Mn^{2+}. What is the equivalent mass of the reducing agent?

48. The freezing point of a solution formed by dissolving 31.17 mg of $CoCl_3 \cdot 5NH_3$ in 50.00 g H_2O is 0.01330°C lower than that of pure water.
a. What is the i factor for this compound?
b. The solution is a good electrical conductor, and the rapid precipitation of AgCl when Ag^+ is added to the solution indicates the presence of Cl^- ions. Further analysis shows that there is no free NH_3 in the solution. Write an equation for the dissociation of this compound into its ions.

49. The Kjeldahl analysis for the nitrogen content of biological material consists of the conversion of all the biological nitrogen to NH_4^+. The solution containing the ammonium ion is then made basic and the ammonia that forms is distilled.

$$NH_4^+(aq) + OH^-(aq) \longrightarrow NH_3(g) + H_2O$$

The ammonia vapor reacts with a carefully measured volume of standard HCl solution.

$$NH_3(g) + H^+(aq) \longrightarrow NH_4^+(aq)$$

The excess HCl is titrated with standard NaOH solution to calculate the amount of acid that has reacted with the ammonia. Use the following data to calculate the percentage of nitrogen in a protein sample.

Mass of protein sample = 0.1057 g

Volume of HCl = 10.00 mL

Concentration of HCl = 0.2045 N

Volume of NaOH = 4.71 mL

Concentration of NaOH = 0.1987 N

50. The Volhard method for analysis of chloride ion in a solution involves adding a carefully measured amount of $AgNO_3$ solution (known to be in excess of the Cl^-) to precipitate all the Cl^- as AgCl.

$$Ag^+(aq) + Cl^-(aq) \longrightarrow AgCl(s)$$

After filtering to remove this precipitate, we determine the moles of excess Ag^+ by titrating with KSCN.

$$Ag^+(aq) + SCN^-(aq) \longrightarrow AgSCN(s)$$

In a particular determination, 25.00 mL of 0.1037 M $AgNO_3$ was added to a chloride solution, and it took 15.42 mL of 0.09873 M KSCN to titrate the excess Ag^+ in the filtrate. What mass of chloride did the solution contain?

chapter 12
Chemical Kinetics

Molecular collisions are necessary for chemical reactions to occur—for example, the reaction pictured here:
$H_2(g) + I_2(g) \longrightarrow 2\,HI(g)$

Any chemist who wants to apply a chemical reaction to some practical purpose faces two important questions: "How far will the reaction go before it reaches equilibrium?" and "How long will it take for the reaction to reach equilibrium?" There is no correlation between a reaction's **kinetic characteristics**—the rate at which it approaches equilibrium and the factors influencing this rate—and the extent of reaction at equilibrium. For example, even though the equilibrium for the formation of water from its elements is very favorable, the reaction can be so slow that a mixture of hydrogen and oxygen can exist indefinitely without any water forming. On the other hand, the dissociation of acetic acid (a reaction with an unfavorable equilibrium) reaches its equilibrium within a millisecond.

Each reaction has its own kinetic behavior pattern. Its rate of approach to equilibrium responds in a unique and characteristic way to changes in *concentration* and *temperature* and to the presence of a *catalyst*. Some knowledge of this kinetic behavior is necessary before we can put a reaction to practical use. For example, one way to analyze for iron is to titrate Fe^{2+} with permanganate solution to an end point consisting of a persistent pink color of excess MnO_4^-.

$$8\,H^+(aq) + MnO_4^-(aq) + 5\,Fe^{2+}(aq) \longrightarrow 5\,Fe^{3+}(aq) + Mn^{2+}(aq) + 4\,H_2O$$

If the reaction were slow, the color of unreacted permanganate would accumulate before the Fe^{2+} was consumed, and the end point would be unreliable. Fortunately, the reaction is fast enough under the titration conditions to produce a clear end point.

Kinetic factors are important in influencing the stabilities of substances. If it were not for the slowness of their decomposition reactions, many known substances

Figure 12.1
Kinetic and equilibrium control of a reaction.

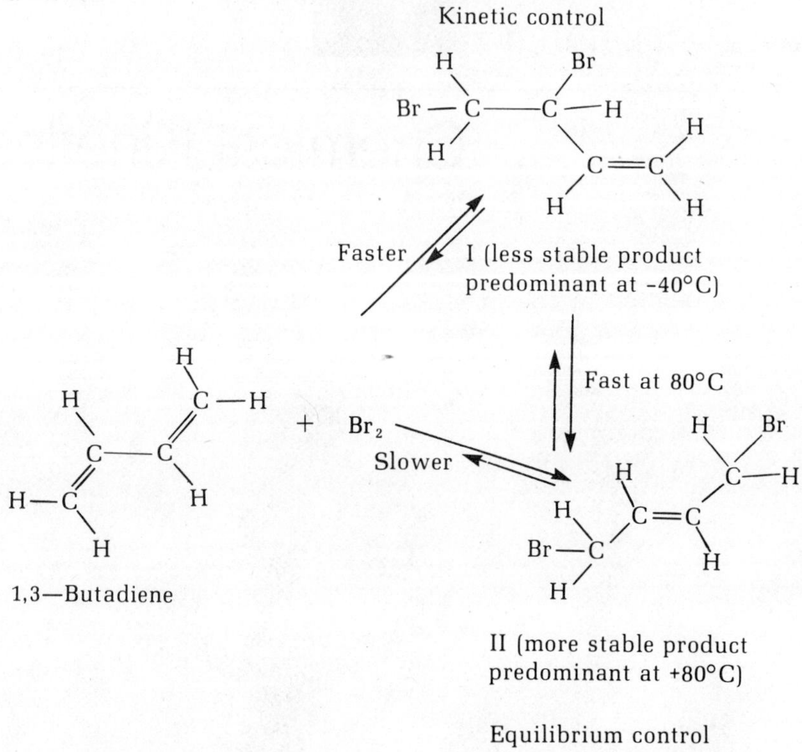

could not exist. For example, the decomposition of acetylene (C_2H_2, a fuel used in welding) has a negative free energy change and its equilibrium is favorable.

$$C_2H_2(g) \rightleftharpoons 2\ C(s) + H_2(g) \qquad \Delta G° = -209\ \text{kJ}$$

Nevertheless, we can store acetylene indefinitely by maintaining conditions for which its decomposition is immeasurably slow. Isomerism is possible only when the reactions that convert the less stable isomers to the most stable isomers are slow.

Relative reaction rates can also influence the course of a reaction and the nature of its products. In many cases (see Figure 12.1 for an example), the reaction produces the less stable of two possible products because the less stable substance is formed more rapidly. This situation is called **kinetic control**. The situation known as **equilibrium control** is the one in which the more stable substance is the major product.

Since kinetic considerations are so important, it would be convenient to formulate general principles for predicting reaction rates. Unfortunately, broad generalizations about reaction rates do not work well, and we must determine the kinetic behavior for each reaction by careful experimental study. In this chapter, we will look at the way chemists gather and interpret kinetic data.

12.1
The Measurement of Reaction Rates

A **reaction rate** is a rate of change of the concentration of a reactant or product. (Reactant concentrations *decrease* during the reaction, product concentrations *increase*.) We can define the rate for the reaction

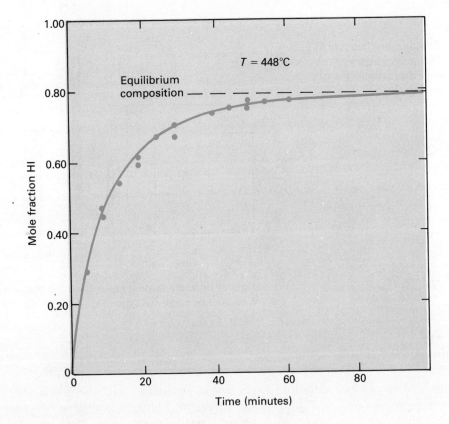

Figure 12.2
Composition change during the formation of HI from an equimolar mixture of H_2 and I_2. *Source:* Data are from M. Bodenstein, *Zeitschrift für Physikalische Chemie* **13**, 57(1894).

$$2\ NO(g) + 2\ H_2(g) \longrightarrow N_2(g) + 2\ H_2O(g)$$

as the rate of formation of nitrogen.*

$$\text{Rate} = \frac{d[N_2]^\dagger}{dt}$$

To measure a reaction rate, we must have (1) a device for measuring elapsed time and (2) a method for estimating the concentration of at least one of the reactants or products. Time measurement is relatively simple. We can use a stop watch, a chart recorder whose paper moves at a known speed, or an oscilloscope with a known sweep speed. (The last device is used to measure the rates of very fast reactions.)

The concentration can be measured by removing a portion of the reaction mixture at a measured time and quickly changing conditions (such as acidity or temperature) to stop the reaction. Having quenched the reaction in the sample, we can analyze its composition at leisure. The graph of concentration versus time shown in Figure 12.2 was prepared from data obtained in this way. Some of the best methods for monitoring the concentration changes in a reacting system involve the continuous measurement of a physical property that depends on the chemical composition of the system. For example, we can observe the extent of the reaction between NO and H_2 by measuring the decrease in the gas pressure (Figure 12.3) since the number of moles of gas decreases as the reaction proceeds. If the products of a reaction absorb

---

*We could also define the rate of this reaction in terms of any of the other reactants or product. The mathematical relationship between different definitions for the rate of the same reaction depends on the reaction's stoichiometry. For example, the rate of decrease of NO concentration is exactly twice the rate of formation of N_2 because two NO molecules are consumed for each N_2 molecule formed.

†The symbol dx/dt stands for the rate of change of the quantity x with time. The brackets around the formula indicate the concentration in moles per liter, so the symbol $d[N_2]/dt$ indicates the rate of increase in the concentration (in moles per liter) of N_2 with time. A negative sign in front of the symbol stands for a rate of decrease in concentration.

Figure 12.3
Following a reaction rate by pressure change. The pressure in this system decreases as the reaction proceeds.

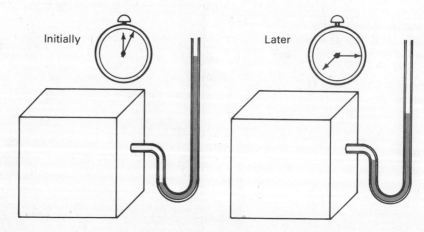

Reaction: $2 NO(g) + 2 H_2(g) \rightarrow N_2(g) + 2 H_2O(g)$

light at different wavelengths than the reactants, the progress of the reaction can be followed by recording the change in the ability of the system to absorb light (Figure 12.4).

12.2
Reaction Mechanisms

The Effects of Concentration

We cannot generalize about the relationship between reaction rates and reactant concentrations. Although increasing the concentration of a reactant raises the rate proportionately in some cases, there are other instances in which a reaction rate is independent of the concentration of a particular reactant. Some reaction rates are proportional to the square of a reactant concentration, others to the square root of concentration. The balanced equation for a reaction tells us nothing about the effect of concentration on its rate. The only way to obtain this information is by systematic experimental observation.

Figure 12.4
Following a reaction rate by light absorption.

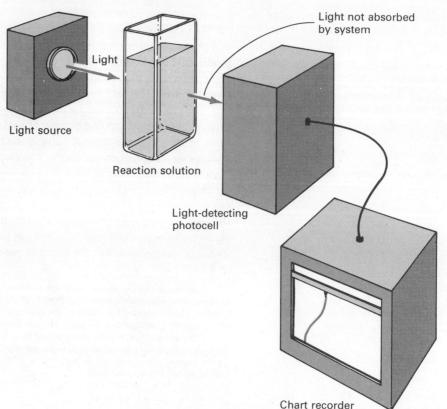

12.2 REACTION MECHANISMS

Table 12.1 Initial Rates of the Reaction: $2\,H_2(g) + 2\,NO(g) \longrightarrow 2\,H_2O(g) + N_2(g)$

Temperature: 1100 K   Rate$_{H_2}$ = $-d[H_2]/dt$ Concentrations in moles per liter.

Experiment	[NO]	[H$_2$]	Relative [H$_2$]	Rate (M/sec)	Relative rate	k (M^{-2}sec^{-1})
Series I: Initial [H$_2$] varies; initial [NO] is constant						
1	5.92×10^{-3}	2.18×10^{-3}	1.00	1.17×10^{-5}	1.00	153
2	5.92×10^{-3}	3.04×10^{-3}	1.39	1.63×10^{-5}	1.39	152
3	5.92×10^{-3}	4.28×10^{-3}	1.97	2.37×10^{-5}	2.03	158

Experiment	[H$_2$]	[NO]	Relative [NO]	Relative [NO]^2	Rate (M/sec)	Relative rate	k (M^{-2}sec^{-1})
Series II: Initial [NO] varies; initial [H$_2$] is constant							
4	5.92×10^{-3}	2.25×10^{-3}	1.00	1.00	0.37×10^{-5}	1.00	123
5	5.92×10^{-3}	4.44×10^{-3}	1.98	3.98	1.52×10^{-5}	4.12	130
6	5.92×10^{-3}	5.32×10^{-3}	2.38	5.57	2.22×10^{-5}	6.00	132

Source: These concentrations and rates are calculated from data of C. N. Hinshelwood and T. E. Green, *Journal of the Chemical Society* (London) **1926**, 730.

An example of an investigation to determine the concentration dependence of a reaction rate is the study of the reaction

$$2\,NO(g) + 2\,H_2 \longrightarrow N_2(g) + 2\,H_2O(g)$$

The initial rates of this reaction (Table 12.1) are observed for a number of experiments in which the initial reactant concentrations are varied. Experiments 1 through 3 have the same NO concentration but different H$_2$ concentrations. Note that the rates for these experiments are proportional to [H$_2$] (the molar concentration of hydrogen). Experiments 4 through 6, which differ in their initial NO concentrations but have the same [H$_2$], show the reaction rate is proportional to [NO]^2.

We summarize these observations in the form of a **rate law**—an equation showing the relationship between the rate of a particular reaction and its reactant concentrations.

$$\text{Rate} = k[H_2][NO]^2$$

where k is a proportionality constant called the **rate constant**. If we have found the correct form of the rate law, the value of the rate constant will be the same for any experiment run at a particular temperature. Although the values for k shown in Table 12.1 do vary somewhat from experiment to experiment, the variation is due to experimental uncertainties in measuring the rates, not to the incorrectness of the rate law.

Each reaction has its own characteristic rate law and rate constant. The forms of many rate laws are similar to the rate law for the reaction between H$_2$ and NO in that the rate is proportional to the product of the reactant concentrations, each concentration being raised to an exponential power known as its **order**.

$$m[X] + n[Y] + p[Z] \longrightarrow \text{Products}$$
$$\text{Rate} = k[X]^a[Y]^b[Z]^c \tag{12.1}$$

The kinetic orders a, b, and c are either zero, whole numbers (usually positive), or simple fractions such as $1/2$ or $3/2$. They bear no obvious relationship to the stoichiometric coefficients m, n, or p and cannot be predicted from the form of the balanced equation. The H$_2$–NO reaction is a case in which the kinetic orders do not match the stoichiometric coefficients. Even though two hydrogen molecules are consumed in the balanced equation, the reaction is only first order in H$_2$.

Determining Kinetic Orders

The initial rate method (the one we used for the reaction between H_2 and NO, see Table 12.1) is a straightforward way of finding the kinetic orders of a reaction. Comparison of the initial rates for several experiments in which one initial concentration varies while the others remain constant reveals the order of the reactant whose concentration does change.

Example 12.1

Use the following data to find the rate law for the decomposition of N_2O_5 at 65°C.

$$2\ N_2O_5(g) \longrightarrow 2\ N_2O_4(g) + O_2(g)$$

Initial concentration of N_2O_5 (M)	Initial rate of decomposition (M/sec)
0.100	2.4×10^{-4}
0.050	1.2×10^{-4}
0.020	4.8×10^{-5}

Solution

Cutting the reactant concentration in half (from 0.100 M to 0.050 M) decreases the concentration by half. A fivefold decrease in concentration (0.100 M to 0.020 M) causes the rate to slow down by a factor of five. Since concentration and rate are proportional, the reaction is first order in N_2O_5.

$$\text{Rate} = k[N_2O_5]$$

Once we have determined the form of the rate law, we obtain the value of the rate constant by taking a single experiment and dividing its rate by the reactant concentrations, each raised to its appropriate power.

Example 12.2

The decomposition of acetaldehyde (CH_3CHO) is second order.

$$CH_3CHO(g) \longrightarrow CH_4(g) + CO(g)$$

The rate of the reaction at 800°C when the concentration of acetaldehyde is 0.020 M is 3.6×10^{-8} M/sec. What is the rate constant?

Solution

The rate law for this reaction is

$$\text{Rate} = k[CH_3CHO]^2$$

which we can rewrite in the form

$$k = \frac{\text{Rate}}{[CH_3CHO]^2} = \frac{3.6 \times 10^{-8}\ \text{M/sec}}{(0.020\ \text{M})^2} = 9.0 \times 10^{-5}\ M^{-1} \text{sec}^{-1}$$

As shown by the dimensional analysis in this problem, the rate constant of a second-order reaction has the unit $M^{-1}\text{sec}^{-1}$, but the unit for a rate constant depends on the order of the reaction. The unit for the rate constant of a first-order reaction is sec^{-1}, and the rate constant for a third-order reaction (such as the reaction between H_2 and NO) has the unit $M^{-2}\text{sec}^{-1}$.

We can also obtain the kinetic rate law from the relationship between concentration and time during a single kinetic experiment. A rate law is a differential equation—it describes how the rate of change of a quantity (concentration) depends on that quantity. Using calculus, we integrate the differential equation to obtain an equation relating the concentration to time. For example, a first-order reaction

$$X \longrightarrow Products$$

$$Rate = -\frac{d[X]}{dt} = k[X]$$

has the integrated rate equation

$$kt = 2.30 \log\left(\frac{[X]_0}{[X]}\right) \tag{12.2}$$

or

$$kt = 2.30 \,(\log [X]_0 - \log [X]) \tag{12.3}$$

where $[X]_0$ is the initial reactant concentration and $[X]$ is the concentration at time t. We can recognize a first-order reaction because its graph of log $[X]$ versus time is a straight line whose slope equals $-k/2.30$ (Figure 12.5).

Another feature of the integrated rate law for a first-order reaction is that it has a constant **half-life**. The half-life of a reaction is the time required for the concentration of the reactant to decrease by half. As can be seen from Figure 12.6, the half-life for the decomposition of N_2O_5 at 55°C is 7.8 min. It takes 7.8 min for the concentration of N_2O_5 to fall to half its original value. Another 7.8 min pass before the concentration decreases from half the original value to ¼ of this concentration, and an additional 7.8 min is required to bring the concentration down to ⅛ its original value. The half-life of a first-order reaction is independent of the reactant concentration. In contrast, the half-life of a reaction with a different kinetic order varies with the

Figure 12.5
Concentration–time relationship for a first-order reaction. Source: Data are from F. Daniels and E. H. Johnston, *Journal of the American Chemical Society* **43**, 53(1921).

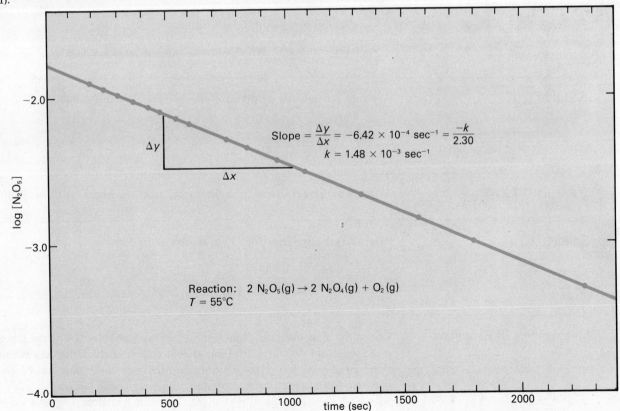

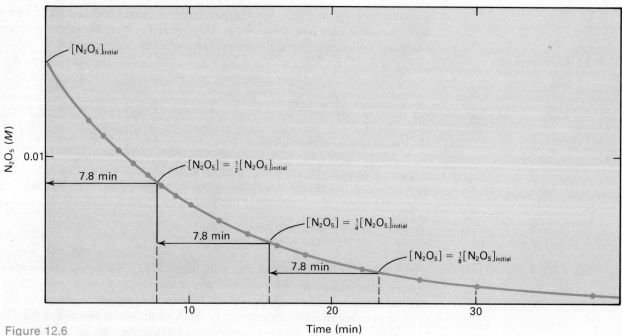

Figure 12.6
Half-life for a first-order reaction.

reactant concentration. The half-life ($t_{1/2}$) of a first-order reaction is related to its rate constant by equation (12.4).

$$t_{1/2} = \frac{2.30 \, (\log 2)}{k} = \frac{0.695}{k} \tag{12.4}$$

Example 12.3

The first-order rate constant for the decomposition of ethyl chloride (C_2H_5Cl)

$$C_2H_5Cl(g) \longrightarrow C_2H_4(g) + HCl(g)$$

is 0.29 sec^{-1} at 600°C. What is the half-life of C_2H_5Cl at 600°C?

Solution

We find the half-life by dividing the rate constant into 0.695.

$$t_{1/2} = \frac{0.695}{0.29 \text{ sec}^{-1}} = 2.4 \text{ sec}$$

Example 12.4

What is the rate constant for a first-order reaction whose half-life is 480 sec^{-1}?

Solution

We can rearrange equation (12.4) to obtain the form

$$k = \frac{0.695}{t_{1/2}} = \frac{0.695}{480 \text{ sec}} = 1.45 \times 10^{-3} \text{ sec}^{-1}$$

Mechanisms of Reactions

One of the most intriguing questions about reaction rates is why the kinetic orders of many reactions do not match their stoichiometric coefficients. Answering this question can give us important insights into the way a reaction occurs. Discrepancies between the kinetic orders and stoichiometric coefficients arise from the way in which we describe chemical changes. We write a chemical equation describing the change that occurs between the time we mix the reactants and the time we analyze for

12.2 REACTION MECHANISMS

products, a description which does not necessarily tell the whole story. In many cases, the observed change is the net result of several simpler reactions occurring in sequence.

The **mechanism** of a reaction is the sequence of molecular events producing the overall chemical change. A reaction mechanism dictates the form of the rate law. For example, the mechanism of the reaction

$$(CH_3)_3CCl + OH^- \longrightarrow (CH_3)_3COH + Cl^-$$

$$\text{Rate} = k[(CH_3)_3CCl]$$

consists of two distinct steps.

Step 1 $(CH_3)_3CCl \longrightarrow (CH_3)_3C^+ + Cl^-$ (slow)

Step 2 $(CH_3)_3C^+ + OH^- \longrightarrow (CH_3)_3COH$ (fast)

The individual steps of a mechanism, called **elementary reactions**, are either **monomolecular** (involving only one reactant molecule), **bimolecular** (a reaction between two molecules) or, in rare cases, **termolecular** (three reactant molecules). The kinetic order of a single elementary reaction matches its molecularity; a monomolecular process is first order, and a bimolecular process is second order. In the example above, the first step, a monomolecular dissociation, has the rate law

$$\text{Rate}_1 = k_1[(CH_3)_3CCl]$$

and the rate of the second step (a bimolecular process) is

$$\text{Rate}_2 = k_2[(CH_3)_3C^+][OH^-]$$

The rate law of the overall reaction depends on the relative rates of the elementary reactions in its mechanism. As far as kinetics are concerned, the most significant elementary reaction is the slowest step, also called the **rate-determining step**. Once the kinetic bottleneck at the rate-determining step is passed, the reaction proceeds rapidly to the final products. A fast elementary reaction following the rate-determining step has no influence on the rate law.

A visible analogy for a rate-determining process (Figure 12.7) consists of an

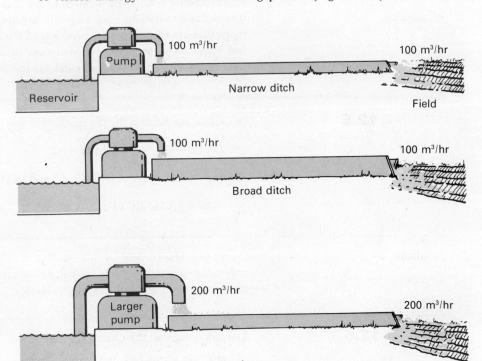

Figure 12.7
A visible analogy for a rate-determining step. Pumping is the rate-determining step. The dimensions of the ditch do not influence the rate of flow of water to the field.

irrigation system in which water is pumped from a reservoir into a ditch that carries the water to a field. In this particular irrigation system, the rate-determining step is the rate of pumping, the pump delivering 100 m³ of water per hour. Since this is not enough to fill the ditch, there is no advantage in enlarging the existing ditch. The water flow reaching the field would still be only 100 m³ per hour. In this system, the flow of water through the ditch is analogous to a fast reaction following the rate-determining step. The only way to speed up the water flow to the field is to improve the rate-determining step, in other words, to pump faster.

Using the concept of the rate-determining step, we can explain the zero kinetic order of hydroxide ion in its reaction with $(CH_3)_3CCl$. Step 1 is the rate-determining step, and since hydroxide ion does not take part in the reaction until step 2 it has no effect on the overall rate of the reaction.

The slow spontaneous decomposition of $(CH_3)_3CCl$ in step 1 forms an **intermediate** (the positive ion) that is rapidly consumed in the subsequent step. Most intermediates are unstable species existing in minute concentrations during the reaction. Since the positive ion forms in a slow reaction and is consumed in a rapid one, its concentration remains too low to detect easily.

Once we discover the rate law of a particular reaction, we try to develop a mechanistic hypothesis consistent with the observed kinetics. It is impossible to prove the validity of any one mechanism by kinetics alone, and we can often think of several possible mechanisms, each agreeing with the rate law. Kinetic analysis provides us with a first step in narrowing the range of possible mechanisms by eliminating those that clearly do not fit the observed rate law.

The major principle that we use in proposing a reaction mechanism based on kinetic data is that the observed rate law depends on the rate-determining step. The kinetic order of a reactant equals the number of molecules of that reactant consumed in the rate-determining step or in previous elementary reactions. If the kinetic order of a reactant is less than its stoichiometric coefficient, then a certain number of reactant molecules equal to the kinetic order is consumed in or before the rate-determining step, and the remainder enter the reaction after the rate-determining step. The reaction between $(CH_3)_3CCl$ and OH^- is an example of this. No hydroxide ions are consumed in the rate-determining step, so the kinetic order of OH^- is zero.

There are many reactions in which the order of a species is greater than its stoichiometric coefficient. In such a case, the number of molecules consumed in or before the rate-determining step still equals the kinetic order, but some of these molecules are regenerated as products of the rate-determining step or of subsequent fast elementary reactions. We will deal in detail with examples of these sorts of mechanisms when we discuss catalysis in the next section.

Example 12.5

Propose a mechanism for the reaction

$$CH_3Cl + OH^- \longrightarrow CH_3OH + Cl^-$$

The rate is first order in both CH_3Cl and OH^-.

$$\text{Rate} = k[CH_3Cl][OH^-]$$

Solution

Since the kinetic orders for this reaction match its stoichiometry, and since there are only two reactant molecules, we can easily explain the rate law by proposing a single step mechanism without any intermediates.

Example 12.6

Remembering that the observed rate law is

$$\text{Rate} = k[H_2][NO]^2$$

propose a mechanism for the reaction

$$2\ H_2(g) + 2\ NO(g) \longrightarrow N_2(g) + 2\ H_2O(g)$$

Solution

Any mechanism in which one of the two hydrogen molecules and both NO molecules react in or before the rate-determining step is consistent with the rate law. It is necessary to propose that a second hydrogen molecule reacts in a fast reaction following the rate-determining step to account for the stoichiometry. One such mechanism is

Step 1 $2\ NO(g) \rightleftharpoons N_2O_2(g)$ (fast)*

Step 2 $N_2O_2(g) + H_2(g) \longrightarrow H_2O_2(g) + N_2(g)$ (slow, rate-determining)

Step 3 $H_2(g) + H_2O_2(g) \longrightarrow 2\ H_2O(g)$ (fast)

When proposing a mechanism, we should check that the net result of the individual steps is the same as the total observed reaction. The proposed mechanism is consistent with the reaction stoichiometry because it shows the net consumption of two molecules each of H_2 and NO and the production of one N_2 and two H_2O molecules.

An alternate mechanism is

Step 1 $NO(g) + H_2(g) \rightleftharpoons H_2NO(g)$ (fast)

Step 2 $H_2NO(g) + NO(g) \longrightarrow H_2O(g) + N_2O(g)$ (slow, rate-determining)

Step 3 $N_2O(g) + H_2(g) \longrightarrow H_2O(g) + N_2(g)$ (fast)

Both mechanisms are consistent with the observed rate law, and a choice between them requires further experimental study. These experiments might include identifying the reactive intermediates by spectroscopic or other physical means.

Heterogeneous Reactions

The rate of a reaction occurring in a heterogeneous system depends on the surface area between the two phases; a reaction between a solid and a liquid or gas is much faster if the solid is finely divided. It is easy to observe this effect by comparing the rate of hydrogen evolution produced by the action of acid on a zinc bar and on zinc powder (Figure 12.8, p. 314). The powder, having a much larger ratio of surface to mass, reacts more rapidly. Rapid oxidation reactions of powders present a hazard because substances that usually oxidize slowly may ignite readily when finely divided. Explosions have resulted from the accidental ignition of coal dust, flour, and powdered metal.

12.3 Catalysis

Homogeneous Catalysis

A **catalyst** increases the rate of a reaction without being consumed. An example of **homogeneous catalysis** (catalysis is the action of a catalyst), where the catalyst and reactants are present in the same phase, is the acid-catalyzed reaction between ethyl acetate ($CH_3CO_2C_2H_5$) and water.

$$CH_3CO_2C_2H_5 + H_2O \rightleftharpoons CH_3CO_2H + C_2H_5OH$$

This reaction, slow in neutral solutions, proceeds rapidly in acid solutions, and its rate is proportional to $[H^+]$.

---

*It is necessary to propose a reversible first step. Otherwise we would predict that large concentrations of N_2O_2 would accumulate in the system. These large concentrations do not occur.

Figure 12.8
The effect of surface area on a heterogeneous reaction:

$2 H^+(aq) + Zn(s) \longrightarrow H_2(g) + Zn^{2+}(aq)$

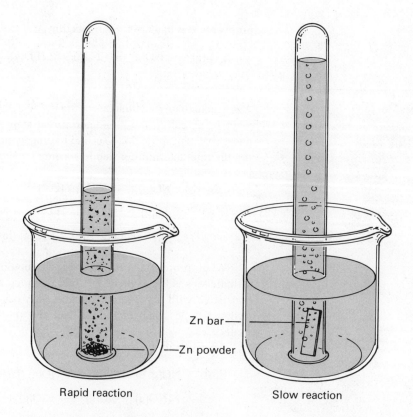

Rapid reaction   Slow reaction

$$\text{Rate} = k[CH_3CO_2C_2H_5][H_2O][H^+]$$

Homogeneous catalysis is another example of a difference between the kinetic order and the stoichiometric coefficient. The stoichiometric coefficient of hydrogen ion in this reaction is zero, but the reaction is first order in $[H^+]$.

The Theory of Homogeneous Catalysis

The kinetic order of a homogeneous catalyst exceeds its stoichiometric coefficient (zero). The catalyst does undergo a chemical change, either in the rate-determining step or in an earlier rapid step, but later it is transformed back into its original molecular state. Because this recycling process enables a single catalyst molecule to aid the reaction of many molecules, catalysts are effective even when their concentrations are low.

In the example of the acid-catalyzed reaction between ethyl acetate and water, the H^+ ion adds to an ethyl acetate molecule to form an intermediate that reacts with water much more rapidly than the neutral ethyl acetate molecule. After this intermediate has reacted, the H^+ ion is expelled from the acetic acid molecule and is free to repeat its catalytic action.

$$CH_3CO_2C_2H_5 + H^+ \rightleftharpoons CH_3CO_2C_2H_5 \cdot H^+ \quad \text{(fast)}$$

$$H_2O + CH_3CO_2C_2H_5 \cdot H^+ \rightleftharpoons CH_3CO_2H_2^+ + C_2H_5OH \quad \text{(slow, rate-determining)}$$

$$CH_3CO_2H_2^+ \rightleftharpoons CH_3CO_2H + H^+ \quad \text{(fast)}$$

Example 12.7

The reduction of Fe(III) by V(III) using copper(II) ion as a catalyst

$$Fe^{3+}(aq) + V^{3+}(aq) \longrightarrow V^{4+}(aq) + Fe^{2+}(aq)$$

is first order in V^{3+} and Cu^{2+} but zero order in Fe^{3+}.

$$\text{Rate} = k[V^{3+}][Cu^{2+}]$$

Propose a mechanism to explain this rate law.

Solution

The rate law indicates that both V^{3+} and Cu^{2+} are involved in or before the rate-determining step and that Fe^{3+} is not reduced until after the rate-determining step. Furthermore, Cu^{2+} must be regenerated in any mechanism we write. One mechanism containing all these features is

$$V^{3+}(aq) + Cu^{2+}(aq) \longrightarrow V^{4+}(aq) + Cu^+(aq) \text{ (slow, rate-determining)}$$

$$Cu^+(aq) + Fe^{3+}(aq) \longrightarrow Cu^{2+}(aq) + Fe^{2+}(aq) \text{ (fast)}$$

Heterogeneous Catalysis

Finely divided palladium metal catalyzes the reaction

$$H_2(g) + C_2H_4(g) \longrightarrow C_2H_6(g)$$

This is an example of **heterogeneous**, or **surface**, **catalysis**, where the catalyst is not in the same phase as the reactants. Although we cannot present detailed mechanisms for heterogeneous catalyses, we are fairly sure they involve chemical reactions between reactant molecules and the catalytic surface (Figure 12.9). Reactant molecules (H_2 in the example above) are **chemisorbed** at active centers on the catalyst's surface. Because they are exothermic processes with heats comparable to heats of reactions, it is

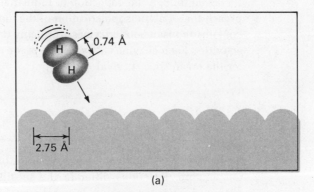

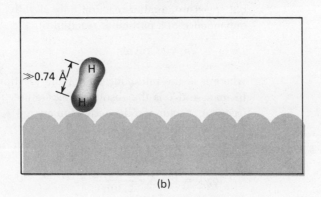

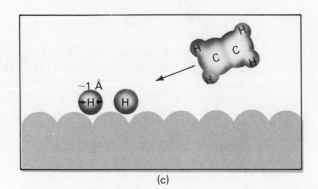

Figure 12.9
Surface catalysis. (a) An H_2 molecule approaches the Pd surface. (b) The H_2 molecule forms a weak bond with Pd surface atoms, weakening the H—H bond and lengthening it. (c) The H_2 molecule breaks apart forming reactive H atoms at the Pd surface, ready to combine with other molecules colliding at the surface. Low-energy electron-diffraction (LEED) studies have verified gas molecule dissociation (H_2 and O_2) at metal surfaces such as nickel and palladium, and observed ordering of the atoms on the metal surface. These metal surfaces have also been observed to change structurally after being used as catalysts for hydrogenation.

likely that chemisorptions form chemical bonds between the reactant molecules and the active centers. Because of distortions in their structures, the chemisorbed molecules react rapidly, and the reaction products diffuse from the surface of the catalyst. The active center, having returned to its original state, is ready to repeat the catalytic process.

Surface catalysis, one of the major foundations of modern chemical technology, allows us to carry out industrial processes at relatively low temperatures, improve product yields, and avoid undesirable side reactions. Different catalysts can cause different reactions of the same substance. In the presence of alumina (Al_2O_3) catalyst, ethanol decomposes to water and ethylene

$$C_2H_5OH(g) \xrightarrow{Al_2O_3} C_2H_4(g) + H_2O(g)$$

but it forms acetaldehyde and hydrogen in the presence of copper.

$$C_2H_5OH(g) \xrightarrow{Cu} CH_3CHO(g) + H_2(g)$$

The surface catalysts already discovered make possible the conversion of crude petroleum into fibers, plastics, solvents, and pharmaceuticals.

Even though the collection of industrial catalysts is impressive, it does not approach the catalytic sophistication of the most primitive organism. Each living cell contains thousands of **enzymes**—proteins with very effective and specific catalytic activities. Each enzyme catalyzes one of the myriad biochemical reactions necessary for the organism's survival.

12.4
Collision Theory

Temperature Effects

Almost every reaction, whether exothermic or endothermic, proceeds faster at higher temperature. Reactions differ in the sensitivities of their rates to a temperature change. A given temperature increase may cause a large rate increase for one reaction but speed up another reaction only slightly. Arrhenius found that the temperature dependence of a particular reaction fits the exponential equation

$$k = Ae^{-E_a/RT} \tag{12.5}$$

where k is the rate constant for the reaction, R is the gas constant expressed in J/K·mol, and T is the absolute temperature. The quantities A and E_a, known as the **frequency factor** and the **activation energy**, respectively, are characteristic constants for each reaction.

We can restate the Arrhenius equation in either of the two following forms,

$$\log k = \log A - \frac{E_a}{2.30RT} \tag{12.6}$$

$$\log\left(\frac{k_2}{k_1}\right) = \frac{E_a}{2.30R}\left(\frac{1}{T_1} - \frac{1}{T_2}\right) \tag{12.7}$$

where k_1 and k_2 are the values of the rate constant at two temperatures T_1 and T_2, respectively. According to equation (12.6), the activation energy of a reaction can be measured by measuring k at several different temperatures and using these data to prepare a graph of $\log k$ versus $1/T$ (Figure 12.10). The result is a straight line whose slope equals $-E_a/2.30R$. Example 12.8 illustrates the use of equation (12.7).

Example 12.8

The first-order rate constant for the decomposition of N_2O_5

Figure 12.10
Estimation of activation energy.

Reaction: $H_2(g) + I_2(g) \rightarrow 2\, HI(g)$

Source: Prepared from the data of M. Bodenstein, *Zeitschrift für Physikalische Chemie* **29**, 295 (1898).

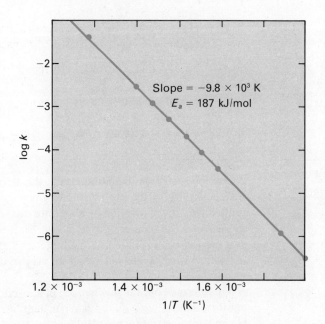

$$2\, N_2O_5(g) \longrightarrow 2\, N_2O_4(g) + O_2(g)$$

is 1.72×10^{-5} sec^{-1} at 25°C and 2.40×10^{-3} sec^{-1} at 65°C. Find the activation energy for this reaction.

Solution

Let $k_1 = 1.72 \times 10^{-5}$ sec^{-1} and $k_2 = 2.40 \times 10^{-3}$ sec^{-1}. Be careful to make sure that the right temperature corresponds to the right rate constant.

$$T_1 = 25°C = 298\text{ K} \qquad T_2 = 65°C = 338\text{ K}$$

$$\log\left(\frac{2.40 \times 10^{-3}\text{ sec}^{-1}}{1.72 \times 10^{-5}\text{ sec}^{-1}}\right) = \log(1.40 \times 10^2) = 2.14$$

$$2.14 = \frac{E_a}{2.30(8.31\text{ J/K-mol})}\left(\frac{1}{298\text{ K}} - \frac{1}{338\text{ K}}\right)$$

$$= \frac{E_a}{19.1\text{ J/K-mol}}(3.36 \times 10^{-3}\text{ K}^{-1} - 2.96 \times 10^{-3}\text{ K}^{-1})$$

$$= \frac{E_a}{19.1\text{ J/K-mol}}(4.0 \times 10^{-4}\text{ K}^{-1})$$

$$E_a = \frac{2.14(19.1\text{ J/K-mol})}{4.0 \times 10^{-4}\text{ K}^{-1}} = 1.02 \times 10^5\text{ J/mol, or } 102\text{ kJ/mol}$$

Example 12.9

Use the answer from Example 12.8 to calculate the rate constant for the decomposition of N_2O_5 at 100°C.

Solution

Let k_2 be the rate constant at 100°C. Therefore $T_2 = 373$ K. We can choose either of the known rate constants and its corresponding temperature for k_1 and T_1, respectively, and we will choose 2.40×10^{-3} sec^{-1} at 338 K. The value of the activation energy should be in J/mol rather than kJ/mol.

$$\log\left(\frac{k_2}{k_1}\right) = \frac{1.02 \times 10^5\text{ J/mol}}{19.1\text{ J/K-mol}}\left(\frac{1}{338\text{ K}} - \frac{1}{373\text{ K}}\right)$$

$$\log\left(\frac{k_2}{k_1}\right) = 5.34 \times 10^3 \text{ K}(2.96 \times 10^{-3} \text{ K}^{-1} - 2.68 \times 10^{-3} \text{ K}^{-1})$$

$$= 5.34 \times 10^3 \text{ K}(2.8 \times 10^{-4} \text{ K}^{-1}) = 1.50$$

$$\frac{k_2}{k_1} = \text{antilog } 1.50 = 31$$

$$k_2 = 31(k_1) = 31(2.40 \times 10^{-3} \text{ sec}^{-1}) = 7.5 \times 10^{-2} \text{ sec}^{-1}$$

As can be seen from equation (12.7), the activation energy of a reaction determines the sensitivity of its rate constant to changes in temperature. Since E_a is positive (with a very few exceptions we will not discuss here), the rate constant for a reaction increases as the temperature is raised. An increase in temperature will cause a large rate increase in a reaction with a large activation energy. Those reactions with small activation energies do not speed up very much as temperature is raised.

Collision Theory

A hypothesis about the mechanism of a reaction can explain the form of the rate law, but it cannot tell us why the rate constant has a particular value or what is responsible for the temperature sensitivity of the rate. To answer these questions, we must try to visualize the structural changes that take place as molecules collide and distort each other until they break apart and form new molecules. This very difficult job involves a detailed analysis of the ways in which the potential and vibrational energies of the molecules change during a productive collision. There is a successful theory, called the **collision theory**, that explains reaction rates on the basis of the structural and energetic changes occurring during collisions, but its details are too complex to discuss in this book. We will present a simplified (and less accurate) version of the collision theory that accounts for the general features of rate behavior and gives us important insights into the way reactions occur.

Two molecules, in order to react with each other, must collide. The collision must be forceful enough to distort the bonds in the colliding molecules beyond their elastic limits. In addition, the reacting molecules must collide at an angle that permits efficient use of their translational kinetic energy to distort the bonds. Molecular rearrangement occurs when the collision has enough energy and the proper orientation. Reactant molecules bounce away unchanged from a collision that is too soft or lined up the wrong way.

Each elementary reaction has a characteristic, required orientation. Consider the reaction

$$H_2(g) + Br(g) \longrightarrow HBr(g) + H(g)$$

This particular reaction occurs when the bromine atom approaches the hydrogen molecule along its axis (Figure 12.11). If the molecules collide in other orientations, their kinetic energy will not be as effective in distorting the covalent bond in H_2, and reaction is less likely to occur.

When a bromine atom and hydrogen molecule approach each other in the proper orientation, repulsive forces between their outer-shell electrons distort hydrogen's covalent bond, and molecular potential energy increases at the expense of kinetic energy. No reaction occurs when the degree of distortion is too weak to rupture the bond—the bond snaps back to its undistorted position, and the molecules regain their kinetic energy. If, on the other hand, the molecules collide at high enough speeds, the bond distortion is large enough to form H and HBr.

Figure 12.12 is a diagram of the potential energy changes occurring when Br and H_2 react. Potential energy increases due to bond distortion as the molecules come closer, reaches a maximum value as bond rearrangement commences, and falls as HBr and H form. The molecular structure associated with the potential-energy maximum is called the **activated complex** (or **transition state**), and the activation

Figure 12.11

Required collision orientation for Br(g) + H$_2$(g) → HBr(g) + H(g).

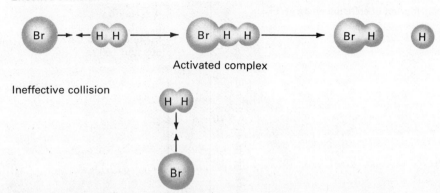

energy is the difference between the potential energy of the activated complex and the undistorted reactant molecules.

An activated complex is not the same an an intermediate in a multistep mechanism. An intermediate's lifetime may be short, but it is stable enough to survive weak bond vibrations. Because the activated complex is the least stable molecular arrangement occurring during the reaction, any vibrational distortion of its reacting bonds (in the direction required for reaction) decomposes it, forming product molecules in some instances and reactants in others. While sensitive analytic techniques can sometimes detect the presence of an intermediate molecule, the lifetime of an activated complex is so brief that none has ever been observed.

Every elementary reaction has its characteristic activation energy. Activation energies of elementary reactions are usually positive—even for exothermic ones—because some distortion of the reactant molecules must precede formation of the chemical bonds in the products. The size of the activation barrier depends on the degree of bond distortion required for reaction. A large activation energy indicates a very distorted activated complex.

Figure 12.12

Changes in molecular potential energy during the reaction between Br and H$_2$.

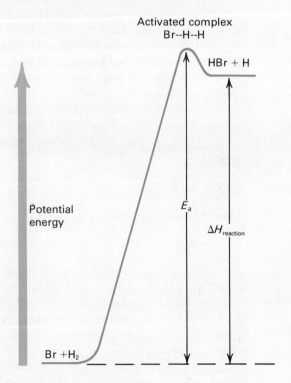

E_a = 73.6 kJ/mol

$\Delta H_{reaction}$ = 67.7 kJ/mol

Figure 12.13
Distribution of collision energies.

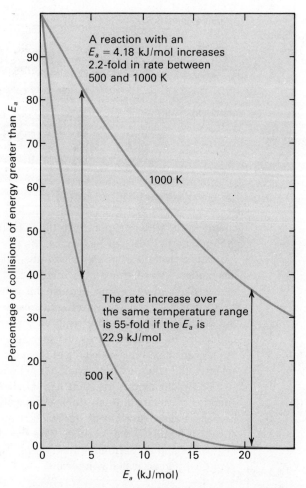

Collisions lead to reaction only if their energy exceeds the activation energy. The fraction of collisions with this required energy equals $e^{-E_a/RT}$ [or log (fraction) = $-E_a/2.30RT$], and a temperature increase raises this fraction (Figure 12.13). A reaction with a low activation energy is not very temperature-dependent because, even at a low temperature, a large percentage of the collisions are energetic enough for reaction, and there is little to be gained from a further temperature increase. If the activation energy for a reaction is large, raising the temperature causes a large rate increase because the number of collisions with the necessary energy rises sharply.

According to the collision theory, the frequency factor, A, in the Arrhenius equation

$$k = Ae^{-E_a/RT}$$

is the **collision frequency**, z, multiplied by the **steric factor**, p.

$$A = pz \tag{12.8}$$

The collision frequency is the number of collisions between reactant molecules per second when the reactant concentrations are each 1 mol/L (that is, 1 M). Calculations based on the kinetic-molecular theory of gases indicate this factor does not vary much from reaction to reaction—it is about 10^{10} collisions M^{-1} sec^{-1}. Large molecules move more slowly than small ones, but they are larger collision targets, and the two effects cancel each other.

Large differences between the frequency factors of different reactions are due to variations in the steric factor. In the earliest form of the collision theory, the steric factor represented the percentage of collisions with the orientation needed for reaction. A low frequency factor was associated with a very stringent limitation on the collision geometry. This is an oversimplification because some reactions have p values greater than one. According to the old collision theory, a reaction that has no limits on how the reactant molecules approach each other would have a steric factor of one, and larger p values would be impossible. This conflict with observation arises because the simplified theory neglects the influences of bond vibrations and molecular rotations on both the activation energy and the frequency factor.

Special Resource Section: Absolute Rate Theory

The **absolute rate theory** treats the formation of the activated complex (AB‡) as an equilibrium process.

$$A + B \rightleftharpoons AB\ddagger \longrightarrow \text{Products}^*$$

Since any vibrational distortion of the reacting bond in the direction required for reaction destroys an activated complex, the activated complex's rate constant for decomposition is its bond vibration frequency. We can use quantum theory to estimate this vibration frequency and we can then estimate the concentration of AB‡. The relationships between the concentrations of A, B, and AB‡ tell us the free energy difference between the activated complex and the reactants. The free energy difference is known as the **Gibbs free energy of activation** ($\Delta G\ddagger$). The free energy of activation in turn depends upon the **enthalpy of activation** ($\Delta H\ddagger$) and the **entropy of activation** ($\Delta S\ddagger$).

$$\Delta G\ddagger = \Delta H\ddagger - T\Delta S\ddagger \tag{12.9}$$

The enthalpy of activation is nearly the same as the Arrhenius activation energy and is also an estimate of the change in potential, vibrational, and rotational energy accompanying a successful collision.

The value of the entropy of activation tells us whether the activated complex is more or less ordered than the reacting molecules. As might be expected, most monomolecular reactions have positive activation entropies—a molecule becomes more disordered as it begins to break apart. Bimolecular reactions tend to have lower activation entropies because joining two molecules together in an activated complex tends to restrict their freedom of motion.

The value of $\Delta S\ddagger$ determines the value of the Arrhenius frequency factor. Reactions with positive activation entropies have large A values. Conversely, a negative activation entropy indicates a reaction with a restricted activated complex and a small frequency factor.

Knowing the values of $\Delta S\ddagger$ and $\Delta H\ddagger$, we can calculate the rate constant for a reaction at any temperature, but the difficulty of estimating these reaction parameters by quantum mechanical calculations limits this application of the absolute rate theory. However, calculations of $\Delta S\ddagger$ and $\Delta H\ddagger$ have been done for simple reactions such as

$$H_2(g) + D(g) \longrightarrow HD(g) + H(g)$$

The more common application of the absolute rate theory is to compute $\Delta H\ddagger$, $\Delta G\ddagger$, and $\Delta S\ddagger$ from observed kinetic data. The enthalpy of activation is estimated from the slope of a graph of log k versus $1/T$, we can calculate $\Delta G\ddagger$ from the value of the rate constant at a given temperature, and the equation

$$\Delta S\ddagger = \frac{\Delta H\ddagger - \Delta G\ddagger}{T} \tag{12.10}$$

gives the value of the entropy of activation.

---

*A similar explanation applies to monomolecular reactions. In these cases we consider the equilibrium between an undistorted reactant molecule and its activated complex

$$A \longrightarrow A\ddagger \longrightarrow \text{Products}$$

Although differing in their details, the collision theory and the absolute rate theory do not conflict with each other. They both explain the effect of temperature on rates in terms of a minimum collision energy required for the reaction. Both theories relate the value of the Arrhenius frequency factor to the degree of restriction of molecular motion in the activated complex.

The two theories share the same limitation—it is very difficult to predict the structure and properties of an activated complex from theoretical considerations. Consequently, our picture of an activated complex is rarely precise. We usually are not sure of the exact geometric arrangement of the atoms or the bond distances, energies, and polarities. Nevertheless, the concept of the activated complex is one of the most widely used ideas in chemistry. Chemists formulate an approximate picture of an activated complex and interpret kinetic effects, such as the influence of solvent polarity on the reaction rate, in terms of changes in the free energy of this complex.

Summary

1. The rate at which a reaction approaches equilibrium and the kinetic factors influencing this rate are characteristic features of the reaction. There is no predictable correlation between the kinetic characteristics of a reaction and the extent of reaction at equilibrium.

2. When a reaction that can produce several products occurs under conditions of kinetic control, it produces the more rapidly formed product in preference to the more stable product. The more stable product is formed under conditions of equilibrium control.

3. A reaction rate is the rate of change of concentration of a reactant or product. We can define a reaction rate in terms of the concentration change of any of the reactants or products, and we can relate these various definitions through the stoichiometry of the reaction.

4. To measure a reaction rate, we must measure elapsed time and be able to estimate the concentration of a reactant or product. The method for measuring concentration must either be rapid compared to the rate of reaction, or we must stop the reaction at desired times by a sudden change in conditions.

5. The balanced equation for a reaction tells us nothing about the effect of concentration on its rate. We obtain this information by a systematic study in which we observe the effects of various concentration changes on the reaction rate.

6. A rate law is an equation showing the relationship between the rate of a particular reaction and its reactant concentrations. Many rate laws have the form

    Rate = $k[X]^a[Y]^b[Z]^c$

    where k is the rate constant, $[X]$, $[Y]$, and $[Z]$ are reactant concentrations, and a, b, and c the orders of the reactants. The order of a reactant is usually a whole number (including zero) or a simple fraction such as ½. Kinetic orders bear no obvious relationship to the stoichiometric coefficients of the reactants.

7. The mechanism of a reaction is the sequence of molecular events producing the overall chemical change. The individual steps of the mechanism, called elementary reactions, are either monomolecular, bimolecular, or termolecular. The kinetic order of an elementary reaction matches its molecularity.

8. As far as kinetics are concerned, the most significant elementary reaction of a mechanism is the slowest step, also called the rate-determining step.

9. An intermediate is an unstable species existing in minute concentrations during a reaction.

10. The order of a reactant equals the number of molecules of that reactant consumed in the rate-determining step or in previous fast elementary reactions. If a reactant's order is less than its stoichiometric coefficient, some reactant molecules react in fast steps following the rate-determining step.

11. The rate of a reaction in a heterogeneous system depends on the surface area between the two reacting phases.

12. A catalyst increases the rate of a reaction without being consumed. The concentration of a catalyst in a homogeneous system appears in the rate law. In heterogeneous, or surface, catalysis, the catalyst is in a different phase than the reactants. Enzymes are proteins with very effective and specific catalytic activities.

13. A homogeneous catalyst exerts its kinetic influence by reacting either in the rate-determining step or in an earlier fast step, but it is not consumed in the total reaction because it is regenerated in a subsequent fast step.

14. In heterogeneous catalysis, the reactant molecules are first chemisorbed at active centers on the catalyst's surface. The molecules then react, and the product molecules diffuse from the surface, allowing other reactant molecules to be adsorbed.

15. The Arrhenius equation,

$$k = Ae^{-E_a/RT}$$

describes the effect of temperature on the reaction rate constant. Each reaction has a characteristic frequency factor (A) and activation energy (E_a). The Arrhenius equation can also be written as

$$\log\left(\frac{k_2}{k_1}\right) = \frac{E_a}{2.30R}\left(\frac{1}{T_1} - \frac{1}{T_2}\right)$$

16. Collision theory: In order to react, molecules must collide with sufficient energy and the proper orientation. During a successful collision, the potential energy of the reacting molecules rises to a maximum associated with a transient molecular form called the activated complex and then falls as the activated complex decomposes to form product molecules. The Arrhenius activation energy is the difference between the energy of the activated complex and the energy of the reactants. The fraction of the collisions with enough energy to form the activated complex is $e^{-E_a/RT}$.

17. The absolute rate theory treats the formation of the activated complex as an equilibrium process.

$$A + B \rightleftharpoons AB\ddagger \longrightarrow \text{Products}$$

The rate of a reaction is the concentration of the activated complex times the vibration frequency of the reacting bond in the activated complex. We can relate a reaction rate constant to the value of the free energy of activation ($\Delta G\ddagger$), the heat of activation ($\Delta H\ddagger$), and the entropy of activation ($\Delta S\ddagger$).

Vocabulary List

kinetic characteristic	bimolecular	activation energy
kinetic control	termolecular	collision theory
equilibrium control	elementary reaction	activated complex
reaction rate	rate-determining step	transition state
rate law	intermediate	collision frequency
rate constant	catalyst	steric factor
order	chemisorb	absolute rate theory
half-life	enzyme	enthalpy of activation
mechanism	frequency factor	entropy of activation
monomolecular		

Exercises

1. Although the equilibrium of the reaction

   $$H_2(g) + D_2(g) \rightleftharpoons 2\ HD(g)$$

   favors the formation of HD, we can prepare mixtures of H_2 and D_2 that do not contain detectable amounts of HD. Explain the role that kinetics plays in this phenomenon.

2. Even though the Gibbs free energy of 1-butene is higher than that of its isomer, 2-butene, a sample of 1-butene retains its purity at room temperature. Explain this in terms of kinetics.

   $$CH_3CH_2CH=CH_2 \qquad CH_3CH=CHCH_3$$
   $$\text{1-Butene} \qquad\qquad \text{2-Butene}$$

3. Which of the following statements about catalysis are false? Explain the nature of the error in each case.
   a. The concentration of a homogeneous catalyst appears in the rate law.
   b. We can measure the rate of a reaction by observing the rate of consumption of its catalyst.
   c. The rate of a reaction catalyzed by a heterogeneous catalyst depends on the surface area of that catalyst.
   d. A catalyst cannot alter a reaction mechanism.
   e. A catalyst may be altered in one step of a reaction, but a subsequent step restores it to its original molecular form.
   f. A catalyst does not change the free energy of a reaction but it does lower the free energy of activation.

4. Which burns faster, sawdust or a block of wood? Why?

5. Fats, which are water insoluble compounds, react with water in the lower intestine. The mechanical action of the intestine, coupled with the action of emulsifying agents called bile salts, forms a colloidal suspension of the fats in water to aid in the reaction process. Explain why forming the emulsion speeds the reaction between the fats and water.

6. What are the kinetic orders for the following reactions?
   a. $O_2(g) + 2\ NO(g) \longrightarrow 2\ NO_2(g)$ Rate = $k[O_2][NO]^2$
   b. $2\ H_2O_2(aq) \longrightarrow 2\ H_2O + O_2(g)$ Rate = $k[H_2O_2][OH^-]$
   c. $3\ ClO^-(aq) \longrightarrow 2\ Cl^-(aq) + ClO_3^-(aq)$ Rate = $k[ClO^-]^2$

7. Write the rate laws for the following reactions.
   a. $2\ S_2O_8^{2-}(aq) + 2\ H_2O \longrightarrow 4\ HSO_4^-(aq) + O_2(g)$
      The reaction is first order in $S_2O_8^{2-}$ and first order in Ag^+.
   b. $C_6H_5NHNHC_6H_5 \longrightarrow NH_2C_6H_4C_6H_4NH_2$
      The reaction is first order in $C_6H_5NHNHC_6H_5$ and second order in hydrogen ion.
   c. $NH_2CN(aq) + H_2S(aq) \longrightarrow (NH_2)_2CS(aq)$
      The reaction is first order in each reactant.

8. Which of the following reactions are catalyzed? Identify the catalyst in each case.
   a. $S_2O_8^{2-}(aq) + 2\ I^-(aq) \longrightarrow 2\ SO_4^{2-}(aq) + I_2(aq)$ Rate = $k[S_2O_8^{2-}][I^-]$
   b. $2\ C_6H_5CHO \longrightarrow (C_6H_5CHO)_2$ Rate = $k[C_6H_5CHO]^2[CN^-]$
   c. $2\ H_2O_2(aq) \longrightarrow 2\ H_2O + O_2(g)$ Rate = $k[H_2O_2][I^-]$

9. Which of the following reaction rates can be estimated by observing the pressure change within the system?
   a. $4\ PH_3(g) \longrightarrow P_4(g) + 6\ H_2(g)$
   b. $2\ N_2O(g) \longrightarrow 2\ N_2(g) + O_2(g)$
   c. $2\ NO(g) \longrightarrow N_2(g) + O_2(g)$

10. Older reference books often give first-order rate constants in units of min^{-1}. The modern style is to use sec^{-1}. Convert a rate constant equal to 0.886 min^{-1} to units of sec^{-1}.

11. Propose a possible mechanism for the first-order reaction

$$SO_2Cl_2(g) \longrightarrow SO_2(g) + Cl_2(g)$$

12. Iodine catalyzes the isomerization of *cis*-1,2-dichloroethene to *trans*-1,2-dichloroethene at 150°C.

    [structure: cis-1,2-dichloroethene → trans-1,2-dichloroethene]

    From the following initial rate data, decide whether the reaction is first order, second order, or half order in I_2. The initial concentration of *cis*-1,2-dichloroethene is the same in all experiments.

Experiment	$[I_2]$ (M)	Initial rate (M/hr)
1	0.0109	0.056
2	0.0295	0.089
3	0.0785	0.146
4	0.1584	0.210

13. Write the rate laws for the following elementary reactions.
    a. $Br_2(g) \longrightarrow 2\ Br(g)$
    b. $H(g) + Br_2(g) \longrightarrow HBr(g) + Br(g)$
    c. $CH_3CHO(g) \longrightarrow CH_3(g) + CHO(g)$

14. What is the rate law for each of the following elementary reactions?
    a. $CH_3(g) + I_2(g) \longrightarrow CH_3I(g) + I(g)$
    b. $HBr(g) \longrightarrow H(g) + Br(g)$
    c. $NO(g) + Cl(g) \longrightarrow NOCl(g)$

15. Raising the concentration of a particular reactant by a factor of 4 increases the reaction rate 16 times. What is the order of this reactant?

16. A fivefold decrease in reactant concentration lowers the reaction rate to 20% of its original value. What is the kinetic order of the reactant?

17. What would be the effect on the rate of a reaction of a threefold increase in the concentration of a reactant if the reaction were the following?
    a. First order in this reactant b. Second order
    c. Zero order

18. What would be the change in rate due to a fourfold decrease in a reactant concentration if the order of this reactant were the following?
    a. First order b. Third order c. ½ order

19. The rate of the reaction between acetone (C_3H_6O) and Br_2

    $$C_3H_6O(aq) + Br_2(aq) \longrightarrow C_3H_5OBr(aq) + H^+(aq) + Br^-(aq)$$

    was studied by removing 25.0-mL portions of the reacting solution at various times and measuring the bromine concentration by reacting the bromine with excess iodine ion,

    $$Br_2(aq) + 2\ I^-(aq) \longrightarrow I_2(aq) + 2\ Br^-(aq)$$

    and titrating the iodine produced with 0.0500 M $Na_2S_2O_3$.

    $$I_2(aq) + 2\ S_2O_3^{2-}(aq) \longrightarrow 2\ I^-(aq) + S_4O_6^{2-}$$

    Prepare a graph of $[Br_2]$ versus time from the following data.

Time (min)	Volume of $Na_2S_2O_3$ (mL)	Time (min)	Volume of $Na_2S_2O_3$ (mL)
0	19.25	70	3.00
10	17.05	75	1.60
20	14.80	80	0.65
60	5.00	83	0.05
67	3.50		

20. The total concentration of gases in a system, expressed in moles per liter, is given by the following equation.

    $$\frac{n}{V} = \frac{P}{RT}$$

    The following data were obtained during the thermal decomposition of sulfuryl chloride at constant volume and 284°C.

    $$SO_2Cl_2(g) \longrightarrow SO_2(g) + Cl_2(g)$$

Time (sec)	Pressure (torr)	Time (sec)	Pressure (torr)
0	180	360	269
120	215	600	302
240	244	900	330

    Prepare a graph of $[SO_2]$ versus time.

21. Find the rate law and estimate the rate constant for the decomposition of diacetone alcohol ($C_6H_{12}O_2$) in basic solution at 25°C.

$$C_6H_{12}O_2(aq) \longrightarrow 2\ C_3H_6O(aq)$$

Experiment	Initial molarities		Rate of decrease of $C_6H_{12}O_2$ (M/sec)
	$[C_6H_{12}O_2]$	$[OH^-]$	
1	0.88	0.044	3.87×10^{-4}
2	0.56	0.046	2.35×10^{-4}
3	0.43	0.046	1.80×10^{-4}
4	0.43	0.094	3.60×10^{-4}
5	0.43	0.0188	7.3×10^{-5}
6	0.43	0.0094	3.7×10^{-5}

23. What is the rate constant for a first-order reaction whose half-life is 480 sec?

25. The reaction

$$HCO_2H \longrightarrow H_2O + CO(g)$$

in 90% H_2SO_4 solution at 25°C is first order and its rate constant is 1.48×10^{-4} sec^{-1}. Draw a graph of $[HCO_2H]$ versus time for this reaction. The initial concentration of HCO_2H is 0.100 M. The time scale should run to 10,000 sec, and you should calculate the concentrations at thousand second intervals.

27. The reaction

$$3\ ClO^-(aq) \longrightarrow 2Cl^-(aq) + ClO_3^-(aq)$$

is second order in ClO^-. The following two-step mechanism was proposed to explain the rate law. Which step is the fast one and which is the slow one?

$$2\ ClO^- \longrightarrow Cl^- + ClO_2^-$$
$$ClO^- + ClO_2^- \longrightarrow Cl^- + ClO_3^-$$

29. The rate of the reaction

$$3\ I^-(aq) + S_2O_8^{2-}(aq) \longrightarrow I_3^-(aq) + 2\ SO_4^{2-}(aq)$$

is proportional to both $[I^-]$ and $[S_2O_8^{2-}]$. Propose a mechanism for this reaction.

31. The reaction

$$2\ NH_3(aq) + ClCH_2CO_2^-(aq) \longrightarrow$$
$$NH_2CH_2CO_2^-(aq) + NH_4^+(aq) + Cl^-(aq)$$

is first order in $ClCH_2CO_2^-$ and zero order in NH_3. Propose a mechanism that is consistent with both the rate law and the stoichiometry. Be sure to indicate which steps are fast and which are slow.

22. a. Find the rate law for the following reaction.

$$IO_3^-(aq) + 8\ I^-(aq) + 6\ H^+(aq) \longrightarrow 3\ I_3^-(aq) + 3\ H_2O$$

Experiment	Initial molarities			Rate of formation of I_3^- (M/sec)
	$[IO_3^-]$	$[I^-]$	$[H^+]$	
1	2.25×10^{-3}	1.8×10^{-3}	1.4×10^{-4}	2.38×10^{-7}
2	2.25×10^{-3}	3.6×10^{-3}	1.4×10^{-4}	9.07×10^{-7}
3	4.5×10^{-3}	1.8×10^{-3}	1.4×10^{-4}	4.77×10^{-7}
4	2.25×10^{-3}	1.8×10^{-3}	0.7×10^{-4}	0.60×10^{-7}
5	2.25×10^{-3}	1.8×10^{-3}	2.8×10^{-4}	8.95×10^{-7}

b. Check the validity of your rate law by calculating the value of the rate constant for each experiment.
c. Calculate the rate if $[IO_3^-]$ is 2.5×10^{-4} M, $[I^-]$ is 8.85×10^{-3} M, and $[H^+]$ is 7.0×10^{-5} M.

24. What is the half-life of a first-order reaction whose rate constant is 5.78×10^{-5} sec^{-1}?

26. The reaction

$$CH_2(CO_2H)_2 \longrightarrow CH_3CO_2H + CO_2(g)$$

in 80% sulfuric acid at 98°C is first order with a rate constant of 4.90×10^{-6} sec^{-1}. Draw a graph of log $[CH_2(CO_2H)_2]$ versus time, starting with an initial concentration of 0.20 M. The scale should run to 5×10^5 sec in intervals of 1×10^5 sec.

28. We believe that the iodination of acetone in basic solution

$$OH^-(aq) + C_3H_6O(aq) + I_2(aq) \longrightarrow$$
$$C_3H_5OI(aq) + I^-(aq) + H_2O$$

proceeds by the following mechanism.

$$C_3H_6O(aq) + OH^-(aq) \longrightarrow C_3H_5O^-(aq) + H_2O\ (slow)$$
$$C_3H_5O^-(aq) + I_2(aq) \longrightarrow C_3H_5OI(aq) + I^-(aq)\ (fast)$$

What rate law is consistent with this mechanism?

30. The rate law for the iodide-catalyzed decomposition of hydrogen peroxide

$$2\ H_2O_2(aq) \longrightarrow 2\ H_2O + O_2(g)$$

is Rate = $k[H_2O_2][I^-]$. Propose a mechanism for this reaction.

32. The reaction

$$2\ Ce^{4+}(aq) + Tl^+(aq) \longrightarrow Tl^{3+}(aq) + 2\ Ce^{3+}(aq)$$

occurs in the presence of Ag^+. Doubling the concentration of Ag^+ doubles the reaction rate. Doubling $[Ce^{4+}]$ also doubles the rate, but changes in $[Tl^+]$ have no effect on the rate.
a. Write the rate law for this reaction.
b. Propose a mechanism for this reaction.

EXERCISES

33. The rate of the reaction

$$H_2O + 2\, S_2O_4^{2-}(aq) \longrightarrow 2\, HSO_3^-(aq) + S_2O_3^{2-}(aq)$$

is second order in $S_2O_4^{2-}$ (rate = $k[S_2O_4^{2-}]^2$), and its rate constant at 10°C is $6.7 \times 10^{-7}\, M^{-1}\, sec^{-1}$. Compute the rate of this reaction when the concentration of $S_2O_4^{2-}$ is $0.250\, M$.

34. The initial rate for the decomposition of phosgene at 715 K

$$COCl_2(g) \longrightarrow CO(g) + Cl_2(g) \qquad \text{Rate} = k[COCl_2][Cl_2]^{1/2}$$

is $2.98 \times 10^{-8}\, M/sec$ when the initial concentrations are $[COCl_2] = 7.76 \times 10^{-4}\, M$ and $[Cl_2] = 1.66 \times 10^{-4}\, M$.
  a. Calculate the rate constant for this reaction at 715 K.
  b. Predict the rate at 715 K if $[Cl_2]$ is initially $3.35 \times 10^{-4}\, M$ and $[COCl_2]$ is $4.94 \times 10^{-4}\, M$.

35. What percentage of the initial reactant will remain after 5.0 hr of a first-order reaction whose rate constant is $5.78 \times 10^{-5}\, sec^{-1}$?

36. One way of defining the rate of the reaction

$$4\, PH_3(g) \longrightarrow P_4(g) + 6\, H_2(g)$$

is as the rate of increase in hydrogen concentration.

$$\text{Rate} = \frac{d[H_2]}{dt}$$

An alternate way is as the decrease in PH_3 concentration.

$$\text{Rate} = -\frac{d[PH_3]}{dt}$$

Use the stoichiometry of this reaction to derive the mathematical relationships between these two rates.

37. Treatment of naphthalene with sulfuric acid at 80°C yields predominantly α-naphthalenesulfonic acid and lesser amounts of its isomer, β-naphthalenesulfonic acid. The predominant product at 160°C is the β-isomer, and the α-isomer changes to the β-isomer at this temperature.
  a. Which of the two isomers is the more stable?
  b. Which reaction, the one producing α-naphthalenesulfonic acid or the one producing the β-isomer, is faster?
  c. Is the product mixture formed at 80°C the result of kinetic control or equilibrium control?

38. The first-order rate constant for the decomposition of N_2O_5

$$2\, N_2O_5(g) \longrightarrow 2\, N_2O_4(g) + O_2(g)$$

is $7.87 \times 10^{-7}\, sec^{-1}$ at 0°C and $4.98 \times 10^{-4}\, sec^{-1}$ at 45°C. Find the activation energy for this reaction.

39. The activation energy for the decomposition of nitrosyl chloride

$$2\, NOCl(g) \longrightarrow 2\, NO(g) + Cl_2(g)$$

is 100 kJ/mol, and its second-order rate constant is $6.0 \times 10^{-4}\, M^{-1}\, sec^{-1}$ at 500 K. Estimate the value of the rate constant at 1000 K.

40. The decomposition of nitrous oxide

$$2\, N_2O(g) \longrightarrow 2\, N_2(g) + O_2(g)$$

has the rate law

$$\text{Rate} = k[N_2O]^2$$

The rate constant is $0.86\, M^{-1}\, sec^{-1}$ at 1030 K and $3.69\, M^{-1}\, sec^{-1}$ at 1085 K. What is the activation energy of this reaction?

41. The conversion of cyclopropane to propene

$$CH_2-CH_2-CH_2(g) \longrightarrow CH_3CH=CH_2(g)$$
(with CH$_2$ forming the triangle)

is a first-order reaction with an activation energy of 272 kJ/mol and a frequency factor of 1.5×10^{15} sec^{-1}. Estimate the rate constant for the conversion of cyclopropane to propene at 25°C.

42. The reaction

$$CO(g) + NO_2(g) \longrightarrow CO_2(g) + NO(g)$$

Rate = $k[CO][NO_2]$

has a frequency factor of 5.0×10^8 M^{-1} sec^{-1} and an activation energy of 116 kJ/mol.
   a. What is the rate constant for this reaction at 25°C?   b. At 100°C?

43. The activation energy for the formation of hydrogen iodide

$$H_2(g) + I_2(g) \rightleftharpoons 2\ HI(g)$$

is 163 kJ/mol H$_2$, and the heat of the reaction is +20 kJ/mol H$_2$. What is the activation energy for the following reaction?

$$2\ HI(g) \rightleftharpoons H_2(g) + I_2(g)$$

[*Hint*: Consider Figure 12.12.]

44. a. What fraction of the collisions at 300 K will have enough energy to bring about a bimolecular reaction with an activation energy of 40 kJ/mol?
   b. 100 kJ/mol?
   c. What fraction of the collisions have at least 40 kJ/mol at 400 K?
   d. 100 kJ/mol at 400 K?

45. The calculated collision frequency for the reaction

$$\beta\text{-}C_{10}H_7O^- + C_2H_5I \longrightarrow \beta\text{-}C_{10}H_7OC_2H_5 + I^-$$

in methanol solution is 2.2×10^{11} M^{-1} sec^{-1}, and the observed Arrhenius frequency factor is 1.0×10^{10} M^{-1} sec^{-1}. What is the value of the steric factor (p) for this reaction?

46. Using the integrated rate equation for a first-order reaction, show to your own satisfaction that the half-life of a first-order reaction is independent of the value of the initial concentration, $[X]_0$.

47. In the vicinity of room temperature, the rate of the reaction

$$O_2(g) + 2\ NO(g) \longrightarrow 2\ NO_2(g)$$

decreases with increasing temperature. The proposed mechanism for this reaction is

$$2\ NO(g) \rightleftharpoons N_2O_2(g) \text{ (fast)}$$

$$N_2O_2(g) + O_2(g) \longrightarrow 2\ NO_2(g) \text{ (slow)}$$

Can you propose a theoretical explanation for this reaction's unusual temperature behavior?

48. Pure PH$_3$, originally present at a pressure of 1.50 atm in a rigid container at 500 K, slowly decomposes according to the equation

$$4\ PH_3(g) \longrightarrow P_4(g) + 6\ H_2(g)$$

After a period of time, the reaction has raised the pressure to 1.85 atm. What percentage of the PH$_3$ has reacted?

chapter 13
Chemical Equilibrium

Mass action and equilibrium concern dissociation of molecules, such as $COCl_2$ splitting into Cl_2 and CO

13.1
The Law of Mass Action

General Principles

When hydrogen reacts with iodine to form hydrogen iodide,

$$H_2(g) + I_2(g) \rightleftharpoons 2\ HI(g)$$

the concentrations of hydrogen and iodine decrease from their initial values, and the hydrogen iodide concentration (initially zero) increases. As the reaction proceeds, the changes in concentration slow down and cease before the reactants are completely consumed (Figure 13.1, curve A). The system reaches a state of **equilibrium**; no further net chemical change occurs, and the composition of the system remains constant.

The reverse reaction (the decomposition of hydrogen iodide to its elements, curve B) also reaches an equilibrium state. The concentration of hydrogen iodide falls as it decomposes, and the concentrations of hydrogen and iodine rise until the equilibrium mixture forms. No further net reaction occurs.

This behavior is typical for **reversible reactions**, incomplete reactions that cease before the maximum possible amount of product (as predicted by stoichiometry) forms. Equilibrium states can be approached from either direction, by the consumption of reactants and the formation of products, or by the reverse process. (The terms reactant and product for a reversible reaction are arbitrary; they depend on the direction in which we choose to write the equation for the reaction.)

Figure 13.1

Composition changes during the reversible reaction

$$H_2(g) + I_2(g) \rightleftharpoons 2\ HI(g)$$

Curve A: Formation of HI from an equimolar mixture of H_2 and I_2.
Curve B: Decomposition of pure HI.
Source: Data are from M. Bodenstein, *Zeitschrift für Physikalische Chemie* **13**, 57(1894).

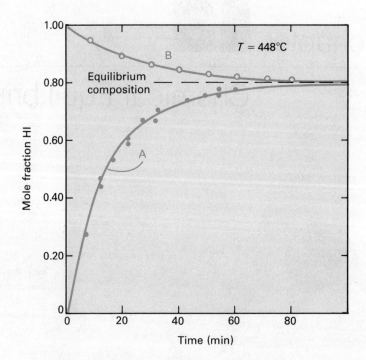

A system at equilibrium is in dynamic balance; the rate of the backward reaction just matches that of the forward reaction.

$$\text{Rate}_{\text{forward}} = \text{Rate}_{\text{reverse}}$$

In the case of the hydrogen iodide equilibrium, hydrogen and iodine molecules continually react to form hydrogen iodide molecules while HI decomposes to H_2 and I_2 at exactly the same rate. Dynamic balance occurs when the Gibbs free energy of the products exactly equals the free energy of the reactants.

$$G_{\text{products}} - G_{\text{reactants}} = \Delta G_{\text{reaction}} = 0$$

Equilibrium in the Gas Phase

We will start the discussion of equilibrium by defining a quantity known as the **mass-action ratio** (Q). For a reaction occurring in the gas phase, the numerator of the mass-action ratio is a multiple of the partial pressures of the products, and each pressure is raised to an exponential power equal to the stoichiometric coefficient of the products. The denominator of the ratio is a similar multiple of reactant pressures. For the hypothetical reaction

$$m\ A(g) + n\ B(g) \rightleftharpoons p\ C(g) + q\ D(g)$$

the mass-action ratio has the form

$$Q_P = \frac{P_C^p P_D^q}{P_A^m P_B^n} \tag{13.1}$$

(The subscript P in Q_P shows that the mass-action ratio is based on partial pressures rather than concentration.) A real example is the mass-action ratio for the formation of hydrogen iodide.

$$Q_P = \frac{P_{HI}^2}{P_{H_2} P_{I_2}}$$

Table 13.1 shows the mass-action ratios of several other gas-phase reactions.

Example 13.1

Write the mass-action ratio for the reaction

$$N_2(g) + 3\ H_2(g) \rightleftharpoons 2\ NH_3(g)$$

13.1 THE LAW OF MASS ACTION

Table 13.1 Mass-Action Ratios

Reaction	Q_P
$H_2(g) + CO_2(g) \rightleftharpoons H_2O(g) + CO(g)$	$\dfrac{P_{H_2O} P_{CO}}{P_{H_2} P_{CO_2}}$
$COCl_2(g) \rightleftharpoons CO(g) + Cl_2(g)$	$\dfrac{P_{CO} P_{Cl_2}}{P_{COCl_2}}$
$B_2H_6(g) \rightleftharpoons 2\,BH_3(g)$	$\dfrac{P_{BH_3}^2}{P_{B_2H_6}}$
$O_2(g) + 2\,SO_2(g) \rightleftharpoons 2\,SO_3(g)$	$\dfrac{P_{SO_3}^2}{P_{O_2} P_{SO_2}^2}$
$P_4(g) + 6\,H_2(g) \rightleftharpoons 4\,PH_3(g)$	$\dfrac{P_{PH_3}^4}{P_{P_4} P_{H_2}^6}$

Solution

The stoichiometric coefficients of N_2, H_2, and NH_3 are one, three, and two, respectively, so their exponents in the mass-action expression are one, three, and two, respectively. Since NH_3 is the only reaction product, its partial pressure appears alone in the numerator and the partial pressures of N_2 and H_2, the reactants, are in the denominator of the mass-action ratio.

$$Q_P = \frac{P_{NH_3}^2}{P_{N_2} P_{H_2}^3}$$

Depending on how much of each of the different components we add to a given system, Q_P can have a variety of values. For example, in the formation of HI, a mixture containing H_2 and I_2 but practically no HI will have a very small mass-action ratio. On the other hand, Q_P will be very large if there is a lot of HI and only low pressures of H_2 and I_2. Although we can start with any mass-action ratio imaginable, as the reaction proceeds the changes in partial pressures will change Q_P, and its value will approach a particular value known as the **equilibrium constant** of the reaction. The composition of the mixture at equilibrium is described by the **law of mass action**: the mass-action ratio of an equilibrium mixture equals the equilibrium constant (K_P) of the reaction.

$$Q_P = K_P \tag{13.2}$$

The value of the equilibrium constant depends on the nature of the reaction and the temperature. We can determine the value of an equilibrium constant by measuring the partial pressures of the system at equilibrium. (Although it is possible to compute K_P from thermodynamic data, these data are usually harder to obtain than the equilibrium pressures.) Table 13.2 shows experimental estimates of the compositions of equilibrium mixtures for the hydrogen iodide reaction. Different initial reactant pressures produce different equilibrium mixtures, but their mass-action ratios are the same within experimental uncertainty.

Example 13.2

The partial pressures of NH_3, N_2, and H_2 in an equilibrium mixture (see Example 13.1 for the reaction) at 450°C are 0.204 atm, 2.45 atm, and 7.35 atm, respectively. What is the value of K_P at 450°C?

Solution

Substituting the partial pressure values into the mass-action ratio (Example 13.1) yields the value of K_P.

13 CHEMICAL EQUILIBRIUM

Table 13.2 **Compositions of Equilibrium Mixtures**

Reaction: $H_2(g) + I_2(g) \rightleftharpoons 2\,HI(g)$ T = 699 K   Pressures are in atmospheres.

$P_{H_2, \text{initial}}$	$P_{I_2, \text{initial}}$	$P_{HI, \text{initial}}$	$P_{H_2, \text{eq}}$	$P_{I_2, \text{eq}}$	$P_{HI, \text{eq}}$	Q_P at equilibrium
0.640	0.571	0	0.167	0.0980	0.946	54.7
0.652	0.519	0	0.204	0.0717	0.895	54.6
0.651	0.430	0	0.262	0.0423	0.777	54.4
0.613	0.617	0	0.129	0.134	0.967	54.0
0.612	0.687	0	0.105	0.180	1.014	54.5
0	0	0.259	0.0275	0.0275	0.204	54.4
0	0	0.614	0.0655	0.0655	0.483	54.3

Source: Pressures are calculated from data of A. Taylor and R. Crist, *Journal of the American Chemical Society* **63**, 1377 (1941).

$$K_P = \frac{P_{NH_3}^2}{P_{N_2} P_{H_2}^3} = \frac{(0.204\ \text{atm})^2}{(2.45\ \text{atm})(7.35\ \text{atm})^3} = 4.28 \times 10^{-5}\ \text{atm}^{-2}$$

Once we know the value of the equilibrium constant, it is easy to predict the direction of the chemical reaction for any mixture. If Q_P is less than K_P, the reaction goes forward, forming more product and lowering the reactant pressures. This raises the value of Q_P until it reaches K_P. A reaction runs backward when Q_P is greater than K_P.

Example 13.3

In which direction will the reaction

$$N_2(g) + 3\,H_2(g) \rightleftharpoons 2\,NH_3(g)$$

go if a mixture of 0.50 atm N_2, 2.0 atm H_2, and 0.050 atm NH_3 is raised to 450°C? Use the K_P value from Example 13.2.

Solution

The mass-action ratio is

$$Q_P = \frac{P_{NH_3}^2}{P_{N_2} P_{H_2}^3} = \frac{(0.050\ \text{atm})^2}{(0.50\ \text{atm})(2.0\ \text{atm})^3}$$

$$= 6.5 \times 10^{-4}\ \text{atm}^{-2}$$

Since this value is larger than the value of K_P (4.28×10^{-5} atm^{-2}), ammonia will decompose into nitrogen and hydrogen. The reaction will decrease the value of Q_P until it equals K_P.

The law of mass action is consistent with Le Chatelier's principle. Consider the equilibrium mixture described in Example 13.2. Now suppose we add more ammonia to it to momentarily raise the ammonia partial pressure above 0.204 atm. According to Le Chatelier's principle, some of this ammonia will decompose to nitrogen and hydrogen (in other words, the equilibrium will shift to the left) in order to relieve the stress of the added ammonia. The law of mass action predicts the same thing since the added ammonia raises the value of Q_P above that of K_P. The decomposition of ammonia and the formation of nitrogen and hydrogen decrease Q_P until it equals K_P.

The difference between Le Chatelier's principle and the law of mass action is that Le Chatelier's principle is purely qualitative. In other words, it tells us the direction of the chemical change but does not predict how far the reaction will go. As we will see in the next section, the law of mass action is useful for quantitative predictions.

13.1 THE LAW OF MASS ACTION

Equilibrium Calculations

We can use the equilibrium constant to calculate the partial pressures of all the components (both reactants and products) of a system at equilibrium provided we know the initial pressure of each component. The law of mass action provides one of the mathematical equations needed to solve this problem. The rest of the required equations are derived from the initial partial pressures and the stoichiometry of the reaction. Examples 13.4–13.6 illustrate the general method of equilibrium calculations.

Example 13.4

Hydrogen and carbon dioxide, each at an initial partial pressure of 20.0 atm, react at 986°C to form water and carbon monoxide.

$$H_2(g) + CO_2(g) \rightleftharpoons H_2O(g) + CO(g)$$

The equilibrium constant at 986°C is 1.59. What is the composition of the equilibrium mixture?

Solution

Since H_2O and CO are produced in equimolar quantities, their partial pressures are equal.

$$P_{H_2O} = P_{CO}$$

Each mole of H_2O formed requires the consumption of a mole each of H_2 and CO_2. Therefore,

$$P_{H_2} = P_{H_2, \text{initial}} - P_{H_2O}$$
$$= 20.0 \text{ atm} - P_{H_2O}$$

and

$$P_{CO_2} = P_{CO_2, \text{initial}} - P_{H_2O}$$
$$= 20.0 \text{ atm} - P_{H_2O}$$

Combining these three stoichiometric equations with the law of mass action, we obtain

$$K_P = \frac{P_{H_2O} P_{CO}}{P_{H_2} P_{CO_2}} = \frac{P_{H_2O}^2}{(20.0 - P_{H_2O})^2} = 1.59$$

The algebraic solution for this equation is easier than for most quadratic equations because we can factor it by taking the square root of both sides.

$$\frac{P_{H_2O}}{20.0 - P_{H_2O}} = \sqrt{1.59} = 1.26$$

The solutions are

$$P_{H_2O} = P_{CO} = 11.2 \text{ atm}$$
$$P_{H_2} = P_{CO_2} = 8.8 \text{ atm}$$

These answers can be easily checked by inserting the concentration values into the mass-action expression to see if Q_P equals K_P.

Example 13.5

What is the composition of an equilibrium mixture of H_2, CO_2, H_2O, and CO at 986°C if the initial partial pressures are as follows?

$$P_{H_2} = 10.0 \text{ atm} \quad P_{CO_2} = 20.0 \text{ atm} \quad P_{H_2O} = 5.0 \text{ atm} \quad P_{CO} = 0$$

(See Example 13.4 for the value of the equilibrium constant.)

Solution

A mole of H_2O is formed for each mole of CO. However, since the initial pressures of H_2O and CO are different, the relationship between the partial pressures of the two products is more complicated than in Example 13.4. The equation is

$$P_{H_2O} = 5.0 \text{ atm} + P_{CO}$$

The partial pressures of H_2 and CO_2 equal their initial pressures minus the pressure of CO.

$$P_{H_2} = 10.0 \text{ atm} - P_{CO}$$
$$P_{CO_2} = 20.0 \text{ atm} - P_{CO}$$

Substituting these three equations into the law of mass action gives an equation in one unknown, P_{CO}.

$$\frac{P_{CO}(5.0 + P_{CO})}{(10.0 - P_{CO})(20.0 - P_{CO})} = 1.59$$

This quadratic equation cannot be factored so we must rearrange it to a form that is convenient for solution (Appendix A.7).

$$\frac{5.0 P_{CO} + P_{CO}^2}{200 - 30.0 P_{CO} + P_{CO}^2} = 1.59$$

$$5.0 P_{CO} + P_{CO}^2 = 318 - 47.7 P_{CO} + 1.59 P_{CO}^2$$

$$0.59 P_{CO}^2 - 52.7 P_{CO} + 318 = 0$$

$$P_{CO} = \frac{52.7 \pm \sqrt{(52.7)^2 - 4(0.59)(318)}}{2(0.59)}$$

There are two solutions to this quadratic equation.

$$P_{CO} = 6.5 \text{ atm} \quad \text{and} \quad P_{CO} = 82.8 \text{ atm}$$

We reject the second solution as unreasonable because it is impossible to obtain 82.8 atm CO from only 10.0 atm H_2. The other partial pressures are

$$P_{H_2O} = 5.0 \text{ atm} + 6.5 \text{ atm} = 11.5 \text{ atm}$$
$$P_{H_2} = 10.0 \text{ atm} - 6.5 \text{ atm} = 3.5 \text{ atm}$$
$$P_{CO_2} = 20.0 \text{ atm} - 6.5 \text{ atm} = 13.5 \text{ atm}$$

Example 13.6

The value of K_P for the reaction

$$2 \text{ SO}_2(g) + \text{O}_2(g) \rightleftharpoons 2 \text{ SO}_3(g)$$

is 1.12×10^3 atm^{-1} at 800 K. Find the partial pressure of the equilibrium mixture formed when SO_2, initially at a partial pressure of 0.200 atm, and O_2, initially at 0.800 atm, are mixed in a rigid container.

Solution

Since each mole of SO_3 formed requires a mole of SO_2,

$$P_{SO_2} = 0.200 \text{ atm} - P_{SO_3}$$

13.1 THE LAW OF MASS ACTION

Each mole of SO_3 requires a half mole of oxygen.

$$P_{O_2} = 0.800 \text{ atm} - \frac{1}{2} P_{SO_3}$$

Application of the mass-action law gives

$$K_P = \frac{P_{SO_3}^2}{(0.200 - P_{SO_3})^2 (0.800 - \frac{1}{2} P_{SO_3})} = 1.12 \times 10^3 \text{ atm}^{-1}$$

Having reduced the problem to a single cubic equation with one unknown quantity, we are faced with the difficult task of solving it. One of the best methods is a process of repeated approximations: We make a guess about the approximate pressure of one of the components and use this value to estimate the pressures of the other components. We check these approximate answers to see if they are consistent with the original guess. If there is an inconsistency, we revise our guess so that it agrees with the approximate answers and continue refining the guess until there is agreement with the approximate estimates.

The large equilibrium constant and the excess of oxygen indicate that most of the SO_2 is converted to SO_3. Let us make a preliminary guess that

$$P_{SO_3} = 0.200 \text{ atm}$$

and

$$P_{O_2} = 0.800 \text{ atm} - \frac{1}{2}(0.200 \text{ atm}) = 0.700 \text{ atm}$$

Using these approximate pressures in the mass-action equation, we find

$$P_{SO_2}^2 = \frac{P_{SO_3}^2}{P_{O_2}(1.12 \times 10^3)} \cong \frac{(0.200)^2}{(0.700)(1.12 \times 10^3)}$$

$$\cong 5.10 \times 10^{-5} \text{ atm}^2$$

$$P_{SO_2} \cong 7.1 \times 10^{-3} \text{ atm}$$

This approximate answer indicates that only 3.5% of the SO_2 remains unchanged, a result that is close to our original guess of nearly complete conversion. We can now revise our estimates for the partial pressures of SO_3 and O_2 to values consistent with a partial pressure of SO_2 of about 7×10^{-3} atm.

$$P_{SO_3} = 0.200 - 0.007 = 0.193 \text{ atm}$$

$$P_{O_2} = 0.800 - \frac{1}{2}(0.193) = 0.703 \text{ atm}$$

The close agreement between the original guesses for these partial pressures and the revised values gives us confidence in the validity of our approximations. A recalculation of the SO_2 pressure based on the revised estimates for SO_3 and O_2 gives a value of 6.9×10^{-3} atm. Note how a sensible guess about approximate equilibrium pressures can quickly lead to the correct solution to a difficult equation. Mathematical approximations are very important in equilibrium calculations. Without them, mathematical complexities would interfere with our understanding of many equilibrium systems.

Other Forms of the Mass-Action Law

We can also express the law of mass action in terms of concentrations such as molarities and mole fractions. If the gas mixture is ideal, the molarities of its components are proportional to their partial pressures. Using $PV = nRT$, we can derive equation (13.3).

$$[A] = \frac{n_A}{V} = \frac{P_A}{RT} \tag{13.3}$$

The mole fraction of a component of an ideal gas mixture is also proportional to its partial pressure.

$$X_A = \frac{P_A}{P_{total}} \tag{13.4}$$

Because of these proportionalities, we can substitute molar concentrations or mole fractions into the mass-action ratio and still maintain the validity of the law of mass action.

An equilibrium constant based on pressure (K_P) does not always have the same numerical value as one based on molarities (K_C) or mole fractions (K_X). These three constants are equal in the case of hydrogen iodide formation because the number of reactant molecules in the balanced equation equals the number of product molecules.

$$K_C = \frac{[HI]^2}{[H_2][I_2]} = \frac{\left(\frac{P_{HI}}{RT}\right)^2}{\left(\frac{P_{H_2}}{RT}\right)\left(\frac{P_{I_2}}{RT}\right)} = \frac{P_{HI}^2}{P_{H_2}P_{I_2}} = K_P$$

$$K_X = \frac{X_{HI}^2}{X_{H_2}X_{I_2}} = \frac{\left(\frac{P_{HI}}{P_{total}}\right)^2}{\left(\frac{P_{H_2}}{P_{total}}\right)\left(\frac{P_{I_2}}{P_{total}}\right)} = \frac{P_{HI}^2}{P_{H_2}P_{I_2}} = K_P$$

In the case of the dissociation of N_2O_4 to NO_2, a reaction producing more molecules than are consumed, this equality does not apply.

$$N_2O_4(g) \rightleftharpoons 2\,NO_2(g)$$

$$K_P = \frac{P_{NO_2}^2}{P_{N_2O_4}}$$

$$K_C = \frac{[NO_2]^2}{[N_2O_4]} = \frac{\left(\frac{P_{NO_2}}{RT}\right)^2}{\left(\frac{P_{N_2O_4}}{RT}\right)} = \frac{P_{NO_2}^2}{P_{N_2O_4}RT} = \frac{K_P}{RT}$$

$$K_X = \frac{X_{NO_2}^2}{X_{N_2O_4}} = \frac{\left(\frac{P_{NO_2}}{P_{total}}\right)^2}{\left(\frac{P_{N_2O_4}}{P_{total}}\right)} = \frac{P_{NO_2}^2}{P_{N_2O_4}P_{total}} = \frac{K_P}{P_{total}}$$

The general relationships between K_P, K_C, and K_X for gas-phase reactions are

$$K_C = \frac{K_P}{(RT)^{\Delta n}} \tag{13.5}$$

$$K_X = \frac{K_P}{P_{total}^{\Delta n}} \tag{13.6}$$

where Δn is the difference between the number of product molecules and reactant molecules in the balanced equation ($\Delta n = +1$ in the dissociation of N_2O_4).

Nonideality

The law of mass action for gases is an ideal law, subject to the same errors as the ideal gas law. Fortunately nonideal effects are small at moderate pressures. The value of K_P for the reaction

$$N_2(g) + 3\ H_2(g) \rightleftharpoons 2\ NH_3(g) \text{ at } 450°C$$

stays reasonably constant (within 25%) below 100 atm (Table 13.3). More serious errors appear at very high compression; K_p for the formation of ammonia increases ten times as the total pressure rises from 100 to 1000 atm.

Although

$$K_p = \frac{P_{NH_3}^2}{P_{N_2} P_{H_2}^3}$$

varies at high pressures, the **thermodynamic equilibrium constant**

$$K_a = \frac{a_{NH_3}^2}{a_{N_2} a_{H_2}^3}$$

is constant over the entire range of pressure. The thermodynamic equilibrium constant is based on the **activities** of the reactants and products. The activity (a) of a substance is a unitless quantity that describes the real free energy of the substance rather than its presumed ideal free energy. It compares the thermodynamic state of a component of a real system to an arbitrarily defined **standard state** at which the component's activity is equal to one.

We are not going to use activities, as such, in this book, but we will present equilibrium constants in a form that approximates the thermodynamic equilibrium constant as closely as possible. Since the activities of gases approximate their partial pressures in atmospheres, a gas will be represented in a mass-action ratio by its pressure in atmospheres. The activity of a solute in a dilute solution is approximately its molarity, so we will write molar concentrations for all solutes in dilute solutions.

One feature of mass-action ratios that often confuses students is that we appear to leave the concentration of water out of the mass-action ratio for a reaction in a dilute aqueous solution even if water is involved in the reaction. For example, the equilibrium constant for the reaction

$$OI^-(aq) + 2\ I^-(aq) + 2\ H^+(aq) \rightleftharpoons I_3^-(aq) + H_2O$$

is written

$$K = \frac{[I_3^-]}{[OI^-][I^-]^2[H^+]^2} \quad \text{not} \quad K = \frac{[I_3^-][H_2O]}{[OI^-][I^-]^2[H^+]^2}$$

Table 13.3 Variation in K_p Due to Nonideality

$N_2(g) + 3\ H_2(g) \rightleftharpoons 2\ NH_3(g)$ at 450°C	
Total pressure (atm)	K_P (atm^{-2})
10	4.34×10^{-5}
30	4.57×10^{-5}
50	4.76×10^{-5}
100	5.26×10^{-5}
300	7.81×10^{-5}
600	16.74×10^{-5}
1000	54.20×10^{-5}

Source: Data are from W. J. Moore, *Physical Chemistry*, 3rd ed. (Englewood Cliffs, N.J.: Prentice-Hall), p. 195. Reprinted by permission of Prentice-Hall, Inc.

The concentration of water does not appear in the numerator of the mass-action ratio because the activity of a pure liquid is, by the definition of the standard state, equal to one, and water in a dilute aqueous solution has nearly the same concentration—and activity—as pure water. In other words, there is very little difference between the water concentrations of various dilute solutions, so the water concentration has very little influence on the equilibrium concentrations of the other reactants and products.

When dealing with reactions in heterogeneous systems, we represent each reactant and product by its activity in its own predominant phase. For example, in the reaction

$$O_2(g) + 2\ Cu(s) + 4\ H^+(aq) \rightleftharpoons 2\ H_2O + 2\ Cu^{2+}(aq)$$

we are interested in the concentration of copper metal in the solid copper phase. This concentration depends on the density of copper but is independent of the total amount of the copper present. Since the amount of copper has no influence on the equilibrium concentrations of the other participants in the reaction, we say that the activity of pure solid copper equals one and write the mass-action law as

$$K = \frac{[Cu^{2+}]^2}{P_{O_2}[H^+]^4}$$

By generalizing this reasoning, we say that the activity of any substance within its own pure solid phase equals one. A simpler example of the insensitivity of an equilibrium system to the total amount of solid present is a saturated solution of I_2 in water. The equilibrium process is

$$I_2(s) \rightleftharpoons I_2(aq)$$

and the equilibrium constant is

$$K = [I_2(aq)]$$

Once the solution is saturated—in other words, the system is in equilibrium—adding more solid I_2 has no effect on the concentration of the solution.

One aspect of the way we write equilibrium constants involves their units. Since activities are unitless, so is any thermodynamic equilibrium constant. Even though we are using molarities and atmospheres instead of activities, the equilibrium constants presented from here on will not have any units. Remember that an answer for an equilibrium concentration will be a molarity if the substance is a solute and a pressure in atmospheres if the substance is a gas.

The value of an equilibrium constant based on molarities and atmospheres, rather than activities, is not truly constant and often changes as the concentrations change. While these nonideal effects are most severe in concentrated solutions, the changes can also be considerable in dilute solutions, particularly for reactions involving ions. Nevertheless, approximate answers derived from nonideal equilibrium constants are very useful. It is often sufficient to estimate an equilibrium concentration within an order of magnitude (a factor of ten) to determine whether a certain separation procedure is feasible, or whether a particular indicator is suitable for a titration.

Having accepted errors due to nonideality, we are free of the necessity of obtaining exact mathematical solutions for equilibrium problems. A mathematical approximation causing an error of less than 10% is acceptable because it is probably less than the nonideality error. In addition, the rather low precision of the published values for most equilibrium constants does not justify the extra work required to obtain exact mathematical solutions for equilibrium concentrations.

13.2
Equilibrium, Thermodynamics, and Kinetics

The Thermodynamics of Equilibrium

As in the examples of equilibrium processes we have discussed earlier—melting, evaporation, and solubility—each chemical reaction occurs with a particular change in enthalpy, entropy, and Gibbs free energy. A reversible reaction proceeds in the direction that decreases the free energy of the system. As the reaction continues in this direction, the resulting changes in the reactant and product concentrations alter the free energy until, at equilibrium, the reactants and products have the same free energy. Since the free energy of the reaction at equilibrium is zero, there is no further net reaction (Figure 13.2).

Another aspect of the change of free energy with the extent of reaction is that the total free energy of the system reaches the minimum value possible for that system as the reaction reaches equilibrium. Any further chemical change in either direction would raise the free energy, producing a less stable mixture, so all net chemical changes cease.

There is an important relationship between the equilibrium constant for a reaction and the **standard free energy** ($\Delta G°$)—the free energy change of the reaction when the activity of each reactant and product equals one. The relationship is

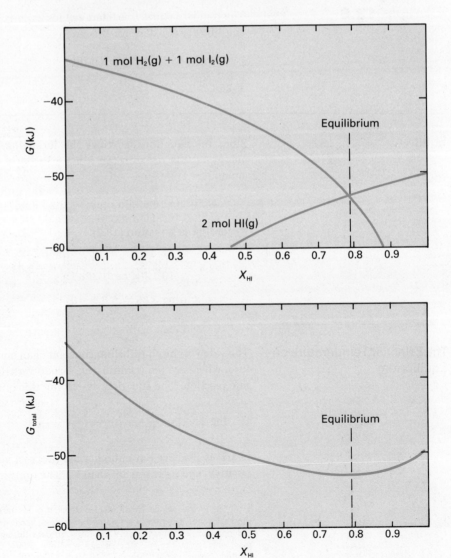

Figure 13.2
Free energy and equilibrium.
Reaction: 1.00 mol H_2 and 1.00 mol I_2 react at 700 K and 2.00 atm.

$$H_2(g) + I_2(g) \rightleftharpoons 2\,HI(g)$$

$$\log K = \frac{-\Delta G°}{2.30RT} = \frac{-\Delta G°}{(19.1 \text{ J/K-mol})T} \qquad (13.7)$$

This equation, which is applicable to all reversible processes, allows us to estimate K from the standard free energy of the reaction. If we know the value of the equilibrium constant, we can evaluate $\Delta G°$.

Example 13.7

The equilibrium constant for the dissociation of N_2O_4 into NO_2 is 0.155 at 25°C. What is the standard free energy change for this reaction?

Solution

Equation (13.7) can be rewritten as

$$\Delta G° = -(19.1 \text{ J/K-mol})T \log K_p$$

$$\log 0.155 = -0.810$$

$$\Delta G° = -(19.1 \text{ J/K-mol})(298 \text{ K})(-0.810) = +4.61 \times 10^3 \text{ J/mol, or } +4.61 \text{ kJ/mol}$$

Example 13.8

The standard Gibbs free energy of formation of ammonia is -16.6 kJ/mol. What is the equilibrium constant for the reaction

$$3 H_2(g) + N_2(g) \rightleftharpoons 2 NH_3(g)$$

at 25°C?

Solution

Since the equation describes the formation of two ammonia molecules, the free energy of this reaction (on a mole basis) is twice the free energy of formation of ammonia.

$$\Delta G° = 2(-16.6 \text{ kJ/mol}) = -33.2 \text{ kJ/mol} = -3.32 \times 10^4 \text{ J/mol}$$

According to equation (13.7),

$$\log K = \frac{-(-3.32 \times 10^4 \text{ J/mol})}{(19.1 \text{ J/K-mol})(298 \text{ K})} = +5.83$$

$$K = \text{antilog } 5.83 = 6.8 \times 10^5$$

The Effect of Temperature on Equilibrium

The value of the equilibrium constant of an endothermic reaction rises with temperature, whereas a temperature increase lowers the equilibrium constant of an exothermic reaction. The equation describing this effect is

$$\log \left(\frac{K_2}{K_1}\right) = \frac{\Delta H°}{19.1 \text{ J/K-mol}} \left(\frac{1}{T_1} - \frac{1}{T_2}\right) \qquad (13.8)*$$

If $\Delta H°$ is positive (an endothermic process) and T_2 is greater then T_1, $\log (K_2/K_1)$ is positive, and K_2 is greater than K_1. The opposite situation occurs if $\Delta H°$ is negative.

---

*As long as we deal with ideal systems, it is not necessary to use a superscript zero to indicate that the heat of the reaction is a standard enthalpy change because enthalpy changes in ideal systems are independent of concentration. Concentration does influence heats of reactions in real systems, and we must specify that the enthalpy change is estimated for a standard state—one in which all reactants and products are at unit activity.

Figure 13.3

The effect of temperature on K_P.

Reaction: $N_2O_4(g) \rightleftharpoons 2\,NO_2(g)$

Source: K. Schreiber, *Zeitschrift für Physikalische Chemie* **24**, 660 (1897).

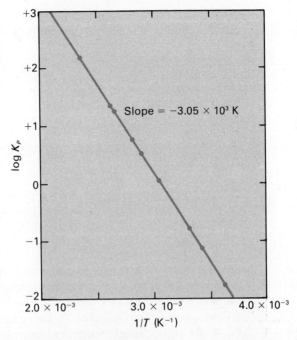

$\Delta H° = (-19.1\ J/K)(-3.05 \times 10^3\ K) = +58.3\ kJ$

The slope of a graph of $\log K$ versus $1/T$ equals $-\Delta H°/19.1$, where $\Delta H°$ is expressed in J/mol (Figure 13.3). Estimates of the slopes of these graphs agree with direct calorimetric measurements of heats of reactions.

Example 13.9

The heat of formation of ammonia is -46.2 kJ/mol. What is the equilibrium constant for the reaction

$$3\,H_2(g) + N_2(g) \rightleftharpoons 2\,NH_3(g)$$

at 500°C? The equilibrium constant at 25°C is 6.8×10^5.

Solution

Since two NH_3 molecules are formed in the reaction, the heat of the reaction is twice the heat of formation of ammonia.

$$\Delta H° = 2(-46.2\ kJ/mol) = -92.4\ kJ/mol,\ \text{or}\ -9.24 \times 10^4\ J/mol$$

$$\log\left(\frac{K_2}{K_1}\right) = \frac{-9.24 \times 10^4\ \cancel{J/mol}}{19.1\ \cancel{J/K\text{-mol}}}\left(\frac{1}{298\ K} - \frac{1}{773\ K}\right)$$

The solution has been arranged so that K_2 is the equilibrium constant at 500°C and K_1 the constant at 25°C.

$$\log\left(\frac{K_2}{K_1}\right) = -4.84 \times 10^3\ K(3.36 \times 10^{-3}\ K^{-1} - 1.29 \times 10^{-3}\ K^{-1})$$

$$= -4.84 \times 10^3\ \cancel{K}(2.06 \times 10^{-3}\ \cancel{K^{-1}})$$

$$= -9.98$$

$$\frac{K_2}{K_1} = \text{antilog}\ -9.98 = 1.1 \times 10^{-10}$$

$$K_2 = 1.1 \times 10^{-10}\ K_1 = (1.1 \times 10^{-10})(6.8 \times 10^5) = 7.5 \times 10^{-5}$$

Kinetics and Equilibrium

One of the theoretical explanations frequently used to explain the law of mass action uses the following kinetic argument. In the hypothetical reaction,

$$m\text{A} + n\text{B} \rightleftharpoons p\text{C} + q\text{D}$$

equilibrium occurs when the rate of the reverse reaction just balances the rate of the forward reaction.

Forward rate = $k_f[\text{A}]^m[\text{B}]^n$

Reverse rate = $k_r[\text{C}]^p[\text{D}]^q$

Forward rate = Reverse rate

$k_f[\text{A}]^m[\text{B}]^n = k_r[\text{C}]^p[\text{D}]^q$

$$\frac{[\text{C}]^p[\text{D}]^q}{[\text{A}]^m[\text{B}]^n} = \frac{k_f}{k_r} = K_{eq} \tag{13.9}$$

Equation (13.9) is a valid equation, and it is true that the rates of the forward and reverse reactions are balanced at equilibrium. However, the theoretical explanation contains a fallacy—it assumes the orders of the forward and reverse reactions equal their stoichiometric coefficients. Yet we know of many cases, such as the formation of $COCl_2$, for which kinetic orders and stoichiometric coefficients differ.

$$CO(g) + Cl_2(g) \rightleftharpoons COCl_2(g)$$

Forward rate = $k_f[CO][Cl_2]^{3/2}$

Equation (13.9) works in spite of this logical error because the kinetic orders of the forward and reverse reactions are related. The rate law for the decomposition of phosgene ($COCl_2$)

Reverse rate = $k_r[COCl_2][Cl_2]^{1/2}$

is related to the rate law for formation of phosgene—the kinetic orders of chlorine exceed its stoichiometric coefficient by 1/2 in both cases. At equilibrium,

$k_f[CO][Cl_2]^{3/2} = k_r[COCl_2][Cl_2]^{1/2}$

$$\frac{[COCl_2]}{[CO][Cl_2]} = \frac{k_f}{k_r} = K_{eq}$$

The rate laws of the forward and reverse reactions for any reversible reaction are related in a way consistent with equation (13.9).

The requirement that the rate laws of the forward and reverse reactions be related in a way consistent with the law of mass action and the dynamic balance at equilibrium also applies to catalyzed reactions. A catalyst raises the rates of the forward and reverse reactions to the same extent, but does not affect the equilibrium concentrations. For example, both the dehydration of ethanol (C_2H_5OH) and its reverse reaction, the hydration of ethylene (C_2H_4), are catalyzed by acid to the same extent.

$$C_2H_5OH \rightleftharpoons C_2H_4 + H_2O$$

Forward rate = $k_f[C_2H_5OH][H^+]$

Reverse rate = $k_r[C_2H_4][H_2O][H^+]$

$$K_{eq} = \frac{k_f}{k_r} = \frac{[C_2H_4][H_2O]}{[C_2H_5OH]}$$

The relationship between the kinetic orders of forward and reverse reactions results from the phenomenon of **microscopic reversibility**—the reverse reaction proceeds via the same intermediates and activated complexes as the forward reaction, but in reverse order. The mechanism for the acid-catalyzed dehydration of ethanol (C_2H_5OH) to form ethylene (C_2H_4) appears to be

$$H^+ + C_2H_5OH \underset{}{\overset{fast}{\rightleftharpoons}} C_2H_5OH_2^+$$

$$C_2H_5OH_2^+ \underset{}{\overset{slow}{\rightleftharpoons}} C_2H_5^+ + H_2O$$

$$C_2H_5^+ \underset{}{\overset{fast}{\rightleftharpoons}} C_2H_4 + H^+$$

A chemical perpetual-motion machine.

We could make a perpetual motion machine if we could find a catalyst that speeds a reaction in one direction without affecting the reverse reaction. The machine shown in the figure consists of a piston driven by a gas-phase reaction that causes a volume change—the formation of phosgene from carbon monoxide and chlorine would do nicely. The reaction proceeds from left to right in the presence of our imaginary catalyst, and the piston moves down the cylinder until a rod from the piston (A) closes a trap door (B) and seals off the catalyst (C). The forward reaction slows down in the absence of the catalyst while the reverse reaction goes on at its usual speed. This shifts the equilibrium to the left, and the gas mixture expands as phosgene decomposes. The piston rises until it pulls open the trap door with the string (D). This exposes the reaction mixture to the catalyst and again reverses the reaction's direction. The cycle repeats, producing work without consuming energy.

Our only problem is finding such a catalyst. This requires getting around microscopic reversibility—we must open a low-energy pathway for formation of phosgene without accelerating its decomposition. This is a job for Maxwell's demon, who could sit on top of the rate-determining energy barrier for the catalyzed reaction and block the path of any decomposing phosgene molecule that tries to get over it.

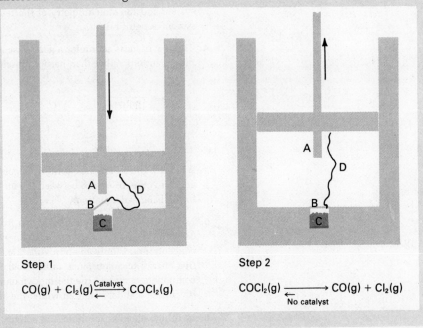

Step 1
$CO(g) + Cl_2(g) \underset{}{\overset{Catalyst}{\rightleftharpoons}} COCl_2(g)$

Step 2
$COCl_2(g) \underset{\text{No catalyst}}{\longrightarrow} CO(g) + Cl_2(g)$

The hydration of ethylene (the reverse reaction) involves protonation, combination of the conjugate acid with water, and the dissociation of a proton from the conjugate acid of ethanol. The forward and reverse rate laws are related because both reactions have the same activated complex for the rate-determining step.

Microscopic reversibility is a consequence of the randomness of molecular behavior. The fastest (and therefore dominant) mechanism for any reaction is the one with the most stable activated complexes. If a series of activated complexes provides the lowest energy pathway leading from reactants to products, it also forms the easiest route from products back to reactants. When a catalyst opens a favorable mechanistic route for the forward reaction, it also provides an easy path for the reverse reaction. Whatever the direction of the reaction, random molecular collisions lead to reaction most readily via these low-energy mechanisms.

Summary

1. When a system is in a state of chemical equilibrium, no further net chemical change occurs, and the composition remains constant. The forward and reverse rates of the reversible reaction are exactly balanced. The free energy change of a reaction at equilibrium is zero.

2. The law of mass action: The mass-action ratio (Q_P) of an equilibrium gas mixture equals the equilibrium constant (K_P)

$$Q_P = K_P$$

The numerator of Q_P is a multiple of the partial pressures of the products, each pressure being raised to an exponential power equal to the stoichiometric coefficient of the substance. The denominator is a similar multiple of reactant pressures. For the hypothetical reaction,

$$m \ A(g) + n \ B(g) \rightleftharpoons p \ C(g) + q \ D(g)$$

$$Q_P = \frac{P_C^p P_D^q}{P_A^m P_B^n}$$

3. Using the law of mass action and equations based on the initial partial pressures of the system and the stoichiometry of the reaction, we can calculate the composition of a gas system at equilibrium.

4. The law of mass action for a gas-phase reaction can be expressed in terms of molarities or mole fractions rather than partial pressures,

$$K_C = \frac{K_P}{(RT)^{\Delta n}}$$

$$K_X = \frac{K_P}{P_{total}^{\Delta n}}$$

where K_C and K_X are the equilibrium constants for molarity and mole fraction, respectively, and Δn is the difference between the number of product molecules and reactant molecules in the balanced equation.

5. Nonideal effects produce deviations from the law of mass action. A law of mass action based on the activities of the components applies at all concentrations, but one based on concentration is only an approximation. Activity is a unitless quantity that tells us the real free energy of a substance rather than its presumed ideal free energy. The activity of a component compares the thermodynamic state in a particular system to that in an arbitrarily chosen standard state in which the activity equals one.

6. In dilute aqueous solutions the activity of water is approximately equal to one and the activities of solutes are approximately equal to their molarities.

7. In heterogeneous systems, the activity of an insoluble solid or liquid equals one.

EXERCISES

8. The standard free energy ($\Delta G°$) of a reaction is the free energy change when each reactant and product has an activity of one. The relationship between the standard free energy and the equilibrium constant is

$$\log K = \frac{-\Delta G°}{2.30\, RT}$$

9. The following equation describes the effect of temperature on an equilibrium constant.

$$\log\left(\frac{K_2}{K_1}\right) = \frac{\Delta H°}{19.1\ \text{J/K-mol}}\left(\frac{1}{T_1} - \frac{1}{T_2}\right)$$

10. A catalyst speeds up the forward and reverse direction of a reversible reaction to the same extent, but does not affect the equilibrium concentrations.

11. Microscopic reversibility is a phenomenon in which the reverse reaction proceeds via the same intermediates and activated complexes as the forward reaction, but in reverse order.

Vocabulary List

equilibrium
reversible reaction
mass-action ratio
equilibrium constant

law of mass action
thermodynamic equilibrium constant
activity

standard state
standard free energy
microscopic reversibility

Exercises

1. Describe the difference between the mass-action ratio and the equilibrium constant for a reaction.

2. Choose the answer that makes the statement correct. The free energy change due to a reaction is zero when:
   a. The reactants are initially mixed
   b. The reactants are completely consumed
   c. The system is at equilibrium
   d. A catalyst is added

3. In order to produce ammonia at a suitable rate, the reaction

$$3\ H_2(g) + N_2(g) \rightleftharpoons 2\ NH_3(g)$$

must be run at high temperatures. At these temperatures, the equilibrium constant for the reaction is much less than 1. Can you suggest a procedure, based on Le Chatelier's principle, for obtaining a good yield of ammonia from hydrogen?

4. In which of the following reactions do K_P, K_C, and K_X have the same value?
   a. $N_2(g) + 3\ H_2(g) \rightleftharpoons 2\ NH_3(g)$
   b. $2\ H_2S(g) + 3\ O_2(g) \rightleftharpoons 2\ SO_2(g) + 2\ H_2O(g)$
   c. $Br_2(g) + Cl_2(g) \rightleftharpoons 2\ BrCl(g)$
   d. $P_4(g) + 6\ Cl_2(g) \rightleftharpoons 4\ PCl_3(g)$

5. Bromine vapor and chlorine gas, each at an initial pressure of 0.500 atm at 25°C, react to form BrCl.

$$Br_2(g) + Cl_2(g) \rightleftharpoons 2\ BrCl(g)$$

The equilibrium mixture contains 0.292 atm each of Br_2 and Cl_2 and 0.416 atm BrCl. What is the value of the equilibrium constant for this reaction at 25°C?

6. The reaction

$$PCl_5(g) \rightleftharpoons PCl_3(g) + Cl_2(g)$$

is at equilibrium in a particular system. Will the mass-action ratio increase, decrease, or stay the same if the following occur?
   a. PCl_5 is added to the system at constant volume
   b. Cl_2 is added at constant volume
   c. PCl_3 is removed at constant volume
   d. The volume of the system is doubled
   e. A catalyst is added

7. Write mass-action ratios for the following reactions
   a. $C_2H_2(g) + 2\ H_2(g) \rightleftharpoons C_2H_6(g)$
   b. $2\ N_2O(g) + O_2(g) \rightleftharpoons 4\ NO(g)$
   c. $F_2(g) + 2\ HCl(g) \rightleftharpoons 2HF(g) + Cl_2(g)$
   d. $2\ Cl_2(g) + 2\ H_2O(g) \rightleftharpoons 4\ HCl(g) + O_2(g)$

8. Write the mass-action ratio for each of the following reactions.
   a. $2\ CO(g) + O_2(g) \rightleftharpoons 2\ CO_2(g)$
   b. $C_2H_6(g) \rightleftharpoons 2\ CH_3(g)$
   c. $2\ O_3(g) \rightleftharpoons 3\ O_2(g)$
   d. $N_2H_4(g) + H_2(g) \rightleftharpoons 2\ NH_3(g)$

9. Write approximate mass-action laws for the following reactions, using appropriate approximations for dilute aqueous solutions and for heterogeneous systems.
   a. $3\ H_2S(g) + 2\ Fe^{3+}(aq) \rightleftharpoons Fe_2S_3(s) + 6\ H^+(aq)$
   b. $H_2O + Mg^{2+}(aq) \rightleftharpoons MgOH^+(aq) + H^+(aq)$
   c. $H^+(aq) + HCO_3^-(aq) \rightleftharpoons CO_2(g) + H_2O$
   d. $Mn(OH)_2(s) + OH^-(aq) \rightleftharpoons Mn(OH)_3^-(aq)$

10. Write mass-action ratios for the following reactions. Use approximations for dilute aqueous solutions and for heterogeneous systems.
    a. $H_2O + BeO(s) \rightleftharpoons Be^{2+}(aq) + 2\ OH^-(aq)$
    b. $Ba_3(PO_4)_2(s) \rightleftharpoons 3\ Ba^{2+}(aq) + 2\ PO_4^{3-}(aq)$
    c. $2\ I^-(aq) + Ag_2SO_4(s) \rightleftharpoons 2\ AgI(s) + SO_4^{2-}(aq)$
    d. $CaCO_3(s) \rightleftharpoons CaO(s) + CO_2(g)$

11. $SO_2Cl_2(g) \rightleftharpoons SO_2(g) + Cl_2(g)$

    $K_P = 3.76$ atm at $110.6°C$ $\Delta H = +46.7$ kJ/mol

    A system at equilibrium at $110.6°$ and 1.82 atm contains 0.18 atm SO_2Cl_2 and 0.82 atm each SO_2 and Cl_2. Will the equilibrium shift to the right, shift to the left, or stay the same if the following occur?
    a. Another 2.00 atm Cl_2 is added to the system
    b. The system is warmed to 200°C
    c. The system is compressed to 10.0 atm
    d. N_2 (inert in this system) is added while the total pressure remains at 1.82 atm
    e. N_2 is added while the system is held rigid
    f. SO_2 is selectively removed from the system

12. The reaction

    $$CO_2(g) + H_2(g) \rightleftharpoons CO(g) + H_2O(g)$$

    is endothermic. According to Le Chatelier's principle, would a system in which this reaction is at equilibrium react from left to right, react from right to left, or stay the same if the following occur?
    a. H_2 is added to the system at constant volume
    b. CO is added to the system at constant volume
    c. Water vapor is removed from the system at constant volume
    d. The system is compressed to half its volume
    e. A catalyst is added
    f. The system is heated

13. What is the mass-action ratio for the reaction

    $$2\ NO(g) + Cl_2(g) \rightleftharpoons 2\ NOCl(g)$$

    when the partial pressures are: NO, 0.300 atm; Cl_2, 0.0500 atm; NOCl, 0.150 atm?

14. Find the mass-action ratio for the reaction

    $$4\ NH_3(g) + 5\ O_2(g) \rightleftharpoons 4\ NO(g) + 6\ H_2O(g)$$

    in a system in which the partial pressure of ammonia is 0.200 atm, the partial pressure of oxygen is 0.400 atm, the partial pressure of nitric oxide is 0.300 atm, and the partial pressure of water is 0.0500 atm.

15. A solution is 0.200 M in $S_2O_8^{2-}$, 0.150 M in I^-, 0.0500 M in SO_4^{2-}, and 0.0010 M in I_3^-. What is the mass-action ratio for the following reaction?

    $$S_2O_8^{2-}(aq) + 3\ I^-(aq) \rightleftharpoons 2\ SO_4^{2-}(aq) + I_3^-(aq)$$

16. A solution is 0.100 M in Tl^+, 0.0100 M in Tl^{3+}, 0.0200 in Ce^{4+}, and 0.200 M in Ce^{3+}. What is the mass-action ratio of the following reaction?

    $$Tl^+(aq) + 2\ Ce^{4+}(aq) \rightleftharpoons 2\ Ce^{3+}(aq) + Tl^{3+}(aq)$$

17. What is the mass-action ratio for the reaction

    $$2\ NaHCO_3(s) \rightleftharpoons Na_2CO_3(s) + H_2O(g) + CO_2(g)$$

    in a system consisting of water vapor and carbon dioxide, each at a partial pressure of 350 torr above a mixture of 5.0 g $NaHCO_3$ and 0.100 g Na_2CO_3?

18. What is the mass-action ratio for the decomposition of hydrogen peroxide

    $$2\ H_2O_2(aq) \rightleftharpoons 2\ H_2O + O_2(g)$$

    when a 0.0500 M solution of hydrogen peroxide is in contact with a gas mixture whose partial pressure of oxygen is 500 torr?

EXERCISES

19. When a sample of PCl_5 is heated to 189°C while the total pressure is held at 1.00 atm, some of the PCl_5 decomposes.

    $$PCl_5(g) \rightleftharpoons PCl_3(g) + Cl_2(g)$$

    The partial pressure of chlorine in the equilibrium mixture is 0.244 atm. What is the equilibrium constant for the decomposition reaction?

20. When bromine is held at 1000 K and 0.500 atm, 2.1% of the Br_2 molecules dissociate into atoms. What is the equilibrium constant for the dissociation of bromine under these conditions?

21. $CO(g) + 2 H_2(g) \rightleftharpoons CH_3OH(g) \quad K_P = 36 \text{ atm}^{-2}$ at 700 K

    What reaction, if any, will occur in a system containing 15 atm CO, 30 atm H_2, and 50 atm CH_3OH at 700 K?

22. $2 Cl_2(g) + 2 H_2O(g) \rightleftharpoons 4 HCl(g) + O_2(g)$

    The equilibrium constant for this reaction is 1.0 atm at 600°C. A system contains 2.0 atm Cl_2, 2.0 atm H_2O, 4.0 atm HCl, and 1.0 atm O_2 at 600°C. What reaction, if any, will occur?

23. $2 HIO_3(s) \rightleftharpoons I_2O_5(s) + H_2O(g) \quad K = 0.012$ at 40°C

    What reaction, if any, will occur in a system containing 15 torr water vapor at 40°C in the presence of a 3:1 mixture (by weight) of solid HIO_3 and solid I_2O_5?

24. The reaction

    $$2 BrI(g) \rightleftharpoons Br_2(g) + I_2(g)$$

    has a K_P value of 32 at 578 K. A system at 578 K contains 0.100 atm BrI, 0.500 atm Br_2, and 3.00 atm I_2. Will the reaction go from left to right, from right to left, or is it at equilibrium?

25. The value of K_P for the reaction

    $$2 SO_2(g) + O_2(g) \rightleftharpoons 2 SO_3(g)$$

    is 1.12×10^3 atm^{-1} at 800 K?
    a. What is the value of K_C at 800 K?
    b. What is the value of K_X at 800 K and 5.00 atm?
    c. What is the value of K_X at 800 K and 50.0 atm?
    d. Are your answers for parts b and c consistent with Le Chatelier's principle?

26. a. The equilibrium constant for the formation of C_2H_6 (ethane) from hydrogen and C_2H_4 (ethylene)

    $$H_2(g) + C_2H_4(g) \rightleftharpoons C_2H_6(g)$$

    is 5×10^{17} atm^{-1} at 25°C. What is the value of K_P for the decomposition of ethane into hydrogen and ethylene at 25°C?

    b. The value of K_P for the formation of ethylene from hydrogen and C_2H_2 (acetylene)

    $$H_2(g) + C_2H_2(g) \rightleftharpoons C_2H_4(g)$$

    is 5×10^{24} atm^{-1} at 25°C. What is the value of K_P for the reaction between hydrogen and acetylene to form ethane at 25°C?

27. The equilibrium constant for the reaction

    $$H_2(g) + Br_2(g) \rightleftharpoons 2 HBr(g)$$

    is 1.6×10^5 at 1024 K. What is the partial pressure of Br_2 in an equilibrium mixture containing 2.00 atm HBr and 10.0 atm H_2 at 1024 K?

28. The equilibrium constant for the reaction

    $$2 HCN(g) \rightleftharpoons H_2(g) + C_2N_2(g)$$

    is 4.0×10^{-4} at 500 K.
    a. What is the equilibrium partial pressure of C_2N_2 if pure HCN, originally present at 25.0 atm, reacts at 500 K?
    b. What is the partial pressure of C_2N_2 at equilibrium if the initial mixture contained 25.0 atm HCN and 10.0 atm H_2?
    c. Are your answers for parts a and c consistent with Le Chatelier's principle?

29. $N_2(g) + O_2(g) \rightleftharpoons 2 NO(g) \quad K_P = 2.70 \times 10^{-4}$ at 1750°C

    What is the equilibrium pressure of NO if:
    a. Pure NO is heated to 1750°C at 15.0 atm?
    b. A mixture of 15.0 atm NO and 15.0 atm N_2 is heated to 1750°C?
    c. A mixture of 7.50 atm N_2 and 7.50 atm O_2 reacts at 1750°C?
    d. A mixture of 22.5 atm N_2 and 7.50 atm O_2 reacts at 1750°C?

30. The value of the equilibrium constant for the following reaction at 936 K is 4.54×10^3.

    $$CO(g) + NiO(s) \rightleftharpoons Ni(s) + CO_2(g)$$

    a. What will be the equilibrium pressures of CO and CO_2 if CO, initially at a pressure of 5.00 atm, reacts with an excess of NiO at 936 K?
    b. How much Ni will be formed if 10.0 mol CO, initially at 5.00 atm, is brought into contact with excess NiO at 936 K?

31. $PCl_5(g) \rightleftharpoons PCl_3(g) + Cl_2(g)$ $K_P = 40.4$ atm at 340°C

   What will be the partial pressures of the equilibrium system if 1.50 atm PCl_5 react at 340°C?

32. What are the partial pressures in the equilibrium system formed by the reaction of 5.00 atm O_2 and 2.00 atm NO at 592 K to form NO_2?

   $O_2(g) + 2\ NO(g) \rightleftharpoons 2\ NO_2(g)$ $K_P = 226$ atm^{-1} at 592 K

33. The equilibrium constant for the reaction

   $3\ H_2(g) + SO_2(g) \rightleftharpoons H_2S(g) + 2\ H_2O(g)$

   is 3.2×10^3 at 1200°C. What is the standard Gibbs free energy change for this reaction at 1200°C?

34. $2\ SO_2(g) + O_2(g) \rightleftharpoons 2\ SO_3(g)$ $K_P = 1.2 \times 10^{22}$ atm^{-1} at 25°C

   What is the standard free energy for this reaction at 25°C?

---

35. The standard free energy change for the reaction

   $CH_4(g) + Cl_2(g) \rightleftharpoons CH_3Cl(g) + HCl(g)$

   is -103.3 kJ/mol at 25°C. What is the equilibrium constant for this reaction at 25°C?

36. The standard Gibbs free energy change for the reaction

   $2\ NOBr(g) \rightleftharpoons 2\ NO(g) + Br_2(g)$

   is $+11.70$ kJ/mol at 25°C. What is the equilibrium constant for this reaction at 25°C?

37. The value of K_P for the dissociation of chlorine

   $Cl_2(g) \rightleftharpoons 2\ Cl(g)$

   is 4.8×10^{-16} at 323°C and 1.29×10^{-3} at 1323°C. What is the heat of dissociation of chlorine?

38. The equilibrium constant of the reaction

   $2\ NO(g) \rightleftharpoons N_2(g) + O_2(g)$

   is 12.4 at 4057 K and 6.35 at 4727 K. What is the enthalpy of this reaction?

39. $SO_2Cl_2(g) \rightleftharpoons SO_2(g) + Cl_2(g)$

   $K_P = 3.76$ atm at 110.6°C $\Delta H = +46.7$ kJ

   Estimate the equilibrium constant for this reaction at 138.2°C.

40. The equilibrium constant for the reaction

   $O_2(g) + 2\ NO(g) \rightleftharpoons 2\ NO_2(g)$

   is 1.6×10^{12} at 25°C and the enthalpy is -113 kJ/mol. What is the value of the equilibrium constant at 500°C?

41. The value of K_P for the dissociation of N_2O_4

   $N_2O_4(g) \rightleftharpoons 2\ NO_2(g)$

   is 1.7×10^{-4} atm at 25°C. What is the partial pressure of NO_2 in an equilibrium mixture of N_2O_4 and NO_2 at 25°C and 2.00 atm?

42. When 1.00 mol CH_3CO_2H reacts with 1.00 mol C_2H_5OH

   $CH_3CO_2H + C_2H_5OH \rightleftharpoons CH_3CO_2C_2H_5 + H_2O$

   0.667 mol $CH_3CO_2C_2H_5$ is formed at equilibrium. Although this reaction takes place in a homogeneous liquid phase, it is not a dilute aqueous solution—the only water in the system is that formed by the reaction.
   a. Calculate K_X for this reaction
   b. Calculate the moles of $CH_3CO_2C_2H_5$ formed when 2.00 mol C_2H_5OH and 1.00 mol CH_3CO_2H react.
   c. Calculate the moles of $CH_3CO_2C_2H_5$ remaining at equilibrium when 1.00 mol $CH_3CO_2C_2H_5$ and 2.00 mol H_2O react.

43. The reaction

   $2\ NO(g) \rightleftharpoons N_2(g) + O_2(g)$

   has an equilibrium constant of 5.4×10^4 at 1620 K, and the rate law for the forward reaction is

   $$\text{Forward rate} = -\frac{d[NO]}{dt} = k_f[NO]^2$$

   $k_f = 192\ M^{-1}\ \text{sec}^{-1}$ at 1620 K

   Write the rate law and calculate the rate constant for the reverse reaction.

44. The rate law for the forward reaction

   $O_2(g) + 2\ NO(g) \rightleftharpoons 2\ NO_2(g)$

   is first order in oxygen and second order in NO, the rate constant (for consumption of O_2) is $1.12 \times 10^4\ M^{-2}\ \text{sec}^{-1}$ at 592 K, and the value of K_C at this temperature is $1.10 \times 10^4\ M^{-1}$. What is the rate law and the value of the rate constant for the reverse reaction?

EXERCISES

45. $CH_3CH_2CH_2CH_3 \rightleftharpoons CH_3\underset{\underset{CH_3}{|}}{C}HCH_3$

    Butane   Isobutane

$$\text{Forward rate} = -\frac{d[\text{Butane}]}{dt} = k_f[\text{Butane}][\text{HAlBr}_4]$$

$k_f = 2.9 \times 10^{-4} M^{-1} sec^{-1}$ at 65°C $K_{eq} = 2.5$ at 65°C

Write the rate law and the rate constant for the reverse reaction at 65°C.

46. The reaction

$$C_6H_6 + H_2SO_4 \longrightarrow C_6H_5SO_3^- + H_3O^+$$

proceeds by the following mechanism

$$H_2SO_4 \longrightarrow SO_3 + H_2O$$

$$SO_3 + C_6H_6 \longrightarrow C_6H_6SO_3 \text{ (rate-determining step)}$$

$$C_6H_6SO_3 + H_2O \longrightarrow C_6H_5SO_3^- + H_3O^+$$

Write the mechanism for the reaction

$$C_6H_5SO_3^- + H_3O^+ \longrightarrow C_6H_6 + H_2SO_4$$

47. **a.** Write the rate law for each elementary reaction, both forward and reverse, involved in the interconversion of C_2H_5OH and C_2H_4. (See page 342.)
    b. Write the mass-action law for each step of this reaction.
    c. Are these equilibrium expressions consistent with the condition that the forward rate equal the reverse rate at equilibrium?
    d. Use the mass-action laws of these individual steps to prove that

$$\frac{[C_2H_4][H_2O]}{[C_2H_5OH]} = K$$

chapter 14

Aqueous Equilibria

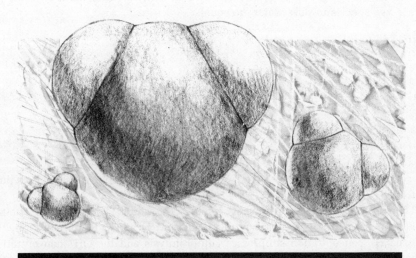

Aqueous equilibria involve reactions of substances with water molecules, H_2O, to give H^+ and OH^-

Whether or not a solution is acidic, basic, or neutral has a profound influence on many reactions that occur in aqueous solutions. The equilibrium state of any reaction consuming or producing either H^+ or OH^- will be sensitive to the concentrations of these ions. Many reactions are catalyzed by H^+ or OH^-, and their rates depend on the solution's acidity. The significance of acidity for reactions occurring in aqueous solutions means that we must know how to measure and control acidity if we are to control the kinetics and equilibrium of a reaction in solution. Because of this, a theory of acid–base equilibria was essential to the development of modern chemical technology. (A major part of this theory was worked out in order to understand the chemistry of brewing.) Acid–base equilibria are extremely important in biochemical reactions since most physiological systems can function only within a narrow acidity range.

The equilibria governing the solubilities of salts are also very important to chemists. An understanding of solubility equilibria helps us to predict the conditions needed for dissolving a salt or for precipitating it from solution. We can use this information to develop a procedure for separating two metal ions or to estimate the sensitivity of a laboratory test for a particular ion.

14.1 Acid–Base Equilibria

The Autoprotolysis of Water

The **hydrated hydrogen ion** (H_3O^+, also called the **hydronium ion**) and hydroxide ion are always present in aqueous solutions and even in pure water because one water molecule can donate a proton to another, a reaction called **autoprotolysis**.

$$H_2O + H_2O \rightleftharpoons H_3O^+ + OH^-$$

A shorthand equation for this reaction is

$$H_2O \rightleftharpoons H^+(aq) + OH^-(aq)*$$

and the process is often called either the dissociation or the ionization of water. Because the equilibrium for autoprotolysis of water is very unfavorable, the concentrations of H^+ and OH^- in pure water are very low. The equilibrium constant for autoprotolysis in water (K_W) is 1.0×10^{-14} at 25°C.

$$K_W = [H^+][OH^-] = 1.0 \times 10^{-14} \tag{14.1}$$

Since autoprotolysis produces equal numbers of H^+ and OH^- ions, the concentrations of these two species in pure water are equal.

$$[H^+] = [OH^-]$$

Therefore,

$$[H^+][OH^-] = [H^+]^2 = 1.0 \times 10^{-14}$$

and

$$[H^+] = \sqrt{1.0 \times 10^{-14}} = 1.0 \times 10^{-7} M = [OH^-]$$

This condition that $[H^+]$ is $1.0 \times 10^{-7} M$ exists in neutral solutions as well as pure water.

There is always an equilibrium between H^+, OH^-, and H_2O regardless of whether the solution is acidic, basic, or neutral. If an acid is added to the solution, the increased H^+ concentration suppresses autoprotolysis and the OH^- concentration decreases. In a similar way, the high OH^- concentration in a basic solution lowers the H^+ concentration. The following two examples illustrate the relationship between the concentrations of hydrogen and hydroxide ions.

Example 14.1

What is the hydroxide-ion concentration in a solution that is $2 \times 10^{-2} M$ in H^+?

Solution

According to equation (14.1), the hydroxide-ion concentration is inversely proportional to the hydrogen-ion concentration.

$$[OH^-] = \frac{1.0 \times 10^{-14}}{[H^+]} = \frac{1.0 \times 10^{-14}}{2 \times 10^{-2}} = 5 \times 10^{-13} M$$

Example 14.2

A solution is $0.0040 M$ in OH^-. What is its H^+ concentration?

Solution

In this case, we divide K_W by the hydroxide-ion concentration to find H^+.

$$[H^+] = \frac{1.0 \times 10^{-14}}{[OH^-]} = \frac{1.0 \times 10^{-14}}{0.0040} = 2.5 \times 10^{-12} M$$

Hydrogen-Ion Concentration; pH

Hydrogen-ion concentrations in dilute solutions (solutions in which no solute exceeds $1 M$) can range from $1 M$ to $1 \times 10^{-14} M$. Because of this large range of possible

---

*For the rest of this chapter, we will use H^+ as the symbol for the hydrated hydrogen ion.

It might seem that beer and pH would not have much in common (other than the fact that students sometimes seek solace in the first when they cannot master the second). In fact, the definition of pH was developed around 1910 at the Carlsberg Laboratory, a Danish laboratory involved with the chemistry and biology of brewing as well as more theoretical studies. Søren Peter Lauritz Sørensen, the head of the laboratory's chemistry department, recognized that acidity has a profound effect on rates of fermentation as well as many other biochemical processes. Without a reliable acidity scale, it would be difficult to perform systematic studies of many technologically important reactions or to make much progress in understanding the theory of these reactions. Sørensen's answer to this need was the pH scale. He did not stop there but also did pioneering work on the preparation of buffer solutions and the use of both indicators and electrodes to measure pH. He was the first scientist to show that pH influences the catalytic effectiveness of enzymes.

Although the Carlsberg Laboratory received its funds from the Carlsberg Breweries, it was not an applied research lab for the breweries. According to the laboratory's charter, no secrets were kept and all experimental and theoretical results were published. Because of this freedom from secrecy, Sørensen's work spread quickly and laid the foundations for much of the technological progress of the twentieth century.

H^+ concentrations, it is usually more convenient to deal with the logarithm of the hydrogen-ion concentration. Since $[H^+]$ cannot be greater than one in dilute solutions, its logarithm will be negative. (It can be positive in concentrated solutions of strong acids such as 3 M HCl.) In order to work with positive numbers, we express the acidity of a solution as the negative base-ten logarithm of the hydrogen-ion activity—**pH**. (The letter p is a general symbol for a negative logarithm of a quantity.) Since we are neglecting nonideality, we will use the approximate definition of pH based on hydrogen-ion concentration rather than activity.

$$pH = -\log [H^+] \qquad (14.2)$$

In pure water and in neutral solutions,

$$pH = -\log (1.0 \times 10^{-7}) = 7.0$$

Hydrogen-ion concentration in an acid solution is greater than $1.0 \times 10^{-7} M$, and the pH is less than 7.0. Basic solutions have lower hydrogen-ion concentrations than 1.0×10^{-7} and pH values higher than 7.0.

Example 14.3

What is the pH value of a solution that is $3.1 \times 10^{-4} M$ in H^+?

Solution

This is really just a problem in evaluating a logarithm. If you do not use a pocket calculator with a log key, it is best to take the logarithms of 3.1 and 10^{-4} separately and add them together. (See Appendix A.4.)

$$pH = -\log (3.1 \times 10^{-4}) = -(\log 3.1 + \log 10^{-4})$$
$$= -(0.49 - 4) = 3.51$$

14 AQUEOUS EQUILIBRIA

Table 14.1 **Common Strong Acids and Strong Bases**

Strong acids

$HCl(aq) \longrightarrow H^+(aq) + Cl^-(aq)$
$HBr(aq) \longrightarrow H^+(aq) + Br^-(aq)$
$HI(aq) \longrightarrow H^+(aq) + I^-(aq)$
$HNO_3(aq) \longrightarrow H^+(aq) + NO_3^-(aq)$
$HClO_4(aq) \longrightarrow H^+(aq) + ClO_4^-(aq)$
$H_2SO_4(aq) \longrightarrow H^+(aq) + HSO_4^-(aq)$ (Only the first dissociation is strong.)

Strong bases

$NaOH(s) \longrightarrow Na^+(aq) + OH^-(aq)$
$KOH(s) \longrightarrow K^+(aq) + OH^-(aq)$

Other common ionic hydroxides are not as soluble as NaOH or KOH. However, most dissolve by a process of complete dissociation. For example, $Ba(OH)_2$ produces 2 mol OH^- for every mole of $Ba(OH)_2$ dissolved.

$$Ba(OH)_2(s) \rightleftharpoons Ba^{2+}(aq) + 2\,OH^-(aq)$$

Example 14.4

What is the hydrogen-ion concentration of a solution whose pH is 11.78?

Solution

In this case, rewrite pH in terms of the next larger whole number.

$$pH = 11.78 = 12 - 0.22$$
$$\log [H^+] = -12 + 0.22$$

Now take the antilog of the two numbers (the whole number and the fractional number) and multiply them together.

$$[H^+] = (\text{antilog } 0.22)(\text{antilog } -12) = 1.7 \times 10^{-12}\, M$$

Since logarithms are useful for calculations, you may find it convenient to state mass-action equations in terms of p values. For example, the autoprotolysis mass-action law can be stated as

$$pH + pOH = pK_W = 14.0 \tag{14.3}$$

where pOH is the negative logarithm of the hydroxide-ion concentration and pK_W is the negative logarithm of the equilibrium constant.

Strong Acids

The dissociations of strong acids (Table 14.1) are always essentially complete in dilute aqueous solutions. The small concentrations of undissociated acid that should, in principle, exist are below the limit of experimental detection. A monoprotic strong acid produces a hydrogen-ion concentration equal to its molarity. The small additional H^+ concentration produced by autoprotolysis of water is negligible unless the molarity of the strong acid is very low (less than $10^{-6}\,M$).*

Example 14.5

What is the pH of 0.048 M HCl?

---

*In such very dilute solutions, autoprotolysis of water is a significant (and often the dominant) contributor of H^+ to the solution. Because of autoprotolysis, a solution of an acid, no matter how dilute, cannot have a pH above 7.00. Dilution cannot convert an acid into a basic solution. Instead, the acid solution begins to resemble pure water as its concentration decreases.

14.1 ACID–BASE EQUILIBRIA

Solution

The hydrogen-ion concentration equals the molarity of HCl (represented by C_{HCl}).

$$[H^+] = C_{HCl} = 0.048$$
$$pH = -\log(0.048) = 1.32$$

Weak Acids

A weak acid such as acetic acid dissociates to a slight extent and has a small equilibrium constant for dissociation (K_A). We can estimate the pH of a solution of a weak acid from its molarity and its **dissociation constant** (see Appendix H).

Example 14.6

What is the pH of 0.20 M acetic acid?

$$CH_3CO_2H(aq) \rightleftharpoons H^+(aq) + CH_3CO_2^-(aq)$$

$$K_A = \frac{[H^+][CH_3CO_2^-]}{[CH_3CO_2H]} = 1.8 \times 10^{-5}$$

Solution

Assuming that dissociation of acetic acid is the only significant source of H^+, we can write the stoichiometric relationship

$$[H^+] = [CH_3CO_2^-]$$

The formation of $CH_3CO_2^-$ lowers the CH_3CO_2H concentration below the 0.20 M added to the solution.

$$[CH_3CO_2H] = 0.20 - [CH_3CO_2^-]$$

We combine these equations with the law of mass action to obtain an equation whose only unknown is $[H^+]$.

$$\frac{[H^+]^2}{(0.20 - [H^+])} = 1.8 \times 10^{-5}$$

In many cases, including this one, we can simplify the solution of the equation by assuming that the degree of dissociation is too small to make a difference in the concentration of undissociated acid.

$$0.20 - [H^+] \cong 0.20$$

The approximate equation is then

$$\frac{[H^+]^2}{0.20} \cong 1.8 \times 10^{-5}$$

$$[H^+] \cong \sqrt{(1.8 \times 10^{-5})(0.20)} = 1.9 \times 10^{-3} \, M$$

$$pH = -\log(1.9 \times 10^{-3}) = 2.72$$

This approximate answer agrees with our assumptions. There is no significant decrease in CH_3CO_2H concentration ($0.20 - 0.0019 \cong 0.20$), and the amount of H^+ produced by autoprotolysis of water is negligible.

We can generalize the approximate method used in Example 14.6 to fit many weak acid solutions. As long as dissociation of the acid is the only important source of H^+, the concentration of the conjugate base of the weak acid is equal to the hydrogen-ion concentration. That is, for the reaction

$$HA(aq) \rightleftharpoons H^+(aq) + A^-(aq)$$

$$[H^+] = [A^-]$$

where HA represents a weak acid. If the decrease in [HA] due to dissociation is negligible, then

$$[HA] = C_A$$

and

$$[H^+] = \sqrt{K_A C_A} \tag{14.4}$$

where C_A equals the molarity of the weak acid solution. For moderately strong acids or for very dilute solutions of weak acids, the degree of dissociation becomes so large that we must use the more exact quadratic equation to find pH. (See Appendix A.7.)

$$\frac{[H^+]^2}{C_A - [H^+]} = K_A \tag{14.5}$$

Bases

A strong base such as NaOH (Table 14.1) is completely dissociated in dilute aqueous solution. The hydroxide-ion concentration of a solution of a monoprotic strong base equals its molarity (C_B).

$$[OH^-] = C_B \tag{14.6}$$

Almost all water-soluble weak bases function by a process known as **hydrolysis**—they receive a proton from water and release a hydroxide ion.

$$B(aq) + H_2O \rightleftharpoons BH^+(aq) + OH^-(aq)$$

$$K_B = \frac{[BH^+][OH^-]}{[B]}$$

Two examples of the hydrolysis of weak bases are

$$NH_3(aq) + H_2O \rightleftharpoons NH_4^+(aq) + OH^-(aq)$$

$$K_B = \frac{[NH_4^+][OH^-]}{[NH_3]}$$

and

$$CH_3CO_2^-(aq) + H_2O \rightleftharpoons CH_3CO_2H(aq) + OH^-(aq)$$

$$K_B = \frac{[CH_3CO_2H][OH^-]}{[CH_3CO_2^-]}$$

Using reasoning similar to that applied to weak acids, we can derive an approximate equation showing the hydroxide-ion concentration in a solution of a weak base. Since the hydrolysis reaction forms one OH^- for every molecule of the conjugate acid (BH^+), the concentrations of these two species are equal.

$$[OH^-] = [BH^+]$$

If the base is weak enough, the concentration of B will be approximately the same as the molarity of the weak base added to the solution (C_B).

$$[B] = C_B - [OH^-] \cong C_B$$

Substituting these relationships into the mass-action law yields

$$K_B = \frac{[OH^-][BH^+]}{[B]} \cong \frac{[OH^-]^2}{C_B}$$

The hydroxide-ion concentration is

$$[OH^-] \cong \sqrt{K_B C_B} \tag{14.7}$$

Example 14.7

What is the pH of 0.020 M ammonia?

Solution

The value of K_B for ammonia is 1.8×10^{-5} (see Appendix H). Use this value and equation (14.7) to find the hydroxide-ion concentration.

$$[OH^-] = \sqrt{K_B C_B} = \sqrt{(1.8 \times 10^{-5})(0.020)} = 6.0 \times 10^{-4}\, M$$

The hydrogen-ion concentration is

$$[H^+] = \frac{1.0 \times 10^{-14}}{[OH^-]} = \frac{1.0 \times 10^{-14}}{6.0 \times 10^{-4}} = 1.7 \times 10^{-11}$$

$$\text{pH} = -\log(1.7 \times 10^{-11}) = 10.77$$

The value of K_B for a weak base is related to the value of the dissociation constant of its conjugate acid, BH^+.

$$K_B = \frac{K_W}{K_A} = \frac{1.0 \times 10^{-14}}{K_A} \tag{14.8}$$

(You can verify this equation by writing out the complete expression for K_W/K_A and canceling terms that appear in both the numerator and the denominator.) Most tables of K_B values do not include data for the anions of weak acids. Instead of trying to look up K_B for an ion such as acetate, we look up the value of K_A of the conjugate acid (acetic acid in the case of acetate) and apply equation (14.8).

Example 14.8

What is the pH of 5.0×10^{-3} M sodium acetate? Assume complete dissociation of sodium acetate into Na^+ and $CH_3CO_2^-$.

$$NaCH_3CO_2(s) \longrightarrow Na^+(aq) + CH_3CO_2^-(aq)$$

Solution

Since the value of K_A for acetic acid is 1.8×10^{-5}, K_B is

$$K_B = \frac{1.0 \times 10^{-14}}{1.8 \times 10^{-5}} = 5.6 \times 10^{-10}$$

According to equation (14.7),

$$[OH^-] = \sqrt{(5.6 \times 10^{-10})(5.0 \times 10^{-3})} = 1.7 \times 10^{-6}\, M$$

$$\text{pOH} = -\log(1.7 \times 10^{-6}) = 5.77$$

Knowing pOH, we can easily compute pH according to equation (14.3).

$$\text{pH} = 14.0 - \text{pOH} = 14.0 - 5.77 = 8.23$$

Buffer Solutions

A major problem facing chemists is how to maintain a moderate pH, either weakly acidic or weakly basic, in a solution. It is possible to prepare a weakly acidic solution (for example, pH 5.0) just by dissolving a very small concentration of strong acid in water, but the pH could easily be changed by one of several events. Dilution of the solution, addition of more acid, or the addition of base would all cause large pH changes. The pH of a **buffer solution**, on the other hand, is relatively insensitive to these changes. Operations causing large acidity changes in dilute HCl (an unbuffered solution) have very little effect on buffers. Because of the stability of their pH values, buffers are widely used in chemical laboratories and industrial processes. Buffer systems in blood maintain the pH within the narrow limits required for health.

A buffer solution contains a weak acid and its conjugate base (or a weak base and its conjugate acid) in comparable concentrations. (A rule of thumb is that the concentrations of the two components should be within a factor of ten of each other.) Two common examples of buffers are (1) a solution 0.100 M in acetic acid and 0.050 M in sodium acetate and (2) a solution 0.010 M in ammonia and 0.020 M in ammonium chloride.

A buffer solution has a much higher pH than a solution containing only the conjugate acid and a much lower pH than a solution of the pure conjugate base. This intermediate pH value is consistent with Le Chatelier's principle. The presence of the conjugate base decreases the degree of dissociation of the acid, while the conjugate acid suppresses the hydrolysis of the base.

Example 14.9

What is the pH of the buffer solution consisting of 0.100 M acetic acid and 0.050 M sodium acetate?

Solution

Dissociation decreases the acetic acid concentration while acetate ion hydrolysis increases it.

$$CH_3CO_2H(aq) \rightleftharpoons H^+(aq) + CH_3CO_2^-(aq)$$

$$H_2O + CH_3CO_2^-(aq) \rightleftharpoons OH^-(aq) + CH_3CO_2H(aq)$$

$$[CH_3CO_2H] = C_A + [OH^-] - [H^+]$$
$$= 0.100 + [OH^-] - [H^+]$$

Assuming neither dissociation nor hydrolysis is extensive enough to significantly alter the acetic acid concentration,

$$[CH_3CO_2H] \cong 0.100$$

Dissociation raises acetate-ion concentration, and hydrolysis lowers it.

$$[CH_3CO_2^-] = C_B + [H^+] - [OH^-]$$
$$= 0.050 + [H^+] - [OH^-]$$

If the effect of dissociation and hydrolysis on acetate-ion concentration is negligible,

$$[CH_3CO_2^-] \cong 0.050$$

Substituting the approximate concentrations of acetic acid and acetate ion into the mass-action law yields

$$K_A = 1.8 \times 10^{-5} = \frac{[H^+](0.050)}{0.100}$$

$$[H^+] = 1.8 \times 10^{-5} \left(\frac{0.100}{0.050}\right) = 3.6 \times 10^{-5} \, M$$

$$pH = -\log(3.6 \times 10^{-5}) = 4.44$$

This answer is consistent with the approximations.

14.1 ACID—BASE EQUILIBRIA

Table 14.2 The Effect of Dilution on Buffer pH

Buffer: acetic acid–sodium acetate in a 2:1 molar ratio
Ideal pH: 4.44 at 25°C

Concentration of sodium acetate (M)	Observed pH (25°C)
0.0010	4.46
0.0025	4.44
0.0050	4.43
0.0100	4.41
0.0250	4.39
0.0500	4.37
0.100	4.34

The approximate method outlined in Example 14.9 applies to most buffers.

$$[HA] \cong C_A \quad \text{and} \quad [A^-] \cong C_B$$

$$[H^+] = K_A\left(\frac{C_A}{C_B}\right) \tag{14.9}$$

or

$$[OH^-] = K_B\left(\frac{C_B}{C_A}\right) \tag{14.10}$$

The last two equations are equivalent because

$$[H^+] = \frac{K_W}{[OH^-]} \quad \text{and} \quad K_A = \frac{K_W}{K_B}$$

These approximate equations fail if either $[H^+]$ or $[OH^-]$ represents a significant fraction (more than 10%) of either C_A or C_B.

Dilution of the buffer lowers the concentrations of the conjugate acid and the conjugate base by the same factor, leaving the C_A/C_B ratio unchanged. According to equations (14.9) and (14.10) for the approximate $[H^+]$ and $[OH^-]$ of a buffer solution, dilution has no effect on the pH of a buffer. The small changes that do occur (Table 14.2) are due mostly to nonideality. (At very high dilution, the approximate equations fail because either dissociation or hydrolysis becomes significant, and the pH approaches 7.0.)

As illustrated by the following two examples, buffers effectively suppress changes in pH due to either added acid or added base (Figure 14.1, p. 360).

Example 14.10

a. How much does the pH change when 1.0 mL of 1.0 M HCl is added to 1.0 L of 3.6×10^{-5} M HCl?

b. How much does the pH change when 1.0 mL of 1.0 M HCl is added to 1.0 L of the buffer solution in Example 14.9?

Solution

a. The initial pH of the solution is $-\log(3.6 \times 10^{-5})$, or 4.44. The number of moles of added HCl equals

$$(1.0 \times 10^{-3} \text{ L})(1.0 \text{ M}) = 1.0 \times 10^{-3} \text{ mol HCl}$$

Addition of this much acid to the liter of dilute HCl raises the molarity of HCl and lowers the pH of the solution.

$$\text{Moles HCl} = 1.0 \times 10^{-3} \text{ mol} + 3.6 \times 10^{-5} \text{ mol}$$

$$\cong 1.0 \times 10^{-3} \text{ mol}$$

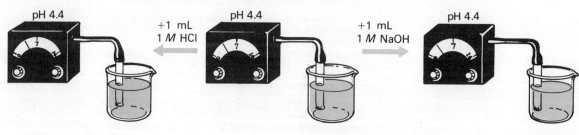

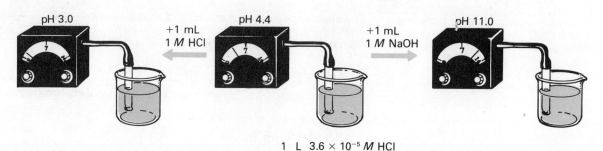

Figure 14.1
The pH stability of a buffered solution.

Solution volume = $1.0 \text{ L} + 1.0 \times 10^{-3} \text{ L} \cong 1.0 \text{ L}$

$[H^+] = \dfrac{1.0 \times 10^{-3} \text{ mol}}{1.0 \text{ L}} = 1.0 \times 10^{-3} \text{ M; pH} = 3.0$

b. As in part a, 1.0×10^{-3} mol HCl is added to the solution. Nearly all of this strong acid reacts with an equivalent amount of acetate ion converting it to an equivalent amount of acetic acid.

$CH_3CO_2^-(aq) + H^+(aq) \rightleftharpoons CH_3CO_2H(aq)$

$C_A = 0.100 + 1.0 \times 10^{-3} = 0.101 \text{ M}$

$C_B = 0.050 - 1.0 \times 10^{-3} = 0.049 \text{ M}$

$[H^+] = 1.8 \times 10^{-5} \left(\dfrac{0.101}{0.049} \right) = 3.7 \times 10^{-5} \text{ M}$

The decrease in pH is barely measurable.

Example **14.11**

a. How much does the pH change when 1.0 mL of 1.0 M NaOH is added to 1.0 L of 3.6×10^{-5} M HCl?

b. How much does the pH change when 1.0 mL of 1.0 M NaOH is added to 1.0 L of the buffer solution in Example 14.9?

Solution

a. The 1.0 mL of 1.0 M NaOH added to the solution comprises 1.0×10^{-3} mol NaOH. This base neutralizes all the HCl, and excess hydroxide ion remains.

Moles excess $OH^- = 1.0 \times 10^{-3} - 3.6 \times 10^{-5} \cong 1.0 \times 10^{-3}$

$[OH^-] = \dfrac{1.0 \times 10^{-3} \text{ mol}}{1.0 \text{ L}} = 1.0 \times 10^{-3} \text{ M; pH} = 11.0$

14.1 ACID–BASE EQUILIBRIA

b. The 1.0×10^{-3} mol NaOH added converts an equivalent amount of acetic acid to acetate ion.

$$CH_3CO_2H(aq) + OH^-(aq) \rightleftharpoons CH_3CO_2^-(aq) + H_2O$$

$$C_A = 0.100 - 1.0 \times 10^{-3} = 0.099\ M$$

$$C_B = 0.050 + 1.0 \times 10^{-3} = 0.051\ M$$

$$[H^+] = 1.8 \times 10^{-5}\left(\frac{0.099}{0.051}\right) = 3.5 \times 10^{-5}\ M$$

There is practically no change in pH.

The ability of a buffer to neutralize acids or bases has its limits. As more acid is added to a buffer, more of the conjugate base is protonated. When nearly all of the conjugate base is neutralized, the pH of the buffer begins to fall sharply. In a similar way, the pH of a buffer will rise sharply when enough base has been added to convert nearly all the conjugate acid to conjugate base.

Buffering is not as effective if the C_A/C_B ratio is very large (greater than 10) or too small (less than 0.10). In such cases, small changes in molarity can have a large effect on the conjugate acid–conjugate base ratio and thus on the acidity. Buffers are most effective within a two-unit pH range bracketed around the pK_A of the conjugate acid (buffer range: $pK_A - 1$ to $pK_A + 1$). We must choose a buffer whose conjugate acid has a pK_A reasonably close to the desired pH. Table 14.3 shows several buffer systems and their effective pH ranges.

Indicators

A weak acid whose color differs from that of its conjugate base can serve as a pH **indicator**. (The indicator color should be so intense that the indicator concentration can be kept low. Otherwise, the indicator's presence might change the pH of the solution.) The pH range in which an indicator changes color depends on its dissociation constant (K_{Ind}).

$$HInd(aq) \rightleftharpoons H^+(aq) + Ind^-(aq)$$

$$K_{Ind} = \frac{[H^+][Ind^-]}{[HInd]}$$

If the hydrogen-ion concentration of the solution is much greater than K_{Ind}, then the color of the conjugate acid of the indicator predominates. The basic color appears when $[H^+]$ is considerably less than K_{Ind}. There is a relatively narrow acidity range (about two pH units) near pK_{Ind} in which both the acid and base forms of the indicator are present in significant concentrations, and the color of the solution is then a mixture of the acid and base colors. It is in this pH range that the indicator is most

Table 14.3 **Buffer Solutions**

Conjugate acid	Conjugate base	K_A	Approximate pH range
H_3PO_4	$H_2PO_4^-$	7.1×10^{-3}	2–3
HCO_2H	HCO_2^-	2×10^{-4}	3–5
CH_3CO_2H	$CH_3CO_2^-$	1.8×10^{-5}	4–6
$H_2PO_4^-$	HPO_4^{2-}	6.3×10^{-8}	6–8
H_3BO_3	$H_2BO_3^-$	6×10^{-10}	8–10
NH_4^+	NH_3	5.6×10^{-10}	8–10
HCO_3^-	CO_3^{2-}	5×10^{-11}	9–11
HPO_4^-	PO_4^{3-}	4.3×10^{-13}	11–12

Table 14.4 Acid–Base Indicators

Indicator	Acid color	Base color	K_{Ind}	pH range
Thymol blue	Red	Yellow	2×10^{-2}	1.2–2.8
Methyl orange	Red	Orange-yellow	3.5×10^{-4}	3.1–4.4
Methyl red	Red	Yellow	1×10^{-5}	4.2–6.3
Bromthymol blue	Yellow	Blue	8×10^{-8}	6.0–7.6
Thymol blue	Yellow	Blue	9×10^{-9}	8.0–9.6
Phenolphthalein	Colorless	Red-violet	a	8.0–9.8

a Two equilibrium reactions occur for phenolphthalein in the color transition range.

effective in enabling us to estimate pH. Table 14.4 lists the colors, K_{Ind} values, and pH transition ranges for several indicators.

Titration Curves

The success of a titration rests on our ability to detect its **equivalence point**—equal numbers of equivalents of acid and base have been added. We use the change in color of an indicator as the **end point** for the titration and assume this end point approximates the equivalence point.

Phenolphthalein might, at first glance, appear to be a poor choice for the end point indicator in the titration of HCl with NaOH. The equivalence point pH for this titration is 7.0, but phenolphthalein does not begin to change color until the pH rises to 8. However, this difference between the pH of the end point and equivalence point does not cause a serious error in the estimation of the equivalent volume of base. To see why this is so, we will examine the changes in pH during the titration.

Consider a titration of 10.00 mL of 0.1000 M HCl with 0.1000 M NaOH. Before the equivalence point is reached, the acid is in excess.

Millimoles excess H^+ = Millimoles HCl − Millimoles NaOH

$$= (0.1000\ M)(10.00\ mL) - (0.1000\ M)V_B$$

where V_B is the volume of base added in milliliters.

Total volume = $V_A + V_B$ = 10.00 mL + V_B

$$[H^+] = \frac{(1.000\ M)(10.00\ mL) - (0.1000\ M)V_B}{10.00\ mL + V_B}$$

The titration solution at the equivalence point is neutral, and its pH is 7.00. There is an excess of base present if the titration is run beyond the equivalence point.

Millimoles excess OH^- = Millimoles NaOH − Millimoles HCl

$$= (0.1000\ M)V_B - (1.000\ M)(10.00\ mL)$$

$$[OH^-] = \frac{(1.000\ M)V_B - (1.000\ M)(10.00\ mL)}{10.00\ mL + V_B}$$

Table 14.5 gives the pH values calculated for different volumes of added base, and Figure 14.2 is a graph of pH versus the volume of base. This sort of graph is called a **titration curve**.

Note the steep rise in the titration curve near the equivalence point. Because this rise is so steep, a sizable error in estimating the end point pH causes a very small error in the measurement of the equivalent volume. Using phenolphthalein as the indicator causes an overestimate of 0.01 mL (a relative error of 0.1%) in the equivalent volume. This error is less than the uncertainty in reading and manipulating most burets. If we use methyl red as the indicator, we underestimate the equivalent volume, but this error is also negligibly small.

We must be more careful in choosing an indicator for the titration of a weak acid by a strong base. A sharp rise in pH does occur at the equivalence point of this

14.1 ACID–BASE EQUILIBRIA

Table 14.5 Calculated pH Values for Acid–Base Titrations

	$C_A = 0.1000\ M$	$V_A = 10.00\ mL$	Base is $0.1000\ M$ NaOH		
		Titration of HCl		Titration of CH_3CO_2H	
V_B (mL)	[H⁺]	pH		[H⁺]	pH
0	0.100	1.0		1.3×10^{-3}	2.9
1.00	0.082	1.1		1.6×10^{-4}	3.8
2.50	0.060	1.2		5.4×10^{-5}	4.3
5.00	0.033	1.5		1.8×10^{-5}	4.7
7.50	0.014	1.8		6.0×10^{-6}	5.2
9.00	5.3×10^{-3}	2.3		2.0×10^{-6}	5.7
9.90	5.0×10^{-4}	3.3		1.8×10^{-7}	6.7
9.99	5.0×10^{-5}	4.3		1.8×10^{-8}	7.7
10.00	1.0×10^{-7}	7.0		1.9×10^{-9}	8.7
10.01	2.0×10^{-10}	9.7		2.0×10^{-10}	9.7
10.10	2.0×10^{-11}	10.7		2.0×10^{-11}	10.7
11.00	2.1×10^{-12}	11.7		2.1×10^{-12}	11.7

titration, but it is not as large as in the case of the titration of a strong acid. The indicator color-transition range must correspond to this smaller pH range.

The calculation of the titration curve for the titration of 10.00 mL of $0.1000\ M$ acetic acid (a typical weak acid) by $0.1000\ M$ NaOH is more complicated than for the titration of HCl. Before the titration begins ($V_B = 0$), we have a solution of a pure weak acid ($K_A = 1.8 \times 10^{-5}$).

$$[H^+] = \sqrt{K_A C_A} = \sqrt{(1.8 \times 10^{-5})(0.1000)} = 1.3 \times 10^{-3}$$

$$pH = 2.89$$

Addition of base converts an equivalent amount of acetic acid to acetate ion and creates a buffer solution.

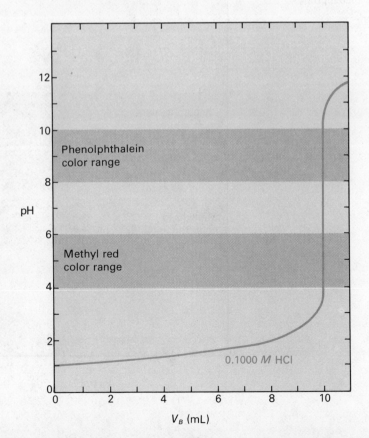

Figure 14.2
Titration curve of a monoprotic strong acid (HCl).

$$CH_3CO_2H(aq) + OH^-(aq) \rightleftharpoons CH_3CO_2^-(aq) + H_2O$$

$$[H^+] = 1.8 \times 10^{-5} \frac{[CH_3CO_2H]}{[CH_3CO_2^-]} = 1.8 \times 10^{-5} \frac{\text{Moles } CH_3CO_2H}{\text{Moles } CH_3CO_2^-}$$

Millimoles excess CH_3CO_2H = Millimoles CH_3CO_2H − Millimoles NaOH

$$= (0.1000\ M)(10.00\ mL) - (0.1000\ M)V_B$$

Millimoles $CH_3CO_2^-$ = Millimoles NaOH = $(0.1000\ M)V_B$

$$[H^+] = 1.8 \times 10^{-5} \frac{(1.000\ M)(10.00\ mL) - (0.1000\ M)V_B}{(0.1000\ M)V_B}$$

(One useful observation about this buffer range of a titration curve is that pH = pK_A when $V_B = {}^1\!/_2\ V_{\text{equiv}}$, a relationship that we can use to estimate the value of K_A for an unknown acid.)

The equivalence point solution is identical to a solution made by dissolving 1.000 mmol of sodium acetate in 20.00 mL water.

$$[OH^-] = \sqrt{K_B C_B}$$

$$K_B = \frac{1.0 \times 10^{-14}}{1.8 \times 10^{-5}} = 5.6 \times 10^{-10}$$

$$C_B = (0.1000\ M)\frac{V_B}{V_{\text{total}}} = \frac{(0.1000\ M)(10.00\ mL)}{10.00\ mL + 10.00\ mL} = 0.0500\ M$$

$$[OH^-] = \sqrt{(5.6 \times 10^{-10})(0.0500)} = 5.3 \times 10^{-6}$$

pOH = 5.28   pH = 8.72

Excess hydroxide ion is the only important base present after the equivalence point is passed, and the titration curves for acetic acid and HCl are similar in this range.

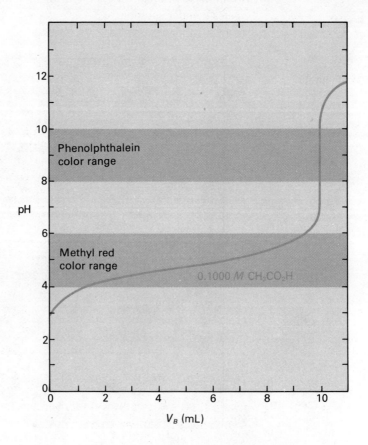

Figure 14.3
Titration curve of a monoprotic weak acid (CH_3CO_2H).

Figure 14.4
Titration curve of maleic acid ($H_2C_4H_2O_4$). A 10.0-mL sample of 0.100 M maleic acid is titrated with 0.100 M NaOH.

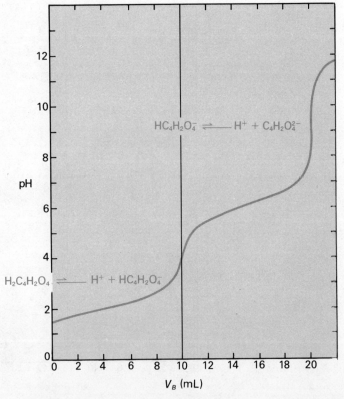

Calculated pH values for several volumes of added base in the acetic acid titration appear in Table 14.5, and the titration curve appears in Figure 14.3. The equivalence point pH lies within the phenolphthalein color-transition range, so this indicator is suitable for the titration. Methyl red, a poor indicator for this titration, starts to change color after only 20% of the acid is titrated, and its color gradually changes during most of the remainder of the titration.

Titrations of Polyprotic Acids

Titration curves of polyprotic acids often contain several distinct buffer regions and several equivalence points characterized by steep pH increases. A diprotic acid, such as maleic acid ($H_2C_4H_2O_4$), whose first proton ($K_{A1} = 1.2 \times 10^{-2}$) dissociates much more readily than the second ($K_{A2} = 6.0 \times 10^{-7}$), has two readily observed equivalence points (Figure 14.4). No significant dissociation of $HC_4H_2O_4^-$ occurs until nearly all the maleic acid molecules lose their first proton, and we can easily distinguish between the neutralizations of the first and second protons.

This is not the case with H_2SO_4, a diprotic acid whose two protons both dissociate readily. The first proton dissociates completely in dilute aqueous solutions, and the dissociation of HSO_4^- to SO_4^{2-} ($K_{A2} = 1.0 \times 10^{-2}$) is significant (10% in 0.1 M H_2SO_4). More HSO_4^- dissociates as the added base neutralizes H^+, and most of the HSO_4^- has dissociated by the time we reach the first equivalence point. There is no clear distinction between the pH ranges at which H^+ and HSO_4^- are neutralized, and the absence of a sharp rise in pH deprives us of a reliable end point for the first equivalence point of sulfuric acid (Figure 14.5).

In the titration of H_3PO_4 (Figure 14.6), a triprotic acid whose three protons have very different dissociation constants ($K_{A1} = 7.1 \times 10^{-3}$; $K_{A2} = 6.3 \times 10^{-8}$; $K_{A3} = 4.3 \times 10^{-13}$), there are three equivalence points, two of which are well-defined by sharp rises in pH. The first, corresponding to the reaction

$$H_3PO_4(aq) + OH^-(aq) \rightleftharpoons H_2PO_4^-(aq) + H_2O$$

occurs at pH 4.7, and methyl red is a suitable indicator for this equivalence point. Continued titration leads to the conversion of $H_2PO_4^-$ to HPO_4^{2-}, and the second

Figure 14.5
Titration curve of H_2SO_4. A 10.0-mL sample of 0.100 M sulfuric acid is titrated with 0.100 M NaOH.

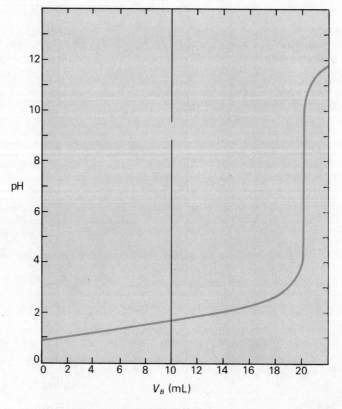

Figure 14.6
Titration curve for phosphoric acid. A 10.0-mL sample of 0.100 M phosphoric acid is titrated with 0.100 M NaOH.

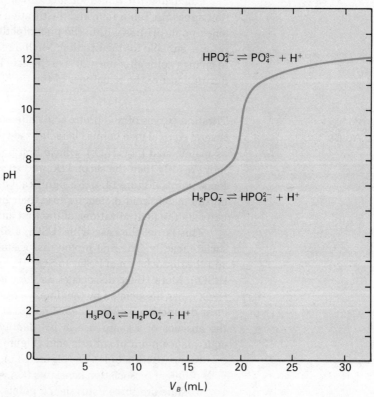

equivalence point occurs near pH 10. This is in the range for phenolphthalein, and the normality of phosphoric acid is twice its molarity when titrated to a phenolphthalein end point. A third mole of hydroxide produces PO_4^{3-}, but since PO_4^{3-} is so basic ($K_B = 2.3 \times 10^{-2}$), excess hydroxide ion does not produce a sharp rise in pH, and there is no clearly defined end point.

14.2 Solubility Equilibria

Solubilities of Salts; Solubility Products

The mass-action law for the solubility of a slightly soluble salt can be expressed in terms of the species (cations and anions) that enter the solution. For example, the law of mass action for the dissolution of $CaSO_4$ in water

$$CaSO_4(s) \rightleftharpoons Ca^{2+}(aq) + SO_4^{2-}(aq)$$

is

$$K_{sp} = [Ca^{2+}][SO_4^{2-}]$$

where K_{sp} is the **solubility product** of calcium sulfate. The solubility product of a salt is the product of the equilibrium concentrations of the ions produced when the salt dissolves, each concentration being raised to a power equal to its stoichiometric coefficient. Values of solubility products for some slightly soluble salts appear in Appendix H.*

The solubility of $CaSO_4$ (in other words, the concentration of its saturated solution) is related to its solubility product. Since each formula unit of $CaSO_4$ that dissolves produces one calcium ion and one sulfate ion, we can express its solubility (s) as

$$\text{Solubility} = s = [Ca^{2+}] = [SO_4^{2-}]$$

and

$$K_{sp} = s^2 \tag{14.11}$$

This relationship holds for any salt, AB, containing equal numbers of cations, A, and anions, B.

Calcium fluoride (a salt of the type AB_2) dissolves to produce one calcium ion and two fluoride ions for each dissolved formula unit.

$$CaF_2(s) \rightleftharpoons Ca^{2+}(aq) + 2\,F^-(aq)$$

The solubility of CaF_2 equals the concentration of Ca^{2+} in a saturated solution because each dissolved CaF_2 unit yields one calcium ion.

$$[Ca^{2+}] = s$$

Since two fluoride ions dissolve for each calcium ion,

$$[F^-] = 2s$$

and

$$K_{sp} = [Ca^{2+}][F^-]^2 = (s)(2s)^2 = 4s^3 \tag{14.12}$$

We can derive the relationship between the solubility of any salt and its solubility product by considering the stoichiometry of its solution process.

Example 14.12   The solubility of PbI_2, which dissolves by the reaction

---

*Solubility products for sodium chloride and other very soluble salts do not appear in Appendix H because they are not of much use. Saturated NaCl solution is so concentrated that its behavior does not come close to ideality. Using a solubility product (an approximate equilibrium constant based on molarities rather than activities) for such a nonideal solution would cause serious errors.

$$PbI_2(s) \rightleftharpoons Pb^{2+}(aq) + 2\,I^-(aq)$$

is 1.18×10^{-3} M. What is the solubility product of PbI_2?

Solution

Since PbI_2 is a salt of the type AB_2, equation (14.12) holds.

$$K_{sp} = [Pb^{2+}][I^-]^2 = 4s^3$$
$$= 4(1.18 \times 10^{-3})^3 = 6.5 \times 10^{-9}$$

Example 14.13

The solubility product of $AgIO_3$

$$AgIO_3(s) \rightleftharpoons Ag^+(aq) + IO_3^-(aq)$$

is 3.0×10^{-8}. What is the solubility of $AgIO_3$?

Solution

We are dealing with a salt of the type AB.

$$K_{sp} = [Ag^+][IO_3^-] = s^2$$

Therefore,

$$s = \sqrt{K_{sp}} = \sqrt{3.0 \times 10^{-8}} = 1.8 \times 10^{-4}\,M$$

The Common-Ion Effect

The solubility of a slightly soluble salt is very sensitive to the presence of another salt containing one of its ions. As predicted by Le Chatelier's principle, a high concentration of a common ion suppresses the solubility of the salt. This phenomenon is called the **common-ion effect**. Example 14.14 illustrates the common-ion effect.

Example 14.14

Predict the solubility of $AgIO_3$ in 0.100 M KIO_3.

Solution

The solubility product used in Example 14.13 applies to this example as well.

$$K_{sp} = [Ag^+][IO_3^-] = 3.0 \times 10^{-8}$$

The concentration of silver ion in 0.100 M KIO_3 saturated with $AgIO_3$ equals the solubility of $AgIO_3$ because this salt is the only source of silver ion in the solution. Although the dissolved silver iodate raises the iodate concentration slightly above its original value of 0.100 M, this increase is small enough to neglect in an approximate treatment.

$$[Ag^+] = s \quad [IO_3^-] = 0.100 + s \cong 0.100$$
$$[Ag^+][IO_3^-] = (s)(0.100) = 3.0 \times 10^{-8}$$
$$s = 3.0 \times 10^{-7}\,M$$

A comparison of the answers for Examples 14.13 and 14.14 confirms that KIO_3 lowers the solubility of $AgIO_3$. The predicted solubility is inversely proportional to the molarity of potassium iodate. The data in Table 14.6 show some deviations, due to nonideality, from strict inverse proportionality but are reasonably consistent with the predicted relationship.

14.2 SOLUBILITY EQUILIBRIA

Table 14.6 Solubilities of $AgIO_3$ in Aqueous Salt Solutions

Solubility in pure water at 25°C = 1.8×10^{-4} M

Solubility in KIO_3 solutions		Solubility in KNO_3 solutions	
C_{KIO_3} (M)	s_{AgIO_3} (M)	C_{KNO_3} (M)	s_{AgIO_3} (M)
0.0010	3.4×10^{-5}	0.0010	1.9×10^{-4}
0.0025	1.4×10^{-5}	0.0025	1.9×10^{-4}
0.0050	7.5×10^{-6}	0.0050	1.9×10^{-4}
0.010	4.0×10^{-6}	0.010	2.0×10^{-4}
0.025	1.8×10^{-6}	0.025	2.1×10^{-4}
0.050	9.9×10^{-7}	0.050	2.2×10^{-4}
0.10	5.6×10^{-7}	0.10	2.4×10^{-4}

Potassium nitrate, having no ion in common with silver iodate, does not decrease its solubility. However, the presence of potassium ion and nitrate ion in the solution create nonideal effects which raise silver iodate's solubility slightly (Table 14.6).

Solubility product calculations are useful for estimating the sensitivities of analytical tests. The following example illustrates one such application.

Example 14.15

The most common test for Ag^+ in a solution is to add a drop of solution containing chloride ion. If sufficient silver ion is present, AgCl ($K_{sp} = 1.8 \times 10^{-10}$) precipitates. We perform such a test by adding 0.05 mL (one drop) of 1 M NH_4Cl to 4 mL of a solution that might contain Ag^+. What is the maximum Ag^+ concentration that can be present without a precipitate forming?

Solution

If no precipitate forms, we can compute the chloride ion concentration in the test solution by calculating the dilution of NH_4Cl.

$$[Cl^-] = (1\ M) \frac{0.05\ mL}{4\ mL} = 0.013\ M$$

Knowing this chloride concentration, we can use the solubility product of AgCl to calculate the maximum Ag^+ concentration. If silver ion were present in higher concentration, a precipitate of AgCl would have formed.

$$K_{sp} = [Ag^+][Cl^-] = 1.8 \times 10^{-10}$$

$$[Ag^+]_{max}(0.013) = 1.8 \times 10^{-10}$$

$$[Ag^+]_{max} = 1.4 \times 10^{-8}\ M$$

The Effect of pH on Solubility

The solubilities of salts of weak acids increase in acidic solutions and decrease as the pH is raised. Hydrogen ion combines with the anion of the salt, decreasing its concentration. According to Le Chatelier's principle, this decrease in anion concentration shifts the solubility equilibrium to the right, allowing more cations into solution. For example, CaF_2 is more soluble in acid solution than in pure water. The equilibrium

$$CaF_2(s) \rightleftharpoons Ca^{2+}(aq) + 2\ F^-(aq)$$

shifts to the right as hydrogen ions combine with fluoride ions.

Table 14.7 Anions Reacting in Acid Solutions

Anion	Reaction
Hydroxide	$OH^- + H^+ \rightleftharpoons H_2O$
Oxide	$O^{2-} + 2H^+ \rightleftharpoons H_2O$
Sulfide	$S^{2-} + H^+ \rightleftharpoons HS^-$
	$HS^- + H^+ \rightleftharpoons H_2S \rightleftharpoons H_2S(g)$
Oxalate	$C_2O_4^{2-} + H^+ \rightleftharpoons HC_2O_4^-$
	$HC_2O_4^- + H^+ \rightleftharpoons H_2C_2O_4$
Carbonate	$CO_3^{2-} + H^+ \rightleftharpoons HCO_3^-$
	$HCO_3^- + H^+ \rightleftharpoons H_2O + CO_2(g)$
Chromate	$2\,CrO_4^{2-} + 2H^+ \rightleftharpoons Cr_2O_7^{2-} + H_2O$
Phosphate	$PO_4^{3-} + H^+ \rightleftharpoons HPO_4^{2-}$
	$HPO_4^{2-} + H^+ \rightleftharpoons H_2PO_4^-$
	$H_2PO_4^- + H^+ \rightleftharpoons H_3PO_4$

$$H^+(aq) + F^-(aq) \rightleftharpoons HF(aq)$$

Many other anions, including hydroxide, oxide, sulfide, oxalate, chromate, carbonate, and phosphate react with hydrogen ion (see Table 14.7) and the solubilities of their salts increase in acidic solution.

This dependence of solubility on pH is useful for separating two metal ions. For example, we can separate iron(III) ion from magnesium ion by selectively precipitating iron(III) hydroxide from an acetate buffer whose pH equals 5.0. Magnesium hydroxide, much more soluble than iron(III) hydroxide, does not precipitate at this weakly acidic pH.

Example 14.16

Compare the solubilities of $Fe(OH)_3$ ($K_{sp} = 4 \times 10^{-38}$) and $Mg(OH)_2$ ($K_{sp} = 1.2 \times 10^{-11}$g) at pH 5.0.

Solution

Let us first consider the solubility of iron(III) hydroxide.

$$[Fe^{3+}][OH^-]^3 = 4 \times 10^{-38}$$

$$[OH^-] = \frac{K_W}{[H^+]} = \frac{1.0 \times 10^{-14}}{1.0 \times 10^{-5}} = 1.0 \times 10^{-9}$$

$$[Fe^{3+}] = s = \frac{4 \times 10^{-38}}{(1.0 \times 10^{-9})^3} = 4 \times 10^{-11}\,M$$

Under these conditions, magnesium ion can be present in high concentrations without a precipitate of its hydroxide forming. Suppose the Mg^{2+} concentration is $1.0\,M$.

$$[Mg^{2+}][OH^-]^2 = (1.0)(1.0 \times 10^{-9})^2 = 1.0 \times 10^{-18} << K_{sp}$$

The Solubilities of Amphoteric Hydroxides

Aluminum hydroxide is an amphoteric substance that is insoluble in neutral solutions. Like other hydroxides, its solubility increases in acid solution.

$$Al(OH)_3(s) \rightleftharpoons Al^{3+}(aq) + 3\,OH^-(aq)$$

$$K_{sp} = [Al^{3+}][OH^-]^3 = 5 \times 10^{-33}$$

Unlike iron(III) hydroxide, aluminum hydroxide dissolves in strongly basic solutions because of its ability to act as an acid.

14.2 SOLUBILITY EQUILIBRIA

$$OH^-(aq) + Al(OH)_3(s) \rightleftharpoons Al(OH)_4^-(aq)$$

$$K_{eq} = \frac{[Al(OH)_4^-]}{[OH^-]} = 10$$

Example 14.17

What is the solubility of $Al(OH)_3$ at:
a. pH 4.0?   b. pH 11.0?

Solution

a. The solubility of $Al(OH)_3$ is the sum of the concentrations of Al^{3+} and $Al(OH)_4^-$.

$$s = [Al^{3+}] + [Al(OH)_4^-]$$

$$[Al^{3+}] = \frac{5 \times 10^{-33}}{[OH^-]^3} = \frac{5 \times 10^{-33}}{(1 \times 10^{-10})^3} = 5 \times 10^{-3} \text{ M}$$

$$[Al(OH)_4^-] = 10\,[OH^-] = 10\,(1 \times 10^{-10}) = 1 \times 10^{-9} \text{ M}$$

$$s = 5 \times 10^{-3} \text{ M} + 1 \times 10^{-9} \text{ M} \cong 5 \times 10^{-3} \text{ M}$$

b. The calculation is carried out for 1.0×10^{-3} M OH^- in the same way as in part a. The concentration of Al^{3+} is 5×10^{-30} M and that of $Al(OH)_4^-$ is 0.01 M, so the solubility is 0.01 M.

Complex Ions and Solubility

Cations, particularly those of the transition metals, can combine with anions or with Lewis bases (such as ammonia) to form polyatomic ions known as complex ions. For example, silver ion combines with two ammonia molecules to form a complex ion.

$$Ag^+(aq) + 2\,NH_3(aq) \rightleftharpoons Ag(NH_3)_2^+(aq)$$

The formation of the complex ion lowers the concentration of the metal ion. This decrease in the concentration of the metal ion applies a stress to the equilibrium for the solubility of a salt of that metal, and more salt dissolves to replace the metal ion that has formed the complex ion. An example of this effect involves the insoluble salt, AgCl ($K_{sp} = 1.8 \times 10^{-10}$). Its solubility in pure water is only 1.3×10^{-5}, but because of the interaction between Ag^+ and NH_3, this salt is considerably more soluble in 1.00 M NH_3.

Another example of the effect of complex-ion formation on solubility involves the influence of chloride-ion concentration on the solubility of AgCl. Chloride ion has a moderate tendency to form a complex ion with Ag^+.

$$Ag^+(aq) + 2\,Cl^-(aq) \rightleftharpoons AgCl_2^-(aq)$$

The initial effect of slowly adding Cl^- to a solution containing Ag^+ is to decrease the solubility of AgCl due to the common-ion effect. But when the chloride ion becomes quite concentrated, the formation of the complex ion becomes the dominant effect and the AgCl begins to redissolve.

$$AgCl(s) + Cl^-(aq) \rightleftharpoons AgCl_2^-(aq)$$

Special Resource Section: Simultaneous Equilibria

We must consider two equilibrium processes to predict the solubility of AgCl in a 1.00 M ammonia solution.

$$AgCl(s) \rightleftharpoons Ag^+(aq) + Cl^-(aq) \quad K_{sp} = 1.8 \times 10^{-10}$$

$$Ag^+(aq) + 2\,NH_3(aq) \rightleftharpoons Ag(NH_3)_2^+(aq) \quad K_{eq} = 1.4 \times 10^7$$

This problem is more complicated than those we considered earlier, but it can be solved by algebraic manipulation of the two equilibrium constants.

To compute the solubility of silver chloride in ammonia solution, we assume most of the dissolved silver is in the form of the complex ion (note the large equilibrium constant for complex-ion formation), and the major dissolving reaction is

$$2\ NH_3(aq) + AgCl(s) \rightleftharpoons Ag(NH_3)_2^+(aq) + Cl^-(aq)$$

The relevant stoichiometric equations are

$$s = [Cl^-] = [Ag(NH_3)_2^+]$$

and

$$[NH_3] = 1.00 - 2[Ag(NH_3)_2^+]$$
$$= 1.00 - 2s$$

Combined with these stoichiometric equations, the law of mass action for complex-ion formation yields the relationship

$$1.4 \times 10^7 = \frac{s}{[Ag^+](1.00 - 2s)^2}$$

or

$$[Ag^+] = \frac{s}{1.4 \times 10^7(1.00 - 2s)^2}$$

By substituting the equations for $[Ag^+]$ and $[Cl^-]$ into the solubility-product expression, we obtain the equation

$$[Ag^+][Cl^-] = \frac{s^2}{1.4 \times 10^7(1.00 - 2s)^2} = 1.8 \times 10^{-10}$$

$$s^2 = 2.5 \times 10^{-3}(1.00 - 2s)^2$$

$$s = 0.050\ (1.00 - 2s) = 0.045\ M$$

Summary

1. The equilibrium constant (K_W) at 25°C for the autoprotolysis reaction

$$H_2O + H_2O \rightleftharpoons H_3O^+ + OH^-$$

or, in simpler form

$$H_2O \rightleftharpoons H^+(aq) + OH^-(aq)$$

is

$$K_W = [H^+][OH^-] = 1.0 \times 10^{-14}$$

2. The pH of an aqueous solution is the negative base-ten logarithm of its hydrogen-ion activity. An approximate definition of p̂H is

$$pH = -\log [H^+]$$

The pH of a neutral solution is 7.0. Acid solutions have pH values lower than 7.0 and basic solutions have higher pH values.

SUMMARY

3. The hydrogen-ion concentration of a monoprotic strong acid equals its molar concentration.

4. The hydrogen-ion concentration of a solution of a weak acid is

    $$[H^+] = \sqrt{K_A C_A} \quad \text{if } C_A >> [H^+]$$

    where K_A is the acid dissociation constant and C_A the molar concentration of the acid.

5. The hydroxide-ion concentration of a monoprotic strong base equals its molar concentration. Most water-soluble weak bases function by a hydrolysis reaction.

    $$H_2O + B \rightleftharpoons BH^+ + OH^-$$

    $$K_B = \frac{[BH^+][OH^-]}{[B]}$$

    For a solution of a weak base,

    $$[OH^-] = \sqrt{K_B C_B} \quad \text{if } C_B >> [OH^-]$$

    The K_B value of a base is related to the acid dissociation constant of its conjugate acid, BH^+.

    $$K_B = \frac{K_W}{K_A}$$

6. A buffer solution contains a weak acid and its conjugate base (or a weak base and its conjugate acid) in comparable concentrations. In a buffer

    $$[H^+] = \frac{K_A C_A}{C_B}$$

    $$[OH^-] = \frac{K_B C_B}{C_A}$$

    A buffer resists changes in pH due to dilution, addition of an acid, or addition of a base.

7. A pH indicator is a weak acid whose color differs from that of its conjugate base. As the pH rises from a value much lower than pK_{Ind} (the negative logarithm of the dissociation constant of the indicator) to a value much higher than pK_{Ind}, the color changes from that of the acid form to the basic form.

8. The pH stays quite low during most of the titration of a strong acid with a strong base. It rises sharply at the equivalence point and levels off at a high pH when the base is in excess. The titration curve for a weak acid is initially steep, then levels off in the buffer region of the acid, rises sharply again at the equivalence point, and then levels off at high pH.

9. The titration curve of a polyprotic acid may contain several equivalence points at which the pH rises steeply and several buffer regions.

10. The solubility product (K_{sp}) of a slightly soluble salt is the equilibrium constant for the dissociation of its ionic crystal lattice in aqueous solution. The K_{sp} is the product of the equilibrium concentrations of the ions produced when the salt dissolves, each concentration being raised to a power equal to its stoichiometric coefficient. For a salt of the type AB,

    $$K_{sp} = [A][B] = s^2$$

    where s is the solubility of the salt. For AB_2 salts,

    $$K_{sp} = [A][B]^2 = 4s^3$$

11. The common-ion effect: A high concentration of a particular ion suppresses the solubilities of salts of that ion.

12. A high concentration of H^+ increases the solubilities of salts of weak acids because the H^+ reacts with their anions to lower the anion concentration.

14 AQUEOUS EQUILIBRIA

13. Amphoteric oxides and hydroxides dissolve in base because they react with OH⁻ to produce anions.

14. Certain anions and Lewis bases (such as NH_3) can increase the solubility of a salt by reacting with the salt's cation to form a complex ion.

Vocabulary List

hydrated hydrogen ion
hydronium ion
autoprotolysis
pH

dissociation constant
hydrolysis
buffer solution
equivalence point

titration curve
solubility product
common-ion effect

Exercises

In order to solve these exercises, it may be necessary to refer to Appendix H for acid and base dissociation constants and for solubility products.

1. Describe the difference between the terms solubility and solubility product.

2. Will the pH of a solution increase, decrease, or stay the same if we dissolve NaOH in it?

3. Classify the following pH values as acidic, basic, or neutral.
   a. 9.5   **b.** 7.0   **c.** 1.5   **d.** 13.0

4. **a.** What is the hydroxide-ion concentration of a solution that is 4.0×10^{-5} M in H^+?
   b. What is the H^+ concentration of a solution that is 0.067 M in OH^-?

5. **a.** What is the pH of a solution that is 6.0×10^{-6} M in H^+?
   b. What is the pH of a solution that is 8×10^{-4} M in OH^-?

6. **a.** What is the H^+ concentration in a solution with a pH of 12.5?
   b. What is the OH^- concentration in a solution whose pH is 8.40?

7. What is the pOH in a solution whose pH is 4.73?

8. Calculate the pH of 0.00385 M HBr (a strong acid).

9. What is the pH of 0.0429 M KOH (a strong base)?

10. What is the pH of a solution that is 0.35 M in HCl and 0.65 M in NaCl?

11. Write out algebraic equations showing the relationships between solubilities and solubility products for the following salts.
    a. $PbCrO_4(s) \rightleftharpoons Pb^{2+}(aq) + CrO_4^{2-}(aq)$
    b. $Hg_2Br_2(s) \rightleftharpoons Hg_2^{2+}(aq) + 2\ Br^-(aq)$
    c. $Ag_2SO_4(s) \rightleftharpoons 2\ Ag^+(aq) + SO_4^{2-}(aq)$
    d. $Cr(OH)_3(s) \rightleftharpoons Cr^{3+}(aq) + 3\ OH^-(aq)$
    e. $MgNH_4PO_4(s) \rightleftharpoons Mg^{2+}(aq) + NH_4^+(aq) + PO_4^{3-}(aq)$

12. Barium hydroxide dissociates completely when dissolved in water.

    $Ba(OH)_2(s) \longrightarrow Ba^{2+}(aq) + 2\ OH^-(aq)$

    Compute the pH of 5.0×10^{-2} M $Ba(OH)_2$.

13. What is the approximate pH of 1.00×10^{-9} M NaOH?

14. What is the pH of 0.50 M HCO_2H (formic acid)?

15. Calculate the pH of 0.020 M HCN.

16. What is the pH of 2.4×10^{-2} M aqueous ammonia?

17. What is the pH of 5.0×10^{-3} M NH_2OH? ($K_B = 1.1 \times 10^{-8}$)

18. The pH of 0.30 M HOBr is 4.6. What is the value of K_A for HOBr?

19. The pH of 1.00 M lactic acid is 1.93.
    a. Compute the dissociation constant of lactic acid.
    b. Compute the pH of 0.050 M lactic acid.

EXERCISES

20. The acid dissociation constant of phenol (C_6H_5OH) is 1.0×10^{-10}. What is the value of K_B for phenylate ion ($C_6H_5O^-$)?

21. The K_B of $(CH_3)_2NH$ is 5.9×10^{-4}. What is K_A for $(CH_3)_2NH_2^+$?

22. Calculate the pH of 0.020 M sodium nitrite, $NaNO_2$.

23. What is the pH of 3.0×10^{-3} M KCN?

24. Calculate the pH of a solution that is 0.010 M in NH_3 and 0.020 M in NH_4Cl.

25. A solution is 0.020 M in imidazole (a base whose K_B is 1.23×10^{-7}) and 0.14 M imidazolium chloride (the salt of HCl and imidazole). What is the pH of the solution?

26. Compute the pH of a solution that is 0.050 M in NaH_2AsO_4 and 0.025 M in Na_2HAsO_4.

27. A solution is prepared by mixing 60.0 mL of 0.0500 M acetic acid, 30.0 mL of 0.0300 M sodium acetate, and 10.0 mL water. What is its pH?

28. The pH of a solution that is 0.050 M in NaH_2PO_3 and 0.0100 M in Na_2HPO_3 is 5.5.
    a. What is the value of K_A for $H_2PO_3^-$?
    b. Calculate K_B for HPO_3^{2-}.

29. Compute the pH values of the following solutions.
    a. A solution that is 0.075 M in HCO_2H and 0.0250 M in $NaHCO_2$
    b. A solution that is 0.750 M in HCO_2H and 0.250 M in $NaHCO_2$

30. Estimate the solubility products of the following salts from their solubilities in pure water.

	Salt	Solubility reaction	Solubility (M)
a.	$PbSO_4(s) \rightleftharpoons$	$Pb^{2+}(aq) + SO_4^{2-}(aq)$	1.26×10^{-4}
b.	$TlCl(s) \rightleftharpoons$	$Tl^+(aq) + Cl^-(aq)$	0.0161
c.	$Hg_2I_2(s) \rightleftharpoons$	$Hg_2^{2+}(aq) + 2\,I^-(aq)$	7×10^{-7}

31. Use the solubilities of the following salts in pure water to find their solubility products.

	Salt	Solubility reaction	Solubility (M)
a.	$Tl_2C_2O_4(s) \rightleftharpoons$	$2\,Tl^+(aq) + C_2O_4^{2-}(aq)$	0.038
b.	$La(IO_3)_3(s) \rightleftharpoons$	$La^{3+}(aq) + 3\,IO_3^-(aq)$	1.03×10^{-3}
c.	$La_2(SO_4)_3 \cdot 9\,H_2O(s) \rightleftharpoons$	$2\,La^{3+}(aq) + 3\,SO_4^{2-}(aq) + 9\,H_2O$	0.045

32. Calculate the solubility of:
    a. AgI in pure water   b. AgI in 0.50 M KI
    c. AgI in 0.50 M KNO_3 d. AgI in 0.10 M $AgNO_3$

33. What are the solubilities of the following?
    a. CaF_2 in pure water   b. CaF_2 in 0.300 M $Ca(NO_3)_2$
    c. CaF_2 in 0.300 M KF

34. Will any precipitate form if equal volumes of 5.0×10^{-3} M $Ba(NO_3)_2$ and 5.0×10^{-2} M $Na_2S_2O_3$ are mixed?

35. Can 10.0 mL of 0.010 M $Ba(NO_3)_2$ be mixed with 5.0 mL of 0.010 M KF without a precipitate forming?

36. A chemist mixes 10 mL of 0.10 M $Ca(NO_3)_2$ and 1.0 mL of 0.10 M Na_2SO_4. Does a precipitate form?

37. A test for the presence of Ba^{2+} in 4.0 mL of solution involves adding 0.10 mL of 0.50 M Na_2SO_4. If no precipitate forms, the conclusion is that Ba^{2+} is absent. What is the maximum concentration of Ba^{2+} that could escape detection because of barium sulfate's slight solubility?

38. What is the maximum possible concentration of Cr^{3+} in a solution whose pH is 6.5?

39. A solution has a pH of 5.7. What is the maximum concentration of Cu^{2+} that it can contain?

40. Draw a graph of pH versus volume of base added for the titration of 20.0 mL of 0.250 M HNO_3 (a strong acid) with 0.500 M KOH (a strong base). Calculate the pH values for the following volumes of added base: 0 mL; 5.0 mL; 9.0 mL; 9.9 mL; 10.0 mL; 10.1 mL; 11.0 mL.

41. Prepare a titration curve for the titration of 20.0 mL of 0.100 M HOCl with 0.200 M NaOH. Compute the pH for each of the following volumes of added base: 0 mL; 1.0 mL; 5.0 mL; 9.0 mL; 9.9 mL; 10.0 mL; 10.1 mL; 11.0 mL.

42. Exactly 25.0 mL of 0.0500 M NH_3 is titrated with 0.100 M HCl.
    a. Calculate the pH values for the following volumes of added acid: 0 mL; 1.0 mL; 5.0 mL; 10.0 mL; 12.0 mL; 12.5 mL; 13.0 mL; 15.0 mL.
    b. Choose an indicator from Table 14.4 that would be suitable for this titration.

43. Prepare a graph for the titration of 25.0 mL of 0.200 M KOH with 0.100 M HCl. Calculate the pH values for the following volumes of added acid: 0 mL; 25.0 mL; 49.0 mL; 49.9 mL; 50.0 mL; 50.1 mL; 51.0 mL; 60.0 mL.

44. Solutions containing the indicator cresol red are yellow below pH 7.2, red above pH 8.8, and orange between pH 7.2 and 8.8.
    a. What is the approximate acid dissociation constant of cresol red?
    b. What is the color of the conjugate base of cresol red?

45. The solubility of $Pb(IO_3)_2$ in 5×10^{-3} M $Pb(NO_3)_2$ is 9.6×10^{-6} M.
    a. Calculate the value of K_{sp} for $Pb(IO_3)_2$.
    b. Calculate the solubility of $Pb(IO_3)_2$ in 1.0×10^{-3} M KIO_3.

46. Which of the following salts will increase in solubility as the solvent is made more acidic?
    a. AgI   b. $PbCO_3$ c. $Mn(OH)_2$ d. $PbSO_4$ e. Ag_3PO_4

47. Zinc hydroxide is an amphoteric substance.

    $$Zn(OH)_2(s) + OH^-(aq) \rightleftharpoons Zn(OH)_3^-(aq) \qquad K_{eq} = 0.005$$

    What are the concentrations of Zn^{2+}, $Zn(OH)_3^-$, and the total dissolved zinc in a saturated solution at the following pH values?
    a. pH 6.0   b. pH 7.0   c. pH 13.0

48. a What must be the molarity of NH_3 in a buffer containing 0.0400 M NH_4Cl and whose pH is 9.50?
    b. What is the molarity of a NaOH solution whose pH is 9.50?
    c. How much will the pH of the ammonia buffer system (pH 9.50) in part a change if it is diluted tenfold with water?
    d. How much will the pH of the NaOH solution (pH 9.50) in part b change if it is diluted tenfold with water?
    e. How much will the pH change if 0.2 mL of 0.50 M NaOH is added to 10.0 mL of the solution in part a?
    f. How much will the pH change if 0.2 mL of 0.50 M NaOH is added to 10.0 mL of the solution in part b?
    g. How much will the pH change if 0.2 mL of 0.50 M HCl is added to 10.0 mL of the solution in part a?
    h. How much will the pH change if 0.2 mL of 0.50 M HCl is added to 10.0 mL of the solution in part b?

---

49. A chemist wishes to prepare 1 L of a pH 7.50 buffer. Suggest a suitable buffer system and calculate the composition of the buffer if the conjugate acid concentration is 0.050 M.

50. What is the pH of a solution formed by mixing 20.0 mL of 0.0500 M HF and 5.0 mL of 0.0500 M NaOH?

51. After an unknown acid has been titrated halfway to its equivalence point, the pH of the titration solution is 6.70. What is the pK_A of the unknown acid?

52. What is the concentration of Zn^{2+} in a solution that is 0.0050 M in $Zn(NO_3)_2$ and 0.050 M in NH_3?

    $$Zn^{2+}(aq) + 4\,NH_3(aq) \rightleftharpoons Zn(NH_3)_4^{2+}(aq) \qquad K_{eq} = 5.0 \times 10^8$$

53. The pH for the first equivalence point in the titration of H_2SO_4 shown in Figure 14.5 is 1.23. Compute the concentrations of HSO_4^- and SO_4^{2-} at this equivalence point.

54. A solution that is 0.025 M in $H_2PO_4^-$ has a pH of 4.50. Estimate the concentrations of H_3PO_4, HPO_4^{2-}, and PO_4^{3-}.

55. What is the solubility of CuI in the following?
    a. Pure water   b. 0.10 M KCN

    $$Cu^+(aq) + 2\,CN^-(aq) \rightleftharpoons Cu(CN)_2^-(aq) \qquad K_{eq} = 1 \times 10^{24}$$

56. Iodide ion displaces chloride ion from a silver chloride precipitate.

    $$I^-(aq) + AgCl(s) \rightleftharpoons AgI(s) + Cl^-(aq)$$

    a. Write the mass-action law for this reaction.
    b. Estimate the value of the equilibrium constant for this reaction from the solubility products of AgCl and AgI.

57. What is the solubility of AgCN in the following?
    a. Pure water   b. 0.02 M HNO_3

58. One way to determine the concentration of Ag^+ in a solution is to titrate it with a standard solution of thiocyanate ion.

$$Ag^+(aq) + SCN^-(aq) \rightleftharpoons AgSCN(aq)$$

Iron(III) ion acts as an end point indicator by forming a red complex ion with SCN^-.

$$Fe^{3+}(aq) + SCN^-(aq) \rightleftharpoons FeSCN^{2+}(aq) \qquad K_{eq} = 1.0 \times 10^3$$

  a. Assuming the titration solution is $0.010\ M$ in $Fe(NO_3)_3$ and that $FeSCN^{2+}$ must be at least $10^{-4}\ M$ before you can see the red color, estimate the concentration of Ag^+ still in solution at the end point.
  b. What is the concentration of SCN^- at the equivalence point?
  c. Would the end-point volume for this titration be greater or less than the equivalence-point volume?
  d. What is the approximate percentage error due to the end point for the titration of 15.00 mL of $0.1000\ M$ $AgNO_3$ with $0.1000\ M$ $KSCN$?

59. Prepare a titration curve for titrating both acidic protons of malonic acid $[CH_2(CO_2H)_2]$ using 10.0 mL of $0.100\ M$ acid with $0.100\ M$ NaOH.

$$K_{A1} = 1.4 \times 10^{-3} \qquad K_{A2} = 2.0 \times 10^{-6}$$

[*Hint*: Except near the first equivalence point, you can neglect one of the two dissociations. To plot the curve near the first equivalence point, you may find it easier to compute the volume of base needed to reach a particular pH value rather than try to estimate the pH from the volume of base added.]

chapter 15
Electrochemical Reactions

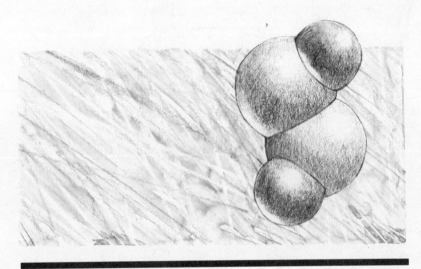

Hydrogen peroxide, H_2O_2, is both an oxidant and a reductant

All chemical reactions involve changes in electric potential energy because electrons move from one energy level to another as molecular or ionic structures change. In many cases, we can extract this energy in a usable form (a current flowing through a circuit) by means of a *galvanic cell*. Electric energy changes to chemical energy in other reactions via the process of *electrolysis*. This chapter deals with the nature of galvanic cells and electrolysis, their theoretical significance, and some of their practical applications.

15.1 Reduction Potentials

Galvanic Cells

A redox reaction such as

$$Cu^{2+}(aq) + Zn(s) \rightleftharpoons Cu(s) + Zn^{2+}(aq)$$

involves electron transfer from the reducing agent (solid zinc in this case) to the oxidizing agent (Cu^{2+}). If Cu^{2+} comes into contact with the zinc surface, electrons pass from zinc atoms to copper(II) ions, losing potential energy in the process. There is no way to put these directly transferred electrons to useful work; to extract work, we must make the electrons flow through an external circuit. Contact between the oxidizing agent and the reducing agent "shorts out" any external circuit attached to the chemical system and deprives us of usable energy.

We prevent the chemical short circuit by keeping the oxidizing agent and reducing agent apart. The flow of electrons from the reducing agent to the oxidizing agent

Figure 15.1 The Daniell cell.

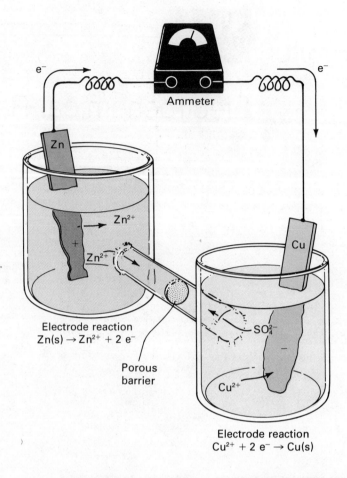

occurs when we supply an external path (a wire) between them. The **Daniell cell** (Figure 15.1) is a typical application of this principle. It consists of two electrodes; one is zinc immersed in $ZnSO_4$ solution, the other copper in $CuSO_4$ solution. A permeable barrier slows down the diffusion of Cu^{2+} toward the zinc electrode while at the same time allowing passage of an internal current—ions migrating due to electrostatic attractions. Since Cu^{2+} does not reach the zinc surface, the redox reaction can occur only if a wire connects the electrodes. Zinc is oxidized

$$Zn(s) \rightleftharpoons Zn^{2+}(aq) + 2\ e^-$$

as its electrons flow through the wire to the copper electrode and reduce Cu^{2+}.

$$2\ e^- + Cu^{2+}(aq) \rightleftharpoons Cu(s)$$

Sulfate ions migrate toward the zinc electrode and Zn^{2+} toward the copper electrode, preventing the accumulation of large charges anywhere within the cell.

The Daniell cell is one example of a **galvanic cell**. Each galvanic cell uses a particular redox reaction and contains two electrodes, or **half-cells**. One half-cell, called the **anode**, is the electrode from which electrons leave the cell and flow into the external circuit. The **cathode** is the half-cell that receives electrons from the external circuit. When an external circuit connects the two half-cells, oxidation occurs in the anode and reduction occurs in the cathode. Ions must be able to migrate between the two half-cells, but the reducing agent and oxidizing agent must not come into contact with each other. When the solutions in the two half-cells differ, we usually place a permeable barrier at the **liquid-liquid juncture** between the two half-cells. Figure 15.2 shows one type of permeable barrier—a **salt bridge**—frequently used in the laboratory.

15.1 REDUCTION POTENTIALS

Figure 15.2
A salt bridge.

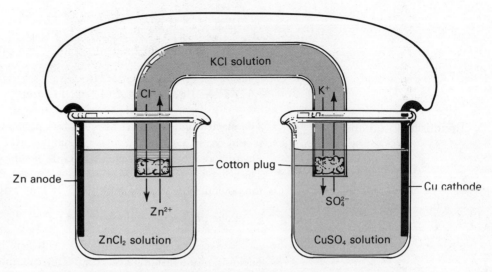

Cell Notation

A shorthand notation for the Daniell cell (Figure 15.1)

$$Zn(s)|ZnSO_4(aq) \vdots CuSO_4(aq)|Cu(s)$$

saves us the trouble of drawing detailed illustrations. In this notation, the different phases and solutions in the cell are listed, starting with the anode on the left and ending with the cathode on the right. Single vertical lines stand for boundaries between phases, while a dashed line stands for a liquid–liquid juncture. (A double vertical line means that a salt bridge is used at the liquid–liquid juncture.)

Some cells do not require any liquid–liquid juncture. The cell shown in Figure 15.3

$$Pt(s)|H_2(g)|HCl(aq)|Hg_2Cl_2(s)|Hg(l)$$

relies on the reaction between hydrogen gas and solid mercury(I) chloride.

$$H_2(g) + Hg_2Cl_2(s) \rightleftharpoons 2\,H^+(aq) + 2\,Cl^-(aq) + 2\,Hg(l)$$

Since neither the oxidant nor the reductant is water soluble, there is no difficulty in preventing their contact. Hydrogen gas passing over a catalytic platinum surface yields electrons

Figure 15.3
A cell without a liquid–liquid juncture.

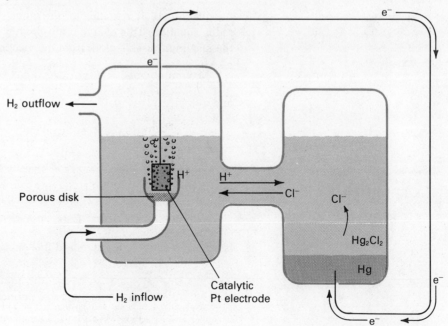

$$H_2(g) \rightleftharpoons 2\,H^+(aq) + 2\,e^-$$

that pass through the external circuit to the cathode and reduce solid mercury(I) chloride.

$$2\,e^- + Hg_2Cl_2(s) \rightleftharpoons 2\,Cl^-(aq) + 2\,Hg(l)$$

Reaction Potentials

The **electric potential** (E) between two points is the potential energy change that would occur if 1 coul of positive charge moved from the first point to the second. Electric potentials are measured in units of **volts** (V). One volt equals 1 J per coulomb. The potential energy decrease in joules due to a flow of charge between two points is the potential (E) in volts multiplied by the charge (Q) in coulombs.

$$\text{Potential energy decrease} = E\mathbf{Q} \tag{15.1}$$

The electric potential difference between the two electrodes of a galvanic cell depends on the reaction occurring in the cell, the concentrations of the reactants and products, and the temperature. To compare potentials for different reactions, we measure their cell potentials at 25°C and at standard concentration conditions: The activity of each reactant and product within its half-cell equals one.* The **standard reaction potential** ($E°$) for the Zn–Cu^{2+} reaction is the potential of a Daniell cell containing 1 M Zn^{2+} in its anode and 1 M Cu^{2+} in its cathode. This potential is 1.10 V at 25°C.

The standard potential of a reaction tells us the maximum amount of usable work it can yield when run under standard conditions. Although this theoretical maximum work yield cannot be obtained in practice, it is an important quantity because it is identical to the decrease in Gibbs free energy in the chemical system within the cell.

$$-\Delta G = \text{Maximum work} \tag{15.2}$$

We calculate the maximum work, and thus the free energy change, by computing the charge transferred between the reducing agent and the oxidizing agent when the reaction is run on a mole scale. Faraday's law (see Section 4.2) provides the equation needed to calculate this charge.

$$\mathbf{Q} = n\mathscr{F} = n(9.65 \times 10^4 \text{ coul/mol}) \tag{15.3}$$

In equation (15.3), n stands for the number of electrons transferred in the balanced redox equation. This charge multiplied by the standard potential of the cell equals the maximum usable work from the reaction under standard conditions.

$$\text{Maximum work} = E°\mathbf{Q} = E°n(9.65 \times 10^4 \text{ coul/mol})$$

and

$$-\Delta G° = E°n(9.65 \times 10^4 \text{ coul/mol}) \tag{15.4}$$

Example 15.1

The standard potential of the reaction

$$Cu^{2+}(aq) + Zn(s) \rightleftharpoons Cu(s) + Zn^{2+}(aq)$$

is +1.10 V. What is the standard free energy change for this reaction?

---

*We will continue to assume that the activities of dissolved species approximately equal their molarities, and the activity of a gas is its partial pressure in atmospheres.

15.1 REDUCTION POTENTIALS

Solution

This reaction involves the transfer of two electrons in the balanced equation, so $2\mathscr{F}$ charge are transferred for each mole of Cu^{2+} reduced. According to equation (15.4),

$$-\Delta G° = (1.10 \text{ V})(2)(9.65 \times 10^4 \text{ coul/mol})$$
$$= (1.10 \text{ J/coul})(2)(9.65 \times 10^4 \text{ coul/mol})$$
$$= 2.12 \times 10^5 \text{ J/mol}$$
$$\Delta G° = -212 \text{ kJ/mol}$$

If zinc metal were to react directly with copper(II) ion (as opposed to reacting via charge transfer through an electric cell), 217 kJ/mol heat would be released ($\Delta H° = -217$ kJ/mol). The 5 kJ/mol difference between the free energy and enthalpy reflects the entropy of the reaction. Even when a Daniell cell is run at such low currents that the charge flow does not generate any heat, 5 kJ/mol heat is used to increase the entropy of the cell's surroundings, and this heat can never do work at 25°C. We compute the standard entropy of the reaction from the equation

$$\Delta S° = \frac{\Delta H° - \Delta G°}{T} \tag{15.5}$$

Example 15.2

Compute the standard entropy of the Zn-Cu^{2+} reaction at 25°C.

Solution

The standard enthalpy is -217 kJ/mol (see above) and the standard free energy is -212 kJ/mol (see Example 15.1). The absolute temperature is 298 K. Inserting this information into equation (15.5), we obtain

$$\Delta S° = \frac{-217 \text{ kJ/mol} - (-212 \text{ kJ/mol})}{298 \text{ K}} = \frac{-5 \text{ kJ/mol}}{298 \text{ K}}$$
$$= -\frac{5 \times 10^3 \text{ J/mol}}{298 \text{ K}} = -17 \text{ J/K-mol}$$

Half-Cell Potentials

It is possible to assign numerical values to the potentials of each half-reaction occurring in a cell—the oxidation half-reaction taking place at the anode and the reduction half-reaction at the cathode. The potential of a cell is the sum of these two **half-cell potentials**.

$$E_{cell} = E_{ox} + E_{red} \tag{15.6}$$

It is not possible to measure the potential of an isolated half-cell; we can only observe the potential of a cell constructed from two half-cells. In order to assign numerical values to half-cell potentials, we arbitrarily assign a value of 0 V to the reduction potential of the **standard hydrogen electrode** at all temperatures. The standard hydrogen electrode consists of hydrogen gas at 1 atm pressure passing through a solution whose hydrogen-ion activity is one and over the surface of a catalytic platinum electrode that is immersed in the solution.

$$H^+(aq)(a = 1) | H_2(g)(1 \text{ atm}) | Pt(s)$$

Half-reaction: $2\ e^- + 2\ H^+(aq) \rightleftharpoons H_2(g) \qquad E° = 0.00 \text{ V}$

We obtain other half-cell potentials by measuring the potentials of cells containing the standard hydrogen electrode.

Figure 15.4
Half-cell potentials.

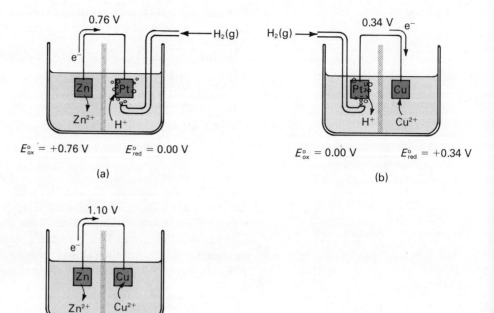

Figure 15.4 shows two examples of how the potential of a half-reaction can be measured relative to the standard hydrogen electrode. A cell consisting of a standard zinc electrode for the anode

$$Zn(s) \rightleftharpoons Zn^{2+}(aq) + 2\ e^-$$

and a standard hydrogen electrode for the cathode has a potential of 0.76 V. Since zinc metal is being oxidized at the anode, the oxidation potential of zinc is +0.76 V [see equation (15.6)]. We can easily convert this number into the reduction potential for Zn^{2+} because the potential of the reverse half-reaction (in this case reduction of Zn^{2+} to zinc metal) is the negative of the potential of the forward reaction. Therefore, the standard reduction potential of Zn^{2+} is −0.76 V.

The relationship between the potentials of forward and reverse half-reactions can be generalized to give an important relationship between oxidation potentials and reduction potentials. The reduction potential of a species equals the negative value of the oxidation potential of its reduced form.

$$E°_{red} = -E°_{ox} \qquad (15.7)$$

According to this rule, the oxidation potential of a standard hydrogen electrode is 0.00 V.

Figure 15.4, part (b), shows a standard copper cathode connected to a standard hydrogen anode. The potential of this cell is 0.34 V and represents the sum of the reduction potential of Cu^{2+}

$$2\ e^- + Cu^{2+}(aq) \rightleftharpoons Cu(s)$$

and the oxidation potential of hydrogen (0.00 V). According to equation (15.6), the reduction potential of Cu^{2+} is +0.34 V.

When we connect a standard copper electrode to a standard zinc electrode (Figure 15.4, part (c)), the zinc electrode is the anode and the copper electrode is the cathode. The measured potential for the cell and for the reaction

$$Cu^{2+}(aq) + Zn(s) \rightleftharpoons Cu(s) + Zn^{2+}(aq)$$

15.1 REDUCTION POTENTIALS

is +1.10 V. This observed value confirms equation (15.6).

$$E°_{cell} = E°_{ox, Zn} + E°_{red, Cu^{2+}}$$
$$= +0.76 \text{ V} + (+0.34 \text{ V}) = +1.10 \text{ V}$$

It is not necessary to measure every half-cell potential against a standard hydrogen electrode. Once the potential of an electrode has been measured against hydrogen, it can serve as a reference for measuring other half-cell potentials. One popular reference electrode is the standard calomel electrode. (Calomel is an old common name for Hg_2Cl_2.)

$$KCl(1\ M)|Hg_2Cl_2(s)|Hg(l)$$

$$2\ e^- + Hg_2Cl_2(s) \rightleftharpoons 2\ Hg(l) + 2\ Cl^-(aq) \qquad E°_{red} = +0.268 \text{ V}$$

Example 15.3

The standard Fe^{2+}-Fe electrode acts as an anode toward the calomel electrode, and the cell potential is 0.71 V. What is the reduction potential for the following half-reaction?

$$2\ e^- + Fe^{2+}(aq) \rightleftharpoons Fe(s)$$

Solution

Since the cell consists of a standard calomel cathode and a Fe-Fe^{2+} anode, the measured cell potential (0.71 V) is

$$E°_{cell} = E°_{red,calomel} + E°_{ox,Fe} = 0.71 \text{ V}$$

The standard reduction potential of calomel is +0.27 V.

$$0.27 \text{ V} + E°_{ox, Fe} = 0.71 \text{ V}$$
$$E°_{ox,Fe} = 0.71 \text{ V} - 0.27 \text{ V} = +0.44 \text{ V}$$

The reduction potential of Fe^{2+} equals the negative value of the oxidation potential of iron [equation (15.7)].

$$E°_{red,\ Fe^{2+}} = -0.44 \text{ V}$$

Appendix I is a table of standard reduction potentials, arranged in decreasing order, for many half-reactions. Since equation (15.7) holds for any pair of reverse reactions, there is no need to list separate oxidation potentials. To find the oxidation potential of a species, locate this species on the right hand side of the appropriate half-reaction in Appendix I and reverse the sign of the reduction potential associated with that half-reaction. Using the half-cell potentials in Appendix I, we can predict the values of many standard reaction potentials by adding the appropriate reduction potential and oxidation potential [equation (15.6)].

Example 15.4

Using the data in Appendix I, calculate the standard potential of the following reaction.

$$Br_2(l) + 5\ Cl_2(g) + 6\ H_2O \rightleftharpoons BrO_3^-(aq) + 10\ Cl^-(aq) + 12\ H^+(aq)$$

Solution

The main problem in any question of this sort is to identify the half-reactions involved. In this case Cl_2 is reduced to Cl^-,

$$2\ e^- + Cl_2(g) \rightleftharpoons 2\ Cl^-(aq) \qquad E°_{red} = +1.36\ V$$

and Br_2 is oxidized to BrO_3^-.

$$Br_2(l) + 6\ H_2O \rightleftharpoons 2\ BrO_3^-(aq) + 12\ H^+(aq) + 10\ e^-$$

To find the oxidation potential, scan Appendix I to find the potential for the half-reaction in which BrO_3^- is reduced to Br_2 (1.52 V). The oxidation potential is the negative value of this number or -1.52 V.

$$E°_{reaction} = E°_{red} + E°_{ox} = +1.36\ V + (-1.52\ V)$$
$$= -0.16\ V$$

The negative value of the reaction potential in Example 15.4 means that the reaction will not proceed spontaneously to the right—$\Delta G°$ is positive. Starting at standard conditions, the reaction proceeds instead to the left, and BrO_3^- oxidizes Cl^-. The Br_2-BrO_3^- half-cell is the cathode of a standard cell based on this reaction.

Although we must multiply the stoichiometric coefficients in the Cl_2-Cl^- half-reaction by five to balance the equation, we do not multiply the reduction potential by five. This is because a volt is a unit of energy per charge (1 V = 1 J/coul) rather than a unit of energy, and half-reaction potentials are a measure of the free energy change *per electron transferred*.

Example 15.5

The following reaction occurs in basic solution.

$$ClO_4^-(aq) + SO_3^{2-}(aq) \rightleftharpoons ClO_3^-(aq) + SO_4^{2-}(aq)$$

What is the standard potential of this reaction?

Solution

Since the reaction occurs in base, the appropriate half-reactions and their potentials appear in the section of Appendix I containing half-reactions in basic solutions. The reduction half-reaction is

$$2\ e^- + H_2O + ClO_4^-(aq) \rightleftharpoons ClO_3^-(aq) + 2\ OH^-(aq) \qquad E°_{red} = +0.36\ V$$

and the oxidation half-reaction is

$$SO_3^{2-}(aq) + 2\ OH^-(aq) \rightleftharpoons SO_4^{2-}(aq) + H_2O + 2\ e^-$$

Since the reduction potential of SO_4^{2-} in base is -0.93 V, the oxidation potential of SO_3^{2-} in base is $+0.93$ V.

$$E°_{reaction} = E°_{red} + E°_{ox} = +0.36\ V + (+0.93\ V) = +1.29\ V$$

Concentration Effects

The concentrations of the components of a half-cell influence its potential, the half-cell potential varying with the mass-action ratio according to **Nernst's equation**,

$$E = E° - \frac{2.303\ RT}{n\mathscr{F}} \log Q \qquad (15.8)$$

where E is the potential at the specified mass-action ratio Q, and n is the number of electrons transferred in the balanced equation. Since R is 8.31 J/K-mol, $\mathscr{F}$ is 9.65×10^4 coul per mole of electrons, and 1 V equals 1 J/coul, we can write the

15.1 REDUCTION POTENTIALS

Nernst equation as

$$\frac{2.303\,RT}{\mathscr{F}} = \frac{2.303\left(8.314\,\frac{J}{K\text{-mol}}\right)}{9.649\times 10^4\,\frac{coul}{mol}} = 1.984\times 10^{-4}\,V/K$$

$$E = E° - \frac{1.984\times 10^{-4}\,T}{n}\log Q \tag{15.9}$$

At 25°C (298.2 K), this relationship is

$$E = E° - \frac{0.0592\,V}{n}\log Q \tag{15.10}$$

Example 15.6

What is the reduction potential of a Cu^{2+}-Cu electrode at 25°C when $[Cu^{2+}] = 5.0\times 10^{-4}$ M?

$$2\,e^- + Cu^{2+}(aq) \rightleftharpoons Cu(s) \qquad E° = +0.34\,V$$

Solution

The mass-action ratio for this half-reaction is $1/[Cu^{2+}]$, and the half-reaction involves a two-electron change. Therefore,

$$\begin{aligned}
E &= +0.34\,V - \frac{0.0592\,V}{2}\log\frac{1}{[Cu^{2+}]}\\
&= +0.34\,V - (0.0296\,V)\log\frac{1}{5.0\times 10^{-4}}\\
&= +0.34\,V - (0.0296\,V)\log(2.0\times 10^3)\\
&= +0.34\,V - (0.0296\,V)(+3.3)\\
&= +0.34\,V - 0.097\,V = +0.24\,V
\end{aligned}$$

The Nernst equation applies to reaction potentials and cell potentials as well as to half-reaction and half-cell potentials. In such cases, $E°$ is the standard reaction potential, n is the number of electrons transferred in the balanced equation, and Q is the mass-action ratio for the whole reaction.

Example 15.7

The standard potential of the reaction

$$2\,Ga(s) + 6\,H^+(aq) \rightleftharpoons 2\,Ga^{3+}(aq) + 3\,H_2(g)$$

is +0.56 V. What is the reaction potential at pH 3.0 in the presence of 2.0 atm H_2 and 0.10 M Ga^{3+} at 25°C?

Solution

The balanced equation involves the transfer of six electrons. The mass-action ratio is

$$Q = \frac{[Ga^{3+}]^2\,P_{H_2}^3}{[H^+]^6} = \frac{(0.10)^2(2.0)^3}{(1.0\times 10^{-3})^6} = 8.0\times 10^{16}$$

According to Nernst's equation,

$$E = +0.56 \text{ V} - \frac{0.0592 \text{ V}}{6} \log (8.0 \times 10^{16})$$

$$= +0.56 \text{ V} - \frac{0.0592 \text{ V}}{6} (16.90) = +0.56 \text{ V} - 0.17 \text{ V}$$

$$= +0.39 \text{ V}$$

The Nernst equation can be used to estimate the equilibrium constant from the standard potential of a reaction. The potential and the Gibbs free energy of a reaction are both zero at equilibrium and the mass-action ratio equals the equilibrium constant. For a reaction at 25°C,

$$0 = E° - \frac{0.0592 \text{ V}}{n} \log K_{eq}$$

and

$$\log K_{eq} = \frac{nE°}{0.0592 \text{ V}} \tag{15.11}$$

Example 15.8

Calculate the equilibrium constant for the reaction

$$Cu^{2+}(aq) + Zn(s) \rightleftharpoons Cu(s) + Zn^{2+}(aq) \quad E° = +1.10 \text{ V}$$

Solution

The balanced equation represents a two-electron transfer. Therefore,

$$\log K_{eq} = \frac{2(1.10 \cancel{V})}{0.0592 \cancel{V}} = 37.3$$

$$K_{eq} = \frac{[Zn^{2+}]}{[Cu^{2+}]} = \text{antilog } 37.3 = 2 \times 10^{37}$$

Concentration Cells

The cell in Figure 15.5 is an example of a **concentration cell**. The electrode on the right is a standard hydrogen half-cell, and the other electrode is a hydrogen half-cell in which the hydrogen pressure is 1 atm and the pH is unknown. We can measure the pH of this left-hand electrode by observing the potential of the concentration cell. The reduction potential of the standard electrode is 0 V, and the reduction potential of the other half-cell is

$$E_{red} = 0.000 \text{ V} - \frac{(0.0592 \text{ V})}{2} \log \frac{P_{H_2}}{[H^+]^2}$$

Since $P_{H_2} = 1$ atm,

$$E_{red} = - \frac{0.0592 \text{ V}}{2} \log \frac{1}{[H^+]^2} = +0.0592 \text{ V (pH)} \tag{15.12}$$

If the pH of the solution is greater than zero, the half-cell has a negative reduction potential and a positive oxidation potential, and the left-hand cell acts as the anode. The cell potential is proportional to the pH of the solution in the left-hand cell.

$$E_{cell} = E_{red} + E_{ox} = 0.000 \text{ V} + 0.0592 \text{ V (pH)}$$

$$pH = \frac{E_{cell}}{0.0592 \text{ V}} \tag{15.13}$$

Figure 15.5
A concentration cell.

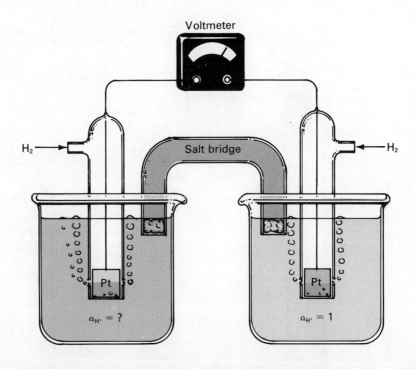

Example 15.9

The potential of the cell shown in Figure 15.5 is 0.336 V and the left-hand half-cell is the anode. What is the pH of the solution in the left-hand electrode?

Solution

According to equation (15.13),

$$\text{pH} = \frac{0.336 \text{ V}}{0.0592 \text{ V}} = 5.69$$

Because hydrogen gas is difficult to handle, the cell in Figure 15.5 is cumbersome. Most pH meters use a **glass electrode** (Figure 15.6) in place of hydrogen electrodes. A glass membrane in this electrode separates the solution being tested from a reference electrode, such as Ag-AgCl immersed in 0.10 M HCl. The potential across the glass membrane depends on the difference in pH between the unknown solution outside the electrode and the reference solution within it. A voltmeter measures the potential difference between the glass electrode and a reference electrode (usually calomel) and displays this potential on a scale calibrated in pH units.*

Other types of glass electrodes show specific responses to certain ions, some being sensitive to differences in NH_4^+ concentration, others responding to barium ion. Using the available selection of glass electrodes, we can measure the concentrations of many different ions potentiometrically (in other words, by measuring the voltage of the electrode).

Cells as Practical Energy Sources

Galvanic cells are useful portable sources of energy. A car's ignition system runs on the direct current from a **lead storage battery**—a series of cells (Figure 15.7) using the reaction

$$PbO_2(s) + Pb(s) + 2\,H^+(aq) + 2\,HSO_4^-(aq) \rightleftharpoons 2\,PbSO_4(s) + 2\,H_2O$$

The anode of a lead cell consists of porous lead held by a grid of inert lead-antimony

---

*A pH difference of one unit corresponds to a hydrogen-ion activity change by a factor of 10, which is equivalent to 0.0592 V from the Nernst equation.

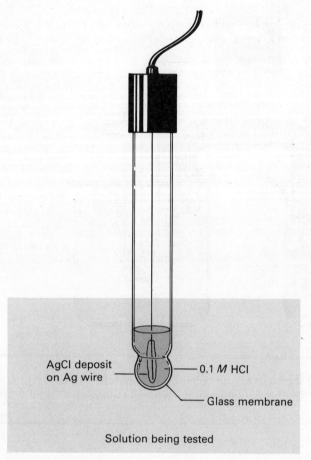

Figure 15.6
A glass electrode.

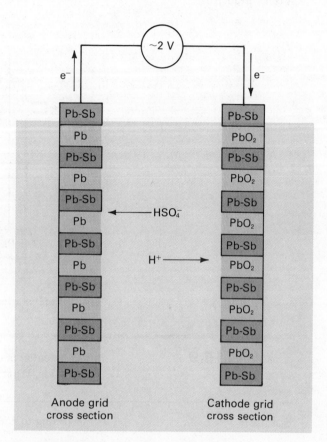

Figure 15.7
A cell of a lead storage battery.

alloy, and the cathode is PbO_2 supported by a similar grid.* The electrolyte is a fairly concentrated solution of sulfuric acid (about 30% by weight). Lead sulfate forms at both electrodes as the battery operates.

$$\text{Pb-Sb(s)}|\text{Pb(s)}|\text{PbSO}_4(s)|\text{H}_2\text{SO}_4(aq)|\text{PbSO}_4(s)|\text{PbO}_2(s)|\text{Pb-Sb(s)}$$

The anode half-reaction is

$$\text{Pb(s)} + \text{HSO}_4^-(aq) \rightleftharpoons \text{PbSO}_4(s) + \text{H}^+(aq) + 2\,e^-$$

and lead(IV) oxide is reduced at the cathode.

$$2\,e^- + \text{PbO}_2(s) + 3\,\text{H}^+(aq) + \text{HSO}_4^-(aq) \rightleftharpoons \text{PbSO}_4(s) + 2\,\text{H}_2\text{O}$$

The potential of a cell is about 2 V. Three cells form a 6-volt battery, and a 12-volt battery contains 6 cells.

The sulfuric acid concentration of a battery decreases and its potential falls as it yields current. Because the density of sulfuric acid indicates its concentration, we

---

*Kinetics play an important part in the success of this cell. The reaction between the anode and the sulfuric acid

$$\text{Pb(s)} + \text{H}^+(aq) + \text{HSO}_4^- \rightleftharpoons \text{PbSO}_4(s) + \text{H}_2(g)$$

has a favorable equilibrium but is too slow to hurt the operation of the cell. Similarly, the lead in the Pb-Sb cathode grid does not react with PbO_2 at an appreciable rate.

can determine if a lead storage battery is discharged by drawing some of its electrolyte into a calibrated hydrometer (a device for rapidly estimating a liquid's density). If a battery is discharged, the sulfuric acid concentration will be too dilute, and the density of the electrolyte will be low.

We can recharge a lead storage battery by attaching the negative pole of a direct-current source to the lead electrode and the positive pole to the PbO_2 electrode. The current electrolyzes the battery, that is, it reverses the cell reaction by reducing $PbSO_4$ at the negative pole to porous lead and oxidizing $PbSO_4$ at the positive pole to PbO_2. When a car's motor is running, some of its mechanical energy generates the electric current needed for recharging.

In spite of its name, a **dry cell** (Figure 15.8) is not dry—it needs water to function—but its electrolyte is a moist paste rather than a fluid. Because of this, we can hold a dry cell in any position without its contents running out. (Do not try this with a lead storage battery!) An inert graphite cathode dips into a paste containing MnO_2, graphite, and NH_4Cl. The cell's zinc outer casing is its anode. A thin layer of paste containing NH_4 and $ZnCl_2$ keeps MnO_2 (the oxidizing agent) from reaching the zinc anode. We do not understand all the details of the complicated chemical reactions in the cell, but we know the zinc anode is oxidized to $ZnCl_2$ and the cathode reduces MnO_2 to Mn_2O_3. Possible reactions are

Anode: $Zn(s) \rightleftharpoons Zn^{2+}(aq) + 2\ e^-$

Cathode: $2\ e^- + 2\ NH_4^+(aq) + 2\ MnO_2(s) \rightleftharpoons Mn_2O_3(s) + 2\ NH_3(aq) + H_2O$

The limited amount of oxidizing agent and reducing agent in an ordinary galvanic cell limits the energy it can deliver before it discharges. A **fuel cell** avoids this disadvantage by having the reactants flow into their appropriate electrodes as the cell operates. Batteries of the fuel cell shown in Figure 15.9 have generated electric power (and drinking water) in space craft. Hydrogen gas passing over the anode is oxidized,

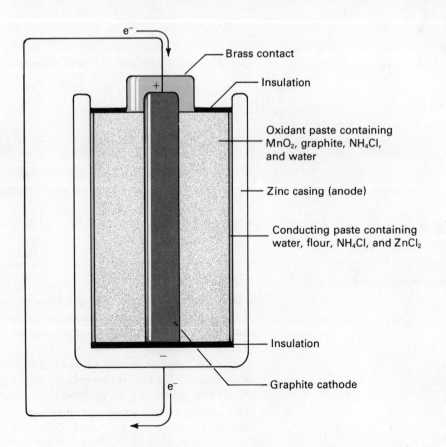

Figure 15.8
A dry cell.

Figure 15.9
A fuel cell.

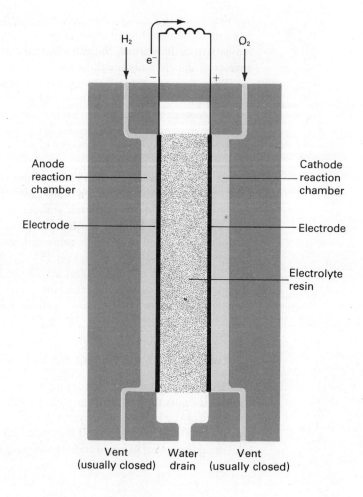

$$H_2(g) \rightleftharpoons 2\ H^+(aq) + 2\ e^-$$

while the cathode reduces a stream of oxygen gas.

$$4\ e^- + O_2(g) + 4\ H^+(aq) \rightleftharpoons 2\ H_2O(l)$$

The total reaction is the formation of water from its elements.

$$2\ H_2(g) + O_2(g) \rightleftharpoons 2\ H_2O(l) \qquad E° = +1.23\ V$$

The electrolyte connecting the two electrodes is a resin containing mobile H^+ ions. The migration of H^+ ions from the anode through the resin to the cathode maintains charge balance within the cell.

15.2 Electrolysis

Decomposition Potentials

Electrolysis—the use of an external potential to produce a thermodynamically unfavorable chemical reaction—is the basis of many important industrial processes. Sodium, aluminum, chlorine, and other elements are prepared by electrolyzing their compounds. Electroplating (electrolysis of solutions of metal ions) is used to produce coatings of metals on cathodes.

We described electrolyses and discussed their stoichiometry in Section 4.2. Now let us consider the factors determining the electric potential needed for a particular electrolysis. An electrolysis reaction has a negative reaction potential, a positive free energy change, and a small equilibrium constant. For example, the free energies in the system

15.2 ELECTROLYSIS

$$Cu(s) + 2 H^+(aq) \rightleftharpoons Cu^{2+}(aq) + H_2(g) \qquad E° = -0.34 \text{ V}$$

favor the reverse reaction (reduction of copper(II) ion by hydrogen gas).

$$\log K_{eq} = \frac{2(-0.34\ \cancel{V})}{0.0592\ \cancel{V}} = -11.5 \qquad K_{eq} = \frac{P_{H_2}[Cu^{2+}]}{[H^+]^2} = 3 \times 10^{-12}$$

Although electrons flow spontaneously from a hydrogen electrode to a copper half-cell, we can reverse the direction of electron flow by applying enough external potential in the opposite direction. This potential forces electrons from the copper electrode to the hydrogen electrode, driving the reaction from left to right.

There is no need to use a system as complicated as a galvanic cell. An electrolysis cell (Figure 15.10) could consist of two copper plates dipping in dilute sulfuric acid. Neither plate reacts with the sulfuric acid until we apply sufficient potential between them. Hydrogen gas then forms at the negative pole of the applied potential (the cathode), and the copper anode corrodes to form Cu^{2+}.

If we could apply the principles of equilibrium to electrolysis, we would predict that the minimum potential required for an electrolysis, called the **decomposition potential**, would be the reverse of the reaction potential. But electrolyses are not run under equilibrium conditions (otherwise they would take too long), and kinetic effects often raise decomposition potentials well above their reverse reaction potentials. For example, the standard potential for the decomposition of water is -1.23 V, but it takes at least 1.7 V (and often more) to make large currents flow through an electrolysis cell and produce hydrogen and oxygen. The slowness of the electrode reactions creates electric resistances at the electrode surfaces, and we need extra voltage called **overpotential** to overcome this resistance. The size of an overpotential (the difference between the actual decomposition potential and the theoretical electrode potential for the reaction) depends on many subtle factors: the nature of the electrode reactions, the electrode surfaces, current densities at the electrodes, and impurities in the solution. It is hard to predict overpotentials, but we know the overpotentials for plating metals tend to be small while those for gas formation can exceed 1 V.

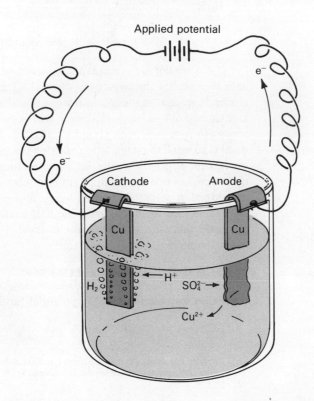

Figure 15.10
Electrolysis.

Concentration Effects in Electrolysis

As the electrolysis in Figure 15.10 proceeds, hydrogen bubbles form at the cathode, and Cu^{2+} forms at the anode. Because this system is not homogeneous (there is no Cu^{2+} near the cathode nor H_2 at the anode), the concentration differences create a galvanic cell whose potential opposes the applied potential.

$$Cu(s)|H_2(g)|H^+(aq) \vdots Cu^{2+}(aq)|Cu(s)$$

As continued electrolysis raises the Cu^{2+} concentration near the anode and depletes the H^+ concentration near the cathode, both the back potential and the required decomposition potential increase. Electrolysis ceases when the decomposition potential equals the applied potential,

$$E_{decomp} = \text{Overpotential} - E° + \frac{0.0592 \text{ V}}{n} \log Q \qquad (15.14)$$

where Q is the mass action ratio for the electrolysis. In the electrolysis in Figure 15.10

$$E_{decomp} = \text{Overpotential} + 0.34 \text{ V} + \log \frac{P_{H_2}[Cu^{2+}]}{[H^+]^2} \text{ *}$$

Competing Electrolyses

There are usually several possible electrolytic reactions that might occur within a particular cell. For example, two possible cathode reactions for the electrolysis of 1 M $ZnSO_4$ at pH 5.0 are

$$2 \text{ e}^- + Zn^{2+}(aq) \rightleftharpoons Zn(s)$$

$$2 \text{ e}^- + 2 H^+(aq) \rightleftharpoons H_2(g)$$

The cathode reaction that occurs is the one with the higher **effective reduction potential** ($E_{red,eff}$).

$$E_{red,eff} = E_{red} - \text{Overpotential} \qquad (15.15)$$

Zn: $E_{red,eff} = -0.76 \text{ V} - \text{Zn overpotential}$

H_2: $E_{red,eff} = 0.00 \text{ V} - \frac{0.0592 \text{ V}}{2} \log \frac{1}{(1.0 \times 10^{-5})^2} - H_2 \text{ overpotential}$

$\qquad = -0.30 \text{ V} - H_2 \text{ overpotential}$

If it were not for overpotentials, hydrogen would form at the cathode in preference to zinc, but the large overpotential for the formation of hydrogen at a smooth electrode surface is enough to reverse this situation, and a smooth cathode reduces zinc instead of hydrogen. Hydrogen will eventually form as the Zn^{2+} is depleted and zinc's effective reduction potential falls, but the residual Zn^{2+} concentration at this point is so small that the plating reaction is essentially quantitative. The high overpotentials for hydrogen evolution allow us to electroplate many metals—nickel and chromium are two other important examples—from aqueous solutions even though the reduction potentials of their ions are negative.

Differences in anodic overpotentials enable us to obtain chlorine by electrolysis of nearly saturated salt solutions (brines). Under the electrolysis conditions, the oxidation potential for the reaction

$$4 \text{ OH}^-(aq) \rightleftharpoons O_2(g) + 2 H_2O + 4 \text{ e}^-$$

is considerably higher than the potential for the desired half-reaction

$$2 \text{ Cl}^-(aq) \rightleftharpoons Cl_2(g) + 2 \text{ e}^-$$

---

*Because copper(II) ion diffuses slowly and the solution is not homogeneous, the mass-action ratio refers to the Cu^{2+} concentration near the anode and the H^+ concentration near the cathode.

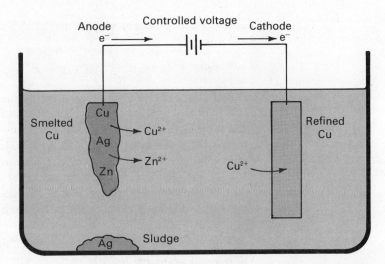

Figure 15.11
Electrorefining of copper.

but the overpotential for oxygen on the graphite anode is about 1 V compared to 0.2 V for chlorine. This difference is enough to give chlorine formation the higher effective potential.

Commercial Electrolysis

Electrolysis, in addition to providing a way to obtain many elements from their compounds (**electrowinning**), is used to purify metals (**electrorefining**), deposit metals on surfaces (**electroplating**), and coat metals with oxide films (**anodizing**). The electrowinning of zinc by electrolyzing solutions of pure $ZnSO_4$ is one of the commercial methods for producing zinc. We will discuss other important electrowinning processes when we survey the chemistry of the individual elements.

Copper produced by smelting is nearly pure but still contains a small amount of impurities, most notably silver (about 4 oz per ton, or 100 g per 1000 kg). We electrorefine (Figure 15.11) smelted copper by making it the anode in an electrolysis whose potential is enough to oxidize copper and the more active impurities such as zinc but not the less active ones such as silver. Copper(II) ions diffuse to the cathode and, being the most easily reduced ions in solution, plate out to form very pure copper. An attractive aspect of this process is the sludge remaining at the anode—it is rich in silver and gold.

The purpose of electroplating is to form a uniform layer of a metal on the surface of a cathode. The cathode could be an automobile bumper in the case of chrome plating or a spoon in the case of silver plating. In principle, electroplating is a simple electrolysis, but so many things can go wrong in the process, and the factors governing success or failure are so subtle that it is as much a craft as a science. If the conditions are not just right, the metal might deposit unevenly or not adhere to the cathode surface, the metal coating might be brittle and crack easily, or it might have a dull, discolored appearance.

Anodizing aluminum creates a thin film of aluminum oxide on its surface. This film is desirable because it resists corrosion and wear, is abrasive, and can absorb paints and dyes. The film is prepared by using aluminum as the anode in an acidic solution.

$$2\,Al(s) + 3\,H_2O \rightleftharpoons Al_2O_3(s) + 6\,H^+(aq) + 6\,e^-$$

We can vary the thickness of the film and emphasize one surface property over another (perhaps abrasiveness over paint absorption) by controlling temperature, current density, and the composition of the acidic solution.

Coulometric Titrations

An electrode can replace a buret in a titration. For example, instead of using a silver nitrate solution to titrate chloride ion,

$$Ag^+(aq) + Cl^-(aq) \rightleftharpoons AgCl(s)$$

we can insert a silver anode into the chloride solution and generate silver chloride by electrolysis.

$$Ag(s) + Cl^-(aq) \rightleftharpoons AgCl(s) + e^-$$

A detecting electrode, sensitive to Ag^+, signals the end point of the titration.

In order to calculate the moles of Cl^- in the solution, we must know how many moles of electrons have been released into the solution at the anode, and this requires a knowledge of the total charge that has passed through the titration cell up to the end point. We compute this charge by measuring the current flowing through the cell and the time required to reach the end point. The **ampere** (amp), the unit of electric current, represents a flow of 1 coul per second. Therefore, the charge equals the current (i) in amperes multiplied by the titration time in seconds.

$$\mathbf{Q} = it \tag{15.16}$$

In the case of the Cl^- titration,

$$\text{Moles } Cl^- = \text{Faradays charge} = \frac{it}{9.65 \times 10^4 \text{ coul}}$$

For any coulometric titration, the moles of titrant generated is

$$\text{Moles titrant} = \frac{it}{n\mathscr{F}} \tag{15.17}$$

where n is the number of electrons transferred per titrant molecule formed.

Example 15.10

It takes 475 sec at a current of 0.0350 amp to precipitate all the chloride ion in 5.00 mL of solution by anodic oxidation of silver metal.

$$Ag(s) + Cl^-(aq) \longrightarrow AgCl(s) + e^-$$

What is the molarity of chloride ion in the solution?

Solution

The total charge passing through the cell is

$$\mathbf{Q} = it = (0.0350 \text{ amp})(475 \text{ sec}) = 16.6 \text{ coul}$$

$$\text{Faradays charge} = \frac{16.6 \text{ coul}}{9.65 \times 10^4 \text{ coul}/\mathscr{F}} = 1.72 \times 10^{-4}\mathscr{F}$$

Since the precipitation of chloride involves a one-electron change,

$$\text{Moles } Cl^- = \text{Faradays} = 1.72 \times 10^{-4} \text{ mol}$$

$$M_{Cl^-} = \frac{1.72 \times 10^{-4} \text{ mol}}{5.00 \times 10^{-3} \text{ L}} = 0.0344 \text{ M}$$

The advantage of this coulometric titration over one using a buret is that we can measure current and time more precisely than we can measure solution volume. This allows the measurement of smaller amounts of Cl^- than could be estimated by a solution titration. Another advantage of coulometric titrations is that electrode reactions can generate unstable titrants such as Ag^{2+} or Mn^{3+}. These unstable ions react quantitatively in the titration reaction immediately after their formation. We cannot

15.2 ELECTROLYSIS

perform equivalent titrations with burets because solutions of these unstable titrants decompose before we get a chance to use them.

Electrolytic Corrosion

Kinetic effects are important in retarding metal corrosion. Aluminum has a large oxidation potential, and its reaction with oxygen

$$3\ O_2(g) + 4\ Al(s) \rightleftharpoons 2\ Al_2O_3(s)$$

has a very large equilibrium constant. But aluminum, a widely used building material, resists corrosion. Initially, a piece of aluminum does react with oxygen in the air, but the oxide forms a thin, impervious layer that protects the rest of the metal from contact with air and prevents further oxidation.

Although iron has a lower oxidation potential than aluminum, it oxidizes faster.

$$3\ O_2(g) + 4\ Fe(s) \rightleftharpoons 2\ Fe_2O_3(s)$$

Iron(III) oxide (the red stuff on rusty iron) is porous and allows the air to continue its corroding action. Rusting rates depend on many factors—moisture, acidity, and the nature of the iron alloy. The last factor is very important; stainless steel (an alloy of iron containing chromium, nickel, and small amounts of manganese) effectively resists rusting, but other iron alloys corrode rapidly.

Because the reaction between dissolved oxygen in water and the surface of cast iron is slow, plumbing pipes can last a long time before extensive corrosion sets in. However, competent plumbers carefully avoid connecting brass or copper directly to iron pipes. This contact catalyzes iron's corrosion by an electrolytic mechanism. The reduction of oxygen,

$$4\ e^- + 4\ H^+(aq) + O_2(aq) \rightleftharpoons 2\ H_2O$$

which is slow at the iron surface, is relatively rapid at the copper surface. The copper acts as a cathode, drawing electrons from the iron pipe and oxidizing it (Figure 15.12).

$$3\ H_2O + 2\ Fe(s) \rightleftharpoons Fe_2O_3(s) + 6\ H^+(aq) + 6\ e^-$$

Figure 15.12
(a) Electrolytic corrosion occurs when an iron pipe is connected directly to a copper pipe. (b) The zinc coating acts as a "sacrificial anode" to prevent corrosion in galvanized iron.

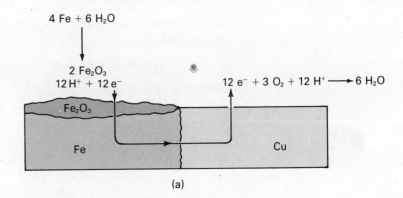

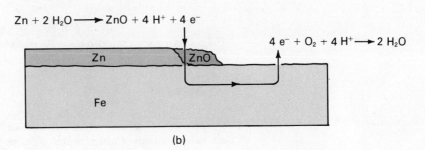

Electrolytic corrosion tends to occur when two different metals are in contact. The one with the higher oxidation potential corrodes rapidly while the one with the lower oxidation potential is protected against corrosion. This is the idea behind galvanized iron—iron coated with a zinc surface. At first, the zinc prevents contact between oxygen and iron, but its protecting effect lasts after oxygen has penetrated to the iron. Since zinc has the higher oxidation potential, it acts as a "sacrificial anode"; it supplies the electrons to reduce oxygen, and no iron corrodes as long as zinc is present.

Summary

1. A galvanic cell consists of two half-cells—a cathode and an anode. A reduction half-reaction occurs in the cathode and an oxidation reaction occurs in the anode when an external circuit connects the two half-cells. A porous barrier prevents contact between the oxidizing agent and reducing agent in the cell while allowing the migration of ions between the half-cells.

2. Cell notation: A single vertical line stands for an interface between two phases. A dashed line represents a liquid–liquid juncture. The anode appears on the left and the cathode on the right.

3. A standard reaction potential ($E°$) is the potential of a cell when the activity of each reactant and product within its half-cell equals one. The standard free energy of a reaction is

$$\Delta G° = -nE°\mathscr{F}$$

where n is the number of electrons transferred in the balanced equation and $\mathscr{F}$ is 9.65×10^4 coul.

4. The cell potential is the sum of the oxidation potential of the anode half-reaction and the reduction potential of the cathode half-reaction. The oxidation potential of the anode half-reaction is the negative of the reduction potential of the product of this reaction.

$$E°_{ox} = -E°_{red}$$

All half-cell potentials are measured against the reduction potential of the standard hydrogen electrode.

$$2\,e^- + 2\,H^+(aq) \rightleftharpoons H_2(g) \qquad E° = 0.000\ V$$

5. The Nernst equation describes the effect of concentrations of reactants and products, expressed as the mass-action (Q) ratio of a half-reaction or reaction, on half-reaction or reaction potentials.

$$E = E° - \frac{1.984 \times 10^{-4}\,T}{n} \log Q$$

or

$$E = E° - \frac{0.0592\ V}{n} \log Q \text{ at } 25°C$$

The Nernst equation can also relate the equilibrium constant to the standard potential of a reaction.

$$\log K_{eq} = \frac{nE°}{0.0592\ V}$$

6. The potential of a concentration cell depends on the difference in concentration between two half-cells, both of which utilize the same half-reaction. The measured potential of a concentration cell can be used to estimate concentrations.

7. Electrolysis is the use of an external potential to produce a thermodynamically unfavorable reaction. The decomposition potential is the minimum potential required for electrolysis. The decomposition potential equals $-E$ for the reaction plus the overpotential for the electrolysis. The size of the overpotential depends on the nature of the electrode reactions, the electrode surfaces, and the composition of the solution.

8. The Nernst equation applies to electrolyses.

$$E_{decomposition} = \text{Overpotential} - E° + \frac{0.0592 \text{ V}}{n} \log Q$$

9. A coulometric titration is based on the fact that the amount of titrant generated by an electrolysis is proportional to current (i) and time.

$$\text{Moles titrant} = \frac{it}{n\mathscr{F}}$$

By measuring the current and time needed to reach an end point, we can estimate the amount of unknown substance in a sample.

10. Electrolytic corrosion tends to occur when two different metals are in contact. The one with the higher oxidation potential corrodes rapidly while the one with the lower oxidation potential is protected against corrosion.

Vocabulary List

Daniell cell
galvanic cell
half-cell
anode
cathode
liquid–liquid juncture
salt bridge
electric potential
volt

standard reaction potential
half-cell potential
standard hydrogen electrode
Nernst equation
concentration cell
glass electrode
dry cell
fuel cell
electrolysis

decomposition potential
overpotential
effective reduction potential
electrowinning
electrorefining
electroplating
anodizing
electrolytic corrosion

Exercises

1. Define the following terms.
   a. Anode   b. Overpotential   c. Standard potential   d. Reduction potential
   e. Electrolysis   f. Fuel cell

2. Which of the following half-reactions could occur at the cathode of a cell?
   a. $2 \text{ Br}^-(aq) \longrightarrow \text{Br}_2(l) + 2 \text{ e}^-$
   b. $\text{Fe}^{2+}(aq) \longrightarrow \text{Fe}^{3+}(aq) + \text{e}^-$
   c. $\text{NO}_3^-(aq) + 3 \text{ H}^+(aq) + 2 \text{ e}^- \longrightarrow \text{HNO}_2(aq) + \text{H}_2\text{O}$
   d. $\text{AgI}(s) + \text{e}^- \longrightarrow \text{Ag}(s) + \text{I}^-$

3. Why do copper, silver, and gold occur naturally in their metallic states whereas potassium, magnesium, and aluminum do not?

4. Why are different products obtained when aqueous solutions of sodium chloride are electrolyzed instead of molten sodium chloride?

5. Describe by appropriate chemical equations, the corrosion that occurs when a copper and a steel hot-water pipe are connected.

6. a. Why is it a poor idea to use iron nails to hold aluminum roofing?
   b. Boating enthusiasts clamp pieces of magnesium to underwater brass fittings. What is the purpose of this procedure?

7. Electrolysis of CdSO₄ solution using a smooth cathode produces cadmium metal at the cathode. If the cathode surface is coated with finely divided platinum, hydrogen is the product even though all other conditions are identical to the first electrolysis. Explain.

8. Consider the metals Ag, Au, Ca, Cr, Cu, and Fe.
   a. Which of these metals is the most active toward oxidation?
   b. Which is the least active toward oxidation?
   c. Which of these metals react with H^+ to form H_2?
   d. Which of these metals can be formed by reducing their ions with H_2?

9. Hydrochloric acid and sulfuric acid solutions do not react with copper, but concentrated nitric acid corrodes it. Write an equation for a possible corrosion reaction involving copper and nitric acid and estimate its standard potential.

10. Why is the voltage available from an automobile battery less in cold weather than in hot?

11. Design galvanic cells based on the following reactions and write shorthand notations for them. Identify the cathode and anode for each cell and show the direction of electron flow and ion migrations.
    a. $2\ Ag^+(aq) + Fe(s) \rightleftharpoons Fe^{2+}(aq) + 2\ Ag(s)$
    b. $Zn(s) + Hg_2SO_4(s) \rightleftharpoons Zn^{2+}(aq) + SO_4^{2-}(aq) + 2\ Hg(l)$
    c. $H_2(g) + 2\ AgCl(s) \rightleftharpoons 2\ H^+(aq) + 2\ Cl^-(aq) + 2\ Ag(s)$

12. Write shorthand notations for galvanic cells based on the following reactions. Show the cathode, anode, direction of electron flow, and directions of ion migrations for each cell.
    a. $2\ Fe^{3+}(aq) + 3\ I^-(aq) \rightleftharpoons I_3^-(aq) + 2\ Fe^{2+}(aq)$
    b. $H_2(g) + Cl_2(g) \rightleftharpoons 2\ H^+(aq) + 2\ Cl^-(aq)$
    c. $4\ Cr^{2+}(aq) + O_2(g) + 4\ H^+(aq) \rightleftharpoons 4\ Cr^{3+}(aq) + 2\ H_2O$

13. What reaction occurs during the operation of each of the following cells?
    a. $Sn(s)|SnCl_4(aq)\ \vdots\ AuCl_3(aq)|Au(s)$
    b. $Fe(s)|Fe(OH)_2(s)|KOH(aq)|NiO_2(s)|Ni(OH)_2(s)|C(graphite)$
    c. $Pt(s)|H_2(g)|H_2SO_4(aq)|O_2(g)|Pt(s)$

14. Write an equation for the reaction occurring during the operation of each of the following cells.
    a. $Pt(s)|H_2(g)|HI(aq)\ \vdots\ HI(aq), I_2(aq)|Pt(s)$
    b. $Ag(s)|AgIO_3(s)|KIO_3(aq)||AgNO_3(aq)|Ag(s)$
    c. $Ag(s)|AgCl(s)|HCl(aq)|Cl_2(g)|Pt(s)$

15. What is the standard free energy change for the following reaction?

$$PtCl_4^{2-}(aq) + 2\ Ag(s) \rightleftharpoons Pt(s) + 2\ AgCl(s) + 2\ Cl^-(aq)$$

$E° = 0.504\ V$ Temperature $= 25°C$

16. The standard potential of the reaction

$$I_2(aq) + H_2SO_3(aq) + H_2O \rightleftharpoons 2\ I^-(aq) + 4\ H^+(aq) + SO_4^{2-}(aq)$$

is +0.37 V. What is the standard free energy change at 25°C?

17. The standard potential of a cell using the reaction

$$2\ MnO_4^-(aq) + 3\ Hg(l) + H_2O \rightleftharpoons 2\ MnO_2(s) + 3\ HgO(s) + 2\ OH^-(aq)$$

is 0.489 V at 25°C. The heat of this reaction is −450.6 kJ.
   a. What is the standard free energy for this reaction?
   b. What is the standard entropy change at 25°C?

18. The heat of the reaction

$$2\ Ni(s) + O_2(g) + 2\ H_2O \rightleftharpoons 2\ Ni(OH)_2(s)$$

is −504.2 kJ and its standard potential is 1.12 V at 25°C. Calculate:
   a. The standard free energy for this reaction
   b. The standard entropy change at 25°C

19. Estimate the standard *oxidation* potentials for the following half-reactions. (See Appendix I.)
    a. $Ni(s) \rightleftharpoons Ni^{2+}(aq) + 2\ e^-$
    b. $H_2SO_3(aq) + H_2O \rightleftharpoons SO_4^{2-}(aq) + 4\ H^+(aq) + 2\ e^-$
    c. $H_2(g) \rightleftharpoons 2\ H^+(aq) + 2\ e^-$

20. What are the standard *oxidation* potentials for the following half-reactions? (See Appendix I.)
    a. $Fe(CN)_6^{4-}(aq) \rightleftharpoons Fe(CN)_6^{3-}(aq) + e^-$
    b. $Ga(s) \rightleftharpoons Ga^{3+}(aq) + 3\ e^-$
    c. $2\ OH^-(aq) + Br^-(aq) \rightleftharpoons BrO^-(aq) + H_2O + 2\ e^-$

21. The cell

$$Pt(s)|H_2(g)|H^+(aq)\ \vdots\ H^+(aq), MoO_2^{2+}(aq), MoO^{3+}(aq)|Pt(s)$$

has a standard potential of 0.483 V. What is the standard reduction potential for the following half-reaction?

$$e^- + MoO_2^{2+}(aq) + 2\ H^+(aq) \rightleftharpoons MoO^{3+}(aq) + H_2O$$

22. A cell consisting of a standard calomel electrode as the anode and a standard cathode based on the half-reaction

$$e^- + 2\ H^+(aq) + ReO_4^-(aq) \rightleftharpoons ReO_3(s) + H_2O$$

has a potential of 0.54 V. What is the standard reduction potential of ReO_4^-?

EXERCISES

$\Delta G = -nFE_0$

23. Compute the standard potentials of the following reactions.
    a. $2\ Cr(s) + 3\ Cu^{2+}(aq) \rightleftharpoons 3\ Cu(s) + 2\ Cr^{3+}(aq)$
    b. $Fe(CN)_6^{3-}(aq) + Fe^{2+}(aq) \rightleftharpoons Fe(CN)_6^{4-}(aq) + Fe^{3+}(aq)$
    c. $Hg(l) + SO_4^{2-}(aq) + 4\ H^+(aq) \rightleftharpoons$
       $Hg^{2+}(aq) + H_2SO_3(aq) + H_2O$

24. Calculate the standard potentials for each of the following reactions.
    a. $6\ Fe^{2+}(aq) + Cr_2O_7^{2-}(aq) + 14\ H^+(aq) \rightleftharpoons$
       $6\ Fe^{3+}(aq) + 2\ Cr^{3+}(aq) + 7\ H_2O$
    b. $I_3^-(aq) + 2\ S_2O_3^{2-}(aq) \rightleftharpoons 3\ I^-(aq) + S_4O_6^{2-}(aq)$
    c. $IO_3^-(aq) + 5\ I^-(aq) + 6\ H^+(aq) \rightleftharpoons 3\ I_2(aq) + 3\ H_2O$
    d. $Br_2(l) + 2\ OH^-(aq) \rightleftharpoons OBr^-(aq) + Br^-(aq) + H_2O$

25. What are the reduction potentials of the following half-cells under the specified conditions? The temperature is 25°C in all cases.
    a. $2\ e^- + Cl_2(g) \rightleftharpoons 2\ Cl^-(aq)$
       $[Cl^-] = 0.10\ M;\ P_{Cl_2} = 0.10\ atm$
    b. $2\ e^- + Zn^{2+}(aq) \rightleftharpoons Zn(s)$ $[Zn^{2+}] = 0.050\ M$
    c. $6\ e^- + BrO_3^-(aq) + 3\ H_2O \rightleftharpoons Br^-(aq) + 6\ OH^-(aq)$
       $[BrO_3^-] = 5.0 \times 10^{-3}\ M;\ [Br^-] = 5.0 \times 10^{-3}\ M;\ pH = 9.50$

26. Calculate the potentials for the following half-cells. The temperature is 25°C in all cases.
    a. $2\ e^- + O_2(g) + 2\ H^+(aq) \rightleftharpoons H_2O_2(aq)$
       $P_{O_2} = 1.0\ atm;\ [H_2O_2] = 0.0100\ M;\ pH = 1.0$
    b. $2\ e^- + Hg^{2+}(aq) \rightleftharpoons Hg(l)$ $[Hg^{2+}] = 0.0020\ M$
    c. $2\ e^- + NO_3^-(aq) + H_2O \rightleftharpoons NO_2^-(aq) + 2\ OH^-(aq)$
       $pH = 11.50;\ [NO_3^-] = 0.200\ M;\ [NO_2^-] = 0.0500\ M$

27. Estimate the potentials of the following reactions under the given conditions. The temperature is 25°C in all cases.
    a. $2\ H_2(g) + O_2(g) \rightleftharpoons 2\ H_2O$
       $P_{H_2} = 0.0040\ atm;\ P_{O_2} = 0.0020\ atm;\ pH = 7.00$
    b. $2\ H^+(aq) + 2\ MnO_4^-(aq) + 3\ H_2O_2(aq) \rightleftharpoons$
       $2\ MnO_2(s) + 3\ O_2(g) + 4\ H_2O$
       $[H_2O_2] = 0.0030\ M;\ [MnO_4^-] = 0.0020\ M;\ pH = 3.0;\ P_{O_2} = 1.00\ atm$

28. What are the potentials of the following reactions at the specified conditions? The temperature is 25°C in both cases.
    a. $14\ H^+(aq) + Cr_2O_7^{2-}(aq) + 6\ Cl^-(aq) \rightleftharpoons$
       $3\ Cl_2(g) + 2\ Cr^{3+}(aq) + 7\ H_2O$
       $[Cr_2O_7^{2-}] = 0.100\ M;\ [Cr^{3+}] = 0.100\ M;\ [Cl^-] = 0.100\ M;$
       $pH = 2.0;\ P_{Cl_2} = 1.00\ atm$
    b. $Sn^{4+}(aq) + 2\ Fe^{2+}(aq) \rightleftharpoons Sn^{2+}(aq) + 2\ Fe^{3+}(aq)$
       $[Sn^{4+}] = 0.500\ M;\ [Fe^{2+}] = 0.500\ M;$
       $[Sn^{2+}] = 2.0 \times 10^{-3}\ M;\ [Fe^{3+}] = 2.0 \times 10^{-3}\ M$

29. Calculate the equilibrium constant at 25°C for
    $3\ HNO_2(aq) + 5\ H^+(aq) + Cr_2O_7^{2-}(aq) \rightleftharpoons$
    $2\ Cr^{3+}(aq) + 3\ NO_3^-(aq) + 4\ H_2O$

30. Calculate the equilibrium constant for the following reaction at 25°C.
    $2\ K(s) + 2\ H_2O \rightleftharpoons 2\ K^+(aq) + 2\ OH^-(aq) + H_2(g)$

31. The cell
    $Pt(s)|H_2(g)(1\ atm)|pH\ 7.00\ buffer\ |aqueous\ solution|H_2(g)(1\ atm)|Pt(s)$
    has a potential of 0.065 V. What is the pH of the solution in the cathode?

32. The potential of the following cell is −0.032 V. What is the pH of the solution in the anode?
    $Pt(s)|H_2(g)(1\ atm)|aqueous\ solution|pH\ 4.00\ buffer|H_2(g)(1\ atm)|Pt(s)$

33. What is the potential and direction of electron flow in the following cell?
    $Cu(s)|Cu^{2+}(aq)(2.0 \times 10^{-4}\ M)|Cu^{2+}(aq)(0.050\ M)|Cu(s)$

34. Calculate the potential and show the direction of flow for the following cell.
    $Ni(s)|Ni^{2+}(aq)(0.200\ M)|Ni^{2+}(aq)(3.00 \times 10^{-3}\ M)|Ni(s)$

35. Identify the cathode and anode half-reactions in each of the following electrolyses.
    a. $2\ NaCl(l) \longrightarrow 2\ Na(l) + Cl_2(g)$
    b. $2\ Cu^{2+}(aq) + 2\ H_2O \longrightarrow 2\ Cu(s) + O_2(g) + 4\ H^+(aq)$

36. Write equations for the half-reactions occurring at each electrode in the following electrolyses.
    a. $2\ H^+(aq) + 3\ I^-(aq) \longrightarrow H_2(g) + I_3^-(aq)$
    b. $2\ NaH(l) \longrightarrow 2\ Na(l) + H_2(g)$

37. Use reduction potentials to decide whether PbO_2 or MnO_2 can be used to prepare chlorine gas from chloride ion in acid solution.

38. a. Using reduction potentials, predict whether H_2O_2 in an acid solution is a strong enough oxidant to convert HI to I_3^-.
    b. Can Zn^{2+} bring about the same reaction?

39. How many grams of lead are consumed when a car battery discharges at a rate of 5.00 amp for 15.0 min? (Assume 100% efficiency.)

40. How many grams of zinc are used up when a standard Daniell cell is run at a current of 0.300 amp for 4.00 min? Assume an excess of $CuSO_4$ and that the cell operates at 100% efficiency.

41. Predict the products, if any, for the following reactions.
    a. $Cu(s) + Fe^{3+}(aq) \longrightarrow$
    b. $Al(s) + Mg^{2+}(aq) \longrightarrow$
    c. $Sn^{4+}(aq) + Cu^{2+}(aq) \longrightarrow$

42. What products, if any, form in the following reactions?
    a. $Hg_2^{2+}(aq) + Sn^{2+}(aq) \longrightarrow$
    b. $Ni^{2+}(aq) + Ag(s) \longrightarrow$
    c. $Fe^{3+}(aq) + I^-(aq) \longrightarrow$

43. The equilibrium constant for

$$Cl_2(g) + H_2O \rightleftharpoons H^+(aq) + Cl^-(aq) + HClO(aq)$$

is 4.66×10^{-4} at 25°C.
  a. Compute the standard potential for this reaction.
  b. Calculate the standard reduction potential for HClO.

44. When the electrolysis of a solution that is 1.00 M in Cu^{2+} and 1.00 M in H^+

$$2\ Cu^{2+}(aq) + 2\ H_2O \longrightarrow 2\ Cu(s) + O_2(g) + 4\ H^+(aq)$$

is run at a current density of 0.010 amp/cm² using a copper cathode and graphite anode, the decomposition potential is 1.41 V. What is the overpotential for this electrolysis?

45. The overpotential for hydrogen formation at a smooth iron cathode using a current density of 0.010 amp/cm² is 0.56 V. The overpotential for reduction of Fe^{2+} to iron metal is negligible at these conditions. What will be the dominant cathode reaction if 0.10 M Fe^{2+} solution whose pH is 2.0 is electrolyzed with a current density of 0.010 amp/cm² using a smooth iron cathode?

46. The mass of $K_2Cr_2O_7$ in a sample is determined by titrating it with Cu^+ generated coulometrically by reducing Cu^{2+} at a cathode.

$$e^- + Cu^{2+}(aq) \longrightarrow Cu^+(aq) \text{(cathode reaction)}$$

$$Cr_2O_7^{2-}(aq) + 14\ H^+(aq) + 6\ Cu^+(aq) \longrightarrow 2\ Cr^{3+}(aq) + 6\ Cu^{2+}(aq) + 7\ H_2O$$

It takes 550 sec using a current of 0.0400 amp to complete the titration. What is the mass of $K_2Cr_2O_7$ in the sample?

47. a. Find the standard free energy for each of the following reactions.

$$H_2(g) + Fe^{2+}(aq) \rightleftharpoons Fe(s) + 2\ H^+(aq)$$

$$\frac{1}{2}H_2(g) + Fe^{3+}(aq) \rightleftharpoons Fe^{2+}(aq) + H^+(aq)$$

$$\frac{3}{2}H_2(g) + Fe^{3+}(aq) \rightleftharpoons Fe(s) + 3\ H^+(aq)$$

(The standard potential for the Fe^{3+}-$Fe(s)$ half-reaction is not in Appendix I.)
  b. Calculate $E°$ for the Fe^{3+}-$Fe(s)$ half-reaction.
  c. Does $E°$ for the Fe^{3+}-$Fe(s)$ half-reaction equal the sum of the standard potentials of the Fe^{3+}-Fe^{2+} and Fe^{2+}-$Fe(s)$ half-reactions?

48. Calculate the maximum amount of useful work produced when 2.00 kg of oxygen gas is converted into water in a fuel cell. Consider the fuel cell to be 75% efficient in converting free energy into electrical energy.

49. The standard reduction potential for the half-reaction

$$4\ e^- + 4\ H^+(aq) + O_2(g) \rightleftharpoons 2\ H_2O$$

is +1.229 V. Calculate the standard reduction potential for

$$4\ e^- + 2\ H_2O + O_2(g) \rightleftharpoons 4\ OH^-(aq)$$

[Hint: $[H^+][OH^-] = 1.0 \times 10^{-14}$.]

50. It is possible to estimate the solubility product of a salt from the standard potentials of two half-cells, one involving a free ion and the other its slightly soluble salt. Find the solubility product of mercury(I) iodide

$$Hg_2I_2(s) \rightleftharpoons Hg_2^{2+}(aq) + 2\ I^-(aq)$$

from the following potentials.

$$2\,e^- + Hg_2^{2+}(aq) \rightleftharpoons 2\,Hg(l) \quad E° = +0.789\text{ V}$$

$$2\,e^- + Hg_2I_2(s) \rightleftharpoons 2\,Hg(l) + 2\,I^-(aq) \quad E° = -0.041\text{ V}$$

[*Hint:* The standard potential of the second half-reaction equals the potential of the first when $[Hg_2^{2+}] = K_{sp}/[I^-]^2$.]

51. We can measure the amount of iron in a sample by converting it to Fe^{2+} ion and titrating it with a standard dichromate solution.

$$Cr_2O_7^{2-}(aq) + 14\,H^+(aq) + 6\,Fe^{2+}(aq) \longrightarrow 2\,Cr^{3+}(aq) + 7\,H_2O + 6\,Fe^{3+}(aq)$$

The end point indicator is colorless in its reduced form, violet in its oxidized form, and its reduction potential is 0.87 V.

$$2\,e^- + 2\,H^+(aq) + Ind(aq) \rightleftharpoons H_2Ind(aq)$$
$$\text{Violet} \text{Colorless}$$

a. What will be the $[Fe^{3+}]/[Fe^{2+}]$ ratio when the $[Ind]/[H_2Ind]$ is 0.10? The titration solution is 1.0 M in H^+.
b. What will be the iron(III)/iron(II) ratio when the Ind/H_2Ind ratio is 10?
c. According to your results from parts a and b, is this indicator suitable for the titration? (It does work in practice because phosphoric acid is added to complex Fe^{3+} and lowers its reduction potential.)

chapter 16
Molecular Spectroscopy

Vibrations of molecules such as CO_2 may be used to characterize and identify them

One of the most useful analytic tools available to chemists is **molecular spectroscopy**, the measurement of the wavelengths of electromagnetic radiation absorbed or emitted by molecules. Since each type of molecule has its own characteristic set of energy levels, it absorbs photons with a unique set of wavelengths. This collection of wavelengths distinguishes the molecule from all other types of molecules and provides a sort of spectroscopic fingerprint that allows chemists to identify the substance unambiguously. If a particular substance is present in a mixture, the spectrum of absorbed or emitted light will tell us so. This is a very powerful analytic technique—even a drop of an unknown liquid or a few grains of an unknown solid can be successfully analyzed, both qualitatively and quantitatively, using modern spectroscopic instruments.

The molecular spectrum of a substance provides us with detailed information about its molecules—their shapes, bond lengths and angles, the ways they vibrate, and the energy needed to distort their bonds. Since no one has ever seen the atoms within a molecule, its description in so much detail is a feat of deduction worthy of Sherlock Holmes. The clues are the frequencies of the light absorbed by the molecule, and the deductions involve sophisticated theories about the way light interacts with matter. We will not discuss these theories in detail, but we will point out the sort of information that can be gained from each type of spectroscopy.

16.1 The Measurement of Spectra

When a molecule *absorbs* a quantum of electromagnetic radiation (a photon) its energy *increases* by the energy of the photon.

$$\Delta E_{molecular} = h\nu \tag{16.1}$$

Depending on its energy, the photon can change the electron arrangement of the molecule, make it vibrate more vigorously, increase the rate at which it spins around its axes of rotation, and even break its bonds if the energy is sufficiently great.

$$\Delta E_{molecular} = \Delta E_{electronic} + \Delta E_{vibration} + \Delta E_{rotation}$$

All of these energies—electronic, vibrational, and rotational—are quantized, and the molecule can absorb only those photons whose energies match the difference between two molecular energy levels. The absorption spectrum of a substance tells us the spacings between its molecular energy levels.

Molecular **electronic energy levels** are relatively far apart (Figure 16.1) and each electronic level has associated with it a number of closely spaced **vibrational energy levels**. Likewise each vibrational level contains a large number of **rotational energy levels**.

Different energies of electromagnetic radiation produce different sorts of molecular energy transitions (Figure 16.2). Photons of **microwaves**—radiation ranging in wavelength from about 1 mm to 1 cm ($\nu \simeq 3 \times 10^{10}$–$3 \times 10^{11}$ Hz)—are not energetic enough to excite a molecule vibrationally or electronically. They can, however, increase the rotational energy of a molecule. **Infrared** radiation ($\nu \simeq 3 \times 10^{12}$–$10^{14}$ Hz) excites molecular rotations and vibrations, and photons in the **visible–ultraviolet** range ($\nu \simeq 10^{14}$–10^{16} Hz) can jolt an electron into a higher molecular orbital as well as change the vibrational and rotational energies of the molecule. It is convenient for us to divide the electromagnetic spectrum into the three frequency ranges, microwave, infrared, and visible-ultraviolet, because of the different types of information yielded by absorptions in each range and because the techniques for measuring each type of spectrum differ.

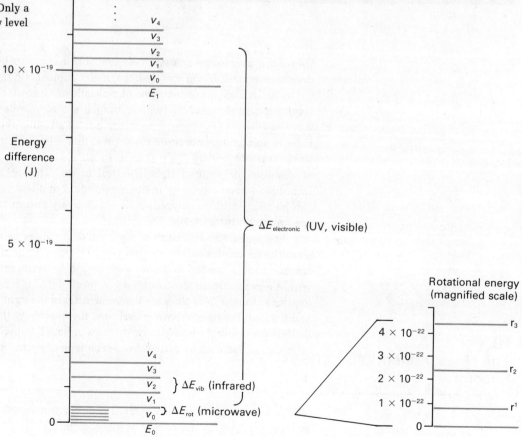

Figure 16.1
Molecular energy levels of CO (approximately to scale). Only a few of each type of energy level appear in this figure.

16.1 THE MEASUREMENT OF SPECTRA

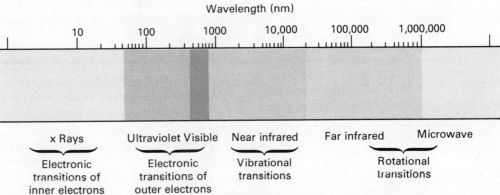

Figure 16.2
Electromagnetic radiation corresponding to the different types of energy transitions.

Spectrophotometers

A **spectrophotometer**, a device for measuring the ability of a substance to absorb light at various frequencies, consists of the following elements: (1) a light energy source, (2) a means of selecting a particular frequency from the source, and (3) a detector to measure the intensity of light passing through a sample of the substance. The nature of these elements and the details of their arrangement in the spectrophotometer vary for each frequency range and from manufacturer to manufacturer.

The Beer–Lambert Law

The absorbance at a particular wavelength depends upon the path length of the light beam through the sample, the concentration of the absorbing substance, and the intrinsic properties of this substance. The **Beer–Lambert law** describes the mathematical relationship between the **absorbance** (A), path length (l), concentration (c), and the **molar absorptivity** (ϵ), also called the **extinction coefficient**, of the absorbing substance.

$$A = \epsilon c l \tag{16.2}$$

If the substance under examination is contained in a solution or in a particular sample cell, then we must compare the intensity of light passing through the substance, its solvent, and cell (I) with the light intensity passing through an identical reference cell containing only the solvent (I_0). The negative log ratio of these two intensities is called the absorbance (A) of the substance at the particular wavelength.

$$A = -\log \frac{I}{I_0} \tag{16.3}$$

Example 16.1

The intensity of light passing through a dichromate solution (often called **transmittance**) is 38%. What is the absorbance?

Solution

Referring to equation (16.3), the ratio of I to I_0 is given as 38%, therefore converting percentage to a decimal gives

$$\frac{I}{I_0} = 0.38$$

Substituting into equation (16.3), we can solve for the absorbance.

$$A = -\log 0.38 = -\log(3.8 \times 10^{-1}) = -(0.58 - 1) = 0.42$$

The molar absorptivity, a characteristic property of the substance at a particular temperature and wavelength, is much more useful to us than the absorbance of any particular solution of that substance.

Example 16.2

The absorbance of 3.12×10^{-5} M myoglobin in a 1.00-cm cell is 0.9245 at 277 nm. What is the molar absorptivity of myoglobin at this wavelength?

Solution

According to the Beer–Lambert law, equation (16.2),

$$\epsilon = \frac{A}{cl} = \frac{0.9245}{(3.12 \times 10^{-5} M)(1.00 \text{ cm})}$$

$$= 2.96 \times 10^4 \, M^{-1} \, cm^{-1}$$

Once we know the value for the molar absorptivity of a substance at a particular wavelength, we can use the value to calculate the absorbance under different conditions of path length and concentration.

Example 16.3

Estimate the absorbance at 277 nm of 2.00×10^{-5} M myoglobin in a 0.50-cm cell.

Solution

Reapplication of the Beer–Lambert law gives

$$A = (2.96 \times 10^4 \, M^{-1} \, cm^{-1})(2.00 \times 10^{-5} M)(0.50 \text{ cm})$$

$$= 0.296$$

In other words,

$$-\log \frac{I}{I_0} = 0.296$$

$$\frac{I}{I_0} = 0.507$$

Thus 50.7% of the light goes through the sample.

Another useful application of the Beer–Lambert law is the estimation of the concentration of a substance from a knowledge of its molar absorptivity and absorbance.

Example 16.4

The molar absorptivity of permanganate ion at 520 nm is $2.50 \times 10^3 \, M^{-1} cm^{-1}$. The absorbance of a permanganate solution in a 0.50-cm cell is 0.742. Estimate the MnO_4^- concentration of this solution.

Solution

The form of the Beer–Lambert law that applies to this problem is

$$c_{MnO_4^-} = \frac{A}{\epsilon l} = \frac{0.742}{(2.50 \times 10^3 \, M^{-1} \, cm^{-1})(0.50 \text{ cm})} = 5.9 \times 10^{-4} M$$

Selection Rules

Absorption of radiation is possible whenever the energy of a photon matches the difference between two molecular energy levels, but photons with the proper energies often escape absorption by molecules they encounter. The molar absorptivity of a particular substance at a particular wavelength depends on the *probability* that a molecule will absorb a photon colliding with it. We can predict this absorption probability by considering the way the oscillating electromagnetic field of light perturbs the molecule. The theory of this interaction (time-dependent perturbation theory), which is much too complex to discuss now, indicates there are **selection rules** governing spectroscopic transitions. Some transitions are highly probable and produce intense absorptions. Others are called **forbidden transitions**, an awesome term that just means there is no effective way for the electromagnetic field of light to interact with the molecule and induce these energy jumps. As a result, these absorption probabilities are practically nil, and no corresponding absorption lines appear in the spectrum of the substance.

16.2 Rotational–Vibrational Spectra

When a polar molecule such as carbon monoxide absorbs a microwave photon, its *rotational* kinetic energy jumps to a higher quantum level.* Likewise, when a polar molecule absorbs a photon of infrared light (a more energetic photon than a microwave photon), the molecule jumps to a higher *vibrational* quantum level. A diatomic molecule, the simplest example of a molecular rotator–vibrator, can be likened to a barbell rotating about its center of mass [Figure 16.3, part (a)]; however, the real molecule is not rigid like the barbell, but can deform and vibrate back and forth. A better likeness of the diatomic molecule is a pair of spherical masses connected by a spring [Figure 16.3, part (b)], but this spring can vibrate only with certain fixed frequencies, and the molecule can rotate only at certain fixed speeds. Both the vibrations and rotations of a molecule are quantized.

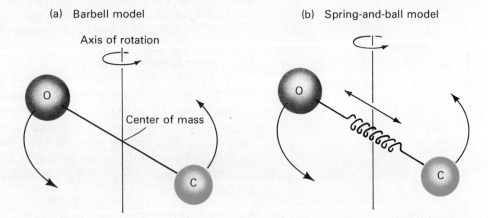

Figure 16.3
A diatomic rotor–vibrator, CO.

Molecular Rotations and Microwaves

The microwave spectrum of a polar diatomic molecule such as CO consists of a series of absorption lines with evenly spaced frequencies. Each of these lines represents a jump in the rotational energy of a molecule from a particular quantized level to the next higher level. The line frequencies depend on only two intrinsic molecular properties—the masses of its constituent atoms and its bond distance. Since we know the atomic masses, we can interpret the microwave absorption frequencies to determine the bond lengths with great accuracy.

A nonpolar diatomic molecule such as H_2 or Cl_2 does not absorb microwaves effectively. Although these molecules also rotate in quantized energy levels, they cannot interact readily with a photon to increase their rotational energies. This is because a rotating molecule must produce an oscillating electromagnetic field in

---

*The rotational changes of many molecules require higher energies than provided by microwaves, and their rotational absorptions appear in the far infrared range (3×10^{11}–10^{13} Hz).

order to be altered easily by light. A rotating polar molecule produces this field, but a nonpolar molecule does not.

More complicated molecules can have more complicated microwave spectra because there can be several axes around which rotation can occur, each axis giving rise to a series of absorption lines. A detailed analysis of such a complicated spectrum can yield detailed information about the shape and size of a molecule. Since the spectrum is the characteristic property of a particular substance, it can also be used to identify the substance. Microwaves from other planets have revealed the existence of many of the substances in the atmospheres of these planets.

Microwaves have many practical applications including radar (the location of distant objects through their reflection of microwaves) and that culinary revolution, the microwave oven. A microwave oven produces microwaves with some of the precise frequencies absorbed by water. This is done by making the oven's interior just the right size so that reflection and interference produce standing waves similar to those for a vibrating string (see Section 5.3) with the desired frequencies. Water molecules in the food absorb the microwaves, rotate faster, and then transfer this rotational energy to other molecules, such as protein molecules, by colliding with them. Since the microwaves penetrate deep into the food sample, the food is cooked evenly throughout. Water-free, nonmetallic objects such as a plastic wrapper or a paper plate do not absorb microwaves and are not burned. However metal foils or plates cannot be used in a microwave oven because the delocalized conduction-band electrons of the metal (see Chapter 9) absorb the microwaves and then quickly re-radiate them, damaging the oven.

Molecular Vibrations and Infrared Radiation

When a molecule absorbs infrared radiation, its bonds vibrate more violently. Molecules can, in this manner, convert infrared radiation to heat (molecular kinetic energy). This is why infrared lamps are the best lamps for transmitting heat. Since molecular vibrations are quantized, a substance absorbs infrared light whose frequency matches the energy difference between the ground state and first excited state of one of its molecular vibrations. For example, the infrared spectrum of CO (Figure 16.4) consists of a single line whose photon energy equals the difference between the two lowest vibrational states of CO.

The frequency of infrared absorption depends on the resistance of a bond to distortion, measured as a quantity called the **force constant**. It is easiest to get a feeling for what the force constant represents by considering a spring. A stiff spring has a large force constant—in other words, it responds to even a slight stretching or compression with a strong opposing force. Flexible springs are easy to distort and have small force constants. By measuring the force constant of a bond, we gain important information concerning the relationship between bond distance and its potential energy.

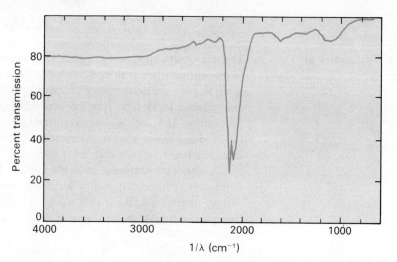

Figure 16.4
The infrared spectrum of CO.

Nonpolar diatomic molecules such as N_2 do not absorb infrared light although they too have quantized vibrational energies and their bonds have characteristic force constants. This is because radiation cannot excite the vibration of a molecule unless the vibrational change alters the dipole moment. Both the vibrational ground state of N_2 and its excited state are nonpolar, so the molecule cannot absorb infrared light. This is an important factor for our environment since the major components (N_2 and O_2) of our atmosphere do not absorb infrared light and so cannot prevent radiant heat from escaping the earth. The effectiveness with which our atmosphere traps radiant heat and thus helps control the earth's mean temperature depends on the concentrations of some of its minor components, in particular CO_2 and water vapor.

One feature of molecular vibrations may conflict with your common sense about motion and temperature—a molecule can never stop vibrating! No matter how much a sample is cooled, its molecules still retain some vibrational energy, called **zero-point energy**. The translational kinetic energy of a molecule falls to zero as the temperature approaches 0 K (the molecule does not move around), but its vibrational energy remains at the zero-point energy. This is one more example of the difference between the apparent behavior of visible objects and the quantized behavior of molecules.

Because triatomic and larger molecules have a variety of ways in which to vibrate (Figure 16.5), their infrared spectra (Figures 16.6 and 16.7) contain more bands than the single band of a diatomic molecule. The number of bands in the spectrum of a triatomic molecule can tell us whether the molecule is linear or bent. Molecules with complicated structures have very complicated spectra (Figure 16.8). These complex spectra provide an important analytic tool—a sort of fingerprint file for

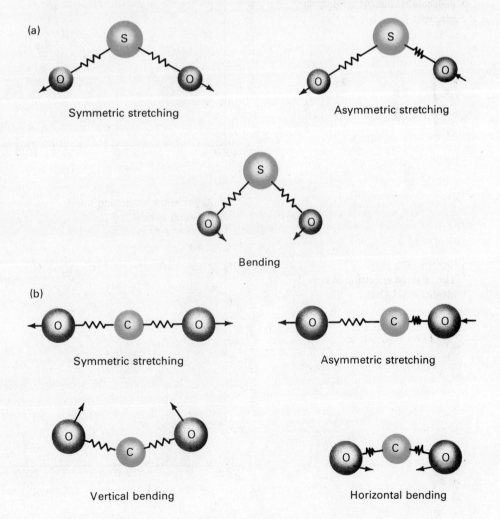

Figure 16.5
Fundamental vibration modes.
(a) In SO_2, a bent triatomic molecule, all vibrations absorb in the infrared because the molecule always has a net dipole moment.
(b) In a linear triatomic molecule such as CO_2, symmetric stretching does not absorb in the infrared because it does not give the linear molecule a net dipole moment. The two bending vibrations have the same energies.

Figure 16.6
The infrared spectrum of SO_2.

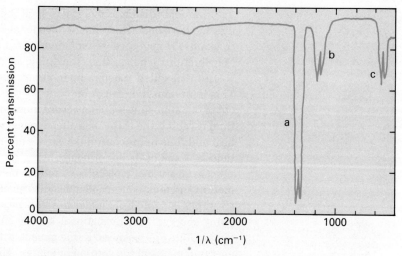

a: Asymmetric stretching band
b: Symmetric stretching band
c: Bending band

Figure 16.7
The infrared spectrum of CO_2. The symmetric stretching motion of CO_2 does not absorb because the molecule retains its zero dipole moment.

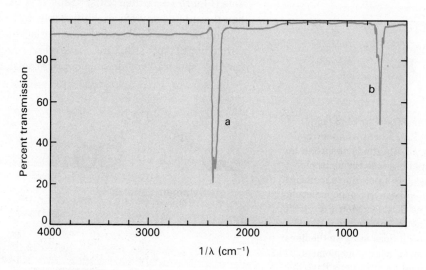

a: Asymmetric stretching band
b: Bending band

Figure 16.8
The infrared spectrum of benzyl alcohol (C_7H_7OH).

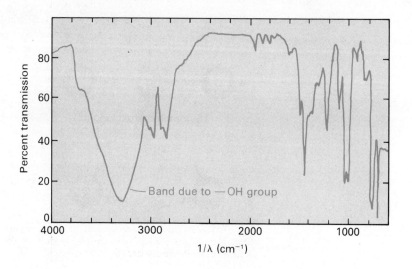

Figure 16.9
The effect of physical state on electronic spectra. (a) Benzene vapor. (b) Benzene dissolved in hexane.

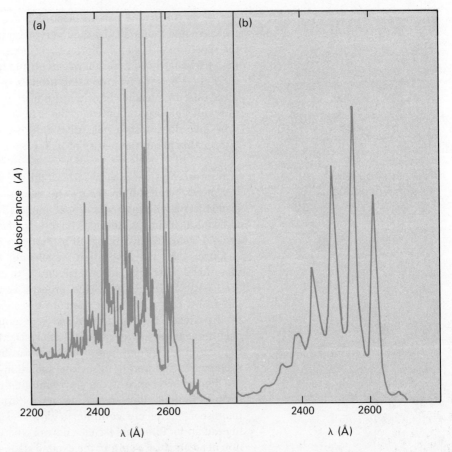

molecules—for identifying substances and deducing molecular structures. Because the infrared energy absorbed by a vibrating chemical bond is characteristic of that type of bond, a particular absorption frequency can reveal the presence of a particular group of atoms within the molecule. For example, an O—H group always absorbs near 1.0×10^{14} Hz* due to the stretching vibration of the O—H bond. The spectrum in Figure 16.7 also contains bands corresponding to O—H bending and both the stretching and bending vibrations of the C—O bond and the various types of C—C and C—H bonds in the molecule. Thus the infrared spectrum of an unknown substance can be used to identify characteristic groups present in the molecule by their particular absorption frequencies. Furthermore, each substance has its own characteristic infrared spectrum by which we can distinguish it from any other substance. Infrared spectroscopy has proven to be a most useful and rapid means of chemical identification, particularly for organic molecules.

16.3
Electronic Spectra

Ultraviolet–Visible Spectra

The absorption of visible or ultraviolet light by a molecule kicks one of its electrons to a higher energy level. Because each excited state has many vibrational and rotational energy levels, ultraviolet–visible spectra of gases consist of complex clusters of absorption lines. In the liquid or in solution, however, these unique bands are merged together because any one molecule now has neighboring molecules that interact by applying changing local electric fields, and damp out specific rotational bands (Figure 16.9).

*It is standard practice to report infrared absorption frequencies in terms of their *wave numbers*—the number of waves per cm. To compute a band's frequency in Hz, multiply its wave number (in cm^{-1}) by the speed of light in cm/sec (3.00×10^{10} cm/sec).

Every substance has an electronic absorption spectrum. Dyes, pigments, and other colored materials, by their very nature, absorb energies in the visible region of the spectrum, but many colorless substances absorb in the near ultraviolet (180–400 nm), particularly compounds of carbon and hydrogen containing multiple bonds. For example, electrons in π bonding orbitals in these compounds absorb characteristic ultraviolet frequencies and jump to a π^* antibonding orbital. The absence of near ultraviolet absorption indicates that a hydrocarbon contains single bonds only. If the substance does absorb near ultraviolet light, the absorption wavelength gives us information about the nature of its multiple bonds.

Fluorescence and Phosphorescence

A **fluorescent** substance absorbs light at one wavelength and then rapidly emits longer wavelength light. This event occurs when absorption produces an electronic excited state with high vibrational energy (E_1^*). The excited state collides with other molecules, losing vibrational energy in the process (E_1), and then returns to the ground state (E_0) by giving off a photon (Figure 16.10). The emission wavelength is *longer* than the absorption wavelength because the energy loss from the low vibrational energy level is less than the original excitation energy, $(E_1^* - E_0) > (E_1 - E_0)$. This process is relatively rare, though, because most molecules can give up *all* their excess energy ($E_1^* - E_0$) to other molecules and return directly to their ground states without emitting a photon. Fluorescence is a useful phenomenon for chemical analysis. Many rocks—for example those containing uranium, thorium, and tungsten—can be identified by their particular fluorescence when exposed to ultraviolet radiation (also called "black light").

Phosphorescence is the slow emission of long wavelength light by an excited state. Although fluorescence occurs within a millisecond after excitation, phosphorescent matter can continue to glow several minutes after exposure to the activating radiation. Phosphorescence occurs when molecular collisions change the direction of an electron's spin in the excited state. In order for the molecule to return to its ground state, this electron spin must be reversed, but a reversal of an electron spin is a very improbable event and thus the decay of the excited state is delayed (Figure 16.11).

Chemical reactions can produce light by a process known as **chemiluminescence**. In chemiluminescent reactions, the initial product is an excited state of a product molecule, and light is emitted as this excited state decays to its ground state.

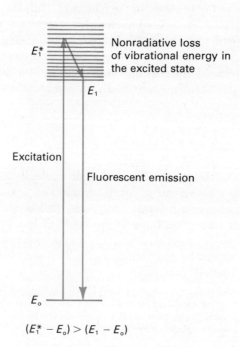

Figure 16.10
Fluorescence.

Figure 16.11
Phosphorescence.

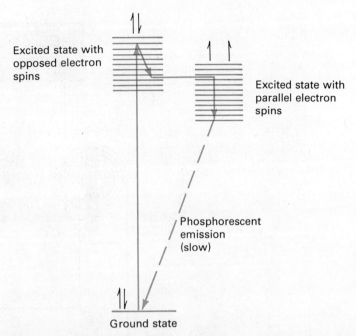

Some of the most fascinating examples of chemiluminescence are the complicated biochemical reactions used by fireflies, glowworms, certain marine animals, and some bacteria to produce light.

16.4 Nuclear Magnetic Resonance

Just as a negatively charged electron possesses spin and a resulting magnetic moment, many nuclei (which are positively charged) also possess spin and magnetic moments; however, these moments are much smaller in magnitude than those of electrons. A nucleus with a net spin can thus have **nuclear spin states** analogous to the spin states of an electron. In the absence of an external magnetic field these nuclear spin states have the same energy, but if a magnetic field is imposed upon the nucleus, the spin states will have different energies.

Consider protium, which is, you recall, a hydrogen atom with one proton and no neutrons in its nucleus. It has a nuclear spin value of ½ and in a strong external magnetic field the nuclei can either line up their magnetic moments with the field, $+½$, or opposed to it, $-½$. When a substance containing hydrogen is placed in a magnetic field and exposed to radiation whose energy corresponds to the difference between its nuclear spin states, it absorbs photons as its nuclei reverse their spins. Because the energy difference between the spin states is small, the frequency of the absorbed radiation is low (in the range for radio waves).

A **nuclear magnetic resonance (nmr)** spectrometer (Figure 16.12) measures absorptions due to changes in nuclear spins. A tube containing a sample of a hydrogen compound lies between the poles of a strong magnet. Two coils, one transmitting radio waves of a particular frequency (often 6.0×10^7 Hz) and the other receiving the radiation, surround the sample tube. The magnetic field strength is changed gradually until it produces an energy difference between the spin states corresponding to the radiofrequency.* When this happens, the nuclear spins flip over, the molecules absorb the radio waves, and a recorder displays the decrease in energy reaching the receiving coil.

At a particular radiofrequency, all protons require the same magnetic field strength for nuclear magnetic resonance, but the *applied* field needed to induce

---

*Some nmr spectrometers maintain a constant magnetic field and vary the radiofrequency.

Figure 16.12

Nuclear magnetic resonance spectrometer.

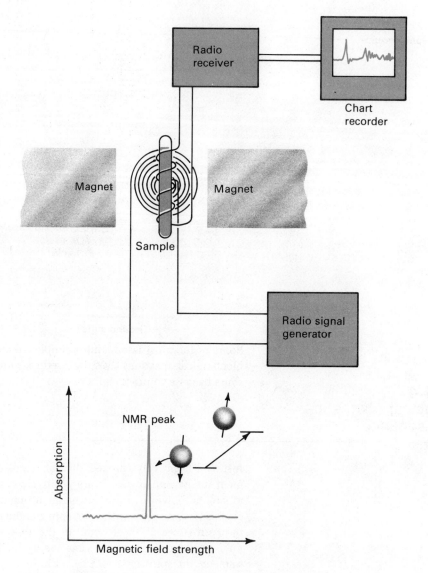

resonance depends on the molecular environment of the proton. This is because the immediate surroundings modify the magnetic field acting on the proton. The electrons surrounding a nucleus insulate it against the applied field, and we must apply stronger magnetic fields to reverse the spins of the better-insulated nuclei. The other nuclei in the immediate vicinity also influence the magnetic field. Those nuclei spinning parallel to the field increase it, while those with opposed spins decrease it.

The dependence of nuclear magnetic resonance on the chemical environment of a proton makes nmr a powerful tool for elucidating molecular structures. Figure 16.13 shows how the nmr spectrum of isopropyl chloride (C_3H_7Cl) reflects its molecular structure. The presence of two absorption peaks in the spectrum shows that there are two types of hydrogen atoms, those in the CH_3 groups (A) and the single hydrogen on the carbon atom that has the chlorine atom attached (B). The relative intensities of these two peaks—six for the A peak to one for the B peak—corresponds to a structure containing six hydrogen atoms of the first type and only one of the second. The detailed structures of the peaks also match the molecular arrangement of isopropyl chloride. The A peak consists of two portions, one representing molecules whose B proton spins with the applied field, the other for molecules for which the B proton opposes the field. The complex appearance of the B peak arises from the various possible combinations of spins for the six A hydrogen atoms. The isomer of isopropyl chloride, n-propyl chloride, has a completely different nmr spectrum that is consistent with its different structure.

Figure 16.13
NMR spectrum of *iso-* and *n*-propyl chloride.

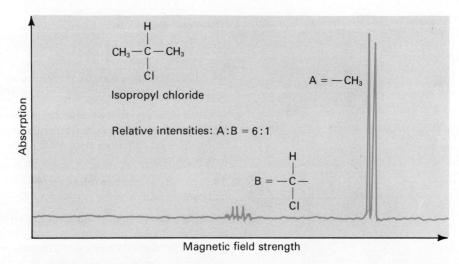

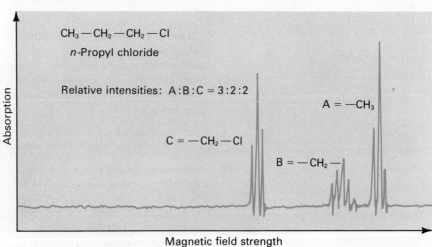

Many other nuclei beside ^{1}H can be studied by nmr—^{19}F, ^{13}C, ^{2}H, ^{17}O are examples—but they give weaker signals than protons and consequently more sophisticated and expensive instrumentation is required.

Summary

1. When a molecule absorbs a photon, its energy increases by the energy of the photon. The energy increase can change the electronic, vibrational, or rotational energies of the molecule, all of which are quantized.
2. Microwaves ($\nu \sim 3 \times 10^{10}$–$3 \times 10^{11}$ Hz) excite the rotational energy of a molecule. Infrared light ($\nu \sim 3 \times 10^{12}$–$10^{14}$ Hz) excites molecular vibrations, and visible–ultraviolet light ($\nu \sim 10^{14}$–10^{16} Hz) excites electrons from lower to higher molecular energy levels.
3. A spectrophotometer is a device for measuring the ability of a substance to absorb light at various frequencies. We record the light-absorbing ability of a solution as its absorbance (A)

$$A = -\log \frac{I}{I_0}$$

where I is the intensity of light passing through the solution and the cell that contains it, and I_0 is the intensity of light through an identical reference cell containing the solvent.

4. The Beer–Lambert law,

$$A = \epsilon c l$$

gives the relationship between the absorbance (A), molar absorptivity (ϵ) of the absorbing substance, concentration (c), and path length (l) of the light beam through the sample.

5. Selection rules govern the probabilities of spectroscopic transitions. Some transitions are highly probable and produce intense absorptions. Forbidden transitions have very low probabilities for photon absorption and no lines corresponding to these transitions appear in the spectrum.

6. There are several selection rules for rotational–vibrational spectra. A most important rule is that photons cannot excite rotations or vibrations of nonpolar diatomic molecules.

7. Microwave spectra supply information about bond distances as well as molecular shapes. Infrared spectra can be used to measure the force constants of bonds and to obtain information about the shapes of molecules. They also provide information about the groups present within molecules and help to identify molecules.

8. Absorption of ultraviolet–visible light by a molecule in the vapor phase results in changes in its electronic energy levels as well as its vibrational and rotational energies. In solution, much of the ultraviolet–visible spectrum's fine structure, which is due to vibrational and rotational changes, is lost. Carbon–hydrogen compounds containing double or triple bonds absorb light whose wavelength is longer than 180 nm as an electron jumps from a π bonding orbital to a π^* antibonding orbital. Carbon–hydrogen compounds containing single bonds only do not absorb above 180 nm.

9. A fluorescent substance absorbs light and reemits it at a longer wavelength. Phosphorescence is the slow emission of long wavelength light by an excited state.

10. Nuclear magnetic resonance (nmr) is the absorption of electromagnetic radiation in the radiofrequency range when a nucleus in a strong magnetic field jumps from one quantized spin state to another. At a particular radiofrequency, nuclear magnetic resonance for the various hydrogen atoms in a molecule may occur at several slightly different applied magnetic fields. These differences occur because the environment about each nucleus alters the net magnetic field acting on that nucleus. Important information about molecular structure can be obtained by careful analysis of the nmr spectrum.

Vocabulary List

molecular spectroscopy	Beer–Lambert law	force constant
vibrational energy level	absorbance	zero-point energy
rotational energy level	molar absorptivity	fluorescence
microwave	extinction coefficient	phosphorescence
infrared	transmittance	chemiluminescence
visible light	selection rules	nuclear spin states
ultraviolet radiation	forbidden transitions	nuclear magnetic resonance
spectrophotometer		

Exercises

1. Explain why infrared radiation is so effective in warming objects it strikes.

2. Explain the difference between fluorescence, phosphorescence, and chemiluminescence.

3. Radar sets function by sending out microwaves. Any object reflecting these microwaves shows up on the radar screen. Why does not microwave absorption by the air render this procedure ineffective?

4. Some scientists are concerned about the "greenhouse effect"—a possible warming of the earth's climate due to increasing concentrations of carbon dioxide in the atmosphere. Why should carbon dioxide, a trace component of our atmosphere, have such a profound effect on the earth's temperature?

EXERCISES

5. Electromagnetic radiation with a photon energy of 5.0×10^{-20} J has what wavelength? To what region of the electromagnetic spectrum does this radiation belong?

6. What is the wavelength of a photon whose energy is 8.0×10^{-19} J? What sort of light (IR, visible, or UV) is this?

7. What is the photon energy of electromagnetic radiation with the following wavelengths? What sort of molecular processes occur on absorption of these radiations?
   a. 3.0×10^{-4} m   b. 500 nm

8. Calculate the photon energies for the following wavelengths and describe what happens to molecules when they absorb these radiations.
   a. 7000 nm   b. 250 nm

9. The lowest frequency band in the rotational spectrum of hydrogen iodide occurs at 3.85×10^{11} Hz. What is the difference in energy between the two rotational energy levels involved in this absorption?

10. Carbon monoxide has a rotational spectrum whose lowest frequency band occurs at 1.15×10^{11} Hz. How much does the rotational energy of carbon monoxide increase when it absorbs this frequency?

11. What is the absorbance of a solution that transmits:
    a. 100% of the light?   b. 20% of the light?

12. Calculate the absorbance of a solution that transmits:
    a. No light   b. 75% of the light

13. The absorbance of a solution at 500 nm is 0.70. What percentage of 500 nm light does the sample transmit?

14. The absorbance of a sample is 1.25 at 350 nm. What percentage of the incident light does the sample transmit at 350 nm?

15. The absorbance at 370 nm of a solution of K_2CrO_4 (0.0400 g/L) in a 1.00-cm cell is 0.987. Chromate ion is the absorbing species. Estimate the value of the molar absorptivity of CrO_4^{2-}.

16. A solution of 1.75 mg of styrene (C_8H_8) in an ether solution whose total volume is 100 mL has an absorbance of 2.02 at 244 nm. The path length of the spectrophotometer cell is 1.00 cm. What is the value of the molar absorptivity of styrene at 244 nm?

17. The molar absorptivity of ferricyanide ion, $Fe(CN)_6^{3-}$, at 436 nm is 743 $M^{-1}cm^{-1}$. What percentage of light does $Fe(CN)_6^{3-}$ transmit at 436 nm if its concentration is 3.00×10^{-3} M, and the path length is 0.200 cm?

18. The molar absorptivity of aniline (C_6H_7N) at 280 nm is $8.60 \times 10^3 M^{-1} cm^{-1}$. A 1.50×10^{-4} M solution of aniline in water is placed in a 0.500-cm cell. What percentage of 280 nm light will this solution transmit?

19. A 1.00-cm thick sample of a $KMnO_4$ solution has an absorbance of 0.470 at 520 nm. What will be the absorbance and the percent transmission of 520 nm light if:
    a. The solution is diluted threefold but the measurement is still made in a 1.00-cm cell?
    b. The original solution is measured in a 3.00-cm cell?

20. What is the effect on the absorbance of a solution if
    a. The path length of the cell is halved?
    b. The concentration of the solution is doubled?

21. How many nmr peaks would you expect to see in a spectrum of each?
    a. Ethanol, CH_3CH_2OH b. Methyl ether, CH_3OCH_3

22. How many peaks would you expect to see in the nmr spectrum of each?
    a. CH_3CH_2Cl b. $CH_3CH_2CO_2CH_3$

23. There are two peaks in the nmr spectrum of ethyl ether, $CH_3CH_2OCH_2CH_3$. What are the relative intensities of these peaks?

24. Trimethylacetic acid, $(CH_3)_3CCO_2H$, has two nmr peaks. What are their relative intensities?

---

25. A sensitive method for detecting traces of iron(II) ion in solution is to add 1,10-phenanthroline (abbreviated phen), a reagent that reacts with iron(II) to form an intensely colored product.

    $$Fe^{2+} + 3\ phen \longrightarrow Fe(phen)_3^{2+}$$

    The molar absorptivity of the product is $1.11 \times 10^3 M^{-1} cm^{-1}$ at 508 nm. When 1.00 mL of an unknown solution is treated with excess 1,10-phenanthroline solution and enough water to bring the total volume to 100 mL, the absorbance at 508 nm is 0.453 measured in a 1.00-cm cell. What is the iron(II) concentration in the unknown solution?

26. Hydrogen peroxide forms a yellow complex with TiO^{2+}. The molar absorptivity of this complex is 643 $M^{-1} cm^{-1}$ at 400 nm, based on the assumption that the complex contains one titanium atom. When 2.00 mL of a solution is treated with enough of an excess of H_2O_2 to convert TiO^{2+} to its complex and then diluted to 25.0 mL, the resulting solution has an absorbance of 0.254 at 400 nm in a 1.00-cm cell. What mass of titanium was present in the 2.00 mL of solution?

27. The ultraviolet spectrum of benzene contains intense bands at 183 nm, 208 nm, and 263 nm. Make an energy level diagram showing the ground state of the benzene molecule and the excited states involved in these absorptions.

28. A microwave oven operates as a resonance cavity in much the same manner as does a vibrating string or an organ pipe. The frequency of microwaves emitted is usually about 2.5×10^9 Hz. Do you think a microwave oven could be made with an internal dimension of about 14 in.? If so, what precise dimension would you choose?

29. The rotational spectrum of HCl prepared from ordinary chlorine consists of two sets of equally spaced bands. Analysis of these absorption frequencies indicates that the sample contains two types of molecules, both with the same bond distances but differing in mass. Explain this phenomenon.

30. Draw the fundamental vibration modes for

   a. N_2 b. $\ddot{S}=C=\ddot{S}$ c. H_2O

31. Carbon disulfide has two intense infrared bands, a carbon–sulfur stretching band at 4.57×10^{13} Hz and a bending band at 1.19×10^{13} Hz. Is CS_2 bent or linear? Explain your answer.

32. How many intense infrared absorption bands would you expect for SCl_2 if its structure were the following? [*Hint:* Refer to Figure 16.5.]

   a. S with two Cl bonded (bent)   b. :Cl̈—S̈—Cl̈:

33. The absorbances at 592 nm of a series of solutions of bromphenol blue (a pH indicator) were measured at various pH values. Each solution had the same total concentration of dissolved bromphenol blue, and the absorbances were measured in identical cells. The following data were obtained.

pH	A	pH	A
2.00	0.00	4.00	0.98
3.00	0.18	4.40	1.43
3.60	0.58	9.00	2.02

   a. Which species, the conjugate acid or the conjugate base of bromphenol blue, absorbs at 592 nm?
   b. Estimate the value of the dissociation constant (K_{Ind}) for bromphenol blue.
   c. Predict the absorbance at pH 5.00.

34. The rate of decomposition of *p*-nitrophenyl acetate ($CH_3CO_2C_6H_4NO_2$) in alkaline solution

   $$CH_3CO_2C_6H_4NO_2(aq) + 2\ OH^-(aq) \longrightarrow CH_3CO_2^-(aq) + NO_2C_6H_4O^-(aq) + H_2O$$

   can be measured because *p*-nitrophenylate ion ($NO_2C_6H_4O^-$) absorbs strongly at 400 nm ($\epsilon = 1.9 \times 10^4\ M^{-1}\ cm^{-1}$). In a buffered solution, the rate of the reaction is first order with respect to *p*-nitrophenyl acetate. When $1.00 \times 10^{-4}\ M$ *p*-nitrophenylacetate has been exposed to a pH 9.72 buffer for 120 seconds, the absorbance in a 1.00-cm cell is 0.256.
   a. What concentration of *p*-nitrophenyl acetate remains after 120 seconds?
   b. Calculate the first-order rate constant for this reaction at pH 9.72.

chapter 17
Hydrogen, Oxygen, and Water

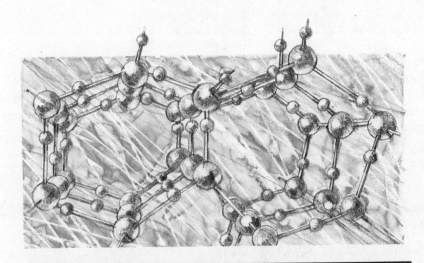

Ice is a unique, open-structured solid; it is more open (less dense) than the liquid and is held together by hydrogen bonds

One of the reasons we devote so much attention to hydrogen and oxygen is that these two elements form binary compounds with almost every other element. Consequently, a study of the hydrides and oxides of the various elements illustrates the periodic law. In addition, hydrogen and oxygen react with each other to form water, a deceptively simple molecule whose unusual properties are necessary for the development and maintenance of life.

17.1 Hydrogen

Scientists estimate hydrogen to be the most abundant element in our vast universe. The hydrogen atoms in "empty space" alone outnumber all other types of atoms combined. Fifteen percent of the atoms on the earth's crust* are hydrogen. However, because the atomic mass of hydrogen is so small, its contribution to the total mass of the earth's crust is slight—less than 1%. Since the mass of the stable elemental form of hydrogen (H_2) is so small, all the hydrogen gas in the earth's atmosphere either escaped the earth's gravitational pull or was chemically combined during the early development of our planet. Almost all the hydrogen present on earth today is present as compounds, in particular such essential substances as water, petroleum and natural gas, and the organic substances that comprise plant and animal matter.

The importance of hydrogen in our biosphere needs little emphasis—all living organisms require water as well as the more complicated hydrogen-containing

---
*The earth's crust is considered to be the cold portion near the earth's surface, the surface waters, and the lower atmosphere.

molecules such as proteins, carbohydrates, fats, and nucleic acids. Life as we know it would not be possible if it were not for the special properties of these molecules, properties that depend on the presence of hydrogen atoms (notably, hydrogen bonds).

The behavior of hydrogen illustrates some of the theoretical concepts we have discussed earlier. We will see how the unique chemical behavior of hydrogen relates to its simple electron configuration, $1s^1$, and we will examine the different properties of hydrogen compounds.

The Preparation of H_2

The stable form of elemental hydrogen, the diatomic molecule H_2, can be prepared in a number of ways. One process involves passing steam over hot coke to produce water gas, a mixture of carbon monoxide and hydrogen.

$$H_2O(g) + C(s) \xrightarrow{800°C} \underbrace{CO(g) + H_2(g)}_{\text{Water gas}}$$

Very little hydrogen is still prepared in this manner because it is quite impure (only 50% H_2) and it is very difficult to separate mixtures of carbon monoxide and hydrogen. Water gas was used at one time as an industrial fuel; now large quantities are converted into methanol, CH_3OH.*

$$CO(g) + 2\ H_2(g) \xrightarrow[\substack{3\text{–}500\text{ atm,} \\ \text{Catalyst}}]{400°C} CH_3OH(g)$$

In another commercial preparation, reasonably pure hydrogen forms when steam is passed over hot iron,

$$3\ Fe(s) + 4\ H_2O(g) \longrightarrow Fe_3O_4(s) + 4\ H_2(g)$$

The purest hydrogen is made by electrolysis of water.

$$2\ H_2O(l) \xrightarrow{\text{Electrolysis}} 2\ H_2(g) + O_2(g)$$

The commercial process uses concentrated sodium hydroxide or potassium hydroxide as the electrolyte and iron or nickel electrodes (Figure 17.1). Hydrogen forms at the cathode and oxygen at the anode. Other industrial electrolyses form large quantities of hydrogen gas as a by-product. For example, when brine solutions are electrolyzed to form chlorine gas, hydrogen is produced at the cathode as a reduction product.

$$2\ NaCl(aq) + 2\ H_2O \xrightarrow{\text{Electrolysis}} Cl_2(g) + H_2(g) + 2\ NaOH(aq)$$

The current major industrial source of hydrogen is petroleum refining. Large amounts are produced by steam cracking of natural gas (which is mostly methane)

$$CH_4(g) + 2\ H_2O(g) \xrightarrow[\text{Catalyst}]{\text{Heat}} CO(g) + 4\ H_2(g)$$

and by cracking of crude oil, a group of reactions in which large molecules of carbon and hydrogen are broken down into smaller fragments to make gasoline, industrial

---

*Methanol is a combustible fuel, currently used in racing and high-performance cars, that could possibly replace or augment gasoline for automobiles. It would pollute the atmosphere less, but its current price is excessive. Recent developments in homogeneous catalysis indicate that certain metal–organic substances catalyze the conversion of CO to CH_3OH at relatively low temperatures and pressures. These processes could dramatically lower methanol's price in the near future.

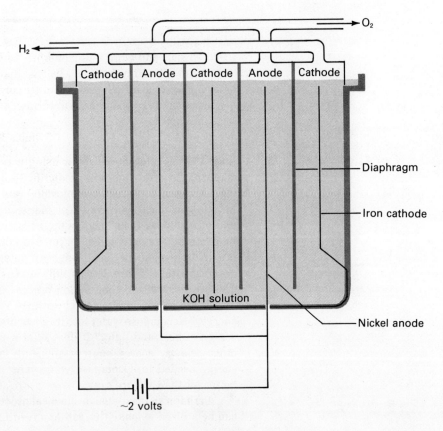

Figure 17.1
Commercial electrolysis cell for H_2 and O_2 preparation from H_2O.

fuels, and other important substances. We will represent a typical cracking process with the following unbalanced equation.

$$C_{14}H_{30}(g) \xrightarrow[\text{Catalyst}]{400\text{--}500°C} C_5H_{10}(g), C_4H_8(g), \text{etc.} + H_2(g)$$

Most of the hydrogen prepared in this way is used for industrial syntheses, in particular the formation of ammonia from nitrogen and hydrogen.

The attractiveness of hydrogen as a fuel is growing steadily as petroleum deposits become depleted. One very desirable feature of hydrogen fuel is that its combustion product (water) does not pollute the atmosphere. The prospect that our technology may one day be based upon hydrogen rather than oil has stimulated research on more effective ways to obtain hydrogen from sources other than petroleum.

Small-scale preparations of hydrogen in the laboratory include reactions of water or acids with active metals such as sodium or zinc,

$$2\ Na(s) + 2\ H_2O \longrightarrow 2\ Na^+(aq) + 2\ OH^-(aq) + H_2(g)$$

$$Zn(s) + 2\ H^+ \longrightarrow Zn^{2+}(aq) + H_2(g)$$

or reaction of aluminum with strongly basic solutions.

$$6\ H_2O + 2\ Al(s) + 2\ OH^- \longrightarrow 3\ H_2(g) + 2\ Al(OH)_4^-$$

Another convenient source of small quantities of hydrogen is the reaction of an ionic hydride salt with water.

$$CaH_2(s) + 2\ H_2O \longrightarrow H_2(g) + Ca(OH)_2(s)$$

Properties

Hydrogen gas is a colorless, odorless, and highly combustible substance. Since it takes 434 kJ to dissociate 1 mol of hydrogen into its atoms (see Table 17.1),

> "... and what will men burn when there is no coal?"
> "Water. Yes, my friends, I believe that one day water will be employed as a fuel, that hydrogen and oxygen which constitute it, used singly or together, will furnish an inexhaustible source of heat and light."
>
> Jules Verne, *The Mysterious Island*
>
> Jules Verne's suggestion about using water as a fuel may be one more example of his shrewdness in predicting technological trends. We are consuming petroleum deposits at an alarming rate, and the millions of years needed for the geological conversion of plant and animal remains into petroleum and other "fossil fuels" makes their replenishment a practical impossibility. In contrast, water can be recycled quickly—the decomposition of water into hydrogen and oxygen produces a potent fuel mixture that reforms water on combustion. The water rapidly makes its way back to streams, lakes, and oceans, and the combustion process does not form any pollutants.
>
> There is, of course, a catch to Jules Verne's prediction. It takes a lot of energy to decompose water into its elements, so water is not "an inexhaustible source of heat and light." It does offer a potentially useful way of utilizing other primary energy sources efficiently and without pollution, but if the energy needed to decompose water must come from petroleum, we are no better off than we are now.
>
> Perhaps an inexpensive chemical process, solar energy, or nuclear fusion can be utilized to make this rather adventuresome idea a practical, workable one.

monatomic hydrogen occurs only at very high temperatures, in an electric discharge, or at reactive surfaces. Because diatomic hydrogen has such a large bond-dissociation energy, reactions of hydrogen, even the exothermic ones, tend to have large activation energies and are slow. This combination of high activation energy and a large negative heat of reaction explains why a hydrogen–oxygen mixture does not react until we ignite it, then it explodes. The high temperature in the ignition spark is enough to cause some reaction,

$$2\ H_2(g) + O_2(g) \longrightarrow 2\ H_2O(g) \qquad \Delta H° = -485\ kJ$$

and the heat released raises the temperatures of the surrounding mixture, causing more reaction and generating still more heat. This snowballing effect releases an unpleasant amount of energy in a dangerously short time—an explosion.

Hydrogen usually acts as a reducing agent although in aqueous solutions it is only a moderate reductant.

Table 17.1 Some Physical Properties of Hydrogen

Molecular hydrogen		Atomic hydrogen	
Melting point	14.1 K	Ionization potential	1305 kJ/mol
Boiling point	20.3 K	Electron affinity	72.7 kJ/mol
Covalent bond radius	0.37 Å	Bohr radius	0.53 Å
Bond energy	434 kJ/mol		

17.1 HYDROGEN

$$H_2(g) + 2\ H_2O \rightleftharpoons 2\ H_3O^+(aq) + 2\ e^- \qquad E°_{oxidation} = 0.00\ V$$

A stream of hydrogen gas can reduce copper(II) ion to copper metal,

$$H_2(g) + Cu^{2+}(aq) \rightleftharpoons 2\ H^+(aq) + Cu(s) \qquad E° = +0.34\ V$$

but not aqueous solutions of zinc ion.

$$H_2(g) + Zn^{2+}(aq) \rightleftharpoons 2\ H^+(aq) + Zn(s) \qquad E° = -0.76\ V$$

However, at higher temperatures a stream of hydrogen can reduce the solid oxides of many active metals.

$$H_2(g) + ZnO(s) \xrightarrow{Heat} Zn(s) + H_2O(g)$$

This last reaction occurs because hydrogen's oxidation product (water vapor) is blown away in the gas stream, and the removal of the product shifts the reaction equilibrium to the right (Le Chatelier's principle).

Hydrogen can oxidize very active metals to form ionic hydrides.

$$2\ Na(s) + H_2(g) \longrightarrow 2\ NaH(s)$$

$$Ca(s) + H_2(g) \longrightarrow CaH_2(s)$$

The crystal lattices of these ionic hydrides contain negative hydride ion, $H:^-$

Atomic hydrogen, unhindered by bond-dissociation energy, is much more reactive than H_2. It can be formed in small concentrations by a direct-current arc or high-energy irradiation of hydrogen gas, but it has a very short life time before recombining—less than half a second. It reduces all oxides except those of the most electronegative elements, and reacts with most metals to form hydrides.

Hydrogen Isotopes

There are three isotopes of hydrogen: **protium** (^1H), **deuterium** (^2H or D), and **tritium** (^3H or T).* Protium, whose nucleus is a proton, is by far the most abundant of the three isotopes, deuterium (one proton and one neutron in the nucleus) accounts for about 0.015% of the atoms, but only one out of every 10,000,000 hydrogen atoms is a tritium atom (one proton and two neutrons). Tritium is radioactive, decaying to ^3He by emitting a beta particle (a high-speed electron) from its nucleus.

$$^3_1H \longrightarrow\ ^3_2He +\ ^0_{-1}\beta$$

We can form compounds containing deuterium in place of ordinary hydrogen. For example, deuterium burns to form D_2O, an analog of H_2O called **heavy water**. It is also possible to synthesize molecules containing both protium and deuterium (for example, HOD, CH_3CO_2D, and CH_2DCO_2H).

The chemical and physical differences between hydrogen compounds and their deuterium analogs are subtle but significant. (See Table 17.2 for a comparison of the properties of H_2O and D_2O.) These differences, termed **isotope effects**, are due to *mass differences* in the two isotopes, and these effects are relatively large for hydrogen compounds. Other elements show isotope effects, but they are considerably smaller because the percentage difference between the masses of two heavy isotopes is small. (Compare the mass ratio of deuterium and protium with the mass ratio of ^{13}C and ^{12}C.)

One particularly important effect of nuclear mass is the **kinetic isotope effect**. Because a covalent bond formed by deuterium is slightly stronger than the corresponding bond with protium, an elementary reaction that breaks a deuterium covalent bond is *slower* than the same reaction involving a bond to protium. For example, the reaction

---

*The assignment of different letter symbols to isotopes of an element is an exception for hydrogen and is not the usual practice. Thus isotopes of oxygen are all called "oxygen."

17 HYDROGEN, OXYGEN, AND WATER

Table 17.2 Physical Properties of D_2O and H_2O

Property	H_2O	D_2O
Normal boiling point (°C)	100.0	101.4
Normal melting point (°C)	0.0	3.8
Density at 25°C (g/cm³)	0.997	1.105
Ionization constant (25°C)	1.0×10^{-14}	0.2×10^{-14}

Figure 17.2
Isotope effects, H_2 and D_2.

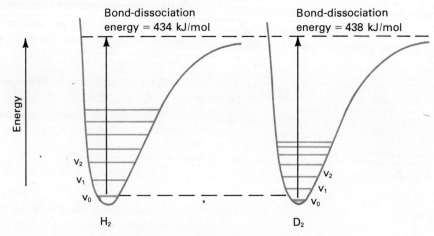

Molecular potential energy diagrams

$$Cl(g) + D_2(g) \longrightarrow DCl(g) + D(g)$$

at 25°C is about seven times slower than the reaction with H_2.

$$Cl(g) + H_2(g) \longrightarrow HCl(g) + H(g)$$

One source of the kinetic isotope effect is the influence of the mass of an atom on the vibrational energy levels of its covalent bonds (Figure 17.2). The relationship between energy and internuclear distance depends only on the nuclear charges and is the same for H_2 and D_2, but the vibrational energies of D_2 are lower than those of H_2 because molecular vibrational energy levels relate *inversely* to the masses. (See Section 16.2.) It is harder to dissociate a D_2 molecule than a H_2 molecule because the deuterium molecule is vibrating less vigorously.

There are several useful applications of isotope effects. One method for separating the isotopes of hydrogen relies on the fact that the electrolysis of heavier D_2O is slower than the electrolysis of normal water. For example, the electrolysis of 999 out of 1000 L of water to H_2 and O_2 leaves 1 L that is rich in deuterium. Pure heavy water is a valuable by-product of the large-scale electrolysis of water, since large quantities are required in nuclear reactors to moderate the rate of the uranium fission reaction (Chapter 22).

17.2 Compounds of Hydrogen

The two common oxidation states for hydrogen are +1 and −1. Hydrogen attains the +1 state by combining with a more electronegative element; its compounds with metals represent the −1 state. Hydrogen interacts chemically with most elements, and the nature of hydrogen compounds—some are covalent, others ionic, and some involve unusual chemical bonds—illustrates periodic relationships, group trends, and the influence of electronegativity differences on the properties of compounds.

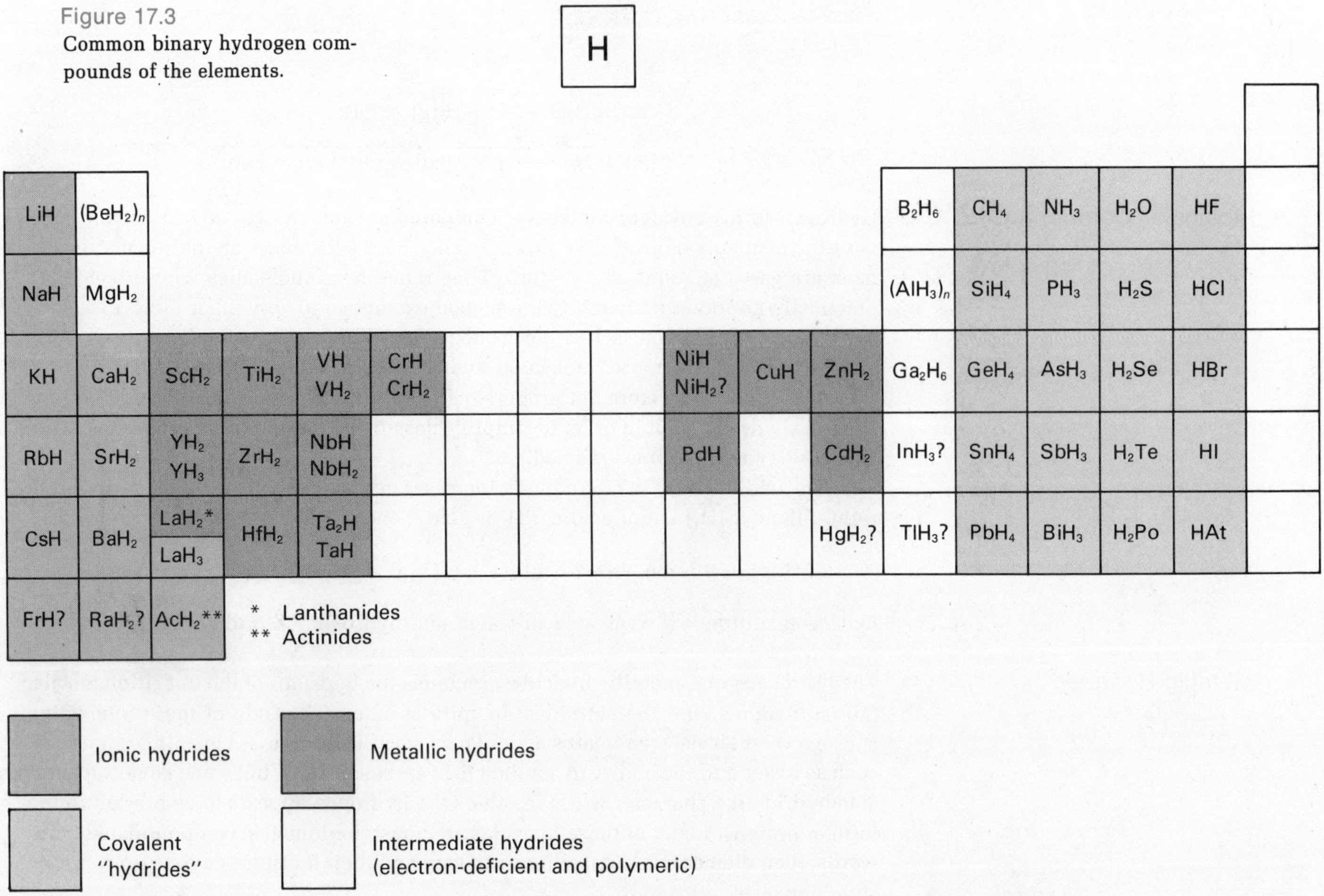

17.2 COMPOUNDS OF HYDROGEN

Figure 17.3
Common binary hydrogen compounds of the elements.

																H
LiH	(BeH$_2$)$_n$										B$_2$H$_6$	CH$_4$	NH$_3$	H$_2$O	HF	
NaH	MgH$_2$										(AlH$_3$)$_n$	SiH$_4$	PH$_3$	H$_2$S	HCl	
KH	CaH$_2$	ScH$_2$	TiH$_2$	VH / VH$_2$	CrH / CrH$_2$			NiH / NiH$_2$?	CuH	ZnH$_2$	Ga$_2$H$_6$	GeH$_4$	AsH$_3$	H$_2$Se	HBr	
RbH	SrH$_2$	YH$_2$ / YH$_3$	ZrH$_2$	NbH / NbH$_2$				PdH		CdH$_2$	InH$_3$?	SnH$_4$	SbH$_3$	H$_2$Te	HI	
CsH	BaH$_2$	LaH$_2$* / LaH$_3$	HfH$_2$	Ta$_2$H / TaH						HgH$_2$?	TlH$_3$?	PbH$_4$	BiH$_3$	H$_2$Po	HAt	
FrH?	RaH$_2$?	AcH$_2$**														

* Lanthanides
** Actinides

- Ionic hydrides
- Metallic hydrides
- Covalent "hydrides"
- Intermediate hydrides (electron-deficient and polymeric)

Figure 17.3 shows the formulas of the common binary compounds of hydrogen with the various elements. (Some of these formulas are oversimplified, but they give a reasonable picture of how the various elements combine with hydrogen.) These compounds fall into four main classes:

1. Ionic hydrides
2. Molecular compounds ("molecular hydrides")
3. Metallic hydrides
4. Polymeric and electron-deficient hydrides

Ionic Hydrides

Hydrogen gas reacts readily with the more electropositive elements—group IA, the lower portion of group IIA (calcium through radium), europium, and ytterbium—at high temperature to form **ionic** (saltlike) **hydrides**. Hydrogen gains an electron and has a charge of −1. The resulting hydride ion has the same electron configuration as helium, but since the nuclear charge of the hydride ion is only +1, the radius is very large, perhaps 2–3 Å. The observed radius of the H$^-$ ion in ionic solids is somewhat smaller, 1.4 Å compared to 0.6 Å for Li$^+$ (an isoelectronic ion, that is, an ion with the same electron configuration).

Alkali metal hydrides such as LiH, NaH, and KH are all white, high-melting solids and have a face-centered cubic structure like NaCl. The electrolysis of any of these hydrides in the molten state produces hydrogen gas at the *anode*,

$$2\ H^- \longrightarrow H_2(g) + 2\ e^-$$

a clear indication of the presence of negative hydrogen ions. Ionic hydrides are also strong reducing agents and react vigorously with water, reducing its +1 hydrogen.

$$\text{NaH(s)} \longrightarrow \text{Na}^+ + \frac{1}{2}\text{H}_2(g) + e^-$$

$$e^- + \text{H}_2\text{O} \longrightarrow \frac{1}{2}\text{H}_2(g) + \text{OH}^-$$

Net reaction: $\text{NaH(s)} + \text{H}_2\text{O} \longrightarrow \text{Na}^+(aq) + \text{OH}^-(aq) + \text{H}_2(g)$

Molecular Compounds

Hydrogen forms covalent **molecular compounds**, such as CH_4, NH_3, H_2S, and HCl, with the elements of groups IV through VII. These substances are quite volatile and most are gases at room temperature. They range from molecules whose bonds are practically nonpolar (PH_3 and CH_4 are good examples) to very polar molecules such as HF. Since hydrogen is less electronegative than most of the elements of these groups (Figure 17.4), most "molecular hydrides" represent hydrogen in its +1 oxidation state and are therefore not properly called hydrides. (The name hydr*ide* implies negative hydrogen, but in order to simplify classification of periodic properties, strict terminology is sometimes relaxed.)

The acidity of these compounds increases across the periodic table from left to right. Silane (SiH_4) is not acidic, PH_3 is acidic enough to react with active metals,

$$2\ PH_3(g) + 6\ Na(s) \longrightarrow 2\ Na_3P(s) + 3\ H_2(g)$$

hydrogen sulfide is a weak acid in water, and hydrogen chloride is a strong acid.

Metallic Hydrides

The third category, **metallic hydrides**, includes the hydrides of the transition metals, the lanthanides, and the actinides. In spite of extensive study of these interesting substances, their nature remains a bit of a mystery. They possess metallic properties, such as luster and the ability to conduct heat and electricity, but some have considerable hydridelike character while in others the hydrogen appears to be present in the form of protons. Many of these hydrides are **nonstoichiometric compounds**. In other words, their elemental compositions can vary and their formulas deviate from simple stoichiometric ratios.

The less hydridic metallic hydrides are often called interstitial hydrides because hydrogen atoms apparently give up their electrons to the conduction bands of the metal (the molecular orbitals responsible for electric conduction) and the resulting protons occupy random interstitial positions in the metal lattice. This proposed structure helps to explain some of the more interesting properties of interstitial hydrides such as the diffusion of hydrogen through hot palladium (Figure 17.5) and the effectiveness of metals such as nickel, platinum, and palladium as catalysts for many reactions of hydrogen. Apparently, the breaking of the H—H covalent bond when the metal hydride is formed removes a major part of the activation energy barrier for reactions of hydrogen. The hydrogen present in the metal hydride is therefore very reactive.

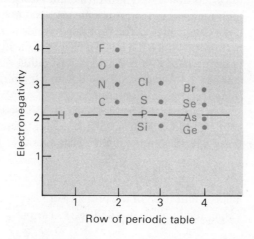

Figure 17.4
Electronegativities of hydrogen and some nonmetallic elements.

Figure 17.5
Representation of H_2 diffusion through palladium.

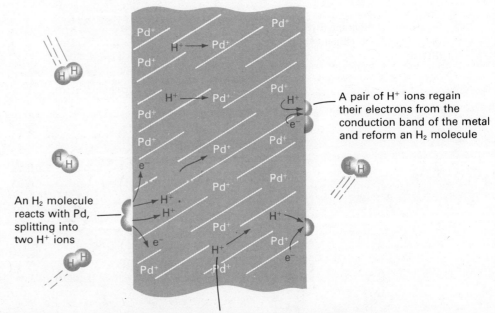

An H_2 molecule reacts with Pd, splitting into two H^+ ions

A pair of H^+ ions regain their electrons from the conduction band of the metal and reform an H_2 molecule

H^+ ions migrate through the crystal lattice of Pd

Polymeric and Electron-Deficient Hydrides

The fourth category includes the hydrides of beryllium, magnesium, the group IIIA elements, and some of the transition metals. These hydrides are **electron-deficient** in the sense that some atoms lack octet electron configurations. Most of these hydrides are **polymeric**—their structures contain repeating structural units. For example, the simplest boron hydride, borane (BH_3), exists only at very high temperatures; at lower temperatures it polymerizes to B_2H_6 (diborane) and larger molecules (B_4H_{10}, B_5H_9, B_5H_{11}). Diborane contains two **three-center bonds** (Figure 17.6, p. 430) in which an orbital from each boron atom combines with a hydrogen orbital to form a bonding orbital that covers all three atoms and contains an electron pair. The resulting *bridged* boron–hydrogen bonds are weaker, longer, and more reactive than the *terminal* B—H bonds, which have typical covalent properties. Three-center bonds occur in the large boron hydride molecules, and hold the polymeric hydrides of beryllium, aluminum, and gallium together.

The group III elements form anions such as BH_4^- and AlH_4^-. Salts of these anions ($NaBH_4$, $LiAlH_4$) are very useful reductants in organic chemistry.

17.3 Hydrogen Bonds

A most important and unique property of hydrogen is the **hydrogen bond.** It is the rather strong attraction between a hydrogen atom covalently bound to a strongly electronegative element of the second period (N, O, or F) and an unshared electron pair on another electronegative second-row element (Figure 17.7, p. 430). The energies of hydrogen bonds range from 4 to 40 kJ/mol. They are *weaker* and *longer* than covalent bonds, but they resemble covalent bonds in that they have a preferred bonding direction. They are considerably stronger than van der Waals attractions.* Although bent hydrogen bonds do exist, there is a strong tendency for the three atoms involved in the bond to lie in a straight line.

Hydrogen bonding is responsible for the unusually high boiling points of HF, H_2O, and NH_3 (Figure 17.8). In general, the boiling points of the covalent hydrides of a periodic family *increase* with molecular mass (the trend for group IV illustrates this), but the boiling point of water is more than 100° higher than that of H_2Te.

---

*Do not confuse hydrogen bonds with the three-centered bonds in the boron hydrides. The three-centered bond is an attraction between a hydrogen atom and two electron-deficient atoms.

Figure 17.6
Structures of diborane (B_2H_6) and the polymeric beryllium hydride $(BeH_2)_x$.

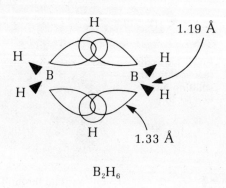

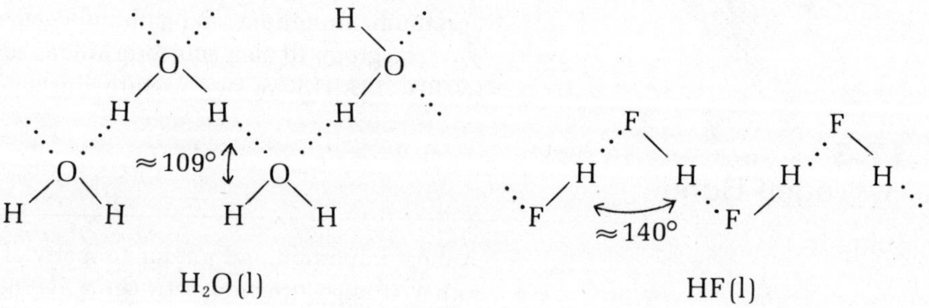

Lewis representation

B_2H_6

$(BeH_2)_x$

Figure 17.7
Hydrogen bonding in H_2O and HF.

$H_2O(l)$ HF(l)

Because of hydrogen bonds, HF, H_2O, and NH_3 form large aggregates of molecules, and it takes extra energy to break up these clusters of associated molecules and form a monomolecular gas. Hydrogen bonds also influence crystal structures and solubilities and are responsible for many of the unusual properties of water. They play an extremely important role in biological systems by maintaining the three-dimensional structures of proteins and nucleic acid molecules.

Example 17.1

Which of the following molecules can form a hydrogen bond with an identical molecule?

17.3 HYDROGEN BONDS 431

Figure 17.8
Normal boiling points of covalent hydrides.

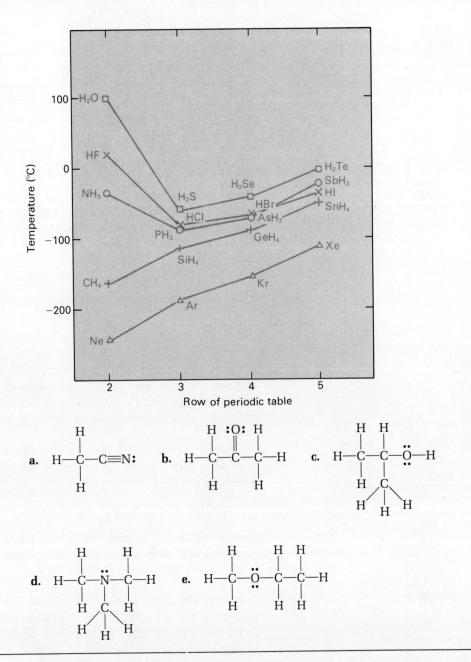

| Solution | All of these structures have one of the requirements for hydrogen bonding in that they have unshared electron pairs on either nitrogen or oxygen. However, only structure c meets the other condition by having a hydrogen atom covalently bonded to nitrogen, oxygen, or fluorine. Therefore, c is the only molecule that can form hydrogen bonds with the same type of molecule. |

It is not easy to explain the nature of hydrogen bonds without getting involved in very complicated quantum mechanical descriptions. Like all other chemical bonds, they are electrostatic attractions, but we cannot classify them as simple dipole–dipole attractions because molecules with larger dipole moments than water are not as strongly attracted to each other. Hydrogen bonding does depend on the small size of the hydrogen atom and on a particular feature of its atomic structure—its lack of inner-shell electrons. A hydrogen atom at the positive end of a polar bond is nearly stripped of its surrounding electrons, and the intense electric field of the exposed proton can exert a strong pull on an electron pair in the vicinity.

17.4
Molecular Oxygen

Of the 105 known elements, oxygen has a special significance for life on earth. Its elemental form (O_2) is the ultimate oxidant for metabolism in most animal organisms, and water, its most important compound, is an essential ingredient of the living cell and provided the environment in which life evolved. Oxygen accounts for approximately 20% by weight of the earth's atmosphere, mostly as O_2, 89% of its water, and 50% of its crust, primarily as oxides, silicates, carbonates, and sulfates.

Properties

The atomic number of oxygen is 8 and the electron configuration is $1s^2 2s^2 2p^4$. Oxygen has six valence electrons and can complete its octet by forming oxide ion (O^{2-}), by participating in two covalent bonds,

$$:\overset{|}{\underset{..}{O}}-$$

or by forming one covalent bond and taking on one extra electron (see Section 7.2).

$$-\underset{..}{\overset{..}{O}}:^-$$

Since oxygen is the second most electronegative element, 3.5 (only fluorine, 4.0, surpasses its electronegativity), covalently bound oxygen atoms usually bear a partial negative charge and the most common oxidation state is -2. Naturally occurring oxygen consists of three stable isotopes, ^{16}O (the most abundant) with traces of ^{17}O and ^{18}O.

As mentioned earlier (Section 7.2), the structure of O_2 does not fit the octet rule—O_2 is paramagnetic and has two parallel electron spins.* The molecular orbital representation of O_2 (Section 6.2)

$$\sigma_{1s}^2 \sigma_{1s}^{*2} \sigma_{2s}^2 \sigma_{2s}^{*2} \sigma_{2p_x}^2 \pi_{2p_y}^2 \pi_{2p_z}^2 \pi_{2p_y}^{*1} \pi_{2p_z}^{*1} \sigma_{2p_x}^{*0}$$

does account for its paramagnetism, and the molecular orbital bond order of two is consistent with molecular oxygen's bond-dissociation energy (493 kJ/mol compared to 941 kJ/mol for N_2 and 150 kJ/mol for F_2).

Preparation

Oxygen is obtained commercially by liquefaction of air and subsequent fractional distillation of the liquid (Figure 17.9) to separate O_2 from N_2. Very pure oxygen is produced commercially by electrolysis of water, a process that is economically worthwhile because of the value of the other product, hydrogen gas.

Oxygen is prepared in small amounts in the laboratory by heating potassium chlorate in the presence of MnO_2 catalyst or by decomposing a solution of a peroxide salt.

$$2\ KClO_3(s) \xrightarrow[MnO_2]{200°C} 2\ KCl(s) + 3\ O_2(g)$$

$$2\ Na_2O_2(aq) + 2\ H_2O \longrightarrow O_2(g) + 4\ Na^+(aq) + 4\ OH^-(aq)$$

The major industrial use of oxygen is as an oxidizer in the manufacture of steel (Chapter 21). It is also used in the oxyacetylene torch and to enrich the oxygen

---

*The first excited state of oxygen is a diamagnetic molecule with a bond order of two and fits the Lewis octet structure.

$$\overset{..}{\underset{..}{O}}=\overset{..}{\underset{..}{O}}$$

This excited state, known as *singlet oxygen*, is almost as stable as the paramagnetic ground state and engages in many interesting and potentially useful reactions.

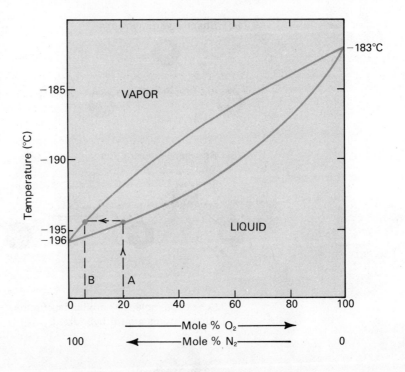

Figure 17.9
Distillation graph for O_2–N_2 mixtures. When liquid air (composition A: 20% O_2, 80% N_2) boils, the vapor has the composition B. Since B is richer in nitrogen than the liquid, the liquid becomes enriched in oxygen as the distillation continues. Repeated distillation produces pure samples of O_2 and N_2.

content of blood. (Patients with respiratory diseases, athletes, and mountain climbers breathe pure oxygen.) Liquid oxygen (LOX) is a fuel oxidant in rocket engines.

Allotropes of Oxygen

Oxygen occurs in two allotropic forms—the common diatomic form (O_2) and ozone (O_3) (Figure 17.10, p. 434). **Allotropes** are forms of the same element differing in the nature of their chemical bonds (as is the case with O_2 and O_3), in their molecular masses (S_6 and S_8), or in their crystal structures (the several allotropes of tin).

Since ozone is considerably less stable than O_2, its formation requires an intense source of energy. The action of ultraviolet radiation produces traces of ozone in the stratosphere.

$$\frac{3}{2} O_2(g) \xrightarrow{UV} O_3(g) \qquad \Delta H = +142 \text{ kJ}$$

The environmental significance of this reaction is that it protects us from excessive and perhaps lethal exposure to the sun's ultraviolet rays because the ozone formed strongly absorbs ultraviolet wavelengths (200–300 nm), and the subsequent decomposition of the ozone into O_2 heats the upper atmosphere. Scientists have recently been evaluating a potential problem due to certain atmospheric pollutants such as the fluorocarbons used as propellants in aerosol spray cans. These compounds might destroy the ozone layer by decomposing in the upper atmosphere to form reactive products which would then consume ozone.

Ozone also occurs in the air of many cities as a constituent of *photochemical smog*—the product of the interaction of the by-products of combustion of gasoline and fuel oil with intense sunlight. Light-induced reactions of the nitrogen oxides formed by cars produce ozone as well as other irritating pollutants. In fact, we measure the concentration of ozone in the air to measure the intensity of smog.

Electric discharges also convert O_2 to O_3. You can smell ozone's penetrating, sweet odor after a lightning storm or near an electric motor's arcing brushes. Ozone mixtures, 5–10% O_3 in O_2, can easily be prepared by passing oxygen gas through a silent electric discharge tube. Ozone has a slight blue color and condenses to a deep blue liquid at −112°C. Ozone is an extremely powerful oxidant,

$$2 e^- + 2 H^+(aq) + O_3(g) \rightleftharpoons O_2(g) + H_2O \qquad E° = +2.07 \text{ V}$$

Figure 17.10
Ozone.

(a) Shape and dimensions

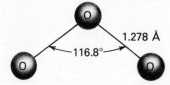

(b) Resonance description

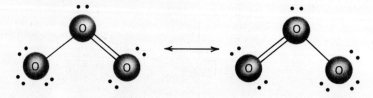

(c) Schematic molecular orbital description. All the atoms are sp^2 hybridized.

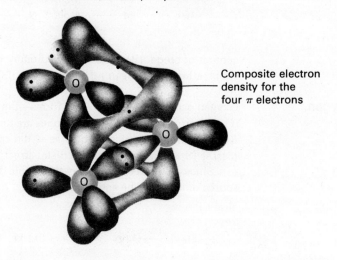

Composite electron density for the four π electrons

and is used commercially as a germicide for treating drinking water and as a bleach in the production of paper from wood pulp.

17.5 Compounds of Oxygen

Molecular oxygen, in spite of its stability and large bond-dissociation energy, reacts with all other elements except He, Ne, and Ar. Some of these reactions occur spontaneously at room temperature, whereas others have high activation energies and require elevated temperatures or catalysts.

Oxides

The nature of the bonds in an oxide depends on the position of the other element in the periodic table. (See Table 17.3 and Figure 17.11.) Electropositive elements—those in groups IA, IIA (except Be), IIIB, and the lower portion of IIIA—form high-melting **ionic oxides** containing O^{2-} ions in their crystal lattices, whereas the oxides of the more electronegative elements are covalent. Some elements of intermediate electronegativity form **polymeric oxides** (for example, SiO_2) that are held together by polar covalent bonds. Nonmetal oxides are molecular in nature and many are gases such as CO, CO_2, NO_2, and ClO_2.

17.5 COMPOUNDS OF OXYGEN

Table 17.3 **Properties of Metal Oxides**

Substance	Oxide type	Melting point (°C)	Solubility of the hydroxide (M)
BeO (beryllia)	Amphoteric	2450	5×10^{-9}
SrO (strontia)	Ionic	2700	6.5×10^{-2}
BaO (baryta)	Ionic	2000	0.22
Li_2O	Ionic	1800	Very soluble
SiO_2	Polymeric covalent	1700	—
SO_2	Molecular	−75	—

Metallic oxides are formed by some transition metals in their lower oxidation states—NbO, $FeO_{0.95}$, NiO, and $TiO_{0.95}$ are some examples. Their crystal structures are close-packed arrays of oxide ions in which the metal ions occupy interstitial positions. Like the metallic hydrides, they retain much of the character of their parent metals—luster, thermal conductivity, and electric conductivity. Many of these oxides have nonstoichiometric formulas, and the crystal defects responsible for the deviation of the formula from the stoichiometric ideal impart interesting and useful optical and electrical properties to these compounds. They are used as photodetectors and semiconductors.

Numerous **mixed-metal oxides**—compounds of two different metals with oxygen—occur in minerals such as $FeTiO_3$ (ilmenite, a titanium ore), spinels such as $MgAl_2O_4$, and perovskites such as $CaTiO_3$.

Figure 17.11
Common oxides of the elements.

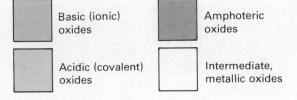

Hydroxides and Oxyacids

Most oxides react readily with water to form compounds containing *hydroxyl* (OH) groups. The acid–base properties of the hydroxyl compounds run the gamut from strongly basic **hydroxides** such as NaOH through amphoteric hydroxides such as $Zn(OH)_2$

$$Zn(OH)_2(s) \rightleftharpoons Zn^{2+}(aq) + 2\ OH^-(aq)$$
$$2\ OH^-(aq) + Zn(OH)_2(s) \rightleftharpoons ZnO_2^{2-}(aq) + 2\ H_2O$$

to strongly acidic **oxyacids**—compounds containing acidic hydroxyl groups—such as H_2SO_4. The corresponding oxides are **anhydrides**, that is, dehydrated forms, of their respective hydroxyl compounds and, as such, can be classified as either basic, amphoteric, or acidic according to the acid–base behavior of the corresponding hydroxyl compound. Thus sodium oxide (Na_2O), a very basic compound, is the anhydride of NaOH, zinc oxide (ZnO) is the amphoteric oxide corresponding to $Zn(OH)_2$,* and SO_3, an acidic gas, is the anhydride of H_2SO_4.

Elements in the lower left of the periodic table form basic oxides, elements located in a diagonal band running through the chart from Be at the upper left through Bi at the lower right form amphoteric oxides, and the acidic oxides are those of the elements in the upper right hand region. Oxide basicity tends to decrease and acidity tends to increase as the periodic group number increases, and the largest atom within a particular group usually forms the most basic oxide (BaO is more basic than MgO). Conversely, the smallest atom in a group forms the most acidic oxides—HNO_3 (the hydrated form of N_2O_5) is more acidic than H_3PO_4 (the hydrated form of P_4O_{10}). The oxidation state of an oxide also influences its acidity, the higher oxide being more acidic than the lower oxide. For example, the acidities of the oxyacids of chlorine increase in the sequence HClO, $HClO_2$, $HClO_3$, $HClO_4$.

The intensity of the charge of a metal ion is the prime factor in the basicity of its oxide and hydroxide. Because large cations with small charges have relatively weak attractions for hydroxide ions, their hydroxide salts dissolve readily to produce strongly alkaline solutions. Hydroxides of small, intensely charged cations are not as soluble and are less effective bases in water. Consequently, $Ca(OH)_2$, the hydroxide of a doubly charged cation, is less basic than KOH, the salt of a larger cation bearing only a single charge. Barium hydroxide is more basic than $Ca(OH)_2$ because Ba^{2+} ions are considerably larger than Ca^{2+} ions.

Example 17.2

Which of each of the following pairs is the more acidic? Why?
a. H_4SiO_4 or H_3PO_4 b. H_2CO_3 or H_4SiO_4 c. H_3AsO_4 or H_3AsO_3

Solution

a. Since silicon is a member of group IVA and phosphorus is the corresponding group VA element, H_3PO_4 is more acidic.
b. Both carbon and silicon belong to group IVA, but carbon is a smaller atom than silicon so its oxyacid, H_2CO_3, is stronger.
c. Since H_3AsO_4 represents arsenic in the +5 oxidation state as opposed to +3 in H_3AsO_3, H_3AsO_4 is the stronger acid.

A related aspect of the effect of the charge intensity of a cation on its acid–base behavior involves the acid hydrolysis of cations. Acid hydrolysis of a metal ion occurs when one or more protons dissociate from adjacent water molecules (Figure 17.12). An intensely charged cation such as Mg^{2+} undergoes hydrolysis because its charge polarizes the O—H bonds in the surrounding water molecules by attracting electrons away from the hydrogen nucleus. This increase in the polarity of an O—H bond enhances the acidity (ability to donate an H^+) of a bound water molecule.

---

*It is hard to distinguish between $Zn(OH)_2$ and the hydrated oxide $ZnO \cdot H_2O$. Because of this ambiguity, which applies to many metal hydroxides and oxides, there is considerable stylistic freedom in writing the formulas of these compounds and the equations for their reactions.

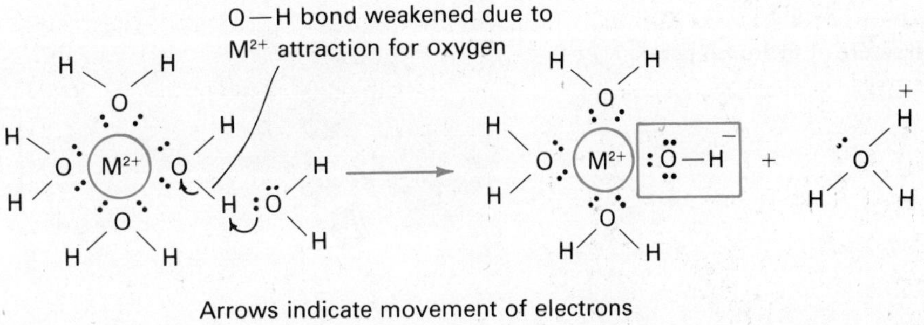

17.5 COMPOUNDS OF OXYGEN

Figure 17.12
Acid hydrolysis of small, highly charged cations.

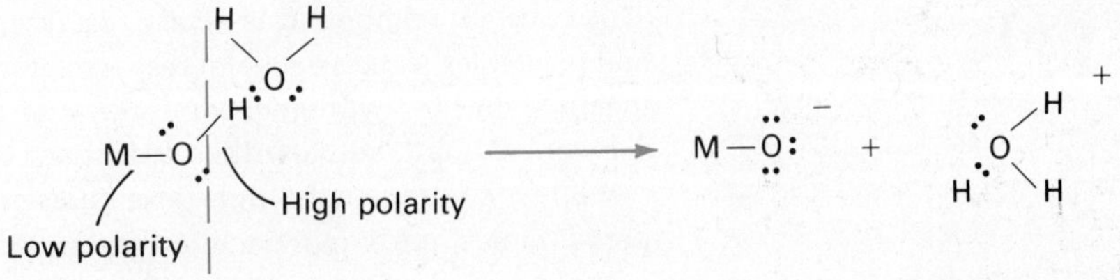

Arrows indicate movement of electrons

Figure 17.13
Acidic hydroxides. Electronegativities of M and O are large; therefore the bond is covalent and the polar O–H bond is easily split by water.

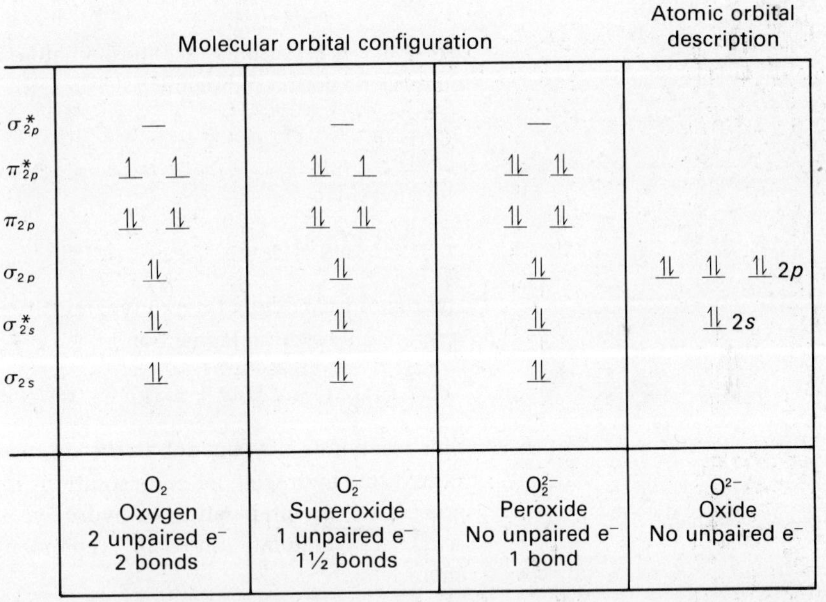

Figure 17.14
Electron configurations of various oxidation states for oxygen.

	Molecular orbital configuration			Atomic orbital description
σ^*_{2p}	—	—	—	
π^*_{2p}	↑ ↑	↑↓ ↑	↑↓ ↑↓	
π_{2p}	↑↓ ↑↓	↑↓ ↑↓	↑↓ ↑↓	
σ_{2p}	↑↓	↑↓	↑↓	↑↓ ↑↓ ↑↓ 2p
σ^*_{2s}	↑↓	↑↓	↑↓	↑↓ 2s
σ_{2s}	↑↓	↑↓	↑↓	
	O_2 Oxygen 2 unpaired e^- 2 bonds	O_2^- Superoxide 1 unpaired e^- 1½ bonds	O_2^{2-} Peroxide No unpaired e^- 1 bond	O^{2-} Oxide No unpaired e^-

Nonmetallic hydroxyl compounds are not at all basic in aqueous solution because the electronegativity of the nonmetallic element is large, and therefore the polarity of the M—O bond is low. There is no tendency for the hydroxyl group to dissociate as OH^- (Figure 17.13).

Peroxides and Superoxides

Ionic **peroxides** are colorless, diamagnetic substances containing the peroxide ion.

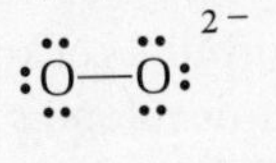

$$:\!\ddot{O}\!-\!\ddot{O}\!:^{2-}$$

Figure 17.14 shows the molecular orbital description of this ion. Ionic peroxides form when some of the more metallic elements (Ba, Sr, Ca) are heated in air.

$$Sr(s) + O_2(air) \longrightarrow SrO_2(s)$$

Peroxide salts react with acidic solutions to produce hydrogen peroxide, H_2O_2 (Figure 17.15).

Figure 17.15
Structure of hydrogen peroxide and peroxydisulfate ion.

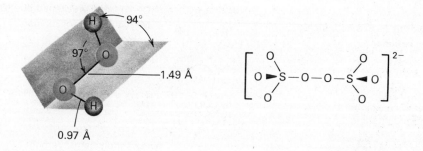

$$Na_2O_2(s) + 2\ H^+(aq) \longrightarrow H_2O_2(aq) + 2\ Na^+(aq)$$

This covalent compound is a very reactive pale blue, sirupy liquid in the pure state and resembles water in several respects. It is highly associated—its molecules cluster together due to hydrogen bonding—and it is both a good insulator and a good solvent for ionic solutes. Like water, its boiling point (152°C) and freezing point (−0.89°C) are unusually high for a substance of its low molecular mass. Hydrogen peroxide is slightly more acidic than water,

$$H_2O_2 \rightleftharpoons H^+(aq) + HO_2^-(aq) \qquad K_A = 1.5 \times 10^{-12}$$

and since it represents an intermediate oxidation state for oxygen (−1), hydrogen peroxide is both a reductant

$$H_2O_2 \longrightarrow O_2(g) + 2\ H^+(aq) + 2\ e^-$$

and an oxidant.

$$H_2O_2 + 2\ H^+(aq) + 2\ e^- \longrightarrow 2\ H_2O$$

Hydrogen peroxide is thermodynamically unstable and decomposes into water and oxygen, an exothermic reaction.

$$2\ H_2O_2 \longrightarrow 2\ H_2O + O_2(g) \qquad \Delta H = -99.0\ kJ/mol$$

This reaction is slow enough in the absence of light, heat, or a catalyst to allow us to store dilute hydrogen peroxide solutions for reasonably long times. The commercial process for the preparation of hydrogen peroxide involves electrolysis of sulfuric acid. Peroxydisulfate ion ($S_2O_8^{2-}$), commonly called persulfate, forms at a platinum anode,

$$2\ HSO_4^-(aq) \longrightarrow S_2O_8^{2-}(aq) + 2\ H^+(aq) + 2\ e^-$$

and the subsequent slow decomposition of this ion

$$2\ H_2O + S_2O_8^{2-}(aq) \longrightarrow 2\ HSO_4^-(aq) + H_2O_2(aq)$$

produces hydrogen peroxide, which is purified by distillation. Both 3% and 30% solutions of H_2O_2 are commercially available, and these solutions are effective bleaches and disinfectants as well as useful laboratory reagents. Pure H_2O_2 was used as the oxidant in early liquid-fuel rockets, but the frequent explosions of this touchy material encouraged rocket developers to seek other propellant systems.

The **superoxide** ion (O_2^-) is paramagnetic—it has an odd number of electrons (Figure 17.14). Its bond distance (1.33Å) is larger than in O_2 (1.21Å) because of the additional electron in a π^* antibonding orbital. This rather unstable ion exists only in crystal lattices stabilized with the cations of large, very electropositive elements such as Rb^+ and Cs^+.

The reactions of the various group IA metals with molecular oxygen follow a trend that is related to the smaller ionization energies and subsequent lattice

17.6 WATER   439

It's hard to think of oxygen, a substance essential to most living things, as a potentially toxic agent, but recent discoveries have implicated oxygen in the irreversible tissue damage that leads to aging. The real culprit appears to be superoxide ion, O_2^-, a species that previously had seemed to be a chemical curiosity with no relevance to the real world. We now know that oxygen molecules react with many compounds in living organisms to form the superoxide ion by a one-electron reduction.

$$e^- + O_2 \longrightarrow O_2^-$$

Superoxide ion is highly reactive and toxic, but most organisms can protect themselves from immediate destruction because they possess a system of enzymes that destroys the superoxide rapidly. The enzyme superoxide dismutase catalyzes the disproportionation of superoxide to oxygen and hydrogen peroxide.

$$2\,H^+ + 2\,O_2^- \longrightarrow O_2 + H_2O_2$$

A second enzyme, catalase, breaks down hydrogen peroxide—also a highly toxic substance—to O_2 and water. The only organisms that lack this enzyme system are those that exist in oxygen-free conditions, and oxygen is a deadly poison for them.

While the superoxide-destroying enzymes enable an organism to survive, a few superoxide ions may last long enough to damage the living cells. One particular area of concern is the attack of superoxide on the fatty material in membranes. It's possible that the superoxide reactions lead to the slow irreversible breakdown of the membranes, a process that may be a major mechanism of aging. One suggested cause of rheumatoid arthritis is a chemical reaction between superoxide and the material that acts as a lubricant for joints.

Another source of membrane damage and aging may be singlet oxygen, the nonmagnetic excited state of O_2. The superoxide dismutase reaction produces this highly damaging form of oxygen, and the body needs some protection against its effects. One possible protector is vitamin E. It helps singlet oxygen lose its excess energy and change into the common paramagnetic form of O_2 and thus perhaps inhibits aging.

stabilities of the larger atoms. Burning in air (a process equivalent to reaction with excess oxygen) converts lithium to its oxide, Li_2O, sodium to its peroxide, Na_2O_2, and potassium, rubidium, and cesium to their superoxides, KO_2, RbO_2, and CsO_2.

Superoxide salts are highly colored solids (paramagnetic substances are usually colored) and have crystal lattice geometries similar to NaCl, like that of CaC_2 (Figure 19.2).

17.6 Water

Properties

Water is so commonplace and so much a part of our lives we tend to take it for granted and perhaps regard it as mundane and uninteresting. On the contrary, its many unusual properties continue to fascinate chemists. Because water molecules form strong hydrogen bonds, it has an unusually high boiling point, freezing point, and heat of vaporization, and it takes a large amount of heat to raise its temperature. Unlike most substances, it expands when it freezes. It is also a very good solvent for ionic substances.

There is nothing unusual about the structure of an individual water molecule, since its bent shape is what we would expect for a molecule with two covalently bonded hydrogen atoms and two unshared electron pairs on oxygen. The unusual properties of water arise from the way its molecules interact with each other. The unshared electron pairs on one water molecule interact with the hydrogen nuclei on neighboring water molecules to form hydrogen bonds. There is a tetrahedral arrangement of bonds, usually two covalent bonds and two hydrogen bonds, around an oxygen atom in water.

These tetrahedra are loosely linked together to form "open" structures with considerable free space (Figure 17.16). Since there is no long-range periodic order, and since the water molecules have a good deal of translational energy, the structure of liquid water is somewhat analogous to the disordered arrangement of the SiO_4 tetrahedra in silicates or glass (Section 19.4). Ice, in contrast, does have periodic order as indicated by its well-defined x-ray diffraction patterns.

There are several polymorphs of ice, each existing in a particular range of temperature and pressure. These polymorphs all contain hydrogen-bonded H_2O tetrahedra but differ in the arrangement of these tetrahedra. In ordinary ice, the tetrahedra link together to form large hexagons, as in diamond, and there is a great deal of empty space in the center of each hexagon.

A somewhat exaggerated phase diagram of water—it overstates the effect of pressure on the freezing point—appears in Figure 17.17. Table 17.4 lists the density of water at various temperatures. Ice is unusual in being *less dense* than its liquid.

Figure 17.16

Geometry and hybridization in water.

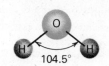

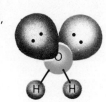

sp^3 hybridization, 2 nonbonded electron pairs

Structures of liquid water and ice.

Ice

Water

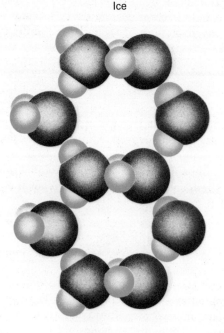

Figure 17.17
Temperature–pressure diagram for H_2O (not drawn to scale).

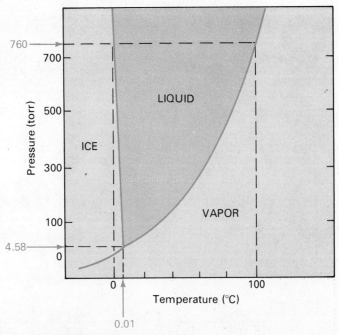

Table 17.4 Density of Liquid Water

Temperature (°C)	Density (g/cm³)
0 (solid)	0.917
0	0.99987
3.98	1.00000
10	0.99970
25	0.99704

This is a reflection of the large amount of free space in ice. When ice melts, many of the hexagonal structures collapse into denser, more disordered arrangements of the molecules, and the volume decreases. At 0°C, a number of the hexagons remain in liquid water, but further heating tends to collapse them and raise the density further. Counteracting this effect is the increase in molecular motion at higher temperature, the effect that causes most substances to expand on heating. At approximately 4°C, the two effects balance each other, and the density of water reaches its maximum value. Water expands above 4° because the increase in thermal motion is more important than the loss of open structure.

Although the volume changes on melting of ice and freezing of water are small (about 10%), they are very significant. The expansion of water when it freezes can burst plumbing and engine blocks. A far more important and beneficial effect is the ecological one—imagine the disastrous effect on aquatic life if ice were denser than water and sank instead of floated. Lacking the insulation of an upper layer of ice, streams and lakes (and perhaps oceans) might freeze solid in winter.

Water's effectiveness as a solvent for ionic substances is directly related to its large polarity. Water molecules, by forming weak electrostatic bonds to ions in the lattice of the salt, partially neutralize the crystal-lattice forces. No longer held as strongly, the ions break away from the crystal surface and are surrounded by water molecules in a manner that insulates their charges, a phenomenon called **solvation** or in the specific case with water, **hydration**. The orientations of the solvating water molecules are different for cations and anions (Figure 17.18). The oxygen atom of a solvating water molecule points to a *cation*; the *hydrogen* atoms are attracted toward *anions*.

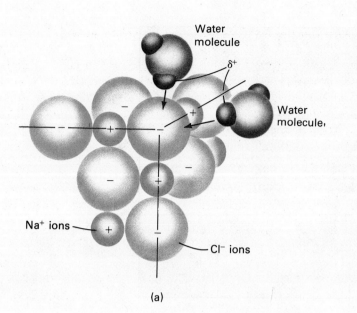

Figure 17.18
Solution and solvation by water. (a) Water molecule interactions neutralize ionic attractions in the crystal and NaCl dissolves. (b) Solvation of cations and anions by water.

Some ionic substances—$BaSO_4$ is a common example—will not dissolve in water. The solubility of an ionic substance depends on the balance of the various thermodynamic factors involved in the solution process. Sodium chloride dissolves in water because its solvation energy balances its lattice energy, and the entropy of solution is positive, that is, it is more disordered. The insolubility of barium sulfate is due to its large positive heat of solution—the relatively weak solvation energies of Ba^{2+} and SO_4^{2-} (large charge but very large ions so that charge density is low) do not compensate for the lattice energy of $BaSO_4$.

Some carbonates, phosphates, and sulfides dissolve to a slight extent because of the hydrolysis of their basic anions.

$$Ca_3(PO_4)_2(s) + 2\,H_2O \rightleftharpoons 3\,Ca^{2+}(aq) + 2\,HPO_4^{2-}(aq) + 2OH^-(aq)$$

$$CaCO_3(s) + H_2O \rightleftharpoons Ca^{2+}(aq) + HCO_3^-(aq) + OH^-(aq)$$

$$ZnS(s) + H_2O \rightleftharpoons Zn^{2+}(aq) + HS^-(aq) + OH^-(aq)$$

Most covalent substances are not very soluble in water. Those that do dissolve are either polar enough to dissociate into ions in solution, engage in a proton transfer reaction with water,

$$HCl(g) + H_2O \rightleftharpoons H_3O^+(aq) + Cl^-(aq)$$

act as a Lewis acid toward water,

$$SO_3(g) + H_2O \rightleftharpoons H_2SO_4(aq) \rightleftharpoons HSO_4^-(aq) + H^+(aq)$$

or form hydrogen bonds with water (CH_3OH, NH_3, and HF for example). Some very large molecules including many proteins are water-soluble because of the interaction of water molecules with hydrogen-bonding groups (such as $-OH$ and $-NH_2$) on their surfaces.

Compounds of Water

Water participates in two interesting classes of compounds, clathrates and hydrates. **Clathrates** (from the Latin *clathratus*—"caged"), or "cage compounds," are solid substances in which small molecules are trapped in the holes of an open crystal structure such as that of water. Water forms clathrates with substances such as $CHCl_3$, Ar, Kr, Xe, Cl_2, and CH_4. Clathrates often have variable and nonstoichiometric formulas such as $Cl_2 \cdot 7.3H_2O$ and $CH_4 \cdot \sim 6H_2O$. The methane–water clathrate, which crystallizes at temperatures up to 15°C, is a particular problem in maintaining natural gas pipelines. If the gas contains too much water, the solid clathrate will clog pipes and valves.

Hydrates are compounds in which water molecules occupy definite positions in a crystal lattice—$CuSO_4 \cdot 5H_2O$, $Na_2CO_3 \cdot 10H_2O$, and $KAl(SO_4)_2 \cdot 12H_2O$ are some examples. The water of hydration can usually be removed by heating. In some cases, the water molecules are coordinated about the metal ions; in others they occupy lattice positions in the overall crystal structure. The hydrated metal ions in a crystal are often very similar to the hydrated (or solvated) ions in aqueous solution. Both $Cu(H_2O)_4^{2+}$—the hydrated cupric ion present in solution—and $CuSO_4 \cdot 5H_2O$ are blue, an observation indicating that cupric sulfate pentahydrate contains the $Cu(H_2O)_4^{2+}$ ion, which is verified by x-ray diffraction. Dehydrating $CuSO_4 \cdot 5H_2O$ removes water from the ion, alters its electronic energy levels, and forms colorless, anhydrous $CuSO_4$.

Summary

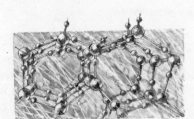

1. Most commercial hydrogen production results from the steam cracking of natural gas

$$CH_4(g) + 2\ H_2O(g) \xrightarrow[\text{Catalyst}]{\text{Heat}} CO(g) + 4\ H_2(g)$$

and catalytic reforming reactions such as

$$C_6H_{14}(g) \xrightarrow[\text{Catalyst}]{\text{Heat}} C_6H_6(g) + 4\ H_2(g)$$

It can also be prepared by the reaction of steam with hot coke, which gives a mixture of carbon monoxide and hydrogen called water gas,

$$H_2O(g) + C(s) \xrightarrow{800°C} \underbrace{CO(g) + H_2(g)}_{\text{Water gas}}$$

and reaction of steam with iron.

$$3\ Fe(s) + 4\ H_2O(g) \xrightarrow{\text{Heat}} Fe_3O_4(s) + 4\ H_2(g)$$

Hydrogen is also prepared on a large scale by electrolysis of water and as a by-product in other industrial processes.

2. Because the bond-dissociation energy of H_2 is 434 kJ/mol, reactions of hydrogen tend to have large activation energies and are often slow, even if the reaction is exothermic.

3. Hydrogen is a moderate reductant toward aqueous solutions and can reduce the oxides of moderately active metals at high temperature. It oxidizes very active metals to form ionic compounds containing hydride ions, H^-.

4. The three isotopes of hydrogen are protium (^1H), deuterium (^2H or D), and tritium (^3H or T). Protium is by far the most abundant isotope. Tritium is radioactive and decays to ^3He by β-particle emission.

$$^3_1H \longrightarrow\ ^3_2He + ^{\ 0}_{-1}\beta$$

5. The mass difference between two isotopes can exert small but significant isotope effects on physical properties, equilibrium constants, and reaction rates. These effects are especially large for the hydrogen isotopes because of their large percentage differences in mass. One important effect is the kinetic isotope effect—an elementary reaction that breaks a covalent bond to deuterium is considerably slower than the corresponding reaction involving covalently bound protium.

6. Hydrogen forms the +1 oxidation state when it reacts with electronegative elements and the −1 state when it reacts with metals. Its various compounds can be classified as ionic

hydrides, molecular compounds, metallic hydrides, or polymeric and electron-deficient hydrides.

The more electropositive elements form ionic hydrides—ionic compounds containing positive metal ions and negative hydride ions. Ionic hydrides are strong reducing agents and react with water to form hydrogen gas and hydroxide ion.

Hydrogen reacts with the elements of groups IV through VII to form molecular compounds—compounds consisting of relatively small covalent molecules.

The transition metals, the lanthanides, and the actinides form metallic hydrides. Some metallic hydrides have considerable ionic character even though they retain many metallic properties. Others appear to be solid solutions of H^+ in metallic crystal lattices.

Beryllium, magnesium, the group IIIA elements, and some transition metals form electron-deficient molecules in which atoms lack the octet configuration. Many of these substances are polymers—large molecules containing repeating structural units. These polymers are held together by three-center bonds in which a bonding orbital covers two metal atoms and a hydrogen atom located between them.

7. A hydrogen bond is a rather strong attraction (4–40 kJ/mol) between a hydrogen atom covalently bound to a strongly electronegative element of the second period (N, O, or F) and an unshared electron pair on another small, second-row element.

8. Oxygen completes its octet by forming the O^{2-} ion, by forming two covalent bonds, or by gaining one extra electron and forming one covalent bond.

9. The ground state of O_2 is a paramagnetic molecule with two unpaired electron spins and a covalent bond order of two.

10. Oxygen is produced commercially by fractional distillation of liquid air and by electrolysis of water.

11. Allotropes are molecular forms of the same element differing in the nature of their chemical bonds, in their molecular masses, or in their crystal structures.

12. The allotropes of oxygen are the common form, O_2, and ozone, O_3. Ozone is produced by the action of ultraviolet light or an electric arc on O_2. Ozone is less stable than O_2 and is a powerful oxidant.

13. Oxides are binary compounds of oxygen that do not contain oxygen–oxygen covalent bonds. The very electropositive metals form ionic oxides such as MgO, more electronegative metals form polymeric oxides such as SiO_2 that are held together by polar covalent bonds, and the oxides of nonmetals are small covalent molecules such as CO_2. Transition metals in their lower oxidation states often form metallic oxides in which the metal ions occupy interstitial positions in a close-packed lattice of oxide ions. Mixed-metal oxides such as $FeTiO_3$ occur in minerals.

14. Most oxides react with water to form either hydroxides or oxyacids, and we can describe an oxide as an anhydride of either a basic, amphoteric, or acidic compound. The elements forming basic oxides lie in the lower left of the periodic table, those forming amphoteric oxides are located in a band running diagonally from the upper left of the periodic table to the lower right, and the elements in the upper right portion of the table form acidic oxides. The higher the oxidation state of an oxide of a particular element, the more acidic it is.

15. Intensely charged cations form weakly basic oxides and hydroxides and tend to undergo extensive acidic hydrolysis because of the strong electrostatic attraction between the cation and hydroxide ions and because the cation's intense charge polarizes the O—H bonds of the surrounding water molecules. This effect is so great in the case of small, highly charged anions such as Al^{3+} that their hydroxides can give up protons and form oxyanions.

16. Nonmetals form covalent bonds to the hydroxyl group and their hydroxyl compounds do not dissociate to form OH^-. These elements do form oxyacids that readily lose protons to form oxyanions. The acidity of the oxyacids increases with oxidation state because the additional oxygens allow for more dispersion of charge in the oxyanion.

17. Peroxides are salts containing the peroxide ion, O_2^{2-}. Several active metals (Ba, Sr, Ca) react with air to form peroxides. Reaction of an ionic peroxide with acid forms hydrogen peroxide, H_2O_2. Hydrogen peroxide is both an oxidizing agent

$$2\ e^- + 2\ H^+ + H_2O_2 \rightleftharpoons 2\ H_2O$$

and a reducing agent.

$$H_2O_2 \rightleftharpoons 2\ H^+ + O_2(g) + 2\ e^-$$

18. The superoxide ion, O_2^-, exists in superoxide salts of large, electropositive cations such as Rb^+ and Cs^+. Superoxide salts are colored, paramagnetic, and very strong oxidants.

19. Water's unusual properties are due to the ability of its molecules to engage in hydrogen bonds. The low density of ice compared to water is due to the free space maintained in the crystal structure by hydrogen bonding. Water is an effective solvent for ionic substances because water molecules surround ions in a manner that insulates against their charges. This process is known as solvation.

20. Clathrates are nonstoichiometric compounds formed when small molecules are trapped in the holes in the open crystal structure of ice. Hydrates are compounds in which water molecules occupy definite positions in an ionic crystal lattice.

Vocabulary List

protium	nonstoichiometric compound	hydroxides
deuterium	electron-deficient hydride	oxyacid
tritium	three-center bond	anhydride
heavy water	hydrogen bond	peroxide
isotope effect	allotropes	superoxide
kinetic isotope effect	ionic oxide	solvation
ionic hydride	polymeric oxide	hydration
molecular hydride	metallic oxide	clathrates
metallic hydride	mixed-metal oxide	hydrates

Exercises

1. Write balanced equations for the formation of hydrogen by the following processes.
   a. Formation of water gas
   b. Reaction of steam and iron
   c. By-product of chlorine in the electrolysis of brine
   d. Electrolysis of water
   e. Reaction of sulfuric acid solution and magnesium

2. Write balanced equations for the chemical reactions, if any, involved in the preparation of oxygen by the following processes.
   a. Liquefaction of air
   b. Electrolysis of water
   c. Decomposition of $KClO_3$
   d. Decomposition of mercuric oxide to its elements
   e. Decomposition of hydrogen peroxide

3. Write balanced equations for:
   a. The reduction of solid NiO in a stream of hot hydrogen
   b. The reduction of Ag^+ solution by hydrogen gas
   c. The oxidation of lithium by hydrogen

4. Write the simplest formula of:
   a. An ionic oxide
   b. A polymeric oxide
   c. An oxide that consists of small covalent molecules
   d. A metallic oxide

5. Write the Lewis formula for a substance in which oxygen completes its octet by:
   a. Gaining two electrons
   b. Forming one covalent bond and taking on an extra electron
   c. Forming two covalent bonds

6. Show by writing the Lewis octet structure and assigning electrons to the appropriate atom that the oxidation state of oxygen is:
   a. -1 in O_2^{2-}
   b. -1 in H_2O_2
   c. $+2$ in F_2O
   d. -2 in SO_3

7. a. Would you expect BH_3 to be a Lewis acid or a Lewis base?
   b. Write an equation for the reaction you would expect between BH_3 and H^-.

8. Write the formula of:
   a. A compound in which hydrogen has the +1 oxidation state
   b. A compound in which hydrogen is in the −1 state

9. Which of the following hydrogen compounds:
   a. Are solids?   b. Are gases or volatile liquids?   c. Is a strong reductant?
   d. Are acids?   e. Is nonstoichiometric?   f. Have metallic character?
   g. Contain three-centered bonds?

   $HI \quad RbH \quad PdH \quad B_2H_6 \quad SiH_4 \quad H_2S \quad LaH_2 \quad BeH_2$

10. Why does HF have a higher boiling point than HCl?

11. Which of the following molecules can form a hydrogen bond with an identical molecule?

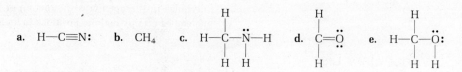

12. Which of the following reactions will occur?
    a. $H_2(g) + Ca^{2+}(aq) \longrightarrow 2 H^+(aq) + Ca(s)$
    b. $H_2(g) + 2 Ag^+(aq) \longrightarrow 2 H^+(aq) + 2 Ag(s)$
    c. $6 H^+(aq) + 2 Cr(s) \longrightarrow 2 Cr^{3+}(aq) + 3 H_2(g)$
    d. $2 H^+ + Hg(l) \longrightarrow Hg^{2+}(aq) + H_2(g)$

13. Describe the volume changes occurring when:
    a. Water is cooled from 3°C to 1°C   b. Water is warmed from 5°C to 8°C
    c. Water freezes

14. Write the simplest formula of the anhydride of:
    a. $Ca(OH)_2$ b. KOH c. $Cr(OH)_3$ d. $Sn(OH)_4$

15. Write the simplest formula for the anyhydride of:
    a. H_2CrO_4 b. H_3PO_4 c. H_2SO_3 d. $HClO$

16. Write the formula of the oxyacid formed from each of the following anhydrides.
    a. Cl_2O_5 b. N_2O_3 c. CO_2

17. Write the formula of the oxyacid formed by reaction of water with:
    a. As_2O_5 b. SeO_3 c. I_2O

18. Write the formula of the hydroxide formed by addition of water to:
    a. SrO b. Li_2O c. Ga_2O_3

19. Write the formula of the hydroxide of each of the following anhydrides.
    a. Fe_2O_3 b. PbO c. Ag_2O

20. Which is more acidic?
    a. H_3PO_3 or H_3PO_4 b. H_2SO_4 or H_2SeO_4
    c. H_3AsO_4 or H_2SeO_4

21. Which is the stronger acid?
    a. H_2CO_3 or HNO_3 b. $HClO$ or $HBrO$ c. HIO_3 or HIO

22. Which is more basic?
    a. $Sr(OH)_2$ or $RbOH$ b. $Sr(OH)_2$ or $Mg(OH)_2$
    c. ScO or Sc_2O_3

23. Which is the stronger base?
    a. TiO or TiO_2 b. KOH or $Ca(OH)_2$
    c. $Al(OH)_3$ or $Ge(OH)_3$

24. Which cation hydrolyzes to the greater extent?
    a. Li^+ or K^+ b. Cs^+ or Ba^{2+} c. Mg^{2+} or Ba^{2+}
    d. Fe^{2+} or Fe^{3+}

25. Which ion hydrolyzes more readily?
    a. Na^+ or Mg^{2+} b. Cr^{2+} or Cr^{3+} c. Al^{3+} or Tl^{3+}

26. What is the charge on the cation in SrO_2?

27. a. Assign oxidation states to each atom in $S_2O_8^{2-}$

    [Lewis structure of $S_2O_8^{2-}$ showing two SO_3 groups connected by an O—O bridge]

    by assigning electrons to the appropriate atoms.
    b. Is the decomposition of $S_2O_8^{2-}$ in water to form H_2O_2 a redox reaction?

EXERCISES

28. The density of solid hydrogen at its freezing point is 0.076 g/cm³. What volume of H_2 gas at standard temperature and pressure can 1.00 cm³ of solid H_2 form?

29. Strontium carbonate is slightly soluble because of the hydrolysis of the carbonate ion. Write a balanced equation for the dissolving of strontium carbonate in water.

30. The density of lithium hydride is 0.82 g/cm³. What volume of hydrogen gas (STP) can be formed by the reaction of 1.00 cm³ of lithium hydride with excess water?

31. Ethyl alcohol, CH_3CH_2OH, has a considerably higher boiling point (79°C) than its isomer, dimethyl ether, CH_3OCH_3 (−23°C). Can you think of a reason for this difference in physical properties?

32. How would you expect the oxygen–oxygen bond energy in ozone to compare with the oxygen–oxygen bond energy in **a.** O_2? **b.** H_2O_2?

33. A student, citing the fact that the oxidation potential of hydrogen is 0.000 V, objects to its description as a mild reductant. The student believes that a substance with a zero oxidation potential cannot be oxidized and therefore cannot reduce anything. What is wrong with this reasoning?

34. What is the pH of 1.00 M H_2O_2?

35. What is the standard potential of each of the following reactions?
    a. $5\ H_2O_2(aq) + 2\ MnO_4^-(aq) + 6\ H^+(aq) \rightleftharpoons 2\ Mn^{2+}(aq) + 5\ O_2(g) + 8\ H_2O$
    b. $2\ Fe^{2+}(aq) + 2\ H^+(aq) + H_2O_2(aq) \rightleftharpoons 2\ Fe^{3+}(aq) + 2\ H_2O$
    c. $2\ H_2O_2(aq) \rightleftharpoons 2\ H_2O + O_2(g)$

36. A 25.00-mL sample of an aqueous solution of H_2O_2 is titrated with 17.50 mL of 0.01850 M $KMnO_4$. What is the molarity of H_2O_2?

    $5\ H_2O_2(aq) + 2\ MnO_4^-(aq) + 6\ H^+(aq) \longrightarrow 2\ Mn^{2+}(aq) + 5\ O_2(g) + 8\ H_2O$

---

37. It is possible to prepare salts of HF_2^-. Can you propose a structure for this ion?

38. The unit-cell edge length of LiH is 4.08 Å, and its crystal lattice is the NaCl type. The ionic radius of Li^+ is 0.60 Å. Calculate the ionic radius of H^-.

39. **a.** Assuming perfect efficiency, use the reaction potential to calculate the minimum amount of work needed to form a mole of hydrogen by electrolyzing water.
    b. What is the maximum amount of usable work that can be obtained by burning a mole of hydrogen gas?
    c. Is hydrogen fuel generated by electrolysis a solution to the energy crisis?

40. **a.** The vapor pressure of pure N_2 at −183°C is 3.73 atm and the vapor pressure of oxygen is 1.00 atm at the same temperature. What is the composition of the vapor that boils off from a liquid mixture of 79 mol % N_2 and 21 mol % O_2 at −183°C?
    b. In which component will the liquid become richer as boiling continues?

41. One method for analyzing for ozone in a gas sample is to bubble the gas into an excess of KI solution. Ozone rapidly oxidizes iodide to I_3^-.

    $2\ H^+(aq) + 3\ I^-(aq) + O_3(g) \longrightarrow O_2(g) + I_3^-(aq) + H_2O$

    The I_3^- is then titrated with $S_2O_3^{2-}$.

    $2\ S_2O_3^{2-}(aq) + I_3^-(aq) \longrightarrow 3\ I^-(aq) + S_4O_6^{2-}(aq)$

    Bubbling 0.250 L (STP) of an ozone–oxygen mixture through a potassium iodide solution generates enough I_3^- to react with 33.5 mL of 0.0200 M $Na_2S_2O_3$. What is the volume percentage of ozone in the gas?

42. a. By referring to the reduction potentials of O_3 and O_2, estimate the standard potential for the decomposition of ozone to O_2.

$$2\ O_3(g) \longrightarrow 3\ O_2(g)$$

b. Estimate $\Delta G°$ for this reaction at 25°C.
c. Compute the value of the equilibrium constant for this reaction at 25°C.
d. What is $\Delta S°$ for this reaction at 25°C?
e. What is the partial pressure of O_3 in an equilibrium mixture containing 1.00 atm O_2 at 25°C?

43. Compute the heat of formation of CaO from the following data: heat of sublimation of Ca = 192 kJ/mol; bond-dissociation energy of O_2 = 493 kJ/mol; first ionization potential of Ca = 587 kJ/mol; second ionization potential of Ca = 1144 kJ/mol; ΔH for $O(g) + 2\ e^- \longrightarrow O^{2-}(g)$ = 640 kJ/mol; crystal-lattice energy of CaO = 3566 kJ/mol.

44. a. The heats of formation of water and CO are −242 kJ/mol and −110 kJ/mol, respectively. What is the heat of reaction for the formation of water gas?
b. During the formation of water gas, some CO reacts with steam to form CO_2 and hydrogen. The heat of formation of CO_2 is −393 kJ/mol. What is the heat of this reaction?
c. Would you improve the ratio of CO to CO_2 in the water gas mixture by raising the temperature of the reacting system? Explain.
d. How much heat is generated by burning 1.00 L (STP) of water gas that is 60 mol % H_2, 20 mol % CO, and 20 mol % CO_2?
e. What volume of oxygen (STP) is needed to burn this mixture?

45. Singlet oxygen is the first excited state of O_2. Unlike the ground state, it has no unpaired electron spins. Its bond order is two. Write a molecular orbital representation for the electron configuration of singlet oxygen.

46. Oxygen compounds show kinetic isotope effects due to the presence of ^{18}O rather than ^{16}O, but these effects are much smaller than the isotope effects for hydrogen. Why are they so much smaller?

47. Which has more vibrational energy near its ground state, HBr or DBr?

48. How does the three-center bond in B_2H_6 differ from a hydrogen bond between two water molecules?

49. How many peaks should there be in the NMR spectrum of B_2H_6? What are their relative intensities?

chapter 18

The Metallic Elements: Periodic Groups IA, IIA, and IIIA

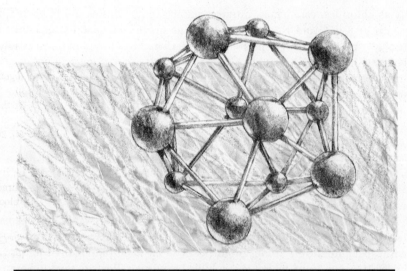

Crystalline boron (a group IIIA element) contains B_{12} icosahedra

The elements of the first three major families of the periodic table are metallic, that is, they lose electrons readily to form cations, and their compounds are predominantly ionic. These elements have typical metallic physical properties—high luster, excellent thermal and electric conductivity, and malleability. Because it is relatively easy to strip the outer-shell electrons from their atoms, most elements in these groups are powerful reducing agents. In this chapter, we will discuss the physical and chemical properties of these elements and see how they relate to the electron configurations and charge-to-size ratio of the atoms.

18.1 The Group IA Elements (Li, Na, K, Rb, Cs, and Fr)

Each of the group IA elements, also called the **alkali metals**, has a single electron in its outermost (valence shell) s orbital (Table 18.1). This electron is easily lost to form the +1 ion, which is quite stable because it has the electron configuration of the preceding noble gas (for example, Na^+ has the Ne configuration). Since the ionization potential for this outer electron is quite low, the metals are highly reactive. Even in the pure metals, these electrons are given up to the metal lattice—they become mobile electrons—and are responsible for the metallic properties, luster, and conductivity of these elements.

Occurrence

Sodium (Na, Latin, *natrium*, for soda), potassium (K, Latin, *kalium*, for potash), and to a lesser extent, lithium are abundant elements in the earth's crust and in its waters. Being reactive, these elements are always found in the chemically combined state,

18 THE METALLIC ELEMENTS: PERIODIC GROUPS IA, IIA, AND IIIA

Table 18.1 Properties of the Group IA Elements

Element	Atomic number	Electron configuration	Melting point (°C)	Boiling point (°C)	Ionization potential (kJ/mol) First	Ionization potential (kJ/mol) Second	Density (g/cm^3)
Li	3	[He] $2s^1$	180.5	1326	520	7300	0.534
Na	11	[Ne] $3s^1$	97.8	883	496	4560	0.971
K	19	[Ar] $4s^1$	63.7	756	420	3050	0.862
Rb	37	[Kr] $5s^1$	39.98	688	403	2630	1.532
Cs	55	[Xe] $6s^1$	28.59	690	375	2290	1.873
Fr	87	[Rn] $7s^1$	27	—	—	—	—

primarily as chlorides, carbonates, nitrates, sulfates, and silicates. As indicated by the presence of K_2CO_3 and Na_2CO_3 in ashes, live organisms contain considerable amounts of K^+ and Na^+. The oceans are rich in NaCl and KCl (3% and 0.1%, respectively) because these salts are soluble. Underground salt mines also occur but only in special rock formations, called salt domes, that protect the salt deposits from water. There are large salt deposits of this type in Oklahoma. Very old, stagnant bodies of water such as the Dead Sea and the Great Salt Lake contain very large concentrations (20–25%) of these salts. These concentrated salt solutions, called brines, are so dense (about 1.2 g/cm^3) that a person can float in them even when sitting up.

Preparation and Properties

The pure alkali metals are prepared by electrolysis of their molten halides or hydroxides. (Sir Humphry Davy first used this technique to isolate sodium in 1807.) The present commercial process for preparing sodium uses an electrolytic reactor called a **Downs cell** (Figure 18.1). Electrolysis of molten mixtures of NaCl with small amounts of Na_2CO_3 or $CaCl_2$ (these mixtures melt at about 600°C compared to 800°C for pure NaCl) at graphite electrodes produces Cl_2 at the anode and molten sodium at the cathode. Liquid sodium, which is insoluble in the melt, floats to the top because of its low density (0.5 g/cm^3) and is drawn off. A wire screen prevents sodium metal from coming in contact with Cl_2 but allows ions to migrate through.

$$NaCl \xrightarrow[\substack{Na_2CO_3 \text{ or } CaCl_2 \\ 600°C}]{\text{Electrolysis}} Na(l) + \frac{1}{2} Cl_2(g)$$

An alternative procedure for preparing alkali metals is to reduce their hydroxides or carbonates with calcium or carbon. This method works in spite of the low

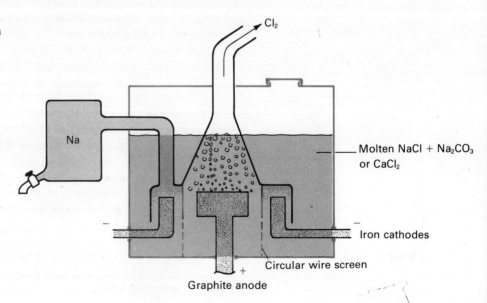

Figure 18.1
Downs cell for preparation of sodium metal from sodium chloride.

reduction potentials of the alkali metal ions because the volatile metals distill out of the reaction mixture. The removal of the product shifts the equilibrium to the right.

$$2\text{NaOH(s)} + \text{Ca(s)} \xrightarrow{\text{Heat}} \text{Ca(OH)}_2\text{(s)} + 2\,\text{Na(g)}$$

(This reduction method can be applied to any low-boiling metal, such as Sc, Eu, Yb.)

The alkali metals have a brilliant luster when pure but they are so reactive that they tarnish after a few seconds in air. To protect against this reactivity they are submerged in an inert organic liquid, such as mineral oil, for storage. They are soft enough to be cut with a knife, easy to melt, malleable (easily deformed), and ductile (easily drawn into wires). They have particularly low densities because they tend to crystallize in a body-centered lattice rather than the more close-packed arrangements, such as face-centered cubic or hexagonal closest packed.

One successful model for the structure of metals describes them as regular arrays of positive metal ions held together by an electron cloud or "sea of electrons." These "valence electrons" are in a conduction band that extends throughout the entire positive ion lattice, like a giant molecular orbital. The freedom of motion of electrons in conduction bands explains many of the observed physical properties of the alkali metals and metals in general. Their mirrorlike metallic luster results from the interaction of electromagnetic radiation with conduction-band electrons. Momentary absorption of light increases the oscillation of these electrons; since the energy levels within the conduction band are very close together, the excited electrons can easily release their extra energy as light. The net result of this process of temporary light absorption followed by rapid emission is the efficient reflection of light from the metal's surface.

Light with sufficiently high frequency (and thus photon energy) can eject electrons from the metal's surface, a phenomenon called the **photoelectric effect**. Alkali metals are particularly susceptible to the photoelectric effect. Having a loose grip on their conduction-band electrons, these elements give up electrons when exposed to low frequency light (visible light compared to the ultraviolet light required by other metals). Cesium, which is very responsive to visible light, is used in the manufacture of photocells, devices that measure light intensity.

The mobile conduction electrons in alkali metals are also responsible for the electrical and thermal conductivity of these elements. These electrons conduct electricity by moving through the crystal lattice under the influence of an electric potential.* The major resistance to electron motion in metallic crystals is the vibrations of the nuclei about their lattice positions—these vibrations disrupt the conduction band—and the conductivity of a metal *decreases* with increasing temperature because of increased thermal vibrations.

Since the alkali metals have relatively few valence electrons in a conduction band, there are no localized electron bonds holding the metal cations together, and it is relatively easy to displace the ions and deform the lattice. The metals of the other periodic groups have stronger lattice forces and are harder and higher melting because they have more conduction-band electrons to hold their lattices together.

The reactivity of the actual metals somewhat limits their commercial use—electric wires made of sodium would quickly corrode; however, large quantities are used as heat-transfer agents in engine valves (sodium) and as coolants in nuclear reactors (liquid sodium and sodium–potassium alloys).

These metals are important commercial reactants because they are such effective reducing agents—their oxidation potentials are about 3 V. Thousands of tons of

---

*The picture of electrons running freely through a wire, although a useful model for electron transport phenomena, may be misleading to beginning students. A more accurate description involves a cascade of displacements of electrons from one atom to the next. Under the influence of an electric potential, electrons move toward the positive pole and displace electrons from adjacent atoms. These electrons in turn displace others, and the process continues as electrons flow into the wire from the negative pole and out the wire at the positive pole.

sodium are consumed annually in the preparation of synthetic organic compounds, detergents, and the antiknock additive for gasoline, tetraethyl lead, $(C_2H_5)_4Pb$. Tetraethyl lead is prepared by the reaction of a sodium–lead alloy, Na_4Pb, with C_2H_5Cl to give $(C_2H_5)_4Pb$.

Reactions

Alkali metals are such strong reducing agents that they react vigorously with even mild oxidants. An example of this is their reaction with water to liberate hydrogen and form a solution of the metal hydroxide.

$$2\ Na(s) + 2\ H_2O \longrightarrow H_2(g) + 2\ Na^+(aq) + 2\ OH^-(aq)$$

The larger alkali metals are so reactive* that the hydrogen gas catches fire from the heat of the reaction, and the resulting explosion showers the surrounding area with molten metal and hot, caustic material. It is prudent to keep alkali metals away from water.

Of the alkali metals familiar to us (francium is so highly radioactive that it has not been prepared in large enough amounts for detailed chemical studies), cesium has the lowest ionization potential. This is because it is a large atom whose outermost electron is far from the nucleus and well-shielded by 54 inner-shell electrons. Lithium, whose ionization potential is considerably larger than cesium's has, however, a higher oxidation potential (Table 18.2). In other words, lithium is more easily oxidized than cesium in solution even though it is easier to remove an electron from gaseous cesium than from gaseous lithium. The ionization potentials of the alkali metals decrease steadily from Li to Cs, but there is no clear trend for the oxidation potentials of these elements. Lithium has the highest oxidation potential, sodium the least.

The ionization potentials and the oxidation potentials of the alkali metals do not correlate with each other because the oxidation potentials *include* the influence of the **sublimation energy** of the metal and the **hydration** (or **solvation**) **energy** of its ion. For example, the first ionization potential of lithium is the energy needed to remove an electron from a lithium atom in the gas phase to produce a gaseous Li^+ ion.

$$Li(g) \longrightarrow Li^+(g) + e^-$$

The oxidation potential of lithium

$$Li(s) \longrightarrow Li^+(aq) + e^- \qquad E° = 3.045\ V$$

is a measure of the free energy change occurring when an atom *in a lithium crystal* loses an electron and becomes a lithium ion *in solution*. Since energy is a state property (Chapter 8), we can evaluate the energy involved in the oxidation half-reaction by examining the thermodynamics of three separate steps—sublimation, ionization, and hydration—that would lead to the overall oxidation process.

Step 1 $M(s) \longrightarrow M(g)$ (sublimation energy)

Step 2 $M(g) \longrightarrow M^+(g) + e^-$ (ionization potential)

Step 3 $\underline{M^+(g) \longrightarrow M^+(aq)}$ (hydration energy)

Step 4 $M(s) \longrightarrow M^+(aq) + e^-$ (oxidation energy)

Table 18.2 lists the energies for these steps for the group IA elements.

Sublimation and ionization *require* energy; hydration *releases* energy because of the strong attraction between ions and polar water molecules. Both sublimation

---

*The term reactivity is not a precise one, particularly when used for comparisons of different elements. Because of subtle kinetic effects, one particular element may be more reactive than another in some reactions, less so in others. For example, compared to the other alkali metals, lithium is slow to react with water, but it is the only element of this group that reacts with N_2.

18.1 THE GROUP IA ELEMENTS

Table 18.2 Energy Changes and Oxidation Potentials for Group IA Elements

	Li	Na	K	Rb	Cs
Sublimation energy (kJ/mol) $M(s) \longrightarrow M(g)$	155	109	90	86	80
Ionization potential (kJ/mol) $M(g) \longrightarrow M^+(g) + e^-$	519	496	419	402	377
Hydration energy (kJ/mol) $M(g)^+ \longrightarrow M^+(aq)$	−519	−406	−322	−301	−276
Energy change (kJ/mol) $M(s) \longrightarrow M^+(aq) + e^-$	+155	+199	+187	+187	+181
Experimental oxidation potentials (volts)	3.045	2.714	2.925	2.925	2.923

energy and ionization energy decrease in the sequence Li to Cs, so it takes less energy to produce a gaseous Cs$^+$ ion from the metal than Li$^+$(g) from Li(s) (457 kJ/mol compared to 674 kJ/mol). However, the large hydration energy of Li$^+$ (519 kJ/mol compared to 276 kJ/mol for Cs$^+$) more than makes up for this difference. The calculated oxidation energies of the alkali metals are consistent with the relative values of the oxidation potentials.*

Ion hydration energies are important factors in ionic reactions and in the behavior of ions in aqueous solutions. Lithium ion, because of its very small size, can form relatively strong electrostatic bonds to the surrounding water molecules, and its hydration releases large amounts of energy. The more diffuse charge of cesium is not nearly as effective in attracting water molecules, and it has a relatively low hydration energy.

Lithium ion, because of its large charge-to-size ratio and the resulting large hydration energy, differs considerably from the other group IA ions. This is a general feature of the chemistry of the first-member elements of a periodic group. In many of their properties—reactivity, solubility of their compounds, and the acid–base properties of their oxides—they bear a closer resemblance to the second element of the family to the right than to the other members of their own family. Thus, lithium and magnesium are quite similar chemically, and there are marked similarities between beryllium and aluminum, and between boron and silicon. These **diagonal similarities** in the periodic table are due to the similar charge intensities of the ions. For example, many lithium and magnesium salts (carbonates, fluorides, and phosphates) are slightly soluble compared to the corresponding soluble salts of Na, K, Rb, and Cs. Lithium, like magnesium, reacts with molecular nitrogen to form a nitride (stable due to its large lattice energy),

$$6\ Li(s) + N_2(g) \longrightarrow 2\ Li_3N(s)$$

and it is the only group IA element that reacts with oxygen in the air to form the oxide, Li$_2$O, rather than a peroxide or superoxide (Section 17.5). Other diagonal similarities appear in Table 18.3.

Compounds

The compounds of the group IA elements are ionic and, except for a few salts, very soluble in water. The alkali metal ions are colorless and do not usually form complex ions. Because their reduction potentials are much less than that of water, they are not

---

*This argument is only approximate since an accurate thermodynamic analysis requires a comparison of the free energies (ΔG) of the various steps. However, entropy effects are not large enough to upset our conclusions in this case, or in the other thermodynamic analyses later in the chapter.

Table 18.3 Diagonal Similarities between Second-Row and Third-Row Elements

Element	Charge-to-size ratio	Oxide type	Halide type	Comparisons		
Mg^{2+}	3.1	MgO Ionic	MgF_2 } Insoluble	Li and Mg react with N_2 to form nitrides		
Li^+	1.7	Li_2O Ionic	LiF	Li and Mg salts more stable to thermal decomposition		
Na^+	1.0	Na_2O Ionic	NaF Soluble			
Al^{3+}	6.0	Al_2O_3 } Covalent polymeric,	Al_2Cl_6 } Covalent polymeric	Oxidation potentials similar; 1.85(Be), 1.66(Al)		
Be^{2+}	6.4	BeO } amphoteric	$BeCl_2$	Hydroxides dissolve in acids and bases		
Mg^{2+}	3.1	MgO Ionic, basic	$MgCl_2$ Ionic	Al_4C_3 and Be_2C give CH_4 on hydrolysis		
Si^{4+}	10	SiO_2 } Acidic oxides	$SiCl_4$ } Molecular gaseous	Si } Semimetallic	SiH_4 } Gaseous	
B^{3+}	15	B_2O_3	BCl_3	B	BH_3	
Al^{3+}	6.0	Al_2O_3 Amphoteric	Al_2Cl_6 Covalent polymeric	Al Metallic	$(AlH_3)_x$ Polymeric	

reduced in dilute aqueous solutions. Since it is hard to precipitate alkali metal compounds or design titrations for alkali metal ions based on redox reactions or complex-ion formation, it would be hard to analyze for these elements if it were not for their characteristic emission spectra. The simple **flame test** (burning a sample in a Bunsen flame) shows the presence of alkali metals by their emission colors—yellow for Na and various shades of violet for K, Rb, and Cs.* The **flame spectrophotometer**, a device for measuring the intensities of various emission lines, allows us to estimate the quantitative content of these elements in a sample. We use sodium's yellow light (590 nm), its only visible emission, as a convenient source of monochromatic light. High-intensity sodium lamps (yellow) and potassium lamps (blue-violet) are used to light streets and airports.

Many alkali metal compounds, particularly those of sodium and potassium, are commercially important: NaCl (table salt), Na_2CO_3 (commonly called washing soda or soda ash), K_2CO_3 (potash), $NaHCO_3$ (baking soda), and NaOH (caustic soda or lye) among others. (The desirable properties of these substances are usually due to the anions, not Na^+ or K^+.)

Most sodium compounds are prepared from sodium chloride or brine solutions, although this approach is apparently changing. For example, until relatively recently most of the world production of $NaHCO_3$, approximately 40 million tons, was by the **Solvay process**. This process involves saturation of a brine solution with ammonia to make it basic, introduction of carbon dioxide to form bicarbonate ion, and then cooling the concentrated solution to precipitate sodium bicarbonate.

$$\underbrace{Na^+ + Cl^-}_{Brine} + NH_3(g) + CO_2(g) \xrightarrow{Cool} NaHCO_3(s) + NH_4^+(aq) + Cl^-(aq)$$

Since the cost of ammonia has risen so high the past few years, this method is no longer competitive. Approximately 60% of the $NaHCO_3$ produced commercially is now mined from large, natural alkali deposits, and then purified.

Brine solutions are also used in another key industrial process. Electrolysis of aqueous NaCl solutions produces sodium hydroxide (Figure 18.2). The cell reactions are

$$2\ Cl^-(aq) \longrightarrow Cl_2(g) + 2\ e^- \quad \text{(anode)}$$

$$2\ e^- + 2\ Na^+(aq) \xrightarrow{Hg} 2\ Na(Hg) \quad \text{(cathode)}$$

*Since the ionization potentials are small for group IA metals, the heat from the flame is sufficient to excite valence electrons that give off the characteristic light upon returning to their ground states.

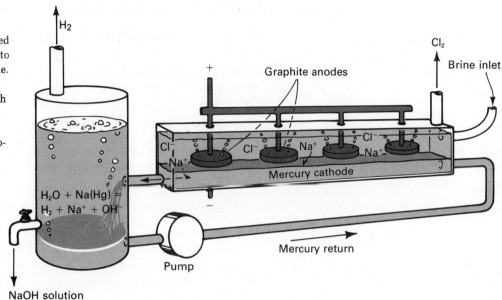

Figure 18.2
Schematic of a de Nora cell. Brine solution is continuously electrolyzed to chlorine at graphite anodes and to sodium metal at a mercury cathode. The sodium dissolves to form an amalgam that is circulated through a reaction vessel. The amalgam reacts with water in the vessel to form sodium hydroxide and hydrogen; the mercury is returned to the cell.

The sodium metal forming at the cathode dissolves in mercury to form a liquid mercury–metal alloy called an **amalgam**. The amalgam is pumped into a separate container where it reacts with water to produce NaOH solution,

$$2\,Na(Hg) + 2\,H_2O \longrightarrow 2\,Na^+(aq) + 2\,OH^-(aq) + H_2(g)$$

and then recycled back to the electrolysis cell. This procedure prevents mixing of the sodium hydroxide solution with the brine and is a very profitable process, because all its products, NaOH, Cl_2, and H_2, are valuable.

Sodium hydroxide and sodium carbonate are both inexpensive bases and are widely used in the manufacture of soap, glass, rayon, cellophane, paper, cleansers, detergents and many other materials necessary to our technological society.

Sodium and potassium ions are present in considerable concentrations in our bodies. In spite of their chemical inertness, these ions play important physiological roles. Their concentrations have a strong influence on osmotic pressure and fluid flow through membranes, and these ions are directly involved in the transmission of nerve impulses. Potassium is particularly important in plant biochemistry and is therefore an important constituent of fertilizers.

18.2 The Group IIA Elements (Be, Mg, Ca, Sr, Ba, and Ra)

The group IIA elements, known as the **alkaline earth metals**, are quite similar to the group IA elements. The members of both families are metallic, good reducing agents, and with the exception of most beryllium compounds and a few magnesium compounds, their compounds are ionic. The alkaline earth metals are less reactive than their group IA counterparts. In all their known compounds, the group IIA elements are in the +2 oxidation state. Some of their important reactions are

$$2\,Ca(s) + O_2(g) \longrightarrow 2\,CaO(s)$$

$$2\,Sr(s) + CO_2(g) \longrightarrow 2\,SrO(s) + C(s)$$

$$3\,Mg(s) + N_2(g) \longrightarrow Mg_3N_2(s)$$

Table 18.4 lists some of the physical properties of the group IIA elements.

18 THE METALLIC ELEMENTS: PERIODIC GROUPS IA, IIA, AND IIIA

Table 18.4 Properties of the Group IIA Elements

Element	Atomic number	Electron configuration	Melting point (°C)	Boiling point (°C)	Ionization potential (kJ/mol) First	Second	Third	Standard oxidation potentials (V)
Be	4	[He] $2s^2$	1278	2770	899	1,760	14,850	1.85
Mg	12	[Ne] $3s^2$	651	1107	738	1,450	7,730	2.37
Ca	20	[Ar] $4s^2$	843	1440	590	1,140	4,840	2.87
Sr	38	[Kr] $5s^2$	769	1380	550	1,060	4,210	2.89
Ba	56	[Xe] $6s^2$	725	1640	503	965	—	2.90
Ra	88	[Rn] $7s^2$	700	1730	—	—	—	—

Properties and Occurrence

All group IIA elements have two valence electrons that they lose easily in chemical reactions. As can be seen from the ionization potentials in Table 18.4, the third (inner-shell) electron is hard to remove, and chemical reactions never produce the +3 ions of these elements.

Because beryllium atoms are so small compared to the other atoms in the group, beryllium's physical properties such as melting point and boiling point (see Table 18.4) differ substantially from those of the other group IIA elements. All these elements are metals and conduct heat and electricity well, but the structural properties of these metals differ considerably. For example, beryllium is strong as steel, hard enough to cut glass, and very brittle, whereas barium is soft and malleable. Only beryllium and magnesium are unreactive enough to be machined—the heat generated in the machining process would ignite the other group IIA metals.

Beryllium–copper alloys, which are highly conducting and resist fatigue (they can be bent back and forth many times without breaking), are used in special electrical components and springs. The windows in x-ray tubes are beryllium because the atoms of this element, having only four electrons, do not interact strongly with x rays. (These very desirable properties are offset though because beryllium and its salts are extremely poisonous.) Because magnesium is light (1.74 g/cm³), its alloys are used in large quantities in aircraft and rockets. Alloying with metals such as aluminum (2.71 g/cm³) is necessary to compensate for the inherent structural weakness of magnesium. Many flares are made from magnesium because it burns with an intense white flame.

Like the group IA elements, the alkaline earth metals are too reactive to occur in the free state in nature. Since most of their naturally occurring compounds (usually silicates and carbonates) are not very soluble, they occur in mineral deposits rather than in the oceans. Only magnesium is present in sufficient concentration in seawater (0.4% by weight) to make its extraction from the ocean economically feasible. Some of the more important alkaline earth minerals are beryl, $Be_3Al_2(SiO_3)_6$, which is called emerald when colored a brilliant green by traces of chromium; dolomite, $MgCO_3 \cdot CaCO_3$; asbestos, $CaMg_3(SiO_3)_4$; and limestone, marble, and chalk—all essentially $CaCO_3$. Calcium is also present in bones and seashells as $CaCO_3$, $Ca_5(PO_4)_3OH$, and $CaSO_4$. Strontium and barium occur as oxides and sulfates, and small amounts of radium occur in uranium ores from radioactive decay of uranium atoms (see Chapter 22).

Preparation

The major commercial preparations of the alkaline earth metals involve electrolysis of their chlorides or fluorides. In the **Dow process** for producing magnesium, lime is added to seawater to precipitate magnesium hydroxide,

$$Mg^{2+}(aq) + CaO(s) + H_2O \longrightarrow Ca^{2+}(aq) + Mg(OH)_2(s)$$

which is then converted to a partial hydrate of magnesium chloride by treating it with HCl and gently heating the $MgCl_2 \cdot 6H_2O$ that forms. Electrolysis of a molten mixture of the partial hydrate with other salts yields magnesium metal at the cathode.

18.2 THE GROUP IIA ELEMENTS

The alkaline earth metals, including magnesium, can all be prepared by chemical reduction of their oxides or halides.

$$MgO(s) + C(s) \xrightarrow{2300°C} Mg(g) + CO(g)$$

$$2\,MgO(s) + Si(s) + 2\,CaO(s) \xrightarrow{Fe} 2\,Mg(g) + Ca_2SiO_4(s)$$

The Stability of the +2 Oxidation State

The first ionization potentials of the alkaline earth metals are all larger than the ionization potentials of the adjacent alkali metals (for example, 738 kJ/mol for Mg versus 496 kJ/mol for Na; 550 kJ/mol for Sr versus 403 kJ/mol for Rb). These increases in ionization potential result from the larger nuclear charges of the alkaline earth metals.

The second ionization potential of a group IIA element is nearly twice as large as its first ionization potential because removal of an electron from a positive ion

$$M^+(g) \longrightarrow M^{2+}(g) + e^-$$

is more difficult than ionization of a neutral atom.

It is important to examine why group IIA elements form +2 ions in their chemical reactions even though it takes extra energy to remove the second electron from an alkaline earth atom. The answer is that ions formed by chemical reactions do not exist in a vacuum, that is, they are either part of a crystal lattice, or they are stabilized by hydration in aqueous solutions. The small, intensely charged +2 ions have greater crystal-lattice and hydration energies than the larger, less charged +1 ions, and the extra hydration and crystal-lattice energies of the +2 ions more than compensate for the required extra ionization energy. The Born—Haber cycles (Figure 18.3, p. 458) show how one might calculate the relative heats of formation for a hypothetical SrCl (assuming a NaCl-type crystal lattice) and for $SrCl_2$. All values are for 1 mol of material.

At first glance, it might seem that both SrCl and $SrCl_2$ could be stable since their formation reactions are both strongly exothermic. However, SrCl is unstable because of its **disproportionation** reaction—a reaction in which one particle of a particular species (in this case Sr^+) oxidizes an identical particle and is itself reduced.

$$2\,Sr^+ \longrightarrow Sr^{2+} + Sr$$

For SrCl the disproportionation reaction is

$$2\,SrCl(s) \longrightarrow SrCl_2(s) + Sr(s)$$

Example 18.1

Use Hess's law (Section 8.1) to predict the energy for the disproportionation of hypothetical SrCl. Use the data from Figure 18.3.

Solution

According to the results in Figure 18.3, the hypothetical energy of formation of SrCl is −226 kJ and the energy of formation of $SrCl_2$ is −780 kJ.

$2\,SrCl(s) \longrightarrow 2\,Sr(s) + Cl_2(g)$	$\Delta E = 2(+226\text{ kJ}) =$	452 kJ
$Sr(s) + Cl_2(g) \longrightarrow SrCl_2(s)$	$\Delta E =$	−780 kJ
$2\,SrCl(s) \longrightarrow Sr(s) + SrCl_2(s)$	$\Delta E =$	−328 kJ

This strongly exothermic reaction would have a very favorable equilibrium constant.

The increasing trend in the oxidation potentials of the alkaline earth metals in the sequence Be–Ba is consistent with the decreasing trend in ionization potentials. The oxidation potentials of Ca, Sr, and Ba are nearly identical because the decreasing

Figure 18.3
Born–Haber cycles for $SrCl_2$ and hypothetical SrCl.

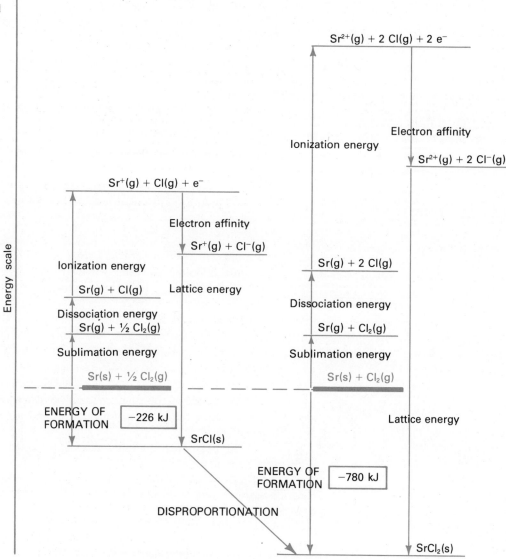

trend in hydration energies (Table 18.5) almost exactly balances the decreasing trend in ionization potentials. The hydration energies of Be^{2+} and Mg^{2+}, although larger than those of the other alkaline earth ions, only partly offset the high ionization potentials of these elements. Consequently, the oxidation potentials of these two elements are considerably lower than those of the larger group IIA elements. In general, beryllium and magnesium are less reactive than the larger alkaline earth metals. For example, Ba, Sr, and Ca all react readily with cold water, magnesium reacts very slowly, and beryllium forms a thin coating of oxide that resists further corrosion.

Carbonates and Bicarbonates

Alkaline earth carbonates, particularly those of the larger cations, are insoluble in water but form soluble bicarbonates in weakly acidic solutions.

$$BaCO_3(s) \rightleftharpoons Ba^{2+}(aq) + CO_3^{2-}(aq)$$

$$BaCO_3(s) + H^+ \rightleftharpoons Ba^{2+}(aq) + HCO_3^-(aq)$$

Water that contains dissolved carbon dioxide is acidic enough to dissolve significant concentrations of $CaCO_3$.

$$CaCO_3(s) + CO_2(g) + H_2O \rightleftharpoons Ca^{2+}(aq) + 2\ HCO_3^-(aq)$$

18.2 THE GROUP IIA ELEMENTS

Table 18.5 Relative Size, Charge, and Hydration Energy Values for the Alkaline Earth Elements

	+1 Ions Size (Å)	+1 Ions Charge/size	+2 Ions Size (Å)	+2 Ions Charge/size	Hydration energies, +2 ions (kJ/mol)
Be	0.44	2.27	0.31	6.45	2385
Mg	0.82	1.22	0.65	3.08	1941
Ca	1.18	0.85	0.99	2.02	1598
Sr	1.32	0.76	1.13	1.77	1464
Ba	1.53	0.65	1.35	1.48	1322

The occurrence of this reaction in limestone caves produces spectacular arrangements of iciclelike stalactites (hanging shafts of $CaCO_3$) and stalagmites (standing shafts). A solution of carbon dioxide in water seeps through the roof of the cave and dissolves some limestone. Evaporation of drops of this solution reverses the solubility equilibrium and redeposits the limestone. Over thousands of years, this drop by drop process of dissolution and reprecipitation of limestone creates the stalagmites and stalactites. In the southwestern United States, the action of carbon dioxide, water, and wind has carved limestone hills into many beautiful shapes.

The presence of dissolved bicarbonates of Ca^{2+}, Mg^{2+}, and Fe^{2+} in water is a condition known as **temporary hardness**. One problem associated with temporary hardness is the precipitation of a coating of metal carbonates, called scale, when the water is heated.

$$Ca^{2+}(aq) + 2\,HCO_3^-(aq) \xrightarrow{\text{Heat}} CaCO_3(s) + CO_2(g) + H_2O$$

Scale clogs boiler valves and heating pipes, and its insulating properties inhibit heat transfer. The second disadvantage of hard water is its adverse effect on washing—the metal ions form an insoluble scum with soap.* This cuts down the amount of soap available for washing and leaves a drab scum on laundry. It is fairly easy to remove temporary hardness by heating the water before using it in order to precipitate scale.

Permanent hardness is the presence of dissolved Ca^{2+}, Mg^{2+}, and Fe^{2+} salts other than the bicarbonates (usually chlorides and sulfates). Because heating cannot remove permanent hardness, this condition requires **water softening** based on a chemical process or ion exchange. Chemical processes involve treating the hard water with Na_2CO_3 (washing soda) to precipitate carbonates

$$Ca^{2+}(aq) + CO_3^{2-}(aq) \rightleftharpoons CaCO_3(s)$$

or adding excess soap to precipitate the scum. A more recent procedure is to add polyphosphate salts such as $Na_6P_6O_{18}$ that form complex ions with the metallic cations.

$$P_6O_{18}^{6-}(aq) + 2\,Ca^{2+}(aq) \rightleftharpoons Ca_2P_6O_{18}^{2-}(aq)$$

These "phosphates" have gained a bad name with ecologists because their presence in waste water appears to stimulate algae growth in lakes and rivers.

Ion exchange involves letting hard water percolate through an insoluble substance, usually a synthetic polymer, that has "soft" ions (Na^+ and K^+) bound to its surface (Figure 18.4). The "hard" ions displace Na^+ and K^+ from their positions on the insoluble surface and are themselves bound to it.

---

*Soaps are usually the sodium salts of organic acids with fairly large molecular masses, for example, stearic acid, $C_{17}H_{35}CO_2H$. Soap scum forms in hard water because calcium salts such as calcium stearate $Ca(C_{17}H_{35}CO_2)_2$ are insoluble. Synthetic detergents are salts of acids whose calcium salts are soluble, so they are not adversely affected by hard water.

Figure 18.4
Ion exchange process to soften water.

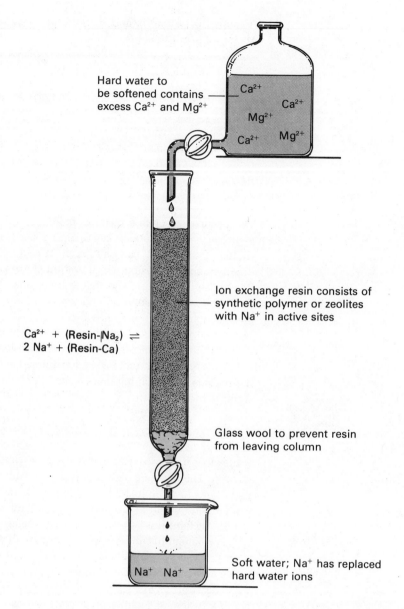

$$Ca^{2+}(aq) + Na_2(\text{ion exchanger})(s) \rightleftharpoons 2\,Na^+(aq) + Ca(\text{ion exchanger})(s)$$

Zeolites are naturally occurring ion exchangers. They are polymeric aluminum silicates in which Na^+ and K^+ are held in the large silicate network by electrostatic attraction. In the synthetic ion exchange resins, ions lie on the surface of organic polymers. Some of these synthetic ion exchange resins can replace cations in solution with hydrogen ions (**cation exchange resins**), and others replace anions with hydroxide ions (**anion exchange resins**). Passage of tap water through a mixture of these particular cation–anion exchange resins produces water that is very pure, so pure that it does not conduct electricity ("deionized water"). (Overall salt content of water is often monitored by measuring its conductance.)

Other Compounds

The small size of the beryllium atom sets its compounds apart from those of the other alkaline earth metals. Many beryllium compounds, including the halides, are polymeric covalent substances rather than ionic; for example, molten $BeCl_2$ (see Figure 7.2 for its structure) is a very poor electrical conductor. Evaporation of beryllium chloride, however, produces gaseous, linear $BeCl_2$ molecules.

$$:\!\ddot{\underset{..}{Cl}}\!-\!Be\!-\!\ddot{\underset{..}{Cl}}\!:$$

18.2 THE GROUP IIA ELEMENTS

Because the beryllium atom in this molecule lacks an octet of electrons, the molecule is an effective Lewis acid.

$$BeCl_2(g) + 2\,Cl^- \longrightarrow BeCl_4^{2-}$$

Of the group IIA hydroxides, $Be(OH)_2$ is the only one that is amphoteric. Consistent with the trends described in Chapter 17, the basicities of the alkaline earth oxides and hydroxides increase in the sequence Be–Ba (Table 18.6). The oxide with the greatest commercial importance, lime (CaO), is prepared by decomposition of limestone.

$$CaCO_3(s) \xrightarrow{\text{Heat}} CaO(s) + CO_2(g)$$

Lime is an inexpensive, commercial base and is an important constituent of mortar. Magnesium oxide (MgO) is a refractory material (melting point > 2500°C) and an excellent electrical insulator, therefore it is widely used to insulate electric wires in heating elements and furnaces. Barium oxide reacts with oxygen to form a peroxide

$$BaO(s) + \frac{1}{2}O_2(g) \xrightleftharpoons{500°C} BaO_2(s)$$

and since this reaction is reversible at higher temperatures, it was once used to extract oxygen from the air.

The sulfates $MgSO_4 \cdot 7H_2O$ (epsom salts), $CaSO_4 \cdot 2H_2O$ (gypsum), and $BaSO_4$ (barite) are all commercially important. Gypsum can be dehydrated to form plaster of Paris ($CaSO_4 \cdot \frac{1}{2}H_2O$). Adding water to plaster of Paris reverses the reaction.

$$2\,[CaSO_4 \cdot \tfrac{1}{2}H_2O(s)] + 3\,H_2O \rightleftharpoons 2\,[CaSO_4 \cdot 2H_2O]$$
$$\text{Plaster of Paris} \hspace{4cm} \text{Gypsum}$$

Since gypsum occupies more volume than plaster of Paris, the hydration process is used to make plaster molds. The swelling that occurs during hydration insures that the gypsum will fill any mold whatever its shape.

Doctors administer suspensions of barium sulfate to patients before taking x-ray pictures of their digestive tracts. This insoluble substance is harmless (although it has a foul taste!) and its opacity toward x rays (because of the 54 electrons in Ba^{2+}) provides the contrast needed to see the otherwise transparent intestines and stomach. Aside from the questionable notoriety strontium has received because of the dangers associated with the radioactive strontium-90 isotope it is most noted for the characteristic red color its salts give to flames; it is used in highway flares and pyrotechnics. Radium salts, because of their radioactivity, glow in the dark; they are used primarily to treat cancers by radiotherapy.

Table 18.6 Some Properties of Group IIA Oxides

Substance	Melting point (°C)	Solubility of the hydroxide (M)	Comments
BeO (beryllia)	2450	5×10^{-9}	Hard as corundum (Al_2O_3)
MgO (magnesia)	2642	3×10^{-4}	Refractory—used in furnace liners, ceramic materials
CaO (lime)	2705	0.022	Industrial use as an inexpensive base and also as a cement
SrO (strontia)	2700	0.065	
BaO (baryta)	2000	0.22	Quite soft

Many of the alkaline earth salts have a strong tendency to hydrate and are useful drying agents, also called desiccants. They maintain dry atmospheres in containers called desiccators and in moisture sensitive equipment. Desiccants can be added to organic liquids to remove traces of water. Anhydrous calcium chloride absorbs so strongly it **deliquesces**—it absorbs so much water that it turns into a moist paste after standing in air. It is sprinkled on dirt roads to wet them down and suppress dust.

18.3 The Group IIIA Elements (B, Al, Ga, In, and Tl)

In keeping with the periodic trend, the group IIIA elements are less metallic than those of the first two groups. Boron, the smallest atom in this group, is **semimetallic**—its electric conductivity is low at room temperature but increases with increasing temperature. (We will discuss semimetallic properties, particularly semiconductance, in more detail in Chapter 19.) With increasing size and numbers of electrons in the sequence Al, Ga, In, Tl, the elements lose electrons more readily, are more basic, and have a greater tendency to form ionic rather than covalent compounds. All these elements form +3 ions but Tl^+ compounds are stable as well.

A striking feature of the smaller group IIIA atoms (B, Al, Ga) is that their first ionization potentials, although larger than those of the corresponding group IA elements, are slightly less than those of group IIA (Table 18.7). This is due to the s^2p^1 valence electron configurations of the group IIIA elements. The outer-shell p electron of boron, which is the one removed in the first ionization step, is farther from the nucleus and more effectively shielded against its positive charge than the s electron in beryllium. Consequently, the p electron in boron, with a slightly higher potential energy, is easier to remove than the stable pair of s electrons in beryllium in spite of the larger nuclear charge of boron.

The larger nuclear charge of the group IIIA elements gives them larger second ionization potentials than the group IIA elements. They have smaller third ionization potentials though, because their third electrons are still in the valence shell and are easier to remove than the inner-shell electrons of the alkaline earth metals. It does take considerable energy to strip all three valence electrons off a group IIIA atom (about 5500 kJ/mol in the case of gallium), but the hydration and lattice energies of the +3 ions (Table 18.8) are more than enough to compensate for the ionization energy. Consequently, all group IIIA elements except boron form some ionic compounds containing +3 ions. Boron, and to some extent, aluminum tend to form covalent bonds.

Occurrence

There are large natural deposits of both boron and aluminum ores, aluminum being the earth's most abundant metal (8% of its crust). Boron occurs primarily in large deposits of borax, $Na_2B_4O_7 \cdot 10H_2O$ and kernite, $Na_2B_4O_7 \cdot 4H_2O$—formed by the evaporation of ancient inland seas—and as the aluminosilicate mineral tourmaline. Aluminum is present in aluminum–silicate minerals, feldspars ($KAlSi_3O_8$), and the ore bauxite ($Al_2O_3 \cdot xH_2O$). Sapphire (Figure 18.5) is α-aluminum oxide (corundum)

Table 18.7 Some Properties of the Group IIIA Elements

Element	Atomic number	Electron configuration	Melting point (°C)	Boiling point (°C)	Ionization potentials (kJ/mol)			
					First	Second	Third	Fourth
B	5	[He] $2s^2 2p^1$	2250		801	2,430	3,660	25,030
Al	13	[Ne] $3s^2 3p^1$	660	2,450	577	1,820	2,745	11,580
Ga	31	[Ar] $3d^{10} 4s^2 4p^1$	29.8	2,237	579	1,980	2,960	6,200
In	49	[Kr] $4d^{10} 5s^2 5p^1$	156.2	2,000	558	1,820	2,700	5,200
Tl	81	[Xe] $4f^{14} 5d^{10} 6s^2 6p^1$	303	1,457	589	1,970	2,880	—

Figure 18.5
Sapphire.

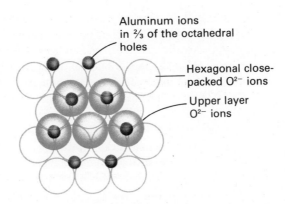

containing small amounts of cobalt and ruby is aluminum oxide colored red by traces of chromium. Gallium, indium, and thallium are present in minute amounts in lead-bearing minerals.

Properties and Preparation

Aluminum may be the most important metal of the twentieth century—its only competitor for the title is iron. Because of its lightness, the strength of its alloys, its ability to conduct heat and electricity, its luster, and its resistance to corrosion, it is used in aircraft, rockets, kitchen utensils, roofing and siding for buildings, wrapping foil, mirror coatings, and paints. Although it is not as good an electric conductor as copper, wires are made of aluminum whenever weight is a factor, such as in electromagnets, transformers, and transmission lines.

It is a very reactive metal (some explosives contain aluminum), but the thin impervious oxide coating that quickly forms on an aluminum surface protects the remaining metal from further corrosion. This oxide coating can be specially formed by electrolysis, a process called **anodizing** (Section 15.2). As the name implies, the piece of aluminum being anodized is part of an anode dipping into a special electrolyte solution. The recipes for anodizing solutions, arrived at more by trial and error than by theory, are usually well-kept secrets. They usually contain acids such as boric, citric, tartaric, chromic, or HSO_4^- whose anions form complexes *within* the oxide film as it grows on the surface of the aluminum anode.

$$H_2O + Al(\text{metal}) \xrightarrow{\text{Electrolysis}} Al_2O_3 \cdot xH_2O \; (+ \text{complexes})$$

The different anodizing bath solutions produce a great variety of coatings with different colors—some are red, some black, and many have a beautiful velvety sheen—and special properties such as corrosion resistance, hardness, abrasiveness, flexibility, and the ability to absorb paints and dyes. One recent anodizing procedure gives aluminum wire its own thin (about 5×10^{-6} m thick), flexible, and electrically insulating coating.

Although aluminum resists corrosion in neutral and moderately acidic solutions, it corrodes rapidly in strong base.

Table 18.8 Comparison of Charge-to-Size Ratios and Hydration Energies of the Metallic Elements

	Li^+	Be^{2+}	B^{3+}	Na^+	Mg^{2+}	Al^{3+}	K^+	Ca^{2+}	Ga^{3+}
Size (Å)	0.60	0.31	0.20	0.95	0.65	0.50	1.33	0.99	0.62
Charge/size	1.67	6.45	15.5	1.05	3.08	6.0	0.75	2.02	4.84
Hydration energy (kJ/mol)	−518	−2,490		−406	−1,921	−4,687	−322	−1,598	

$$2\ OH^-(aq) + 6\ H_2O + 2\ Al(s) \longrightarrow 2\ Al(OH)_4^-(aq) + 3\ H_2(g)$$

This is why aluminum pots should never be washed in strongly alkaline cleansers.

The **Hall–Heroult process** is the commercial preparation of aluminum by electrolysis of its oxide. Charles Hall and Paul Heroult were both university students (Hall in the United States and Heroult in France) and the same age (23) when they quite independently discovered an economical method for producing aluminum from its ore. The discovery was that Al_2O_3 (melting point > 2000°C) could be dissolved in cryolite (Na_3AlF_6) to produce a lower-melting (about 1000°C), conducting solution from which aluminum metal could be produced electrolytically. Figure 18.6 depicts a typical Hall electrolysis cell. The cell is first filled with cryolite or a mixture of fluoride salts (CaF_2 and NaF or AlF_3). Electric current is passed through graphite electrodes. The cryolite melts, and purified bauxite ($Al_2O_3 \cdot xH_2O$) is slowly added. Molten aluminum forms at the cathode

$$Al^{3+} + 3\ e^- \longrightarrow Al(l)$$

and sinks to the bottom of the cell where it is drained off at regular intervals.

Elemental boron can be prepared by electrolysis or by reduction of its compounds with more active elements. For example,

$$B_2O_3(s) + 3\ Mg(s) \xrightarrow{Heat} 3\ MgO(s) + 2\ B(s)$$

$$2\ BCl_3(g) + 3\ H_2(g) \xrightarrow{Heat} 2\ B(s) + 6\ HCl$$

Although elemental boron is usually amorphous and difficult to crystallize, some of its polymorphs are well characterized. These polymorphs contain B_{12} units (Figure 18.7) in which the boron atoms lie at the twelve vertices of a regular icosahedron (a three-dimensional figure with twenty triangular faces). A boron atom's small size (its radius is 0.82 Å) and its tendency to form covalent bonds are responsible for these tightly bound B_{12} units.

Elemental boron is light gray and extremely hard. It is a very poor conductor and is almost inert chemically. Like aluminum, it has a positive oxidation potential, but the strong covalent bonds of the B_{12} units retard its reactions. It is resistant to hot concentrated acids such as HCl and HF, and reacts only slowly with concentrated HNO_3. The usual method for dissolving elemental boron is to fuse it with alkalis such as KOH.

$$2\ B(s) + 6\ KOH(s) \xrightarrow{Heat} 2\ K_3BO_3(s) + 3\ H_2(g)$$

$$2\ H_2O + K_3BO_3(s) \longrightarrow 3\ K^+(aq) + 2\ OH^-(aq) + H_2BO_3^-(aq)$$

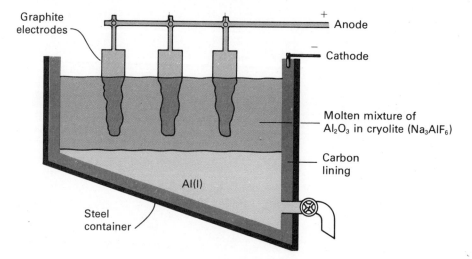

Figure 18.6
Hall process for production of aluminum.

Figure 18.7

Elemental boron. α-Boron consists of B_{12} icosahedra roughly arranged in cubic face-centered positions, joined to each other by weaker bonds. Other allotropes also contain boron icosahedra but are packed together differently.
Source: B. F. Decker and J. S. Kasper, *Acta Crystallographica* **12**, 503(1959).

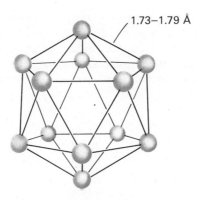
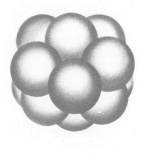

1.73–1.79 Å

B_{12} unit (icosahedron)   space-filling structure

Aluminum, gallium, indium, and thallium are soft metals. In fact indium and thallium are so soft they can be cut with a knife, and their softness makes them useful materials in vacuum gaskets. The melting points of the larger group IIIA metals are low, and gallium (melting point 29.8°C or 86°F) melts on hot days or in your hand. Since the boiling point of gallium is extremely high (2237°C), it remains a liquid in a temperature range extending over 2200°C. Like mercury, liquid gallium does not wet glass and is used in special thermometers.

Compounds

Except for boron, the group IIIA elements are quite reactive and combine readily when heated with electronegative elements such as oxygen, halogens, and sulfur.

$$2\,Al(s) + 3\,O_2(g) \xrightarrow{Heat} Al_2O_3(s)$$

$$2\,In(s) + 3\,Cl_2(g) \xrightarrow{Heat} 2\,InCl_3$$

$$16\,Ga(s) + 3\,S_8(s) \xrightarrow{Heat} 8\,Ga_2S_3(s)$$

Boron does react with nitrogen, however, at very high temperatures to form covalent boron nitrides.

$$2\,B(s) + N_2(g) \xrightarrow{1200°C} 2\,BN(s)$$

There are two boron nitride polymorphs (two crystalline forms), each resembling a polymorph of carbon; a BN unit is isoelectronic with two carbon atoms. The crystal structure of one of the boron nitride polymorphs (Figure 18.8) is similar to that of

Figure 18.8
Boron nitrides.

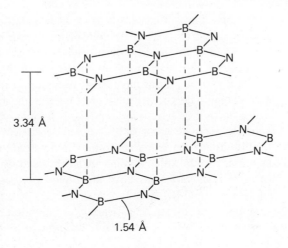
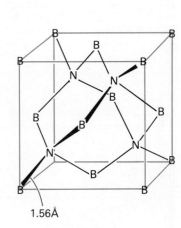

3.34 Å

1.54 Å   1.56 Å

Graphitelike BN   Diamondlike BN

diamond (Figure 9.19)—the boron and nitrogen atoms are all in the sp^3 state of hybridization, and each atom is covalently bound to four tetrahedrally arranged nearest neighbors. This diamondlike polymorph is about as hard as diamond, but the polarity of its bonds makes it more reactive.

Boron, nitrogen, and hydrogen form covalent molecules whose complicated structures resemble those of organic compounds. For example, borazine, $B_3N_3H_6$,

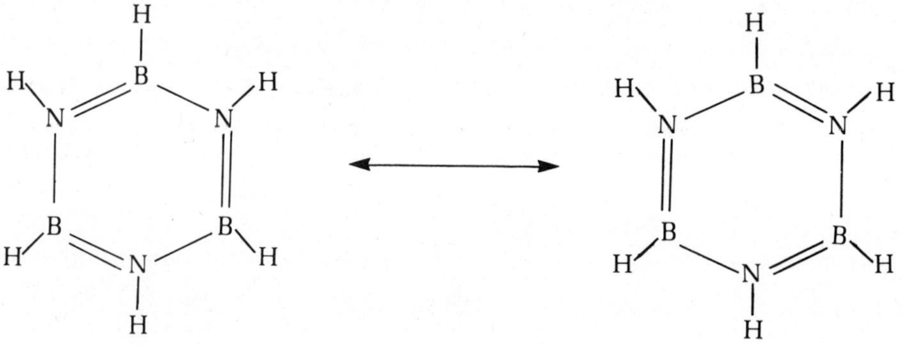

resembles benzene

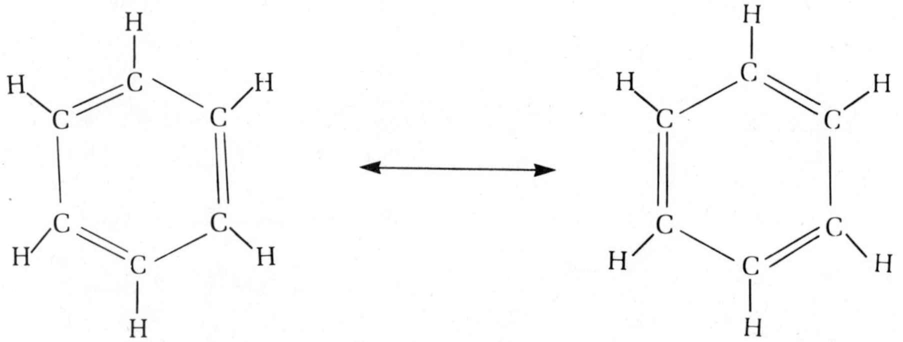

Boron forms **borides** with electropositive elements—groups IA through IIIA, the transition elements, lanthanides, and actinides—but the covalent nature of boron persists in these compounds. They show little or no ionic character. (Mg_3B_2 is a possible exception.) Borides are all hard, refractory, and rather inert materials. Most have unusual optical and electric properties. For example, the electric conductivity of TiB_2 is more than ten times that of titanium itself. The usual method for preparing borides is direct combination of the elements at high temperature.

$$2\ B(s) + Ti(s) \xrightarrow{Heat} TiB_2(s)$$

Aside from a few saltlike borides, they have metallic properties similar to those of the interstitial hydrides. There are four distinct types of boride compounds differing in the bonding of the boron atoms—those in which the boron atoms are (1) isolated, (2) in chains, (3) two-dimensional networks, or (4) three-dimensional clusters.

The boron hydrides have been discussed in the previous chapter. Many of the more complex boron hydride molecules contain mixtures of boron–boron covalent bonds, as well as boron–hydrogen and bridged (three-center) hydrogen bonds, as seen in diborane (Figure 18.9).

Boron–oxygen compounds resemble silicon–oxygen compounds in that both the boron and the silicon atoms are surrounded by tetrahedra of oxygen atoms. The boron–oxygen molecules have disordered polymeric structures consisting of BO_4 tetrahedra joined by shared oxygen atoms. The many different glassy borate structures that occur incorporate a variety of transition elements to form different colors. Consequently, borax, a tetraborate, is used to prepare ceramic glazes and colored glasses, and is a welding flux for dissolving metal oxides.

Figure 18.9
Structures of some boron hydrides.

B₅H₉   B₅H₁₁

The hydroxides of the group IIIA elements show a transition from acidic properties through amphoteric to basic as atomic size increases. Boron hydroxide B(OH)₃ (called boric acid, H_3BO_3) is a weak hydroxy acid, Al(OH)₃ and Ga(OH)₃ are amphoteric, and the hydroxides of indium and thallium are basic. The group IIIA elements give us our first example of another important trend in the middle families (IIIA–VA) of the periodic table—the lower oxidation states of a family become more stable as atomic size increases. Thallium, the largest atom in group IIIA, is the only member of the group to have a stable +1 state, and this state is more stable than Tl(III).

Many of the covalent compounds of boron and aluminum, particularly the halides, are electron deficient and effective Lewis acids.

Summary

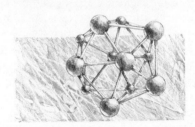

1. The group IA elements, also called the alkali metals, are highly reactive metals containing one valence electron per atom. These metals all form the +1 oxidation state, and their compounds are ionic.

2. Alkali metals are prepared commercially by electrolysis of their molten salts.

3. The metallic properties of the group IA elements—heat and electric conductivity, luster, and malleability—are due to their metallic crystal structures. These structures consist of body-centered cubic arrays of positive metal ions through which the electrons occupying the conduction bands of the crystals can move freely. The alkali metals are particularly susceptible to the photoelectric effect—the release of electrons from a metal's surface when exposed to light of high enough frequency.

4. Alkali metals are strong reducing agents. There is no particular trend in the oxidation potentials of the alkali metals because the decreasing trend in the sublimation energies and ionization potentials as atomic size increases is offset by a decreasing trend in the hydration energies of the cations.

5. Hydration energy and chemical behavior are strongly influenced by the charge-to-size ratio of the alkali metal ion. Because of this effect, there are diagonal similarities between the

properties of the first member of a periodic group and those of the second element of the family to the right.

6. With the exception of a few lithium salts, alkali metal compounds are quite soluble in water. Alkali metal ions do not form complexes readily, are not easy to precipitate, and cannot be reduced in aqueous solutions, but they can be detected by flame tests that display the characteristic colors of their emission spectra.

7. There are many commercially important alkali metal salts, many of which are prepared from NaCl or its concentrated solutions (brines). For example, sodium hydroxide is prepared by the electrolysis of brine.

8. Sodium and potassium ions play important physiological roles, particularly in the transmission of nerve impulses.

9. The elements of group IIA, also called the alkaline earth metals, are metals, good reducing agents, and with the exception of beryllium tend to form ionic compounds. They are less reactive than the group IA elements. Having two outer-shell electrons, an alkaline earth metal forms the +2 oxidation state.

10. Of the alkaline earth metals, only Be and Mg are unreactive enough to be used as metals. Magnesium metal is prepared by the Dow process—the electrolysis of a molten partial hydrate of magnesium chloride obtained by precipitation of $Mg(OH)_2$ from seawater and treatment of the $Mg(OH)_2$ with HCl.

11. The group IIA elements form the +2 oxidation state because the increase in crystal lattice energy associated with the small, intensely charged +2 ions more than compensates for the ionization energy needed to remove the second electron.

12. Temporary hardness is the presence of dissolved bicarbonates of Ca^{2+}, Mg^{2+}, and Fe^{2+} in water. Heating temporarily hard water in boilers and pipes leaves a scale of carbonates,

$$Ca^{2+} + 2\ HCO_3^- \xrightarrow{Heat} CaCO_3(s) + CO_2(g) + H_2O$$

and the metal ions interfere with the action of soap. Permanent hardness is the presence of Ca^{2+}, Mg^{2+}, and Fe^{2+} salts other than the bicarbonates. The permanently hard water can be "softened" by adding Na_2CO_3 to precipitate the carbonates or by adding polyphosphate salts to form complexes with the cations. Softening by ion exchange involves letting the water percolate through an insoluble substance that has Na^+ or K^+ bound to its surface. The calcium ions and other "hard" ions replace Na^+ and K^+ at the surface of the ion exchanger and are themselves bound to it. Zeolites are naturally occurring cation exchangers. Synthetic cation exchange resins and anion exchange resins are also widely used.

13. Beryllium compounds tend to be covalent and polymeric. Monomeric $BeCl_2$ is a strong Lewis acid. $Be(OH)_2$ is the only amphoteric group IIA hydroxide. Many group IIA salts absorb water so effectively that they act as drying agents, also called desiccants. Anhydrous calcium chloride deliquesces—it absorbs so much water that it turns into a moist paste after standing in air.

14. The group IIIA elements are less metallic than those of group IA and IIA, boron being semimetallic—its electric conductivity is low at room temperature but increases with increasing temperature. The larger atoms of group IIA are increasingly more metallic, have a greater tendency to form ionic compounds, and their hydroxides are more basic. The group IIIA elements form the +3 oxidation state, and thallium also forms the +1 state.

15. Aluminum is a light metal with good heat and electric conductivity. Although reactive, aluminum is protected against extensive corrosion by the oxide film that forms on its surface. We can produce oxide coatings with various characteristics by oxidizing aluminum at an electrolytic anode, a process called anodizing.

16. Aluminum is produced commercially by the Hall–Heroult process: purified bauxite ($Al_2O_3 \cdot xH_2O$) is dissolved in molten cryolite (Na_3AlF_6), and the molten solution electrolyzed to form aluminum at the cathode.

17. Elemental boron consists of many polymorphs, most containing B_{12} units arranged in the forms of regular icosahedra. Gallium, indium, and thallium are all soft metals.

EXERCISES   469

18. Boron, the least reactive of the group IIIA elements, reacts with nitrogen at high temperature to form a boron nitride (BN) polymorph whose crystal structure resembles diamond. Boron reacts with the group IA and IIA metals and the lanthanides and actinides to form metal borides of various structures. It forms boron hydrides such as B_2H_6, B_5H_9, and B_5H_{11}, molecules containing three-center bonds. Boron oxides have disordered polymeric structures in which each boron atom is surrounded by a tetrahedron of oxygen atoms.

19. The group IIIA hydroxides show a transition from acidic properties, through amphoteric to basic as atomic size increases.

20. Many compounds of B and Al are electron deficient and are good Lewis acids.

Vocabulary List

alkali metals	Solvay process	zeolite
Downs cell	amalgam	cation exchange resin
photoelectric effect	alkaline earth metal	anion exchange resin
sublimation energy	Dow process	deliquesce
hydration energy	disproportionation	semimetallic
diagonal similarity	temporary hardness	anodizing
flame test	permanent hardness	Hall–Heroult process
flame spectrophotometer	water softening	

Exercises

1. Write the formula for:
   a. Washing soda   b. Limestone   c. Lime
   d. Gypsum   e. Plaster of Paris   f. Borax

2. What is the formula for each?
   a. Bauxite   b. Beryl   c. Asbestos   d. Cryolite
   e. Diborane   f. Anhydrous calcium chloride

3. Write balanced equations for the reaction of each of the following metals with water.
   a. Cesium   b. Beryllium   c. Calcium   d. Lithium

4. Write balanced equations for the commercial methods of producing:
   a. Sodium   b. Sodium hydroxide
   c. Sodium bicarbonate   d. Magnesium   e. Lime
   f. Aluminum

5. Write the outer-shell electron configuration of:
   a. Aluminum   b. An alkaline earth metal
   c. An alkali metal

6. Write the outer-shell electron configuration of:
   a. In^+ b. An alkali metal ion   c. Be^{2+}

7. a. Which of the following solutes will produce temporary hardness?
   b. Which will produce permanent hardness?

   $NaHCO_3$ $MgCl_2$ $NaCl$ $Mg(HCO_3)_2$
   KNO_3 $Ca(NO_3)_2$

8. a. Which of the following compounds will produce permanent hardness when dissolved in water?
   b. Which will produce temporary hardness?

   $MgSO_4$ $Fe(HCO_3)_2$ $Mg(NO_3)_2$ $KHSO_4$
   $NaNO_3$ $CsCl$

9. What can you conclude from the observation that a sample produces a bright yellow flame when placed in the flame of a Bunsen burner?

10. A simple salt imparts a red color to a Bunsen burner flame when placed in it. What can you conclude from this observation?

11. Why is it necessary to electrolyze molten salts of sodium rather than aqueous solutions of these salts in order to obtain sodium metal?

12. a. If light has a frequency below a certain threshold value, which is a characteristic of the metal involved, it cannot produce a photoelectric effect. Explain why this is so.
    b. Why do the alkali metals have particularly low threshold frequencies?

13. a. Would you expect the ionization potential of francium to be greater or less than that of Cs^+?
    b. Would you expect the hydration energy of Fr^+ to be greater or less than that of Cs^+?
    c. Explain both of your predictions in terms of the charge-to-size ratio of the atoms and ions.

14. a. To what group IIIA element does Be show the greatest resemblance?
    b. In what ways does Be resemble this element?
    c. In what ways is Be a typical group IIA element?

15. Limestone dissolves in hydrochloric acid with the rapid evolution of a gas. Write a balanced net ionic equation for this reaction.

16. a. Write the equation for the process that occurs when a brine solution is passed through a cation exchange resin with H^+ bound to its surface.
    b. After treatment with this cation exchange resin, the solution is passed through an anion exchange resin with OH^- bound to its surface. Write the equation for the reaction that occurs.
    c. Why is it that the solution that is passed through these two resins is deionized even though each resin replaces one ion with another?

17. a. Which group IIA oxide is more soluble in a strongly basic solution than in a neutral solution? Why?
    b. Does this oxide dissolve in acid?

18. In the formation of barium peroxide by the reaction between BaO and O_2, which element is oxidized and which is reduced? Balance the equation for this reaction.

19. Describe the trends in the following properties going down the group IA column of the periodic table.
    a. Atomic radius   b. Melting point   c. Electronegativity

20. List four important products derived from common salt (NaCl). Write equations to show how each is obtained.

21. How might you easily distinguish between samples of NaCl and CsCl?

22. What is an easy way to tell the difference between $MgSO_4$ and $BaSO_4$?

23. Why is the oxidation potential of lithium the largest of any group IA element?

24. By considering a Born–Haber cycle, show why the formation of $CaCl_3$ is very unlikely.

25. Briefly explain each of the following.
    a. Anodizing   b. A Downs cell   c. A three-center bond
    d. A covalent aluminum compound acting as a Lewis acid

---

26. The standard method for measuring the amount of barium in a sample of soluble salts is to dissolve the sample and add an excess of a soluble sulfate salt.
    a. What reaction occurs in this process?
    b. If the SO_4^{2-} concentration after completion of the reaction is 0.0100 M, what is the concentration of Ba^{2+} remaining in solution?

27. When a 0.500-g sample that contains soluble barium salts is treated with excess sodium sulfate, 0.485 g of precipitate forms. What is the percentage of barium in the soluble salt sample?

28. The method for quantitative estimation of Ca^{2+} is to precipitate the calcium with excess oxalate ion.

$$Ca^{2+}(aq) + C_2O_4^{2-}(aq) \longrightarrow CaC_2O_4(s)$$

After the calcium oxalate precipitate has been washed free of soluble impurities, it is treated with sulfuric acid,

$$CaC_2O_4(s) + 2\ H^+(aq) \longrightarrow Ca^{2+}(aq) + H_2C_2O_4(aq)$$

and titrated with potassium permanganate.

$$6\ H^+(aq) + 2\ MnO_4^-(aq) + 5\ H_2C_2O_4(aq) \longrightarrow 2\ Mn^{2+}(aq) + 10\ CO_2(g) + 8\ H_2O$$

A 0.800-g sample forms enough calcium oxalate to reduce 30.4 mL of 0.100 M KMnO$_4$. What is the percentage of calcium in the sample?

29. The density of sodium metal is 0.9712 g/cm^3. Calculate the radius of a sodium atom.

30. Use the values of the solubility product of CaCO$_3$, the dissociation constants for H$_2$CO$_3$, and the equilibrium constant for the formation of carbonic acid from CO$_2$

$$CO_2(g) + H_2O(l) \rightleftharpoons H_2CO_3(aq) \quad K = \frac{[H_2CO_3]}{P_{CO_2}} = 0.0338$$

to predict the equilibrium constant for the following reaction.

$$CaCO_3(s) + CO_2(g) + H_2O \rightleftharpoons Ca^{2+}(aq) + 2\ HCO_3^-(aq)$$

31. The chemistry of the lightest element of a periodic group is not typical of the rest of the elements in that group. Cite evidence to support this statement.

32. The density of beryllium is 1.86 g/cm^3, and its atomic radius is 1.125 Å. What type of crystal lattice does beryllium have?

33. The density of lithium metal is 0.53 g/cm^3. Calculate the atomic radius of lithium. Lithium forms a body-centered cubic lattice.

34. Calculate the solubility of strontium fluoride in:
    a. A solution saturated with CaF$_2$ b. A solution buffered at pH 2.0

35. Lithium bromide crystallizes with a NaCl-type unit cell and a cell edge length of 5.501 Å. Assume that Li$^+$ is so small that Br$^-$ ions are in contact. Calculate the radius of the bromide ion. What is the maximum radius of Li$^+$ based on the bromide ion radius calculation? Compare this value to the literature value.

36. Extrapolate the data in Table 18.2 to estimate the sublimation energy and ionization potential of francium and the hydration energy of Fr$^+$. Use these extrapolated data to estimate the standard oxidation potential of francium.

37. Refer to Appendix H for the appropriate K_{sp} values and devise a method for separating Ba^{2+} from Ca^{2+} using either K$_2$CO$_3$ or K$_2$CrO$_4$, whichever is suitable.

38. The enthalpy for the formation of lime from calcium carbonate

$$CaCO_3(s) \rightleftharpoons CaO(s) + CO_2(g)$$

is +178 kJ and the standard entropy change is +159 J/K.
    a. Calculate the equilibrium pressure of CO$_2$ at 25°C.
    b. Calculate the temperature at which the pressure of CO$_2$ is 1.00 atm. (Assume $\Delta H°$ and $\Delta S°$ do not change with temperature.)

39. Calculate the density of BaO (NaCl-type crystal lattice). The unit-cell edge length is 5.539 Å. Compare this to the literature value for the density of BaO.

40. Solid potassium superoxide, KO$_2$, is used to remove CO$_2$ from impure air and release O$_2$. Write a balanced equation for this reaction.

41. Potassium metal crystallizes with a body-centered cubic unit cell whose edge length is 5.321 Å. Assume that the metal is composed of K$^+$ ions (radius = 1.33 Å) and delocalized electrons. What fraction of the space in the lattice is filled by K$^+$ ions? Compare this value with the percent of filled space calculated by assuming that potassium atoms are in contact.

chapter 19

Carbon and the Group IVA Elements

CHFClBr is an example of a simple organic molecule that is chiral

Carbon, the most notable element of group IVA, plays a fundamental role in our existence. Its compounds are the stuff of life—the nutrients, the building blocks of plant and animal tissue, and the vast number of complex substances that catalyze, moderate, and control biochemical reactions. Other carbon compounds, petroleums and petrochemicals, are the primary fuels for our technological society (at least for the time being) and the raw material for plastics and synthetic fibers.

Although we will direct most of our attention to carbon, the other elements of group IVA are also very important. Compounds of silicon comprise a large portion of the earth's minerals. Tin and lead, in addition to their importance to our modern economy, played a major role in our early technological development. Metallurgy took a giant leap forward with the discovery that small amounts of tin alloyed with copper produced bronze, an alloy hard enough for forging tools (and weapons). Tin was so important to the Bronze Age cultures of the Middle East that, more than 3000 years ago, Phoenician sailors braved the Atlantic Ocean to reach the tin mines of Cornwall. The use of lead predates recorded history. Water for the Roman baths flowed through lead pipes, many of which are still intact. There is considerable evidence, in fact, that the lead used by the wealthier Romans for pipes, bath and tub floors, and water containers, contaminated the water sufficiently to cause health problems, birth defects, and sterility, which contributed significantly to their decline.

As indicated by their central position in the periodic table, the group IVA elements tend to be intermediate in character—neither strongly metallic nor nonmetallic. Their s^2p^2 outer-shell electron configurations, half an octet, allow them to form networks of strong covalent bonds; however, this tendency is less pronounced for the

Table 19.1 Some Properties of the Group IVA Elements

Element	Atomic number	Electron configuration	Atomic covalent radius (Å)	Melting point (°C)	Boiling point (°C)	First electron Ionization potential (kJ/mol)
C	6	[He] $2s^2 2p^2$	0.77	3727	4830	1090
Si	14	[Ne] $3s^2 3p^2$	1.17	1410	2680	790
Ge	32	[Ar] $3d^{10} 4s^2 4p^2$	1.22	937	2830	780
Sn	50	[Kr] $4d^{10} 5s^2 5p^2$	1.41	232	2270	710
Pb	82	[Xe] $4f^{14} 5d^{10} 6s^2 6p^2$	1.54	327	1725	715

heavier members of the group and consequently their metallic nature is greater. Although carbon, silicon, and germanium are high-melting, polymeric, covalent substances, tin and lead are low-melting, metallic, and tend to form +2 and +4 ions (See Table 19.1).

Silicon, the most abundant element of the group (and second only to oxygen in the earth's crust), occurs extensively as SiO_2 in the form of either sand or quartz and as silicates in rocks. Elemental carbon occurs as graphite and as diamond, and large deposits of carbon compounds occur as coal, petroleum, natural gas, and mineral carbonates. Germanium, tin, and lead are much less abundant elements and are found mainly as sulfides such as PbS (galena) and oxides such as SnO_2 (cassiterite).

19.1 Atomic and Molecular Carbon

Allotropes

Carbon has two polymorphic forms, **graphite** and **diamond**. (**Polymorphs** are different crystalline forms of the same substance.) In other words, graphite and diamond are crystalline **allotropes**. The contrast in properties between these two forms of the same element is striking. Graphite, a black, soft, high-melting substance, is a good electric conductor and an effective lubricant (it has a slippery, greasy feel), and its conductivity and inertness make it a favorite material for constructing electrodes. Diamond, however, is one of the hardest substances known. When pure, it is colorless, extremely high melting, and does not conduct electricity. These contrasting properties of diamond and graphite arise from the different types of bonding within their respective crystals.

The structure of graphite consists of layers in which the carbon atoms are arranged in hexagons (Figure 19.1). Each carbon atom is in the sp^2 hybridization state (the hybrid orbitals are formed from the s orbital and the two horizontal p orbitals, p_x and p_y), and the planar trigonal arrangement of covalent bonds connects each atom to its three nearest neighbors, 1.42 Å away. There is one unhybridized p orbital (the p_z orbital), perpendicular to the hexagonal plane, for each carbon atom, and these orbitals merge to form π molecular orbitals extending over the entire layer. The freedom of motion of electrons in these delocalized π orbitals gives graphite its metallic character—electric conductivity, opacity, and luster. Its melting point is high because of the strength of the covalent bonds *within* the layers of atoms. The distance between parallel layers is relatively large (3.35 Å), and the weak van der Waals forces between them cannot prevent one plane of atoms from sliding over the next. This loose bonding between the atomic layers, significantly enhanced by the presence of foreign molecules such as H_2O, O_2, and N_2, is responsible for graphite's lubricating ability.

Diamond's hardness, high melting point, transparency and brilliance, and its inability to conduct electricity all arise from its crystal structure. (See Figure 9.19 for the unit cell of diamond.) A perfect diamond crystal is a giant molecule (a mac-

Figure 19.1
Structures of graphite and diamond. In graphite, carbons in layers have sp^2 hybridization. All carbons in diamond have sp^3 hybridization.

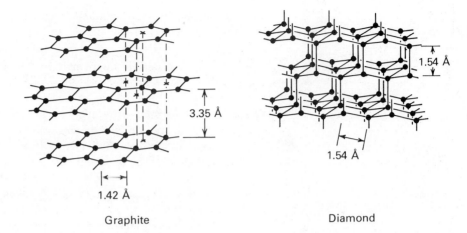

romolecule) in which each carbon atom forms four covalent bonds to its four nearest neighbors, 1.54 Å away and arranged at the corners of a tetrahedron. This tetrahedral arrangement of carbon bonds is consistent with the sp^3 state of hybridization. Since the covalent bonds holding the crystal together are very strong (the dissociation energy of each bond is 343 kJ/mol), they resist chemical and physical attack.

In spite of its inertness, diamond is thermodynamically less stable than graphite at room temperature and pressure, and the conversion of diamond to graphite is an exothermic process that decreases the free energy of the system.

$$\text{C(diamond)} \xrightarrow[\text{1 atm}]{\text{298 K}} \text{C(graphite)} \qquad \Delta H° = -1.88 \text{ kJ/mol}$$

The interconversion of diamond and graphite has, however, such a large activation energy that we do not have to worry about this economically disastrous reaction occurring at room temperature.

Considerable effort and research has been given to the successful preparation of synthetic diamonds from graphite. Increased pressure favors diamond formation because diamond is more dense than graphite (3.51 g/cm³ compared to 2.22 g/cm³), and the stress of the pressure drives the allotropic equilibrium toward the state that occupies the smaller volume (Le Chatelier's principle). This is also consistent with the dependence of enthalpy on pressure.

$$\Delta H = \Delta E + P\Delta V$$

Since ΔV for the conversion of graphite to diamond is negative, the reaction becomes exothermic at high pressure.

Even at high temperatures, the conversion of graphite to diamond is so slow that it requires a suitable metal catalyst (Fe or Cr). Graphite reacts in the presence of a catalyst at 2000°C and 10^5 atm to form small diamond crystals (a few millimeters in size). These synthetic diamonds, usually gray due to contamination by amorphous particles of carbon, are worthless as gems, but they are valuable for preparing cutting and grinding tools.

The Isotopes of Carbon

Naturally occurring carbon contains two stable isotopes, ^{12}C (98.9% of natural carbon) and ^{13}C (~ 1.1%). As we learned earlier, the atomic mass unit (symbol amu) is based on the mass of a ^{12}C atom as a standard. According to the definition of the atomic mass unit, one ^{12}C atom has a mass exactly equal to twelve atomic mass units. A third naturally occurring isotope, ^{14}C, is radioactive and decays by emitting a high-energy electron (a beta particle, see Section 22.2).

$$^{14}_{6}\text{C} \longrightarrow \, ^{14}_{7}\text{N} + \, ^{0}_{-1}\beta$$

19.2
Compounds of Carbon

Carbon atoms, because of their small size, their resistance to polarization, and their intermediate electronegativity (2.5) usually form covalent bonds. The only compounds in which carbon has appreciable ionic character are those formed with very metallic elements, such as the alkali metals and the alkaline earth metals. Rather than make a detailed study of carbon combined with all other elements, as we did with hydrogen and oxygen, we shall only indicate briefly the general types of compounds formed and some of the representative features.

Carbides

The compounds of carbon combined with elements of equal or lower electronegativity* can be divided into several classes, the two most important being the **ionic carbides** and the **polymeric covalent carbides**.

Carbides are usually prepared by the direct combination of the elements at high temperature,

$$2\ C(graphite) + 2\ Na(s) \xrightarrow{2500°C} Na_2C_2(s)$$

or by the reaction of metal oxides with carbon.

$$3\ C(graphite) + CaO(s) \xrightarrow{3000°C} CaC_2(s) + CO(g)$$

Ionic carbides, when pure, are colorless, transparent, and easily decomposed by water (or acid) to yield the metal hydroxide and a hydrocarbon, usually acetylene (C_2H_2). Calcium carbide, for example, is an ionic carbide that is used extensively for the commercial preparation of acetylene. The crystal lattice contains metallic cations and C_2^{2-} (acetylide) ions (Figure 19.2). The acetylide ion is a doubly negative ion composed of two carbon atoms joined by a triple bond (isoelectronic with N_2).

$$:C\equiv C:^{2-}$$

Treatment of an acetylide with water results in two successive Brønsted–Lowry reactions and the production of acetylene.

$$C_2^{2-}(s) + H_2O \longrightarrow C_2H^-(aq) + OH^-(aq)$$

$$C_2H^-(aq) + H_2O \longrightarrow C_2H_2(g) + OH^-(aq)$$

Polymeric covalent carbides are formed between carbon and elements of similar size and electronegativity, such as boron and silicon. Boron carbide, B_4C, and silicon carbide, SiC, closely resemble diamond; they are extremely hard, high melting, and chemically inert. The resemblance to diamond reflects a similarity in crystal structure—the crystal lattice of the polymorph of silicon carbide used in cutting tools (carborundum) is diamondlike with alternate atoms of silicon and carbon.

Oxides of Carbon

The most important oxide of carbon is carbon dioxide (CO_2), a stable, odorless, and colorless gas. It is the product of the complete combustion of carbon or hydrocarbons,

$$C(s) + O_2(g) \longrightarrow CO_2(g)$$

$$CH_4(g) + 2\ O_2(g) \longrightarrow CO_2(g) + 2\ H_2O(g)$$

and it is also a major product of animal metabolism. Carbon dioxide can be prepared by heating carbonates

---

*Hydrocarbons—compounds of carbon and hydrogen—are notable exceptions. They are, like the binary compounds of carbon with electronegative elements [such as $(CN)_2$, CO, CO_2, CS_2, CF_4, and CCl_4] all covalent, volatile substances.

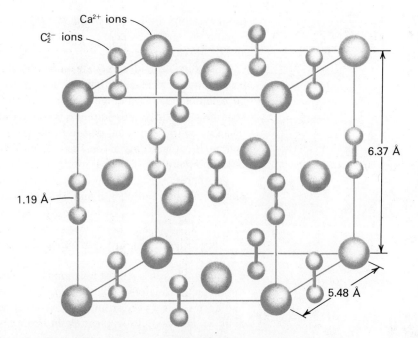

Figure 19.2
Ionic carbide structure. Calcium carbide has the NaCl-type structure but actually is tetragonal due to the shape of the C_2^{2-} ions. Many ionic carbides, such as CeC_2, MgC_2, and UC_2, have this structure.

$$CaCO_3(s) \xrightarrow{\text{Heat}} CaO(s) + CO_2(g)$$

or treating them with acid.

$$CaCO_3(s) + 2\,H^+(aq) \longrightarrow Ca^{2+}(aq) + H_2O + CO_2(g)$$

Although carbon dioxide accounts for only a small portion of the atmosphere (0.035% by volume), it is an extremely important component of our biosphere. When animals metabolize vegetable matter, they convert carbohydrates into CO_2, which is then absorbed by plants and used to resynthesize carbohydrates. Carbon dioxide thus forms a key link in the *carbon cycle* that maintains a balance of material between plants and animals.

An ecologically important property of carbon dioxide is its ability to absorb infrared radiation by excitation of its bending and asymmetric stretching vibration modes (Section 16.2). The major components of the atmosphere, N_2 and O_2, being nonpolar diatomic molecules, do not absorb infrared radiation. Consequently, the concentration of CO_2 in the atmosphere is a critical factor in determining how much of the sun's heat the earth receives and retains.

Carbon dioxide dissolves in water to form the weak diprotic acid, carbonic acid (H_2CO_3),* which forms bicarbonate ion (HCO_3^-) and carbonate ion (CO_3^{2-}) on neutralization.

$$H_2CO_3 + OH^- \rightleftharpoons H_2O + HCO_3^- \qquad K = 4.5 \times 10^7 \; (K_{A1}/K_W)$$

$$HCO_3^- + OH^- \rightleftharpoons H_2O + CO_3^{2-} \qquad K = 4.7 \times 10^3 \; (K_{A2}/K_W)$$

These dissociation reactions have great biochemical significance; carbon dioxide dissolved in our blood establishes a HCO_3^-–CO_3^{2-} buffer system that helps to maintain

---

*Although carbon dioxide is quite soluble in water (about 0.3 M at room temperature and normal pressure), most of the dissolved CO_2 molecules retain their identity, and the concentration of H_2CO_3 is very low.

$$H_2O + CO_2(aq) \rightleftharpoons H_2CO_3$$

Like a solution of $NH_3(g)$ in water, which we usually write as $NH_3(aq)$ not NH_4OH, H_2CO_3 has never been isolated and therefore we might prefer $CO_2(aq)$ or $CO_2 \cdot H_2O$ to H_2CO_3.

the pH within the narrow limits, 7.34–7.42, required for life. If the pH of the blood were to change by ± 0.5 pH unit, and remain there for a short period, collapse, coma, and death would result.

Bicarbonate and carbonate salts are typical ionic substances. Bicarbonates are water soluble, whereas carbonates, except for those of the group IA elements, tend to be insoluble. The insoluble carbonates include many important minerals, such as limestone ($CaCO_3$) and dolomite ($MgCO_3 \cdot CaCO_3$). The soluble carbonates, sodium bicarbonate (baking soda) and sodium carbonate (soda ash), are important industrial commodities, millions of tons being consumed annually in the baking industry and in the preparation of such important products as glass and soap.

Carbon monoxide, a colorless, odorless and stable gas is the product of *incomplete* combustion of carbon and hydrocarbons.

$$2\ C(s) + O_2(g) \longrightarrow 2\ CO(g)$$

$$2\ CH_4(g) + 3\ O_2(g) \longrightarrow 2\ CO(g) + 4\ H_2O(g)$$

Carbon monoxide is a most poisonous gas because of its reaction with hemoglobin in the blood. Hemoglobin takes up O_2 in the lungs and deposits it in the body tissues, but since CO is a very strong Lewis base (it can donate an electron pair to electron-deficient species), it displaces O_2 and complexes very tightly to the iron atom in the heme group. The CO binds so strongly to heme (200 times better than O_2) that even though there might be only a small concentration of CO in the air, after a few minutes it effectively replaces large quantities of O_2 because it is not released easily by the heme. A person can be poisoned by CO without realizing it until it is too late. Even exposure to relatively modest CO concentrations—heavy city traffic or cigarette smoking—for short periods of time effectively decreases the O_2 carrying efficiency of the blood.

Carbon monoxide is a good reducing agent and can be prepared by dehydrating formic acid, a pungent, irritating substance found in ants and insect bites.

$$HCO_2H \xrightarrow{H_2SO_4} H_2O + CO(g)$$

Although isoelectronic with N_2,

$$:C\equiv O: \quad :N\equiv N:$$

carbon monoxide is polar, more reactive, and combines readily with transition metal elements to form metal carbonyls.

$$Ni(s) + 4\ CO(g) \xrightarrow[\text{Pressure}]{\text{Heat}} Ni(CO)_4(l)$$

Hydrogen Cyanide and Cyanide Ion

Hydrogen cyanide, HCN, is an extremely poisonous, colorless, sweet-smelling gas whose aqueous solution (hydrocyanic acid) is weakly acidic.

$$HCN(aq) \rightleftharpoons H^+(aq) + CN^-(aq) \quad K_A = 5.0 \times 10^{-10}$$

Cyanide ion, CN^-, is isoelectronic with carbon monoxide

$$:C\equiv N:^-$$

and resembles this molecule in that it is a Lewis base and forms complexes with transition metal ions.

$$Ag^+(aq) + 2\ CN^-(aq) \rightleftharpoons Ag(CN)_2^-(aq)$$

This complex-forming tendency is responsible for the toxicity of cyanide solutions and HCN. Like CO, cyanide ion complexes with metal-containing proteins involved in key biochemical reactions and renders them ineffective.

19.3 Organic Chemistry

In the early nineteenth century, all substances were believed to belong to one of two classes: inorganic or organic. Inorganic substances were those from mineral sources, while organic substances composed and derived from all *living* systems. Organic compounds were thought to be fundamentally different from inorganic compounds, and required "vital forces" for their synthesis. This theory collapsed in 1828 when Friedrich Wöhler synthesized a totally organic substance, urea (a major constituent of animal urine) from inorganic materials. Wöhler, while attempting to prepare ammonium cyanate (NH_4OCN) from silver cyanate and ammonium chloride, heated the solution to evaporate the water and obtained urea.

$$[NH_4OCN] \xrightarrow{\text{Heat}} H_2N-\underset{\text{Urea}}{\overset{\overset{\displaystyle O}{\|}}{C}}-NH_2$$

A detailed distinction between organic and inorganic chemistry is not particularly useful and might obscure the important fact that there is no fundamental difference between the bonds in an organic molecule and those in inorganic molecules, and that an organic reaction proceeds according to the same thermodynamic principles as any other reaction. The division between organic and inorganic chemistry persists, however, because the chemistry of carbon compounds is so complicated—there are well over a million known organic substances but only 100,000–200,000 inorganic compounds.

Functional Groups

Carbon's ability to form such an immense number of compounds with such varied properties is due to its intermediate electronegativity, four valence electrons, and its ability to form strong double and triple bonds (a feature of atoms in the second period). The formation of one bond by a carbon atom has little distorting effect on the other electrons of the atom, and they can participate in other covalent bonds. These bonds can join carbon atoms together (called **catenation**), or connect carbons to hydrogen or any other nonmetallic element (such as O, N, P, S, or a halogen) in a molecule. The four bonds of a carbon atom can all be single bonds, or they may be some combination of single and multiple bonds. Organic molecules can be extremely large—there are stable molecules containing thousands of carbon atoms joined by strong covalent bonds. Given all these ways to form bonds, carbon has ample opportunity to form its bewildering array of compounds.

The early recognition by Mendeleev and Meyer that there are families of elements was a major step in understanding chemistry. It is just as useful to divide organic compounds into families, groups of substances that exhibit similar chemical reactions, and relate the typical properties of a family to a structural feature common to all molecules in that family. This common feature is known as a **functional group**—a portion of a molecule responsible for the reactions that characterize it as a member of a particular family. Table 19.2 lists some of the major families of organic compounds and their functional groups. For example, all alcohols contain the hydroxyl (—OH) group, all amines contain a nitrogen atom bound by three single bonds, and all carboxylic acids contain the carboxyl group (—CO_2H). Some compounds are **polyfunctional**—they contain two or more functional groups and show reactions typical of each. For example, molecules of amino acids contain both an amino group and a carboxyl group.

Table 19.2 **Functional Groups**

Family	Functional group	Example
Alkane	C—C and C—H bonds only	Methane, CH_4
Alkene	$\searrow$C=C$\swarrow$	Ethylene, CH_2=CH_2
Alkyne	—C≡C—	Acetylene, HC≡CH
Aromatic hydrocarbon	Aromatic ring	Benzene, C_6H_6
Alcohol	—Ö—H	Methanol, CH_3OH
Ether	—C—Ö—C—	Ethyl ether, $CH_3CH_2OCH_2CH_3$
Aldehyde	—C(=Ö)—H	Acetaldehyde, CH_3CH=O
Ketone	—C—C(=Ö)—C—	Acetone, $CH_3\overset{O}{\overset{\|}{C}}CH_3$
Carboxylic acid	—C(=Ö)—Ö—H	Acetic acid, CH_3CO_2H
Ester	—C(=Ö)—Ö—C—	Methyl acetate, $CH_3CO_2CH_3$
Amine	—C—N̈—H, —C—N̈—C—, or —C—N̈—C—C— with H	Methylamine, CH_3NH_2; Dimethylamine, $(CH_3)_2NH$; Trimethylamine, $(CH_3)_3N$
Amide	—C(=Ö)—N̈—H, —C(=Ö)—N̈—C— with H	Acetamide, $CH_3\overset{O}{\overset{\|}{C}}$—$NH_2$

Alkanes

Alkanes are compounds of carbon and hydrogen (hydrocarbons) **that contain only single bonds.** Because they lack a functional group other than C—C and C—H bonds, alkanes do not engage in many reactions, but their few reactions are very important to our technology. The major source of alkanes is petroleum deposits; their combustion is our present major energy source, and their thermal decomposition (cracking) provides the simple molecules we need for synthesizing larger molecules.

All alkanes, being nonpolar, are insoluble in water. Their melting and boiling points depend on their molecular sizes and shapes: the smaller alkanes (CH_4 through C_4H_{10}) are the constituents of natural gas; the molecules found in gasoline contain from 5 to 10 carbon atoms; lubricating oils are composed of larger, less volatile alkanes; alkanes containing 20 or more carbon atoms are present in solids such as paraffin wax and asphalt.

The three simplest alkane molecules (Table 19.3) are methane (CH_4), ethane (C_2H_6), and propane (C_3H_8). The presence of four or more carbon atoms in an alkane allows for **isomerism** (Section 7.2); the four carbon atoms can be arranged in a single chain, or they can form a branched arrangement. The possibilities for different branched isomers increases with the number of carbon atoms. There are two butane isomers (C_4H_{10}), three pentanes (C_5H_{12}), and five hexanes (C_6H_{14}). Isomerism is a major reason for the complexity of organic chemistry. Since the unique properties of each substance depend on the molecular structure, the molecular formula is not an adequate description of the substance, and we must discover the arrangement of atoms within the molecule before we can understand its chemistry.

The presence of atoms other than carbon and hydrogen also increases the possibilities for isomerism. For example, propane, the only substance with the formula C_3H_8, forms two isomeric chlorination products with the formula C_3H_7Cl.*

$$\begin{array}{cc}
\text{H} \quad \text{H} \quad \text{H} & \text{H} \quad \text{H} \quad \text{H} \\
| \quad | \quad | & | \quad | \quad | \\
\text{H}-\text{C}-\text{C}-\text{C}-\text{Cl} & \text{H}-\text{C}-\text{C}-\text{C}-\text{H} \\
| \quad | \quad | & | \quad | \quad | \\
\text{H} \quad \text{H} \quad \text{H} & \text{H} \quad \text{Cl} \quad \text{H} \\
\text{1-Chloropropane} & \text{2-Chloropropane}
\end{array}$$

Enantiomers

A rather different and very subtle type of isomerism occurs in the case of the chlorobutanes (C_4H_9Cl). There are two isomers that fit the structural formula for 2-chlorobutane:

$$\begin{array}{c}
\text{CH}_3\text{CH}_2\text{CHCH}_3 \\
| \\
\text{Cl}
\end{array}$$

In each isomer, the carbon atoms lie in an unbranched chain, and the chlorine is bound to an interior atom of the chain. The two isomers are very similar—they have the same melting point and boiling point, and most of their other chemical and physical properties are identical. Their most obvious difference is the way they interact with **plane-polarized light**—radiation whose electromagnetic waves all lie in the same plane (Figure 19.3). When plane-polarized light passes through a sample of one of these isomers, its plane of polarization twists in a clockwise direction; the other isomer rotates the plane of polarization to an equal degree in the counterclockwise direction. **Optical activity**—the ability to change the plane of polarization of light—always occurs in pairs of isomers known as **enantiomers**. The *dextrorotatory* enantiomer rotates the polarized light clockwise, and the *levorotatory* enantiomer has an equal counterclockwise effect.

Enantiomers are **mirror images** of each other—one molecule held up to a mirror would produce the image of its enantiomer (Figure 19.4, p. 484). All objects, including molecules, have mirror images, but in many cases there is no difference between the object and its reflection; the mirror image is the original object seen from a different perspective. A cube or any other symmetric object is identical with its mirror image. It is only an *asymmetric* object—an object we can describe as right-handed or left-handed—that differs from its reflection. (The hand is a common example.) This property of being either right-handed or left-handed is called **chirality**, and molecules possessing it are **chiral**. All optically active molecules are chiral, and one enantiomer is right-handed, the other left-handed.

The chirality of carbon compounds is usually due to the tetrahedral arrangement of single bonds around a carbon atom. Enantiomerism occurs when a carbon atom binds four *dissimilar* atoms or groups of atoms. One such *chiral center* is present in a 2-chlorobutane molecule—the carbon atom that is joined to a hydrogen atom, a chlorine atom, a methyl group (—CH_3), and an ethyl group (CH_3CH_2—). If any two

---

*The name 1-chloropropane means that a chlorine atom is bound to the terminal carbon (carbon number 1) of the propane carbon chain. In 2-chloropropane the carbon atom bearing the chlorine is the *second* one, counting from the end of the chain. A similar numbering system locates the positions of the methyl groups in alkanes (see Table 19.3).

Table 19.3 **Some Alkanes**

Molecular formula	Name	Structural formula
CH_4	Methane	$H-CH_2-H$ with H above and below (tetrahedral)
C_2H_6	Ethane	H_3C-CH_3
C_3H_8	Propane	$H_3C-CH_2-CH_3$
C_4H_{10}	Butane	$H_3C-CH_2-CH_2-CH_3$
	2-Methylpropane (Isobutane)	$H_3C-CH(CH_3)-CH_3$
C_5H_{12}	Pentane	$H_3C-CH_2-CH_2-CH_2-CH_3$
	2-Methylbutane (Isopentane)	$H_3C-CH(CH_3)-CH_2-CH_3$
	2,2-Dimethylpropane (Neopentane)	$C(CH_3)_4$
C_6H_{14}	Hexane	$H_3C-CH_2-CH_2-CH_2-CH_2-CH_3$

19.3 ORGANIC CHEMISTRY

Molecular formula	Name	Structural formula

	2-Methylpentane	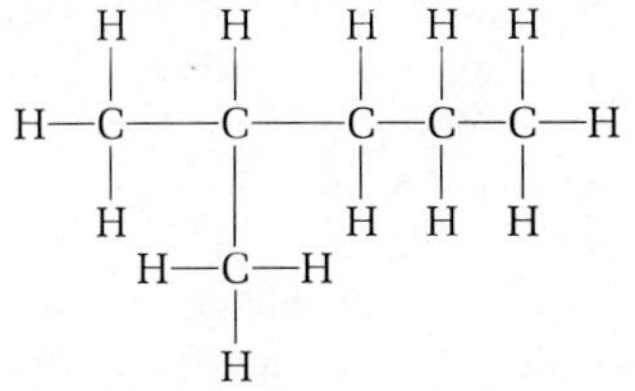 H H H H H
		H—C—C—C—C—C—H
		H H H H
		    H—C—H
		        H

	3-Methylpentane	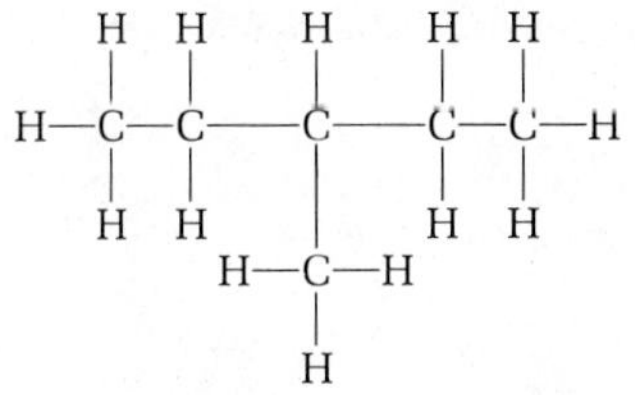 H H H H H
		H—C—C—C—C—C—H
		H H H H
		     H—C—H
		       H

	2,2-Dimethylbutane	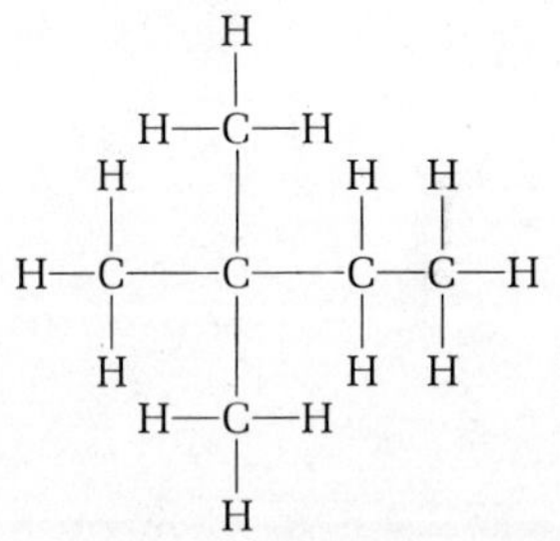 H
		   H—C—H
		H H H
		H—C—C—C—C—H
		H H H
		   H—C—H
		     H

	2,3-Dimethylbutane	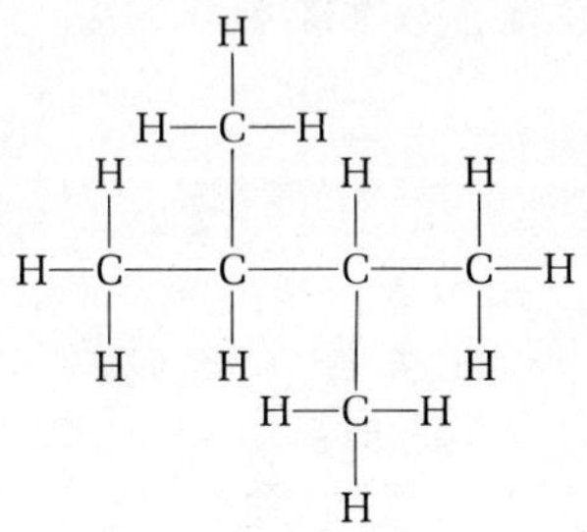 H
		   H—C—H
		H H H
		H—C—C—C—C—H
		H H H
		    H—C—H
		      H

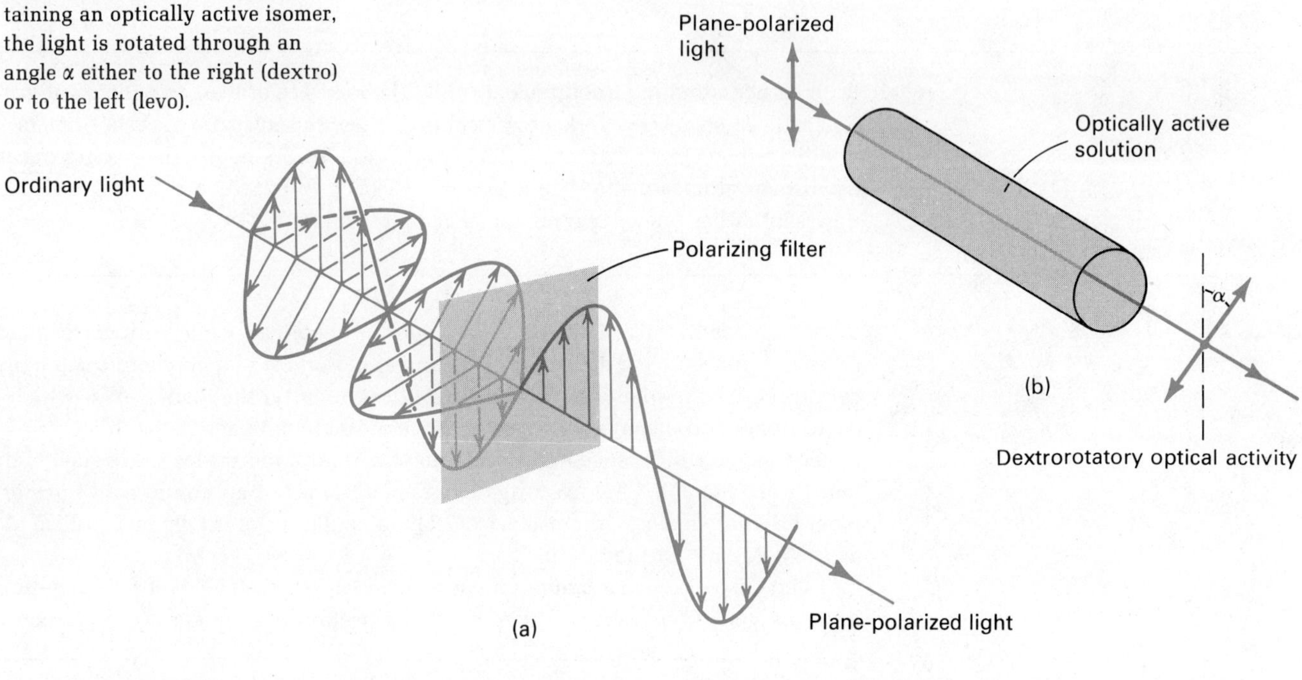

Figure 19.3
(a) Ordinary, unpolarized light consists of many waves all perpendicular to the direction of propagation of the beam. After passing through a polarizing filter, the wave disturbance is in a single plane. (b) When plane-polarized light passes through a solution containing an optically active isomer, the light is rotated through an angle α either to the right (dextro) or to the left (levo).

Figure 19.4
Enantiomers. (a) Two enantiomers are nonidentical mirror images. A pair of hands and a pair of 2-chlorobutane molecules are examples of enantiomers. (b) A symmetric object or molecule is identical with its mirror image and does not have an enantiomer. A cube and a molecule of 2-chloropropane are each symmetric.

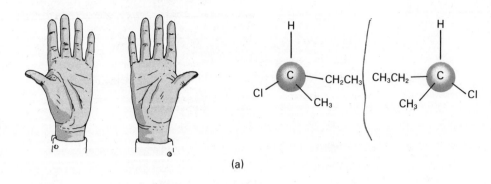

(a)

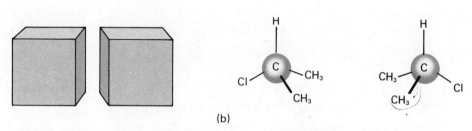

(b)

atoms (or groups of atoms) attached to a carbon atom are identical (as in the case of the central carbon atom in 2-chloropropane), the molecule is not chiral and does not have an enantiomer.

Example 19.1

Which of the following structures can exist as enantiomers?

a. CH$_3$CHCH$_2$Cl
      |
      Cl

b. CH$_3$CH$_2$CHCl$_2$

c. ClCH$_2$CH$_2$CH$_2$Cl

d.   Cl
      |
  CH$_3$CCH$_3$
      |
      Cl

Solution

Each carbon atom in structures b, c, and d is bound to at least two identical groups. Since these structures lack chiral centers, they cannot exist as enantiomers. The middle carbon atom of structure a is bound to four different groups—a chlorine atom, a hydrogen atom, a methyl group, and a —CH$_2$Cl group. Structure a, as written, could represent either one of two mirror-image molecules.

Since two enantiomers have the same types of chemical bonds with corresponding polarities, internuclear distances, and bond angles, most of their chemical and physical properties are identical. Except for optical activity, the instances in which two enantiomers do behave differently involve their interaction with other chiral molecules. This difference in reactivity to other chiral molecules is extremely important in biochemistry. We can digest one enantiomer of an amino acid but not the other because enzymes, themselves chiral molecules, catalyze the metabolism of one amino acid but not the other.

The clearest visible analogy for the difference in behavior of two enantiomers involves your own hands. Either hand fits around a symmetric object such as a

Figure 19.5
Ethylene. All atoms lie in the same plane.

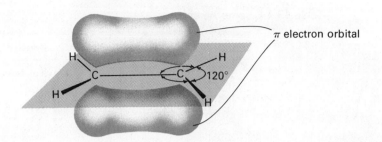

hammer equally well. Now fit your right hand into a right-hand glove and then try to put the same glove on your left hand. The fit is not as good. The same sort of difference in interaction occurs when enantiomers react with a pure sample of asymmetric molecules.

Alkenes and Alkynes

Alkenes, such as ethylene, C_2H_4, are hydrocarbons containing a carbon–carbon double bond. This bond involves two carbons in the sp^2 state of hybridization. One carbon–carbon bond is formed by σ overlap between two hybrid orbitals (see Figure 19.5) and the second by π overlap between the remaining unhybridized p orbitals. Each carbon atom forms two more covalent bonds to hydrogen atoms by σ overlap with its remaining two hybrid orbitals. The π bond prevents rotation of the double bond and holds the ethylene molecule in a *rigid* arrangement in which all four hydrogen atoms lie in the same plane as the two carbon atoms. The bond angles in ethylene are all close to 120°, consistent with sp^2 hybridization.

There is only one alkene with the formula C_3H_6 (propene), but there are four alkene isomers with the formula C_4H_8 (Table 19.4). In addition to chain branching, there are two other sources of isomerism specifically associated with the double bond. One is the location of the double bond in the chain; it can either connect an outer carbon atom to an interior one (1-butene), or it can lie between two interior atoms (as in the 2-butene isomers).

The second source of double-bond isomerism is the geometric arrangement of the groups attached to the doubly bound carbon atoms. There are two **geometric isomers** of 2-butene: *cis*-2-butene, a molecule whose two methyl groups lie on the *same* side of the double bond, and *trans*-2-butene, the isomer that would be produced by rotating the double bond in *cis*-2-butene (Figure 19.6). Since double-bond rotation, unlike single-bond rotation, is extremely difficult in the absence of high temperature, light, or a catalyst, there is no difficulty in maintaining pure samples of geometric isomers.

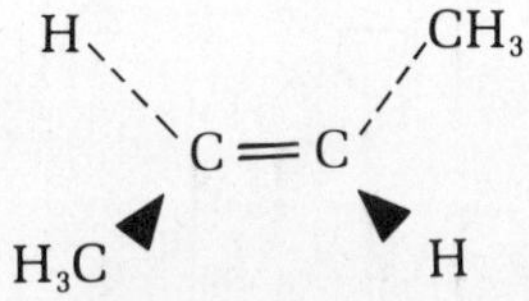

Geometric isomerism cannot occur if either of the two doubly bound carbon atoms bears two identical groups—rotation of the double bond in 1-butene would not change the structure of the molecule. (Do not confuse geometric isomers with enantiomers; neither *cis*- nor *trans*-2-butene is chiral and neither is optically active.)

Figure 19.6
Cis and *trans* isomers of 2-butene.

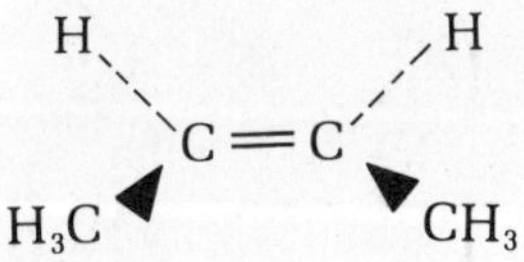

[*cis*-2-Butene structure]

[*trans*-2-Butene structure]

Example 19.2

Which of the following alkenes has a geometric isomer? Write its structure.

a. [ClCH₂ / H C=C H / H]   b. [CH₃ / H C=C Cl / H]   c. [CH₃ / H C=C Cl / H]

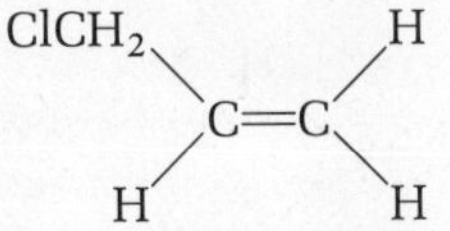

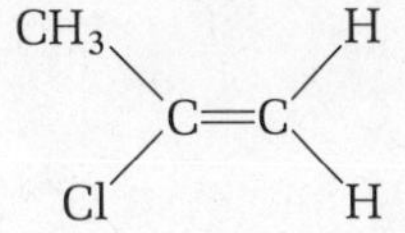

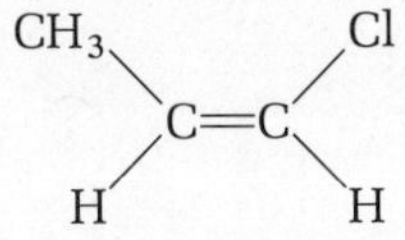

Solution

Since the right-hand carbon atoms of both a and b are bound to two hydrogen atoms, rotation of these double bonds would not change their structures. Therefore, they do

Table 19.4 **Some Alkenes**

Molecular formula	Name	Structural formula
C_2H_4	Ethylene (ethene)	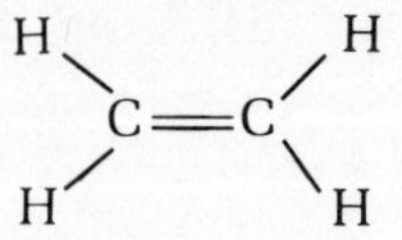
C_3H_6	Propene	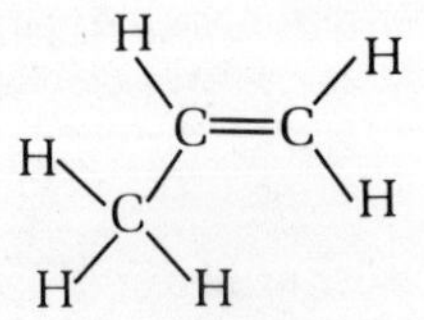
C_4H_8	1-Butene	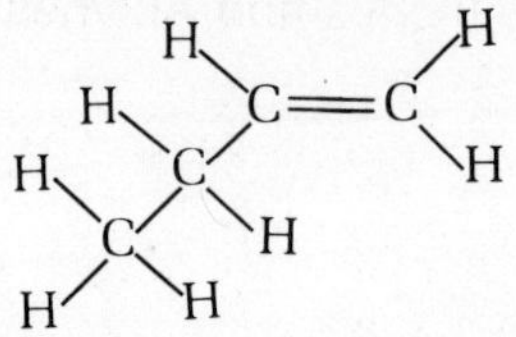
C_4H_8	cis-2-Butene	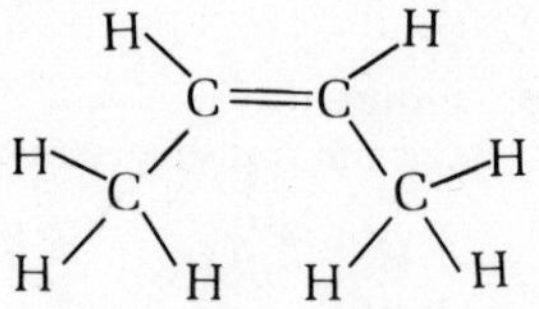
C_4H_8	trans-2-Butene	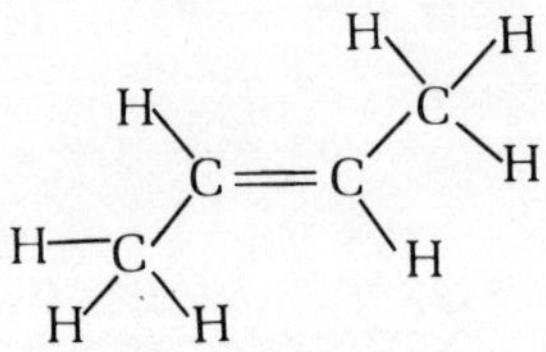
C_4H_8	2-Methylpropene (Isobutylene)	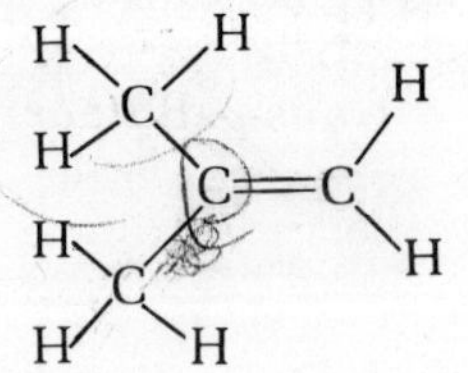

not have geometric isomers. Neither double-bonded carbon atom in c is bound to two identical groups. Rotation of this double bond would produce the geometric isomer.

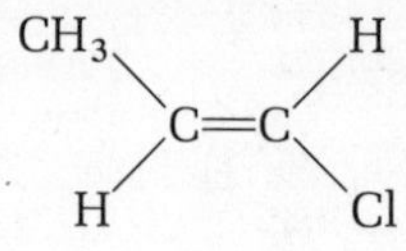

Alkynes contain carbon–carbon triple bonds. Each triply bonded carbon atom is in the sp state of hybridization, and the triple bond consists of one σ bond between a pair of hybrid orbitals and two π bonds between unhybridized orbitals. The linear

Figure 19.7
Acetylene, an alkyne.

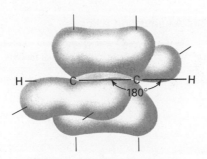
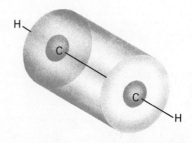

Acetylene is a linear molecule with 2 π_p bonds

The resulting π molecular orbital is cylindrically symmetric

shape of acetylene, C_2H_2, fits this description of the triple bond (Figure 19.7). The linearity of triple bonds prevents geometric isomerism in alkynes because triple bond rotation would not alter the molecular shape.

Molecules containing double or triple bonds are susceptible to addition reactions to form molecules containing only single bonds (*saturated molecules*).

$$H_2C{=}CH_2 + H_2 \xrightarrow{\text{Catalyst (Pt)}} CH_3CH_3$$

$$HC{\equiv}CH + 2\,H_2 \xrightarrow{\text{Pt}} CH_3CH_3$$

$$H_2C{=}CH_2 + Br_2 \longrightarrow BrCH_2CH_2Br$$

$$HC{\equiv}CH + 2\,Br_2 \longrightarrow Br_2CHCHBr_2$$

If an organic substance readily engages in addition reactions, it is a clear indication that its molecules contain either double or triple bonds (or both). A simple laboratory test for alkenes and alkynes is to add bromine to the organic compound. If bromine's characteristic reddish color fades out rapidly, it is probably due to addition to a multiple bond.

Cyclic Compounds

Some organic molecules contain rings of carbon atoms and are known as **cyclic compounds**. Cyclopropane, C_3H_6, consists of a triangle of carbon atoms, each carbon atom being bound to two hydrogen atoms (Table 19.5). There are rings containing four, five, six, and larger numbers of carbon atoms. Some rings contain double bonds (cycloalkenes), some have carbon chains projecting from the ring (ethylcyclopentane, for example), and others are composed of several rings. The rings in **heterocyclic compounds** contain atoms other than carbon.

Figure 19.8
A puckered, strain-free conformation of cyclohexane.

Cyclopropane rings (and to a lesser extent cyclobutanes) are somewhat unstable because of *ring strain*. The bonds in the cyclopropane ring are forced into 60° angles rather than the 109.5° angles characteristic for singly bound carbon atoms. The resulting orbital overlap is less effective than in normal single bonds, and the cyclopropane ring tends to break open to form unstrained molecules. Larger rings are relatively free of strain. The bond angles required for the pentagonal structure of cyclopentane are nearly equal to the tetrahedral bond angle. The bond angles in cyclohexane are normal because the ring puckers (Figure 19.8) to avoid bond angles greater than 109.5°. Consequently, cyclopentane, cyclohexane, and the larger cycloalkane rings behave like typical alkanes.

Aromatic Compounds

Benzene is a hydrocarbon whose properties sharply differentiate it from the compounds we have considered so far. Although its formula, C_6H_6, seems to indicate a high degree of unsaturation—a six-carbon alkane has eight more hydrogens—benzene does not engage readily in the addition reactions typical of molecules containing multiple bonds. Chemical and physical evidence shows that a benzene molecule consists of a planar hexagon of carbon atoms with each carbon atom bound

Table 19.5 Some Cyclic Alkanes and Alkenes

Name	Structural formula	Shorthand structural formula
Cyclopropane	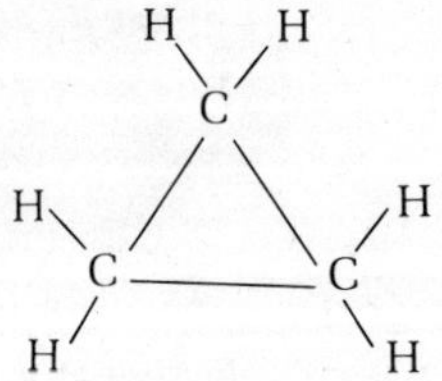	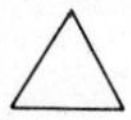 △
Cyclobutane	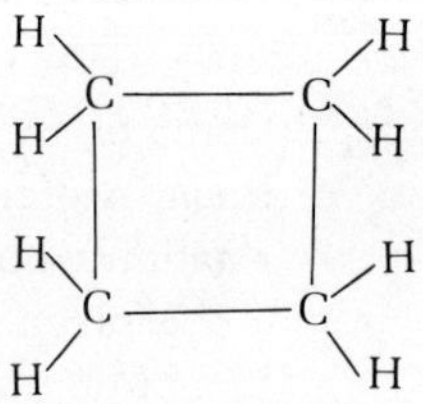	□
Cyclopentane	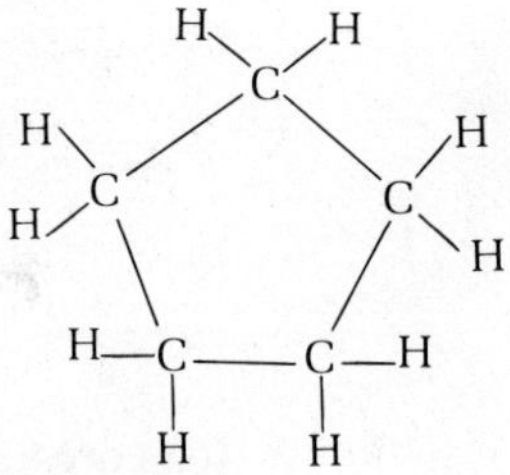	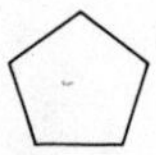 ⬠
Cyclohexane	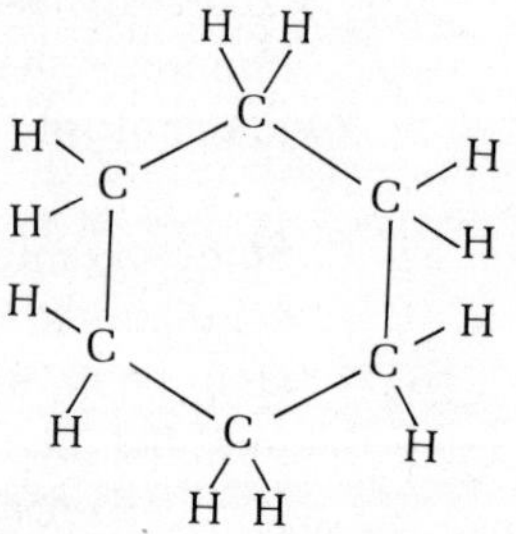	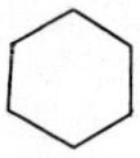 ⬡
Cyclopentene	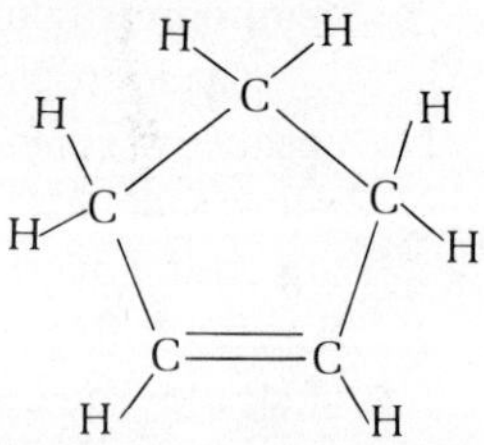	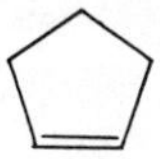 ⬠ (with double bond)
Ethylcyclopentane	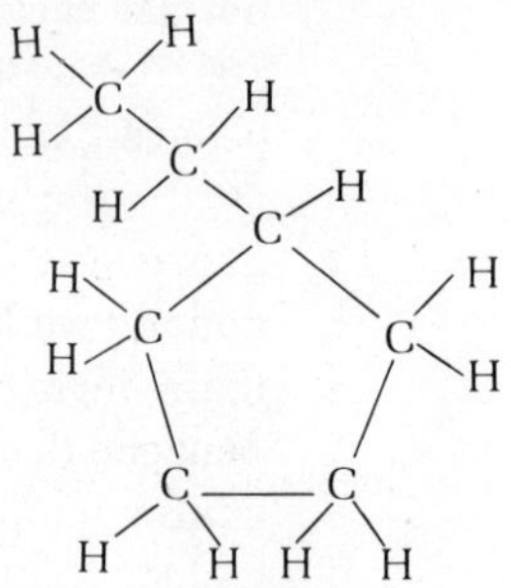	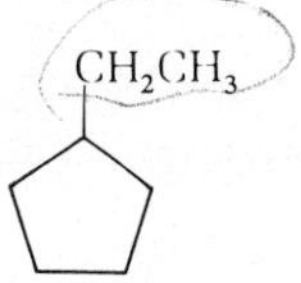 ⬠–CH$_2$CH$_3$

to a hydrogen (Figure 19.9). Each carbon atom is in the sp^2 hybridization state, and its hybrid orbitals form σ bonds with each of the two adjacent carbon atoms and with a hydrogen atom. The six unhybridized p orbitals (one on each carbon atom) interact in a π fashion to create a set of molecular orbitals that cover the entire ring. The composite electron density cloud for the six π electrons consists of two donut-shaped lobes parallel to the ring, one above it and one below it. This very stable electronic arrangement resists addition reactions, processes that would disrupt the circular π orbitals.

It is hard to translate the bonding in benzene into a Lewis structure. The best we can do is to represent benzene as a hybrid of two resonance forms, each containing a ring of carbon atoms held together by alternating double and single bonds. This description is accurate, the carbon–carbon bond lengths are intermediate between those observed for single bonds and double bonds, but it is so cumbersome that chemists usually draw only one of the two resonance forms. This might be misleading because there are no double bonds in benzene; we know, however, that this shorthand notation does not represent the true structure of benzene. A more recent symbol for benzene consists of a circle (representing the six π electrons) within a hexagon.

Many organic molecules have cyclic π electron systems that resist addition reactions. Most of these **aromatic** molecules contain one or more benzene rings. Others have heterocyclic rings (for example, pyridine) whose bonds resemble those in benzene, and a few (not shown here) have unusual sorts of cyclic structures. Molecules lacking an aromatic ring system are classified as **aliphatic** molecules. Alkanes, alkenes, alkynes, and cycloalkanes are all aliphatic substances.

Figure 19.9
Benzene and other aromatic molecules.

(a) Benzene σ bonds

(b) Benzene π electron cloud

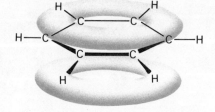

(c) Benzene resonance description

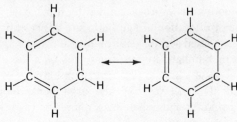

(d) Modern symbolism for benzene

(e) Other aromatic molecules

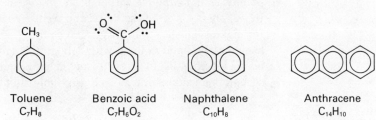

Toluene C_7H_8 Benzoic acid $C_7H_6O_2$ Naphthalene $C_{10}H_8$ Anthracene $C_{14}H_{10}$ Pyridine C_5H_5N

Structure Proof

Discovering the structure of a complicated organic molecule is often a difficult job, one that sometimes takes years of work. Although it is impossible to predict the precise experimental procedure needed to elucidate the structure of a particular compound, there is a general method for attacking structure-proof problems.

The first stages of a structure proof are straightforward; quantitative elemental analysis and molecular mass measurements tell us the molecular formula of the compound. Simple chemical tests, abetted by infrared spectroscopy (see Section 16.2), can tell us what functional groups are present in the molecule. For example, if a compound is a weak acid (K_A about 10^{-5}), its molecule probably contains at least one carboxyl group ($-CO_2H$).

Once the molecular formula of a compound is known and its functional groups identified, we must determine the arrangement of atoms in the molecule. This can be a most demanding task if the molecule is large (see Figure 19.10). The usual method is to decompose the molecule into smaller fragments using specific reaction conditions that produce predictable structural changes. From the structures of the decomposition fragments—they are easier to identify than the original molecule—we can reconstruct the original structure.

Many organic compounds are natural constituents of the biological world; living organisms produce them and can decompose them into simpler substances. Other organic substances—those occurring in petroleum and coal deposits—were formed by the decomposition of buried biological matter due to the intense heat and pressure under the earth's surface. As chemists learned how to synthesize complicated carbon-containing molecules, man-made organic substances began to play increasingly important roles in our lives.

Synthetic organic substances such as dyes, fibers, pesticides, pharmaceuticals, and plastics are such common features of modern life that it's easy to take them for granted and ignore their pervasive social and economic impact, and although it's popular to refer to our era as the nuclear age, synthetics have much more influence on our lives than nuclear reactors. The development of artificial fibers such as nylon frees agricultural land and a large labor force from the growing of plant fibers, allowing greater production of foodstuffs (a process aided by other chemical innovations).

The proliferation of synthetic substances and the bulk in which we use them has created serious environmental problems. Most synthetic substances can't be metabolized by live organisms, and the waste products of our technological society are continually accumulating. Some are absorbed by plants and animals but, resisting the organism's biochemical mechanisms for decomposing foreign substances, remain in the organism and may be passed on to higher organisms in the food chain. Plastics present a solid waste problem. Unlike the vegetable matter they replace, they aren't biodegradable, and the pile of synthetic waste material grows daily.

Another problem with a technology based on synthetic carbon compounds is that although carbon is one of the earth's more abundant elements, its supply is nevertheless *finite*, and we are running out of the raw material for making plastics and fibers. The solution of the limited supply of petroleum may go hand-in-hand with solving the solid waste problem. Since carbon is too valuable to throw away, the recycling of solid waste will preserve carbon in a usable form as well as getting rid of heaps of garbage.

Figure 19.10
Some possible structures for $C_4H_8O_2$.

CARBOXYLIC ACIDS

$CH_3CH_2CH_2CO_2H$ CH_3CHCO_2H
                                  $|$
                                 CH_3

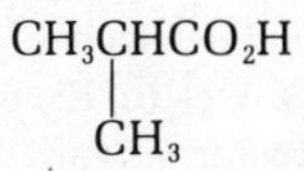

ESTERS

$CH_3CH_2\overset{O}{\overset{\|}{C}}-OCH_3$ $CH_3\overset{O}{\overset{\|}{C}}OCH_2CH_3$

$\overset{O}{\overset{\|}{HC}}-O-\overset{CH_3}{\underset{CH_3}{CH}}$ $\overset{O}{\overset{\|}{HC}}-OCH_2CH_2CH_3$

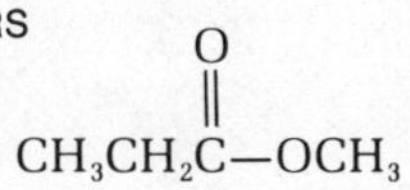

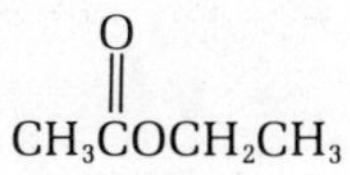

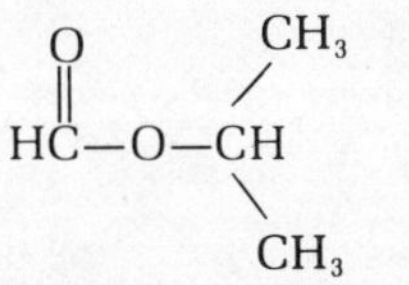

HYDROXY ALDEHYDES SUCH AS

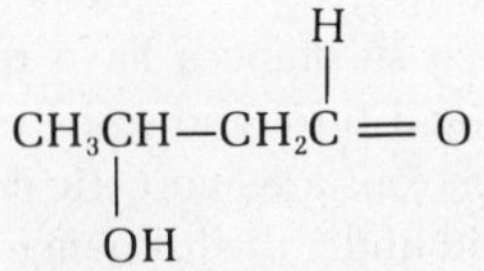

$CH_3CH-CH_2\overset{H}{\overset{|}{C}}=O$
$|$
OH

HYDROXY KETONES SUCH AS

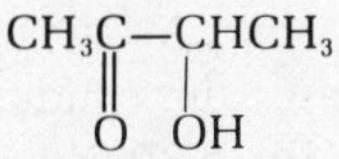

$CH_3\underset{O}{\overset{}{C}}-\underset{OH}{\overset{}{C}}HCH_3$

DIHYDROXY ALKENES SUCH AS

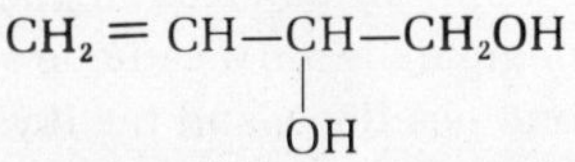

$CH_2=CH-\underset{OH}{\overset{}{C}H}-CH_2OH$

CYCLIC COMPOUNDS SUCH AS

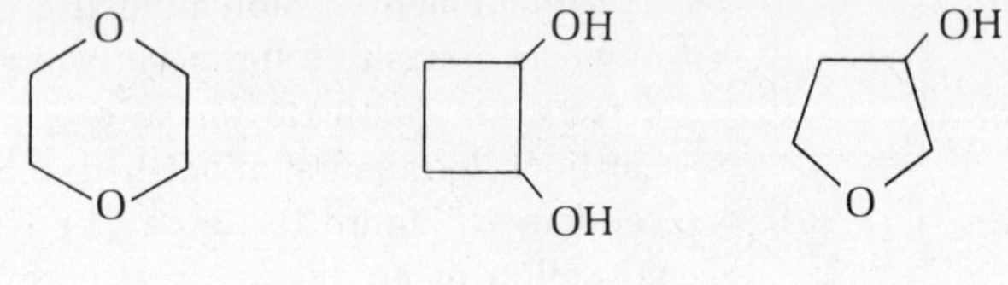

(cyclic structures: 1,4-dioxane; cyclobutane-1,2-diol; 3-hydroxytetrahydrofuran)

19.4
Silicon, Germanium, Tin, and Lead

Properties of the Metals

The four elements following carbon in group IVA have considerably more metallic character than carbon. Metallic silicon and germanium are light gray in color, very hard and brittle materials. The pure metals are only poor electric conductors at room temperature, however, increased temperatures, or the presence of crystal defects or impurities enhances the conduction significantly. Because of these **semiconducting** properties, silicon and germanium are used extensively to prepare electronic components such as transistors, rectifiers, photocells, and detectors.

Figure 19.11
Types of semiconductors. (a) Electron-deficient semiconductor, *p*-type (positive). (b) Excess electron semiconductor, *n*-type (negative).

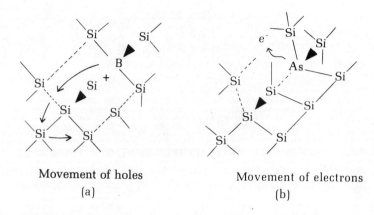

Figure 19.12
Stoichiometric defect semiconductors. (a) Schottky type (for example, CrO and Fe_3O_4). (b) Frenkel type (for example, AgBr).

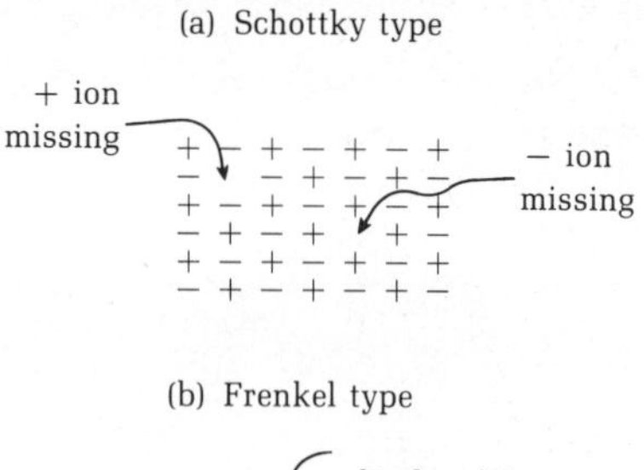

Both silicon and germanium have diamondlike crystal structures in which tetrahedral networks of σ bonds unite sp^3 hybridized atoms. As in diamond, the electrons in the localized σ orbitals are not free to conduct electricity but unlike diamond, these substances have relatively low-energy, unoccupied molecular orbitals that extend throughout the entire crystal. At high temperatures a significant number of electrons are energetic enough to occupy these conduction band molecular orbitals, and under the influence of an external electric potential these conduction band electrons will "flow" through the crystal.

The room temperature conductivity of silicon and germanium can also be increased rather dramatically by the presence of small quantities of elements from periodic groups IIIA or VA—even a few atoms of the impurity for every million silicon atoms increases the conductivity a thousandfold! When the impurity is a group VA atom (As, Sb, or Bi) occupying a silicon atom's lattice site, its fifth valence electron, which is unused in bonding, moves easily through the crystal in this *n*-type semiconductor (*n* for negative). (See Figure 19.11.) Group IIIA atoms (B, Al, Ga), on the other hand, create electron vacancies in silicon and germanium lattices. These "holes" migrate toward a cathode as electrons from neighboring atoms flow into the vacant bond positions, and the crystal is called a *p*-type (positive) semiconductor.

Unusual electrical properties are also exhibited by some pure substances or compounds other than silicon or germanium when they contain certain types of lattice defects. These **defect structures** are of two general types—stoichiometric and nonstoichiometric. Stoichiometric crystals have defects that result from (1) equal numbers of positive and negative ions missing from their lattice positions (Schottky defects, Figure 19.12, part a) or (2) positive or negative ions occupying interstitial locations, tetrahedral or octahedral holes, rather than their regular lattice sites (Frenkel defects, Figure 19.12, part b). Nonstoichiometric defects (Figure 19.13) involve (1) cation or anion excesses, with resulting conductivity by electron or positive "hole" migration (ZnO heated to give $Zn_{>1}O$), (2) cation excess due to different possible oxidation states, or (3) cation excess in alkali metal halides produced by x-ray irradiation or heating of the alkali halide in alkali metal vapors. Glass that has been out in the sun for years turns purple because the ultraviolet radiation produces nonstoichiometric defects in its structure. A glass jar containing a radium salt also turns purple due to the intense x rays from radioactive decay of radium atoms.

Tin and lead, also gray in color, are more typical metals than are silicon and germanium since they are soft, malleable, and good conductors. Both tin and lead are used in many important alloys, such as brass, solder, and motor bearings, and lead is used extensively for storage-battery plates. These metals resist corrosion quite well; the lead linings of large acid tanks and the lead plumbing in laboratories resist oxidation by H_2SO_4 because of lead's large overpotential for hydrogen formation.

Figure 19.13
Nonstoichiometric semiconductors. (a) Cation or anion excess, as much as several percent (for example, $Zn_{>1}O$). (b) Different oxidation states of the cation (for example, $Ni_{>1}O$, $Fe_{>1}S$, and $Cu_{>1}I$). (c) Cation excess resulting in trapped electrons, called "F" centers.

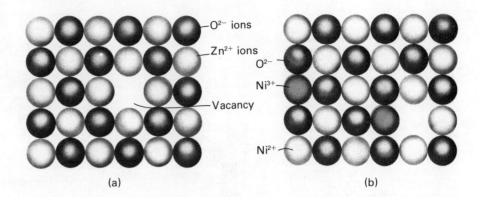

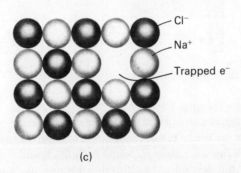

Although tin is too poor a reducing agent to give the galvanic protection afforded by zinc, it is an anticorrosion coating for iron (as in the "tin can"). The heavy nuclei of lead make it an effective shield against x rays and radioactivity.

In contrast to lead, which has only one crystalline modification (cubic-closest packed), tin has three polymorphic forms, α-, β-, and γ-Sn. Gray tin (the α-phase) has the diamond structure, is very brittle, but unlike diamond, fractures easily. White tin (β), the stable form at room temperature, has a close-packed tetragonal structure. White tin changes to γ-tin when heated above 161°C and to α-tin when cooled below 13.2°C. The latter phase transition, accompanied by an enormous decrease in volume (27%), accounts for the so-called "tin disease" that harried Europe several centuries ago. At that time, dinnerware and many musical instruments were made of tin. During severe winters, organ pipes and plates would appear "diseased" and disintegrate as the temperatures in churches and homes fell below the transition temperature (13.2°C is ~56°F) for the conversion of β-tin to α-tin.

Preparation

Relatively impure silicon and germanium are obtained by reduction of the oxides with coke or hydrogen.

$$SiO_2(s) + C(s) \xrightarrow[\text{Furnace}]{\text{Electric}} Si(s) + CO_2(g)$$

$$GeO_2(s) + C(s) \xrightarrow[\text{Furnace}]{\text{Electric}} Ge(s) + CO_2(g)$$

$$GeO_2(s) + 2\,H_2(g) \xrightarrow[\text{Furnace}]{\text{Electric}} Ge(s) + 2\,H_2O(g)$$

It is necessary to process the impure metals by special procedures to obtain these elements in the high purity required for electronic components. The impure metals are converted to volatile halides such as $SiCl_4$ or $SiHBr_3$, and the halides purified by fractional distillation. Thermal decomposition or reduction of the distilled halide produces quite pure samples of the metal, but it still takes a final **zone-refining** process (Figure 19.14) to reduce impurities to about one part per billion. As a rod of

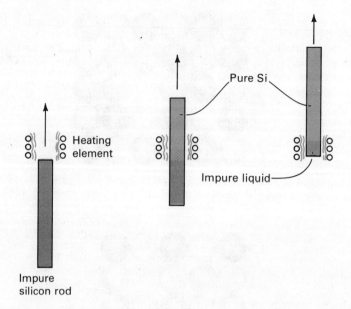

Figure 19.14
Zone refining of Si. Impurities, which are more soluble in the melt than in the solid, collect in the lower part of the rod leaving pure Si in the upper part.

the metal passes slowly through a narrow heating element, the metal in the immediate vicinity of the heater melts, and the narrow liquid zone moves from one end of the rod to the other. Since the impurities are more soluble in the liquid phase than in the solid, they accumulate in the liquid zone and are carried to the end of the rod. After several passes through heating elements, the leading end of the rod is very pure, and the tail end, which contains the impurities, is cut off.

Smelting (the melting and reduction of ores) of SnO_2 (cassiterite) and PbS (galena) produces tin and lead.* It is often necessary to roast the ores in air before smelting in order to remove impurities, such as sulfur and arsenic from SnO_2, and metal impurities from PbS. Electrorefining (Section 15.2) purifies the smelted metals.

Compounds of Silicon

Silicon compounds illustrate two important differences between a third-row element of a periodic family (in this case silicon) and its second-row counterpart (carbon). The first difference is that silicon (as well as the heavier group IVA elements) tends not to form sp^2 or sp hybrid orbitals and does not form π bonds involving its p orbitals. In contrast to CO_2, a linear, multiply bonded molecule, SiO_2 (silica) is polymeric and contains sp^3 hybridized silicon atoms at the centers of tetrahedra of covalently bound oxygen atoms. Silicates also contain these SiO_4 tetrahedra, whereas the carbonate ion is trigonal planar and contains π bonds between carbon and the surrounding oxygen atoms. The second difference between silicon and carbon is the ability of silicon to use unoccupied $3d$ orbitals to form superoctet structures (Section 7.2) such as SiF_6^{2-}. The covalent bonds holding these superoctet species together involve hybrid orbitals from third-shell s, p, and d orbitals (sp^3d^2 orbitals). Lacking corresponding low-energy d orbitals, carbon cannot form stable structures containing more than four covalent bonds to a carbon atom.

Silicon occurs naturally as silica and silicates, and the SiO_4 tetrahedra in silicates are linked in many different ways to give a wide variety of formulas and structures. Silicate structures are often very complex, and the numerous silicates present in minerals differ in the number of oxygen atoms shared between two silicon atoms, the geometric arrangement of the tetrahedra (Figure 19.15), and in the number, type, and arrangement of the metallic cations. The physical properties of some silicates reflect the arrangement of the tetrahedra within; for example, lamellar silicates, such as mica, fracture easily into thin sheets, whereas the silicate chains in pyroxenes produce fibrous minerals such as asbestos.

*The term "smelting" refers to any process by which a metal is obtained by heating its ore in a furnace. This covers several different types of reactions, some involving only the ore and oxygen, others requiring carbon or CO as reducing agents.

Figure 19.15 Silicates.

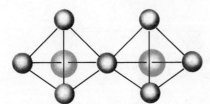

Orthodisilicates, $Si_2O_7^{6-}$

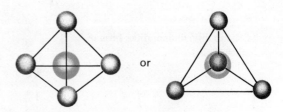

Orthosilicates, SiO_4^{4-}, as in zircon, $ZrSiO_4$

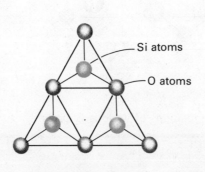

Metasilicates, SiO_3^{2-} ($Si_3O_9^{6-}$ rings)

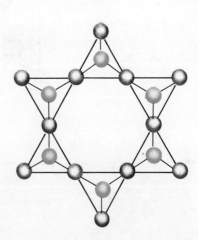

$Si_6O_{18}^{12-}$ rings, as in beryl (emerald), $Be_3Al_2Si_6O_{18}$

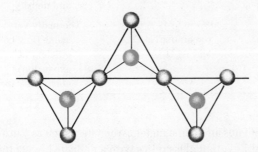

$(SiO_3)_n^{2n-}$ chains (pyroxenes), as in spodumene, $LiAl(SiO_3)_2$; also occurs as $(Si_2O_5)_n^{2n-}$ layers in clays and micas, for example, talc, $Mg_3(OH)_2(Si_2O_5)_2$

Pure silica (SiO_2) occurs in two crystalline modifications, quartz and cristobalite. The SiO_4 tetrahedra in quartz form large helices, spiraling either to the left or to the right; thus there are enantiomers of quartz. The structure of cristobalite resembles diamond; silicon atoms occupy positions corresponding to those of carbon in diamond, and the oxygen atoms lie between silicon atoms (Figure 19.16).

Slow cooling of molten silica produces an amorphous material that we know as glass. The lack of an x-ray diffraction pattern (Section 9.4) shows that this solid lacks long-range, periodic order. (There is some local order in the arrangement of the SiO_4 tetrahedra.) Common glass is prepared by slowly cooling solutions of basic metal oxides in molten silica; varying the composition of the solution, we can produce many types of glass with widely differing properties (see Table 19.6). Colored glass results from the presence of small quantities of metal ions—cobalt for blue, manganese for violet, chromium for green, and uranium for yellow.

The birthplace of the glass industry was apparently Egypt in the Eighteenth Dynasty, about 1500 B.C. Glass was made by heating together silica, lime (CaO),

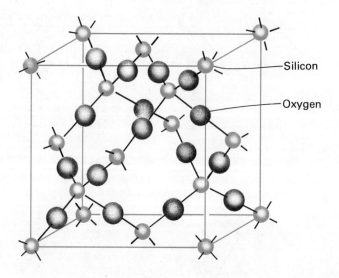

Figure 19.16
Cristobalite, diamondlike form of silica (SiO$_2$).

Table 19.6 Properties and Compositions of Various Glasses

Type of glass	Composition	Properties	Uses
Soft or soda	75% SiO$_2$, 15% Na$_2$O, 5% CaO, remainder Al$_2$O$_3$ and MgO	Inexpensive, common type	Bottles, jars, windows, glass tubing
Flint	45% SiO$_2$, 45% PbO, 10% K$_2$O	Optically clear, high refraction and density	Cut glass, dinner ware, ornamental
Pyrex brand, borosilicate	80% SiO$_2$, 12% B$_2$O$_3$, remainder Na$_2$O and Al$_2$O$_3$	Thermally very stable (low coefficient of expansion), chemically inert	Laboratory glassware, high or low temperature applications

alkaline carbonates and some metal carbonates, such as CuCO$_3$, to impart color to the glass. Ancient Egyptian glass relics with a delicate blue color contain approximately 3% CuCO$_3$ while those with 20% CuCO$_3$ are a deep purplish-blue.

The fusion of silica mixtures containing large quantities of sodium carbonate produces a water-soluble material, called "water glass," containing polymeric species with the general formula [SiO$_2$]$_n$(OH)$_2^{2-}$. Acidification of water glass yields silicic acids, the weak acid counterparts of the polymeric silicates. Since silica is a weak acid, strong alkaline solutions etch glass by converting the silica to soluble silicates.

$$[SiO_2]_n(s) + 2\ OH^- \longrightarrow [SiO_2]_n(OH)_2^{2-}$$

This is why you should not put NaOH solution in a buret with a glass stopcock—the hydroxide ion etches the smooth glass in the stopcock, and the etched surface jams the stopcock. Since silica is not at all basic, glass is resistant to acids.

Compounds of Germanium, Tin, and Lead

All group IVA elements exhibit +4 oxidation states and form stable tetrahalides and dioxides. The +2 states also occur,* and there is a definite trend toward increasing

---

*By playing the game of assigning the +1 oxidation state to hydrogen and −2 to oxygen, we can compute many different oxidation states for carbon in its organic compounds. However, these oxidation states do not tell us much about the nature of these compounds and are not a useful basis for classifying them.

stability of the lower oxidation state in the larger atoms. Lead and tin, in contrast to silicon and germanium, have many stable +2 compounds—substances such as $SnCl_2$, SnO, PbO, and many Pb(II) salts. Although tin(II) compounds, such as stannous chloride ($SnCl_2$), are effective reducing agents and are readily oxidized to stannic compounds, Pb(II) compounds are much harder to oxidize, and PbO_2 is a strong oxidant (Table 19.7).

Neither Sn(II) nor Sn(IV) are stable in dilute aqueous solutions because of their strong tendencies to hydrolyze and form insoluble oxides.

$$Sn^{2+}(aq) + H_2O \rightleftharpoons SnO(s) + 2\,H^+(aq)$$

$$Sn^{4+}(aq) + 2\,H_2O \rightleftharpoons SnO_2(s) + 4\,H^+(aq)$$

Stannous and stannic salts do dissolve readily in HCl solution to form stable chloride complexes, $[SnCl_4]^{2-}$ and $[SnCl_6]^{2-}$, respectively.

The easiest way to prepare stannous chloride solution is to dissolve tin metal in hydrochloric acid.

$$Sn(s) + 2\,H^+(aq) + 4\,Cl^-(aq) \longrightarrow SnCl_4^{2-}(aq) + H_2(g)$$

This solution, usually stored over excess tin to prevent air oxidation to stannic chloride, is used to reduce metal ions such as Fe(III), Hg(II), and Ag(I).

$$2\,Cl^-(aq) + SnCl_4^{2-}(aq) + 2\,Fe^{3+}(aq) \longrightarrow 2\,Fe^{2+}(aq) + SnCl_6^{2-}(aq)$$

$$4\,Cl^-(aq) + SnCl_4^{2-}(aq) + 2\,Hg^{2+}(aq) \longrightarrow Hg_2Cl_2(s) + SnCl_6^{2-}(aq)$$

Most Pb(II) salts are insoluble, the nitrate and acetate being the most common exceptions.

Group IVA oxides and hydroxides tend to be acidic, and all the elements form complex oxy and hydroxy anions.

$$CO_2 + OH^-(aq) \rightleftharpoons HCO_3^-(aq)$$

$$SiO_2(s) + 4\,OH^-(aq) \rightleftharpoons SiO_4^{4-}(aq) + 2\,H_2O$$

$$SnO_2(s) + 2\,OH^-(aq) \rightleftharpoons SnO_3^{2-}(aq) + H_2O$$

$$SnO(s) + 2\,OH^-(aq) \rightleftharpoons SnO_2^{2-}(aq) + H_2O$$

$$PbO(s) + 2\,OH^-(aq) \rightleftharpoons PbO_2^{2-}(aq) + H_2O$$

The properties of these oxides illustrate the trend toward decreasing acidity and increasing basicity with increasing size of the metal atom. Carbon dioxide is the group's most acidic oxide, whereas SnO_2 is amphoteric—it tends to dissolve in strong acid as well as in base. A similar trend holds for SnO and PbO; lead(II) oxide is the more basic and less acidic compound. Tin(II) oxide is more basic and less acidic than SnO_2, an example of the influence of oxidation state on acidity.

Red lead (Pb_3O_4), the red pigment in the undercoating used to protect iron against corrosion, does not represent an unusual oxidation state for lead. Formed by

Table 19.7 **Representative Stable Compounds of Ge, Sn, and Pb**

Element	Reactant		
	O_2	I_2	S
Ge	GeO_2	GeI_4	GeS_2
Sn	SnO_2	SnI_4	SnS_2
Pb	PbO and Pb_3O_4	PbI_2	PbS

roasting PbO in air, this substance contains Pb(II) atoms and Pb(IV) atoms in a 2:1 ratio.

There are many organometallic compounds of the heavier group IVA metals in which the metal is covalently bound to alkyl groups. The most important of these compounds, tetraethyl lead [$Pb(C_2H_5)_4$], is added to some grades of gasoline to assure smooth combustion of the fuel. There is increasing concern, however, over the pollution of our environment by toxic metals such as mercury, cadmium, chromium, and lead—and rightfully so because the actual amounts of lead in our environment have increased at an alarming rate since the introduction of tetraethyl lead additive for gasolines. Literally thousands of tons of lead are released into the atmosphere each day, and when attempts are made to analyze areas or substances for their lead content, chemists must use extreme techniques to avoid undue contamination of the substance being analyzed. Lead is literally everywhere—in our hair, on our clothes, skin, dishes, laboratory apparatus—and we must first have a "clean room" (much, much cleaner than a surgical operating room), free from lead contamination, in which we can carry out experiments to gather meaningful data. The Romans' lead problems seem trivial compared to our massive success at pollution. We are increasingly aware of these problems, though, and scientists have a responsibility to preserve as well as create a beneficial environment—tetraethyl lead is rapidly being removed from gasolines.

Summary

1. Of the group IVA elements, carbon is the most notable because its compounds form the molecular basis for plant and animal life. Group IVA elements, because of their half-filled valence shells (s^2p^2), form chiefly covalent compounds.

2. Silicon is the most abundant element of the group occurring extensively in rocks and earth as silica (SiO_2) and silicates.

3. Carbon has two polymorphs, diamond and graphite. The carbon atoms in each are bonded differently, sp^3 and sp^2 hybridized, respectively, giving diamond and graphite very different properties.

4. The principal isotope of carbon, ^{12}C, is the basis of the relative scale of atomic masses with a mass of exactly 12.

5. Carbides are compounds of carbon combined with elements of equal or lower electronegativity.

6. Carbon dioxide (CO_2) is formed by complete combustion of carbon compounds, whereas carbon monoxide (CO) is formed during incomplete combustion. Carbon dioxide is the common link in the carbon cycle involving plant and animal metabolism. In water, CO_2 dissolves to form carbonic acid. Limestone ($CaCO_3$) and sodium bicarbonate ($NaHCO_3$) are two important salts derived from CO_2. Carbon monoxide is very poisonous because it acts as a strong Lewis base and replaces O_2 in blood hemoglobin so effectively that body tissues starve for oxygen.

7. Cyanide ion, CN^-, is isoelectronic with CO. Gaseous HCN is poisonous and its water solution is a very weak acid, hydrocyanic acid.

8. Silicon, germanium, tin, and lead are considerably more metallic than carbon. Silicon and germanium are semiconductors whose electrical properties change dramatically when a small amount of a group IIIA element (called p-type semiconductors) or a group VA element (called n-type semiconductors) is added.

9. Substances that have defect structures also possess unusual electrical properties; they are of two main types: stoichiometric and nonstoichiometric. Schottky and Frenkel defects are two kinds of stoichiometric defects. Nonstoichiometric defects involve cation or anion excesses; these can be produced chemically or by irradiation.

10. Tin and lead are soft metals, good conductors, and are used chiefly in alloys, platings, and batteries. Tin has three crystalline modifications; one (gray tin) has the diamond structure.

11. Smelting (the melting and reduction) of SnO and PbS produces tin and lead metals. They are electrorefined to increase purity; ultrapure metals are produced by a technique called zone refining.

12. Silicon primarily forms sp^3 bonds and reacts readily with oxygen to form silicates. Tetrahedra of SiO_4 link in a variety of ways to form layered, chain, and three-dimensional network crystals. Slow cooling of silica (SiO_2) produces an amorphous glass whose properties can be extensively varied by addition of other metal ions. These ions occupy positions in the relatively open glass structure.

13. Germanium, tin, and lead form +4 and +2 oxidation states, the latter being more stable with the larger atoms. Tin salts tend to hydrolyze in other than strongly acidic solutions forming insoluble oxides; $SnCl_2$ is a good reducing agent. Oxides and hydroxides tend to be acidic and hydrides are covalent.

14. Organic chemistry, in the modern sense of the word, is the study of carbon compounds. The study of so large a number of substances is made easier by understanding their functional groups, such as —OH in alcohols.

15. Alkanes, alkenes, and alkynes are all hydrocarbons consisting of only carbon and hydrogen. Alkanes contain only single bonds, alkenes have carbon–carbon double bonds, and alkynes have carbon–carbon triple bonds.

16. Enantiomers are mirror-image isomers of the same compound that affect plane-polarized light differently; this effect is called optical activity. Enantiomers have asymmetric, or chiral, carbons (the four groups joined to an sp^3 carbon are all different), giving the molecule a left- or right-handedness.

17. Geometric isomers of carbon compounds are possible if the molecules have double bonds, because the conformation of parts of the molecule is fixed. Cis and trans isomers are geometric isomers.

18. Carbon compounds can be cyclic (rings of carbon atoms) or heterocyclic (some of the atoms are other than carbon atoms).

19. Aromatic carbon compounds such as benzene (C_6H_6) show different reactivities than aliphatic compounds (like alkanes) because they possess π molecular orbitals that extend over many atoms or perhaps the entire molecule.

Vocabulary List

graphite	isomerism	cyclic compound
diamond	plane-polarized light	heterocyclic compound
allotrope	optical activity	aromatic
ionic carbide	enantiomer	aliphatic
polymeric covalent carbide	chirality	semiconductor
catenation	alkene	defect structure
functional group	geometric isomer	zone refining
alkane	alkyne	smelting

Exercises

1. Write balanced equations for the reaction of:
   a. Lead(IV) oxide plus excess 6 M NaOH   b. Tin metal plus 6 M HCl
   c. Tin(II) chloride plus mercury (II) chloride   d. Silica plus concentrated NaOH

2. On the basis of the properties of the group IVA elements:
   a. Which element forms the most basic oxide?
   b. Which element has the highest boiling point?
   c. Which element has the lowest melting point?
   d. Which element is the most metallic?

3. Carbon dioxide is a gas at room temperature, whereas SiO_2 is a high melting solid. Explain this in terms of the structures of the two compounds and the differences between second- and third-row elements.

4. Silicon forms the SiF_6^{2-} ion, but carbon does not form a corresponding ion. Why is this so?

5. Complete and balance the following reactions.
   a. $Si(s) + F_2(g) \longrightarrow$
   b. $Li_2C_2(s) + H_2O \longrightarrow$
   c. $GeO_2(s) + H_2(g) \longrightarrow$

6. A classic test for the presence of CO_2 in a gas is to bubble the gas through a clear solution obtained by shaking lime (CaO) with water ("limewater"). If a white precipitate forms, CO_2 is present. Write equations for:
   a. The preparation of limewater   b. The formation of the precipitate due to CO_2

7. Why is diamond an electric insulator? Would you expect a crystal of graphite to conduct electricity equally well in all directions? Explain.

8. Draw the geometric isomer of

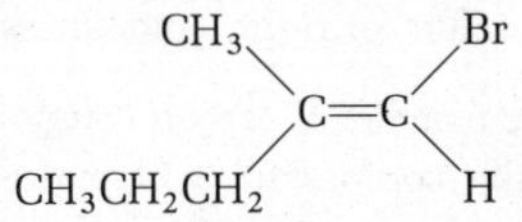

    $$\underset{CH_3CH_2CH_2}{\overset{CH_3}{\diagdown}} C=C \underset{H}{\overset{Br}{\diagup}}$$

9. Draw the enantiomer of

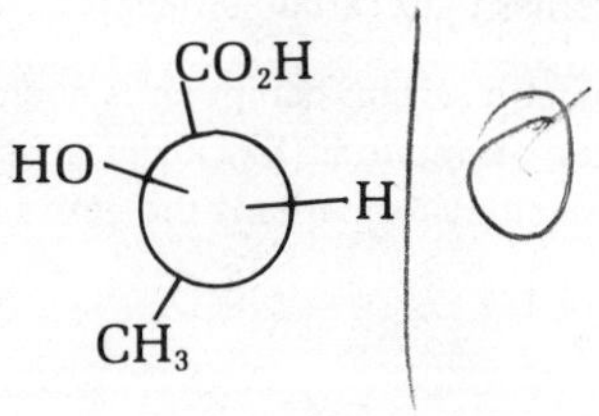

    [structure with CO₂H, HO, H, CH₃ substituents]

10. Predict the products of the following reactions.

    a. $CH_3C\equiv CH + 2 H_2 \xrightarrow{Pt}$
    b. 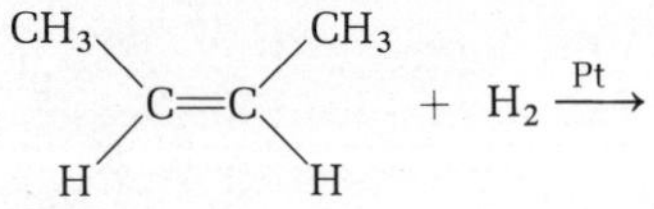 $\underset{H}{\overset{CH_3}{\diagdown}} C=C \underset{H}{\overset{CH_3}{\diagup}} + H_2 \xrightarrow{Pt}$
    c. $CH_3CH=CH_2 + Br_2 \longrightarrow$
    d. $CH_3C\equiv CCH_3 + 2 Br_2 \longrightarrow$

11. Which of the following molecules is:
    a. Aromatic?   b. Heterocyclic?   c. Both aromatic and heterocyclic?
    d. Strained?   e. A cycloalkene?

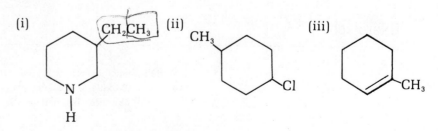

(iv) (v) 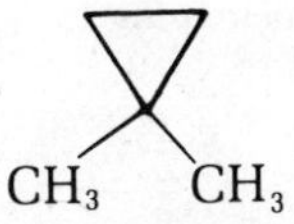 (vi)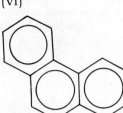

(vii) 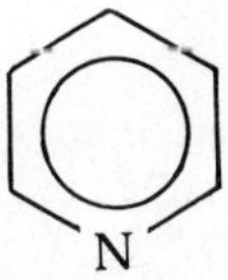 (viii) 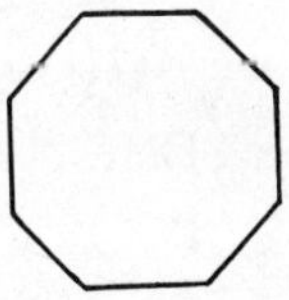 (ix)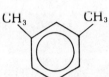

12. Calculate the number of metric tons (1 metric ton = 10^3 kg) of coke (assume pure carbon) needed to produce 2000 kg of silicon.

13. Calculate the mass of CaO produced by heating 1000 kg of dolomite rock. Assume the rock is pure $MgCO_3 \cdot CaCO_3$.

14. Geologists easily identify carbonate rocks, such as limestone, by placing hydrochloric acid on them. How would a carbonate respond to this treatment?

15. Amorphous forms of carbon such as charcoal and lampblack resemble graphite in appearance and yield the same products upon oxidation. Suggest an experiment to distinguish between charcoal and graphite and describe the results you would expect.

16. Write a balanced equation for:
    a. The oxidation of $SnCl_4^{2-}$ by $HgCl_2$
    b. The reduction of Fe(III) using Sn(II) in acid solution

17. Write the Lewis structure for:
    a. CO   b. CO_2 c. CN^- d. CO_3^{2-}

18. Write the Lewis structure for:
    a. H_2CO b. C_2N_2 c. CCl_4 d. C_2H_4

19. Classify each of the following compounds according to its functional group.
    a. $CH_3CH_2NHCH_2CH_3$
    b. $C_6H_5COCH_3$ (phenyl methyl ketone)
    c. $CH_3CH_2CHCH_2CH_3$ with CH_3 branch
    d. CH_3CH_2Br
    e. C_6H_5—CH_2CH_2OH
    f. $(CH_3)_2CHCO_2H$

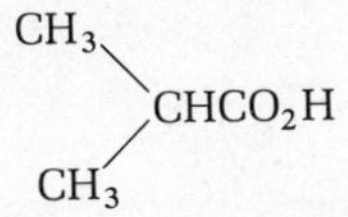

20. To which functional class does each of the following compounds belong? If a compound is polyfunctional, list all appropriate classes.
    a. $CH_3CH_2COCH(CH_3)_2$
    b. $(CH_3)_2CHOCH(CH_3)_2$
    c. $CH_3CH=CHCH=O$
    d. $NH_2CH_2CH_2OH$
    e. $CH_3C\equiv CH$
    f. CH_3CNH_2 (with C=O)

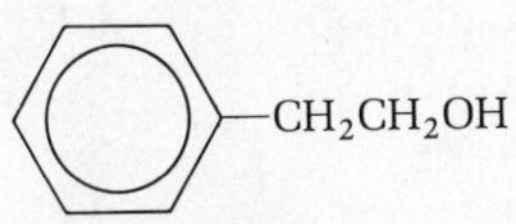

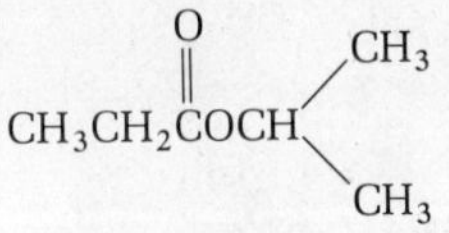

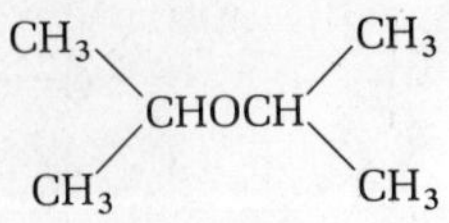

21. Write the structure of an isomer of $CH_3CH_2CH_2OH$.

22. Draw the structure of an isomer of CH_3CHNH_2 with CH_3 branch.

23. Write the structures of two isomers of dimethylformamide that are also amides.

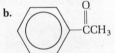

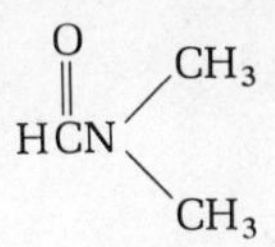

24. Write the structures of five isomers of pentanoic acid, $CH_3CH_2CH_2CH_2CO_2H$. These isomers can be either carboxylic acids or esters.

25. Write the structure of one molecule, other than one of those appearing in Table 19.2, belonging to each of the following functional group classes.
   a. Ketone   b. Alcohol   c. Alkene   d. Ether

26. Write the structure of one molecule for each of the following functional group classes. Do not use structures from Table 19.2.
   a. Carboxylic acid   b. Amine   c. Aldehyde
   d. Alkyne

27. Which of the following molecules are chiral? Draw the enantiomer of each chiral molecule.

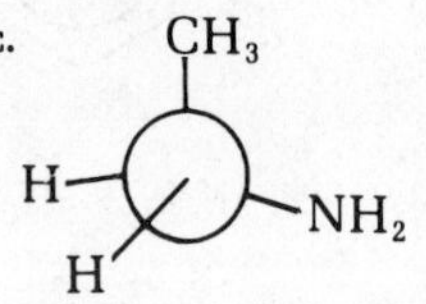

28. Are the following molecules chiral? If so, draw their enantiomers.

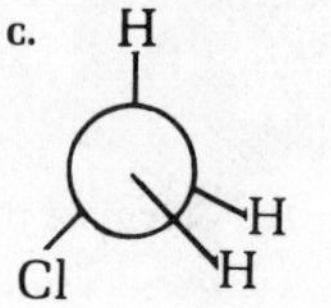

29. Write structures for all the possible isomers of C_4H_9Br, not counting stereoisomers.

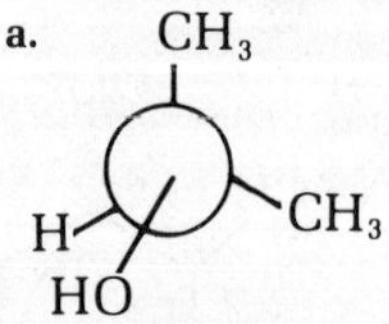

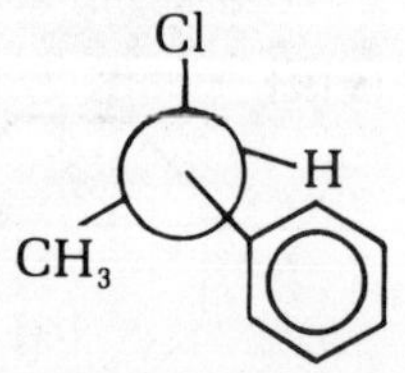

30. Write the structures for all isomers, except stereoisomers, with the formula $C_3H_6Cl_2$.

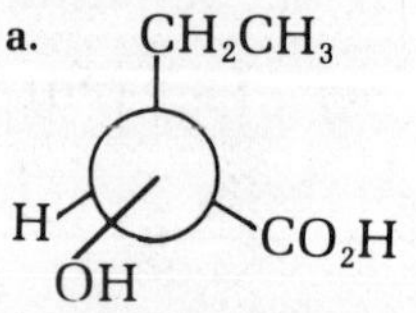

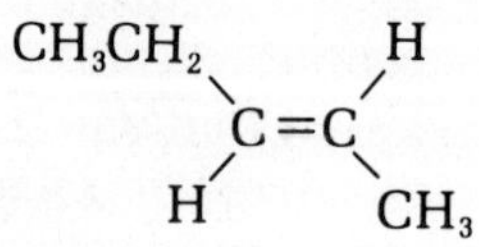

31. Which of the following structural formulas can represent one of two enantiomers? Draw these enantiomers.

   a. 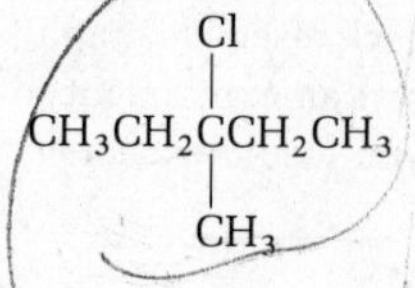 $CH_3CH_2CCH_2CH_3$ with Cl and CH_3 substituents
   b. 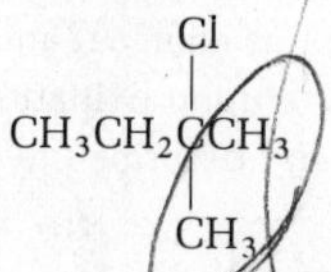 $CH_3CH_2CCH_3$ with Cl and CH_3 substituents
   c. 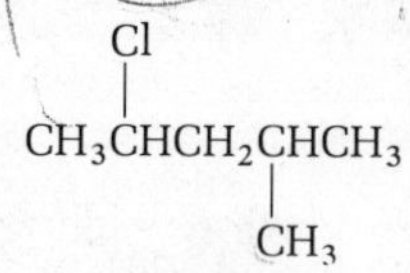 $CH_3CHCH_2CHCH_3$ with Cl and CH_3 substituents

32. Which of the following can exist as one of two enantiomers? Draw these enantiomers.
   a. 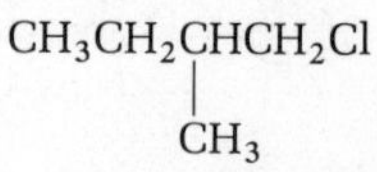 $CH_3CH_2CHCH_2Cl$ with CH_3
   b. 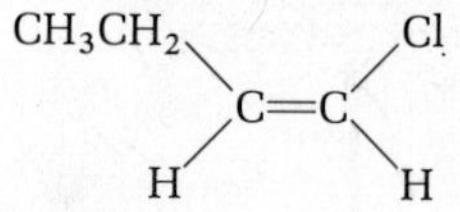 alkene with CH_3CH_2, Cl, H, H
   c. 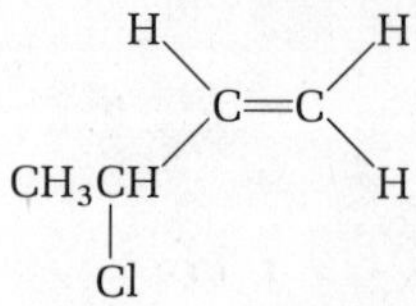 alkene with CH_3CH(Cl), H, H, H

33. Which of the following structures is one of a pair of geometric isomers? For those that are, draw the geometric isomers.
   a. 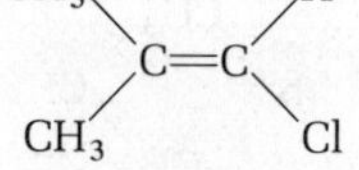 CH_3, CH_3 / H, Cl on C=C
   b. 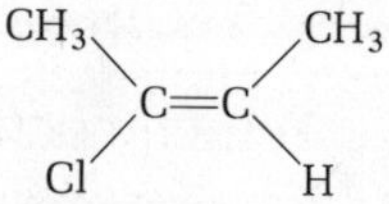 CH_3, Cl / CH_3, H on C=C
   c. CH_3CH_2, CH_3 / H, Cl on C=C

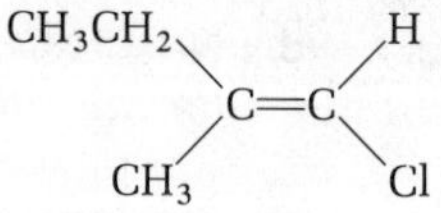

34. Which of the following molecules has a geometric isomer? Draw each geometric isomer.
   a. 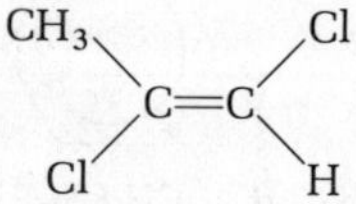 CH_3, Cl / Cl, H on C=C
   b. 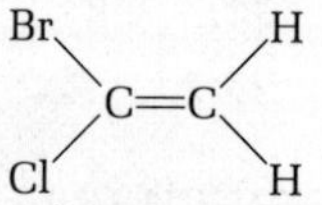 Br, Cl / H, H on C=C
   c. $CH_3CH_2C{\equiv}CH$

35. The diamond crystal lattice is face-centered cubic with a unit-cell edge length of 3.56 Å. The carbon atoms occupy the lattice points and half of the cell's tetrahedral holes. Compute the density of diamond.

36. Tin(IV) chloride ($SnCl_4$), a covalent molecule, reacts with chloride ion to form hexachlorostannate(IV) ion ($SnCl_6^{2-}$). Use the valence-shell electron-pair repulsion method (VSEPR) to predict the geometry of these two species.

37. There are nine isomers with the formula C_7H_{16}. Draw them.

EXERCISES

38. Sketch each of the following molecules and describe the hybridization state for each carbon atom in the molecule.
   a. CH_2Cl_2 **b.** CCl_4 **c.** $CH_3CH{=}CH_2$ **d.** $CH_3\underset{\underset{O}{\|}}{C}CH_3$

39. The approximate formula of soda glass is $Na_2O \cdot CaO \cdot 6SiO_2$. How many kg of Na_2CO_3 are required to make 1 kg of this glass? How many kg of silica?

40. Give a reasonable chemical explanation for each of the following observations.
   a. Ca^{2+} and Mg^{2+} salts commonly occur together or as mixed salts.
   b. Limestone is more soluble in rainwater than in pure, deionized water.
   c. Concentrated sodium hydroxide solutions are stored in plastic containers.

41. Aluminosilicates, as well as silicates, are abundant, common anions in minerals. An example is $AlSiO_5^{3-}$. Why do you suppose these anions are so plentiful?

42. Calculate $\Delta G°$ (standard conditions) for the conversion of graphite to diamond.

   $$C(\text{graphite}) \longrightarrow C(\text{diamond})$$

   The entropy of diamond is 3.26 J/K-mol *less* than that of graphite under these conditions.

43. Assuming that no change in $\Delta E°$ or $\Delta S°$ occurs with pressure, calculate the pressure that would be needed to make diamond thermodynamically stable at 25°C. Use the entropy information from Exercise 42.

44. Air contains approximately 0.04% CO_2 by volume. The solubility of CO_2 in water at a partial pressure of 1.00 atm is 0.03 M. What is the approximate pH of water saturated with CO_2 from the air?

45. What is the pH of a solution obtained by dissolving 2.0 g of CaC_2 in water to make 100 mL of solution?

chapter 20

The Nonmetallic Elements and Noble Gases: Periodic Groups VA–O

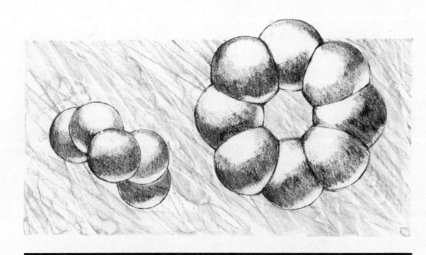

Common sulfur consists of molecules of eight sulfur atoms in a puckered ring, S_8

Most of the elements of groups VA through VII A have nonmetallic character—they tend to accept electrons and form negative ions, their oxides are acidic, and their elemental forms are polyatomic. There is a substantial decrease in electronegativity and an increase in metallic character going from the lightest to the heaviest member of each family, and the heaviest element in each family, Bi, Po, and At respectively, exhibits significant metallic behavior—electrical conductivity, high luster, and basic oxides. Since polonium and astatine are highly radioactive, and very rare, bismuth is the only abundant metal in groups VA–VIIA.

20.1 Group VA Elements (N, P, As, Sb, and Bi)

All the group VA elements play important roles in technology, but nitrogen and phosphorus, being essential ingredients for living systems, are particularly necessary for our well being. Nitrogen occurs in a wide variety of biochemically important compounds—proteins and nucleic acids (see Chapter 23) are just two examples—and the biological role of phosphorus, involving both energy transfer within organisms and the transfer of genetic information, is equally important. Both nitrogen and phosphorus are essential ingredients of fertilizers.

Each member of group VA has five outer-shell electrons in an s^2p^3 arrangement. Both the complete loss of these electrons to form a +5 ion or the gain of three electrons to produce a −3 ion require rather large amounts of energy. As a result, most compounds of these elements involve covalent bonds. Compounds containing −3 ions do occur, but there are no known compounds containing +5 ions. As we

20 THE NONMETALLIC ELEMENTS AND NOBLE GASES

Table 20.1 Properties of the Group VA Elements

Element	Atomic number	Electron configuration	Melting point (°C)	Boiling point (°C)	First electron ionization potential (kJ/mol)	Atomic[a] radius (Å)
N	7	[He] $2s^2 2p^3$	−210	−196	1400	0.73
P	15	[Ne] $3s^2 3p^3$	44.1	280	1010	1.10
As	33	[Ar] $3d^{10} 4s^2 4p^3$	Sublimes		947	1.21
Sb	51	[Kr] $4d^{10} 5s^2 5p^3$	631	1380	834	1.45
Bi	83	[Xe] $4f^{14} 5d^{10} 6s^2 6p^3$	271	1500	703	1.54

[a]Atomic radii in this and following tables are the homonuclear single-bonded values from *Tables of Interatomic Distances and Configurations in Molecules and Ions*, Special Publications No. 11 and 18, *The Chemical Society* (London), 1958 and 1965.

have seen in the earlier discussions of groups IA–IVA, there can be considerable differences in properties among members of the same family, and the variations among the group VA elements are even more pronounced than the earlier cases (Table 20.1). At one end of the scale, nitrogen and phosphorus, being typical nonmetals, can form positive oxidation states only by binding to very electronegative atoms such as oxygen, fluorine, or chlorine. Bismuth, in contrast, forms positive oxidation states in most of its compounds. Compared to the high electronegativity of nitrogen (3.0) and its small size (radius of 0.74 Å), bismuth is rather large (1.51 Å) and it can lose its relatively weakly held valence electrons to form cations.

The oxidation states of a group VA element can vary from −3 to +5. Figure 20.1 displays the relative free energies of these oxidation states for each of the group VA elements. The free energy value of the zero oxidation state for each element is set at zero, and the free energies of the other oxidation states relative to the zero state are plotted on the graph. We compute this free energy difference by multiplying the half-reaction potential for the conversion of the element from a particular oxidation state to its zero state by the number of electrons involved in the process. For example, the reduction potential for NO_3^- going to N_2 in acid solution is 1.24 V. Since this is a five-electron reduction per nitrogen atom, the free energy of a nitrogen atom in NO_3^-

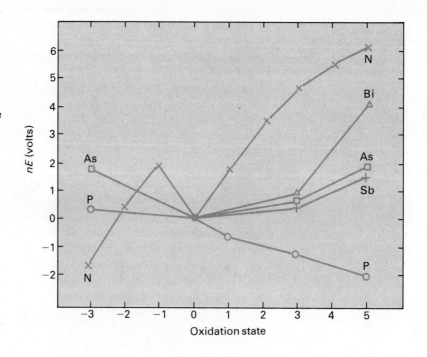

Figure 20.1
Oxidation state–free energy diagram for the group VA elements. Nitrogen is the most individual; it is stable as the element and in the −3 state. All its positive states except +5 tend to disproportionate in acid solution. Phosphorus tends to disproportionate to P(V) and PH_3. As, Sb, and Bi are all quite similar and Bi(V) is a strong oxidizing agent.

20.1 GROUP VA ELEMENTS

Table 20.2 Occurrence and Uses of Group VA Elements

Element	Occurrence	Form	Uses
Nitrogen	Air	N_2	Fertilizers
	Nitrate deposits	KNO_3, $NaNO_3$	Explosives
	Plants and animals	Amino acids, proteins	
Phosphorus	Phosphates	$CaF_2 \cdot 3Ca_3(PO_4)_2$, fluorapatite	Fertilizers
		Apatites containing chiefly $Ca_3(PO_4)_2$	Matches
			Alloyed with copper
	Plants and animals	ATP, ADP (see Chapter 23)	
	Bones, teeth	$Ca_5(OH)(PO_4)_3$, hydroxyapatite	
Arsenic	Sulfide	As_2S_2, realgar	Alloys
		As_2S_3, orpiment	Insecticides
Antimony	Sulfide	Sb_2S_3, stibnite	Alloys, usually with lead (battery plates)
Bismuth	Oxide, sulfide, sometimes elemental	Bi_2S_3, bismuthinite	Low-melting alloys
		$Bi_2O_3 \cdot H_2O$, bisinite	

is 5×1.24 V or 6.20 eV* higher than that of a nitrogen atom in N_2. The steepness of the line connecting two oxidation states indicates the strength of the oxidants and reductants. Bismuth(V), a powerful oxidant, has a free energy far above that of Bi(III), the next lower stable oxidation state. The high free energy of As(III) compared to elemental arsenic shows that it is a strong reductant.

Occurrence

Elemental nitrogen (N_2), an abundant substance, constitutes 78% (by volume) of air. Being part of all living organisms, nitrogen compounds occur in all plants and animals and in biological waste products. Manure and guano (bird droppings) are very effective fertilizers because they are so rich in nitrogen compounds. The only important nitrogen containing minerals are KNO_3 (saltpeter) and $NaNO_3$ (Chile saltpeter), substances that are so water soluble that their deposits occur mainly in deserts such as those in Chile. There are extensive deposits of phosphorus compounds such as phosphate minerals and bone; in fact phosphorus is a more abundant element than nitrogen. The most significant ores of arsenic, antimony, and bismuth are their sulfides and oxides. Table 20.2 summarizes the occurrence of the group VA elements.

The Nitrogen Cycle and Nitrogen Fixation

Although nitrogen is the major constituent of the atmosphere, the N_2 molecule is so unreactive (it has a triple bond) that most organisms cannot use it directly for nutritional purposes. **Nitrogen fixation**, the conversion of elemental nitrogen into biologically useful forms, occurs in nature by the action of lightning on the atmosphere but mostly by the action of certain microorganisms such as blue-green algae and the bacteria located in the root nodules of leguminous plants (for example, peanuts and clover). These nitrogen-fixing organisms contain special enzymes that catalyze the conversion of N_2 to ammonium salts, and other enzymes that speed the biological oxidation of NH_4^+ to NO_2^- and NO_3^-. Plants can then absorb these ions from the soil and use them to synthesize nitrogen-containing organic substances, particularly amino acids and proteins, and animals obtain the amino acids necessary for their normal

---

*An electron volt (eV) is the energy an electron gains by being accelerated by a potential of 1 V. 1 eV = 1.60×10^{-19} J.

growth by eating the plants (or other herbivorous animals). The fixed nitrogen returns to the soil via animal excrement and decaying animal and vegetable matter. The balance between atmospheric nitrogen and fixed nitrogen is maintained by denitrifying bacteria that convert nitrogen-containing ions back into N_2. The interaction of all these processes—nitrogen fixation, the transfer of nitrogen from the soil to plants then to animals and then back to the soil, and denitrification—is called the **nitrogen cycle** (Figure 20.2).

The Haber Process

As long as the human population remained relatively low, the nitrogen cycle provided enough nitrogen fertilizers to meet agricultural needs, but by the beginning of the twentieth century it was clear that natural sources of fixed nitrogen could no longer meet the demands of the world's rapidly expanding population. We escaped that food crisis because of the development in 1913 of the **Haber process**, a method for producing ammonia by the reaction of hydrogen and nitrogen. Fritz Haber's success in finding the conditions suitable for the reaction

$$N_2(g) + 3 H_2(g) \rightleftharpoons 2 NH_3(g)$$

was due to his rational application of thermodynamic theory. Even though no one had ever carried out this reaction successfully, thermodynamic data (the free energy of the reaction is -33 kJ at room temperature) convinced Haber that the reaction had a favorable equilibrium constant and was feasible. The trick was to find conditions at which the reaction occurs at a reasonable rate. He achieved this partly by finding a suitable catalyst (Fe_3O_4 containing some K_2O or Al_2O_3) and partly by working at high temperature (500°C) to overcome the reaction's activation energy. High temperature is, however, a disadvantage; since ammonia formation is strongly exothermic ($\Delta H = -92$ kJ), the equilibrium constant decreases as temperature increases. Haber compensated for this by using high pressures (100–200 atm). Pressure shifts the equilibrium toward ammonia formation—the direction that decreases pressure because the number of molecules of gas decreases—and also speeds the reaction because the effective reactant concentration is increased.

The Haber process provides ammonia for industrial and agricultural use in nearly unlimited quantities (the annual world production is approximately 40 million tons), and the availability of inexpensive ammonia enables us to maximize the yield of foodstuffs from the world's limited supply of arable land. But the initial social effect of the Haber process was military rather than agricultural. The effective development of the Haber process coincided with the start of World War I. Nitrates

Figure 20.2
The nitrogen cycle.

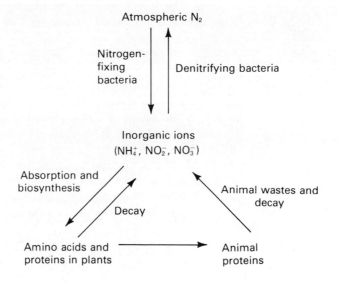

are essential for making explosives* and the world's major source of nitrates was in Chile. Since the British blockade cut Germany's access to Chilean nitrates, the German war effort faced collapse. But the ammonia produced in the Haber process can be converted to nitrates, and by rapidly expanding facilities for ammonia production, Germany was able to provide its own nitrates throughout the long war.

Preparation of the Group VA Elements

Nitrogen is obtained commercially by fractional distillation of liquefied air (Figure 17.9). A trace of argon contaminates this commercial nitrogen and makes it slightly denser than purer forms, an observation that provided one of the first hints of the existence of the noble gases. The decomposition of certain nitrogen compounds such as ammonium nitrite

$$NH_4NO_2(s) \xrightarrow{\text{Heat}} N_2(g) + 2\,H_2O(g)$$

provides pure nitrogen.

Although phosphorus is more abundant than nitrogen, it is difficult to obtain the pure element. The most common method is to reduce phosphate minerals in high-temperature electric furnaces using coke as the reductant. Sand (SiO_2) is also a necessary reactant.

$$2\,Ca_3(PO_4)_2(s) + 6\,SiO_2(s) + 10\,C(s) \longrightarrow 6\,CaSiO_3(s) + 10\,CO(g) + P_4(g)$$

As the effluent gas, containing mainly P_4 and CO, is bubbled through water, the white allotrope of phosphorus forms. The water also keeps air away from the phosphorus, a substance that is so flammable it ignites in air.

The production of the remaining group VA elements involves reduction of their oxides and sulfides using carbon or iron as the reducing agent. For example,

$$As_2S_3(s) + Fe(s) \longrightarrow 2\,As(g) + 3\,FeS(s)$$

The amounts of As, Sb, and Bi produced as by-products from refining Cu and Ag more than meets the world's needs.

Properties and Uses of the Elements

The only elemental form of nitrogen is N_2, a colorless and unreactive gas. (The French word for nitrogen, *azote*, is derived from the Greek word meaning inert.) We have already discussed the chief industrial use of nitrogen, the conversion to ammonia by the Haber process. The ammonia is then used to synthesize fertilizers, nitric acid and nitrates, synthetic fibers such as nylon, and plastics such as polyurethanes.

Phosphorus exists at room temperature in three allotropes (Figure 20.3), the white, red, and black forms. **White phosphorus** (sometimes called yellow phosphorus) is a very poisonous, reactive solid. It consists of tetrahedral P_4 molecules. Heat (over 400°C) or light converts white phosphorus to **red phosphorus**, the more stable form. Although there have been reports of several structurally different modifications of red phosphorus, they are all apparently amorphous (refer to Section 9.4) arrangements of polymers formed by linked P_4 tetrahedra. **Black phosphorus**, formed when white phosphorus is heated at high pressure, is the least reactive allotrope and has a double-layered structure.

*The introduction of gunpowder—a mixture of KNO_3, charcoal, and sulfur—into medieval Europe had a profound effect on the nature of warfare. Shakespeare bears witness to this in *King Henry the Fourth, Part I* Act I, Scene I.

> And that it was great pity, so it was,
> This villanous saltpetre should be digged
> Out of the bowels of the harmless earth
> Which many a good tall fellow had destroyed
> So cowardly; and but for these vile guns,
> He would have himself have been a soldier

Figure 20.3
Allotropes of phosphorus.

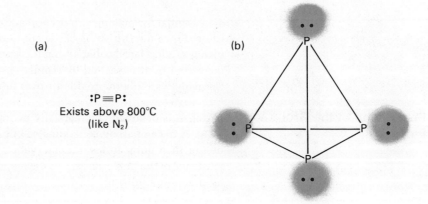

(a)

:P≡P:
Exists above 800°C
(like N_2)

(b)

White phosphorus
(tetrahedral)

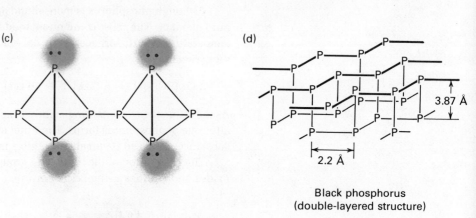

(c)

Red phosphorus
(postulated to consist of
chains of linked tetrahedra)

(d)

Black phosphorus
(double-layered structure)

The properties of these allotropes reflect their inherent structural differences. White phosphorus is a waxy, low-melting solid that dissolves in carbon disulfide (CS_2), a nonpolar solvent. Because of its high molecular mass, red phosphorus is less volatile than the white form and is insoluble. Black phosphorus is unreactive, and its somewhat metallic appearance resembles that of graphite.

The major uses of elemental phosphorus are for synthesizing its oxides, compounds that are then used to produce phosphates for fertilizers and laundry detergents, and P_4S_3, used in preparing matches. Phosphorus is also important because of its flammability, and tracer bullets and incendiary bombs and shells contain white phosphorus.

White phosphorus was once used in matches, but its extreme toxicity makes it unsuitable for this purpose. Modern match heads (Figure 20.4) consist of a paste of lead dioxide, P_4S_3, and powdered glass. Friction ignites the match by initiating an exothermic reaction between P_4S_3 and PbO_2. Safety matches on the other hand do not contain any phosphorus in their heads. Instead, the match box contains a strip of red phosphorus and abrasive. Striking the match against this strip ignites the oxidant–reductant mixture in the match head.

Arsenic and antimony have allotropic forms similar to those of phosphorus. Moderate heating of the yellow forms of either arsenic or antimony or evaporation of the CS_2 solutions of these elements produces the gray colored, semimetallic allotropes. The gray forms of As, Sb, and Bi are stable, and their structures resemble the double-layered structure of black phosphorus.

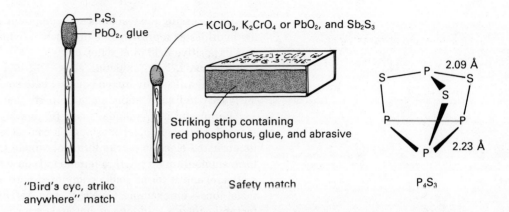

Figure 20.4
Use of phosphorus in matches.

Although important pesticides contain arsenic compounds, the major commercial use of this element is in preparing special alloys. The presence of small amounts of arsenic in lead increases the metal's strength and the surface tension of the molten metal. The latter property is useful in preparing lead shot. Because of their high surface tension, drops of Pb–As alloy retain their spherical shapes when dropped into cold water and solidify to form nearly perfect spheres.

Small quantities of antimony and bismuth increase the hardness of many metals and also promote the expansion of the alloy on solidification. The first property makes Babbitt metal (90% Pb, 7% Sb, 3% Cu) desirable for motor bearings, and the second makes antimony- and bismuth-containing alloys useful for making printing type. As these molten alloys solidify in a type mold, they expand to fill the mold completely and produce an accurate casting. There are several low-melting bismuth alloys such as Wood's metal (50% Bi, 25% Pb, 12.5% each of Sn and Cd, melting point 71°C) that are used in automatic fire sprinklers and fire doors. The fire's heat easily melts the alloy to release the sprinkler valve or door lock.

Reactions and Compounds of Nitrogen

The elements of group VA exhibit a variety of chemical reactions that produce compounds whose oxidation states range from −3 to +5. Nitrogen is the only element of the group to form stable compounds for each oxidation state between −3 and +5 (plus the fractional average oxidation state −1/3). For example, the oxides of nitrogen cover all the oxidation states from +1 to +5, whereas the other group VA elements form primarily +3 and +5 states (Table 20.3).

Nitrogen, a member of the second row of the periodic table, can use only its outer-shell s and p orbitals for bonding but the other elements of group VA can form bonds involving their d orbitals. Consequently, nitrogen differs from the other members of the group in being unable to form superoctet molecules similar to PCl_5. Nitrogen also has a greater tendency than the larger group VA atoms to form strong multiple bonds involving π overlap of atomic orbitals. These striking differences between a second-row element and the larger members of its periodic group are common to all nonmetallic groups. (For example, compare the chemistry of carbon and silicon, Chapter 19.)

Table 20.3 Oxides of Group VA Elements

Oxidation state	Nitrogen	Phosphorus	Arsenic	Antimony	Bismuth
(Elemental)	N_2	P_4	As	Sb	Bi
+1	N_2O				
+2	NO				BiO
+3	N_2O_3 (NO_2^-)	P_4O_6	As_4O_6	Sb_4O_6	Bi_2O_3
+4	NO_2	P_4O_8		Sb_2O_4	Bi_2O_4
+5	N_2O_5 (NO_3^-)	P_4O_{10} (PO_4^{3-})	As_2O_5	Sb_2O_5	Bi_2O_5

In discussing the reactions and compounds of nitrogen, we will concentrate on the hydrides as representative of negative oxidation states and the oxides as examples of positive oxidation states.

Ammonia, NH_3 (oxidation state -3), is a colorless gas with a pungent odor. There is significant hydrogen bonding between NH_3 molecules (Section 17.3) so it is fairly easy to liquefy ammonia, a property that makes it a useful refrigerant.

Because of its pyramidal shape (Figure 20.5) and the polarity of its bonds, ammonia is a strongly polar substance. It can act as an effective Lewis base by donating its unshared electron pair to form coordinate covalent bonds. It accepts a proton to form ammonium ion, NH_4^+, a tetrahedral ion whose structure and bonding resemble methane, and it forms stable complexes with electron acceptors such as transition metal ions—for example, $Cu(NH_3)_4^{2+}$ and $Ag(NH_3)_2^+$. Ammonia is very soluble in water, and aqueous solutions of ammonia are slightly basic.

$$NH_3(aq) + H_2O \rightleftharpoons NH_4^+(aq) + OH^-(aq) \qquad K_B = 1.8 \times 10^{-5}$$

Liquid ammonia, like water, is an effective polar solvent and has the unusual property of dissolving very electropositive metals to form blue solutions containing solvated electrons. Liquid ammonia solutions show acid–base reactions similar to those in water, but in this case the neutralization reaction is

$$NH_4^+ + NH_2^- \rightleftharpoons 2\, NH_3$$

rather than

$$H_3O^+ + OH^- \rightleftharpoons 2\, H_2O$$

Salts of ammonium ion, NH_4^+, resemble alkali metal salts, being crystalline, water-soluble substances, except that they relatively easily decompose into ammonia and an acid.

$$NH_4Cl(s) \xrightarrow{300°C} NH_3(g) + HCl(g)$$

Since NH_4^+ is a reducing agent, ammonium salts of strongly oxidizing anions such as ammonium nitrate (NH_4NO_3) and ammonium perchlorate (NH_4ClO_4) are dangerous materials that may explode with mild heating, mechanical shock, or even spontaneously. A ship full of NH_4NO_3 caught fire and exploded destroying Texas City, Texas in 1947.

Hydrazine, N_2H_4 (oxidation state -2), is a colorless liquid at room temperature (melting point 20°C). Like ammonia, it is hydrogen bonded, polar, and soluble in water. Being isoelectronic with hydrogen peroxide (Figure 20.6) it resembles H_2O_2 in its reactivity and is much more reactive than ammonia. It reacts energetically with oxygen

$$N_2H_4(l) + O_2(g) \longrightarrow N_2(g) + 2\, H_2O(g)$$

Figure 20.5
(a) Structures and inversion of ammonia, NH_3. (b) Structure of ammonium ion, NH_4^+.

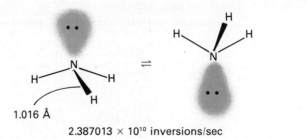

Figure 20.6
Structures of hydrazine, N_2H_4, and H_2O_2.

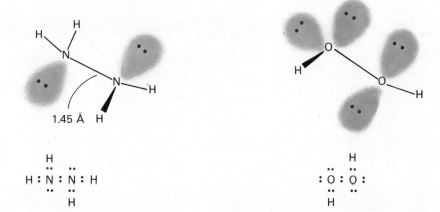

and, like H_2O_2, has been used as a rocket fuel. It is an effective reducing agent. For example,

$$N_2H_4(aq) + 2\ Cl_2(g) \xrightarrow{H^+} 4\ HCl(aq) + N_2(g)$$

A diprotic base, it forms salts of $N_2H_5^+$ (hydrazonium ion) and $N_2H_6^{2+}$ ions. It is a weak base in aqueous solution

$$N_2H_4(aq) + H_2O \rightleftharpoons N_2H_5^+(aq) + OH^-(aq) \qquad K_B = 1.0 \times 10^{-6}$$

Figure 20.7
Structure of hydroxylamine, NH_2OH.

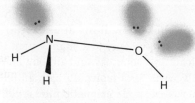

and acts as a Lewis base to form complexes with metal ions.

Hydroxylamine, NH_2OH (oxidation state -1), is an unstable, white solid, but its aqueous solutions are stable. Being weakly basic,

$$NH_2OH(aq) + H_2O \rightleftharpoons NH_3OH^+(aq) + OH^-(aq) \qquad K_B = 6.6 \times 10^{-9}$$

it forms stable salts such as $(NH_3OH^+)Cl^-$ and $(NH_3OH^+)NO_3^-$. Looking at the structure of hydroxylamine (Figure 20.7), we can think of it as an ammonia molecule in which an OH replaces one of the hydrogen atoms. It can act either as an oxidant or reductant. It is prepared by careful reduction of nitrate or nitrite salts with tin or SO_2.

Hydrazoic acid, HN_3 (oxidation state $-1/3$), is a very unstable, colorless liquid. It can give up a proton to form azide ion, N_3^-, an ion with a symmetric and linear structure (Figure 20.8). There are two distinct types of azide compounds, covalent azides formed by heavy metals (PbN_3, AgN_3) and ionic azides formed by electropositive metals (NaN_3, KN_3). Covalent azides are very unstable, and since they explode when struck, they are commonly used in percussion or detonating caps. Ionic azides are quite stable and resemble halide salts. They do decompose at high temperatures.

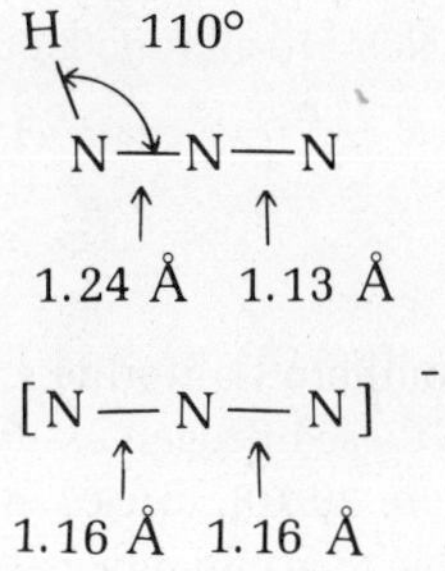

$$2\ KN_3(s) \xrightarrow{350°C} 2\ K(l) + 3\ N_2(g)$$

Figure 20.9 shows the structures of the oxides and oxyanions of nitrogen. Nitrous oxide, N_2O (oxidation state $+1$, also called dinitrogen monoxide), a colorless, unreactive gas, is prepared by the thermal decomposition of ammonium nitrate.

$$NH_4NO_3(l) \xrightarrow{250°C} N_2O(g) + 2\ H_2O(g)$$

Its chief uses are as the propellant gas in whipped-cream spray cans and as an anaesthetic. The mild euphoria accompanying the anaesthetic action of nitrous oxide has earned it the nickname "laughing gas."

Nitric oxide, NO (oxidation state $+2$, also called nitrogen monoxide), is a molecule containing an odd number of electrons and is thus a free radical (Section

Figure 20.8
Structures of HN_3 and N_3^- (in KN_3). In this figure and in Figure 20.9 only the approximate geometry and bond distances are indicated for some nitrogen-containing compounds. Most valence electrons and the various resonance structures possible are omitted for clarity.

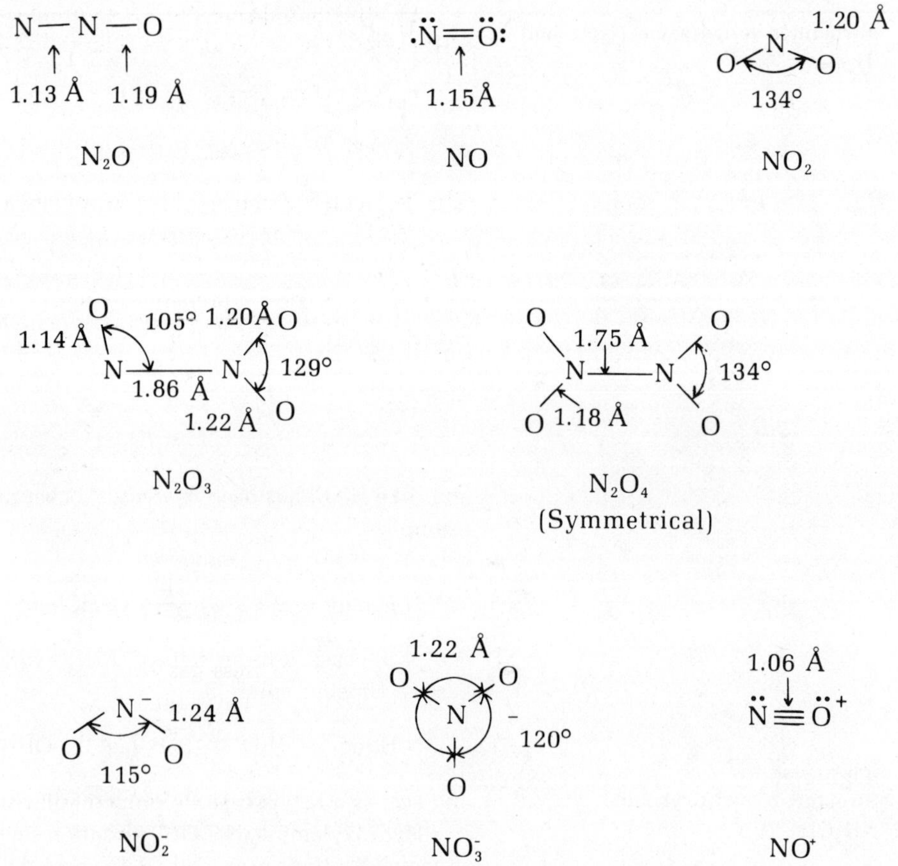

Figure 20.9 Structures of some nitrogen–oxygen compounds.

7.2); however, many of its properties are not typical of free radicals. It has surprisingly little tendency to combine to form N_2O_2 and does so only to a slight extent in the liquid and solid states. Unlike other free radicals, it is colorless, and although it is paramagnetic, its magnetic moment is unusually weak for a molecule with an odd electron. These properties can be explained in terms of subtle magnetic interactions within the molecule, but the details of this explanation are too complex for this course.

Nitric oxide is a common reduction product of dilute nitric acid.

$$8\ H^+ + 3\ Cu(s) + 2\ NO_3^- \longrightarrow 3\ Cu^{2+} + 2\ NO(g) + 4\ H_2O$$

It forms to a slight extent in the atmosphere due to the natural effect of light or lightning (a small contribution to nitrogen fixation).

$$N_2(g) + O_2(g) \xrightarrow[\text{electric discharge}]{\text{Light or}} 2\ NO(g)$$

In modern times, this reaction occurs on an unpleasantly large scale in automobile engines and is a major source of urban air pollution. Although NO is itself innocuous, sunlight rapidly converts it to NO_2,

$$2\ NO(g) + O_2(g) \xrightarrow{\text{Light}} 2\ NO_2(g)$$

a poisonous gas whose brown color often tints the skies of Southern California and other sunny, urban areas. Other harmful substances form from these nitrogen oxides. One of them, peroxyacetyl nitrate (also called PAN—see Figure 20.10), causes millions of dollars damage to crops each year. No one has been able to accurately assess the human damage caused by PAN.

Figure 20.10
Structure of peroxyacetyl nitrate (PAN).

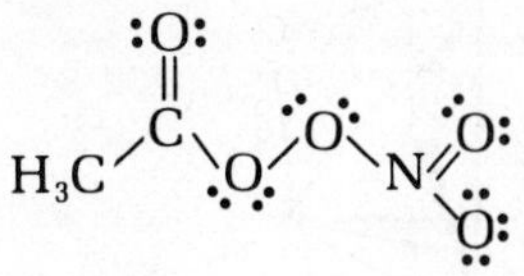

Dinitrogen trioxide, N_2O_3 (oxidation state +3), is the anhydride of nitrous acid, HNO_2. Both the oxide and the acid are unstable, decomposing to NO and NO_2.

$$N_2O_3(g) \longrightarrow NO(g) + NO_2(g)$$

$$2\ HNO_2 \longrightarrow NO(g) + NO_2(g) + H_2O$$

The molecular structure of N_2O_3 looks like a molecule of NO and one of NO_2 held together by a weak N—N bond. Nitrous acid is a weak acid whose slight blue color resembles that of N_2O_3. It is a strong oxidizing agent but can itself be oxidized to nitric acid. It can be prepared by adding sulfuric acid to cold solutions of nitrite (NO_2^-) salts. Nitrites are much more stable than nitrous acid and can be formed by mild reduction of nitrates.

$$2\ NaNO_3(s) \xrightarrow{\text{Heat}} 2\ NaNO_2(s) + O_2(g)$$

Nitrogen dioxide, NO_2, and dinitrogen tetroxide, N_2O_4, represent the +4 oxidation state. The three species NO_2, N_2O_4, and NO are in equilibrium, and the equilibrium is temperature dependent.

$$N_2O_4(g) \rightleftharpoons 2\ NO_2(g) \rightleftharpoons 2\ NO(g) + O_2(g)$$

At $-11°C$, the colorless gas N_2O_4 has practically no tendency to dissociate, but at room temperature there is enough red-brown NO_2 in the gas mixture to see easily. At 140°C the dissociation of N_2O_4 is complete. The red color fades at 650°C as NO_2 breaks down to NO and O_2.

Nitrogen dioxide is an odd-electron molecule. Being highly paramagnetic and colored and showing a reasonable tendency to associate to form N_2O_4, it is a fairly typical free radical.

Dinitrogen pentoxide, N_2O_5 (oxidation state +5), is the anhydride of nitric acid, HNO_3. Quite reactive and somewhat unstable, it is prepared by dehydration of HNO_3.

$$12\ HNO_3(l) + P_4O_{10}(s) \longrightarrow 4\ H_3PO_4(l) + 6\ N_2O_5(g)$$

Nitric acid and nitrate (NO_3^-) salts are important industrial chemicals. The commercial production of nitric acid involves the air oxidation of ammonia in the **Ostwald process**.

$$4\ NH_3(g) + 5\ O_2(g) \xrightarrow[1000°C]{Pt} 4\ NO(g) + 6\ H_2O(g)$$

$$2\ NO(g) + O_2(g) \longrightarrow 2\ NO_2(g)$$

$$3\ NO_2(g) + H_2O \longrightarrow 2\ HNO_3(aq) + NO(g)$$

Before the development of the Ostwald process, the major source of nitrates was the large deposits of $NaNO_3$ in the deserts of Chile.

Reactions and Compounds of P, As, Sb, and Bi

Although the reactions of the two major allotropes of phosphorus, the white and red forms, are generally the same, most phosphorus reactions are carried out with the red form because it is safer to handle. Partial combustion of red phosphorus in air produces phosphorus(III) oxide, P_4O_6 (oxidation state +3). This substance, also called phosphorous oxide, contains tetrahedral molecules with three oxygen atoms and one unshared electron pair around each phosphorus atom (Figure 20.11). Additional oxidation joins oxygen atoms to the unshared electron pairs, forming P_4O_{10} (oxidation state +5), a white solid that is often called phosphorus pentoxide because its formula was previously thought to be P_2O_5. It is the anhydride of orthophosphoric acid, H_3PO_4, and is so reactive toward water that it is extremely deliquescent and will dehydrate HNO_3 to N_2O_5 and H_2SO_4 to SO_3. These dehydration reactions are very exothermic.

Figure 20.11
Phosphorus oxide structures.

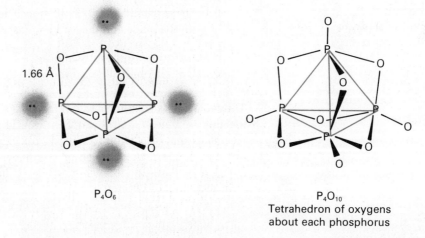

The reaction of white phosphorus in moist air produces P_4O_6 and generates a pale greenish light in the process. (The name phosphorus derives from the Greek word for "light-giving.") This light arises from the direct conversion of some of the reaction's energy to light energy, a phenomenon known as **chemiluminescence** (see Section 16.3). Chemiluminescence occurs because the P_4O_6 molecules are initially formed in their excited states and release photons as they fall to the ground state. The presence of more oxygen quenches the chemiluminescence and forms P_4O_{10}.

The reaction of water with P_4O_6 produces phosphorous acid, $H_2(HPO_3)$. [Be careful about its spelling. The name of the element is phosphorus (accent on first syllable), but the +3 oxidation state substances are named as phosphorous compounds (accent on second syllable).] We write the formula in this way to indicate that one of the three hydrogen atoms is bound directly to the phosphorus atom and the other two are located on oxygens (Figure 20.12). Phosphorous acid is only a diprotic acid; the hydrogen bound to the phosphorus atom is not at all acidic.

There are various acids representing the +5 oxidation state of phosphorus. In addition to orthophosphoric acid (H_3PO_4), there are dehydrated forms including metaphosphoric acid (HPO_3), and polymeric forms such as pyrophosphoric acid ($H_4P_2O_7$) and triphosphoric acid ($H_5P_3O_{10}$). The salts of these polymeric acids can have very complex structures that often resemble silicate structures in that a tetrahedron of oxygen atoms surrounds each phosphorus atom.

Like phosphorus, As, Sb, and Bi form +3 and +5 oxides when heated in air, and also react with sulfur and the halogens to form corresponding sulfides and halides.

Figure 20.12
Lewis structures for some phosphorus acids and ions.

Phosphorous acid, $H_2(HPO_3)$

Orthophosphoric acid

Pyrophosphoric acid

$(P_nO_{3n+1})^{-2-n}$

Polyphosphates

Sodium triphosphate $Na_5P_3O_{10}$

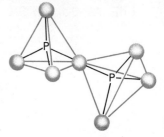

Pyrophosphate ion $P_2O_7^{4-}$

$$16\ Bi(s)\ +\ 5\ S_8(s)\ \longrightarrow\ 8\ Bi_2S_5(s)$$

$$2\ As(s)\ +\ 3\ Cl_2(g)\ \longrightarrow\ 2\ AsCl_3(l)$$

Phosphorus and the remaining group VA elements react with electropositive elements such as Na, K, and Mg to form phosphides, arsenides, antimonides, and bismuthides representing the −3 state. All the group VA elements form hydrides; in addition to the hydrides of nitrogen, there are phosphine (PH_3), arsine (AsH_3), stibine (SbH_3), and bismuthine (BiH_3).

The heavier elements of group VA are harder to oxidize, and as we have seen in group IVA, their higher oxidation states are less stable. For example, it takes a strong oxidizing agent in a basic medium to oxidize a Bi(III) compound to bismuthate ion (the +5 state).

$$3\ OH^-(aq)\ +\ Bi(OH)_3(s)\ +\ Cl_2(g)\ \longrightarrow\ BiO_3^-(aq)\ +\ 2\ Cl^-(aq)\ +\ 3\ H_2O$$

Bismuthate salts such as $NaBiO_3$ are strong oxidants in acid solutions.

All the +3 (-*ous*) and +5 (-*ic*) oxides of As, Sb, and Bi dissolve in water, but only arsenic oxide (As_2O_5) is definitely acidic.

One of the commercially more important and controversial phosphate compounds is sodium triphosphate, $Na_5P_3O_{10}$, a substance that is added to laundry detergents. Although this substance does aid the washing process—it maintains an alkaline pH at which the detergent functions best, and it softens hard water by complexing with Ca^{2+} and Mg^{2+}—its prime function involves market psychology rather than chemistry. The amount of synthetic detergent needed for a wash load is far smaller than the amount of soap needed for the same wash. When synthetic detergents were first introduced, customers followed the habit of adding large amounts of it to the wash, and the excess of detergent caused all sorts of problems such as an overflow of suds. The resulting consumer dissatisfaction nearly destroyed the market for synthetic detergents. That's when sodium triphosphate came in; it was added as a "builder"—an essentially inert substance that provided a reassuring bulk to the detergent. With the addition of builders to detergents, consumer resistance evaporated and synthetic detergents almost completely displaced soaps from the laundry room.

The problem with using sodium triphosphate as a builder is that it drains via sewage into lakes and rivers where it exerts an adverse ecological effect. (Phosphate fertilizers leached out of farmland also add to the problem.) It isn't that the phosphates are poisonous, on the contrary, they play an essential role in biochemical systems. The difficulty is that the increase in the phosphate level in water stimulates plant growth, and the organisms most able to benefit from this bounty are the more primitive ones such as algae. Fish, unable to compete with the rapidly multiplying algae for other natural commodities, are wiped out, and the lake turns into a smelly, weed-choked bog. This process, known as eutrophication, is a natural part of the life cycle of a lake, but the introduction of large quantities of phosphates can accelerate the process tremendously. In protecting our environment, we mustn't limit ourselves to thinking in such oversimplified terms as toxic or nontoxic, biodegradable and nonbiodegradable. We must consider the effect of concentration levels of substances on the delicate balance of ecological systems.

$$As_2O_5(s) + 3\ H_2O \rightleftharpoons 2\ H_3AsO_4(aq)$$

Arsenic acid, H_3AsO_4, is a triprotic acid like H_3PO_4 but is somewhat weaker, and it is a strong oxidant whereas H_3PO_4 is not.

Arsenic(III) oxide As_4O_6, is extremely poisonous and has been used as an insecticide and rodent poison, but it is being replaced by less toxic chemicals.

20.2 Group VIA Elements (O, S, Se, Te, and Po)

The group VIA elements, having s^2p^4 outer-shell electron configurations (Table 20.4), tend to add two electrons to form -2 ions or form two covalent bonds to attain their octets of electrons. Except for oxygen, all these elements can form superoctet species such as SF_6. They are more nonmetallic than the corresponding group VA elements, but as in the case of other periodic groups, there is a marked increase in metallic character with atomic size. Polonium, the largest atom in the group, appears to form some compounds, such as PoO_2, that contain positive polonium ions. The positive oxidation states of the more electronegative elements (S and Se) occur when they form covalent bonds with even more electronegative elements (O, F, and Cl), the $+4$ and the $+6$ being most important. Figure 20.13 is an oxidation state-free energy graph similar to that shown for the group VA elements (Figure 20.1).

Table 20.4 Properties of the Group VIA Elements

Element	Atomic number	Electron configuration	Melting point (°C)	Boiling point (°C)	First electron ionization potential (kJ/mol)	Atomic size (Å)
O	8	[He] $2s^2 2p^4$	-219	-183	1310	0.73
S	16	[Ne] $3s^2 3p^4$	119	445	1000	1.02
Se	34	[Ar] $3d^{10} 4s^2 4p^4$	220	685	941	1.16
Te	52	[Kr] $4d^{10} 5s^2 5p^4$	450	1390	869	1.43
Po	84	[Xe] $4f^{14} 5d^{10} 6s^2 6p^4$	250	960	810	1.67

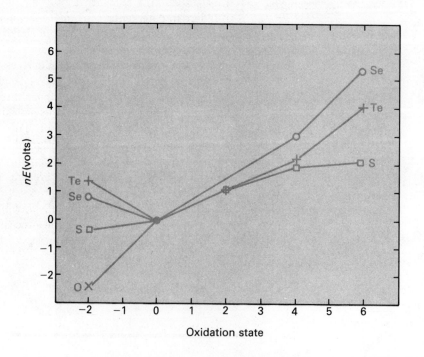

Figure 20.13
Oxidation state–free energy diagram for group VIA elements. Only negative oxidation states commonly occur with oxygen. The stability of higher oxidation states of S, Se, and Te decreases with atomic number. Intermediate $+4$ states are more stable than $+6$ for Te and Se.

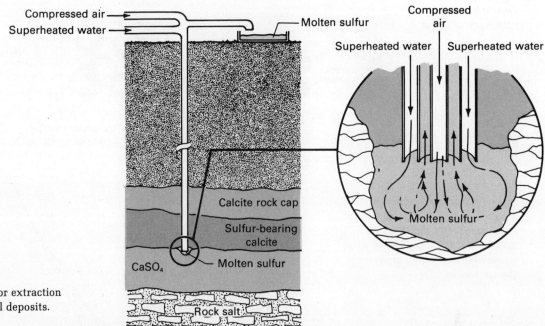

Figure 20.14
The Frasch process for extraction of sulfur from natural deposits.

Occurrence

The discovery of sulfur dates back to prehistoric times, and because of its occurrence near volcanoes, sulfur commanded the awe of ancient man. It is the brimstone of "fire and brimstone," and the devil is often described as "reeking of sulfurous fumes." Although not abundant (about 0.6% of the earth's crust), sulfur is one of the cornerstones of chemical technology, mainly because of the industrial importance of sulfuric acid and the agricultural importance of sulfate fertilizers. It occurs in small amounts in living systems and plays several critical biochemical roles.

Free sulfur occurs in large natural deposits and can be mined without further purification. The largest sulfur deposits occur near the Gulf of Mexico and in Sicily. Since the Gulf of Mexico deposits are in the porous limestone caps that cover deep subterranean rock salt deposits—geological formations that cannot be reached by conventional mine shafts—the sulfur is extracted by the **Frasch process** (Figure 20.14). This method involves the pumping of superheated water (above 116°C) down through concentric pipes drilled into the limestone caps. The hot water melts the sulfur, which is then pumped to the surface through one of the inner pipes. The nearly pure sulfur crystallizes in large bins.

Some metal sulfides such as pyrites (FeS_2, also known as "fool's gold" because of its yellow metallic luster), galena (PbS), and sphalerite (ZnS) are useful sources of sulfur. The sulfur dioxide (SO_2) that forms when these ores are roasted to obtain the metal is used to prepare sulfuric acid. (See the next section.) These metal sulfides also contain selenium and tellurium contaminants, and the recovery of these two elements from the exhaust dust of the roasting process is their major method of production.

Properties

All the group VIA elements except oxygen are solids at room temperature, and all can exist in several allotropic forms. There are perhaps over a dozen different sulfur allotropes. The most common of the well-characterized sulfur allotropes are the orthorhombic and monoclinic forms. Each of these allotropes contains covalently bound, puckered rings of eight sulfur atoms (Figure 20.15), the difference in the allotropes being a difference in the crystal packing of the S_8 units. **Orthorhombic sulfur**, the most stable form at room temperature, changes to the **monoclinic** form above 95.5°C (Figure 20.16). Liquid and gaseous sulfur also contain S_8 rings although these tend to break into smaller fragments at higher temperature. Because the intermolecular attractions between S_8 molecules are weak, sulfur has a low melting point

Figure 20.15
Allotropes of sulfur.

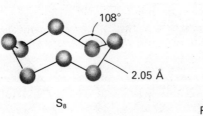

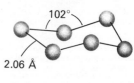

Packing model for orthorhombic sulfur—room temperature stable form

Figure 20.16
Phase diagram for sulfur (not drawn to scale). Dashed lines within the monoclinic phase region indicate transition conditions when nonequilibrium (rapid) phase change occurs.

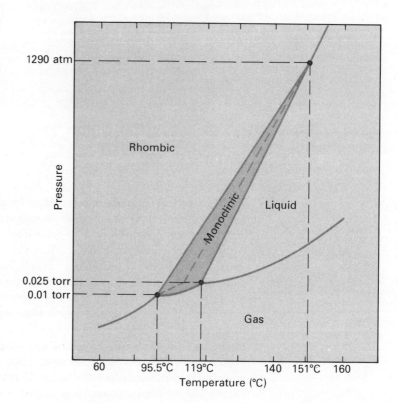

and dissolves in nonpolar solvents such as CS_2 and benzene. The S_8 molecules maintain their integrity in these solutions.

Rhombohedral sulfur forms when HCl is added to cold $Na_2S_2O_3$ solutions.

$$6\ S_2O_3^{2-} \xrightarrow{\text{Conc. HCl}} S_6(\text{rhombohedral}) + 6\ SO_3^{2-}$$

This allotrope, containing S_6 molecules, is less stable than the S_8 form probably because the bond angle (102°) is sufficiently different from the tetrahedral angle (109°) and excessive bond strain occurs within the ring.

Other forms of sulfur containing S_7 and S_9 rings have also been prepared, but they are very unstable.

Plastic sulfur forms when molten sulfur is poured into cold water. This material consists of long chains of sulfur atoms as well as some rings. Its amorphous appearance is consistent with its disordered molecular structure.

At high temperature, S_8 molecules decompose to S_4 and S_2. The latter molecule has a diradical structure (a structure with two parallel unpaired spins) similar to that of O_2.

Although selenium and tellurium also form eight-membered ring allotropes similar to those of sulfur, these allotropes are thermodynamically less stable than the gray semimetallic forms of these elements. The semimetallic allotropes contain long spiral chains of atoms in which the bonding is primarily covalent.

Polonium is quite different from the other elements of the group in having typical metallic electric conductivity and crystal structures. One of the two known allotropes of polonium, the α-phase, is the only reported example of a metallic crystal with a simple cubic structure and one atom per cell.

Reactions and Compounds of Sulfur

Sulfur is very reactive both as an oxidizing and a reducing agent and forms binary compounds with every element except gold, platinum, and the noble gases. Even the reaction with such an unreactive metal as mercury proceeds at room temperature. (That is why we sprinkle sulfur on spilled mercury—the sulfide that forms on the mercury surface prevents evaporation of this toxic metal.)

The compounds of sulfur in its -2 oxidation state include hydrogen sulfide (H_2S), bisulfide (HS^-) salts, sulfides, and many sulfur-containing organic compounds such as methyl mercaptan (CH_3SH). Hydrogen sulfide is a bent molecule (Figure 20.17) similar in structure to water but much less polar. Not being hydrogen bonded, H_2S has a much lower boiling point ($-61°C$) than water. A foul-smelling, toxic gas, it is easily formed by the reaction of hydrogen gas and molten sulfur,

$$8\ H_2(g) + S_8(l) \longrightarrow 8\ H_2S(g)$$

or by the action of acid on metal sulfides.

$$FeS(s) + 2\ H^+(aq) \longrightarrow Fe^{2+}(aq) + H_2S(g)$$

Although it is only moderately soluble in water (a saturated solution at 1 atm and 25°C is about 0.1 M), the weakly acidic properties of H_2S make it very soluble in alkaline solutions.

$$H_2S(g) + OH^-(aq) \rightleftharpoons HS^-(aq) + H_2O$$

$$HS^-(aq) + OH^-(aq) \rightleftharpoons S^{2-}(aq) + H_2O$$

The sulfides of metals fall into two classes, the ionic sulfides formed by electropositive metals (for example, Na_2S, K_2S, and BaS) and the metallic, nonstoichiometric sulfides formed by the transition metals (for example, FeS, NiS, and $Fe_{0.86}S$). The ionic sulfides have simple ionic crystal lattices. The structure of sodium sulfide, Na_2S, resembles the fluorite structure of CaF_2 (see Figure 9.16) except that the sulfide ions, being doubly charged, occupy the face-centered cubic lattice sites, whereas the Na^+ ions lie in the tetrahedral holes. This structure is called the **antifluorite lattice** (Figure 20.18). Ionic sulfides are soluble in water, and their solutions are very basic because of the extensive hydrolysis of the sulfide ion.

$$S^{2-}(aq) + H_2O \rightleftharpoons HS^-(aq) + OH^-(aq)$$

Almost all the metallic sulfides are insoluble in water and have very low solubility products. However, many sulfides dissolve in acid because the sulfide ion is the conjugate base of a weak acid. Moderately soluble sulfides such as MnS ($K_{sp} = 6.3 \times 10^{-12}$) precipitate when a neutral solution of the metal ion is saturated with H_2S but not if the solution is strongly acidic. Other sulfides such as HgS

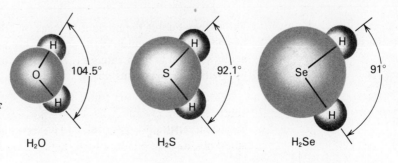

Figure 20.17
Comparison of water, hydrogen sulfide, and hydrogen selenide molecules. There is less hydrogen–hydrogen and nonbonded electron-pair repulsion as the central atom size increases and less tendency for sp^3 hybridization.

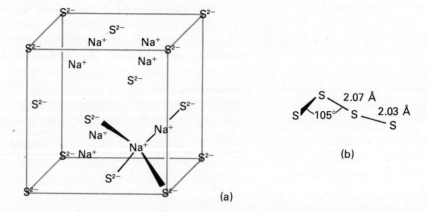

Figure 20.18
(a) Na$_2$S has an antifluorite lattice structure. (b) The S$_4^{2-}$ ion in BaS$_4$ has a geometry similar to H$_2$O$_2$.

($K_{sp} = 1 \times 10^{-53}$) are so insoluble that they precipitate even at low pH. This is the basis for various schemes for analyzing mixtures of metal ions. Treatment of a strongly acidic solution of the metal ions with H$_2$S precipitates only those with very insoluble sulfides. By then raising the pH of the solution, we can precipitate the moderately insoluble sulfides, leaving the metal ions whose sulfides are soluble (the group IA and IIA ions) in solution.

Example 20.1

A solution containing 10^{-3} M each of Hg^{2+} and Mn^{2+} and whose pH is maintained at 1.0 is saturated with H$_2$S gas to give a concentration of dissolved H$_2$S of about 0.1 M. Which ion will precipitate? Estimate its concentration in solution after equilibrium has been attained.

Solution

The concentration of S^{2-} in the solution can be calculated from the concentration of H$_2$S, the pH, and the dissociation constants of H$_2$S ($K_{A1} = 1.0 \times 10^{-7}$; $K_{A2} = 1 \times 10^{-14}$).

$$[\text{HS}^-] = \frac{1.0 \times 10^{-7}[\text{H}_2\text{S}]}{[\text{H}^+]} = \frac{(1.0 \times 10^{-7})(0.1)}{0.1} = 1 \times 10^{-7}$$

$$[\text{S}^{2-}] = \frac{1 \times 10^{-14}[\text{HS}^-]}{[\text{H}^+]} = \frac{(1 \times 10^{-14})(1 \times 10^{-7})}{0.1} = 1 \times 10^{-20}$$

This extremely low sulfide-ion concentration is still sufficient to precipitate HgS. The concentration of Hg^{2+} remaining in solution is

$$[\text{Hg}^{2+}] = \frac{K_{sp}}{[\text{S}^{2-}]} = \frac{1 \times 10^{-53}}{1 \times 10^{-20}} = 1 \times 10^{-33}$$

The ion product for MnS does not exceed its solubility product, so MnS does not precipitate.

$$[\text{Mn}^{2+}][\text{S}^{2-}] = (10^{-3})(1 \times 10^{-20}) = 1 \times 10^{-23} < K_{sp} = 6.3 \times 10^{-12}$$

Rather than bubble the noxious H$_2$S gas into the solution, we can precipitate sulfides by adding thioacetamide to the acidic solution. This compound reacts rapidly in the acidic solution to form enough H$_2$S to saturate the solution without filling the laboratory with this hazardous gas.

$$\underset{\text{Thioacetamide}}{\text{CH}_3\overset{\overset{\text{S}}{\|}}{\text{C}}\text{NH}_2(\text{aq})} + \text{H}_2\text{O} \xrightarrow{\text{H}^+} \text{CH}_3\overset{\overset{\text{O}}{\|}}{\text{C}}\text{NH}_2(\text{aq}) + \text{H}_2\text{S}(\text{aq})$$

Hydrogen persulfide, H_2S_2 (oxidation state -1), is a yellow liquid similar to hydrogen peroxide in molecular structure but less stable. This compound, as well as S_2^{2-} and other polysulfide ions, decompose exothermically to sulfur and hydrogen sulfide.

Sulfur forms a variety of interesting molecules, some with rings and some with chains of atoms (Figure 20.19) combined with moderately electronegative atoms such as Cl and N.

The most important representatives of the positive oxidation states of sulfur (mainly the +4 and +6 states) are the oxides, SO_2 and SO_3, the corresponding oxyacids, H_2SO_3 and H_2SO_4, and the oxyanions, HSO_3^- (bisulfite), SO_3^{2-} (sulfite), HSO_4^- (bisulfate), and SO_4^{2-} (sulfate) (Figure 20.20).

Sulfur dioxide, SO_2, is a gas used to bleach wood pulp in the preparation of paper and to prepare SO_3 and sulfuric acid. Its industrial preparation involves either the roasting of metal sulfides or the reduction of calcium sulfate by carbon at high temperature.

$$CaSO_4(s) + C(s) \xrightarrow{1200°C} 2\ CaO(s) + SO_2(g) + CO_2(g)$$

A major constituent of air pollution, SO_2 is produced by burning petroleum or coal that contains sulfur impurities. When dissolved in water, it forms sulfurous acid, H_2SO_3, a diprotic weak acid.

$$H_2SO_3(aq) \rightleftharpoons H^+(aq) + HSO_3^-(aq) \qquad K_{A1} = 0.017$$

$$HSO_3^-(aq) \rightleftharpoons H^+(aq) + SO_3^{2-}(aq) \qquad K_{A2} = 6.2 \times 10^{-8}$$

Oxidation of SO_2 by oxygen in the presence of a catalyst, a method known as the **contact process**, is the industrial source of sulfur trioxide.

$$2\ SO_2(g) + O_2(g) \xrightarrow[400°C]{V_2O_5\ \text{or Pt}} 2\ SO_3(g)$$

Sulfuric acid, a diprotic strong acid, is prepared from SO_3. Since SO_3 is not rapidly dissolved in water, the SO_3 is dissolved in H_2SO_4 to form oleum (or fuming sulfuric acid) and the oleum is then added to water.

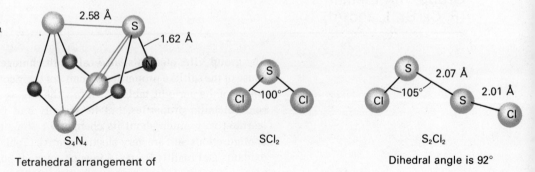

Figure 20.19
Some compounds of sulfur with nitrogen and chlorine.

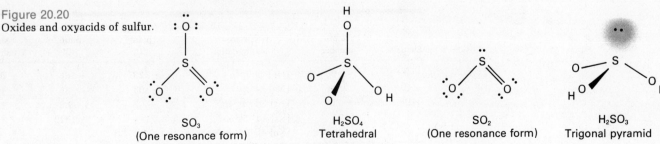

Figure 20.20
Oxides and oxyacids of sulfur.

$$SO_3(g) + H_2SO_4(l) = H_2SO_4 \cdot SO_3(l) \text{ or } H_2S_2O_7(l)$$
Oleum

$$H_2SO_4 \cdot SO_3(l) + H_2O = 2 H_2SO_4(l)$$

Sulfuric acid is one of the most important commodities in the chemical industry (the annual production is over 30 million tons), and the contact process is one of those reactions whose development made modern chemical technology possible. Concentrated commercial sulfuric acid (96–98% by weight) is a syrupy, colorless liquid that acts as a powerful dehydrating agent. For example, it so effectively removes water from sugar that only a black carbon mass remains. Although it is a strong, corrosive acid, H_2SO_4 is only a mild oxidant. With few exceptions, it attacks only those metals active enough to be oxidized by H^+.

Another important sulfur compound is sodium thiosulfate, $Na_2S_2O_3$, the "hypo" used in photographic dark rooms. The thiosulfate ion, $S_2O_3^{2-}$, forms a stable complex with Ag^+

$$Ag^+(aq) + 2 S_2O_3^{2-}(aq) \rightleftharpoons Ag(S_2O_3)_2^{3-}(aq) \qquad K_{eq} = 2.9 \times 10^{13}$$

and thus dissolves excess AgBr from the photographic film, leaving behind the silver image that was formed by the earlier development process. Sodium thiosulfate is also used in an important series of titrations called **iodometric titrations**. The ion under analysis, for example Cu^{2+}, reacts with an excess of iodide ion to form the stoichiometric amount of triiodide ion, I_3^-.

$$5 I^-(aq) + 2 Cu^{2+}(aq) \longrightarrow I_3^-(aq) + 2 CuI(s)$$

The solution is then titrated with a standard thiosulfate solution until the last trace of I_3^- is reduced to I^-.

$$I_3^-(aq) + 2 S_2O_3^{2-}(aq) \longrightarrow 3 I^-(aq) + S_4O_6^{2-}(aq)$$

From the volume and normality of the thiosulfate solution, we can measure the amount of Cu^{2+} in the sample.

20.3 Group VIIA Elements (F, Cl, Br, I, and At)

The group VIIA elements, also called the **halogens** (from *halos*, Greek for salt) because of the saltlike properties of many of their compounds, are the most nonmetallic family of the periodic table. (Astatine, the largest atom of the group, appears to show some metallic properties, but its scarcity and intense radioactivity keep us from learning very much about its chemistry.) Since they have s^2p^5 outer-shell electron configurations and are very electronegative (Table 20.5), the halogens are all reactive oxidants and readily form −1 ions or covalent bonds to attain their octet configura-

Table 20.5 Properties of the Group VIIA Elements

Element	Atomic number	Electron configuration	Melting point (°C)	Boiling point (°C)	Atomic radius (Å)
F	9	[He] $2s^22p^5$	−223	−188	0.71
Cl	17	[Ne] $3s^23p^5$	−103	−35	1.00
Br	35	[Ar] $3d^{10}4s^24p^5$	−7.2	58.8	1.14
I	53	[Kr] $4d^{10}5s^25p^5$	114	184	1.33
At	85	[Xe] $4f^{14}5d^{10}6s^26p^5$	(300)	—	—

Figure 20.21
Oxidation state–free energy diagram for the halogens. As with previous groups, the stability of −1 states decreases with increasing atomic number. Higher oxidation states are less stable.

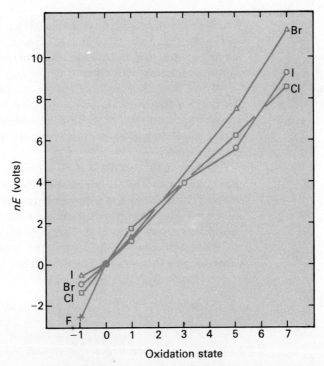

tions. The oxidation states of halogens range from −1 to +7. The free energy-oxidation state graph for the halogens is given in Figure 20.21.

Occurrence

Because of their reactivity, the halogens do not occur free in nature, but are present mainly as salts of alkali metals or alkaline earth metals. Being soluble in water, these salts occur in the oceans (the chloride content of seawater is about 2% and the bromide content 0.01%—see Table 20.6), in stagnant inland bodies of water, in the dry beds of primeval lakes and oceans, and in underground salt deposits that are protected from water by rock formations.

Although many halogen compounds are extracted from salt mines and brine wells, the major source of chlorine and bromine is seawater. Iodide concentrations in seawater are, however, quite low, but some marine animals and plants such as kelp concentrate iodine to a remarkable degree and provide us with an important source of this element. Iodine is an important material for proper health and purifying table salt usually removes this essential iodine. The major source of iodine is from iodate (IO_3^-) salts present in the Chilean nitrate deposits. Fluorine is obtained from the relatively insoluble minerals fluorite (also called fluorspar, CaF_2), cryolite (Na_3AlF_6), and fluorapatite [$CaF_2 \cdot 3Ca_3(PO_4)_2$].

Properties and Preparation of the Elements

Fluorine, F_2, is an extremely reactive, colorless gas. Its relatively low F–F bond energy, the high electron affinity of fluorine atoms, and the large lattice energies of

Table 20.6 Composition of Salt Obtained from Evaporation of Seawater

Substance	Percent of total salt obtained
NaCl	77.8
$MgCl_2$	9.4
$MgSO_4$	6.6
$CaSO_4$	3.4
KCl	2.1
$MgBr_2$	0.2
$CaCO_3$	Trace

fluoride salts all contribute to the reactivity. Its reactions with metals are so vigorous and exothermic that many metals ignite in the presence of pure fluorine. Fluorine atoms also form strong covalent bonds to carbon and other nonmetals, consequently F_2 reacts violently with organic matter.

There is no chemical oxidant capable of oxidizing fluoride compounds to F_2—fluorine has the largest known reduction potential. The preparation of fluorine involves the electrolysis of liquid mixtures of KF and HF using cells made of steel or monel metal (a nickel–copper alloy that resists fluorine corrosion) and special graphite electrodes.

In view of the reactivity of fluorine, one element that even reacts with the noble gases Kr, Xe, and Rn, it might seem that it would be impossible to store this gas because of the absence of a container that can resist it. Fortunately, steel and nickel do the job. Fluorine reacts with these metals, but in doing so coats them with thin, impervious layers of fluorides that stop further corrosion.*

Elemental chlorine, Cl_2, a gas with a greenish tint (it gets its name from *chloros*, Greek for green) is obtained by the electrolysis of brine.

$$2\ NaCl(aq)\ +\ 2\ H_2O\ \xrightarrow{\text{Electrolysis}}\ 2\ NaOH(aq)\ +\ Cl_2(g)\ +\ H_2(g)$$

(We described this electrolysis in Chapter 18 as a method for preparing sodium hydroxide.) This electrolysis is a good illustration of the practical importance of overpotential effects. Considering only the half-cell potentials, we would predict that oxygen should form at the anode in preference to chlorine. Instead, chlorine forms because of the large overpotentials for oxygen formation at graphite electrodes. Even with this overpotential effect, a small amount of O_2 forms with the Cl_2, and the electrolysis of dilute chloride solutions produces a mixture of 20% O_2 and 80% Cl_2. Some chlorine is prepared commercially by the **Deacon process**, the oxidation of HCl gas.

$$4\ HCl(g)\ +\ O_2(g)\ \xrightarrow{450°}\ 2\ Cl_2(g)\ +\ 2\ H_2O(g)$$

The major uses of Cl_2 are for the production of chlorine-containing organic compounds, as an industrial bleach, and as a germicide for water purification. Chlorine gas was the first poison gas used in World War I, but it was quickly replaced by more sophisticated chemical warfare agents, some of which contain chlorine—phosgene, $COCl_2$, and mustard gas, $(C_2H_4Cl)_2S$. None of these agents proved militarily decisive.

Bromine, Br_2, a dark red, volatile liquid, is usually prepared by oxidation of bromide salts with chlorine gas.

$$Cl_2(g)\ +\ 2\ Br^-(aq)\ \xrightarrow{H^+}\ Br_2(l)\ +\ 2\ Cl^-(aq)$$

We use considerable quantities to prepare silver bromide salts for photographic film and bromine-containing organic compounds such as ethylene bromide, $C_2H_4Br_2$. Ethylene bromide is a gasoline additive, along with tetraethyl lead, it helps prevent the depositing of Pb on the engine walls by forming volatile $PbBr_2$, which exits through the exhaust.

Iodine, I_2, a deep violet solid that easily sublimes to form a violet vapor, can be prepared either by oxidizing iodide salts or reducing iodates.

$$2\ IO_3^-(aq)\ +\ 5\ HSO_3^-(aq)\ \longrightarrow\ I_2(aq)\ +\ 5\ SO_4^{2-}(aq)\ +\ 3\ H^+(aq)\ +\ H_2O$$

---

*When liquid fluorine was tested as an oxidant for rocket fuels it was necessary to clean the fuel chambers, fuel lines, and valves scrupulously and then expose them to gaseous F_2 or HF to coat all the surfaces. The liquid F_2 was pumped in after this treatment, and the system was reasonably stable. Serious explosions occurred when traces of grease or organic solvents remained!

Figure 20.22
Iodine and triiodide ion. The I_3^- ion is linear but unsymmetrical. Effectively it is an I_2 molecule whose bond gets weaker (longer) when weakly bonded to an I^- ion.

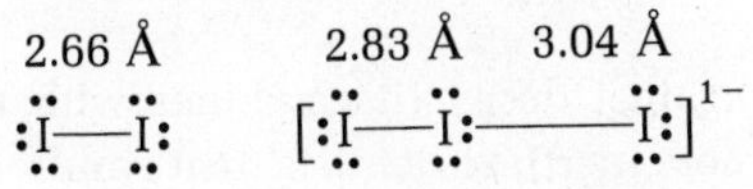

2.66 Å   2.83 Å   3.04 Å

:Ï—Ï: [:Ï—Ï———Ï:]$^{1-}$

$I_2(s)$ I_3^- in $CsI_3(s)$

Although I_2 resembles the other halogens in not being very soluble in water, it dissolves readily in iodide solutions by forming the triiodide ion, I_3^-. (See Figure 20.22.)

$$I_2(aq) + I^-(aq) \rightleftharpoons I_3^-(aq)$$

Iodine reacts with starch to produce an intense blue complex, a sensitive color reaction that is characteristic of I_2. The fading of this color provides the end point for iodometric titrations (see Section 20.2), and its appearance signals the end of iodimetric titrations—redox titrations using I_3^- solutions as the oxidant. For example,

$$I_3^-(aq) + H_2O + H_3AsO_3(aq) \longrightarrow 3\,I^-(aq) + 2\,H^+(aq) + H_3AsO_4(aq)$$

(After all the As(III) has been oxidized excess I_3^- will form the blue complex with the starch.)

All the halogens dissolve in carbon tetrachloride more readily than in water, a phenomenon that is sometimes used to detect the presence of iodide or bromide ions in aqueous solutions. An oxidant such as HNO_2 or chlorine water solution is added to a test tube containing the aqueous solution and CCl_4 (a liquid that is immiscible with water). If I^- is present, the oxidant oxidizes it to I_2, which then dissolves in CCl_4 to produce a violet solution. Stronger oxidants such as MnO_4^- oxidize Br^- to Br_2, and the red color of bromine appears in the carbon tetrachloride.

Halides

The **halides**, compounds of halogens in which the halogen is in the -1 oxidation state, include ionic halides such as NaCl, KBr, and BaI_2 and covalent halides formed with nonmetals and moderately electronegative metals. The halides of the elements of groups IA and IIA (except beryllium), the lanthanides, and some transition metals are ionic. Also, the covalent nature of metal halides increases as the charge-to-size ratio of the metal ion increases. For example, there is a gradual decrease in ionic character from RbCl, a typically high-melting (melting point 715°C) ionic substance, through $SrCl_2$ and YCl_3 to $ZrCl_4$, a solid that sublimes at a relatively low temperature (330°C) to produce tetrahedral $ZrCl_4$ molecules in the gas phase.

The vast majority of the ionic halides are very soluble in water. Fluorides are the least soluble of the ionic halides because their strong lattice energies more than make up for the large hydration energy of the fluoride ion. Calcium fluoride is one of the few slightly soluble ionic halides. The other insoluble metal halides, the chlorides, bromides, and iodides of Ag^+, Pb^{2+} and Hg_2^{2+}, have considerable covalent character.

Halide ions form complexes with many metal ions, and the fluoride ion is a particularly effective ion. It is so powerful a complexing agent that fluoride solutions etch glass by forming complexes with silicon.

$$2\,H_2O + SiO_2(s) + 6\,F^-(aq) \longrightarrow SiF_6^{2-}(aq) + 4\,OH^-(aq)$$

The commercially most important halide salt is and always has been sodium chloride. A large part of its current production is used to synthesize other essential materials such as $NaHCO_3$ (Section 18.1), NaOH, HCl, NaOCl, and Cl_2, but a lot of NaCl is consumed in its traditional role as a food preservative and an essential component of our diets. We have depended on salt as a dietary supplement ever since agriculture developed and grains, vegetables, and cooked meats replaced more primitive, salt-containing foods. The development of trade and long-distance travel would have been impossible without the use of salt for preserving meat and fish. The preparation of salt from seawater is one of the oldest known industries. The pay of Roman soldiers consisted in part of salt rations—the word salary comes from the Latin *salarium*, which is in turn derived from *sal*, Latin for salt—and small cakes of salt were used as currency in Africa and Asia.

Approximately 7 million tons of sodium chloride, about 70 lb per person, are used for human consumption in the United States each year. As you know from your own experience only a small portion of this shows up as table salt; the bulk is used in a variety of ways in the commercial preparation of food. Table salt usually contains a trace of KI for dietary reasons—iodide-deficiency causes disease of the thyroid gland—and small amounts of Na_2CO_3, $MgCO_3$, or Na_3PO_4 to prevent caking of the salt in moist air.

Trace amounts of fluoride protect teeth against decay (it combines with the calcium hydroxyphosphate in teeth to make it more inert), so dentists treat children's teeth with fluorides, many toothpastes contain fluoride salts, and trace concentrations of fluorides are added to many water supplies.

The halides of the more electronegative elements or the halides of a metal in one of its high oxidation states are covalent. These covalent compounds include the hydrogen halides (HF, HCl, HBr, and HI), some metal halides such as $BeCl_2$ and $AlCl_3$, nonmetal halides such as SF_4, OF_2, and PCl_3, and interhalogen compounds such as BrF_5. The formation of covalent halides occurs by direct combination of the elements, and most of these reactions are strongly exothermic.

$$H_2(g) + F_2(g) \longrightarrow 2\ HF(g) \quad \Delta H = -550\ kJ$$

Fluorine is such a strong oxidant that fluorination reactions promote elements to their highest possible oxidation states and fluorine's small size allows for maximum coordination number with the other element. These products are often superoctet species such as IF_7, SF_6, and WF_6.

The hydrogen halides are all typical polar covalent molecules. They all dissolve in water, and all except HF ($K_A = 6.7 \times 10^{-4}$) are strong acids in water. Hydrogen fluoride is a relatively weak acid because its strong covalent bond more than compensates for the hydration energy of F^- (Table 20.7) and because hydrogen bonds with water stabilize HF.

Hydrogen chloride, commercially the most important hydrogen halide, is marketed mainly as concentrated hydrochloric acid, a 36% aqueous solution. A slightly impure form of hydrochloric acid called muriatic acid (from *muria*, Latin for brine) is used to clean surfaces and adjust the pH values of swimming pools. Hydrochloric acid is a nonoxidizing acid, in other words, its anion (Cl^-) cannot act as an oxidant. This is an advantage for many applications. For example, nitric acid does not corrode aluminum very well because nitrate ion reacts with the aluminum surface to form an oxide film that inhibits further corrosion. Hydrogen chloride dissolves aluminum rapidly because the only reaction is the redox reaction between H^+ and Al. Another factor that makes hydrochloric acid corrosive toward metals is the ability of Cl^- to form complexes with metal ions. Aqua regia, a mixture of HCl and HNO_3, dissolves gold because of the combination of the oxidizing power of the nitrate ion and the ability of Cl^- to form complexes with gold ions.

$$Au(s) + 4\ H^+(aq) + NO_3^-(aq) + 4\ Cl^-(aq) \longrightarrow AuCl_4^-(aq) + NO(g) + 2\ H_2O$$

Table 20.7 Energy Values (kJ) for Hydrogen Halide Processes

Reaction	HF	HCl	HBr	HI
$HX(aq) = HX(g)$	23.9	−4.2	−4.2	−4.2
$HX(g) = H(g) + X(g)$	535	405	340	272
$H(g) = H^+(g) + e^-$	1320	1320	1320	1320
$e^- + X(g) = X^-(g)$	−348	−367	−345	−315
$H^+(g) + X^-(g) = H^+(aq) + X^-(aq)$	−1514	−1393	−1364	−1330
$HX(aq) = H^+(aq) + X^-(aq)$	18	−40	−54	−57

The advantages of hydrochloric acid over the other hydrohalic acids (aqueous solutions of hydrogen halides) are (1) it is inexpensive, (2) it is a strong acid compared to HF, and (3) unlike HBr and HI, it is stable against oxidation. Hydrogen chloride is cheap because it is produced by the reaction of sodium chloride and sulfuric acid,

$$2\,Na^+(aq) + 2\,Cl^-(aq) + 2\,H^+(aq) + SO_4^{2-}(aq) \longrightarrow 2\,Na^+(aq) + SO_4^{2-}(aq) + 2\,HCl(g)$$

a reaction that uses inexpensive reactants and produces, in addition to HCl, Na_2SO_4, an important component of fertilizers.

Although HF is a weak acid, a small amount of it mixed with other acids helps them to dissolve metals because of the formation of metal–fluoride complexes. Hydrogen fluoride, which is prepared by treating solid CaF_2 with concentrated H_2SO_4, is a volatile liquid and an excellent polar solvent. Like water, it is extensively hydrogen bonded, and solid HF contains long chains of associated HF molecules (Section 17.3). Hydrogen bonding to fluorine is in fact so strong that there are stable salts of the hydrogen bonded ion HF_2^-, called bifluoride ion.

$$[:\!\ddot{\underset{..}{F}}\!-\!H\!-\!\ddot{\underset{..}{F}}\!:]^-$$

There are many very important organic halogen compounds. These include solvents such as chloroform ($CHCl_3$) and carbon tetrachloride (CCl_4), refrigerants such as Freon-12 (CF_2Cl_2), and polymers such as polyvinyl chloride (PVC), polyvinylidene chloride (Saran wrap), and Teflon (Figure 20.23). Teflon is particularly useful because of its inertness and its smooth, stick-free surface. Many chlorinated hydrocarbons such as DDT were used extensively as pesticides but their use is decreasing because they resist biodegradation and because many animals concentrate them in their fatty tissue.

Oxides, Oxyacids, and Oxyanions

The halogens form a variety of oxides, oxyacids, and oxyanion salts ranging up to the +7 oxidation state (Table 20.8). The predominant positive oxidation states are the odd-numbered ones: +1, +3, +5, and +7. Fluorine forms only two oxides, OF_2 and O_2F_2, and these oxides do not react with water to form oxyacids. (According to the rules for assigning oxidation states, fluorine in these oxides is in the −1 state.) As noted earlier (Section 17.5), the acid strengths of the oxyacids increase with increasing oxidation state; hypochlorous acid, HClO, is a relatively weak acid ($K_A = 3 \times 10^{-8}$), whereas $HClO_4$ is extremely strong. The oxides, oxyacids, and

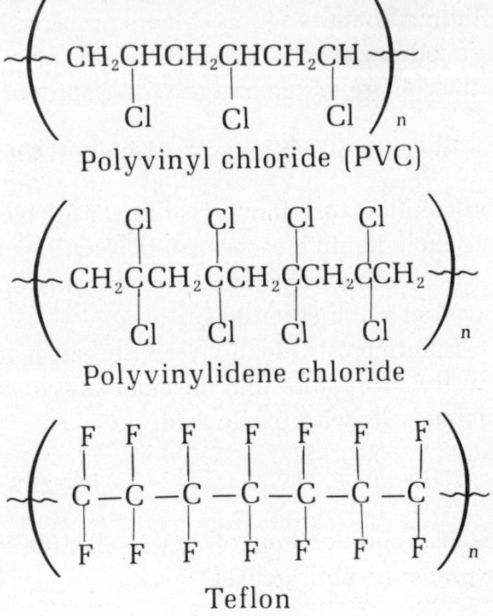

Figure 20.23
Halogen-containing polymers (n represents a large number).

$$\left(\!\!-CH_2\underset{\underset{Cl}{|}}{CH}CH_2\underset{\underset{Cl}{|}}{CH}CH_2\underset{\underset{Cl}{|}}{CH}\!-\!\right)_n$$
Polyvinyl chloride (PVC)

$$\left(\!\!-CH_2\underset{\underset{Cl}{|}}{\overset{\overset{Cl}{|}}{C}}CH_2\underset{\underset{Cl}{|}}{\overset{\overset{Cl}{|}}{C}}CH_2\underset{\underset{Cl}{|}}{\overset{\overset{Cl}{|}}{C}}CH_2\underset{\underset{Cl}{|}}{\overset{\overset{Cl}{|}}{C}}CH_2\!-\!\right)_n$$
Polyvinylidene chloride

$$\left(\!\!-\underset{\underset{F}{|}}{\overset{\overset{F}{|}}{C}}\!-\!\underset{\underset{F}{|}}{\overset{\overset{F}{|}}{C}}\!-\!\underset{\underset{F}{|}}{\overset{\overset{F}{|}}{C}}\!-\!\underset{\underset{F}{|}}{\overset{\overset{F}{|}}{C}}\!-\!\underset{\underset{F}{|}}{\overset{\overset{F}{|}}{C}}\!-\!\underset{\underset{F}{|}}{\overset{\overset{F}{|}}{C}}\!-\!\underset{\underset{F}{|}}{\overset{\overset{F}{|}}{C}}\!-\!\right)_n$$
Teflon

Table 20.8 Halogen Oxides, Oxyacids, and Oxyanions

Element	Oxidation state	Oxide	Oxyacid	Oxyanion
F	−1	F_2O	—	—
Cl	+1	Cl_2O	HOCl Hypochlorous acid	ClO^- Hypochlorite
	+3	—	$HClO_2$ Chlorous acid	ClO_2^- Chlorite
	+4	ClO_2	—	—
	+5	—	$HClO_3$ Chloric acid	ClO_3^- Chlorate
	+7	Cl_2O_7	$HClO_4$ Perchloric acid	ClO_4^- Perchlorate
Br	+1	—	HOBr Hypobromous acid	BrO^- Hypobromite
	+4	BrO_2	—	—
	+5	—	$HBrO_3$ Bromic acid	BrO_3^- Bromate
	+7	—	$HBrO_4$ Perbromic acid	BrO_4^- Perbromate
I	+1	—	HOI Hypoiodous acid	OI^- Hypoiodite
	+4	I_2O_4	—	—
	+5	I_2O_5	HIO_3 Iodic acid	IO_3^- Iodate
	+7	I_2O_7	HIO_4, H_5IO_6 Metaperiodic acid	$IO_4^-, H_3IO_6^{2-}$ Metaperiodate

oxyanion salts are all strong oxidants. Concentrated perchloric acid, $HClO_4$, is a particularly dangerous oxidant that reacts violently with all organic matter—wood, clothing, and flesh!

Commercially, the most important oxyanion salt is sodium hypochlorite, NaOCl. Solutions containing this salt are prepared by electrolyzing brine to form chlorine at the anode and OH^- at the cathode (see Section 18.1). During electrolysis the contents of the cell are mixed and the Cl_2 disproportionates in the alkaline solution.

$$Cl_2(g) + 2\ OH^-(aq) \rightleftharpoons Cl^-(aq) + ClO^-(aq) + H_2O$$

The solution that forms is a powerful bleach, and it is the active constituent of household liquid bleaches. Solid NaOCl is an effective germicide, and its use in place of Cl_2 in home swimming pools avoids the expensive equipment and hazards involved in using chlorine gas.

Chlorine dioxide, a yellowish gas, is a free radical and strong oxidant. Rather than trying to store this unstable substance, we prepare it as needed by reducing chlorate salts with oxalic acid.

$$2\ H^+(aq) + H_2C_2O_4(aq) + 2\ ClO_3^-(aq) \longrightarrow 2\ ClO_2(g) + 2\ CO_2(g) + 2\ H_2O$$

The disproportionation of ClO_2 in alkaline solution provides the commercial method for preparing chlorite (ClO_2^-) salts.

$$2\ ClO_2(g) + 2\ OH^-(aq) \rightleftharpoons ClO_2^-(aq) + ClO_3^-(aq) + H_2O$$

20.4 The Noble Gases
(He, Ne, Ar, Kr, Xe, and Rn)

One of the six noble gases terminates each of the six rows of the periodic table. Except for helium, whose outer-shell configuration is $1s^2$, each noble gas has an s^2p^6 configuration. Because these elements have closed subshell configurations and high ionization potentials, they do not gain or lose electrons easily, and they do not form many compounds. (They were called "inert gases" before their reactions were discovered in 1962.) They are all colorless, monatomic gases at room temperature, and they have very low boiling points (Table 20.9) because their only intermolecular forces are weak van der Waals forces.

Occurrence

Although sometimes called "rare gases," the noble gases are fairly abundant in our atmosphere and, in the case of helium, occur in natural gas deposits. Argon is the most plentiful noble gas (0.9% of air) and radon, which is extremely radioactive, occurs only to a very slight extent. Except for radon and helium, the pure gases are produced by liquefaction and fractional distillation of air. The major source of helium is the fractional distillation of natural gas from certain helium-rich wells.

The Discovery of the Noble Gases

Because of their chemical inertness, noble gases escaped detection until the late nineteenth century. There had been an earlier hint of their presence when Cavendish in 1784 noticed that about 1% of air resisted all attempts at chemical reaction, but this observation was ignored. About a century later, Lord Rayleigh observed that samples of nitrogen obtained from air were slightly denser than nitrogen formed by thermal decomposition of pure NH_4NO_2. Suspecting a denser impurity in the nitrogen from air, he reacted this sample with magnesium

$$N_2(g) + 3\ Mg(s) \xrightarrow{\text{Heat}} Mg_3N_2(s)$$

and was left with the same inert gas mixture that Cavendish had seen.

Rayleigh and Sir William Ramsay, who discovered the noble gases independently, had the benefit of two important tools—one theoretical and one analytical—not available to Cavendish. The theoretical tool was Mendeleev's periodic law. Since, at the time, no such unreactive elements were known, the newly discovered gas had to belong to a previously undiscovered periodic family. Therefore, Rayleigh and Ramsay knew that they had to look for several new elements rather than one, and the periodic law indicated the number of family members and their approximate atomic masses.

The analytic tool was atomic emission spectroscopy. Because the inert gas sample emitted several intense lines that are very weak in the spectrum of air, it was clear that elements that are minor components of air make up the inert portion. In the effort

Table 20.9 Properties of the Noble Gases

Element	Atomic number	Estimated atomic radius (Å)	Normal boiling point (°C)	Melting point (°C)	Ionization potential (kJ/mol)
He	2	0.54	−268.9	−269.7	2370
Ne	10	0.65	−246	−248.6	2080
Ar	18	0.95	−185.8	−189.4	1520
Kr	36	1.10	−152	−157.3	1350
Xe	54	1.30	−108.0	−111.9	1170
Rn	86	1.45	−61.8	−71	1030

to isolate the pure elements, emission spectroscopy was an invaluable aid in monitoring the compositions of samples. The technique is so effective that helium (named after *helios*, Greek for sun) was discovered by its emission lines in the sun's spectrum before it was isolated on earth.

Rayleigh and Ramsay isolated argon ("idle"), krypton ("hidden"), and xenon ("stranger") from air. Helium and radon were discovered as products of the radioactive decay in uranium ores (Section 22.1).

Uses

Helium is commercially the most important noble gas. Its low density makes it useful for inflating weather balloons and blimps. (Although H_2 is more buoyant than He, it has the disadvantage of being dangerously flammable.) Helium's extremely low boiling point (4.18 K) makes liquid helium an effective refrigerant for obtaining extremely low temperatures (cryoscopic temperatures). It provides inert atmospheres for heli-arc welding and other scientific and engineering applications requiring light, inert atmospheres.

Helium is used in place of nitrogen in underwater "breathing atmospheres" because helium has such a low solubility in the blood. Nitrogen is quite soluble and when air is breathed by divers at great depths in the ocean (extreme pressures) nitrogen dissolves in the blood. If the diver surfaces rapidly, N_2 comes out of solution forming gas bubbles in the blood at the joints, causing great pain and injury. A side effect from using helium is that the diver's voice sounds very high—the vocal cords vibrate faster because helium is less dense than air. Argon, which is heavier than air, is more suitable than helium in artificial atmospheres because helium's buoyancy can cause problems in handling equipment.

The atomic emission spectrum of neon has had a great esthetic impact—for better or for worse—on our surroundings. The brightly colored "neon lights" are discharge tubes containing various gases, the color of the lamp depending on the gas present in it. (Red neon lights are the only ones that actually contain neon.)

Argon, which provides the inert atmosphere in incandescent and fluorescent lights, is one of the few substances that is completely recyclable. We get it from the air, use it, and return it to the air in its original state. Although helium is equally inert, we must worry about its available reserves since we obtain helium from natural gas deposits and let used helium escape into the atmosphere. The helium is still around, but its low concentration in the air makes it more expensive to recover than from natural gas.

Noble Gas Compounds

In 1962, Neal Bartlett deprived the noble gases of their reputation for inertness by showing that the larger noble gas elements (Kr, Xe, Rn) can react with fluorine. This discovery of noble gas reactions is an excellent illustration of the application of the principles of chemical bonding. Bartlett knew that O_2 could be oxidized to O_2^+ ions, which then form the ionic solid O_2PtF_6. Since the ionization potential of O_2 is almost the same as that of xenon,

$$O_2(g) \longrightarrow O_2^+(g) + e^- \qquad 1180 \text{ kJ}$$

$$Xe(g) \longrightarrow Xe^+(g) + e^- \qquad 1170 \text{ kJ}$$

Bartlett realized that xenon might undergo a similar reaction, a hunch he confirmed by heating Xe and PtF_6 which formed $XePtF_6$, a red, crystalline compound. Other noble gas compounds, fluorides and oxides, have since been synthesized.*

The larger the noble gas atom, the lower its ionization potential and the more likely it is to form stable compounds. Although radon is the largest and most easily oxidized member of the family, its intense radioactivity limits the study of its chemistry, therefore more xenon compounds are known. Krypton apparently forms some compounds, but no reactivity has been observed for argon, neon, or helium.

*Solid clathrates in which noble gas atoms are trapped in the cagelike lattices of another substance (Section 17.6) had been discovered many years earlier.

20.4 THE NOBLE GASES

Table 20.10 **Some Noble Gas Compounds**

Compound	Physical state	Melting point (°C)	Comment
XeF_2	Crystalline	130	Hydrolyzes in water to Xe and O_2
XeF_4	Crystalline	114	Stable
$XeOF_2$	Crystalline	31	Unstable
XeF_6	Crystalline	50	Stable, forms XeF_8^{2-} ions
$XeOF_4$	Liquid	−46	Stable
XeO_3	Solid		Explosive; only stable in solution
XeO_4	Gas		Very unstable

Xenon fluorides, XeF_2, XeF_4, and XeF_6, form when various xenon–fluorine mixtures are heated for several hours in sealed tubes. These colorless, crystalline compounds react with water to form oxyfluorides ($XeOF_4$ and XeO_2F_4) and oxides (XeO_3 and XeO_4). (See Table 20.10.) For example,

$$XeF_6(s) + 3 H_2O \longrightarrow XeO_3(aq) + 6 HF(aq)$$

Xenon fluorides are very powerful oxidants and fluorinating agents.

$$2 XeF_2(s) + 2 H_2O \longrightarrow 2 Xe(g) + O_2(g) + 4 HF(aq)$$

The molecular geometries of the xenon compounds (Figure 20.24) are good examples of the relationship between molecular shape and the number of groups and unshared electron pairs surrounding an atom (Section 7.3). There are two fluorines and three unshared electron pairs around the xenon atom in XeF_2. The orbital arrangement that minimizes electrostatic repulsion in this system is a trigonal bipyramidal arrangement in which three orbitals point to the corners of an equilateral triangle and the other two are perpendicular to the plane of this triangle, one up and one down. Since unshared electron pairs are more stable in the triangular (equatorial) orbitals (120° apart), the fluorine atoms occupy the vertical (axial) positions, giving XeF_2 its observed linear shape.

Xenon difluoride is a superoctet species containing ten valence electrons around Xe—eight from the Xe atom and one from each F atom—and its bonds involve one of xenon's d orbitals. The five orbitals in the trigonal bipyramidal arrangement are a set of five hybrid sp^3d orbitals constructed from a 5s orbital, three 5p orbitals, and one 5d orbital.

There are twelve valence electrons around the Xe atom in XeF_4, and these twelve electrons occupy six sp^3d^2 hybrid orbitals, orbitals obtained from one 5s orbital, three 5p orbitals, and two 5d orbitals. The octahedral arrangement of these six orbitals in which the two unshared electron pairs occupy opposite vertices minimizes electrostatic repulsions and gives XeF_4 its square planar shape.

Figure 20.24
Molecular geometries of Xe compounds. (Description is that of bonded atoms only, not the orbital hybridization geometry.)

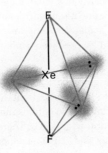
XeF_2
Linear

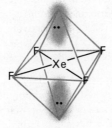
XeF_4
Square planar

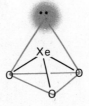
XeO_3
Trigonal pyramid

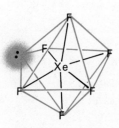
XeF_6
Distorted octahedral

Liquid Helium

Liquid helium is very strange stuff. In the first place, there are two distinct liquid phases. Above 2.2 K, liquid helium behaves fairly normally, but below 2.2 K it changes to a liquid phase known as helium II. This phase is an unusually good heat conductor, but its strangest property is its **superfluidity**—it has essentially *no viscosity*. Because of its lack of viscosity, it can flow up the wall of a container, over the edge and down the other side. It also has unusual optical and acoustical properties—it is not a normal liquid and is called a "quantum mechanical liquid!"

Summary

1. The group VA elements (N, P, As, Sb, and Bi) have s^2p^3 outer-shell electron configurations, and their oxidation states range from -3 to $+5$.

2. Nitrogen occurs as N_2 gas in the atmosphere, as nitrogen-containing compounds in animal and vegetable matter, and as nitrate deposits. Phosphorus occurs as phosphates present in a variety of minerals. Arsenic, antimony, and bismuth occur mainly as sulfides and oxides.

3. The nitrogen cycle is the conversion of N_2 to ammonium ion and then to nitrate by the action of microorganisms, the transfer of these nitrogen-containing species to plants and animals, the return of nitrogen compounds to the soil through plant and animal wastes, and the conversion of nitrate and ammonium ions to N_2 by denitrifying bacteria.

4. The Haber process is the commercial method for preparing ammonia by the reaction of nitrogen and hydrogen in the presence of a catalyst at high temperature and pressure.

5. Elemental nitrogen is prepared commercially by fractional distillation of liquid air. The reduction of phosphates by coke in the presence of sand at high temperature yields elemental phosphorus. Arsenic, antimony, and bismuth are prepared by reducing their sulfides.

6. The elemental form of nitrogen is N_2. Phosphorus exists as three allotropes—white phosphorus (P_4) and the polymeric red and black forms. Arsenic and antimony have yellow allotropes, similar in structure to white phosphorus, and gray, semimetallic forms similar to black phosphorus.

7. Nitrogen, a second-row element, differs from the other group VA elements in its tendency to form strong π bonds and its inability to form superoctet molecules.

8. Ammonia, NH_3, is a polar molecule with a pyramidal shape. Being hydrogen bonded, it is easy to liquefy, and liquid ammonia is a polar solvent. Ammonia dissolves readily in water and is weakly basic. Ammonium (NH_4^+) salts are typical ionic compounds.

9. In addition to ammonia, the representatives of the negative oxidation states of nitrogen are hydrazine, N_2H_4 (-2), hydroxylamine, NH_2OH (-1), and hydrazoic acid, HN_3, and azides ($-1/3$). There are both ionic azides such as KN_3 and covalent azides such as AgN_3.

10. The oxides of nitrogen and their oxidation states are nitrous oxide, N_2O ($+1$), nitric oxide, NO ($+2$), dinitrogen trioxide, N_2O_3 ($+3$), nitrogen dioxide, NO_2, and dinitrogen tetroxide, N_2O_4 (both $+4$), and dinitrogen pentoxide, N_2O_5 ($+5$). Both NO and NO_2 are free radicals, but NO is unusually stable and has an unusually low magnetic moment for a free radical. Dinitrogen trioxide and dinitrogen pentoxide are the anhydrides of nitrous acid (HNO_2) and nitric acid (HNO_3), respectively, and these acids dissociate to form nitrite ion (NO_2^-) and nitrate ion (NO_3^-), respectively. Nitric acid is produced commercially by the Ostwald process—the air oxidation of ammonia.

11. Phosphorus reacts with air to produce the sesquioxide, P_4O_6 (oxidation state $+3$), and P_4O_{10} ($+5$). The oxidation of white phosphorus to P_4O_6 yields light, a phenomenon known as chemiluminescence. The sesquioxide is the anhydride of phosphorous acid, $H_2(HPO_3)$, and P_4O_{10} reacts with water to form orthophosphoric acid, H_3PO_4. There are dehydrated forms of phosphoric acid including metaphosphoric acid, HPO_3, and polymeric forms such as pyrophosphoric acid, $H_4P_2O_7$. These acids form a large variety of salts with complex structures, all of which contain tetrahedra of oxygen atoms surrounding a central phosphorus atom.

12. Arsenic, antimony, and bismuth form the +3 and +5 oxidation states.

13. The group VIA elements (O, S, Se, Te, and Po) have s^2p^4 outer-shell electron configurations and oxidation states ranging from -2 to $+6$.

14. Sulfur is extracted from porous limestone deposits deep below the earth's surface by the Frasch process in which superheated water is pumped into the limestone to melt the sulfur and compressed air forces the molten sulfur up pipes leading to the surface.

15. Sulfur exists as several allotropic forms. Both orthorhombic sulfur, which is the most stable form at room temperature, and monoclinic sulfur, which is stable above 95.5°C, contain rings of eight sulfur atoms. Rhombohedral sulfur, which contains S_6 rings, forms by treatment of $Na_2S_2O_3$ solution with HCl. Plastic sulfur is an amorphous modification formed by quickly cooling molten sulfur. The most stable allotropes of Se and Te are the semimetallic forms containing long, covalently bound, spiral chains of atoms. Polonium is a metallic element.

16. Hydrogen sulfide (H_2S) is a bent molecule similar in structure to water but much less polar. It is not strongly hydrogen bonded. Being weakly acidic, it dissociates to bisulfide ion (HS^-) and sulfide ion (S^{2-}). Metal sulfides are either ionic (Na_2S) or metallic, nonstoichiometric compounds ($Fe_{0.86}S$).

17. We can take advantage of the weak acidity of H_2S to selectively precipitate very insoluble metal sulfides by saturating an acidic solution of the metal ions with H_2S. Moderately insoluble sulfides do not precipitate under these conditions because of the very low concentration of S^{2-} in an acidic solution.

18. The most important positive oxidation states of sulfur are $+4$ and $+6$. The $+4$ compounds include SO_2, its corresponding acid, H_2SO_3 (sulfurous acid), and salts of HSO_3^- (bisulfite ion) and SO_3^{2-} (sulfite). The $+6$ compounds include SO_3, H_2SO_4 (sulfuric acid), and the salts of HSO_4^- (bisulfate) and SO_4^{2-} (sulfate). Sulfuric acid, one of the most important industrial chemicals, is produced by the contact process, the catalytic oxidation of SO_2 to SO_3, which is dissolved in H_2SO_4 to form $H_2SO_4 \cdot SO_3$ (oleum) and then in water to form H_2SO_4.

19. Sodium thiosulfate, $Na_2S_2O_3$ (average oxidation state $+2$), is used as a complexing agent for silver salts in photography and to titrate iodine solutions.

20. The group VIIA elements (the halogens—F, Cl, Br, I, and At) have s^2p^5 outer-shell electron configurations and oxidation states ranging from -1 to $+7$.

21. The major sources of the halogens are Cl^- and Br^- in seawater and salt deposits, IO_3^- (iodate) salts occurring in Chilean nitrate deposits, and fluoride minerals such as CaF_2 (fluorite), Na_3AlF_6 (cryolite), and $CaF_2 \cdot 3Ca_3(PO_4)_2$ (fluorapatite).

22. Fluorine (F_2) is extremely reactive because of its relatively weak covalent bond, its high electron affinity, and the large lattice energies of fluoride salts. Fluorine is prepared by electrolysis of molten mixtures of KF and HF. Chlorine (Cl_2) is synthesized by electrolysis of brine solutions, a reaction that succeeds because large overpotentials for O_2 formation make Cl_2 production the more favorable anode reaction. Bromine (Br_2) is prepared by oxidizing Br^- with Cl_2, and iodine (I_2) is formed by reducing iodates.

23. Iodine forms a characteristic blue complex with starch. I_2 reacts with I^- to form triiodide ion (I_3^-). Solutions of I_3^- are used as oxidants in iodimetric titrations.

24. The halides, the -1 oxidation-state halogen compounds, fall into two classes, ionic and covalent. Most ionic halides are water soluble. Some partially covalent metal halides such as AgCl, Hg_2Cl_2, and $PbCl_2$ are insoluble. Halide ions, particularly F^-, form complexes with metal ions.

25. Covalent halides include the hydrogen halides, some metal halides, particularly those in high oxidation states, and nonmetal halides such as SF_4. The hydrogen halides are typical polar covalent compounds, and all except HF are strong acids. Hydrochloric acid, the commercially most important hydrogen halide, is produced by the action of H_2SO_4 on NaCl. Hydrogen fluoride is strongly hydrogen bonded and forms the hydrogen-bonded anion HF_2^-.

26. The halogens form a variety of oxides, oxyacids, and oxyanions. The higher the oxidation state, the stronger the oxyacid. All the halogen oxyacids are strong oxidants. Sodium hypochlorite, NaOCl, a commercial bleach, is formed by the action of OH^- solutions on Cl_2. Chlorine dioxide, ClO_2, is an unstable free radical.

27. The noble gases (He, Ne, Ar, Kr, Xe, and Rn) have closed subshell configurations—$1s^2$ for helium, and outer-shell s^2p^6 configurations for all the other family members. They are all gases with low boiling points.

28. The noble gases are obtained by fractional distillation of liquid air or, in the case of helium, certain natural gas deposits.

29. Xenon, krypton, and radon react with fluorine to form various fluorides, for example XeF_2, XeF_4, and XeF_6. The fluorides react with water to form oxyfluorides and oxides.

Vocabulary List

nitrogen fixation
nitrogen cycle
Haber process
white phosphorus
red phosphorus
black phosphorus
Ostwald process

chemiluminescence
Frasch process
orthorhombic sulfur
monoclinic sulfur
rhombohedral sulfur
plastic sulfur
antifluorite lattice

contact process
iodometric titration
halogen
Deacon process
halide
superfluidity

Exercises

1. Which element is:
   a. The most metallic in group VA?
   b. The most reactive halogen?
   c. The noble gas with the highest oxidation potential?

2. Which group VA element does not have allotropic forms?

3. a. Which should give the more sharply defined x-ray diffraction patterns, white phosphorus or red phosphorus?
   b. Which has the higher Gibbs free energy, white or red phosphorus?

4. Which of the following species is least likely to be stable? Why?
   a. $N_2H_5^+$ b. NO_2 c. NO_2^+ d. NO_2^- e. NO_3 f. K_3N g. $LiNH_2$ h. $Cu(NH_3)_4^{2+}$

5. Acetic acid acts as strong acid in liquid ammonia. Write the equation for its acid reaction.

6. Which element is oxidized in the formation of N_2O by the thermal decomposition of ammonium nitrate? Which element is reduced?

7. a. When nitric acid oxidizes a metal, a red-brown gas appears above the solution. What species is responsible for this color?
   b. The same red-brown gas appears when $Cu(NO_3)_2 \cdot 3H_2O$ is heated strongly. The residue left after this reaction is CuO. What element must have been oxidized? Write a balanced equation for the decomposition of $Cu(NO_3)_2 \cdot 3H_2O$.

8. Which of the following is:
   a. An ionic halide? b. A covalent halide? c. A halide complex ion?
   d. A free radical?
   SF_2 $CuCl_6^{4-}$ $CaCl_2$ ClO_2 I_2

9. Which substance, NH_3 or PH_3, would you expect to have the higher boiling point? Why?

10. Sodium bismuthate oxidizes Mn^{2+} to permanganate ion in acid solution and is itself reduced to BiO^+ in the process. Write a balanced net ionic equation for this reaction.

11. Briefly describe the following processes.
    a. Haber process   b. Frasch process   c. Ostwald process   d. Contact process
    e. Which of these processes *does not* involve a chemical reaction?

EXERCISES 537

12. What is the anhydride of:
   a. Phosphoric acid?   b. Perchloric acid?   c. Nitric acid?   d. Selenic acid?

13. Draw the Lewis structure for SF_4. Use the valence-shell electron-repulsion method (VSEPR) to predict the shape of this molecule.

14. Name the following compounds.
   a. $Na_2S_2O_3$ b. $HBrO_2$ c. H_2SeO_3 d. KH_2PO_4

15. Name the following compounds.
   a. $Na_4P_2O_7$ b. KN_3 c. AsH_3 d. NH_4NO_2

16. Sodium hydroxide is a weak base in liquid ammonia. Write the equation for its basic reaction.

17. Write the equation for the neutralization of a strongly acidic liquid ammonia solution by NaOH.

18. Using the Lewis structure

$$\ddot{\text{N}}=\text{N}=\ddot{\text{N}}^-$$

assign oxidation states to each atom in N_3^-. Are these values consistent with an average oxidation state of $-1/3$?

19. Draw a Lewis structure for NO^+. Is this structure consistent with the +3 oxidation state of nitrogen?

20. Write a balanced net ionic equation for the reduction of nitrate ion to hydroxylamine by SO_2 in acid solution.

21. Write a balanced net ionic equation for the formation of N_2O_3 by adding sulfuric acid to a nitrite solution. Is this a redox reaction?

22. a. What mass of SO_2 can be prepared by roasting 1000 kg (1 metric ton) of PbS?

$$2\ PbS(s)\ +\ 3\ O_2(g)\ \longrightarrow\ 2\ PbO(s)\ +\ 2\ SO_2(g)$$

   b. What is the STP volume of the SO_2 produced in this reaction?

23. What volume (STP) of SO_2 does it take to produce 2000 kg of 98% H_2SO_4 by the contact process?

---

24. It takes 24.35 mL of 0.05872 M $Na_2S_2O_3$ to titrate a sample containing I_2 until the blue starch–iodine color fades out. What mass of I_2 was in the sample?

25. A method for arsenic analysis involves converting all the arsenic to As_2O_3 and then titrating the As_2O_3 with I_3^- solution until a blue starch–iodine color appears.

$$5\ H_2O\ +\ 2\ I_3^-(aq)\ +\ As_2O_3(aq)\ \longrightarrow$$
$$2\ H_3AsO_4(aq)\ +\ 6\ I^-(aq)\ +\ 4\ H^+(aq)$$

What is the percentage of arsenic in a 0.7352-g sample if it takes 35.84 mL of 0.03274 M I_3^- to complete the titration?

26. What is the maximum concentration of Cu^{2+} that can be present in a solution saturated with H_2S at pH 0.70 at 25°C?

27. What is the maximum concentration of Ni^{2+} in an H_2S saturated solution at pH 3.0 and 25°C?

28. Write a balanced equation for the overall change in the Ostwald process. What volume of ammonia and what volume of oxygen (both at STP) are needed to produce 1 metric ton (1000 kg) of 68% nitric acid assuming no loss of material due to side reactions?

29. Commercial hydrochloric acid is 36% HCl by mass and its density is 1.18 g/cm³.
   a. What is its molality?   b. What is its molarity?

30. The standard reduction potential for sulfur going to H_2S

$$2\ e^-\ +\ H^+(aq)\ +\ S(s)\ \longrightarrow\ H_2S(g)$$

is +0.14 V. What is the free energy difference in kJ/mol between -2 sulfur and elemental sulfur?

31. Use Figure 20.13 to estimate the reduction potential for reducing S(VI) to elemental sulfur.

32. a. What is the density of nitrogen gas at standard conditions?
   b. What is the density at STP of a mixture of 98.8% N_2 and 1.2% Ar?

33. The Kjeldahl method is a procedure for measuring the amount of nitrogen in a protein-containing sample such as wheat. The sample is first heated in sulfuric acid to convert all

the nitrogen to NH_4^+. The solution is then made basic by adding excess sodium hydroxide to convert NH_4^+ to NH_3. When this solution is distilled, the NH_3 evaporates and is trapped in a previously measured excess amount of HCl solution. By titrating the excess HCl left after all the NH_3 has reacted, we can estimate the amount of nitrogen in the sample. A 0.735-g sample of wheat flour is treated according to the Kjeldahl procedure and the ammonia distilled into 25.00 mL of 0.4928 M HCl. It takes 29.37 mL of 0.2528 M NaOH to titrate the excess HCl. What is the percentage of nitrogen in the sample?

34. Predict the freezing point of a solution of 0.500 g of white phosphorus in 100 g of benzene. The freezing point of pure benzene is 5.44°C, and its molal freezing-point-depression constant (K_f) is 5.12°C/m.

35. a. How many sulfur atoms are present in the antifluorite unit cell of Na_2S?
    b. How many Na atoms are present?

36. What is the equilibrium constant for the following reaction?

$$S^{2-}(aq) + H_2O \rightleftharpoons HS^-(aq) + OH^-(aq)$$

37. Predict the pH of 0.10 M Na_2S. [Neglect the effect of the reaction

$$HS^-(aq) + H_2O \rightleftharpoons H_2S(aq) + OH^-(aq)]$$

38. Calculate the standard potentials and equilibrium constants for the following reactions at 25°C.
    a. $2 Cl_2(g) + 4 OH^-(aq) \rightleftharpoons O_2(g) + 4 Cl^-(aq) + 2 H_2O$
    b. $2 IO_3^-(aq) + 5 H_2SO_3(aq) \rightleftharpoons I_2(aq) + 5 SO_4^{2-}(aq) + 8 H^+(aq) + H_2O$
    c. $Cl_2(g) + 2 OH^-(aq) \rightleftharpoons Cl^-(aq) + ClO^-(aq) + H_2O$

39. a. The Gibbs free energy for the formation of ammonia

$$N_2(g) + 3 H_2(g) \rightleftharpoons 2 NH_3(g)$$

is −33 kJ at 25°C. What is the equilibrium constant of this reaction at 25°C?
    b. The heat of this reaction is −92 kJ. What is the Gibbs free energy at 500°C?

40. What is the equilibrium constant at 500°C for the reaction given in Exercise 39?

41. Attempts to measure the molecular mass of sulfur by measuring the density of sulfur vapor run into the problem of the conversion of S_8 to S_6 at high temperature. Sulfur vapor at 350°C and 172 torr has a density of 1.04 g/L.
    a. What is the average molecular mass of the sulfur vapor?
    b. What are the partial pressures of S_6 and S_8? (Neglect the possible presence of other sulfur molecules such as S_4 and S_2.)
    c. What is the value of the equilibrium constant for

$$3 S_8(g) \rightleftharpoons 4 S_6(g)$$

42. The equilibrium for the Deacon process

$$4 HCl(g) + O_2(g) \rightleftharpoons 2 H_2O(g) + 2 Cl_2(g)$$

is

$$K_P = \frac{P_{H_2O}^2 P_{Cl_2}^2}{P_{O_2} P_{HCl}^4} = 256 \text{ at } 350°C$$

Compute the partial pressure of the equilibrium mixture formed at 350°C by the reaction of 1.00 atm of HCl and 0.250 atm O_2 in a rigid container.

43. On the basis of the phase diagram for sulfur (Figure 20.16), decide whether monoclinic or orthorhombic sulfur is more dense. Explain the reasoning behind your answer.

44. There are three triple points (conditions of temperature and pressure at which three different phases are in equilibrium) in the phase diagram for sulfur (Figure 20.16). Describe the phases present at each triple point.

chapter 21

Transition Metals, Lanthanides, and Actinides

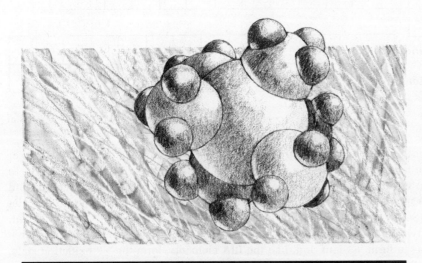

An interesting characteristic of transition elements is their ability to form complexes, such as $Co(NH_3)_6^{3+}$

The 58 metallic elements (over half the known elements) that comprise the transition metals, lanthanides, and actinides (Figure 21.1) include some of the most important metals in our technology. We use enormous quantities of iron–carbon alloys (steels) to build cars, bridges, skyscrapers, railroads, and a vast number of the artifacts of our manufacturing society. Other transition metals such as Ti, Mo, Cr, Mn, Co, and Ni are minor but key constituents of these alloys. Copper carries the electric current we need to light our cities and run our machines. Many of these elements play important biochemical roles—their atoms are essential parts of many enzymes. Hemoglobin, a protein that carries oxygen through our blood streams, contains iron and vitamin B-12 contains cobalt. Several actinide elements such as uranium and plutonium are the fuels of nuclear reactors.

21.1 General Considerations

Electron Configurations

A study of the chemical and physical properties of these elements reveals the ways in which the inner-shell electron configuration of an atom can influence its behavior. The transition metals, lanthanides, and actinides differ from the elements of the major periodic groups (the main-group elements) in that they have similar outer-shell configurations, usually s^2, but differing numbers of electrons in their *inner d* and *f* subshells. The **transition metals**—30 elements in 3 different periods (Sc–Zn, Y–Cd, and La–Hg)—have various *d* electron populations, and the lanthanides and actinides vary in the number of 4*f* and 5*f* electrons, respectively.

21 TRANSITION METALS, LANTHANIDES, AND ACTINIDES

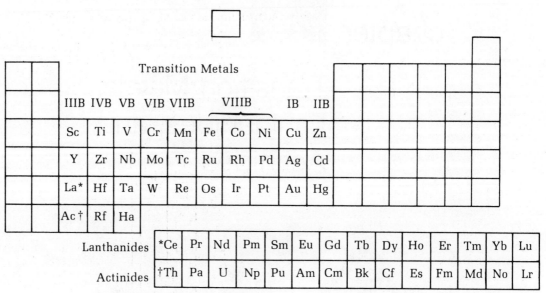

Figure 21.1
Transition metals, lanthanides, and actinides.

Some transition elements strongly resemble the members of one particular major periodic family; others have a more tenuous but still recognizable relationship to major-group elements. An example of a close relationship between a main group and its corresponding transition-metal subgroup is the similarity of the group IIIB elements (Sc, Y, and La) to the IIIA elements (B, Al, Ga, In, and Tl). A IIIB element, having two outer-shell s electrons and one inner-shell d electron, can lose these three electrons to form +3 ions and, like the IIIA elements, form the +3 oxidation state. The similarity between the IIIA and IIIB elements extends to their metallic properties.

As the subgroup number gets higher, the connection with the main group becomes less clear. There is not much in common between the metal manganese, a member of group VIIB, and the halogens of group VIIA, other than that they have the same maximum oxidation state (+7). The properties of iron, cobalt, and nickel and the six elements below them in the periodic table do not resemble the properties of any main group of elements. Groups IB (Cu, Ag, and Au) and IIB (Zn, Cd, and Hg), located at the right-hand end of the transition-metal region of the periodic table, do, however, show many similarities to their corresponding main groups.

The transition metals differ from the main group elements in that they have a greater variety of positive oxidation states (Table 21.1) because there is not much difference between the energies of s and d electrons in the atom of a transition metal, and both types of electrons are accessible for chemical bonding. The oxidation state reached by a transition metal in a particular reaction depends on how many of these accessible electrons are involved in the reaction. Because the outer-shell s electrons are more exposed to the influence of other atoms than are the inner-shell d electrons, almost all transition metals lose their two s electrons to form stable +2 oxidation states. The higher oxidation states require involvement of the d electrons.

The **lanthanide** elements have different numbers of 4f electrons, but most of these are not involved in chemical reactions. The 14 lanthanides (the elements Ce through Lu—although lanthanum is usually included) behave like atoms with a d^1s^2 configuration (this is the ground state of several of them) and resemble the group IIIB elements by forming mainly +3 ions. Other oxidation states (Ce^{4+}, Pr^{4+}, Sm^{2+}, Eu^{2+}, and Yb^{2+}) do occur, and some of these illustrate the stabilities of electron configurations in which a subshell (in this case the 4f) is either empty (as in Ce^{4+}), half-filled (Eu^{2+}), or completely filled (Yb^{2+}).

The d^1s^2 configuration of several lanthanides and the prevalence of the +3 oxidation state do not correspond with the building method (Aufbau principle, Section

21.1 GENERAL CONSIDERATIONS

Table 21.1   Common Oxidation States of the Transition Metal Elements

Sc $3d^1 4s^2$	Ti $3d^2 4s^2$	V $3d^3 4s^2$	Cr $3d^5 4s^1$	Mn $3d^5 4s^2$	Fe $3d^6 4s^2$	Co $3d^7 4s^2$	Ni $3d^8 4s^2$	Cu $3d^{10} 4s^1$	Zn $3d^{10} 4s^2$
+3	+2 +3 +4	+2 +3 +4 +5	+2 +3 +6	+2 +4 +7	+2 +3	+2 +3	+2	+1 +2	+2

Y $4d^1 5s^2$	Zr $4d^2 5s^2$	Nb $4d^4 5s^1$	Mo $4d^5 5s^1$	Tc $4d^5 5s^2$	Ru $4d^7 5s^1$	Rh $4d^8 5s^1$	Pd $4d^{10} 5s^0$	Ag $4d^{10} 5s^1$	Cd $4d^{10} 5s^2$
+3	+4	+3 +5	+3 +4 +5 +6	+4 +6 +7	+2 +3 +4 +6 +8	+2 +3 +4	+2 +4	+1 +2	+2

La $5d^1 6s^2$	Hf $5d^2 6s^2$	Ta $5d^3 6s^2$	W $5d^4 6s^2$	Re $5d^5 6s^2$	Os $5d^6 6s^2$	Ir $5d^7 6s^2$	Pt $5d^9 6s^1$	Au $5d^{10} 6s^1$	Hg $5d^{10} 6s^2$
+3	+4	+5	+4 +5 +6	+3 +4 +6 +7	+3 +4 +6 +8	+3 +4	+2 +4	+1 +3	+1 +2

5.4) for predicting electron configurations and chemical behavior. Admittedly, the building method is an oversimplification, but we are still unable to adequately explain why one, and, except for Ce and Pr, only one inner-shell electron is available for chemical reactions.

The inner-shell electrons of the **actinides** (the series of 14 elements beginning with Th) are more accessible than those in the lanthanides—they are farther out from the nucleus and closer in energy to the outer s electrons, and consequently the actinides show a greater variety of oxidation states. For example, uranium, an atom with three $5f$, one $6d$, and two $7s$ electrons, forms the +6 oxidation state; UF_6 and UO_2^{2+} are two examples. The chemistry of the actinides is very interesting but attempts to study them are often frustrated by their intense radioactivity.

The partially empty d orbitals of the transition metals play an important role in the chemistry of these atoms. The ions of these elements, being electron-pair acceptors, are Lewis acids and form complexes by combining with **ligands** (combining groups) containing unshared electron pairs. These ligands can be simple species such as CO, Cl^-, and NH_3, larger organic molecules, or even giant biologically important molecules such as proteins. (Investigation of metal–protein complexes, a very important aspect of biochemistry, is one of the most active fields of current transition-metal research.) These complexes are often superoctet structures and exhibit large **coordination numbers**—the number of ligands around the metal atom. Transition-metal coordination numbers of 4 and 6 are common, and others such as 2, 5, 7, and 8 occur often.

Metallic Properties

Although transition-metal elements are less metallic than the elements of group IA and IIA in the chemical sense—they tend to share electrons more readily and do not form monatomic ions as easily—they all exhibit metallic physical properties. They are malleable, lustrous, and conduct electricity well. With the exception of mercury (a liquid at room temperature), they are solids, generally with close-packed crystal lattices, either hexagonal or face-centered cubic. Their hardness, brittleness, and melting points vary from element to element in a way that shows the influence of electron configuration on physical properties. The high melting points and hardness of the transition metals is due to their unpaired d electrons that can interact with d orbitals on adjacent atoms to produce covalent bonds that raise the lattice energy and melting point of the crystal. Since the bonds formed by the d electrons are more localized than the metallic bonds of the group IA and IIA elements, most transition metals are harder and more brittle than the group IA and IIA metals. The exceptions are copper, zinc, and the other metals of groups IB and IIB. These elements have no unpaired d electrons and are consequently soft and malleable and have relatively low melting points. Zinc melts at 419°C compared to 1890°C for chromium, and mercury is a liquid at room temperature.

The unpaired d electron spins of transition metals make these atoms *paramagnetic*, and the atoms with the most unpaired electrons are located in the middle of each transition row—elements such as Cr, Mn, and Fe. Because of special features of their crystal structures, iron, nickel, and many transition-metal alloys are **ferromagnetic**. A ferromagnetic metal is attracted by a magnetic field much more strongly than a simple paramagnetic substance and it can form permanent magnets. The crystal structure of a ferromagnetic metal maintains magnetic domains in which the atoms have parallel magnetic moments rather than randomly oriented moments.

The Lanthanide Contraction

Each successive lanthanide atom is slightly smaller than the preceding one because the electrons are in the same shells and subshells but increased nuclear charge attracts the electrons more strongly. Although the size difference between adjacent elements is small (Table 21.2), the cumulative effect over the entire lanthanide series is significant (about 20%). This size decrease, known as the **lanthanide contraction**, has an important effect on transition-metal chemistry. Because of the lanthanide contraction, the transition metals following the lanthanides (Hf through Hg) are nearly the same size as the corresponding elements of the second transition row (Zr

Table 21.2 Names and Sizes of the Lanthanides

Name	Symbol	Electron configuration	Radius of the 3+ ions (Å)[a]
Lanthanum	La	$4f^0 5d^1 6s^2$	1.17
Cerium	Ce	$4f^1 5d^1 6s^2$	1.15
Praseodymium	Pr	$4f^3 5d^0 6s^2$	1.13
Neodymium	Nd	$4f^4 5d^0 6s^2$	1.12
Promethium	Pm	$4f^5 5d^0 6s^2$	1.11
Samarium	Sm	$4f^6 5d^0 6s^2$	1.10
Europium	Eu	$4f^7 5d^0 6s^2$	1.09
Gadolinium	Gd	$4f^7 5d^1 6s^2$	1.08
Terbium	Tb	$4f^9 5d^0 6s^2$	1.06
Dysprosium	Dy	$4f^{10} 5d^0 6s^2$	1.05
Holmium	Ho	$4f^{11} 5d^0 6s^2$	1.04
Erbium	Er	$4f^{12} 5d^0 6s^2$	1.03
Thulium	Tm	$4f^{13} 5d^0 6s^2$	1.02
Ytterbium	Yb	$4f^{14} 5d^0 6s^2$	1.01
Lutetium	Lu	$4f^{14} 5d^1 6s^2$	1.00

[a] R. D. Shannon and C. T. Prewitt, *Acta Crystallographica* **B25**, 925 (1969).

Table 21.3 Relative Sizes of Some Ions

	Radius (Å) for coordination number 6		
2+	3+	4+	5+
Ca 1.00	Sc 0.75	Ti 0.61	V 0.54
Sr 1.16	Y 0.89	Zr 0.72	Nb 0.64
Ba 1.36	La 1.05	Hf 0.71	Ta 0.64

through Cd). Having similar sizes (Table 21.3), the elements of the second and third transition rows resemble each other chemically very closely and differ significantly from the first-row transition elements (Sc through Zn) because these first-row elements are considerably smaller.

21.2 Coordination Compounds

The Work of Alfred Werner

Chemists in the late 1800s had a great deal of trouble understanding the nature of many transition-metal compounds. According to the terminology of the time, most elements had either fixed "valences" (essentially what we now call oxidation states) such as +1 for Na, −1 for Cl, −2 for O, or a characteristic set of simple valences (+2 and +4 for Pb, +1 and +2 for Hg), and the formulas of an element's compounds could be predicted from its characteristic valences. In the case of transition metals, however, the valence of the metal atom seemed to change depending upon the nature of the atoms or molecules that combined with it. For example, cobalt forms a trichloride, $CoCl_3$, consistent with a valence of +3, but this compound reacts with ammonia to form several compounds with *different formulas* (Table 21.4). Because these compounds did not fit the valence theories of the nineteenth century, they were called "complex compounds," a name that survives today in the terms **complex** and **complex ion**. At the end of the nineteenth century, the Swiss chemist Alfred Werner provided the first adequate description of the structures of complex ions.

Werner first dealt with the observation that the various ammonia complexes of $CoCl_3$ react differently with $AgNO_3$, a reagent that rapidly precipitates chloride ion as AgCl. When $CoCl_3 \cdot 6NH_3$ is treated with an excess of Ag^+, 3 mol AgCl precipitate for every mole of complex, but the complex $CoCl_3 \cdot 5NH_3$ yields only 2 mol AgCl. The two isomeric forms of $CoCl_3 \cdot 4NH_3$ act the same way in the presence of excess Ag^+, each yielding 1 mol AgCl, while $CoCl_3 \cdot 3NH_3$ does not react rapidly with $AgNO_3$ at all.

Werner concluded that a chlorine atom in a $CoCl_3$–ammonia complex could be in either one of two chemical states. It could be present as a *chloride ion*, in which case it would react rapidly with $AgNO_3$, or it could be *bound directly* to the metal atom and resist rapid precipitation as AgCl. According to this interpretation of the Ag^+ reaction data, $CoCl_3 \cdot 6NH_3$ has three Cl^- ions, $CoCl_3 \cdot 5NH_3$ two, $CoCl_3 \cdot 4NH_3$ one, and $CoCl_3 \cdot 3NH_3$ none.

Table 21.4 Ammonia Complexes of $CoCl_3$

Formula	Color	Original name	Number of Cl^- precipitated by Ag^+
$CoCl_3 \cdot 6NH_3$	Orange	Luteo cobaltic chloride	3
$CoCl_3 \cdot 5NH_3$	Purple	Purpureo cobaltic chloride	2
$CoCl_3 \cdot 4NH_3$	Green	Praseo cobaltic chloride	1
$CoCl_3 \cdot 4NH_3$	Violet	Violeo cobaltic chloride	1
$CoCl_3 \cdot 3NH_3$			0

Werner's second conclusion was that in addition to its "primary valence" a transition metal has a "secondary valence" (we now call this the *coordination number*), which is characteristic of the metal and its oxidation state and indicates the number of attached groups. In all the ammonia complexes of $CoCl_3$, as well as in all other Co(III) complexes, the cobalt atom has a coordination number of six. In $CoCl_3 \cdot 6NH_3$, the six ammonia molecules occupy the six coordination positions and the three chlorines are present as chloride ions. The formula could then be written as

$[Co(NH_3)_6]Cl_3$

where the cation is $[Co(NH_3)_6]^{3+}$ and the anions are 3 Cl^- ions. The six ligands in $CoCl_3 \cdot 5NH_3$ are the *five* ammonia molecules and *one* chloride ion, and only two Cl^- ions in the formula unit are free to react with Ag^+. The formula is written as $[Co(NH_3)_5Cl]Cl_2$. The cobalt atom in $CoCl_3 \cdot 4NH_3$ is bound to *four* ammonia molecules and two chloride ions; the formula is $[Co(NH_3)_4Cl_2]Cl$. The cation in this case is $[Co(NH_3)_4Cl_2]^+$, because the two bound chloride ions partially neutralize the charge from the cobalt(III).

The measured conductivities of solutions of the various complex ions (Table 21.5) confirmed Werner's theory. Since $[Co(NH_3)_6]Cl_3$ dissociates into four ions, its solutions are better conductors than those of $[Co(NH_3)_5Cl]Cl_2$, a substance that dissociates into only three ions. Werner also studied similar platinum complexes; their formulas, numbers of ions, and conductivities are listed in Table 21.5 along with the cobalt complexes. The colligative properties of these solutions also fit Werner's proposed structures.

Geometric Isomers

Werner extended his theories of these complex compounds by proposing that the coordination positions around the metal atom were arranged in characteristic geometric patterns, and he correctly postulated the shapes of the complexes by examining the *number* of isomers for each formula.

Consider three possible geometric arrangements for the six ligands around a Pt(IV) complex (Figure 21.2): (1) a planar hexagon centered on the metal atom, (2) a trigonal prism, and (3) an octahedron. (We consider only the ligands attached directly to the central metal atom; the two remaining chloride ions occupy lattice positions in the crystalline solid.) The key to discovering the true geometry of this complex is the observation that there are only *two* isomeric forms of the compound $[Pt(NH_3)_4Cl_2]Cl_2$. If the complex ion $[Pt(NH_3)_4Cl_2]^{2+}$ were either a hexagon or a trigonal prism, there would be three possible isomers (Figure 21.3), so Werner rejected the

Table 21.5 Conductivities of Some Complexes

Formula	Number of ions per formula unit	Conductivity (ohm^{-1})	Werner formula
Cobalt complexes			
$CoCl_3 \cdot 6NH_3$	4	430	$\{Co^{(NH_3)6}\}Cl_3$
$CoCl_3 \cdot 5NH_3$	3	260	$\{Co_{Cl}^{(NH_3)5}\}Cl_2$
Platinum complexes			
$PtCl_4 \cdot 6NH_3$	5	523	$\{Pt^{(NH_3)6}\}Cl_4$
$PtCl_4 \cdot 5NH_3$	4	404	$\{Pt_{Cl}^{(NH_3)5}\}Cl_3$
$PtCl_4 \cdot 4NH_3$	3	230	$\{Pt_{Cl_2}^{(NH_3)4}\}Cl_2$
$PtCl_4 \cdot 3NH_3$	2	97	$\{Pt_{Cl_3}^{(NH_3)3}\}Cl$
$PtCl_4 \cdot 2NH_3$	0	~0	$\{Pt_{Cl_4}^{(NH_3)2}\}$

Source: A. Werner and A. Miolati, *Zeitschrift für Physikalische Chemie* **12**, 35(1893) and **14**, 506(1894).
Additional reading: *Journal of Chemical Education* **36**, 521(1959).

Figure 21.2
Possible geometries of six-coordinate complexes. The L stands for any ligand and the bonds are indicated by black lines. Part b of the octahedral complex is an attempt to show the arrangement of bonds in an octahedral complex without showing the outline of the octahedron. Part c is an even simpler representation of the fact that four ligands lie in a single plane while the other two are located perpendicular to this plane, one above and one below. It is important to understand that there is no difference between the vertical and horizontal ligand positions. If we were to turn the octahedron on its side, the two vertical ligands would move into the horizontal plane, and two of the horizontal ligands would replace them in the vertical positions.

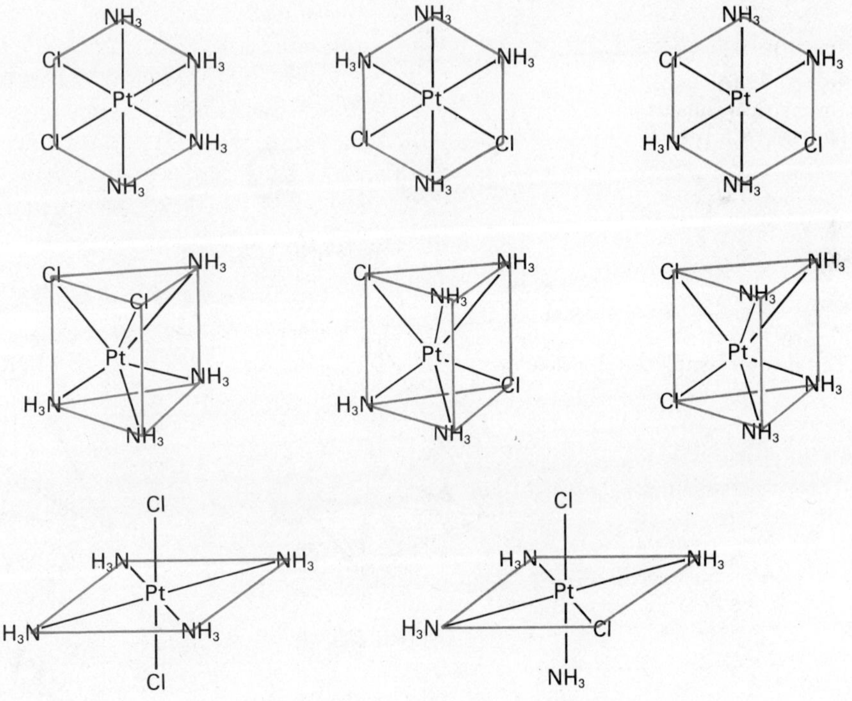

Figure 21.3
Possible isomers for proposed shapes of six-coordinate complexes.

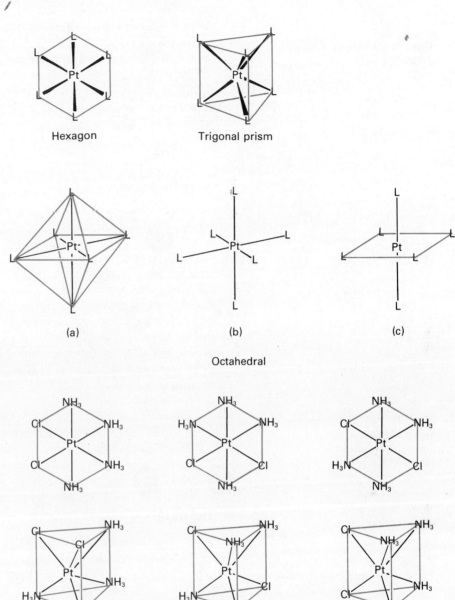

possibility that the complex had either of these shapes. Since there are only two possible octahedral structures for [Pt(NH$_3$)$_4$Cl$_2$]$^{2+}$, the octahedral shape is consistent with the observed number of isomers.

These distinct isomeric forms are called **geometric isomers**. In the *cis* isomer (Figure 21.4), the two Cl$^-$ ligands occupy *adjacent* vertices of the octahedron. (Any vertex of an octahedron is adjacent to four of the other five vertices.) The chloride ligands in the *trans* isomer of [Pt(NH$_3$)$_4$Cl$_2$]$^{2+}$ occupy vertices on *opposite* sides of the octahedron.

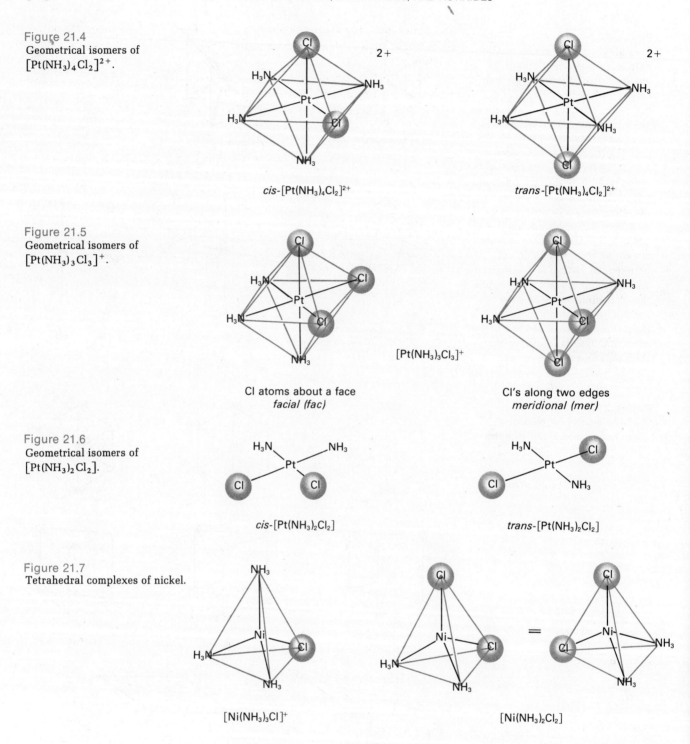

Figure 21.4
Geometrical isomers of [Pt(NH₃)₄Cl₂]²⁺.

Figure 21.5
Geometrical isomers of [Pt(NH₃)₃Cl₃]⁺.

Figure 21.6
Geometrical isomers of [Pt(NH₃)₂Cl₂].

Figure 21.7
Tetrahedral complexes of nickel.

The fact that there are only two isomers of [Pt(NH₃)₃Cl₃]Cl is also consistent with the octahedral shape for the cation. One geometric isomer has three chloride ligands at the three vertices of the same triangular face of the octahedron. The chlorides of the other isomer occupy two opposite ends of the octahedron and one of the intervening ligand positions (Figure 21.5). In all known cases of six-coordinate complexes, the number of isomers fits the hypothesis of an octahedral shape, and this shape has been confirmed by x-ray crystallography.

The four-coordinate complexes of Pt(II) are square planar. This shape explains why there are two geometric isomers of [Pt(NH₃)₂Cl₂] (Figure 21.6). The Cl⁻ ligands in the *cis* form are at adjacent corners of the square whereas those in the *trans* isomer occupy opposite corners.

Other four-coordinate complexes such as some Ni(II) complexes are tetrahedral rather than square planar. Consequently, there is only one form of [Ni(NH$_3$)$_2$Cl$_2$] (Figure 21.7).

We can often decide which isomer is *cis* and which is *trans* by observing its preferred reactivity or by measuring the molecule's dipole moment. For example, *cis*-[Pt(NH$_3$)$_2$Cl$_2$] has a large dipole moment because the electronegative chlorine atoms are on the same side of the molecule. The *trans* isomer has no dipole moment because the two Pt—Cl bond dipole vectors point in opposite directions and cancel each other. The same is true of the Pt—N bonds. The polarity differences give the two isomers differing physical properties, such as their solubilities, and these differences make the separation of the geometric isomers possible.

In addition to the square planar and tetrahedral shapes of four-coordinate complexes and the octahedral shape of six-coordinate complexes, there are trigonal-bipyramidal and square-pyramidal shapes in five-coordinate complexes (Figure 21.8). We will discuss the types of hybrid orbitals and chemical bonds involved in some of these complexes later.

Figure 21.8
Common geometrical structures and their hybrid orbital compositions.

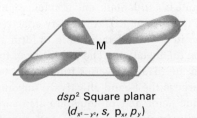
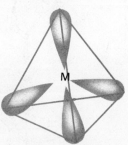

dsp^2 Square planar
($d_{x^2-y^2}$, s, p_x, p_y)

sp^3 Tetrahedral
(s, p_x, p_y, p_z)

(a) Four-coordinate types

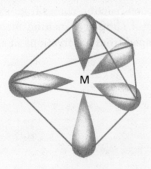
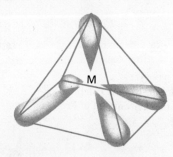

dsp^3 Trigonal bipyramidal
(d_{z^2}, s, p_x, p_y, p_z)

dsp^3 Square pyramidal (uncommon)
($d_{x^2-y^2}$, s, p_x, p_y, p_z)

(b) Five-coordinate types

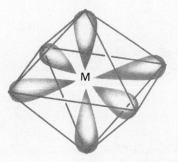

d^{2}sp^3 Octahedral
($d_{x^2-y^2}$, d_{z^2}, s, p_x, p_y, p_z)

(c) Six-coordinate structure

Chelates

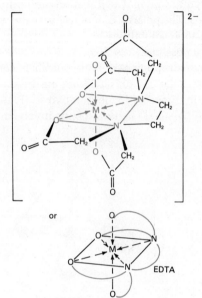

Figure 21.9
EDTA complexing ability (M^{2+} ion).

The ligands attached to the central metal ion of a complex do not have to be simple ions (such as Cl^-) or molecules (such as NH_3); they can be large complicated structures (Table 21.6). Some ligands are **polydentate**. In other words, the ligand molecule has several functional groups capable of donating electron pairs to a metal ion, and a polydentate ligand can occupy several of the metal ion's coordination positions at the same time. For example, a molecule of ethylenediamine has two amine groups (note that *amine* is the name for $-NH_2$ groups while *ammine* is for the NH_3 group in complexes), each of which can donate an electron pair, and the molecule is flexible enough to arrange itself around a Co(III) ion so it occupies *two* of the octahedral positions. Some transition-metal ions do not form stable complexes with ethylenediamine because their sizes and orbital geometries are incompatible with the dimensions and shape of an ethylenediamine ligand.

The process of forming complexes of metal ions with polydentate ligands is called **chelation** (after the Greek word for claw) and polydentate ligands are often referred to as **chelate groups** or **chelating agents**. Depending on the number of Lewis-base groups on the ligand, it is classified either as **monodentate** (one complexing group), **bidentate** (two complexing groups), or **polydentate** (tridentate, tetradentate, and so on). For example, ethylenediamine is a bidentate, diethylenetriamine is a tridentate, and ethylenediamine tetraacetate (EDTA) is a hexadentate ligand (Figure 21.9). EDTA is a very effective ligand and chelates with almost every metal ion in a 1:1 molecular ratio to form an octahedral complex *regardless* of the charge on the metal ion. The stabilities of these complexes do depend on pH, however, and the pH effect varies from metal ion to metal ion. EDTA complexes are so stable because the ligand effectively surrounds the metal ion.

Amino acids and proteins are also very effective chelating agents, and many biologically important proteins bind specific metal ions such as Zn^{2+} and Fe^{2+}. Hemoglobin, the protein that binds O_2 and carries it through the blood stream, contains Fe^{2+} bound to two ligands—a protein, and a porphyrin group, which is a planar tetradentate ligand (Figure 21.10). The O_2 molecule binds to the iron's sixth coordination position. A good deal of our understanding of the mechanism of the hemoglobin–oxygen reaction comes from extending the knowledge we have gained from our studies of simpler transition metal complexes. The theories of complex structures developed from observations of simple ligands enable us to interpret the

Figure 21.10
Hemoglobin–oxygen complex. Schematic environment of Fe in heme.

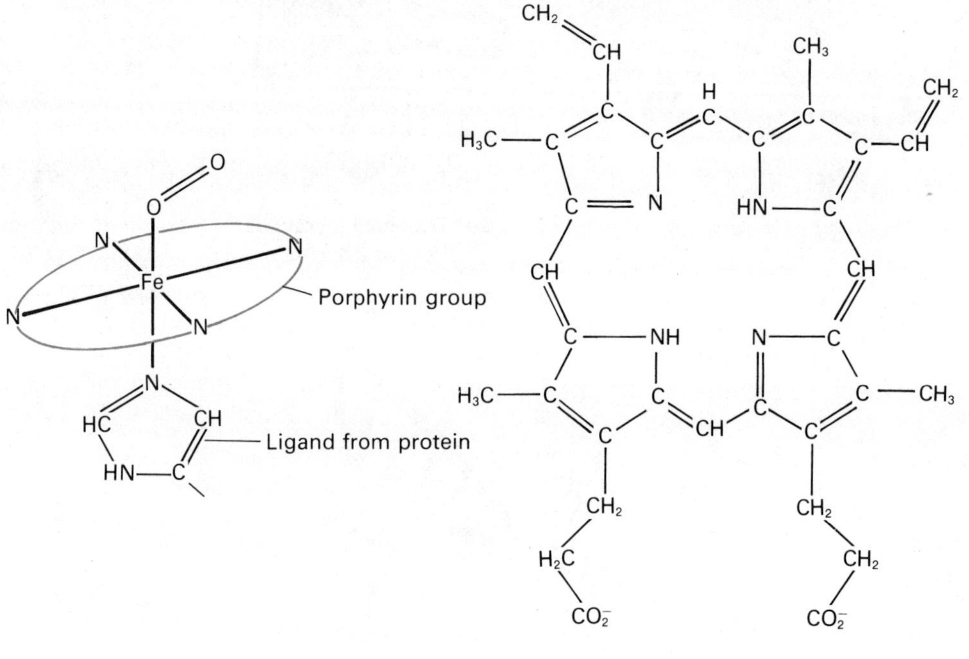

Structure of porphyrin ring

21.2 COORDINATION COMPOUNDS

Table 21.6 **Lewis-Base Ligands**

Simple molecules or ions
(Only the donated electron pair is indicated.)

:NH_3 :X^- (X = any halogen)
:CN^- :CO
H_2O:

More complex molecules (monodentates and polydentates)

Name	Structural formula	Ligand type
Pyridine (py)	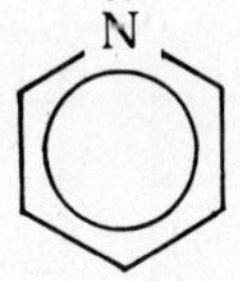 pyridine ring with :N	Monodentate
Ethylenediamine (en)	$H_2\ddot{N}CH_2CH_2\ddot{N}H_2$	Bidentate
Diethylenetriamine (dien)	$H_2\ddot{N}CH_2CH_2\ddot{N}HCH_2CH_2\ddot{N}H_2$	Tridentate
Ethylenediamine-tetraacetate ion (EDTA^{4-})	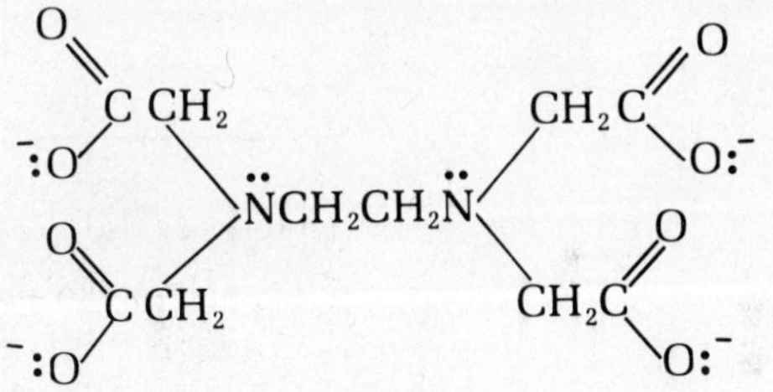 $^-$:Ö—C(=O)—CH$_2$ \ $\ddot{N}$CH$_2$CH$_2\ddot{N}$ / CH$_2$—C(=O)—Ö:$^-$ (four carboxylate arms)	Hexadentate
Oxalate ion	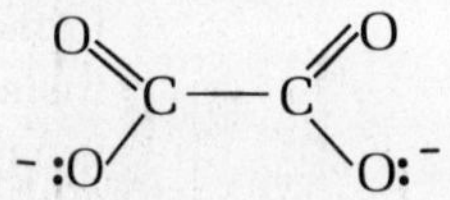 $^-$:Ö—C(=O)—C(=O)—Ö:$^-$	Bidentate
Dimethylglyoxime (dmg)	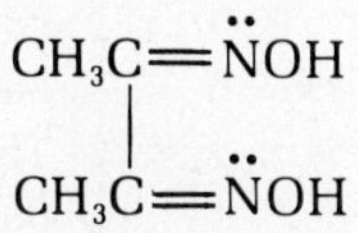 $CH_3C{=}\ddot{N}OH$ / $CH_3C{=}\ddot{N}OH$	Bidentate
Glycine (gly)	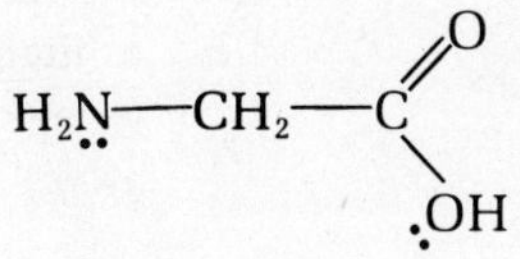 $H_2\ddot{N}{-}CH_2{-}C(=O){-}\ddot{O}H$	Bidentate

Figure 21.11
Optical isomers of $[Co(en)_3]^{3+}$.

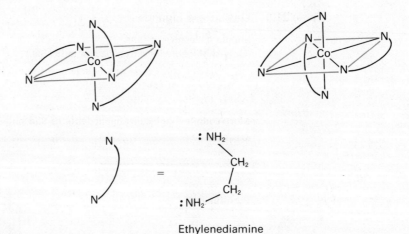

Figure 21.12
Optical isomers of $[Co(en)_2Cl_2]^+$.

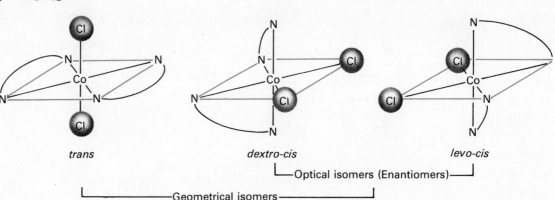

Optical Isomers

spectra and magnetic properties of hemoglobin and its O_2–complex in terms of the subtle configurational arrangements of the metal ion and its ligands. The study of such a complex system as hemoglobin requires the combined efforts of organic, physical, and inorganic chemists.

There are two enantiomers (mirror-image isomers) of the octahedral complex formed by cobalt and three ethylenediamine molecules. They are called tris(ethylenediamine)cobalt(III) ion, $[Co(en)_3]^{3+}$ (Figure 21.11).* Instead of being geometric isomers, these two structures are nonsuperimposable mirror images of each other and, like the enantiomers of chiral organic molecules, their solutions rotate the plane of polarized light (see Section 19.3). Another example of optical isomerism in octahedral complexes occurs with the bis(ethylenediamine)dichlorocobalt(III) ion, $[Co(en)_2Cl_2]^+$ (Figure 21.12). There are cis and trans forms of this complex, but in addition, the cis form exists as two enantiomers. The trans form, having a mirror plane, is superimposable on its mirror image and has no enantiomers. Therefore, there are a total of three $[Co(en)_2Cl_2]^+$ isomers—a pair of enantiomers and one additional geometric isomer.

The Valence-Bond Description of Coordination Complexes

Linus Pauling first attempted to describe complex ions in quantum mechanical terms when he applied the **valence-bond method** to these structures. Pauling assumed that the ligands, which are Lewis bases, donate electron pairs to a set of *empty* hybrid orbitals on the central metal atom to form a covalent bond. This sort of bond, formed by electron-pair donation from a Lewis base to a Lewis acid (in this case the metal atom) is called a **coordinate covalent bond**, hence the name **coordination complex**.

*See Appendix F for notes on nomenclature.

Pauling was able to correlate various complex geometries with different modes of hybridization of the orbitals of the metal atoms. For example, octahedral shapes can be explained by the use of d^2sp^3 hybrid orbitals. These are six identically shaped orbitals formed by averaging two d orbitals, one s orbital, and three p orbitals, and they point to the corners of an octahedron. Tetrahedral shapes are consistent with sp^3 hybrid orbitals, and square-planar shapes are associated with dsp^2 hybrids.

Example 21.1

Describe the bonding and electron configuration of the octahedral complex $Cr(CO)_6$.

Solution

Since the complex is octahedral, the ligands are bound through a set of d^2sp^3 hybrid orbitals. A chromium atom in its ground state electron configuration, $[Ar]\,3d^5 4s^1 4p^0$

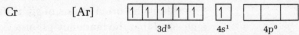

does not have the empty d^2sp^3 orbitals necessary for forming six coordinate bonds, but electron rearrangement to an excited configuration (Cr‡) frees these orbitals.

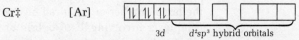

The hybrid orbitals then receive electron pairs from the ligands.

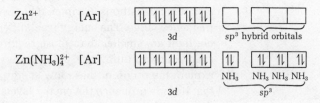

Example 21.2

What is the nature of the bonding and the electron configuration in the tetrahedral complex ion $Zn(NH_3)_4^{2+}$?

Solution

Having lost the two $4s$ electrons, Zn^{2+} has the $[Ar]\,3d^{10}4s^04p^0$ configuration and has the empty sp^3 orbitals needed for a tetrahedral complex.

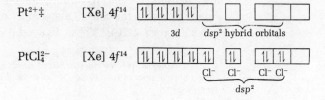

Example 21.3

The complex ion $PtCl_4^{2-}$ is square planar. Describe its bonding and electron configuration.

Solution

The square-planar shape indicates that the ligands are bound by a set of four dsp^2 hybrid orbitals. Rearrangement of the eight d electrons of Pt^{2+} frees the orbitals needed for square-planar bonding.

The valence-bond picture allows us to correlate the magnetic properties of a complex with its electron configuration and state of hybridization. Thus all d^6 complexes* with d^2sp^3 hybrid bond orbitals, for example $Cr(CO)_6$ and $Co(NH_3)_6^{3+}$, are diamagnetic because they do not have unpaired electrons. However, the octahedral Co(III) complex, CoF_6^{3-}, possesses four unpaired electrons and is very paramagnetic. In this case, the hybrid orbitals result from the $4s$ orbital, the three $4p$ orbitals, and two of cobalt's $4d$ orbitals—sp^3d^2 hybrid orbitals.

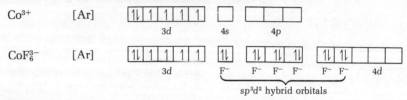

The calculated orientations of sp^3d^2 orbitals are octahedral (just like d^2sp^3). Pauling suggested that the tendency to form CoF bonds is so strong that the F^- ions can form stable bonds with the higher-energy $4d$ orbitals and no electronic rearrangement to free two $3d$ orbitals for bonding is required. Since sp^3d^2 complexes use outer-shell d orbitals, we call them **outer-orbital complexes** (or high-spin complexes because of the large number of unpaired electrons remaining) as opposed to the **inner-orbital complexes** involving d^2sp^3 hybrid orbitals.

The valence-bond method allows us to correlate the shape of a complex with the calculated geometries of hybrid orbitals and with the magnetic properties of the complex. Its weakness is that there are no precise guidelines for predicting in advance when a six-coordinate complex will be an outer-orbital type or an inner-orbital type, or when a four-coordinate complex will be tetrahedral or square planar. Furthermore, since it ignores the excited electronic states of the complexes, the valence-bond method does not deal with the most vivid feature of transition-metal complexes and their solutions—their dramatic variety of characteristic colors.

Special Resource Section: Crystal-Field Theory

A colored species absorbs part of the visible spectrum (light varying in wavelength from 400 to 700 nm). When we look at a colored solution, we see the visible light that is *not* absorbed. For example, solutions of $[Ti(H_2O)_6]^{3+}$ are red-violet because they *absorb* yellow light and *transmit* the red and blue wavelengths (Figure 21.13). There are many colored transition-metal complexes because these complexes can undergo electronic transitions (Chapter 16) corresponding to energy absorption in the visible frequency range. The energy level differences associated with the absorption of visible light are smaller than those required for ultraviolet absorption (most *colorless* species absorb ultraviolet light), so any theory on the colors of complex ions must explain the origin of these low-energy electron transitions. Furthermore, it is clear that ligands influence the energy levels involved in visible-light absorption because different complexes of the same metal ion have a variety of colors. For example, cupric ion in 6 M HCl gives a green solution, whereas in slightly acidic solutions it is blue and in 6 M $NH_3 \cdot H_2O$ it turns a deep blue-violet.

Crystal-field theory (CFT) was developed by Hans Bethe, a physicist, in order to describe the electronic states of transition-metal ions in crystal lattices. It was later adopted by chemists to explain the colors of coordination complexes. The fundamental feature of this theory is that it considers the distorting effect due to the electric field of the ligands on the various *unhybridized* orbitals of the central atom. In its simplest form, crystal-field theory does not try to deal with the *nature* of the bonds in the complex (because it cannot!), but only with the effects resulting from several ions coming close to the central metal ion. Consider the d orbitals of a completely isolated atom (Figure 21.14). These five orbitals are degenerate (they all have the same

---

*Complexes are often classified as d^6, d^7, d^8, and so on, depending on the number of d electrons in the uncomplexed atom or ion. Thus, Co^{3+}, Fe^{2+}, and Cr^0 complexes are d^6, Co^{2+} complexes are d^7, and Ni^{2+} complexes are d^8.

Figure 21.13
Visible-light absorption spectrum of $[Ti(H_2O)_6]^{3+}$.

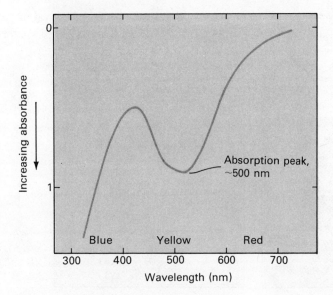

energy), and each has its characteristic orientation in space. If the atom is placed in a spherical and uniform electric field, the d orbital energies change, but they all change to the same degree, and their degeneracy persists. If, however, the electrostatic field is not uniform, it will affect some d orbitals differently than others. A balloon presents a good analogy (Figure 21.15). Placing the balloon in a pressurized tank shrinks the balloon but does not distort its spherical shape because the inward force on the balloon exerts the same strength over its entire surface. In contrast, poking the balloon on opposite sides with two fingers—a nonuniform force—deforms the balloon.

Now consider what happens when six octahedrally oriented ligands approach a central metal atom. The electric fields of the ligands alter the energies of the d orbitals, but the specific effect on each orbital depends upon the relative orientation of the orbital and the ligand. According to calculations based upon a free metal ion–electrostatic field, the energies of the d_{z^2} and $d_{x^2-y^2}$ orbitals increase because their lobes point *directly* at the ligands (Figure 21.16), and electrons in these orbitals would be repelled by the ligands. The other three d orbitals (d_{xy}, d_{xz}, and d_{yz}), pointing *between* the ligands, decrease in energy. This effect, called **orbital splitting**, results in an energy difference between the two sets of orbitals. The magnitude of the splitting, called 10 Dq, depends upon the relative electrostatic strengths of the ligands. The relative strengths of a few common ligands are in the sequence

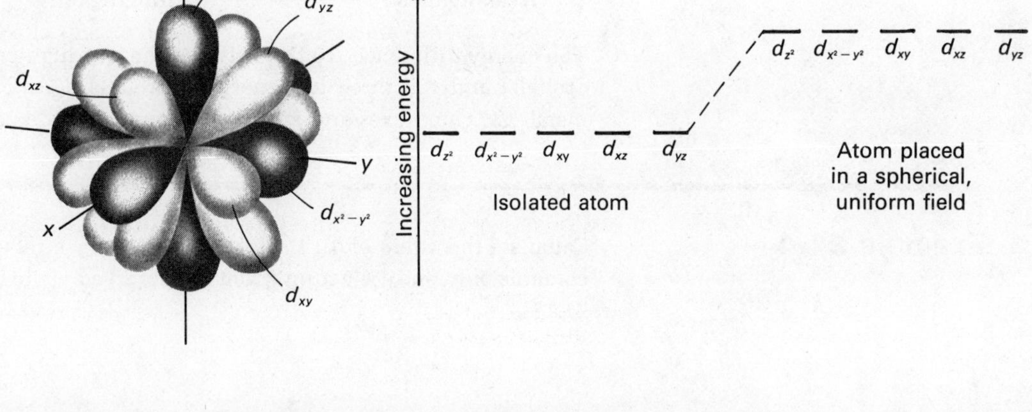

Figure 21.14
Effect of a spherical, uniform electric field on the d orbitals.

554   21 TRANSITION METALS, LANTHANIDES, AND ACTINIDES

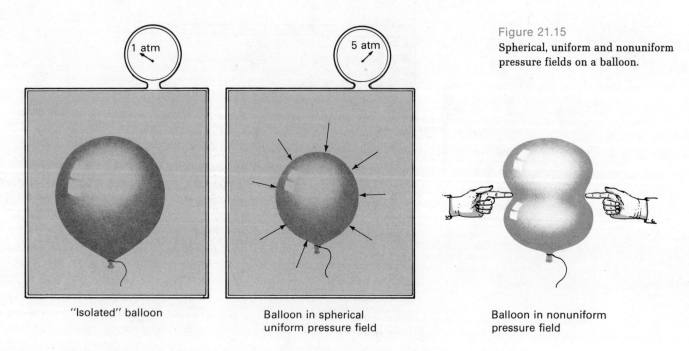

Figure 21.15
Spherical, uniform and nonuniform pressure fields on a balloon.

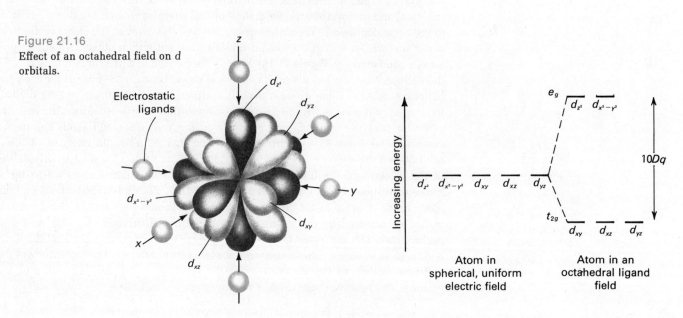

Figure 21.16
Effect of an octahedral field on d orbitals.

I^-, Br^-, Cl^-, F^-, OH^-, H_2O, NH_3, en, CN^-, CO

Weak ligands ⟶ Strong ligands

The energy difference, 10 Dq, between the two high-energy d orbitals (called the e_g orbitals) and the three low-energy d orbitals (t_{2g} orbitals) is therefore largest for metal–CO complexes and smallest for metal–iodide complexes.

Example 21.4   Calculate the value of 10 Dq for $Ti(H_2O)_6^{3+}$ ion (a relatively simple case because it contains only one d electron) from the observed visible absorption spectrum (Figure 21.13).

Solution

Since the absorption peak occurs approximately at 500 nm, the energy of this light and thus the energy of the electron transition is

$$\Delta E = h\nu$$

$$\nu = \frac{c}{\lambda} = \frac{3.00 \times 10^8 \text{ m/sec}}{5.00 \times 10^{-7} \text{ m}} = 6.0 \times 10^{14} \text{ sec}^{-1}$$

$$\Delta E = (6.63 \times 10^{-34} \text{ J-sec})(6.0 \times 10^{14} \text{ sec}^{-1}) = 4.0 \times 10^{-19} \text{ J}$$

On a molar scale, the value of $10\,Dq$ is

$$10\,Dq = (4.0 \times 10^{-19} \text{ J})(6.0 \times 10^{23} \text{ mol}^{-1}) = 2.4 \times 10^5 \text{ J/mol} = 240 \text{ kJ/mol}$$

Because of vibrational effects (Section 16.2), the ion absorbs a broad portion of the yellow-green part of the visible spectrum. Complexes with stronger ligands have *larger* crystal-field splitting and absorb *shorter* wavelength light, and conversely weak ligands form complexes that absorb at *longer* wavelengths. Even though the electronic transitions in complexes containing several d electrons are much more complicated than the Ti(III) d^1 case and their spectra are harder to explain, the influence of ligand strength on the absorption spectrum is still apparent. For example, $[Co(NH_3)_6]^{3+}$ absorbs in the blue-violet region of the visible spectrum and transmits orange-yellow light. Because CN^- is a *stronger* ligand than NH_3, the complex ion $[Co(CN)_6]^{3-}$ absorbs in the ultraviolet region and is colorless (Figure 21.17).

Crystal-field theory was not intended to and cannot deal with the electronic character of *bonded* ligands, but **ligand-field theory** (LFT), which is an attempt to incorporate crystal-field theory and molecular-orbital theory, can. The ligand-field theory model for an octahedral complex is one in which the d_{z^2} and $d_{x^2-y^2}$ orbitals of the metal atom form hybrid orbitals with its outer-shell s and p orbitals, and these hybrid orbitals overlap with the orbitals of the ligands to form bonding and antibonding molecular orbitals. The d_{xy}, d_{xz}, and d_{yz} orbitals of the metal do not take part in bonding and are termed nonbonding molecular orbitals. Ligand-field theory can explain the existence of inner-orbital and outer-orbital complexes; consider Co(III) complexes, for example. In order to form an inner-orbital complex, the two e_g orbitals must be empty so they can accept electron pairs from ligands. This occurs with strong ligands such as NH_3 because the strong ligand field creates such a large energy difference between the t_{2g} and e_g levels that all six d electrons are in the t_{2g} orbitals and the e_g orbitals are free for bonding. However, when the ligand is weak, for example F^- in CoF_6^{3-}, the crystal-field splitting is too small to prevent d electrons from occupying the e_g orbitals (Figure 21.18). Since all the d orbitals are occupied, the ligands can form bonds with the outer-orbitals only.

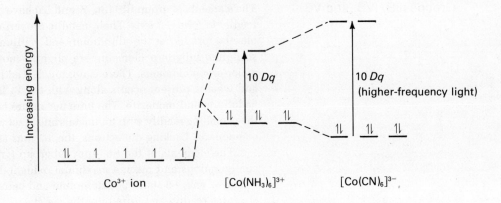

Figure 21.17
Effect of a ligand field on d orbital splitting.

Figure 21.18
Molecular orbitals for $[Co(NH_3)_6]^{3+}$ based on ligand-field theory.

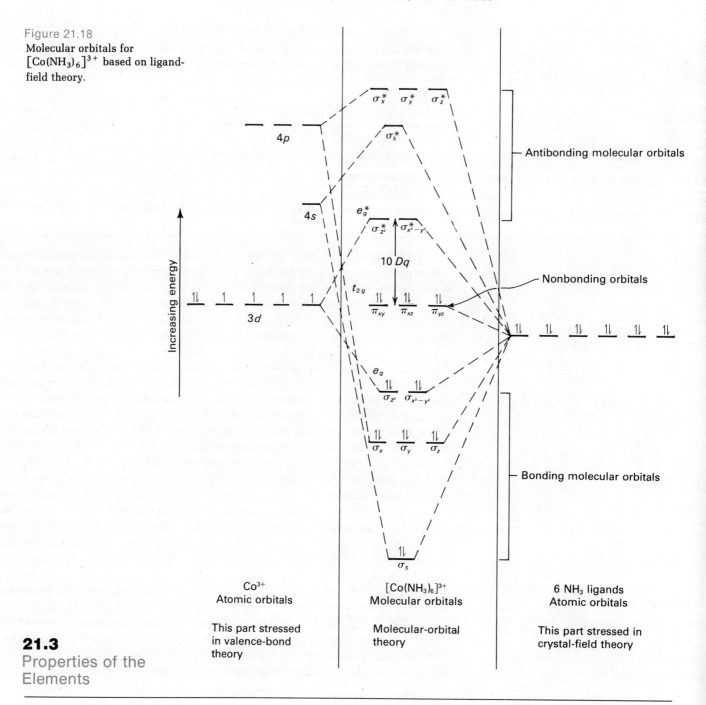

21.3 Properties of the Elements

Groups IIIB, IVB, and VB

The elements of group IIIB (Sc, Y, and La) have d^1s^2 electron configurations and react readily to form +3 ions. Their metallic properties as well as their reactivities resemble the properties of aluminum and gallium, but the compounds formed by scandium-subgroup elements are slightly more ionic than those formed by the major-group elements. There is not much scandium, yttrium, or lanthanum on earth and what is present occurs along with the 14 lanthanides in the complex minerals gadolinite and monazite. The pure metals are hard, good electrical conductors, and they combine readily with nonmetals and react with water to form +3 oxidation-state compounds. Lacking d electrons, the +3 ions and their salts are colorless.

The elements of the titanium subgroup (group IVB—Ti, Zr, and Hf) have d^2s^2 configurations and reach a maximum oxidation state of +4. Titanium, in addition, forms +2 and +3 states but zirconium and hafnium, being larger atoms, lose both d electrons readily and form only the +4 state.

Titanium, the most important element of the subgroup, is an abundant element (the tenth most abundant on earth), occurring mainly as the minerals rutile, TiO_2, and ilmenite, $FeTiO_3$. Titanium metal's lightness, strength, and corrosion resistance—it forms inert oxide and nitride surface coatings that protect the metal from further corrosion—make it a desirable metal for use in aircraft and rockets. The most important compound of titanium is TiO_2, a white pigment that has replaced the more toxic lead oxide pigments in paint. (Art forgeries have been exposed because the paintings, supposedly several centuries old, contained TiO_2 white paint.) The tetragonal crystal lattice of rutile (Figure 21.19) is an alternative to the fluorite structure (see Figure 9.23) for ionic compounds of the type AB_2. Perovskite, $SrTiO_3$, has a crystal lattice that is common for ionic compounds of the type ABX_3. The titanium ions in perovskite occupy the lattice points of a simple cubic lattice while Sr^{2+} occupies the cube centers and O^{2-} the midpoints of the cube edges.

Solutions of Ti(III) ions (also called titanous) are obtained by reducing Ti(IV) solutions with zinc and acid. Because these solutions contain $Ti(H_2O)_6^{3+}$, they are violet. Being strong reducing agents, they rapidly lose their color as they are oxidized to solutions of colorless Ti(IV) species such as TiO^{2+}.

Because of the lanthanide contraction, zirconium and hafnium are nearly the same size (1.45 Å and 1.44 Å, respectively) and are therefore chemically similar to each other. Zircon, $ZrSiO_4$, is a brilliant gemstone but is softer than diamond and not as durable.

The group VB elements (V, Nb, and Ta) are very hard metals and are used in special alloy steel. Because tantalum is inert except at very high temperatures, it is used to make equipment for many large-scale chemical operations. Vanadium and niobium (once called columbium) do react with concentrated acids, and all the group VB metals react with electronegative elements such as chlorine, fluorine, and oxygen when heated. Having d^3s^2 electron configurations, they have maximum oxidation states of +5, but this oxidation state is represented by oxyions such as VO_2^+ (stable in acid) and VO_3^- (stable in base) rather than the intensely charged +5 ions.

Significant quantities of vanadium are recovered from some oil deposits, a situation that may be due to the use of vanadium compounds to transport oxygen in the blood system of prehistoric sea creatures. There is at least one currently living species, the tunicates (also called sea squirts), who possess greenish blood with V(III) as the central metal atom in its oxygen-transporting molecule.

The Chromium and Manganese Subgroups

The chromium VIB subgroup (Cr, Mo, and W) and the manganese VIIB subgroup (Mn, Tc, and Re) contain hard, relatively unreactive metals whose primary uses are for forming special-use alloys and fabricating devices. Large quantities of chromium are used to make stainless steel and for chrome plating. Molybdenum metal is used for x-ray tubes and special high-temperature steels, and tungsten (also called wolfram) is used for filaments in incandescent light bulbs and as tungsten carbide for cutting tools.

Figure 21.19

Structures of rutile, TiO_2, and perovskite, $SrTiO_3$.

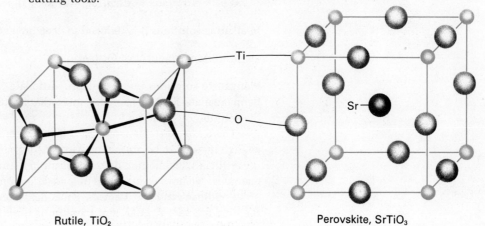

Rutile, TiO_2

Perovskite, $SrTiO_3$

The most important oxidation states of chromium are Cr(III) and Cr(VI). There is a large variety of complex ions and salts of Cr(III), each with its own characteristic color (chromium is derived from the Greek word *chromos* meaning color), but in aqueous solution hydrated Cr^{3+} ion is green. Chromate ion (CrO_4^{2-}), dichromate ion ($Cr_2O_7^{2-}$), and chromium trioxide (CrO_3) represent the +6 oxidation state of chromium. There is a rapid, reversible and pH dependent equilibrium between chromate and dichromate ions in solution.

$$2\ CrO_4^{2-}(aq) + 2\ H^+(aq) \rightleftharpoons Cr_2O_7^{2-}(aq) + H_2O$$

Dichromate is the dominant species in acid solutions, but the equilibrium is reversed at higher pH values (roughly greater than 7), and CrO_4^{2-} forms. It is easy to observe this reaction because CrO_4^{2-} is bright yellow and $Cr_2O_7^{2-}$ deep orange, and we can alter the color of Cr(VI) solutions by changing the pH.

Acidic chromium(VI) solutions are strong oxidants

$$6\ e^- + Cr_2O_7^{2-}(aq) + 14\ H^+(aq) \rightleftharpoons 2\ Cr^{3+}(aq) + 7\ H_2O \qquad E° = 1.33\ V$$

and dichromate solutions are used in redox titrations such as the analysis for iron by oxidation of Fe^{2+} to Fe^{3+}.

$$6\ Fe^{2+}(aq) + Cr_2O_7^{2-}(aq) + 14\ H^+(aq) \rightleftharpoons 2\ Cr^{3+}(aq) + 6\ Fe^{3+} + 7\ H_2O$$

Basic chromium(VI) solutions are much weaker oxidants.

$$3\ e^- + CrO_4^{2-}(aq) + 4\ H_2O \rightleftharpoons Cr(OH)_3(s) + 5\ OH^-(aq) \qquad E° = 0.13\ V$$

Chromium(VI) is an essential trace element. It is necessary for proper human metabolism, but slightly larger quantities—as little as 50 μg per liter of solution—are extremely toxic and can apparently cause cancers. Chromium, in fact, is rapidly becoming one of our most serious metal pollutants, along with mercury, cadmium, and lead, in our modern technological world.

Manganese forms the +2, +3, +4, +6, and +7 oxidation states. The +7 state is represented by permanganate ion, MnO_4^-, a strong oxidant that is used in redox titrations, for example the titration of oxalic acid ($H_2C_2O_4$).

$$5\ H_2C_2O_4(aq) + 2\ MnO_4^-(aq) + 6\ H^+(aq) \longrightarrow 10\ CO_2(g) + 2\ Mn^{2+}(aq) + 8\ H_2O$$

Since the intensely violet permanganate solution is reduced to Mn^{2+}, a pale pink ion (essentially colorless in dilute solutions), the end point of the permanganate titration is the persistence of the violet color of excess MnO_4^-. In neutral solutions, MnO_4^- is reduced to MnO_2, a brown, insoluble substance.

$$3\ e^- + 4\ H^+(aq) + MnO_4^-(aq) \rightleftharpoons MnO_2(s) + 2\ H_2O$$

In alkaline solutions it is reduced to deep green solutions of manganate ion MnO_4^{2-}.

$$e^- + MnO_4^-(aq) \rightleftharpoons MnO_4^{2-}(aq)$$

Manganate solutions quickly disproportionate to MnO_2 and MnO_4^- when acidified. Permanganate solutions are also somewhat unstable and slowly oxidize the water in the solution on standing.

There are no stable isotopes of technetium, and none of its isotopes has a long enough lifetime to be present on earth. As its name implies (Greek, *technetos*, meaning artificial), technetium does not occur in nature but can be synthesized by nuclear reactions. We have been able to prepare kilogram quantities of this metal and in spite of its radioactivity we know a good deal about its chemistry. Technetium and rhenium are very similar, and rhenium is used in small amounts to improve corrosion resistance and electrical properties of alloys.

Iron, Cobalt, Nickel, and the Noble Metals

The three elements following manganese in the first transition row—iron, cobalt, and nickel—bear no obvious relationship to the main-group elements, and there is no easy way to relate their oxidation states to their electron configurations. Iron, with a d^6s^2 configuration, cobalt (d^7s^2), and nickel ($d^{10}s^0$) all form +2 and +3 ions. The six elements below them in the periodic table—Ru, Rh, Pd, and Os, Ir, Pt—are much less reactive than their first-transition-row counterparts. This resistance to reaction, coupled with their attractive appearances, is why these six elements are often called the noble metals. The noble metals also show higher oxidation states than Fe, Co, and Ni; ruthenium ($4d^7 5s^1$) and osmium ($5d^6 6s^2$), each with a total of eight outermost electrons, can form +8 oxidation states.

The Iron Age began around 1200 B.C. in the Hittite empire, situated in what is now eastern Turkey. Iron had been known and used long before that; there are forged iron artifacts dating back to 2700 B.C., and even older jewelry was decorated with iron beads from meteorites. But the Hittites were the first to discover that heating iron in the presence of carbon produces steel, a metal that is harder than bronze. The Hittites guarded their secret jealously and for two centuries had a monopoly on preparing iron tools and weapons, but when their empire collapsed, Hittite blacksmiths carried their knowledge in all directions, and their steel-making skills quickly spread to the Mediterranean lands, Asia, Africa, and northern Europe.

The social, political, and military effects of steel were profound. The steel axe and plowshare cleared forests and brought previously unusable land under cultivation. Although bronze, an expensive metal, provided tools and arms for a powerful elite, it was inaccessible to the average man and Bronze Age societies were highly stratified. The availability of cheap and abundant iron had an equalizing effect that may have been a necessary precondition for Greek democracy. The military effect of steel was overwhelming—a bronze sword is no match for a steel one and nations that possessed iron weapons could dominate their Bronze Age neighbors. Military conquest helped spread Iron Age technology to new areas. For example, Celtic invaders brought the Iron Age to Scandinavia and the British Isles.

The technology of steel making grew in sophistication as it spread. Man learned how to produce harder steel by rapidly quenching the hot metal in water and how to give it a sharp edge by heat tempering. Indeed, it took a skilled smith to make quality steel. If there was not enough carbon in the steel, it would be too soft; if it was not quenched properly, it would be too brittle. The skill (and luck) required to forge a superior sword probably gave rise to the legends of "magic" swords such as King Arthur's Excalibur.

The last major technical problem of iron making to be mastered was the production of cast iron objects. Unable to attain the very high temperatures needed to melt iron, ancient man had to laboriously beat hot iron into the desired shapes. Cast iron was first prepared in seventh-century China, apparently because the high phosphorus content of Chinese iron lowered its melting point considerably. It was not until the fourteenth century and the development of efficient bellow furnaces that Europeans learned to make cast iron, an important step that opened the way to the mass production of iron artifacts.

Iron, because of its desirable metallic properties and the abundance of iron ores, is the most important metal of this group, and it is one of our fundamental commodities. Iron ranks fourth in abundance in the earth's crust and occurs as silicates, aluminosilicates, oxides (hematite, Fe_2O_3, and magnetite, Fe_3O_4), and as iron pyrites, FeS_2. In addition to comprising about 5% of igneous rocks, iron makes up an even higher percentage of inner basaltic rocks, and the earth's core is mostly molten iron. There are several allotropes of iron. Ferrite, the most stable form at room temperature and pressure, has a body-centered cubic lattice.

The preparation of iron from ore usually involves the reduction of iron oxides by CO at very high temperatures,

$$3\ Fe_2O_3(s) + CO(g) \xrightarrow{\text{Low temperature}} 2\ Fe_3O_4(s) + CO_2(g)$$

$$Fe_3O_4(s) + CO(g) \xrightarrow{\text{High temperature}} 3\ FeO(s) + CO_2(g)$$

and finally,

$$FeO(s) + CO(g) \rightleftharpoons Fe(l) + CO_2(g)$$

Because the equilibrium constant for the last step is only 1.68 at 700°C and 4.17 at 1000°C, efficient iron production requires both high temperatures and rapid removal of CO_2 to shift the equilibrium toward iron formation. This is achieved in the modern **blast furnace** (Figure 21.20), a large industrial apparatus that can produce as much as 1500 tons of iron per day. Iron ore, coke, and limestone are fed into the top of the furnace through a pressure cap. Hot air blown in through the bottom of the blast furnace oxidizes the coke to CO, and this reaction generates the heat needed to reduce the iron ore. The purpose of the limestone is to react with silicate impurities in the ore to form calcium silicate, which melts, floats to the top of the molten iron, and is drawn off as **slag** and used to make cement. Iron, whose melting point is very high (1540°C), melts in the blast furnace because carbon impurities lower its melting point significantly. (Iron containing 4.3% C melts at 1130°C). The molten iron is drained into molds called "pigs" and solidifies as **pig iron**. Pig iron is converted to wrought iron and steel by further heating to oxidize and remove objectionable amounts of impurities, mainly carbon, silicon, sulfur, and phosphorus.

Steel is iron containing 0.05–2% carbon. The carbon makes steel much harder than pure iron, and the hardness and flexibility depend upon the precise amounts of carbon and other components present in the alloy. There are several distinct iron–carbon phases: (1) cementite (Fe_3C), (2) austenite, or carbon dissolved in the γ-allotrope (face-centered cubic) of iron, and (3) pearlite, or carbon in α-iron. When steel is annealed and quenched (heated and then quickly cooled), the austenite and pearlite form distinct crystals embedded in a continuous matrix of cementite. The properties of steel such as tensile strength, hardness, malleability, ductility, and corrosion resistance depend primarily on the amount of cementite present as well as its temperature of formation and the particular annealing and quenching process used.

Most steel is manufactured by the **open-hearth** and **basic oxygen processes**. (A minor amount is produced in electric furnaces.) In the open-hearth method, pig iron, scrap iron, and perhaps some Fe_3O_4 are heated together in a blast of oxygen-rich air in a large, open, shallow furnace lined with dolomite or limestone ($MgCO_3 + CaCO_3$) bricks. The air oxidizes most of the carbon impurity to CO and carries it away while the oxides of silicon, manganese, sulfur, and phosphorus that are formed by the oxidation process combine with the basic oxides in the brick lining to form slag. After several hours of this treatment, a hole is punched through the wall of the furnace and the molten steel flows out into collecting molds. In the basic oxygen process, a large cylindrical vessel, lined with basic CaO and MgO bricks, is charged with scrap and pig iron, and calcium and magnesium carbonates. The vessel is

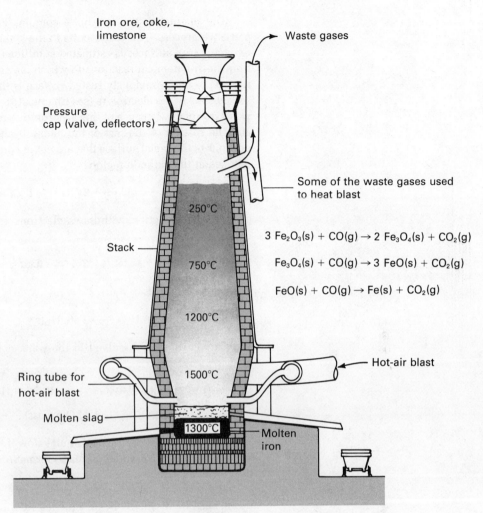

Figure 21.20
Schematic drawing of a blast furnace for production of pig iron.

Overall reactions

$$2\ C\ (coke) + O_2 \rightarrow 2\ CO + Heat$$

$$Fe_2O_3 + 3\ CO \rightarrow 2\ Fe\ (C\text{-containing liquid}) + 3\ CO_2$$

$$CaCO_3\ (limestone) \xrightarrow{Heat} CaO + CO_2$$

$$CaO + SiO_2 \rightarrow CaSiO_3\ (slag)$$

heated and pure oxygen is bubbled through the molten mass using a long, water-cooled tube. The chemical reactions are essentially the same as in the open-hearth process—that is, oxidation of silicon, manganese, sulfur, phosphorus, and carbon impurities. The oxygen converts some of the molten iron to FeO. It mixes throughout the rest of the molten mass and reacts with Si and Mn impurities forming iron.

$$Si + 2\ FeO \longrightarrow 2\ Fe + SiO_2$$

$$Mn + FeO \longrightarrow Fe + MnO$$

This basic oxygen process produces a very high-purity steel in a very short time (approximately 1 hour) and has grown so much in importance that over 50% of commercial steel is currently manufactured by this process.

Iron is a good reducing agent at room temperature and a very powerful one at high temperatures. It is oxidized rather easily by hydrogen ion in weak and strong aqueous acids, but is rendered passive (much less reactive) by powerful oxidants such as dichromate and cold concentrated nitric acid due to the formation of surface oxide films. Iron forms stable +2 and +3 oxidation states, the most common ions being pale green $[Fe(H_2O)_6]^{2+}$ and yellow $[Fe(H_2O)_6]^{3+}$.

The rusting of iron is a quite complex and, needless to say, extremely costly corrosive process. The yearly costs in the United States of protecting, repairing, and replacing iron surfaces is estimated at billions of dollars. Most rusting of iron arises from electrochemical reactions in which the corroding iron surface is the anode, the iron oxide on a previously rusted surface is the cathode, and water containing dissolved salts is the electrolyte (see Chapter 15). Iron will not rust in dry air or in pure water free of air. Because of this electrolytic process, a rusted portion of iron accelerates the rusting of the rest of the metal. It also explains how rusting can occur at portions of the metal surface that are not in contact with air (Figure 21.21). Oxygen is reduced at the cathode region,

$$2\,e^- + O_2(g) + H_2O \longrightarrow 2\,OH^- \qquad E°_{red} = +0.40\text{ V}$$

and this half-reaction withdraws electrons from the anode region at which iron corrosion occurs.

$$Fe(s) \longrightarrow Fe^{2+} + 2e^- \qquad E°_{ox} = +0.44\text{ V}$$

The overall result is

$$Fe(s) + O_2(g) + H_2O \longrightarrow Fe(OH)_2(s)$$

Carbon dioxide dissolved in the water accelerates rusting by acting as a weak acid.

$$Fe(s) + 2\,H_2CO_3 \longrightarrow Fe^{2+} + 2\,HCO_3^- + H_2(g)$$

$$4\,Fe^{2+} + 8\,HCO_3^- + O_2 + H_2O(\text{excess}) \longrightarrow 2\,Fe_2O_3 \cdot H_2O(s) + 8\,H_2CO_3$$

The initial oxidation by H_2CO_3 is quite slow if the iron is very pure because the pure metal has a large overpotential for H_2 formation, but impurities in the metal increase the corrosion rate.

Iron forms many complexes, for example, the hydrated Fe(III) oxide that makes up rust can be dissolved by an aqueous solution of oxalic acid, $H_2C_2O_4$, because the oxalate ion, $C_2O_4^{2-}$, is a bidentate ligand that forms a complex with Fe^{3+}.

$$Fe_2O_3 \cdot xH_2O(s) + 4\,H_2C_2O_4 \longrightarrow 2\,Fe(C_2O_4)_2^-(aq) + 2\,H^+(aq) + (x+3)\,H_2O$$

Some dyes and inks are suspensions of very insoluble, colored cyanide complexes of iron. For example, prussian blue is the pigment formed by the reaction of Fe^{3+} and hexacyanoferrate(II) ion, $[Fe(CN)_6]^{4-}$.

Figure 21.21
Rusting of iron.

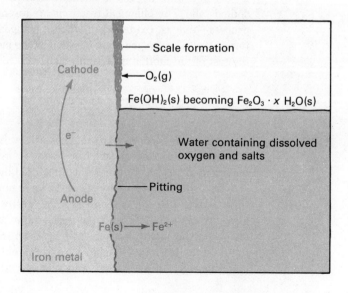

$$4\ Fe^{3+}(aq) + 3[Fe(CN)_6]^{4-}(aq) \longrightarrow Fe_4[Fe(CN)_6]_3(s)$$
Prussian blue

Cobalt and nickel are important constituents of alloys, but are not used as much as the pure metals. A type of steel particularly noted for its corrosion resistance is called stainless steel. It is typically an alloy of iron and carbon with chromium and nickel. A very common stainless steel is called 18-8; it is 18% Cr, 8% Ni, and 74% Fe. The principal sources of cobalt and nickel are the arsenide ores of copper, silver, and lead, and cobalt and nickel are recovered as by-products of smelting for the more precious metals.

Cobalt and nickel, like iron, form +2 and +3 oxidation states, but the +3 oxidation states of these elements are much stronger oxidants than Fe(III). Compounds of Ni(III) are quite rare, and the stable Co(III) species are complex ions with fairly strong ligands. Both elements form a variety of complexes. The ions most commonly encountered are $[Ni(H_2O)_4]^{2+}$, a green species, and the pink ion, $[Co(H_2O)_6]^{2+}$.

The noble metals are important because of their inertness and are used in making jewelry, special laboratory equipment, and are becoming increasingly useful in heterogeneous catalysis. Platinum and palladium are very effective catalysts in many industrial processes such as the conversion of SO_2 to SO_3 for production of sulfuric acid, catalytic converters for unburned hydrocarbons in automobile and industrial exhausts, and in hydrogenation reactions. In the latter application their catalytic action appears to involve the formation of hydrogen atoms by dissociation of H_2 molecules at the metal surface (Section 17.2). It is of interest to note that the three best hydrogenation catalysts—Ni, Pt, and Pd—are all located in the same vertical column of group VIII in the periodic table. Even though they have different ground-state electronic configurations, their ability to dissociate H_2 into H atoms at their surfaces yet not form stable metal hydrides is the key to their success.

Groups IB and IIB

Copper, silver, and gold, the elements of subgroup IB, are among the earliest metals known, primarily because they occur in the free state in nature. The pure metals are too soft to be used for tools but their bright, shiny appearance made them ideal materials for jewelry and decorative artifacts. Because of their inertness, scarcity, and lack of competing practical uses in ancient times, they were used in coins. (The group IB metals are sometimes called the *coinage metals*.) Although we no longer use gold or silver coins (and the days of the copper penny may be numbered), gold and silver are still the foundation for the world's major currencies.

Alloys of the group IB elements are considerably harder than the pure metals themselves. Bronze, an alloy of copper and tin, was the most important metal of ancient times, the era we call the Bronze Age, and its use, both practical and decorative, persisted well into the Iron Age. Today, we use considerable amounts of copper to make brass, a copper–zinc alloy.

The group IB elements all have $d^{10}s^1$ electron configurations and form the +1 oxidation state, a point of similarity with the group IA elements. However, copper, silver, and gold are much less reactive than the alkali metals, they exhibit higher oxidation states, and they form complex ions. Like Na^+, the +1 ions of the group IB elements are colorless, but they differ from Na^+ in that they form many insoluble salts, including the halides. The +1 ions of copper and gold are not particularly stable; copper(I) (cuprous) ion, Cu^+, disproportionates to elemental copper and copper(II) (cupric) ion, Cu^{2+},

$$2\ Cu^+(aq) \rightleftharpoons Cu(s) + Cu^{2+}(aq) \qquad E° = 0.37\ V$$

and Au^+ disproportionates to the metal and Au(III) compounds.

$$6\ Au^+(aq) + 3\ H_2O \rightleftharpoons 4\ Au(s) + Au_2O_3(s) + 6\ H^+(aq)$$

Thus the stable species of Cu(I) and Au(I) are the insoluble salts such as CuI and AuCl or complexes such as $[Cu(CN)_3]^{2-}$. The +2 oxidation state of copper is stable and includes the blue hydrated ion, $[Cu(H_2O)_4]^{2+}$, and complex ions such as $[Cu(NH_3)_4]^{2+}$ (violet), and $[CuCl_4]^{2-}$ (green). Silver and gold have unstable +2 oxidation states. Of the members of group IB, only gold has a stable +3 oxidation state.

Copper occurs in the free state and in ores such as cuprite, Cu_2O, malachite, $CuCO_3 \cdot Cu(OH)_2$ (many houses, buildings, and churches in Europe and the eastern United States used copper sheeting on roofs and the greenish color taken on by the copper is due to formation of this basic carbonate), and the sulfides CuS and Cu_2S. We obtain crude copper metal by smelting the sulfide.

$$Cu_2S(s) + O_2(g) \longrightarrow 2\,Cu(s) + SO_2(g)$$

Electrorefining (Section 15.2) converts the impure smelted product to pure copper. Silver and gold also occur as the free metals and as the sulfides and arsenides.

All the group IB metals have excellent thermal and electrical conductivities, similar to group IA, the alkali metals. The major modern use of copper is as an electric conductor; it has a very low resistance to electric currents, it can be drawn into flexible wires, it does not corrode readily and, until recently, it was inexpensive. However, as the world's reserves of copper dwindle, its price is rising and there have been attempts to replace copper with inferior but less expensive materials such as aluminum.

Silver is used to prepare photographic film. The film contains an emulsion of silver halides (usually AgBr). Exposure to light sensitizes these salts so that they can be reduced more easily than salts not exposed to light. Treatment of the exposed film with a reducing solution leaves a residue of silver metal that forms the photographic image.

Although gold is used mainly in jewelry, gold fillings or teeth, and as a monetary standard, it has specialized industrial uses. Having low electrical resistance and being very corrosion resistant, gold-plated conductors and connectors are used in most modern electronic components.

The group IIB elements (Zn, Cd, and Hg), have $d^{10}s^2$ electron configurations. They have two outer-shell electrons and no partially filled inner subshells, so they are quite similar to the alkaline earth elements. Although less reactive than the main-group elements, the group IIB metals do resemble them in forming colorless +2 ions. Mercury, in addition, forms the +1 oxidation state in the form of mercurous ion, Hg_2^{2+}, and its salts; the ion is diamagnetic, thus one s electron is lost from each mercury atom and they share the other one in a covalent bond, $^+Hg\text{-}Hg^+$. One feature of the zinc subgroup elements not present in the alkaline earth metals is their tendency to form complex ions, such as $Zn(NH_3)_4^{2+}$ and $Cd(CN)_4^{2-}$. The group IIB elements are fairly plentiful and occur primarily as sulfides—ZnS (two allotropes are sphalerite and zinc blende (see Figure 9.24), CdS, and the HgS ore called cinnabar. Some mercury also occurs free or as amalgams of silver and gold in nature. The metals of all three elements are produced by smelting the ores. In this process the sulfides are roasted to convert them to oxides,

$$2\,ZnS(s) + 3\,O_2(g) \xrightarrow{\text{Heat}} 2\,ZnO(s) + 3\,SO_2(g)$$

and the oxides are then reduced with carbon.

$$ZnO(s) + C(s) \xrightarrow{1200°C} Zn(g) + CO(g)$$

Since the reaction temperature exceeds the boiling point of zinc, the metal vapor distills out of the reaction vessel. Roasting cinnabar produces metallic mercury directly, and the pure metal can be distilled from the furnace.

$$HgS(s) + O_2(g) \longrightarrow Hg(g) + SO_2(g)$$

21.3 PROPERTIES OF THE ELEMENTS

The metals are used extensively in alloys—zinc in brass, cadmium in solders and in bearings, and mercury in dental fillings (a Ag-Hg amalgam). Both zinc and cadmium are used to plate iron and steel to give it a protective coating against corrosion. For example, galvanized iron is iron coated with a zinc surface (Section 15.2). Even if the surface is scratched, the zinc still provides protection for the iron because zinc is more easily oxidized ($E°_{ox}$ = 0.76 V) than iron ($E°_{ox}$ = 0.44 V) and iron becomes the cathode in the cell. This process is usually called *cathodic protection*.

Both zinc and cadmium are used extensively in the manufacture of batteries, and large quantities of cadmium are employed as control rods in nuclear reactors. Because mercury is the only liquid metal at room temperature, it is used extensively in thermometers and barometers. It is also used in high-intensity mercury vapor lamps because its emission spectrum contains many strong lines.

The Lanthanides (or Rare Earth Elements) and the Actinides

At the end of the eighteenth century, the Finnish chemist, Johan Gadolin, began the study of the oxide ore (or "earth" as it was then called) that was later to bear his name—gadolinite. From this single source, 17 elements—the group IIIB metals and the 14 lanthanide elements—were eventually isolated.* The older name for the lanthanides, the "rare earths," arose from the incorrect belief that these elements are scarce and the early chemists' additional mistaken identification of the oxides as the elements. In fact, one of the rarer lanthanides, thulium, is more abundant than either mercury, cadmium, or bismuth.

The reason it took so long to characterize these elements was the difficulty in separating them as pure elements. The lanthanides are very similar to each other and to the group IIIB elements—they form +3 oxidation states and their compounds have similar solubilities. Until 1940, the known procedures for purifying the elements were discouragingly tedious, requiring many repeated cycles of precipitation and dissolution to take advantage of the slight solubility differences between lanthanide salts. The expense of these methods limited the commercial use of the elements. The problem of purifying lanthanide compounds was solved by the application of **ion exchange chromatography** techniques. There are many types of **chromatography**,† all involving the distribution of the components of a mixture between a stationary adsorbing phase and a moving solvent phase. The stationary phase (Figure 21.22) is usually packed in a column, the mixture added to the top of the column, and the moving phase allowed to percolate through the stationary adsorbent. Those components of the mixture that are weakly bound to the stationary phase move rapidly down the column while strongly adsorbed substances move slowly. The distribution equilibrium of a substance is reestablished each time it comes into contact with a fresh portion of the column, so the substance goes through the equivalent of many purification cycles during its journey down the column. Consequently, two substances with slightly different adsorption tendencies will be separated from each other and emerge from the base of the column at different times. The stationary phase in ion exchange chromatography is an ion exchange resin (Section 18.2) and the moving phase is a solution with a carefully controlled pH and total ionic strength. Even a mixture of all the +3 lanthanide ions can be separated into its components by a single pass through a long column of a cation exchange resin.

The individual metals can be obtained by electrolysis of the fused chlorides or by reduction of the halide salts using a more active metal.

$$3\ Ca(s) + 2\ YbCl_3(s) \longrightarrow 2\ Yb(g) + 3\ CaCl_2(s)$$

---

*An indication of how lanthanides and group IIIB elements occur in the same mineral deposits is that four elements—yttrium, terbium, erbium, and ytterbium—are all named after the small Swedish mining town of Ytterby. We do not know if this indicates excessive local pride or just lack of imagination on the part of early rare-earth chemists.

†Although the name chromatography seems to imply a relationship to color, the only connection is a historical one. The first known application of chromatography was to separate mixtures of plant pigments.

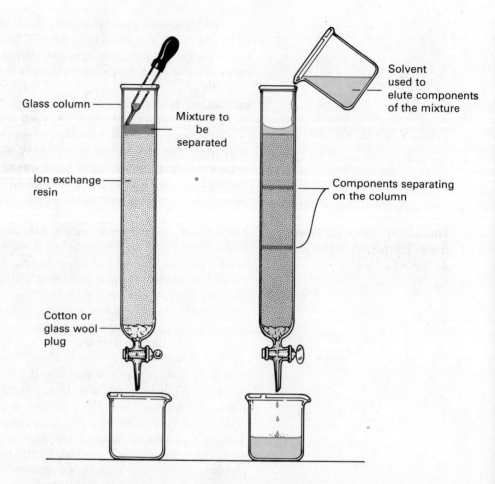

Figure 21.22
Column chromatography.

Because ytterbium has a low boiling point, it distills from the reaction mixture and condenses in a pure state. The pure elements are silvery, not too hard, reactive metals. They combine readily with the more reactive nonmetals and, if their surfaces are clean, will react with water or CO_2 and burn in air if heated slightly. Mechanical friction can initiate combustion, and flints used in cigarette lighters contain a mixture of the lighter lanthanide metals called *mischmetal*.

The lanthanides are characterized by their partially filled $4f$ subshells. Their electron configurations fit the classification $4f^{0-14}$, $5d^{0-1}$, $6s^2$, and they all form +3 oxidation states. Their +3 ions are similar in size (see Table 21.2) and the elements resemble each other more closely than do the transition-metal elements. The colors of the +3 ions vary according to the number of unpaired $4f$ electrons, an effect that is apparent from the similar colors of a lighter lanthanide ion (one with fewer than seven $4f$ electrons) and a heavier one (with more than seven $4f$ electrons) having the same number of unpaired electrons (Table 21.7). Thus Pr^{3+} ($4f^2$, two unpaired electrons) and Tm^{3+} ($4f^{12}$, two unpaired electrons) are both green. Because the colors of the lanthanide ions depend on electronic transitions in an *inner* subshell, ligands do not affect the transition energies very much, and the color of all the complexes of a particular +3 lanthanide remains essentially the same. This lack of influence of environment on color is the reason why lanthanides are sometimes incorporated into glass and colorless crystals to impart particular colors and special optical properties. Several lanthanides, notably europium, phosphoresce with bright colors when bombarded by electrons and these "rare earth phosphors" are used in color television sets and lasers.

Most of the cases in which a lanthanide has an oxidation state other than +3 involve situations where the $4f$ subshell of the ion is either empty, Ce(IV), half-filled, Tb(IV) and Eu(II), or completely filled, Yb(II). Ceric ion, Ce^{4+}, is a strong oxidant used in redox titrations.

21.3 PROPERTIES OF THE ELEMENTS

Table 21.7 Colors of Lanthanide (III) Ions (M^{3+})

Symbol	Number of unpaired electrons	Color	Number of unpaired electrons	Symbol
La	0	Colorless	0	Lu
Ce	1	Colorless	1	Yb
Pr	2	Green	2	Tm
Nd	3	Reddish	3	Er
Pm	4	Pink/yellow	4	Ho
Sm	5	Yellow	5	Dy
Eu	6	Pink/colorless	6	Tb
Gd	7	Colorless	7	Gd

$$e^- + Ce^{4+}(aq) \longrightarrow Ce^{3+}(aq) \qquad E° = 1.61 \text{ V}$$

Europium and ytterbium, in addition to their tendency to form +2 ions, resemble group IIA metals since they dissolve in liquid ammonia to produce blue solutions containing solvated electrons, and like barium and strontium, they form insoluble sulfates and moderately soluble hydroxides. Cerium, terbium, europium, and ytterbium, having unique electronic configurations, can easily be separated from the other lanthanides because they form oxidation states other than +3.

The actinide elements, elements 90–103 (Table 21.8), have partially filled $5f$ subshells, $5f^{0-14}6d^{0-2}7s^2$. Because these energy levels are even closer together than the f, d, and s subshells in the lanthanides, the actinides form a greater variety of oxidation states, and higher oxidation states such as Th(IV) in ThO_2 and U(VI) in UO_2^{2+} are more stable. None of the actinides has stable isotopes; however, thorium and some uranium isotopes are quite abundant in nature, and there are trace amounts of actinium, protactinium, neptunium, and plutonium because they have isotopes with long enough lifetimes to have survived since the formation of the elements. The

Table 21.8 Actinide Elements and New 6d Series Elements

Element	Symbol	Atomic number	
Actinium	Ac	89	
Actinides			
Thorium	Th	90	
Protactinium	Pa	91	
Uranium	U	92	
Neptunium	Np	93	
Plutonium	Pu	94	
Americium	Am	95	
Curium	Cm	96	
Berkelium	Bk	97	
Californium	Cf	98	
Einsteinium	Es	99	
Fermium	Fm	100	
Mendelevium	Md	101	
Nobelium	No	102	
Lawrencium	Lr	103	(called Joliotium by Russian workers)
6d Series (Suggested names)[a]			
Rutherfordium	Rf	104	(called Kurchatovium, Ku, by Russian workers)
Hahnium	Ha	105	(called Nielsbohrium by Russian workers)

[a] Rutherfordium (104) and Hahnium (105) are the currently accepted names.

elements of atomic number 93 and above, called the **transuranium elements**, are prepared by synthetic nuclear reactions. The major interests of these elements are their nuclear reactions, a subject we will consider in the next chapter.

Summary

1. The transition metals are considered as a separate but similar group of elements since they contain electrons in inner-shell d orbitals.

2. The transition metals are termed subgroup elements (B rather than A) because their outermost electron configuration is like that of the corresponding main-group elements. They differ, however, in that they usually exhibit a variety of positive oxidation states, because the energy of the inner d electron is similar to that of the s electrons.

3. The lanthanides and actinides are IIIB subgroup elements, with a corresponding +3 common oxidation state, but they have differing numbers of electrons in a subshell two shells closer to the nucleus, the $4f$ and $5f$ subshells, respectively.

4. The partially filled subshells of the transition metals, lanthanides, and actinides give them special chemical and physical properties. They are good Lewis acids and readily form complexes with Lewis bases, they exhibit metallic physical properties, and they have unpaired electrons so their atoms are paramagnetic and some solids are ferromagnetic.

5. The lanthanide contraction is the relatively small decrease in atomic size of the lanthanides resulting from their increasing nuclear charge but inner-shell filling of electrons. Their similar size and principal oxidation state (+3) result in very similar chemical behavior for the group.

6. Alfred Werner first characterized coordination compounds or complexes as molecules or ions in which there are more atoms (or groups of atoms) bonded to a central atom than expected based on the number of electrovalent or covalent bonds the central atom usually forms. The number of atoms or atom groups attached to the central atom is expressed as the coordination number.

7. Coordination compounds often form geometric isomers, which differ in the spatial arrangements of the same set of ligands about a central atom. *Cis* (adjacent) and *trans* (opposite) isomers are two examples. These complexes also form optical isomers.

8. The common geometrical shapes of coordination complexes are: four-coordinate complex, square planar or tetrahedral; five-coordinate complex, trigonal bipyramidal or square pyramidal; six-coordinate complex, octahedral.

9. A Lewis-base ligand often has more than one atom per molecule that is capable of donating an electron pair. These are termed chelate or polydentate ligands.

10. The Pauling valence bond description of coordination complexes assumes that the atomic orbitals of the central atom form an empty hybrid orbital set, such as sp^3 or d^2sp^3, which then accepts electron pairs from the Lewis-base ligands to form coordinate covalent bonds. The geometry and electronic properties of the complex depend upon which hybrid orbital set is used.

11. Crystal-field theory (CFT) and ligand-field theory (LFT) explain the observed colors, and magnetic and electronic properties of coordination complexes by considering the ligand's effect upon the d (or f) orbitals of the central metal ion. These d orbitals are made nonequivalent in energy to different degrees depending upon the electron-donating strength of the ligand, so electron transitions (absorption of light) are possible between different d orbitals.

12. Group IIIB, IVB, and VB elements are all hard metals, react readily with nonmetals, and form expected maximum oxidation states of +3, +4, and +5. Titanium metal, an aircraft metal, and one form of TiO_2 called rutile, used as a paint pigment, are perhaps most important; all the elements are used to impart special properties to alloys.

13. The chromium (VIB) and manganese (VIIB) subgroup elements are harder and more unreactive than the previous subgroups. Chromium metal is used in very large amounts as a protective plating; all others—technetium only slightly because it must be prepared artificially—are used in alloy preparation. The maximum oxidation state of chromium, +6, represented by chromate ion, CrO_4^{2-}, and dichromate ion, $Cr_2O_7^{2-}$, is a strongly oxidizing state and readily reduces to the stable +3 state. Molybdenum and tungsten (also called wolfram) are inert and very high melting but can be formed by thermal reduction. Manganese's highest oxidation state (+7) is represented by permanganate ion, MnO_4^-, which is also strongly oxidizing and is reduced to the stable, pink Mn(II) ion or brown MnO_2.

14. The group VIII elements—iron, cobalt, nickel, and the noble metals—bear no direct resemblance to any main-group elements. Iron is most important; we use more of it than all other metals combined. It is prepared from iron ores by reduction with carbon in a blast furnace. The resulting pig iron is purified further by either the open-hearth or basic oxygen processes to produce a low-carbon–iron alloy, steel. Iron rust, chiefly $Fe_2O_3 \cdot xH_2O$, results from an electrochemical process that requires the presence of both oxygen and moisture.

15. Cobalt and nickel, also important alloying metals, have stable +2 oxidation states and readily form coordination complexes. The noble metals are becoming increasingly important in catalytic processes.

16. Copper, silver, and gold (group IB) occur free in nature, are soft metals, and are used primarily in photographic films (Ag), electrical wires and connectors, and fabrication of alloys such as bronze. Zinc, cadmium, and mercury (group IIB) are also used as alloying metals—cadmium and zinc in the manufacturing of batteries and as protective-plating metals, and mercury, since it is moderately inert and a liquid, in thermometers and scientific apparatus.

17. The lanthanides (4f elements) are relatively abundant elements. They are difficult to separate chemically, but separation can be accomplished by ion exchange chromatography. Ceric ion, Ce(IV), is a strong oxidant and the stable lanthanide ions, Eu(II) and Yb(II) with half-filled and filled 4f subshells, respectively, are chemically similar to group IIA ions, Ca(II) and Sr(II). Most of the actinides (5f elements) are prepared artificially.

Vocabulary List

transition metal	chelate group	ligand-field theory
lanthanide	chelating agent	blast furnace
actinide	bidentate	slag
ligand	valence-bond method	pig iron
coordination number	coordinate covalent bond	steel
lanthanide contraction	coordination complex	open-hearth process
complex	outer-orbital complex	basic oxygen process
complex ion	inner-orbital complex	ion exchange chromatography
geometric isomer	crystal-field theory	chromatography
polydentate	orbital splitting	transuranium element
chelation		

Exercises

1. Define the following terms.
   a. Geometrical isomer   b. Coordination number   c. Ligand
   d. Lanthanide contraction   e. Ferromagnetism

2. Why do transition metal elements have so many oxidation states while the lanthanides have primarily one oxidation state?

3. What are the most common coordination numbers for transition metal ions?

4. What does the Roman numeral in the name of a complex represent?

5. Describe a simple chemical test that would allow you to distinguish between the two compounds, [Co(NH$_3$)$_5$Br]SO$_4$ and [Co(NH$_3$)$_5$SO$_4$]Br.

6. Write balanced chemical equations for:
   a. The dissolving of silver chloride in aqueous ammonia
   b. The oxidation of ferrous ion using ceric ion
   c. The conversion of a blue copper sulfate solution into a green solution using 6 M HCl
   d. The roasting of cinnabar

7. The visible absorption spectrum of cis-[Co(NH$_3$)$_4$(H$_2$O)$_2$]$^{3+}$ has a maximum at about 500 nm. Predict its color.

8. Most automobile radiator-cleaning substances for removing rust from the cooling system contain oxalic acid. How does the oxalic acid remove the rust?

9. Suppose the element number 108 could be prepared at some future date. Use periodic relationships to predict the following properties.
   a. Electron configuration   b. Highest oxidation state   c. Common oxidation states

10. Which absorbs the higher-energy visible light, an orange compound or a green one?

11. Clearly explain the distinction between
    a. Covalent and coordinate covalent bonds   b. Di- and tridentates
    c. cis and trans isomers of complex ions
    d. Enantiomers and geometrical isomers

12. It is sometimes necessary to remove traces of O$_2$ from nitrogen gas by bubbling the slightly impure N$_2$ through a strongly reducing medium. Using half-cell potentials, decide which would be the best solution for this purpose, Cu$^+$, Ti^{2+}, or Fe^{2+}.

13. Aqueous solutions of copper(II) sulfate are blue and the anhydrous salt is white. If concentrated ammonia is added to aqueous copper(II) sulfate solution, it turns deep blue-violet. Explain the chemical basis for these phenomena.

14. Which atom is larger, Pr or Yb?

15. Of the following three elements, which two are the most similar in atomic size, Ni, Pd, or Pt? Why?

16. What is the coordination number of the metal atom in each of the following species?
    a. KAu(CN)$_4$ b. [Co(NH$_3$)$_5$SO$_4$]$^+$ c. [Cu(en)$_2$]$^{2+}$

17. What is the coordination number of the metal in:
    a. [Pt(NH$_3$)$_2$(H$_2$O)Cl]$^+$?
    b. [Co(C$_2$O$_4$)$_3$]$^{3-}$
    c. The EDTA complex of Cr^{3+}?

18. What is the oxidation state of the metal in:
    a. [Cu(NH$_3$)$_4$]$^{2+}$?   b. NiCl$_4^{2-}$?   c. [Ru(H$_2$O)$_3$Cl$_3$]?

19. Find the oxidation state of the metal in each of the following species.
    a. [Co(NH$_3$)$_5$SO$_4$]$^+$ b. K$_3$[CuF$_6$]   c. Fe(C$_2$O$_4$)$_2^-$

20. How many geometrical isomers are there for:
    a. [Co(NH$_3$)$_2$Cl$_4$]$^-$, octahedral?
    b. [AuCl$_2$Br$_2$]$^-$, square planar?
    c. [CoCl$_2$Br$_2$]$^{2-}$, tetrahedral?

21. How many geometrical isomers are there for each of the following formulas?
    a. [Co(en)$_2$Cl$_2$]$^+$, octahedral
    b. [Cu(en)$_3$]$^{2+}$, octahedral
    c. Fe(CO)$_5$, trigonal bipyramidal

22. Do any of the structures in Exercise 20 exist as optical isomers?

23. Which, if any, of the formulas in Exercise 21 can represent optical isomers?

24. Write the IUPAC name for each of the following. (See Appendix F.)
    a. KAu(CN)$_4$ b. [Cr(NH$_3$)$_6$]Cl$_3$ c. K$_2$[PtCl$_6$]
    d. [Mo(CN)$_6$]$^{4-}$ e. [Fe(CN)$_6$]$^{4-}$ f. Fe$_2$(CO)$_9$

25. Name each of the following according to the IUPAC rules.
    a. [Cu(en)$_2$]$^{2+}$ b. [Co(H$_2$O)$_2$(NH$_3$)$_4$]$^{3+}$
    c. [Al(H$_2$O)$_6$]$^{3+}$ d. [Ni(CN)$_4$]$^{2-}$
    e. [Ni(NH$_3$)$_4$]$^{2+}$ f. Cs$_2$[TaF$_7$]

26. Write chemical formulas for:
    a. Hexachlorovanadate(III) ion
    b. Diamminesilver(I) ion
    c. Tris(oxalato)cobaltate(III) ion

27. Write the formulas for each of the following.
    a. Dichloroaurate(I) ion
    b. Bromochlorodiammineplatinum(II)
    c. Tetraquadichlorochromium(III) bromide

EXERCISES

28. Use the valence-bond method to assign electron configurations to each of the following species.
    a. $[Ag(CN)_2]^-$, linear   b. $[Cu(NH_3)_4]^{2+}$, square planar
    c. $Cr(CO)_6$, octahedral

29. Using the valence-bond method, write the electron configurations for:
    a. $[NiCl_4]^{2-}$, tetrahedral   b. $[Cr(H_2O)_6]^{3+}$, octahedral
    c. $Fe(CO)_5$, trigonal bipyramidal

30. Draw structures showing the geometry of the *cis* and *trans* forms of $[Co(H_2O)_4Br_2]^+$.

31. Draw structures showing the geometry of the two forms of $[Cr(NH_3)_3(H_2O)_3]^{3+}$

32. How many possible isomers, both geometrical and optical, are there for:
    a. $[Coen(NH_3)_2BrCl]^+$, octahedral?
    b. $Ru(H_2O)_3Cl_3$, octahedral?

33. How many geometrical and optical isomers are there for each of the following?
    a. $[Pt(NH_3)_2(H_2O)Cl]^+$, square planar
    b. $[Zn(NH_3)_2(H_2O)_2]^{2+}$, tetrahedral

34. Construct the ligand-field molecular-orbital diagram for $[Fe(CN)_6]^{3-}$. How many unpaired electrons does this complex have?

35. Write the ligand-field molecular-orbital diagram for $[Cr(NH_3)_6]^{3+}$ and use this diagram to determine the number of unpaired electrons in the complex.

36. What does the prefix *tetrakis* in a complex name stand for?

37. What sort of experiment might decide which geometric isomer of $Pd(NH_3)_2Cl_2$ is the *cis* form and which is the *trans*?

38. There is evidence that a few Cu(III) compounds exist. The green substance $K_3[CuF_6]$ is one. Describe the electronic structure of the $[CuF_6]^{3-}$ ion using:
    a. Valence-bond theory   b. Ligand-field theory

39. How could you determine experimentally whether $[Pt(NH_3)_4]^{2+}$ is tetrahedral or square planar?

40. Will magnetic measurements allow you to decide whether $[Cu(CN)_4]^{2-}$ is tetrahedral or square planar?

41. The complex ion $[Co(NH_3)_5SO_4]^+$ is reddish and $[Co(NH_3)_5Br]^{2+}$ is violet. In which complex is the 10 Dq value the smaller?

42. Using standard reduction potentials and the K_{sp} for $Cr(OH)_3$, calculate the equilibrium constant for

$$2\ CrO_4^{2-}(aq) + 2\ H^+(aq) \rightleftharpoons Cr_2O_7^{2-}(aq) + H_2O$$

43. Write the electron configurations for Rh and Rh^{3+} and a valence-bond electron configuration for $RhCl_6^{3-}$ (an inner complex).

44. An orange complex of nickel with the apparent formula $Ni(CN)_2 \cdot 2KCN$ has no unpaired electrons. Can you give a structural explanation for this observation?

45. An octahedral ligand field splits the energies of the d orbitals. Would it also split the energies of the p orbitals? What about a tetrahedral ligand field?

46. An analytic procedure for the determination of iron, cobalt, and nickel in alloys is to separate the ions on a chromatographic column and then titrate each ion with EDTA. If a 0.100-g sample of alloy required 26.5 mL of 0.0140 M EDTA to completely titrate its Co^{2+}, what percentage of cobalt is in the alloy?

47. Use half-cell potentials to predict the products, if any, of the following reactions.
    a. $Ce(s) + H_2O$ b. $Ce^{4+}(aq) + Fe^{2+}(aq)$ c. $Ti^{2+}(aq) + Fe^{3+}(aq)$

48. Calculate $E°$ for the reaction between 1 M $K_2Cr_2O_7$ and water in the presence of 1 M H^+. Since acid dichromate solutions are stable, how do you explain your answer?

49. After an exposed photographic film has been treated with a reducing solution to convert the light-sensitized AgBr to silver metal, it is necessary to treat the film so that future exposure to light will not darken the unreacted AgBr. This is done by bathing the film in a concentrated solution of sodium thiosulfate ("hypo"). What chemical reaction occurs during this process?

50. Group IIB elements show many similarities to group IIA elements. Compare zinc and calcium with regard to:
   a. Oxidation states
   b. Solubilities of their carbonates
   c. Colors of their ions
   d. The reactions, if any, of their ions with aqueous ammonia
   e. The behavior of their oxides when stirred with concentrated aqueous NaOH

chapter 22

Nuclear Properties and Radioactivity

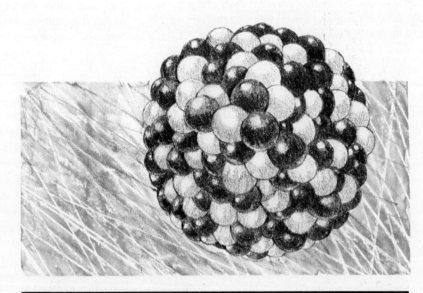

The radioactive uranium-238 nucleus contains 92 protons and 146 neutrons

Nuclear energy is one of the most significant discoveries of the twentieth century. Since the explosion of the first atomic bomb in 1945, the fact that nuclear transformations can release tremendous amounts of destructive energy has dominated world diplomacy and become the focal point of a persistent anxiety about the future of humanity. The use of controlled nuclear reactions as an energy source has generated intense debates about its potential benefits and its possible harmful effects. Both the destructive and the constructive potentials of nuclear energy have stimulated decades of intensive research into the properties of atomic nuclei, their structures, the factors that stabilize them, and the nature of their interconversions.

22.1 Radioactivity

The Discovery of Radioactivity

In 1896, the French physicist Henri de Becquerel discovered that high energy radiation, capable of penetrating several layers of cardboard and exposing a photographic plate, emanates from uranium ores. Since this radiation, called **radioactivity**, is not influenced by temperature, pressure, or chemical environment, it became clear that these rays were due to changes in the atoms themselves rather than changes in molecular structures. Becquerel's co-workers, Pierre and Marie Curie, showed that the radioactivity of the uranium ore was due to several previously undiscovered elements, polonium, francium, and radium, in addition to the uranium. Ernest Rutherford made the surprising discovery that a pure radium sample decomposes to form radon, a radioactive noble gas. This gave the key clue to the nature of radioactive processes—they are spontaneous reactions that change one element into another.

Radioactivity occurs when an unstable nucleus decays and emits smaller particles or energy in the form of very high frequency light. **Natural radioactivity** is due to the presence of unstable nuclei on earth.

There are three major types of radioactivity: alpha rays, beta rays, and gamma rays. **Alpha rays** are streams of $^4_2$He nuclei called **alpha particles** (symbol α) emitted with high energies from radioactive nuclei. Being positive, alpha rays bend toward a negative electric pole (Figure 22.1). **Beta rays** are streams of high-speed electrons, called **beta particles** (symbol β), emitted by a nucleus. (Their source and high energy are the only things that distinguish β particles from the electrons in an atom's orbitals.) Being negative, beta rays bend toward a positive electric pole, and because beta particles are much less massive than alpha particles, they bend more sharply than alpha rays. **Gamma rays** (symbol γ) are light rays of extremely high energy and therefore very short wavelengths (about 0.01 nm). Like other forms of electromagnetic radiation, they are not deflected in an electric field.

The Measurement of Radioactivity

Because of their high energies, radioactive particles or rays *ionize* the matter they pass through. This ionization is the basis of most methods for detecting radioactivity. (It is also why radioactivity is so unhealthy.) For example, the ionization reactions caused by radiation sensitize the silver salts in a photographic film and make it easier to reduce these salts to silver metal. Radiation will also ionize the air in an electroscope, a device in which two pieces of metal foil repel each other due to their similar charges. The ionization of the surrounding air allows the pieces of foil to lose their charges, and they fall toward each other. The exposure of film and the collapse of an electroscope provide inexpensive methods for detecting radioactivity, and both these methods are used to monitor the exposure to radioactivity of workers at nuclear reactors and in laboratories dealing with radioactive materials.

There are more sophisticated and expensive devices that measure radioactivity by detecting changes in the ionization of gases. These instruments—**ionization counters**, **proportional counters**, and **Geiger–Müller counters**—may differ considerably in their details, but all operate on the same basic principle (Figure 22.2). Radiation ionizes gas molecules (usually argon) within the counter to produce pairs of positive and negative ions. Charged collector plates (or wires) attract the positive ions in one direction and the negative charges in the other, and the discharge of these ions at the plates is recorded by the instrument's electrical system as a current pulse. The various types of counters differ primarily in the amount of voltage on the collector plates and in the method used to amplify the electronic signal.

Cloud chambers (Figure 22.3) and **bubble chambers** allow us to observe the paths of individual radioactive particles and the behavior of these particles in magnetic fields. The cloud chamber contains slightly supersaturated vapor, that is, the

Figure 22.1
Alpha, beta, and gamma radioactivity from uranium ores.

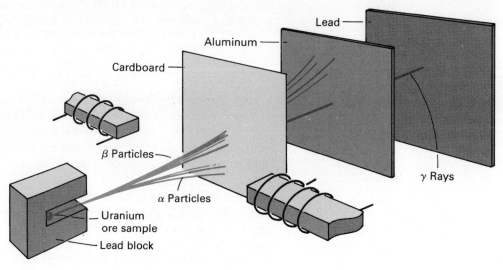

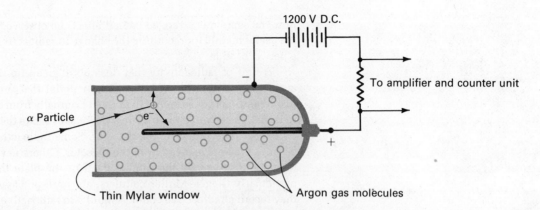

Figure 22.2
Geiger–Müller counter.

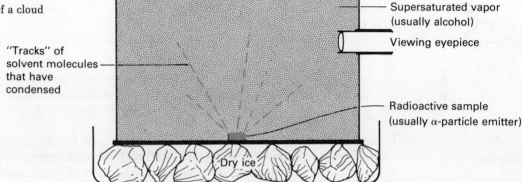

Figure 22.3
Cross-sectional diagram of a cloud chamber.

vapor pressure of the material is slightly greater than the equilibrium vapor pressure, but no condensation has occurred. The passage of a radioactive particle through the cloud chamber leaves a trail of ions, and these ions stimulate the condensation of the vapor to the liquid (just as scratches on a glass beaker stimulate the formation of precipitates in saturated solutions). The particle's path appears as a trail of tiny droplets that can be photographed. Because of the low density of the vapor, a cloud chamber is not much use for studying very high energy particles or particles that do not interact readily with matter because such particles pass through the chamber without interacting with many molecules. The bubble chamber solves this problem by using a slightly superheated liquid (usually liquid H_2) held just above its boiling point. The ion trails left by radioactive particles facilitate evaporation of the liquid and create trails of gas bubbles. The high density of the liquid, compared to the vapor in a cloud chamber, increases the probability that a radioactive particle will interact with molecules.

One common type of radiation detector, the **scintillation counter**, is not based on ionization. In this device, which is particularly useful for detecting gamma rays, the radiation passes through a crystal or liquid solution and excites molecules to higher energy electronic states. As the molecules return from these excited states to their normal ground states, they emit photons of visible or ultraviolet light. (This is the principle of the ZnS detector screen used in Rutherford's scattering experiment, Section 4.4.) A photocell measures the intensity of the light emitted by the detector.

When we use a radioactivity detector—an ionization counter, a proportional counter, a Geiger–Müller counter, or a scintillation counter—we record the number of counter responses in a given time period, the most common unit of radioactivity being **counts per minute** (cpm). This method of reporting radioactivity is valid only if all measurements are made exactly the same way, because the type of counter, the distance of the sample from the detector, the size of the sample, and the details of how the sample is arranged can all affect the observed count. Therefore, cpm units are useful mainly for recording the change in radioactivity of a particular sample over a period of time.

A unit of radioactivity that does not depend on the counting technique is the **curie** (Ci)—a quantity of radioactive material that produces 3.700×10^{10} nuclear disintegrations per second.* (It derived originally from the radiation emitted by 1 g of radium.) To convert cpm to curies, we must estimate the percentage of actual radioactivity that is recorded by the detector. Some radioactive rays will escape detection because they move away from the detector. Others never reach the counter because they are absorbed by the sample itself or the plate that holds it. Some detectors, particularly Geiger–Müller counters, are unresponsive for short time periods after they record a radioactive count. Should two radioactive particles enter the counter at nearly the same time, the counter will fail to detect one of them. We must correct for all of these factors before we can estimate the number of curies in a sample.

22.2 Nuclear Properties

The nucleus itself is an extremely small particle; its radius is 100,000 times smaller than the radius of an atom. It is characterized by its atomic number (Z) and **mass number** (A), which is *the number of neutrons plus protons*. For convenience, particles within the nucleus such as the neutron and proton are called **nucleons**.

Nuclear Binding Energy

Since the nucleus is so small and carries an intense positive charge (the charge on a helium nucleus is about 10^{22} coul per cubic meter, which corresponds to a repulsive energy of almost 10^9 kJ/mol), scientists have considerable difficulty explaining how such intensely charged particles can be so closely confined and still be stable. Although we do not yet fully understand the nature of the forces that keep the electrostatic repulsion between helium's two protons from splitting the nucleus, we do know the source of the required **nuclear binding energy**. The mass of the $^4_2$He nucleus (often expressed as helium-4) is a bit *smaller* than the sum of the masses of its constituent parts—two protons and two neutrons—and somehow the missing mass has been converted into the binding energy of the nucleus. According to Einstein's law of conservation of mass and energy,

$$\Delta E = \Delta m c^2 \tag{22.1}$$

where ΔE and Δm represent changes in energy and mass, and c is the velocity of light, 2.998×10^8 m/sec.

Example 22.1

Find the nuclear binding energy for the $^4_2$He nucleus.

Solution

In order to calculate the nuclear binding energy, we must compute the difference between the sum of the masses of the constituent parts of the nucleus and the mass of the nucleus itself (Table 22.1).

Mass of two protons	$= 2 \times 1.00728$ amu	$= 2.01456$ amu
Mass of two neutrons	$= 2 \times 1.00867$ amu	$= 2.01734$ amu
Total mass of constituent parts		$= 4.03190$ amu
Mass of a helium-4 nucleus		$= 4.00150$ amu
Mass difference $= \Delta m = 4.03190$ amu $- 4.00150$ amu $= 0.03040$ amu		

---

*The curie is a dangerously large amount of radioactivity. The radioactivity of samples encountered in the laboratory is usually under a millicurie and often on the order of a few microcuries (μCi).

22.2 NUCLEAR PROPERTIES

Table 22.1 Mass Differences and Binding Energies of Some Nuclei

	Atomic number	Nuclear mass (amu)	Mass difference (amu)	Binding energy (MeV)	Binding energy per nucleon (MeV)
^{2}H	1	2.01355	0.00240	2.23	1.11
^{4}He	2	4.00150	0.03040	28.32	7.08
^{7}Li	3	7.01436	0.04216	39.27	5.61
^{9}Be	4	9.00999	0.06248	58.20	6.47
^{11}B	5	11.00656	0.08186	76.25	6.93
^{12}C	6	11.99671	0.09899	92.21	7.68
^{14}N	7	13.99923	0.11242	104.72	7.48
^{16}O	8	15.99052	0.13708	127.69	7.98
^{19}F	9	18.99346	0.15876	147.88	7.78
^{27}Al	13	26.97439	0.24163	225.08	8.34
^{35}Cl	17	34.95952	0.32030	298.36	8.52
^{40}Ar	18	39.95250	0.36928	343.98	8.60
^{56}Fe	26	55.92066	0.52872	492.50	8.79
^{64}Zn	30	63.91268	0.60050	559.37	8.74
^{90}Sr	38	89.8864	0.84108	783.47	8.71
^{120}Sn	50	119.8747	1.09620	1021.1	8.51
^{157}Eu	63	156.8914	1.38222	1287.5	8.20
^{186}W	74	185.9107	1.59906	1489.5	8.01
^{206}Pb	82	205.9295	1.74254	1623.2	7.88
^{238}U	92	238.0003	1.93528	1802.71	7.57

m_p = 1.00728 amu   1 atomic mass unit = 931.5 MeV
m_n = 1.00867 amu
m_e = 0.000549 amu

To obtain the binding energy in J/mol, we must first convert this mass into kg/mol of helium nuclei. Since an amu is the same as a gram per mole, the mass loss per mole is

$$\left(0.03040 \frac{g}{mol}\right)\left(\frac{1 \text{ kg}}{1000 \text{ g}}\right) = 3.040 \times 10^{-5} \text{ kg/mol}$$

According to equation (22.1),

$$\Delta E = \left(3.040 \times 10^{-5} \frac{\text{kg}}{\text{mol}}\right)\left(2.998 \times 10^{8} \frac{\text{m}}{\text{sec}}\right)^2 = 2.732 \times 10^{12} \frac{\text{kg-m}^2}{\text{sec}^2\text{-mol}}$$

$$= 2.732 \times 10^{12} \text{ J/mol, or } 2.732 \times 10^{9} \text{ kJ/mol}$$

The energy that binds the nucleons together in a nucleus is analogous to the bond energy that holds atoms together to form a molecule; however, the nuclear binding energy of almost 3×10^9 kJ/mol is considerably greater than the modest energies of chemical bonds, which are usually less than 500 kJ/mol.

Nuclear physicists usually report the binding energy for a single nucleus in units of **electron volts**. One electron volt (eV) is the energy an electron gains as it is accelerated through a potential (V) of one volt. Since the charge of an electron (**Q**) is 1.602×10^{-19} coul, and since

$$E = \mathbf{Q}V$$

1 eV equals 1.602×10^{-19} J. The binding energy of one $^{4}_{2}$He nucleus is therefore

$$\frac{2.732 \times 10^{12} \text{ J/mol}}{(6.022 \times 10^{23} \text{ nuclei/mol})(1.602 \times 10^{-19} \text{ J/eV})} = 2.832 \times 10^{7} \text{ eV/nucleus}$$

This value is also expressed as 28.32 MeV (1 MeV equals 1 million electron volts).

Each nucleus has a characteristic nuclear binding energy (see Table 22.1). Stable nuclei have enough binding energy to maintain their structures, but nuclei with insufficient binding energies undergo spontaneous transformations to more stable nuclei. In the process, the total mass of the system decreases slightly as some mass is transformed into energy. Iron-56 has the greatest binding energy per nucleon and thus represents the most stable nucleus. (During the initial formation of the earth, temperatures were probably not excessive, but the presence of large quantities of unstable isotopes such as uranium and thorium radioactively heated the earth to a molten state. Iron-56, being the most stable nucleus, increased in amount and collected at the earth's center while metallic silicates and oxides "floated" outward to form the mantle. As the radioactivity decayed away, the planet cooled and the crust formed.) Nuclei with lower mass numbers than iron can increase their binding energy by **fusion** (joining together) to form a nucleus with a larger atomic number while elements with higher mass numbers than iron can become more stable by undergoing **fission** (splitting apart).

Nuclear Stability

Many nuclei are stable; others decay by one of several spontaneous nuclear reaction processes. We have not yet gained a complete understanding of the nature of the nucleus and the forces holding it together, but several important trends concerning nuclear stability are apparent. These trends give us some hints as to the nature of the binding forces within the nucleus.

1. There appears to be an upper limit to the number of protons a nucleus can hold and still remain stable. The stable nucleus containing the most protons is $^{209}_{83}$Bi, and none of the elements of atomic number 84 or greater has a stable isotope. As the atomic number increases, so does the total repulsive electrostatic force between the protons; there are compensating attractive forces, but there is an upper limit to the amount of positive charge a nucleus can bear without decomposing.

    Physicists invoke the analogy of a **liquid drop** to explain this upper limit to nuclear size and charge. A drop of liquid is stable because the attractive forces between its molecules maintain the minimum surface and prevent it from splitting into smaller spherical drops that would have a larger total surface. But the larger the drop, the greater the force that is needed to keep it from breaking up, and there is a maximum size to a drop that can resist splitting when subjected to slight distorting forces. A large nucleus is unstable because it is too large for the binding forces acting between the nucleons. Although this liquid-drop model explains some of the behavior of large nuclei, it does not explain the instability of smaller nuclei.

2. With the exception of $^1_1$H and $^3_2$He, every stable nucleus has at least as many neutrons as protons. For the lighter elements, atomic number 20 or less, stable nuclei have approximately the same number of protons and neutrons, but heavier, stable nuclei have many more neutrons than protons. Figure 22.4, a graph of the number of neutrons versus the number of protons in stable nuclei, indicates the role neutrons play in nuclear binding.

3. A majority of the stable nuclei have *even numbers* of protons and neutrons; there are only five stable nuclei ($^2_1$H, $^6_3$Li, $^{10}_5$B, $^{14}_7$N, and $^{180}_{73}$Ta) containing both an odd number of protons and neutrons.

4. There are so-called **magic numbers** of nucleons (2, 8, 20, 50, 82, and 126) that seem to confer added stability to a nucleus. For example, the unusual stability of a nucleus with either 50 protons or 50 neutrons can be seen from the binding energies per nucleon (Figure 22.5) or by the large number of stable nuclei in which this magic number occurs. Tin (50 protons) has ten stable isotopes, an unusually large number, and there are six stable nuclei with 50 neutrons compared to four with 48 and four with 52. These nuclear magic numbers, reminiscent of the numerical relationships in the periodic table, are

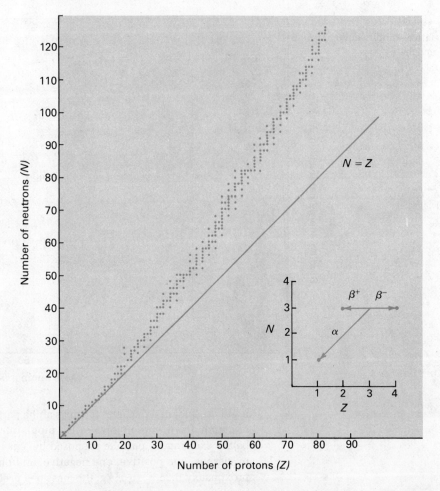

Figure 22.4
Graph of number of neutrons versus number of protons in stable isotopes of naturally occurring nuclei. The insert shows the changes in N and Z that occur when nuclei emit α or β particles or positrons.

explained in the same way: there are **shells of energy levels** within the nucleus. The protons occupy one set of quantum states, and the neutrons occupy a completely independent set of energy levels, and they apparently tend to occupy these states in pairs—hence the preferred occurrence of stable nuclei with even numbers of nucleons. The magic numbers represent cases in which shells are completely filled. Just as the low reactivity of the noble gases arises from the complete filling of their low-energy subshells by electrons (2, 8, 18, etc.), the filling of low-energy shells by nucleons stabilizes the nucleus.

Theories of Nuclear Binding

The way in which binding energy varies with mass number indicates that a nucleus is held together by strong but extremely short-range forces that attract *adjacent* nucleons but have little or no effect on more distant nucleons. Secondly, the particle scattering that occurs when one proton crashes against another, when compared with experiments in which neutrons collide with protons, shows that the attractive forces between nucleons are *independent of charge*. In other words, the binding force between a proton and a neutron is the same as between two protons or between two neutrons. In the case of two protons interacting, the attractive force is opposed by the strong electrostatic repulsions between the two positive charges. Neutrons stabilize a nucleus by providing binding forces without causing any electrostatic repulsions.

According to current theory, first presented by the Japanese physicist H. Yukawa in 1935, the attractive force between adjacent nucleons involves the exchange of extremely small particles between them—analogous to the exchange of electrons between two atoms in chemical forces. Shortly after Yukawa proposed his theory, a group of subatomic particles called **mesons** was discovered (from Greek *mesos* meaning intermediate, because the mass of the meson is intermediate between that of the electron and proton), first in cosmic rays and later as fragments of collisions between

Figure 22.5
Relative stability curve for nuclei.

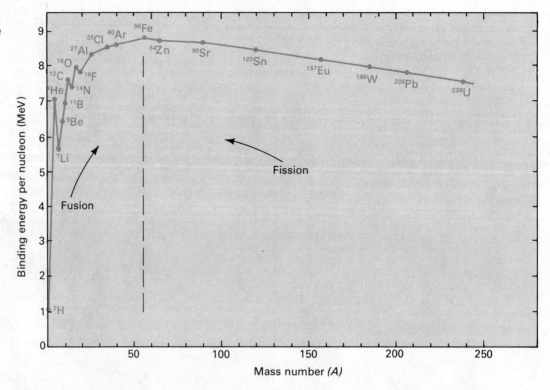

highly accelerated particles (such as alpha particles or protons) and nuclei. One type of meson, the **pi-meson** (also called **pion**), interacts strongly with nucleons and may be the exchanging particle proposed by Yukawa. There are three different types of pi-mesons, one positive, one negative, and one neutral, and all have a mass of approximately 0.25×10^{-27} kg; the mass of a proton is 1.67×10^{-27} kg for comparison.

Nuclear Spins

Nucleons, whether protons or neutrons, have quantized spin states, each producing a weak magnetic field in one of several directions. We have already discussed the two quantized spin states of the $^{1}_{1}H$ nucleus in Chapter 16 when we considered nuclear magnetic resonance of hydrogen compounds. Larger nuclei containing even numbers of each type of nucleon have no spin, because the nucleons pair up with opposing spins; but those nuclei containing either an odd number of protons or an odd number of neutrons have nuclear spins. Nuclear magnetic resonance studies using the spins of nuclei such as $^{19}_{9}F$, $^{13}_{6}C$, and $^{31}_{15}P$ have been extremely useful in elucidating the structure and character of new chemical species.

22.3 Nuclear Transformations

A nucleus that is too large or has the wrong combination of protons and neutrons is unstable and undergoes a nuclear transformation that usually changes its mass and charge. Most heavier nuclei, $A > 140$, transform or decay by α-particle emission, because this process lowers the coulomb repulsive energy of the nucleus—the number of positive charges is decreased by two—but the nuclear binding energy does not change because the α particle is almost as tightly bound as in a heavy nucleus. Whichever decay process occurs, it is an attempt by the unstable nucleus to achieve a more stable configuration.

Charge is conserved in all nuclear transformations, and the sum of the atomic numbers of the products of a nuclear reaction always equals the sum of the reactant atomic numbers. Although mass is not conserved in a nuclear transformation because some of it changes to energy, the mass change is always so much smaller than 1 amu

Alpha Decay

When a nucleus emits an alpha particle, its atomic number decreases by two and its mass number by four.

$$^{238}_{92}U \longrightarrow {}^{234}_{90}Th + {}^{4}_{2}He$$

Note how the atomic numbers and mass numbers balance in this equation. The product from an alpha-decay process is an isotope of the element *two spaces back* in the periodic table from the original element. The product nuclei are easily identified by proper balancing of the sub- and superscripts.

Example 22.2

Write a balanced equation for the alpha decay of neodymium-144.

Solution

The atomic number of neodymium is 60. Alpha decay decreases this to 58, the atomic number for cerium. It also decreases the mass number from 144 to 140.

$$^{144}_{60}Nd \longrightarrow {}^{140}_{58}Ce + {}^{4}_{2}He$$

Beta Decay and Neutrinos

A nucleus whose neutron-to-proton ratio is too large can emit a beta particle (also called a **negatron**). This process effectively converts a neutron into a proton,

$$^{1}_{0}n \longrightarrow {}^{1}_{1}p + {}^{0}_{-1}\beta$$

thereby raising the atomic number by one while leaving the mass number unchanged. The product is an isotope of the element one position beyond the reactant in the periodic table. For example,

$$^{210}_{83}Bi \longrightarrow {}^{210}_{84}Po + {}^{0}_{-1}\beta$$

Careful observations of beta-decay processes indicated, however, that the kinetic energy of the products is usually less than the energy equivalent to the mass lost in the process. To account for this discrepancy as well as apparent changes in the total nuclear spin, a violation of the law of conservation of momentum, Wolfgang Pauli proposed in 1931 that some of the energy is carried off by extremely small, neutral particles called **neutrinos**. Having practically no rest mass (less than 3.3×10^{-34} kg) and no charge, neutrinos do not interact with matter readily and are thus very difficult to detect. After many futile attempts, the neutrino was finally observed using large scintillation counters placed next to a powerful nuclear reactor. A characteristic series of scintillations showed the interaction of a neutrino with a proton to produce a neutron and a positron.

Positron Emission

Nuclei with insufficient neutrons to make them stable can undergo radioactive decay by emitting a **positron** (symbol β^+ or $_{+1}^{0}\beta$), a particle with the same mass as an electron but a +1 charge. The result of this emission is the conversion of a proton to a neutron, a decrease of one atomic number, and no change in mass number.

$$^{99}_{46}Pd \longrightarrow {}^{99}_{45}Rh + {}^{0}_{+1}\beta$$

It is very hard to observe positrons since the positron is the **antiparticle** of the electron. Two antiparticles have identical mass but opposite charges of the same size. When antiparticles collide, both are annihilated. In other words, they are transformed into a pair of photons of equal energy. Since electrons are so common in matter, most positrons are quickly annihilated, and the usual evidence for positron emission is the characteristic frequency of the gamma rays produced by annihilation.

Example 22.3

Compute the frequency of a gamma ray produced by positron–electron annihilation.

Solution

We can calculate the energy released by the annihilation of these two antiparticles by using equation (22.1).

$$E = 2(\text{mass of electron})c^2$$
$$= 2(9.11 \times 10^{-31} \text{ kg})(2.998 \times 10^8 \text{ m/sec})^2 = 1.64 \times 10^{-13} \text{ J}$$

Since two photons are produced, the energy per photon is half of this value.

$$E_{\text{photon}} = \frac{1}{2}(1.64 \times 10^{-13} \text{ J}) = 8.20 \times 10^{-14} \text{ J}$$

According to Planck's law, the frequency associated with this energy is

$$\nu = \frac{E}{h} = \frac{8.20 \times 10^{-14} \text{ J}}{6.63 \times 10^{-34} \text{ J·sec}} = 1.24 \times 10^{20} \text{ sec}^{-1}$$

Positrons occasionally survive long enough to be observed in a cloud chamber. We can recognize positrons because their paths bend in a magnetic field to a degree consistent with a charge-to-mass ratio of an electron, but the direction of bending is consistent with a positively charged particle.

Electron Capture

An alternate (and sometimes competing) mode of radioactive decay of neutron-deficient nuclei is **electron capture**. In this process, the nucleus captures one of the surrounding electrons, usually from the first quantum shell. The change produced is the same as for positron emission, the atomic number decreases by one and the mass number remains constant.

$$^{105}_{47}\text{Ag} + ^{0}_{-1}\text{e} \longrightarrow ^{105}_{46}\text{Pd} + \text{x rays}$$

Since no particle is emitted, the only evidence for electron capture is the formation of the product nuclei and the emission of x rays that are produced as electrons fall from higher energy levels to take up the position left empty by the captured electron.

Gamma Rays

Many radioactive transformations are accompanied by the emission of gamma rays. Gamma rays are streams of high-energy photons formed from part of the energy released by the nuclear reaction. Some of this energy shows up as kinetic energy of the product particles, but some may occur as gamma rays. The gamma rays emitted during a nuclear reaction are often discrete frequencies rather than a broad continuum of frequencies. This is another piece of evidence in favor of the shell model of the nucleus. There are excited states for a nucleus in which the shell arrangement of the nucleons is less stable than the ground state, and the excited nucleus emits a gamma photon to reach the ground state. These excited states are formed in some radioactive processes. For example, $^{137}_{55}\text{Cs}$ emits a beta particle to form an excited state of $^{137}_{56}\text{Ba}$, and this excited nucleus releases a photon whose frequency corresponds to the difference in energy between the excited state and ground state.

The Kinetics of Radioactive Decay

All radioactive decay processes follow a first-order rate law. For example, the rate of alpha-particle emission by $^{210}_{84}\text{Po}$ is proportional to the number of polonium-210 nuclei in the sample. As the number of unstable nuclei decreases over a period of time, the rate of alpha-particle emission decreases proportionately.

Figure 22.6

First-order decay curve of Po-210, half-life of 138.4 days.

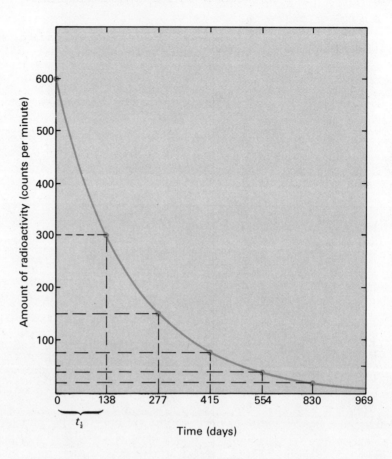

One feature of a first-order process is that it has a constant **half-life**. The half-life of a particular nucleus is the time it takes half the nuclei of that type in a sample to decay. Figure 22.6 illustrates the constant half-life of radioactive decay. It takes 138.4 days for the radioactivity of polonium-210 to decrease to half its original intensity, another period of 138.4 days to reduce the radioactivity to one-fourth, and a third period of 138.4 days reduces the radioactivity to one-eighth of what it was at the start of the observations.* Since decay is a first-order process, a graph of the logarithm of radioactivity versus time is a straight line with a negative slope, Figure 22.7. (See Section 12.2.)

We can express the decay rate mathematically as

$$\ln \frac{X_0}{X} = kt \qquad (22.2)$$

where X_0 represents the number of nuclei originally (time zero) and X is the number of nuclei at time t; k is the rate constant. Since the decay times are usually in half-lives rather than rate constants, we can obtain a relationship for the half-life, $t_{1/2}$ by setting $X = \frac{1}{2} X_0$. Then

$$\ln \frac{X_0}{\frac{1}{2} X_0} = kt_{1/2}$$

and since $\ln (a) = 2.30 \log (a)$

$$2.30 \log 2 = kt_{1/2}$$

$$t_{1/2} = \frac{0.693}{k} \qquad (22.3)$$

---

*Because it is a statistical law based on the likely behavior of a large sample of nuclei, the first-order rate law does not apply when only a few (perhaps 50 or less) radioactive nuclei remain in the sample.

Figure 22.7
Log plot of first-order decay curve for Po-210.

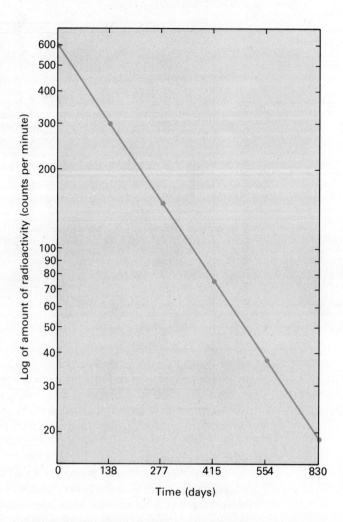

Example 22.4

Polonium-210 has a half-life of 138.4 days. How much of a sample of polonium that originally weighed 1.00 g remains after 60 days?

Solution

According to equation (22.3),

$$k = \frac{0.693}{138.4 \text{ day}} = 5.01 \times 10^{-3} \text{ day}^{-1}$$

We can use equation (22.2) to calculate the amount of polonium-210 remaining at any time. After 60 days,

$$\ln \frac{1.00 \text{ g}}{X} = (5.01 \times 10^{-3} \text{ day}^{-1})(60 \text{ day}) = 0.301$$

$$2.30 \log \frac{1.00 \text{ g}}{X} = 0.301$$

$$\log \frac{1.00 \text{ g}}{X} = 0.131$$

$$\frac{1.00 \text{ g}}{X} = \text{antilog } (0.131) = 1.35$$

$$X = \frac{1.00 \text{ g}}{1.35} = 0.74 \text{ g of } {}^{210}\text{Po remaining after 60 days}$$

Example 22.5

After 52 hours of decay, only 0.060 g of ^{64}Cu remains of an original sample of 1.00 g. What is the half-life for ^{64}Cu?

Solution

Using equation (22.2), we can calculate the value of k, the rate constant for first-order decay.

$$2.30 \log \frac{1.00 \text{ g}}{0.060 \text{ g}} = k(52 \text{ hr})$$

$$2.81 = k(52 \text{ hr})$$

$$k = \frac{2.81}{52 \text{ hr}} = 5.4 \times 10^{-2} \text{ hr}^{-1}$$

According to equation (22.3),

$$t_{1/2} = \frac{0.693}{5.4 \times 10^{-2} \text{ hr}^{-1}} = 12.8 \text{ hr}$$

The product of a radioactive decay is often itself unstable, and the radiation given off by the daughter nuclei (the nuclei formed in the first radioactive process) complicates the change in total radioactivity with time (Figure 22.8). The form of a graph of log of radioactivity (in cpm) versus time is not a straight line if the radiation emitted by the daughter is significant compared to the radioactivity of the parent. The actual form of the decay curve depends on which decays faster, the parent or the daughter. Nonlinear decay curves can also occur if the sample contains several radioactive nuclei, each decaying at a different rate.

Natural Radioactivity

It is believed that the elements were originally formed approximately 5 billion yr ago. If that is true, it is not likely that a nucleus with a half-life of less than 200 million yr would be around in the earth's crust in detectable amounts unless some natural radioactive process replenishes it. That is to say, any nucleus with a half-life less than 2×10^8 yr will have decayed for more than ten half-lives, and less than one out of every thousand original nuclei will have survived. Some naturally occurring radioactive nuclei—$^{40}_{19}$K (half-life = 1.27×10^9 yr), $^{87}_{37}$Rb (5.0×10^{10} yr), and $^{238}_{92}$U (4.51×10^9 yr), for example—still exist on the earth because their decay rate is so slow. Many shorter-lived nuclei do exist on earth, however, because they are daughters of very slowly decaying nuclei that continuously replenish them. For example, although $^{234}_{90}$Th has a half-life of only 24 days, it still exists in minerals because it is the daughter of $^{238}_{92}$U, a long-lived nucleus, and the decay of the uranium is always providing new thorium.

There are several cases in which a long-lived radioactive nucleus is the patriarch of many generations of shorter-lived nuclei, each of which in turn decays until eventually a stable nucleus is formed. Figure 22.9 shows one such natural decay scheme, starting with the alpha decay of U-238 and terminating with the alpha decay of Po-210 to form Pb-206, a stable isotope. In between these two events, many radioactive processes, some of them alpha decays and some beta decays, occur. Some of these steps are rapid, for example,

$$^{234}_{91}\text{Pa} \longrightarrow {}^{234}_{92}\text{U} + {}^{0}_{-1}\beta \qquad t_{1/2} = 1.18 \text{ min}$$

but others such as

$$^{234}_{92}\text{U} \longrightarrow {}^{230}_{90}\text{Th} + {}^{4}_{2}\text{He} \qquad t_{1/2} = 2.48 \times 10^5 \text{ yr}$$

have rather long half-lives. All of the nuclei formed in this decay series exist in nature because the slow decay of U-238 replenishes them.

Figure 22.8
Parent–daughter decay curve.

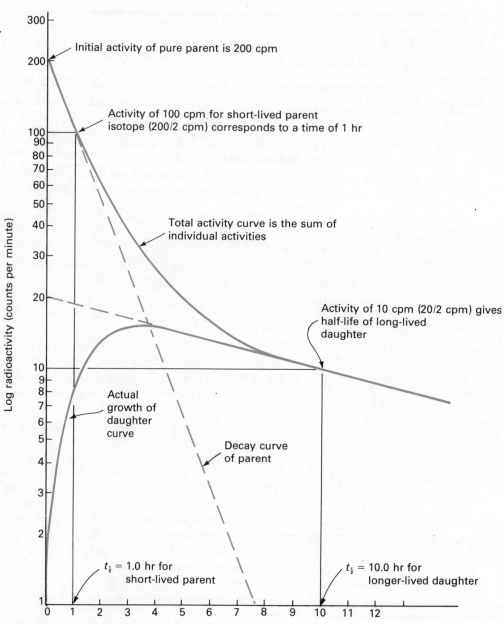

Induced Radioactivity

Some unstable nuclei are formed in the laboratory or in the upper atmosphere by collisions of high-energy particles with stable nuclei. The subsequent decay of the nuclei produced in such a reaction is called **induced radioactivity**.

When the bombarding particle is positively charged, such as deuterons ($_1^2H^+$ ions) or α particles, there are very strong repulsive electrostatic forces between it and the target nucleus as the bombarding particle gets close to the target nucleus. In order for the particle to crash into the target nucleus and disrupt its structure, it must have a very high kinetic energy (about 10 MeV or more). Various techniques are used to impart this energy, principally acceleration using electric fields. In the **linear accelerator** (Figure 22.10), the particles cascade through a sequence of alternating electric fields (the ± fields provide a push–pull effect), each field in turn accelerating the particles and increasing their energy. The **cyclotron** (Figure 22.11) is a circular particle accelerator. As in the linear accelerator, alternating electrostatic fields push–pull the particles to greatly increase their velocities, but the particles are forced to follow a circular path because there is an intense magnetic field perpendicular to their velocity direction. As the particles pick up speed they spiral out from the center

22.3 NUCLEAR TRANSFORMATIONS

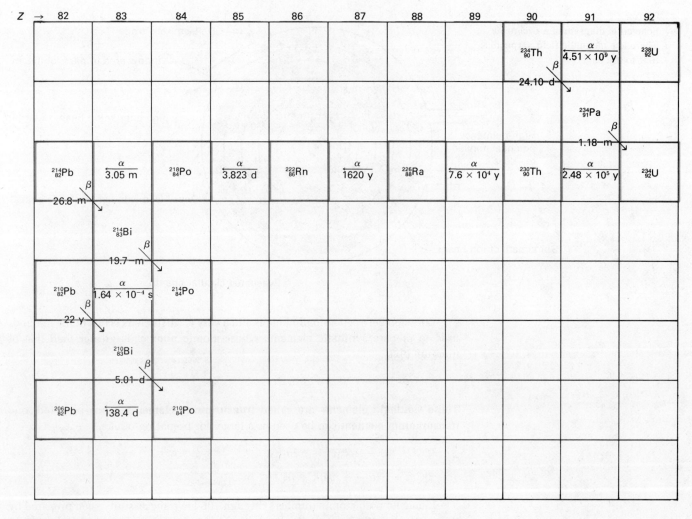

Figure 22.9
Uranium-238 decay scheme: radioactive $^{238}_{92}U$ to stable $^{206}_{82}Pb$. The mode of decay is given along with the half-life for the decay process in seconds (s), minutes (m), days (d), and years (y).

of the cyclotron. Since the strength of the magnetic field is the confining force for the particles, when their kinetic energies exceed the magnetic restraining force they are deflected out of the cyclotron onto the bombardment target.

When an energetic positive particle strikes a nucleus, it often causes a reaction in which smaller particles (protons or neutrons) break away from the target nucleus.

$$^1_1H + ^{63}_{29}Cu \longrightarrow ^{63}_{30}Zn + ^1_0n$$

The $^{63}_{30}Zn$ isotope, being unstable, decomposes by positron emission.

$$^{63}_{30}Zn \longrightarrow ^{63}_{29}Cu + ^{0}_{+1}\beta$$

Nuclear reactions are often written in abbreviated form as $^{63}Cu(p,n)^{63}Zn$, which indicates the reactant isotope, reacting and product particles, and the product isotope.

Figure 22.10
Schematic diagram of a linear accelerator.

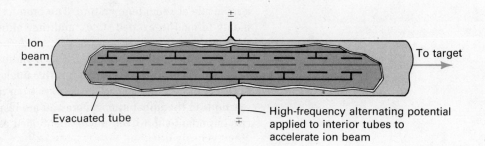

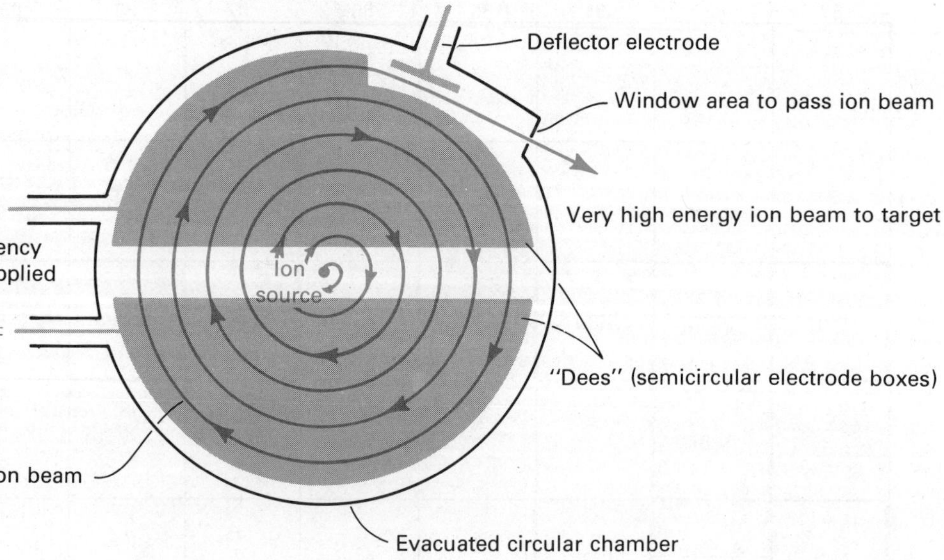

Figure 22.11
Schematic diagram of a cyclotron. There is a magnetic field perpendicular to the page.

- Deflector electrode
- Window area to pass ion beam
- Very high energy ion beam to target
- "Dees" (semicircular electrode boxes)
- Evacuated circular chamber
- Spiral path of ion beam
- High-frequency potential applied to "dees"
- Ion source

Collision between a small nucleus and a very large nucleus is one of the methods used to prepare synthetic elements whose atomic number is greater than that of uranium (92).

$$^{238}_{92}\text{U} + ^{12}_{6}\text{C} \longrightarrow ^{245}_{98}\text{Cf} + 5\,^{1}_{0}\text{n}$$

These synthetic elements are called **transuranium elements**. Once prepared, one transuranium element can be used as a target for preparing others.

$$^{252}_{98}\text{Cf} + ^{10}_{5}\text{B} \longrightarrow ^{257}_{103}\text{Lr} + 5\,^{1}_{0}\text{n}$$

$$^{244}_{96}\text{Cm} + ^{4}_{2}\text{He} \longrightarrow ^{245}_{97}\text{Bk} + ^{1}_{1}\text{H} + 2\,^{1}_{0}\text{n}$$

Elements with atomic numbers through 106 have successfully been prepared by this method—although admittedly only a very few atoms of element number 106 have been produced, and they are very unstable with half-lives on the order of milli- or microseconds. There has recently been considerable discussion by some scientists who predict, on the basis of nuclear-shell theory, that we will encounter stable or modestly stable isotopes of some *superheavy nuclei*, those with atomic numbers 114 or 118, for example.

Another method of inducing radioactivity is by neutron bombardment. Neutrons were first produced by bombarding beryllium or boron with alpha particles,

$$^{4}_{2}\text{He} + ^{11}_{5}\text{B} \longrightarrow ^{14}_{7}\text{N} + ^{1}_{0}\text{n}$$

but the best available source of neutrons is a nuclear reactor. Since neutrons are neutral, they are not repelled by the positive nuclear protons and they can combine with target nuclei even when they are traveling at rather slow speeds. Because there can be more time for interaction, many nuclei readily capture these "slow-moving" neutrons, called **thermal neutrons** because they have about as much kinetic energy as a molecule at room temperature. The product nuclei from these reactions are commonly formed in excited states, and they either emit gamma rays as they return to their ground state, or they undergo other forms of radioactive decay.

Another rather important induced radioactivity process results from **cosmic rays**, which are streams of small positive nuclei, mesons, and neutrons that bombard the earth from outer space. These "rays from the cosmos (heavens)" induce nuclear reactions in the upper atmosphere and are responsible for the natural occurrence of two important short-lived nuclei mentioned earlier, tritium and carbon-14.

$$_0^1 n \text{ (fast)} + {}_7^{14}N \longrightarrow {}_6^{12}C + {}_1^3H$$

$$_0^1 n \text{ (thermal)} + {}_7^{14}N \longrightarrow {}_6^{14}C + {}_1^1H$$

Fission

Nuclear fission is the splitting of a large nucleus into two smaller fragments; several neutrons are usually produced as well. Although some naturally occurring nuclei undergo slow, spontaneous fission, the most important fission reactions are induced by neutron capture. The capture of a thermal neutron by U-235 produces an excited U-236 nucleus that undergoes rapid fission.

$$_0^1 n \text{ (thermal)} + {}_{92}^{235}U \longrightarrow [{}_{92}^{236}U] \longrightarrow {}_{36}^{94}Kr + {}_{56}^{139}Ba + 3\,{}_0^1n$$

This equation represents one typical fission reaction of U-236. In actual fact, there are a variety of fission reactions occurring simultaneously; for example, the fission can also produce Rb and Cs rather than Kr and Ba (see Figure 22.12).

A comparison of nuclear binding energies shows that energy would be released by the fission of any heavy nucleus (mass number 80 or more), but fission is a rather rare phenomenon because a nucleus must be distorted considerably before it splits. The liquid-drop model is particularly useful in describing fission. According to this model, a stable drop has a spherical shape but an outside force can make it vibrate into distorted shapes. If the oscillations are violent enough, the drop becomes so distorted that it splits apart. A similar picture accounts for fission—a nucleus must oscillate into a sufficiently distorted shape before it can split. For most nuclei, the binding energy is strong enough to prevent these oscillations, but a few particularly large nuclei whose binding energies are too weak to maintain their undistorted shapes undergo spontaneous fission, for example, Cf-252. For other large nuclei, the capture of a neutron provides the energy that sets the nucleus into the violent vibrations that split it.

Fission releases large amounts of energy, about 2×10^{10} kJ/mol (200 MeV) in the case of U-235 fission. In addition, the fission of one nucleus produces several neutrons, and these neutrons can in turn stimulate fission in other nuclei. The situation in which an energy-producing reaction also increases the concentration of particles that initiate it leads to a reaction mechanism known as a **branched-chain process**

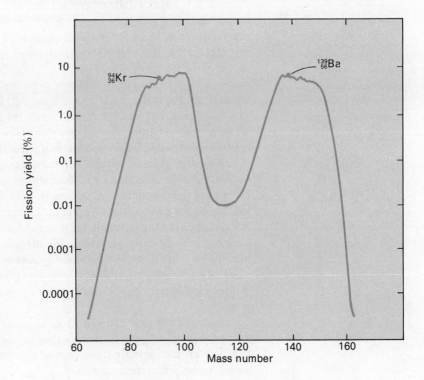

Figure 22.12
Distribution of isotopes resulting from ^{235}U fission process. *Source:* J. M. Siegal (Ed.), "Plutonium Project," *Journal of the American Chemical Society* **68**, 2437(1946). Copyright by the American Chemical Society.

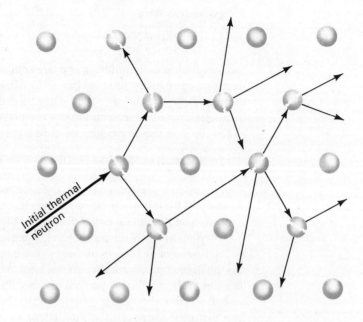

Figure 22.13
Illustrative branched-chain process in which $^{235}_{92}U$ atoms are "split" by neutrons.

(Figure 22.13). The fission of one U-235 nucleus produces approximately 2.5 neutrons, and these can cause fission in 2.5 additional U-235 nuclei, and so on. This snowballing process continues very rapidly and the continually accelerating energy output can lead to tremendous explosions such as those that destroyed Hiroshima and Nagasaki. To achieve self-sustaining fission reactions, the probability for neutron capture must be about 40% or greater, and the mass of fissionable material required to insure this probability is called a **critical mass**. For an explosion, the critical mass must be put together rapidly and forcibly held there or else the mass will fly apart with only a relatively little fizzle.

Nuclear Reactors

Controlled nuclear reactions provide far more socially acceptable ways of utilizing fission reactions than do atom bombs. The challenge to running a controlled nuclear reaction is to exert a delicate control over the probability of neutron capture so that fission runs at a self-sustaining rate but never reaches explosive speeds. A **nuclear reactor** (Figure 22.14) consists of many rods of subcritical amounts of fissionable material (called **fuel elements**) interspersed with control rods made of a substance that absorbs neutrons efficiently, such as cadmium. The fuel rods are usually surrounded by a **moderator** that slows neutrons to thermal speeds without capturing them. Graphite, water, and "heavy water" (D_2O) are all good moderators.

Since one fuel element has a subcritical mass, too many neutrons escape it to allow a self-sustaining fission reaction. However, each fuel element can capture neutrons emanating from the other fuel elements in the reactor, and if the probability of this capture is large enough, the fission reaction will continue briskly. The job of the control rods is to stop enough neutrons to keep fission from running wild. The operator of the reactor adjusts the positions of the control rods to maintain a sensitive balance on the fission rate. Pushing the control rods deeper into the reactor slows the reaction, pulling them out allows it to run faster. The purpose of the moderator is to slow the initially fast neutrons to the thermal speeds at which capture is far more likely. The safe operation of a nuclear reactor is uppermost in any design plan, and many of the special safety features are specifically planned for rapid insertion of the control rods into the fuel element area to prevent any attempt by the reactor to "run away."

Nuclear reactors serve mainly as a source of heat energy to drive turbines and thus produce electricity. Although they are not a panacea to the energy crisis, they will play an increasingly important role in power production in the near future if permitted by public acceptance.

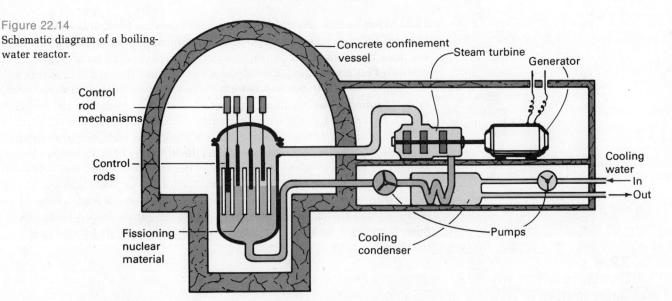

Figure 22.14
Schematic diagram of a boiling-water reactor.

Another use of reactors is as a source of neutrons for producing synthetic nuclei by neutron-capture reaction. **Breeder reactors** use their neutrons to create more fissionable material than they consume. For example, U-238, the most abundant uranium isotope, is not a suitable fission fuel, but it can be converted to Pu-239, an important nuclear fuel, in a breeder reactor. A breeder reactor uses fissionable uranium-235 or plutonium-239 as fuel to produce energy in the normal way, but since on the average 2.5 neutrons are produced per fission reaction, some of the extra neutrons can be captured by the U-238 present. The reactions are

$$^1_0n + {}^{238}_{92}U \longrightarrow {}^{239}_{92}U \longrightarrow {}^{239}_{93}Np + {}^{\ 0}_{-1}\beta \longrightarrow {}^{239}_{94}Pu + {}^{\ 0}_{-1}\beta$$

Fusion

An energy source with far more potential than nuclear fission is **nuclear fusion**—the merging of light nuclei to form heavier ones. For example, the process

$$4\,{}^1_1H \longrightarrow {}^4_2He + 2\,{}^{\ 0}_{+1}\beta$$

would yield about 25 MeV, or about 600 million kJ per gram of hydrogen consumed! (The subsequent annihilation of the positrons would yield about another 2 MeV.) This reaction is apparently a primary source of the sun's energy, but since this reaction has a very high activation energy that exceeds even the energies available in the sun, the fusion of four protons into a He-4 nucleus occurs by a roundabout mechanism. We believe that the mechanism of fusion in the sun's center is the **carbon–nitrogen cycle**. The steps are

$$^{12}_{6}C + {}^1_1H \longrightarrow {}^{13}_{7}N$$

$$^{13}_{7}N \longrightarrow {}^{13}_{6}C + {}^{\ 0}_{+1}\beta$$

$$^{13}_{6}C + {}^1_1H \longrightarrow {}^{14}_{7}N$$

$$^{14}_{7}N + {}^1_1H \longrightarrow {}^{15}_{8}O$$

$$^{15}_{8}O \longrightarrow {}^{15}_{7}N + {}^{\ 0}_{+1}\beta$$

$$^{15}_{7}N + {}^1_1H \longrightarrow {}^{12}_{6}C + {}^4_2He$$

Net change $\quad 4\,{}^1_1H \longrightarrow {}^4_2He + 2\,{}^{\ 0}_{+1}\beta$

Farther out in the sun, but still in its interior, the **proton–proton chain** appears to be the dominant mechanism. The net reaction is

$$4\,{}^1_1H \longrightarrow {}^4_2He + 2\,{}^{\ 0}_{+1}\beta$$

Fusion reactions are attractive energy sources because the fuel is essentially unlimited and the reactions do not produce the radioactive waste products that fission reactions do. Unfortunately, we have not yet mastered the technology of running controlled fusion reactions; the only large-scale production of fusion energy on earth has been in the highly destructive form of the hydrogen bomb. Fusion reactions, also called **thermonuclear reactions**, require temperatures of more than 200 million °C. At these high temperatures, electrons are stripped from atoms, and matter collapses into a jumble of nuclei and electrons called a **plasma**. The main barrier to producing a controlled thermonuclear reactor has been the difficulty of maintaining plasma long enough for fusion to occur. Plasmas are so hot that they would melt any container wall, so we must learn how to hold plasmas in intense magnetic and electrical fields rather than conventional vessels. In thermonuclear bombs such as the hydrogen bomb, a small fission bomb creates the high temperature needed for fusion. Under these conditions, fusion is inefficient but still occurs rapidly enough to release tremendous amounts of destructive force.

22.4 The Uses of Radioactivity

Quite apart from the use of fission or fusion as energy sources, radioactivity has proved extremely useful in chemistry, medicine, biology, geology, and archeology. It provides us with sensitive analytic techniques, acts as a probe for determining the mechanisms of chemical reactions and physiological processes, and has given us clues about the age of the earth and the conditions of its formation. The following examples illustrate some of the useful applications of radioactivity.

Analytic Chemistry

One of the most sensitive analytic applications of nuclear chemistry is **neutron activation analysis**, a nondestructive analytical technique for measuring the elemental composition of a sample. The sample is placed in a nuclear reactor to expose it to intense bombardment by thermal neutrons. The resulting neutron-capture reactions create radioactive nuclei. The unique radioactivities and decay-particle energies of the various products, like fingerprints, indicate the specific types and concentrations of nuclei that were formed by neutron capture. By comparing the radioactivity produced in the sample with that produced in a standard sample, we can estimate the composition of the unknown. The technique is so sensitive that it has detected trace components that are present in concentrations as low as one part per billion. The details of the composition of a sample, as revealed by neutron activation analysis, can provide hints as to the origin of the sample, and many of the most interesting applications of neutron activation involve this sort of detective work. Art historians have been able to check the authenticity of paintings because each old master mixed his own pigments in a characteristic way that influenced the composition of the pigment. Criminologists have used neutron activation to trace soil samples to their exact source, and historians have exposed Napoleon's hair to neutron activation to ascertain the cause of his death. (It seems to have been arsenic poisoning.)

Geology

The most important contribution of radiochemistry to geology is the estimation of the time elapsed since the earth's rocks formed. The various methods that have been used are all based on the idea that if we measure the ratios of concentrations of certain nuclei in rocks, we can use our knowledge of the rates of the nuclear reactions that produce them to compute the length of time the decay has been going on. One method is based on the ratio of Pb-206 to U-238 in uranium ores whose only source of lead is the decay scheme of U-238 (see Figure 22.9). The number of Pb-206 atoms in the rock equals the number of U-238 atoms that have decayed since the rock's formation. A comparison with the number of U-238 atoms still present in the sample tells us its age expressed in terms of numbers of U-238 half-lives. Since this half-life is

4.51 billion yr, we can calculate the age of the rock. Other techniques are based on the Ar/K ratio (the argon, although a gas, remains trapped in the rock's crystal lattice),

$$^{40}_{19}K + {}^{0}_{-1}e \longrightarrow {}^{40}_{18}Ar \qquad t_{1/2} = 1.27 \text{ billion yr}$$

or the $^{87}_{37}Rb/^{87}_{38}Sr$ ratio.

$$^{87}_{37}Rb \longrightarrow {}^{87}_{38}Sr + {}^{0}_{-1}\beta \qquad t_{1/2} = 50 \text{ billion yr}$$

The age of the earth has also been estimated from the ratios of $^{204}_{82}Pb$, $^{206}_{82}Pb$, and $^{207}_{82}Pb$ in ordinary lead ore samples. The principle is that since the earth's formation, the content of the 206-isotope has steadily increased due to U-238 decay and the 207-isotope has grown due to U-235 decay, whereas the 204-isotope, not being the daughter of any decay scheme, has remained constant throughout the earth's lifetime. A knowledge of the rates of the decay scheme coupled with estimates of the primeval isotopic composition of lead (based on the isotopic analysis of lead in uranium-free meteorites) leads to an estimate of the earth's age of 4.5 billion yr.

Archeological Dating

The intensity of C-14 radioactivity in any artifact that contains carbon—in other words, anything made from wood, leather, bone, shell, beeswax, resins, or anything else derived from plant or animal sources—allows archeologists to estimate the age of the object. The age of a grave can be calculated from the carbon-14 activity of the skeleton, and even the charcoal from an ancient fire can tell us when the fire was lit.

The idea behind the **carbon-dating** method, conceived by Willard Libby, is that carbon-14 has been formed at a uniform rate throughout geologic time by cosmic ray bombardment of nitrogen in the upper atmosphere.* The reaction is

$$^{14}_{7}N + {}^{1}_{0}n \longrightarrow {}^{14}_{6}C + {}^{1}_{1}H$$

The carbon-14 atoms are then oxidized to form radioactive carbon dioxide, that is then incorporated by plants and passed on to animals that eat the plants. Anything that is alive, a plant or an animal, is constantly exchanging radioactive carbon with its environment, and the concentration of carbon-14 quickly reaches a steady level in a live organism and maintains this level as long as the organism lives. This radioactivity level corresponds to 15.3 disintegrations per minute per gram of carbon. Death, whether it is a matter of a natural death, a tree being cut down, or an animal being slaughtered, cuts off the source of radioactive carbon, and its concentration slowly decreases due to radioactive decay.

$$^{14}_{6}C \longrightarrow {}^{14}_{7}N + {}^{0}_{-1}\beta \qquad t_{1/2} = 5730 \text{ yr}$$

A comparison of the level of carbon radioactivity in an ancient piece of wood with that present in a freshly cut wood sample reveals the fraction of $^{14}_{6}C$ nuclei that have decayed since the old wood was cut, and this fraction can be used to compute the age of the wood. Some of the gigantic sequoia trees have been determined to be about 3000 yr old, and fire charcoal from caves in France occupied by prehistoric man was dated as being 16,000 yr old.

Medical and Biological Applications

Radiation is quite harmful to living organisms. The high energies of alpha, beta, and gamma rays ionize molecules within tissue and cells and the chemical changes resulting from the ionization can harm the organism. The effect on genetic material (DNA) can be particularly harmful because it leads to mutations that can be passed on to future generations.

In spite of the harm done by radioactivity, it has a very useful place in medicine. Patients with surgically untreatable cancer are exposed to radiation (usually gamma

---

*Apparently the assumed steady rate of formation of carbon-14 nuclei throughout recent geologic time is not correct. Appropriate corrections can be made, however, and this dating technique is quite useful for determining dates from 1–50,000 yr ago with an accuracy of ± 200 yr.

rays from a cobalt-60 source) in the hope of destroying the tumor without doing too much harm to normal tissue. Although there are limits to the amount of this procedure, called *radiotherapy*, a patient can tolerate, it has successfully arrested and even cured cancers. Many radioisotopes are used in the treatment and diagnosis of diseases; iodine-131, for example, when taken internally (in very small amounts, of course) rapidly concentrates in the thyroid gland where it can destroy cancerous cells.

Biologists use radioactivity to create mutations in test organisms and thus learn more about the nature of heredity. Many mutations of the microorganism *neurospora crassa* have been prepared and studies of neurospora mutants lacking metabolic enzymes have elucidated the details of metabolic processes. Since mutation is a random process, radioactive irradiation of grains and seeds is occasionally beneficial and successfully has led to the development of hardier varieties of wheat, corn, and vegetables. Another interesting use of radiation is to sterilize insects, usually the males, as a modern-day pest-control technique.

Summary

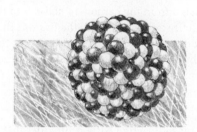

1. Unstable nuclei are radioactive; they change (decay) into other nuclei in an attempt to become more stable. Three major types of radioactive decay processes give α particles (He-4 nuclei), β particles (high-energy electrons), and γ rays (very high energy electromagnetic radiation).

2. Radioactivity is usually detected and measured by its ability to produce ions in ionization-counter devices, such as Geiger–Müller counters, or in cloud or bubble chambers. Scintillation detectors do not measure ionization and are particularly useful for measurement of γ-radiation.

3. Units of radioactivity are usually counts per minute (or disintegrations per minute). The curie (Ci) measures quantity, not rates, and is the amount of material that produces 3.700×10^{10} disintegrations per second.

4. The nucleus of an atom contains protons and neutrons, called nucleons, that represent most of the mass of an atom but only a small fraction of its volume. The different nuclei are represented as $^A_Z X$, where X is the symbol for the element, Z is the atomic number, and A is the mass number (the number of neutrons plus the number of protons).

5. The difference between the actual nuclear mass and the sum of the masses of the nucleons is called the mass difference. Expressed as energy (usually in electron volts) using $\Delta E = \Delta m c^2$, this is called the nuclear binding energy. Iron-56 has the greatest binding energy per nucleon; elements of lower atomic number can become more stable by fusion, whereas elements of higher atomic number must undergo fission to become more stable.

6. The liquid-drop and nuclear-shell models are two theories of nuclear structure. The shell model is patterned after the Bohr electron-shell theory and better explains many of the observed characteristics of nuclei, such as the magic numbers.

7. Current theories attribute exchange of small particles, mesons, to account for the large nuclear attractive forces required to hold nuclei together.

8. Nuclei have quantized spin states—many with net nuclear spin are studied using nuclear magnetic resonance instruments.

9. In nuclear transformations there is no change in the total mass number of the system. Common transformation processes are alpha decay (α), beta decay (β), positron emission (β^+), electron capture, gamma rays (γ), and neutrinos.

10. All radioactive decay processes follow a first-order rate law. The half-life of a particular nucleus is the time it takes half the nuclei of that type to decay. If a radioactive nucleus has a radioactive product (called daughter), the decay scheme and decay-rate curve becomes more complex.

EXERCISES

11. Many elements possess natural radioactivity, and others can acquire induced radioactivity. Linear accelerators and cyclotrons are devices in which various nuclei are accelerated (given more energy) and shot at target nuclei to form new nuclei with greater atomic and mass numbers. Neutrons can also be used as bombarding nuclei, and since they are not charged, less energy is required for them to combine with target nuclei. Thermal neutrons are particularly effective as bombarding nuclei.

12. Nuclear fission is the splitting of a large nucleus into two smaller fragments. Nuclear fusion, the merging of lighter nuclei to form heavier ones, is the process whereby the sun produces energy. One particular process is called the carbon–nitrogen cycle.

13. Nuclear explosions occur when a critical mass of fissionable material undergoes a chain reaction.

14. Nuclear reactors produce heat (and power) by controlling the rate of nuclear fission of a critical mass of material using moderators. Breeder reactors create nuclear fissionable material by neutron capture reactions as they operate.

15. Radioactivity is a useful tool for analysis by neutron activation, dating of rocks and artifacts, and medical diagnosis and radiotherapy.

Vocabulary List

radioactivity	electron volt	transuranium element
natural radioactivity	fusion	thermal neutron
alpha ray	fission	cosmic ray
alpha particle	liquid-drop analogy	nuclear fission
beta ray	magic number	branched-chain process
beta particle	shells of energy levels	critical mass
gamma ray	meson	nuclear reactor
ionization counter	pi-meson	fuel element
proportional counter	negatron	moderator
Geiger–Müller counter	neutrino	breeder reactor
cloud chamber	positron	nuclear fusion
bubble chamber	antiparticle	carbon–nitrogen cycle
scintillation counter	electron capture	proton–proton chain
counts per minute	half-life	thermonuclear reaction
curie	induced radioactivity	plasma
mass number	linear accelerator	neutron activation analysis
nucleon	cyclotron	carbon dating
nuclear binding energy		

Exercises

1. Briefly define the following terms.
   - **a.** Nucleon **b.** Meson **c.** Alpha particle
   - **d.** Positron **e.** Half-life **f.** Critical mass **g.** Millicurie

2. Why is more energy required to bombard target nuclei with protons than with neutrons?

3. Why are α particles deflected less than β particles in an electric field? (See Figure 22.1.)

4. Why is the bubble chamber a more suitable device for observing a neutrino than the cloud chamber?

5. Present some evidence for the shell theory of the nucleus.

6. Write a nuclear equation describing an emission process and a capture process.

7. What is the difference between:
   - **a.** Fusion and fission **b.** A moderator and a control rod?
   - **c.** A nucleon and a neutron? **d.** A β particle and an electron?

8. Why do fission chain reactions not occur in natural uranium ores?

9. Would an unstable platinum isotope present during the formation of the earth tend to engage in fusion reactions or decay into smaller fragments? Why? (See Figure 22.5.)

10. List the numbers on protons and neutrons in each of the following nuclei.
    a. Oxygen-18   b. Tin-118   c. Plutonium-239

11. How many protons and how many neutrons are there in:
    a. Nitrogen-15?   b. Iodine-131?   c. Radium-226?

12. Which of the following nuclei contain either an odd number of protons or an odd number of neutrons?
    a. P-31   b. Ar-40   c. K-38   d. Fe-58

13. Which of the following nuclei contain an odd number of either type of nucleon?
    a. Mg-24   b. B-10   c. Na-23   d. Cd-64

14. Complete and balance the following nuclear reactions.
    a. $^{6}_{4}B \longrightarrow ? + 2\,^{1}_{1}H$
    b. $^{108}_{44}Ru \longrightarrow \,^{0}_{-1}B + ?$
    c. $^{230}_{90}Th \longrightarrow \,^{226}_{88}Ra + ?$

15. Complete and balance the following nuclear reactions.
    a. $^{106}_{46}Pd + ? \longrightarrow \,^{109}_{47}Ag + \,^{1}_{1}H$
    b. $^{252}_{98}Cf + \,^{11}_{5}B \longrightarrow \,^{257}_{103}Lr + ?$
    c. $^{24}_{11}Na \longrightarrow \,^{24}_{12}Mg + ?$
    d. $^{98}_{43}Rh \longrightarrow ? + \,^{0}_{+1}\beta$

16. Write balanced nuclear reactions from the following notations (d = deuteron, $^{2}_{1}H$; p = proton, $^{1}_{1}H$).
    a. $^{23}Na(d, \alpha)^{21}Ne$ b. $^{39}K(p, \alpha)^{36}Ar$
    c. $^{115}In(n, \gamma)?$ d. $^{98}Mo(d, n)^{99}Tc$

17. Complete the following nuclear reactions.
    a. $^{7}_{3}Li(p, n)$_____   b. $^{197}_{79}Au(n, \gamma)$_____
    c. $^{14}_{7}N(n, p)$_____

18. Write nuclear reactions for α decay from:
    a. Bismuth-210   b. Radon-222   c. Thallium-194
    d. Thorium-232

19. Write nuclear reactions for β decay of:
    a. Indium-115   b. Strontium-90   c. Sulfur-35
    d. Cobalt-60

20. Write the nuclear symbol for each of the stable isotopes of:
    a. O   b. Sn   c. Cd
    Refer to Figure 22.4 for the necessary information.

21. Using Figure 22.4 as a source of information, write the nuclear symbol for each stable isotope of:
    a. S   b. Sr   c. F

22. Which stable isotopes contain:
    a. 80 neutrons?   b. 82 neutrons?   c. 84 neutrons?
    See Figure 22.4.

23. Write the nuclear symbol for each stable isotope with:
    a. 18 neutrons   b. 20 neutrons   c. 22 neutrons

---

24. Calculate the total binding energy and the binding energy per nucleon for:
    a. $^{3}_{2}He$ (mass 3.01493 amu)
    b. $^{99}_{43}Tc$ (mass = 98.88264 amu)
    c. $^{254}_{98}Cf$ (mass = 254.03359 amu)

25. Calculate the binding energy per nucleon for:
    a. $^{16}_{8}O$ (mass = 15.99052 amu)
    b. $^{35}_{17}Cl$ (mass = 34.95952 amu)

26. Calculate the energy change in J/mol for the following nuclear reaction.

    $$2\,^{2}_{1}H \longrightarrow \,^{3}_{2}He + \,^{1}_{0}n$$

    (Mass of $^{3}_{2}He$ = 3.01493 amu.)

27. What is the energy change in J/mol for the following process?

    $$4\,^{1}_{1}H \longrightarrow \,^{4}_{2}He + 2\,^{0}_{+1}\beta$$

28. Carbon-14 dating techniques are only useful for periods of time back to 50,000 yr or roughly ten half-lives of the radioactive isotope. What fraction of the original amount of carbon-14 would remain after ten half-lives?

29. The half-life of plutonium-239 is 2.4×10^4 yr. How much plutonium would remain from a 100-kg sample after 2000 yr?

30. Estimate the age of a rock that contains 0.265 g of Pb-206 and 1.650 g of U-238. Assume that all the Pb-206 came from the decay of the U-238.

31. What is the age of a rock whose argon-40 to potassium-40 ratio is 5.2?

32. A charcoal fragment from the Lascaux Caves in France has a carbon-14 to carbon-12 ratio of 2.0 disintegrations per minute per gram of carbon. How old is the charcoal?

33. A sample of wrapping cloth from an Egyptian mummy gives 9.5 disintegrations/min-g C. What is the age of the cloth?

EXERCISES

34. Prepare a graph of the log of radioactivity versus time from the following data. Use the graph to determine the half-life of the isotope (P-32).

Time (hr)	Counts per minute	Time (hr)	Counts per minute
0	950	8	605
2	850	10	540
4	750	15	410
6	675	20	305

35. Radioactivity data for a sample of Rh-98 is shown below. Use it to prepare a graph of the log of radioactivity versus time and determine the half-life of Rh-98.

Time (min)	Counts per minute	Time (min)	Counts per minute
0	1325	6	778
1	1232	7	710
2	1105	8	670
3	1006	9	599
4	950	10	563
5	869		

36. The size of a nucleus can be approximated by the equation $r = 1.3 \times 10^{-15} A^{1/3}$, where r is the nuclear radius in meters and A is the mass number. Calculate the nuclear radii of:
    a. $^{2}_{1}H$ **b.** $^{4}_{2}He$ **c.** $^{56}_{26}Fe$ **d.** $^{238}_{92}U$

37. Approximately one molecule in every 10^{12} molecules of atmospheric CO_2 contains C-14. How many C-14 atoms are there in a 180-pound person whose body is 20% carbon by weight?

38. The sun gives off energy at the rate of approximately 10^{17} J/sec. Calculate the sun's mass loss during a 24-hr day.

39. Calculate the mass loss and the energy released in the α decay of plutonium-239 (mass = 239.05216 amu). The mass of uranium-235 is 235.04393 amu.

40. Find the energy equivalent of the mass of:
    a. A neutron   **b.** A proton

41. Suggest a feasible nuclear reaction pathway for the conversion of a lead isotope into gold.

42. In order to obtain the half-life of a substance, the radioactivity of a fresh sample was measured to be 4260 disintegrations per second (dps). After 5.0 days, the activity had fallen to 540 dps. What is the half-life of the isotope?

43. Compute the total amount of energy released in the decay of a U-238 atom to a Pb-206 atom via the sequence outlined in Figure 22.9.

44. An antiproton (a particle with the same mass as a proton but a negative charge) collides with a proton, annihilating both particles and producing a pair of photons. What is the frequency of the photons?

45. Rhodium-98 decays by positron emission to ruthenium-98. Technetium-98 decays to ruthenium-98 by beta decay. Both processes are accompanied by the emission of a γ ray with the same frequency for each process. Explain this in terms of the shell model for the nucleus.

chapter 23

Biochemistry

Ribose, a five-carbon sugar, is an important ingredient of nucleic acids

Some biochemists believe we will eventually explain all the complex workings of living organisms in terms of the properties of their molecules. This view strikes other observers* as overconfident; they feel the variety and complexity of living systems and the evolutionary process transcend the laws of chemistry and physics. But even if chemical principles should prove inadequate to explain life processes, life does require very complicated chemical systems. Describing these systems is the job of biochemists.

Nineteenth-century chemists believed that inexplicable "vital forces" operated in living systems, and biochemical processes could never occur in the absence of a live organism. They discarded this theory when they accidentally discovered a biochemical transformation (the fermentation of sugar) taking place in a system free of living cells. Rapid progress followed this initial discovery, and biochemists can now produce test-tube versions of many of the reactions a cell uses to derive energy, synthesize the substances it needs, and reproduce itself.

It would be wrong to pay so much attention to reactions in cell-free systems that we ignore one of the most important features of a cell—its highly organized structure (Figure 23.1). A cell is a complex entity containing several types of internal structures, each involved with a different aspect of the cell's chemistry. Oxidation reactions necessary for metabolism occur in the mitochondria and protein synthesis takes place in the ribosomes. The nucleus contains the material (deoxyribonucleic acid—DNA) that stores genetic information. The membranes separating the parts of a cell

---

*One of the most articulate proponents of the view that life is more than a chemical process is Michael Polanyi. See his article, "Life's Irreducible Structure," in *Science* **160**, 1309–1312(1968).

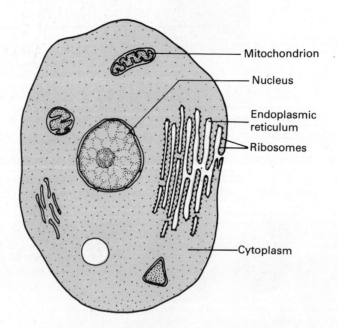

Figure 23.1
A cell.

have highly organized structures and specific biochemical functions. Fully aware of the importance of internal structures and membranes, biochemists are trying to understand their behavior by studying their molecular structures.

We will discuss the biochemistry of a "cell" or "organism," vague designations that might offend readers familiar with the complexity of the biological world. Cells are not all alike; bacteria cells are more primitive than plant or animal cells. The photosynthetic cells of green plants contain chloroplasts (Section 23.7), structures not present in other types of cells. A highly evolved organism contains many different sorts of specialized cells. Ignoring this diversity of cell structure is an oversimplification required by the introductory nature of this chapter.

All cells, regardless of type, resemble each other in fundamental ways. The same classes of substances—proteins, carbohydrates, lipids, and nucleic acids—show up in all cells and serve similar biochemical functions. This underlying unity allows us considerable leeway in generalizing about the molecular nature of living organisms.

23.1 Carbohydrates and Lipids

Carbohydrates (Figure 23.2), commonly known as sugars and starches, are important sources of food energy. Another type of carbohydrate, cellulose, gives many plant tissues their tough, fibrous properties. Carbohydrates fall into two main categories—**monosaccharides** (simple sugars) and **polysaccharides**. A monosaccharide is either an aldehyde (such as glucose) or a ketone (such as fructose) containing several alcohol (−OH) groups (usually one for each carbon atom except for the aldehyde or ketone group). Most monosaccharides have a strong tendency to form cyclic structures, and these cyclic molecules are usually dominant in monosaccharide samples. A polysaccharide is a polymer built up from cyclic monosaccharide units. Sucrose (table sugar) is a dimer containing one glucose unit and one fructose unit. Starch and cellulose are both large polymers of glucose, but they differ in the three-dimensional arrangement of the bonds holding the monosaccharide units together. This subtle structural difference is the reason why cellulose is indigestible but starch is an important nutrient.

Lipids (Figure 23.3) are a class of water-insoluble compounds involved in biochemical systems. The most common lipids are **fats**—molecules containing three ester bonds between the three alcohol groups of glycerol and three straight-chain carboxylic acids (called fatty acids). The main biochemical function of fats is food

23.1 CARBOHYDRATES AND LIPIDS

MONOSACCHARIDES

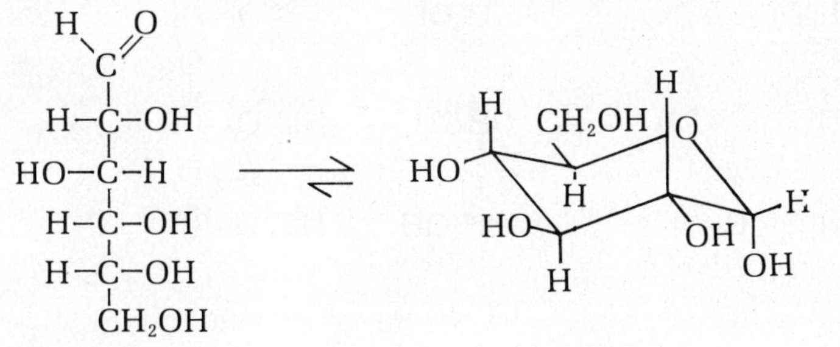

D—Glucose ⇌ A cyclic form

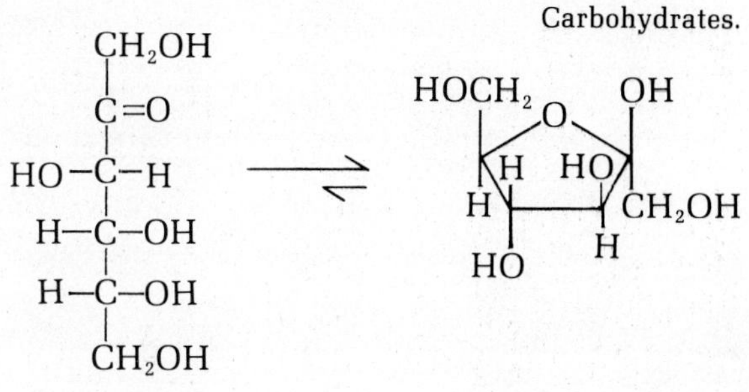

D—Fructose ⇌ A cyclic form

POLYSACCHARIDES

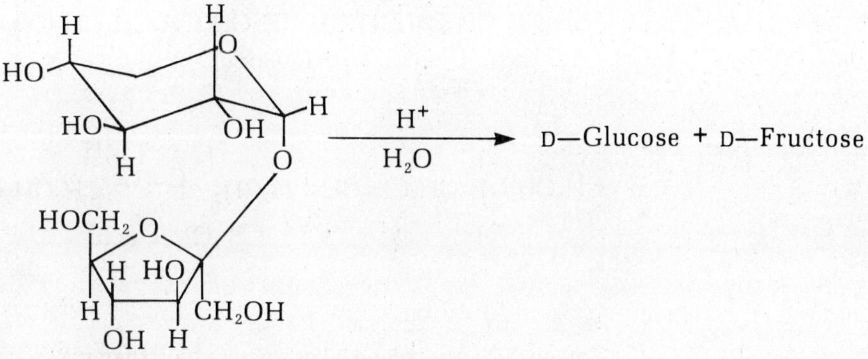

Sucrose $\xrightarrow{H^+, H_2O}$ D—Glucose + D—Fructose

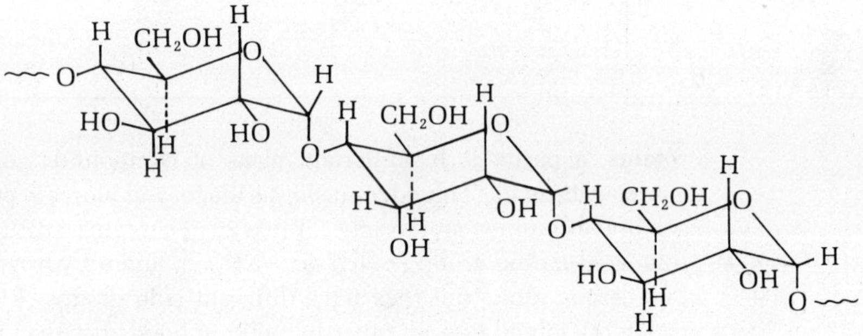

A portion of a starch molecule

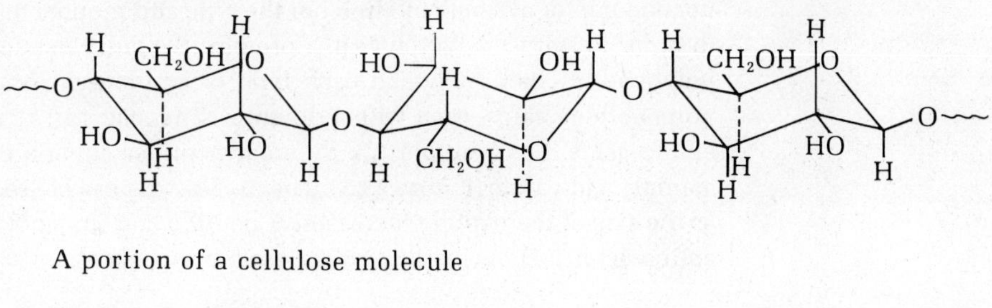

A portion of a cellulose molecule

Figure 23.2
Carbohydrates.

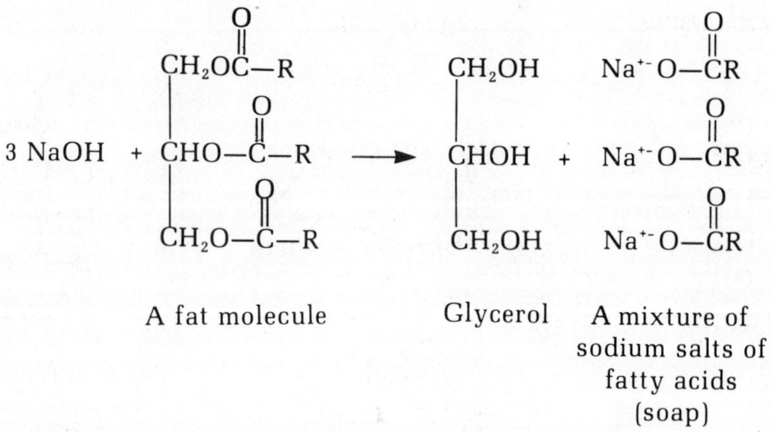

Figure 23.3
Lipids.

$$3\ NaOH + \begin{matrix} CH_2O\overset{O}{\overset{\|}{C}}-R \\ CHO-\overset{O}{\overset{\|}{C}}-R \\ CH_2O-\overset{O}{\overset{\|}{C}}-R \end{matrix} \longrightarrow \begin{matrix} CH_2OH \\ CHOH \\ CH_2OH \end{matrix} + \begin{matrix} Na^+\ ^-O-\overset{O}{\overset{\|}{C}}R \\ Na^+\ ^-O-\overset{O}{\overset{\|}{C}}R \\ Na^+\ ^-O-\overset{O}{\overset{\|}{C}}R \end{matrix}$$

A fat molecule · · · · · · Glycerol · · A mixture of sodium salts of fatty acids (soap)

Typical fatty acids

$$CH_3CH_2CH_2CH_2CH_2CH_2CH_2CH_2CH_2CH_2CH_2CO_2H$$
Lauric acid

$$CH_3CH_2CH_2CH_2CH_2CH_2CH_2CH_2\underset{}{\overset{H}{\diagdown}}C=C\underset{CH_2CH_2CH_2CH_2CH_2CH_2CH_2CO_2H}{\overset{H}{\diagup}}$$
Oleic acid

storage. When someone takes in more food than needed, some of the excess nutrients are converted to fats and stored in the body. If, later, the food intake drops below the person's immediate needs, the fats will be metabolized to supply the needed energy. Lipids, including fats, are important constituents of cell membranes and, as such, can play subtle and significant biochemical roles.

23.2 Proteins

Proteins, a particularly important class of compounds comprising enzymes and major constituents of blood, muscle, cartilage, and hair, are polymers of **amino acids**. Each of the common amino acids (Table 23.1) contains a carboxylate group ($-CO_2^-$), a protonated amino group ($-NH_3^+$ or $-NH_2^+-$), and a hydrogen atom attached to the same carbon atom, but they have different side groups (R). The amino acids in a protein are joined to each other by peptide bonds between the amino group of one amino acid and the carboxylate group of its neighbor in the **polypeptide chain**.

The polypeptide chains of a protein have a very specific sequential arrangement of amino acids. Figure 23.4 shows the sequence of amino acids—also called the **primary structure**—of insulin. Because of the specific nature of this sequence, it is not enough for a chemist to find out the type and number of each amino acid present in order to learn the structure of a protein. We must also locate the positions in the polypeptide chain at which each type of amino acid occurs. A large number of polypeptide chains, each with a different sequence, can be made from a given set of amino acids. For example, six different peptides contain one unit each of glycine, alanine, and valine (Figure 23.5). The number of possible sequences expands rapidly as the size of the peptide increases. A peptide consisting of one each of ten different amino acids has one of 3,628,800 possible sequences.

23.2 PROTEINS

Table 23.1 Amino Acids

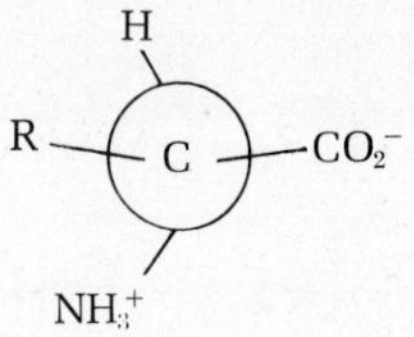

$$\underset{NH_3^+}{\overset{H}{R-C-CO_2^-}}$$

Name	Abbreviation	R (at pH 7)
Glycine	Gly	H—
Alanine	Ala	CH_3-
Valine	Val	$(CH_3)_2CH-$
Leucine	Leu	$(CH_3)_2CHCH_2-$
Isoleucine	Ile	CH_3CH_2CH- $\quad\quad\quad\ \ \|$ $\quad\quad\quad CH_3$
Phenylalanine	Phe	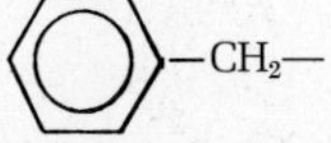 $C_6H_5-CH_2-$
Tryptophane	Try	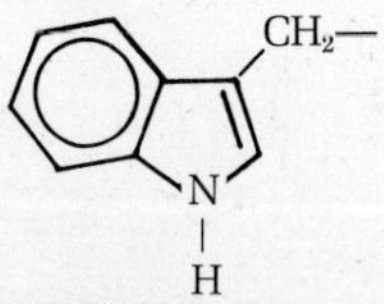 (indole)$-CH_2-$
Methionine	Met	$CH_3SCH_2CH_2-$
Serine	Ser	$HOCH_2-$
Threonine	Thr	CH_3CH- $\quad\ \ \|$ $\quad\ OH$
Cysteine	CySH	$HSCH_2-$
Tyrosine	Tyr	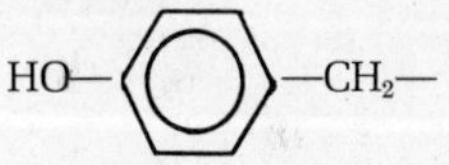 $HO-C_6H_4-CH_2-$
Aspartic acid	Asp	$^-O_2CCH_2-$
Asparagine	Asn	$NH_2\overset{O}{\overset{\|}{C}}CH_2-$
Glutamic acid	Glu	$^-O_2CCH_2CH_2-$
Glutamine	Gln	$NH_2\overset{O}{\overset{\|}{C}}CH_2CH_2-$
Lysine	Lys	$^+NH_3CH_2CH_2CH_2CH_2-$
Arginine	Arg	$NH_2\overset{^+NH_2}{\overset{\|}{C}}NHCH_2CH_2CH_2-$
Histidine	His	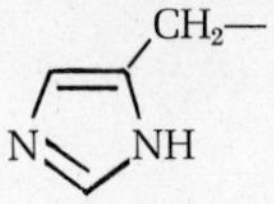 (imidazole)$-CH_2-$
Proline	Pro	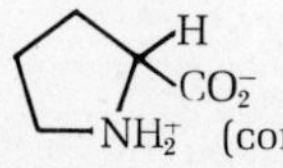 (pyrrolidine structure) (complete structure)

Example 23.1 Write all the possible polypeptide sequences for a peptide containing two alanine residues, one serine, and one aspartic acid. (Use the three letter abbreviations for the amino acids.)

Figure 23.4
(a) Disulfide bridge. (b) Primary structure of insulin.

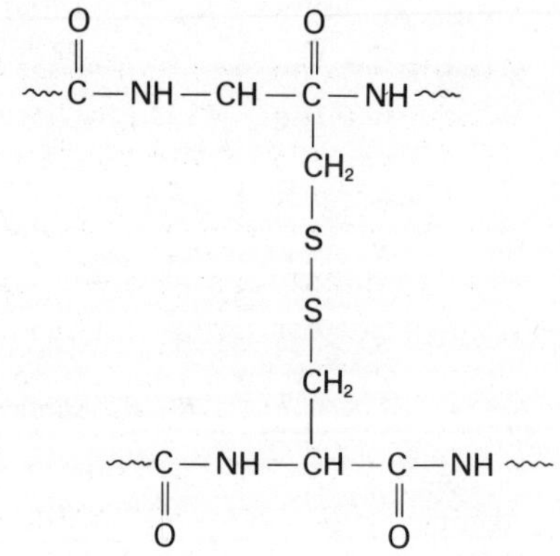

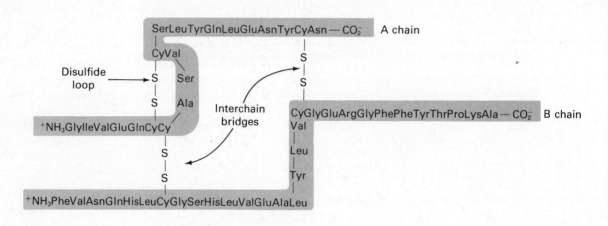

(a) Disulfide bridge

(b) Primary structure of insulin

Figure 23.5
Peptide sequences.

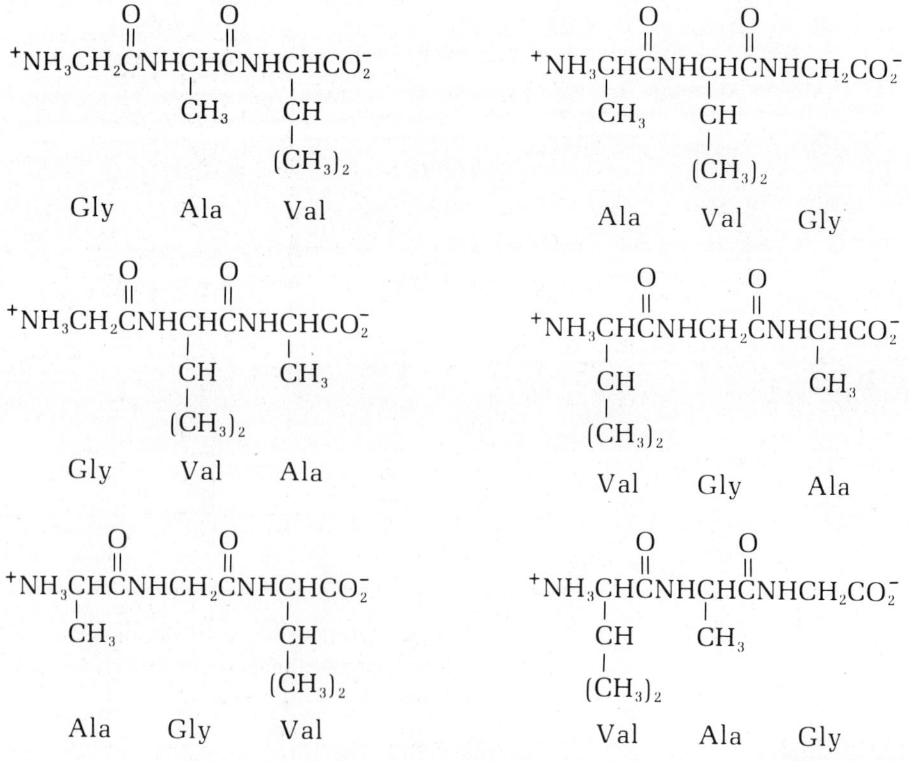

Solution

Ala.Ala.Ser.Asp	Ala.Ala.Asp.Ser	Ala.Ser.Ala.Asp	Ala.Asp.Ala.Ser
Ala.Ser.Asp.Ala	Ala.Asp.Ser.Ala	Ser.Ala.Ala.Asp	Asp.Ala.Ala.Ser
Ser.Ala.Asp.Ala	Asp.Ala.Ser.Ala	Ser.Asp.Ala.Ala	Asp.Ser.Ala.Ala

A polypeptide chain is quite flexible and can take on many different shapes. Each polypeptide chain in a protein twists into a very specific shape, and changes in this shape can destroy its biological activity. A variety of forces—hydrogen bonding, electrostatic attraction, and the tendency of nonpolar groups to cluster together—stabilize the native conformation of a protein. Two types of shapes that occur frequently in proteins are the **alpha helix** (Figure 23.6), a spiral arrangement of the

Figure 23.6
Secondary structure. (a) Alpha helix. (b) Beta structure.

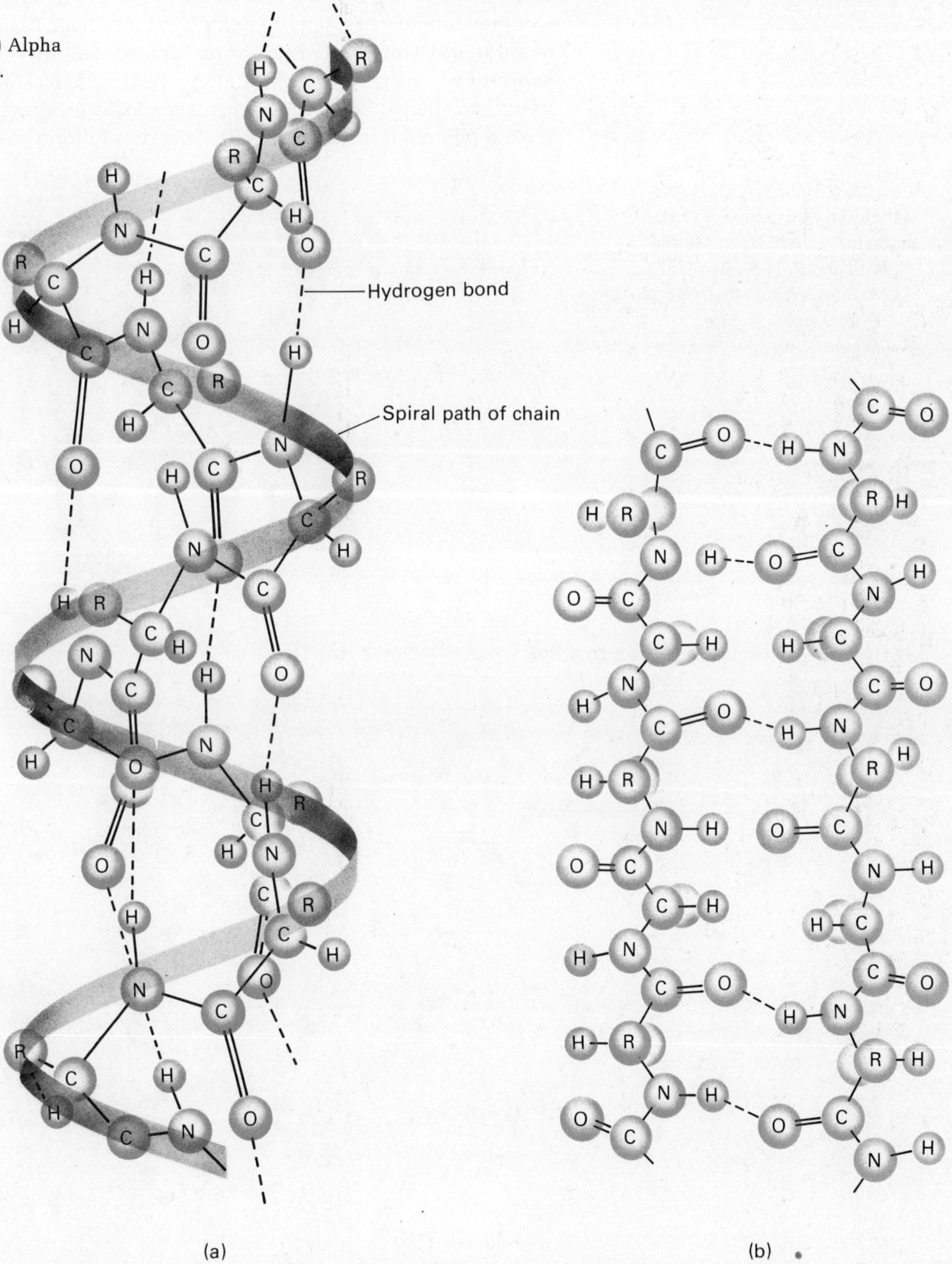

(a)   (b)

peptide chain, and the **beta structure**, an arrangement in which two peptide chains (or two portions of a single-folded chain) lie side by side. The portions of a protein that cannot be described as either alpha helix or beta structure are called **random coils**. A typical protein molecule may contain portions with each of the three different classes of three-dimensional arrangements, and the relative amounts of these three structures constitute the **secondary structure** of a protein.

There is more to the three-dimensional shape of a protein than its secondary structure. The random coil portions of a polypeptide chain and weak points in its alpha helices and beta structures give the chain considerable flexibility. Attractive forces fold the polypeptide chains into characteristic convoluted shapes (the **tertiary structure**—Figure 23.7). Some proteins are aggregates of several polypeptides, and these chains lie in very precise positions and orientations. The spatial arrangement of these polypeptides relative to each other constitute the **quaternary structure**.

Protein molecules tend to unfold and lose their characteristic shapes when exposed to heat, extremes of pH, or nonaqueous solvents. These reactions, called **denaturations**, change the physical properties of proteins and destroy their biological usefulness. Denaturation occurs, for example, when an egg is boiled. A water-soluble protein (albumin) in the egg white denatures to form a water-insoluble amorphous

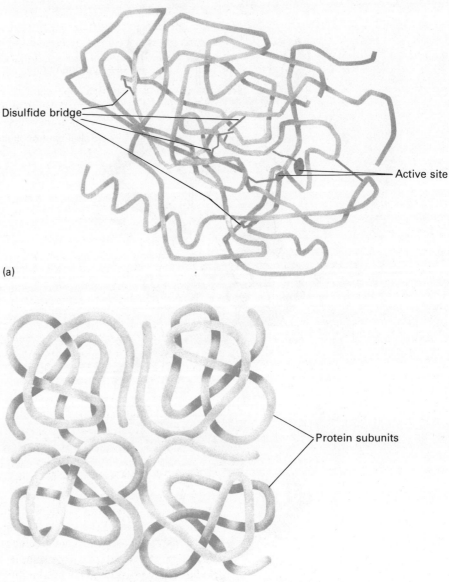

Figure 23.7
(a) Tertiary structure of chymotrypsin. *Source*: B. W. Matthews, P. B. Sigler, R. Henderson, and D. M. Blow, *Nature* **214**, 652(1967).
(b) Quaternary structure.

solid and although the boiled egg white may help meet our nutritional needs, it can no longer perform its primary function of nurturing a chick embryo. Boiling an egg for 3 min is too mild a treatment to break covalent bonds—it is the loss of the secondary and tertiary structures of the protein that cause these profound changes in properties.

23.3 Enzymes

Biochemical systems are far more complex than the most sophisticated industrial chemical plants. A living cell is the site of thousands of different reactions occurring simultaneously under essentially the same mild conditions of pH (nearly neutral) and temperature (about 300 K). The cell's chemical system is remarkable for its precision and efficiency. Every reaction taking place within the cell is necessary for its well-being, and the cell does not waste any energy on useless reactions. Each biochemical reaction runs at the exact speed required for proper cell function.

The amazing degree of kinetic control in biochemical systems is due to the presence of very effective and specific catalysts called **enzymes**. Each biochemical reaction requires a particular enzyme to maintain its speed. The *specificity* of enzymes surpasses that of other catalysts. Most enzymes catalyze one specific reaction of one substance. (Some catalyze the same reaction for several closely related substances.) An ordinary catalyst accelerates similar reactions for a large group of compounds containing the same functional group and often speeds up undesirable reactions as well as the desired one.

An example of the specificity of an enzyme is the catalysis of the hydration of fumarate ion by the enzyme fumarate hydratase. (The name of an enzyme usually describes what the enzyme does and ends in the suffix *-ase*.)

$$\text{Fumarate} + H_2O \xrightleftharpoons[]{\text{Fumarate hydratase}} \text{L-Malate}$$

This is the only reaction catalyzed by this enzyme; other organic molecules containing double bonds are inert in its presence. A further feature of its specificity is the fact that the enzyme-catalyzed reaction produces only one of the two possible stereoisomers of malate (a phenomenon known as *stereospecificity*). Hydrogen ion catalyzes the hydration of fumarate, but the reaction is not stereospecific (both stereoisomers form), and H^+ also hydrates almost every other substance with a carbon–carbon double bond.

Another aspect of the subtlety of enzyme action is that certain enzymes called **regulatory enzymes** change their catalytic properties in response to the cell's needs. A reaction catalyzed by a regulatory enzyme is part of a sequence of reactions forming an essential substance that modulates the activity of the regulatory enzyme. The enzyme, active when the cell is deficient in the essential substance, loses its activity when the substance is present in sufficient amounts. This feedback control (Figure 23.8) keeps the cell from wasting energy by synthesizing more material than it needs.

All enzymes are proteins. Some are simple proteins composed solely of polypeptide chains. Others require a nonprotein molecule or ion, called a **cofactor**, for their activity. Some cofactors are complicated organic molecules called **coenzymes** (Figure 23.9), others are metal ions. For example, every molecule of the peptide-hydrolyzing enzyme carboxypeptidase contains a Zn^{2+} ion bound at a specific site in the protein. Removal of this cofactor produces a catalytically inactive protein.

Regulatory
enzyme
Inhibition
Proline

Figure 23.8
Feedback inhibition of a regulatory enzyme. The final product of the biosynthetic sequence, proline, turns off the regulatory enzyme needed for the first step. Therefore, the synthesis shuts down whenever the cell has enough proline.

Pyridoxal phosphate
Lipoic acid
Thiamine pyrophosphate

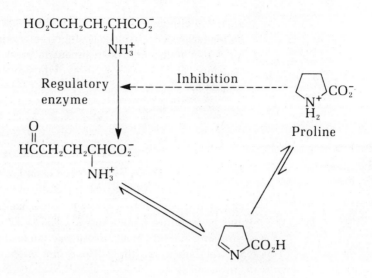

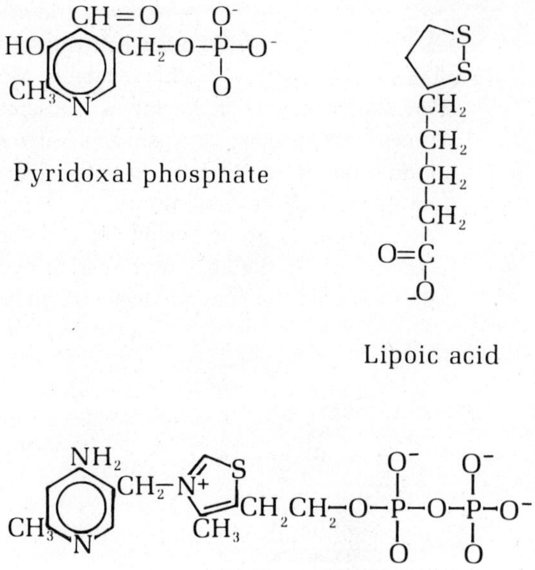

Figure 23.9
Coenzymes.

The first step in an enzyme-catalyzed reaction is the binding of a specific reactant molecule, called the **substrate** (S), to the enzyme (E) to form an **enzyme–substrate complex** (ES). This complex reacts rapidly to form products (P) that diffuse rapidly from the enzyme, leaving it free to deal with other substrate molecules.

$$E + S \rightleftharpoons ES \longrightarrow E + P$$

The rate of the reaction depends partly on how well the enzyme binds the substrate, and partly on how rapidly the enzyme–substrate complex reacts.

Substrate molecules interact with a small portion of the enzyme known as its **active site**. The size, shape, and polarity of an active site and the spatial arrangement of the catalytic groups are responsible for the activity and specificity of an enzyme. Substrate molecules fit into the active site, and their reactive bonds lie near the catalytic groups of the active site. Other molecules either cannot bind to the active site or, if they can bind, do so in a way that precludes catalysis.

An active site consists either of portions of different polypeptide chains or several separate segments of the same chain. The active site of an enzyme that requires a cofactor contains that cofactor. The protein in the absence of the cofactor is inactive because part of its active site is missing. The polypeptide segments (and cofactor, if needed) of an active site are close together in a native enzyme but move apart if the secondary and tertiary structure of the protein unfolds. This disruption of the structure of an active site due to denaturation destroys the catalytic activity.

The groups responsible for the catalytic abilities of an active site are the side groups of certain amino acids and portions of the cofactor (if present). We believe the secret of an enzyme's efficiency and specificity lies in the delicate organization of the catalytic groups in its active site. Each group is in the precise position required for maximum catalytic efficiency, and the juxtaposition of several catalytic groups enhances the effectiveness of all of them. For example, the active site of chymotrypsin (see Figure 23.7) contains a histidine and a serine side group arranged in space so that they can cooperate in catalyzing the substrate's reaction.

23.4 Nucleic Acids

An essential feature of life is an organism's ability to pass genetic characteristics to its offspring. When a cell divides, the parent cell gives its progeny coded information concerning the proteins it must synthesize to maintain its identity. This code resides in molecules of **deoxyribonucleic acids (DNA)**, substances that can **replicate**, that is, direct the synthesis of identical offspring molecules, and carry programs for synthesizing specific polypeptides.

Molecules of DNA contain very long **polynucleotide chains** (Figure 23.10) whose backbones consist of deoxyribose molecules linked to each other by phosphate groups. Each deoxyribose unit carries a nitrogen-containing **base group**. The most frequently occurring base groups in DNA are thymine, cytosine, adenine, and guanine (abbreviated T, C, A, and G, respectively). The specific sequence of base groups in a DNA chain carries its genetic code.

The key to the replicating ability of DNA lies in its three-dimensional structure (Figure 23.11). Two intertwining chains form the **double helix** structure characteristic of native DNA. Hydrogen bonds between pairs of base groups, one from each chain, hold the two chains together. The base groups point inward, perpendicular to the axis of the helix, and form a stack of parallel base pairs. Thymine pairs with adenine and cytosine with guanine. No other pairing combinations are possible because of the distance between the chains and the necessity for hydrogen bonding between the base groups. A pair of large base groups (A or G) cannot fit in the available space, and a pair of small base groups (T or C) would be too far apart to form hydrogen bonds. Therefore, the base pair must include one large and one small base group. Hydrogen bonds connect oxygen atoms on one sort of base group with amino groups on the pair partners. For example, the oxygen of thymine can form a hydrogen bond with the amino group of adenine but cannot bind to the oxygen in guanine.

These strict base-pairing requirements mean that the two intertwined chains must have complementary sequences; the sequence of one chain dictates the sequence of the other chain. For example, if a particular portion of one chain of a double helix has the sequence TCAG, the sequence of the other chain must be AGTC.*

Example 23.2

What base sequence is complementary to the sequence GGACT?

Solution

Since C binds with G, and A with T, the complementary sequence is CCTGA.

*Since two intertwined DNA strands run in opposite directions, the complementary sequences must be read in opposite directions.

Figure 23.10
Deoxyribonucleic acid (DNA).

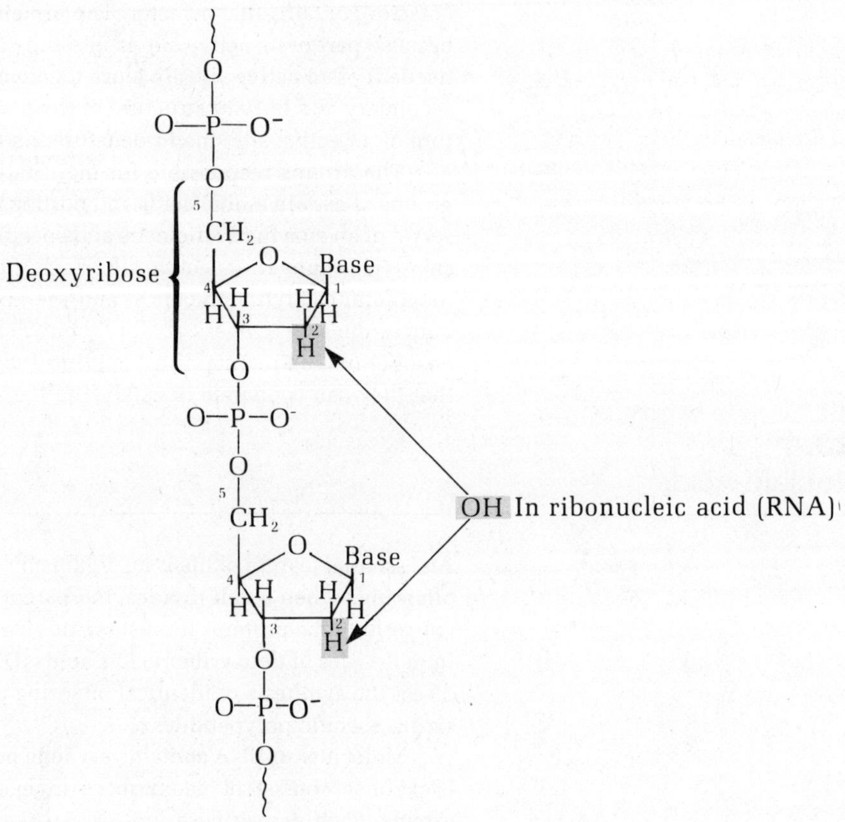
Deoxyribose
Base
OH In ribonucleic acid (RNA)

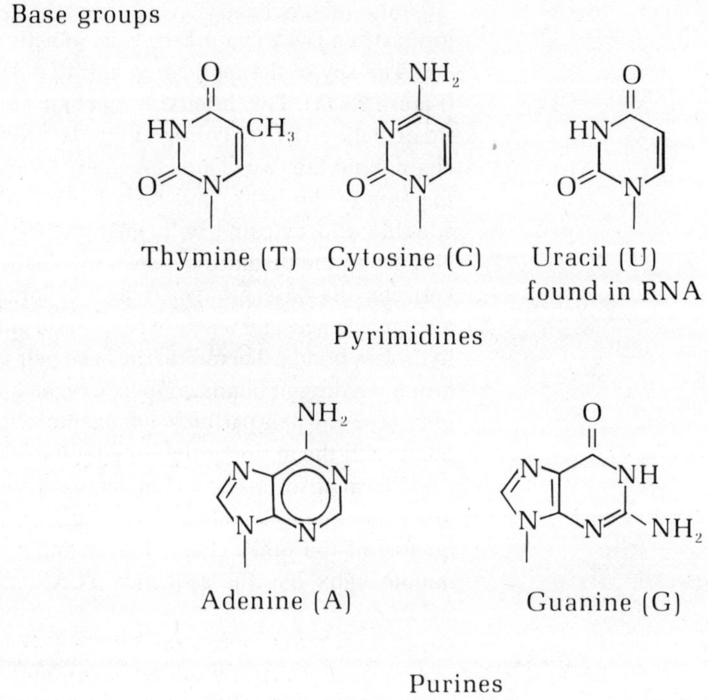
Base groups
Thymine (T)
Cytosine (C)
Uracil (U)
found in RNA
Pyrimidines
Adenine (A)
Guanine (G)
Purines

Watson and Crick, who proposed the double-helix structure of DNA in 1953, showed how such a structure could lead to genetic replication. According to their hypothesis (Figure 23.12), replication begins with the unwinding of the double helix. Each unwound chain serves as a template for synthesizing a new complementary chain.

Figure 23.11
(a) The DNA double helix. (b) Hydrogen bonding between the base pairs, thymine-adenine and cytosine-guanine, holds the two chains together. *Source:* Adapted from P. C. Hanawalt and R. A. Haynes, "The Repair of DNA," *Scientific American* **216**, 38 (Feb. 1967).

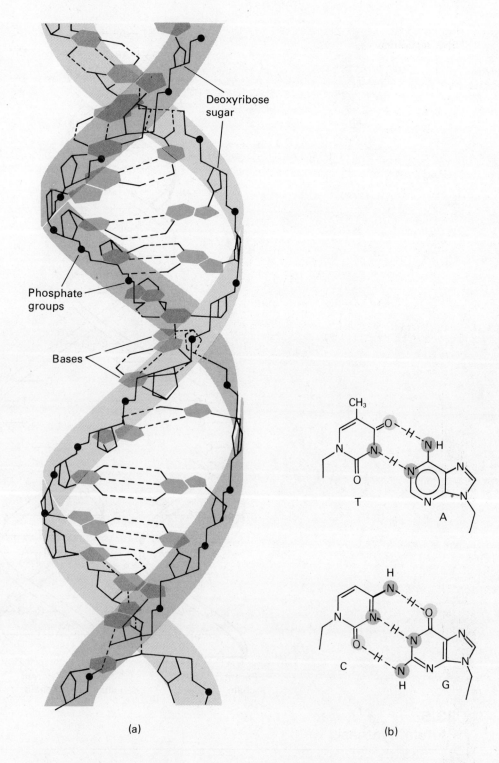

(a)   (b)

Two double-helix molecules, identical to each other and to their parent, form in the process. Each offspring molecule contains one chain from its parent and one newly synthesized chain. The Watson–Crick model is to molecular genetics what Dalton's theory was to chemistry. It provides us with a useful working hypothesis on which to base genetic research, and subsequent experiments have confirmed the hypothesis and given us new insights into the nature of replication. For example, DNA can be synthesized in a test tube provided some single stranded DNA is present to act as a template, and the chain sequence of the newly synthesized DNA is complementary to the added DNA.

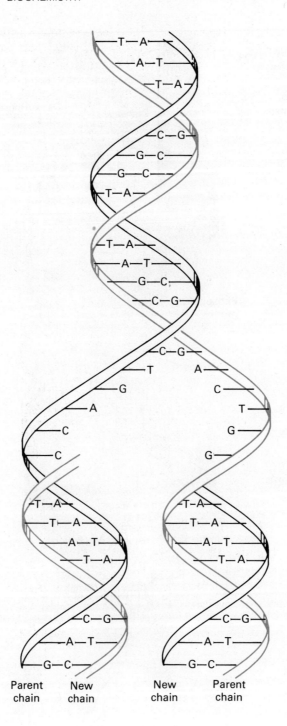

Figure 23.12
DNA replication.

23.5
Protein Synthesis

The DNA in the genes of a cell provides the information the cell needs to synthesize its polypeptides, the base sequences of the DNA dictating the polypeptide sequences. The DNA acts as a template for synthesizing complementary chains of **ribonucleic acid (RNA)**, called **messenger RNA (mRNA)**, which then migrate to the ribosomes and direct protein synthesis. This indirect method for using genetic information prevents exposing DNA to conditions that might damage it and cause genetic mutations, and it also allows a single DNA molecule to send messages to many ribosomes.

Ribonucleic acid chains are similar to DNA chains, the only differences being (1) the presence of a hydroxyl group at carbon number 2 of each ribose unit and (2) the

23.5 PROTEIN SYNTHESIS

replacement of thymine base groups with uracil (U) bases. The structures of uracil and thymine are very similar and both pair with adenine.

Synthesis of proteins in ribosomes is a complex process (Figure 23.13) following the principle of complementary nucleotides. For each amino acid, the cell contains a specific **transfer RNA (tRNA)** and a specific enzyme for joining that amino acid to the end of its tRNA molecule. A tRNA molecule, much smaller than mRNA, coils back on itself to form a double helix structure from a single chain. Some of the base groups in tRNA cannot participate in the double helix, and one segment of three nucleotide units (the **anticodon**) is free to bind to a complementary three-unit sequence (the **codon**) of the mRNA chain. A tRNA molecule bearing the polypeptide's N-terminal amino acid (AA_1) binds the first codon of the mRNA sequence, the tRNA molecule with the next amino acid (AA_2) hooks up to the adjacent codon, and an enzyme forms a peptide bond between the two amino acids. This process continues until the polypeptide chain is complete.

Since each mRNA codon binds a specific amino acid, the base sequence of the mRNA molecule determines the sequence of the polypeptide chain. Three of the 64 (4^3) possible codons have no corresponding tRNA anticodons and serve as periods in

Figure 23.13
Protein synthesis.

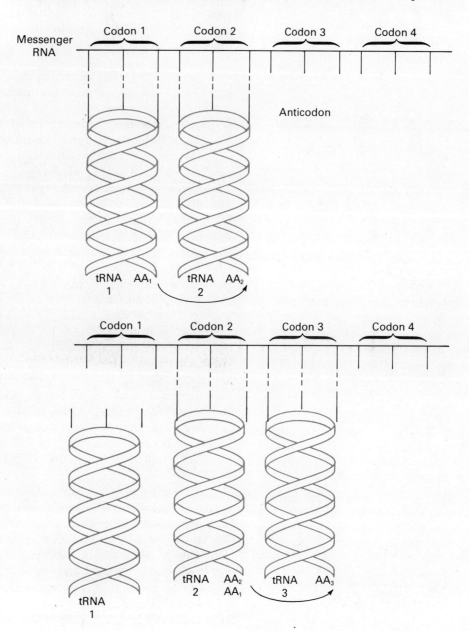

the genetic message—they signal when polypeptide synthesis is complete for a chain. Since there are 61 remaining codons and only 20 amino acids, there must be several codons for a particular amino acid. For example, the codons UUC and UUU code for the insertion of phenylalanine.

Biochemists now know the complete genetic code and could, in principle, add a RNA molecule with a known sequence to a ribosomal system to produce a particular protein. But because it is very difficult to synthesize a particular RNA molecule, we cannot compete with living systems when it comes to directing protein synthesis—not now, in any case.

In 1763 Edward Stone, a rural English clergyman, reported the effectiveness of extracts of willow bark in curing "agues" (malarial fevers). He had investigated the fever-reducing properties of willow bark for two reasons. First, like quinine, it tastes bitter. Second, willows grow in the swampy regions where malaria was prevalent, and Stone believed diseases and their cures resided in the same environment. Although his reasoning was naive, there is no disputing Stone's discovery—willow bark is effective in controlling fevers.

Chemists improved on Stone's discovery by isolating the bark's active ingredient—salicylic acid. An inexpensive synthesis of salicylic acid allowed the wide-spread use of this medication, which also proved useful for relieving migraine headaches and arthritic pains. The only problem was its noxious taste—many patients preferred their aches.

The chemists' answer to this problem was to prepare aspirin, a derivative of salicylic acid. The idea was that the acid in the patient's stomach would decompose aspirin to salicylic acid, allowing the patient to have his salicylic acid without having to eat it. But it turns out a large portion of the ingested aspirin reaches the blood stream intact and is an even better pain-reliever than salicylic acid. Once again, an effective medication was discovered for the wrong reason.

No one knows how aspirin works. We can describe its beneficial effects and the harm done by taking too much, but although we have learned a few details about the chemical changes it causes in the body, we lack a coherent picture of its biochemical role. This fundamental ignorance is the rule in pharmaceutical chemistry—in no case do we know the chemical events associated with a medicine's action. (We do better with poisons—we have quite a complete picture of the biochemical action of cyanide.)

So things really haven't progressed that much since Stone's days. Pharmaceutical companies sometimes discover new products by "screening" vast numbers of compounds for their medicinal effects. The investigation of an old Asian folk remedy led to the discovery of the tranquilizer reserpine. Sometimes a drug is discovered by following a hunch or by sheer accident. (Careless handling of a pipet led to the discovery of the hallucinatory effect of LSD.)

Until we achieve the necessary biochemical knowledge, we will never be sure of the effects of a new pharmaceutical agent. Perhaps one day we will understand enough about enzymes, membranes, and biochemical mechanisms to predict what sort of molecular structure will produce a desired physiological effect, but until then we must rely on scrupulous clinical testing to determine the usefulness and safety of each new drug.

Figure 23.14
Interconversion of adenosine triphosphate (ATP) and adenosine diphosphate (ADP).

ATP^{4-} + H$_2$O $\rightleftharpoons$ ADP^{3-} + HO–P(=O)(O$^-$)–O$^-$ + H$^+$

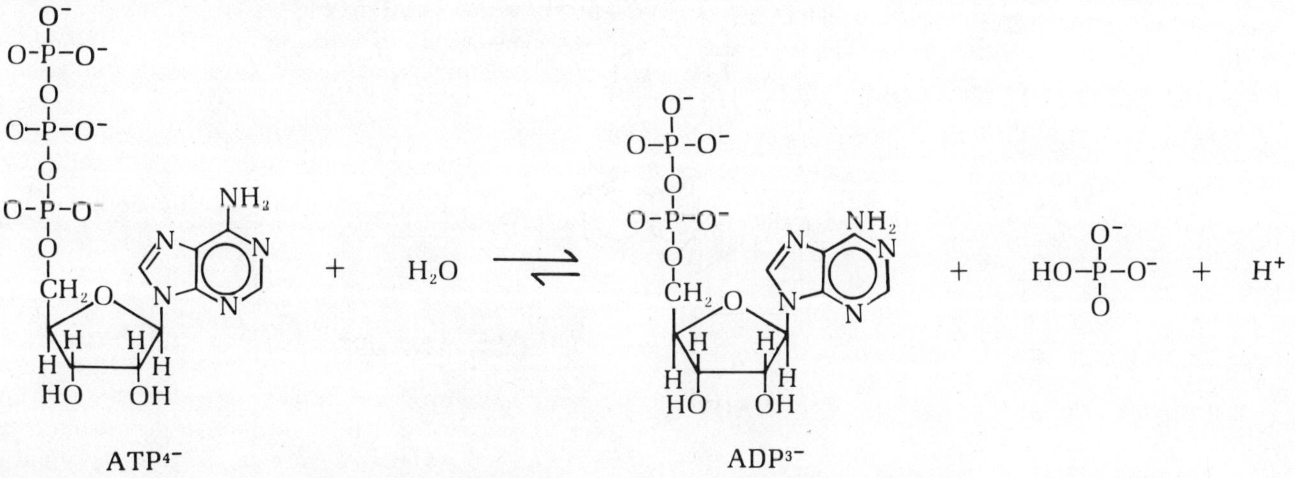

23.6
Bioenergetics

An organism is an efficient isothermal engine. Without requiring any flow of heat, it utilizes a high percentage of the free energy available from its nutrients to perform mechanical work (as in muscle contraction), synthesize molecules (biosynthetic work), and move molecules across membranes from dilute solutions to concentrated solutions (osmotic work). The organism's achievements put man-made engines to shame. No practical man-made machine runs on chemical energy without an intervening heat flow or electric current, and the only ways we know for making a molecule move against a concentration gradient require electrical or mechanical energy.

Adenosine triphosphate (ATP) is the key to the efficient use of free energy in biochemical systems. This anion (Figure 23.14), present in high concentrations in all cells, is decomposed by water to **adenosine diphosphate (ADP)** and inorganic phosphate. Because of the moderate free energy of the reaction of ATP with water (−31 kJ),* it is fairly easy to reverse this reaction and convert ADP to ATP. Many metabolic reactions yield the free energy required for synthesizing ATP from ADP and phosphate.

A large percentage of the free energy released by the metabolism of cell nutrients forms ATP instead of being given off as heat or increased entropy. In this way, a cell uses ATP as an intermediate storehouse of metabolic free energy. When the organism performs work, whether mechanical, biosynthetic, or osmotic, the conversion of ATP to ADP provides the needed energy.

Nutrients are reduced organic compounds, and their metabolism involves their oxidation. Living cells contain several specific biochemical oxidants (Figure 23.15): nicotinamide adenosine dinucleotide (NAD$^+$), nicotinamide adenosine dinucleotide phosphate (NADP$^+$), and the flavin nucleotides (FMN and FAD). These molecules engage in a variety of enzyme-catalyzed redox reactions and are reduced to NADH, NADPH, FMNH$_2$, and FADH$_2$, respectively. While NADPH carries out the reduction reactions necessary for biosynthesis, the reduced flavins and NADH give up their electrons to oxygen.

*Standard free energies for biochemical reactions refer to a standard state at pH 7.0. The concentrations of all reactants and products, other than H$^+$, are 1 M.

Figure 23.15
Biochemical oxidants.

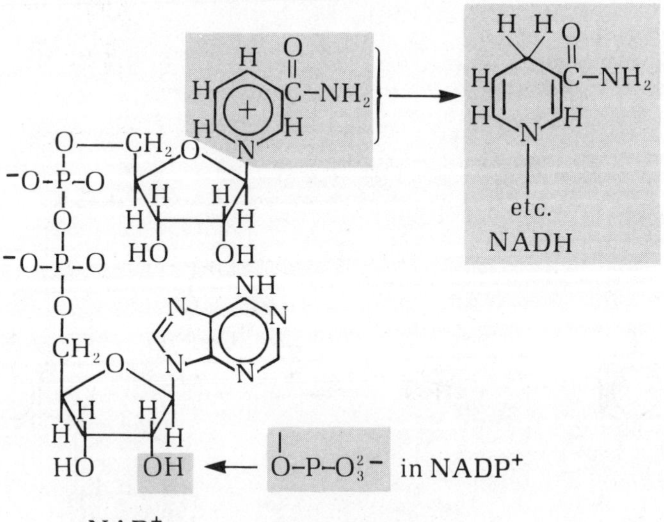

NAD+

Flavin nucleotides

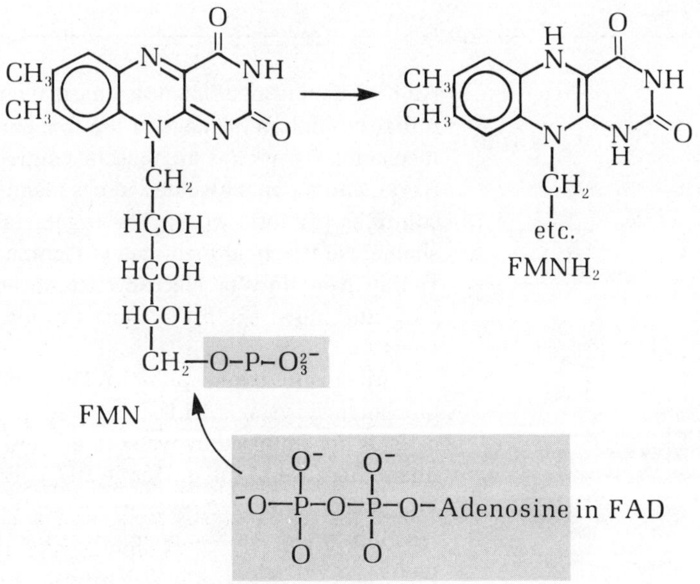

$$NADH + H^+ + \tfrac{1}{2}\,O_2 \longrightarrow H_2O + NAD^+ \qquad \Delta G^\circ = -220\text{ kJ}$$

$$FMNH_2 + \tfrac{1}{2}\,O_2 \longrightarrow H_2O + FMN \qquad \Delta G^\circ = -169\text{ kJ}$$

Reoxidation of the reduced nucleotides by oxygen occurs in the cell's mitochondria via a complex sequence of enzyme-catalyzed redox reactions called the **electron transport chain** (Figure 23.16). The electron transport chain involves a group of complicated molecules called **cytochromes**. These are proteins containing a heme group which, in turn, contains a transition-metal atom that can be in one of two oxidation states. The cytochromes are readily oxidized and reduced, and the electron transport chain proceeds as cytochromes take electrons from reduced nucleotides and then transfer these electrons to other cytochromes until a reduced cytochrome finally reduces O_2 to water.

There is an advantage in having the electrons transferred from the reduced nucleotides to oxygen in so many steps rather than in a single step. If NADH were

Figure 23.16
Electron transport chain.

$$NADH \longrightarrow NAD^+ + H^+ + 2e^- \longrightarrow \boxed{\text{Cytochromes}} \longrightarrow 2e^- + 2H^+ + \tfrac{1}{2}O_2 \longrightarrow H_2O$$

with ATP produced (and ADP + P consumed) at three points along the cytochrome chain.

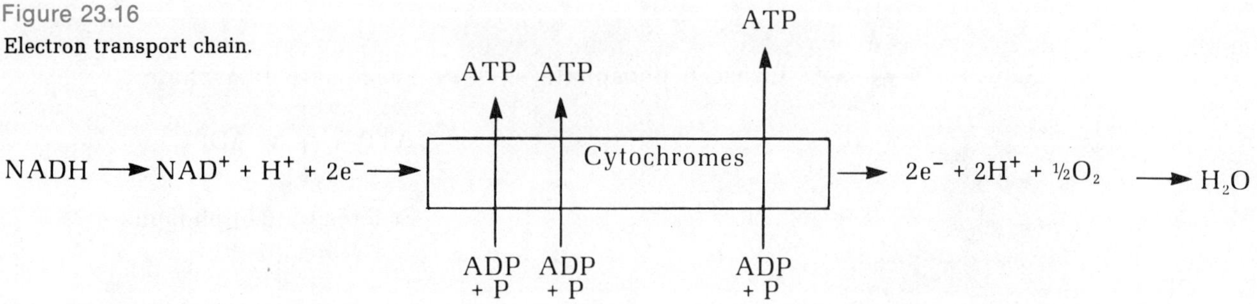

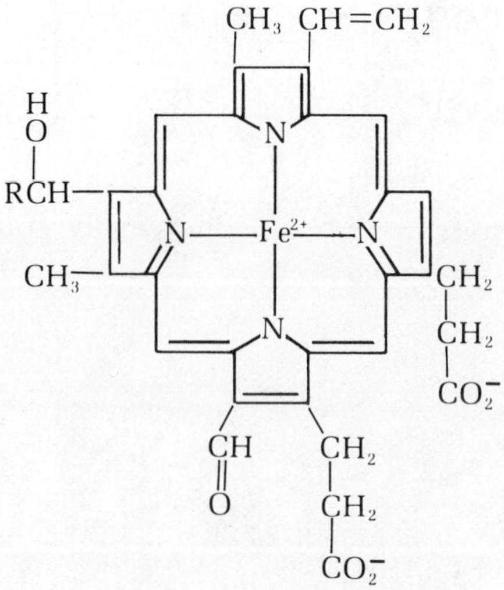

Heme

oxidized by O_2 directly, most of the free energy of the reaction would be released as heat, and the organism would lose the opportunity to store it for useful work. By having the process go in several steps, each releasing a moderate amount of free energy, the mitochondria are able to synthesize several ATP anions for each pair of electrons transferred. This process is known as **oxidative phosphorylation**.

One way of looking at the function of the electron transport chain is to draw an analogy with an electrochemical cell. When an oxidant and a reductant come into contact and react with each other directly, all possibility for converting the free energy of the reaction to usable electrical energy is lost. We get around this problem by preventing direct contact between the reactants, forcing the transfer of electrons from the cathode to the anode to go via an external circuit. In an analogous way, there is no direct reaction between reduced nucleotides and O_2 (due to the lack of a suitable enzyme), and the electrons flowing from NADH or $FMNH_2$ to O_2 are forced to travel by a route that yields useful work in the form of ATP.

One of the major sources of nutritional free energy is the metabolism of glucose. The net chemical result of this process is the oxidation of glucose to form carbon dioxide and water.

$$C_6H_{12}O_6 + 6\,O_2 \longrightarrow 6\,CO_2 + 6\,H_2O \qquad \Delta G° = -2840 \text{ kJ}$$

This oxidation involves a very complex sequence of enzyme-catalyzed reactions that can be divided into two main groups of reactions, **glycolysis** of glucose to yield

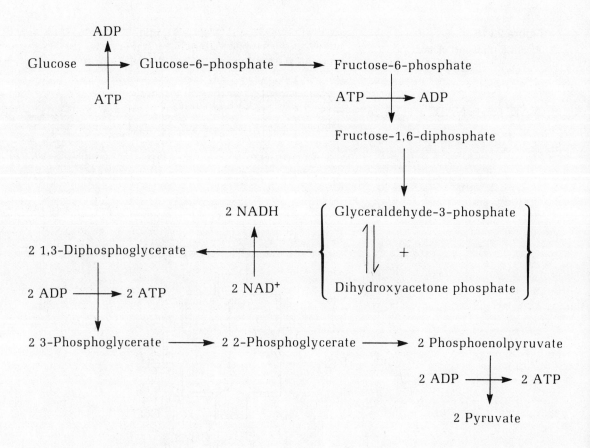

Figure 23.17
Glycolysis.

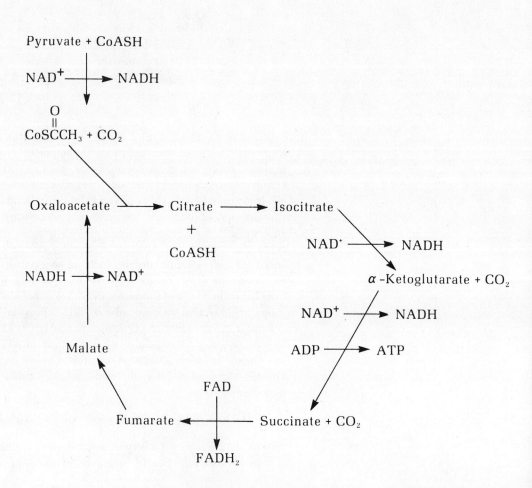

Figure 23.18
Tricarboxylic acid cycle.

23.7 Photosynthesis

pyruvate (Figure 23.17) and the oxidation of pyruvate to CO_2 in a series of reactions known as the **tricarboxylic acid cycle** (Figure 23.18). These processes form both ATP and reduced nucleotides. When the ATP ions produced by electron transport from the reduced nucleotides are taken into account, the metabolism of each mole of glucose is responsible for the synthesis of 36 mol of ATP. The free energy stored in these 36 mol of ATP (−1100 kJ) is about 40% of the free energy of oxidation of glucose, representing a substantial degree of free energy conservation.

There is a limited amount of O_2 and reduced carbon nutrients (such as glucose) in the earth's biosphere. If there were no way to regenerate these substances, our biosphere would quickly reach equilibrium. Since equilibrium in biochemical systems is synonymous with death, it is fortunate that many organisms are capable of **photosynthesis**—the use of the sun's energy to synthesize reduced carbon compounds and form oxygen.* The photosynthesis performed by green plants and many one-celled organisms keeps life on earth going.

Structures called **chloroplasts** contain the highly organized systems of pigments, electron carriers, and enzymes necessary for photosynthesis in green plants. Each chloroplast contains two structurally similar pigments, **chlorophyll** a and b (Figure 23.19). These molecules, which contain magnesium, are responsible for trapping photons of sunlight. Each absorbed photon excites the chlorophyll molecule, raising one of its electrons to a higher energy level. The excited electrons, following circuitous routes to get back to their ground states, use their energy to form O_2, NADPH, and ATP.

Figure 23.19
Chlorophyll.

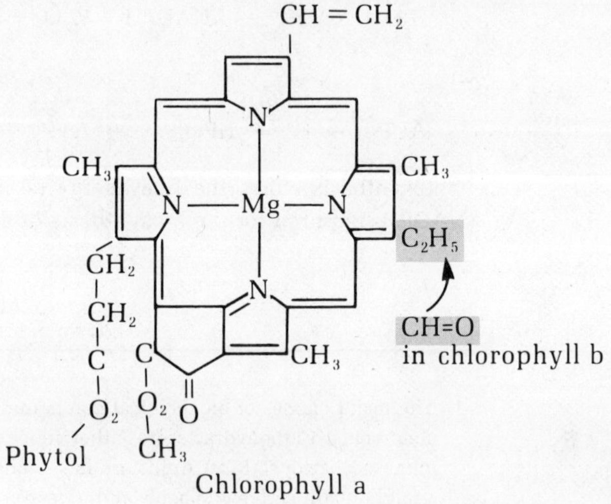

Chlorophyll a

Figure 23.20 shows a simplified version of what happens when a chloroplast absorbs photons. The chloroplast contains two pigment systems: system I, which contains chlorophyll a, and system II, which contains both chlorophyll a and b. When system I absorbs photons, it acts like a cathode; a pair of excited electrons flow through a complex system of electron carriers to reduce $NADP^+$.

$$2 \text{ Chlorophyll} \xrightarrow{\text{Light}} 2 \text{ Chlorophyll}^+ + 2 \text{ e}^-$$

$$2 \text{ e}^- + NADP^+ + H^+ \longrightarrow NADPH$$

*Not all photosynthetic organisms produce oxygen. Some oxidize hydrogen sulfide to sulfur,

$$H_2S \longrightarrow S + 2 H^+ + 2 \text{ e}^-$$

and others use even more exotic sources of electrons.

Figure 23.20
Simplified scheme of photon-induced reactions in photosynthesis.

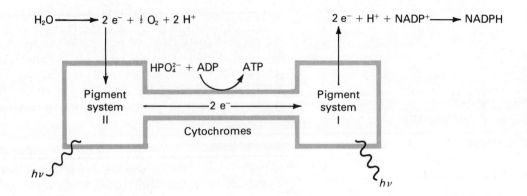

System II is a photosynthetic anode. As its electrons, excited by light, flow to system I, its ionized chlorophylls oxidize water.

$$2 \text{ Chlorophyll} \xrightarrow{\text{Light}} 2 \text{ Chlorophyll}^+ + 2 \text{ e}^-$$

$$H_2O + 2 \text{ Chlorophyll}^+ \longrightarrow \tfrac{1}{2} O_2 + 2 H^+ + 2 \text{ Chlorophyll}$$

The electrons flowing from system II to system I move through a series of cytochromes similar to the electron transport system in mitochondria and, as in mitochondria, oxidative phosphorylation of ADP occurs. The reactions caused by the photons are

$$H_2O + NADP^+ \longrightarrow NADPH + \tfrac{1}{2} O_2 + H^+$$

and

$$ADP^{3-} + H^+ + HPO_4^{2-} \longrightarrow ATP^{4-} + H_2O$$

Photosynthesis stores the free energy (as ATP) and generates the reducing agent (NADPH) required for the biosynthesis of compounds such as glucose.

Summary

1. The major classes of biochemical compounds are carbohydrates, lipids, proteins, and nucleic acids. Carbohydrates are either monosaccharides or polysaccharides (polymers of monosaccharides). Most lipids are fats—molecules with three ester bonds between three straight-chain carboxylic acids and glycerol.

2. Proteins are polymers of amino acids. Each amino acid contains a carboxylate group ($-CO_2^-$), a protonated amine group ($-NH_3^+$ or $-NH_2^+-$), and a hydrogen atom attached to the same carbon atom. The amino acids in a protein are joined together by peptide bonds to form polypeptide chains.

3. The primary structure of a protein is the sequence of its polypeptide chains.

4. Native proteins have specific three-dimensional shapes that are stabilized by hydrogen bonding, electrostatic attractions, and the tendency of nonpolar groups to cluster together. Two types of three-dimensional arrangements found in proteins are the alpha helix and the beta structure. Portions of a protein that are neither of these two types are called random coil structures.

5. The secondary structure of a protein is the relative amount of alpha helix, beta structure, and random coil regions in its polypeptide chains. The tertiary structure of a protein is the characteristic shape of its polypeptide chains. When a protein is an aggregate of several polypeptides, the relative orientation of these polypeptides is called the quaternary structure.

6. Denaturation, the loss of the characteristic shape of a protein molecule, changes the physical properties of the protein and destroys its biological activity.

7. Enzymes are naturally occurring catalysts with a high degree of effectiveness and specificity. Some enzymes are simple proteins and others contain a nonprotein cofactor. Some cofactors are metal ions; others are complicated organic molecules called coenzymes.

8. The mechanism of an enzyme-catalyzed reaction involves the reaction of a specific reactant—the substrate—with the enzyme to form a complex that then reacts rapidly to form products.

9. The small portion of an enzyme molecule directly involved in catalysis is called the active site. A cofactor forms part of the active site of an enzyme that requires the cofactor. Denaturation disrupts the active site and renders it inactive.

10. Genetic information is coded into molecules of deoxyribonucleic acid (DNA) and passed from parent to progeny when DNA molecules from the parents direct the synthesis of identical offspring molecules, a phenomenon called replication. A DNA molecule contains polynucleotide chains that are polymers of nucleotides. Each nucleotide contains a base group, and the sequence of base groups in DNA carries the genetic code. The DNA molecule contains two polynucleotide chains intertwined in a double helix that is held together by hydrogen bonding between base groups on each chain. The sequence of bases on one chain complements the sequence of the other.

11. During replication, the DNA double helix unwinds, and each chain directs the synthesis of its complementary chain. This produces two offspring DNA molecules that are identical to the parent molecule.

12. DNA directs the synthesis of complementary chains of messenger ribonucleic acid (mRNA), which in turn directs the synthesis of polypeptide chains with specific amino acid sequences. Each amino acid is bound by an enzyme-catalyzed reaction to its specific transfer RNA (tRNA) molecule that binds through a three-nucleotide portion of its chain (the anticodon) to a complementary three-nucleotide sequence (the codon) of mRNA. An enzyme forms peptide bonds between amino acids joined to adjacent codons, thus producing a specific amino acid sequence in the polypeptide.

13. A cell stores energy derived from metabolism by using this energy to synthesize adenosine triphosphate (ATP) from adenosine diphosphate (ADP). When it needs energy to perform work, the cell reconverts ATP to ADP and releases its stored energy.

14. Cells contain several biochemical oxidants involved in metabolism. Biochemical redox reactions convert these oxidants to their reduced forms that are reoxidized by oxygen via a complex sequence of enzyme-catalyzed redox reactions called the electron transport chain. Electron transport is accompanied by the formation of ATP by a process called oxidative phosphorylation.

15. A major source of cell energy is the metabolism of glucose. This occurs via a complicated sequence of steps that produces many molecules of ATP and thus conserves a large portion of glucose's free energy.

16. Photosynthesis is the use of the sun's energy to synthesize reduced carbon compounds and form O_2. The green pigment in plants that traps the light energy is chlorophyll. Photons excite electrons in chlorophyll molecules, and as these electrons return to their ground states via circuitous biochemical mechanisms, they use their energy to form O_2, NADPH, and ATP.

Vocabulary List

carbohydrate	amino acid	tertiary structure
monosaccharide	primary structure	quaternary structure
polysaccharide	alpha helix	denaturation
lipid	beta structure	enzymes
fat	random coil	regulatory enzymes
protein	secondary structure	cofactor

coenzymes
substrate
enzyme–substrate complex
active site
deoxyribonucleic acid (DNA)
replication
polynucleotide chain
base group

double helix
ribonucleic acid (RNA)
messenger RNA (mRNA)
transfer RNA (tRNA)
anticodon
codon
adenosine triphosphate (ATP)
adenosine diphosphate (ADP)

electron transport chain
cytochrome
oxidative phosphorylation
glycolysis
tricarboxylic acid cycle
photosynthesis
chloroplast
chlorophyll

Exercises

1. Draw the structure of
   a. A typical amino acid
   b. A portion of a polypeptide chain containing two different amino acid units
   c. A portion of a single strand of DNA
   d. A portion of a starch molecule
   e. A fat

2. Which of the following structures is:
   a. A portion of a polypeptide?
   b. A portion of a polynucleotide chain?
   c. A base group in a nucleic acid?
   d. An amino acid?

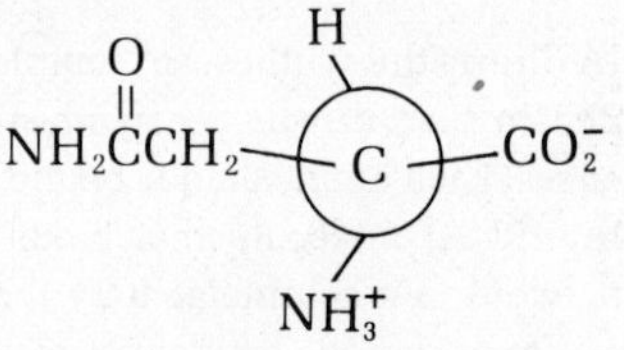

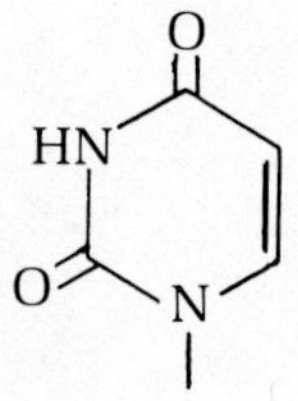

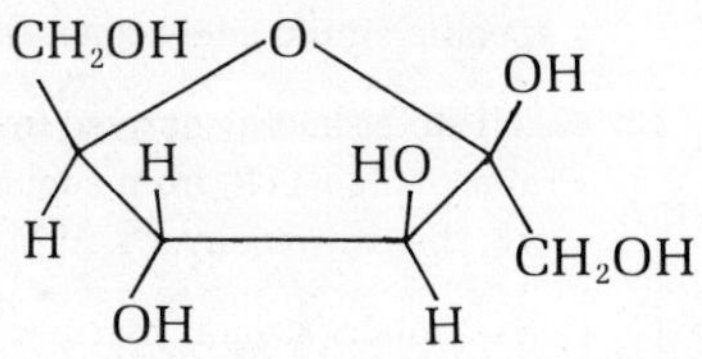

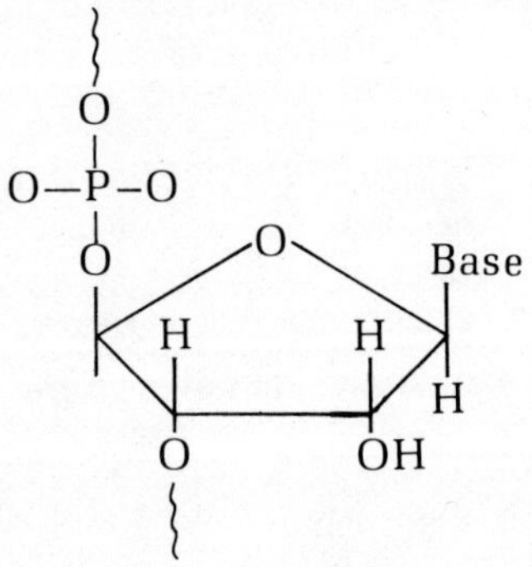

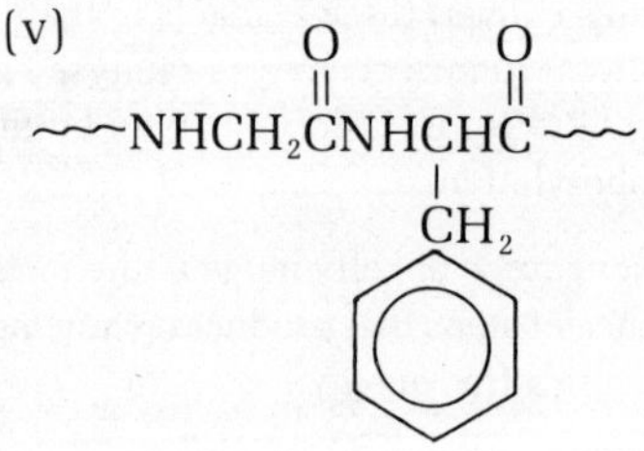

3. Which of the following is:
   a. A monosaccharide in its cyclic form?
   b. A monosaccharide in its open form?
   c. A fat?
   d. A fatty acid?
   e. A portion of a polysaccharide?

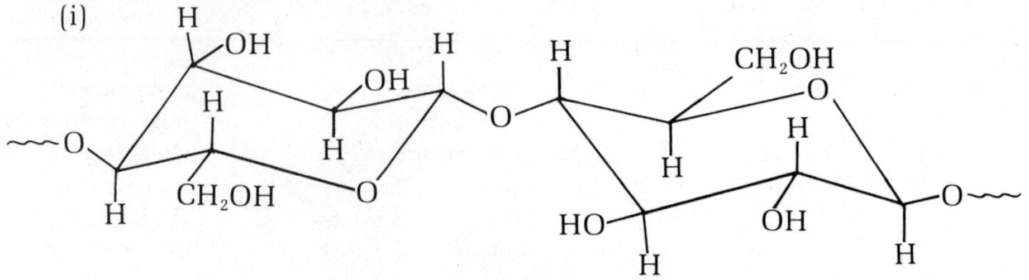

(ii)
```
      H   O
       \ //
        C
        |
   HO—C—H
        |
   HO—C—H
        |
    H—C—OH
        |
    H—C—OH
        |
      CH₂OH
```

(iii) [cyclic sugar structure with CH₂OH, OH, H, HO groups]

(iv) $NH_3^+CH_2CH_2CH_2CH_2CHCO_2^-$
                              |
                             NH_3^+

(v) $CH_3CH_2CH_2CH_2CH_2CH_2CH_2CH_2CH_2CH_2CH_2CH_2CH_2CO_2H$

(vi)
$$\begin{array}{l} O \\ \| \\ RCOCH_2 \\ \quad | \\ O \\ \| \\ RCOCH \\ \quad | \\ O \\ \| \\ RCOCH_2 \end{array}$$

4. Define the following terms.
   a. Enzyme   b. Coenzyme   c. DNA replication   d. Regulatory enzyme

5. Define the following terms.
   a. Denaturation   b. Active site   c. Photosynthesis   d. Electron transport chain

6. Define the following terms.
   a. Messenger RNA   b. Alpha helix   c. Osmotic work
   d. Codon   e. Cytochrome

7. To which structural class of substances do enzymes belong: lipids, monosaccharides, proteins, amino acids, or nucleic acids?

8. Which of the following is:
   a. A biochemical oxidant?   b. A heme group?   c. ADP?

(i) [heme-like porphyrin structure with Fe^{3+} center, substituents: $CH_2CH_2CO_2^-$, $CH_2=CH$, CH_3, $CH_2CH_2CO_2^-$, $CH=O$, CH_3, and R—CH(OH)—]

(ii)

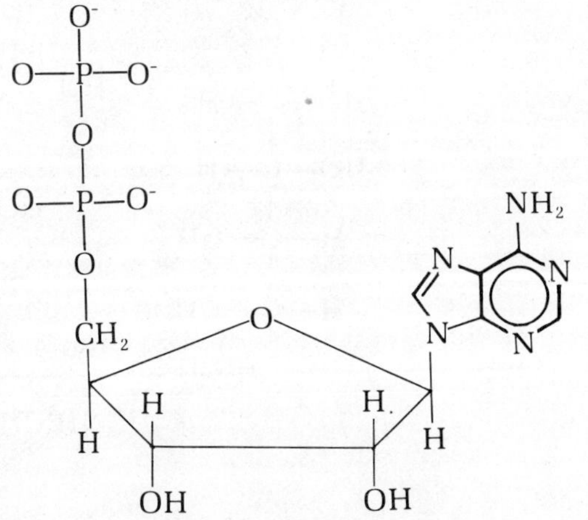

(iii)

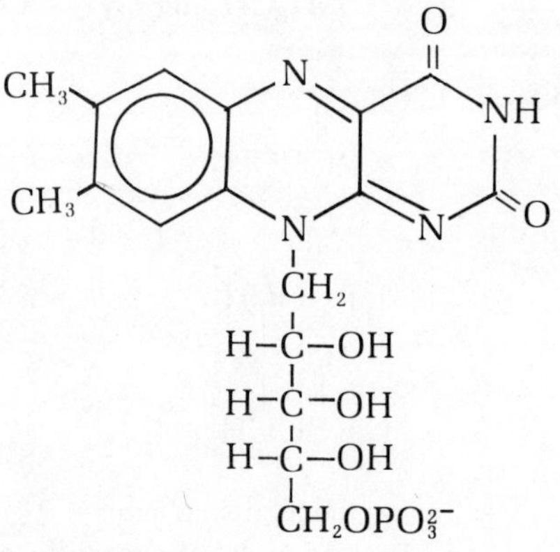

(iv)

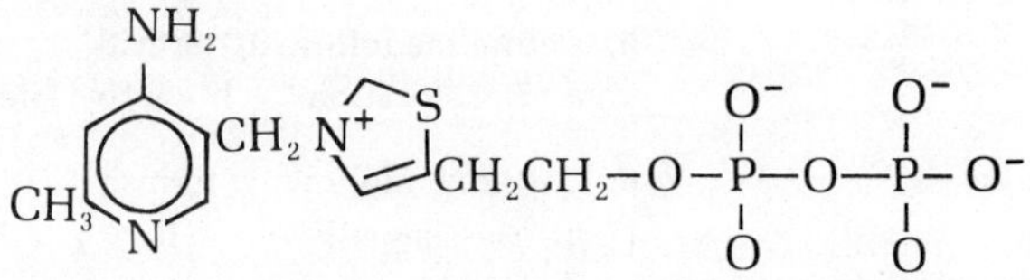

9. Which of the following statements are false? Correct the errors.
   a. The alpha helix is a common feature of the primary structures of many proteins.
   b. Hydrogen bonds stabilize the alpha helix.
   c. Hydrogen bonds stabilize the beta structure.
   d. Hydrogen bonds are the only force stabilizing the native structures of proteins.
   e. Random coils appear only in denatured proteins.
   f. Once we discover the relative amounts of alpha helix, beta structure, and random coil in a protein, we know its complete three-dimensional structure.

10. Do the following statements refer to the primary, secondary, tertiary, or quaternary structure of the protein?
    a. About 70% of myoglobin is in the alpha helix arrangement.
    b. Hemoglobin consists of four nearly identical protein units bound together by noncovalent forces.
    c. Part of the sequence of insulin is Gly.Ileu.Val.Glu.

11. Draw the structure of the tripeptide Ala.Ser.Asp.

12. The standard free energy change for the reaction

    $$NADH + H^+ + \tfrac{1}{2} O_2 \longrightarrow NAD^+ + H_2O$$

    is -220 kJ at pH 7. Electron transport from NADH to O_2 produces three molecules of ATP by oxidative phosphorylation. Assuming standard conditions within mitochondria, calculate the percentage of free energy conserved by oxidative phosphorylation.

13. What sequence of bases in one strand of DNA is complementary to the sequence TGGAC in the other strand?

14. A portion of single-stranded DNA has the sequence GTACA. What is the sequence of the portion of another strand of DNA that will bind with the first portion in a double helix?

15. a. What nucleotide base is complementary to uridine?
    b. Draw a picture of the hydrogen bonding between uridine and its complementary base.
    c. What codon will bind a molecule of transfer RNA with the anticodon UAG?

16. A tripeptide contains one unit each of phenylalanine, tyrosine, and histidine. Draw all the possible different tripeptides with this amino acid composition.

17. There are twelve different peptides containing two glycine units and one each of alanine and valine. Write their sequences.

18. Only one amino acid has no enantiomers. Which one is it?

19. Which of the amino acids in Table 23.1 have more than one asymmetric center?

20. Name an amino acid that has a large nonpolar side group.

21. a. Since amino acids are amphoteric, what reaction would you expect glycine to undergo in a strongly acidic solution? Draw the structure of the product.
    b. What is the product of glycine's reaction in a strongly basic solution?

22. Insulin (Figure 23.4) contains four glutamic acid units, no aspartic acid, and one unit each of arginine and lysine. What would you expect the charge of an insulin molecule to be at pH 7.0?

23. One feature of the tertiary structures of proteins is that amino acids with nonpolar side groups tend to lie in the center of the molecule in order to avoid contact with water while amino acids with hydrogen bonding or ionic side groups are located at the outer surface of the molecule. A segment of the peptide chain in ribonuclease has the sequence —Ileu.Ileu.Val.Ala—. Would you expect this segment to be at the outer surface of the protein or buried within its interior?

24. There is strong evidence that two histidine units in ribonuclease, one at position 12 in the peptide chain and the other at position 119, participate at the active site of the enzyme. Use this observation to explain why denaturation destroys the catalytic activity of ribonuclease.

25. The standard free energy at pH 7 for the hydrolysis of phosphoenolpyruvate

$$\begin{array}{c} CH_2 \\ \| \\ COPO_3^{2-} \\ | \\ CO_2^- \end{array} + H_2O \rightleftharpoons \begin{array}{c} CH_3 \\ | \\ C=O \\ | \\ CO_2^- \end{array} + HPO_4^{2-}$$

is −62 kJ. The standard free energy for hydrolysis of ATP to ADP at pH 7 is −31 kJ. What is the standard free energy of the reaction below at pH 7?

$$\begin{array}{c} CH_2 \\ \| \\ COPO_3^{2-} \\ | \\ CO_2^- \end{array} + H^+ + ADP^{3-} \rightleftharpoons \begin{array}{c} CH_3 \\ | \\ C=O \\ | \\ CO_2^- \end{array} + ATP^{4-}$$

26. a. How many moles of ATP must be converted to ADP to provide enough work to lift a 20.0-kg mass 2.0 m? Assume 25% efficiency in converting the free energy of ATP to mechanical energy. The free energy change for the conversion of ATP to ADP is −31 kJ/mol and the acceleration due to gravity is 9.8 m/sec^2.
    b. What mass of glucose must be metabolized to CO_2 and H_2O to supply this ATP?

27. An enzyme catalyzes the decomposition of a polypeptide by water provided the amino acids are of one enantiomeric type (the L-form) but will not attack polypeptides made from the mirror-image forms. This is an example of:
    a. Oxidative phosphorylation b. Electron transfer c. Stereospecificity
    d. A regulatory enzyme e. Biological oxidation f. A cofactor

28. There are 16 different aldehyde monosaccharides with the formula $C_6H_{12}O_6$. None of these compounds has a branched carbon chain, and each has one —OH group for each carbon except the aldehyde group. Two of these compounds, glucose and galactose, are shown below in their open-chain forms. In general terms, what is the nature of the differences in structure between these isomers?

$$\begin{array}{cc}
\text{H}\diagdown\!\!\diagup\text{O} & \text{H}\diagdown\!\!\diagup\text{O} \\
\text{C} & \text{C} \\
| & | \\
\text{H—C—OH} & \text{H—C—OH} \\
| & | \\
\text{HO—C—H} & \text{HO—C—H} \\
| & | \\
\text{H—C—OH} & \text{HO—C—H} \\
| & | \\
\text{H—C—OH} & \text{H—C—OH} \\
| & | \\
\text{CH}_2\text{OH} & \text{CH}_2\text{OH} \\
\text{Glucose} & \text{Galactose}
\end{array}$$

29. Years ago, ATP was often incorrectly called a "high-energy compound." Why would a compound with a high free energy be unsuitable for the task of storing metabolic energy?

30. The photosynthetic reaction

    $$H_2O + NADP^+ \xrightarrow{\text{Light}} NADPH + H^+ + \tfrac{1}{2} O_2$$

    requires the energy of two photons, one in pigment system I and the other in pigment system II, for each electron transferred.
    a. How much light energy (in kJ) is used to produce a mole of NADPH if 700 nm light is used for photosynthesis?
    b. What percentage of this energy is preserved as increased free energy due to the reduction of $NADP^+$?

    $$NADPH + H^+ + \tfrac{1}{2} O_2 \longrightarrow H_2O + NADP^+ \qquad \Delta G° = -220 \text{ kJ at pH 7}$$

    c. One mole of ATP forms for each pair of electrons flowing from pigment system I to pigment system II. What percentage of the light's energy is preserved by this phosphorylation?

31. The enzymatic conversion of fructose-6-phosphate to fructose-1, 6-diphosphate

    $$ATP^{4-} + C_6H_{11}PO_9^{2-} \rightleftharpoons ADP^{3-} + C_6H_{10}P_2O_{12}^{4-} + H^+$$

    is stimulated by high concentrations of ADP and inhibited by high ATP concentrations.
    a. Is this effect consistent with Le Chatelier's principle?
    b. Suggest an explanation for this effect.
    c. How does this phenomenon aid in the efficient functioning of an organism?

32. Malate dehydrogenase catalyzes the oxidation of L-malate to oxaloacetate,

    $$\begin{array}{c}
    CO_2^- \\
    | \\
    HO—C—H \\
    | \\
    CH_2CO_2^-
    \end{array} + NAD^+ \rightleftharpoons NADH + H^+ + \begin{array}{c}
    CO_2^- \\
    | \\
    C{=}O \\
    | \\
    CH_2CO_2^-
    \end{array}$$

    L-Malate $\qquad\qquad\qquad\qquad\qquad\qquad$ Oxaloacetate

    but the enzyme is not active toward the enantiomer of L-malate (D-malate).
    a. Malate dehydrogenase also catalyzes the reduction of oxaloacetate by NADH. Which enantiomer of malate will be produced in this reaction?
    b. What important kinetic principle does this illustrate?

EXERCISES

33. One technique used to crack the genetic code was to use a synthetic messenger RNA of known base composition in a test-tube protein synthesis. Each test tube contained the synthetic RNA, ribosomes, transfer RNAs, all the necessary enzymes and cofactors for protein synthesis, and all the common amino acids. One amino acid in each test tube carried a radioactive label, and the incorporation of its radioactivity into the protein was evidence that the synthetic RNA carried its codon.
   a. Lysine is the only amino acid incorporated when the messenger RNA is polyadenylic acid (polyA). What is the codon for lysine?
   b. A synthetic messenger RNA was prepared by polymerizing equal quantities of U and G units. What possible codons could occur in this messenger RNA?
   c. Assuming the polymerization to form messenger RNA in part b is completely random, estimate the relative amounts of each possible codon.

34. Sickle-cell anemia is a genetic disease caused by the presence of a mutant gene for the synthesis of hemoglobin. The difference between normal hemoglobin and the mutant hemoglobin (hemoglobin S) is that the mutant contains valine in a position occupied by glutamic acid in the normal protein. The codons for glutamic acid are GAA and GAG, and the valine codons are GUU, GUG, GUA, and GUC. The most likely mutations involve altering a single base unit in DNA. What are the likely possibilities for the codon for valine at the mutant position in hemoglobin S?

35. An enzyme catalyzed reaction is usually first order in substrate at low substrate concentrations but independent of substrate concentration at high substrate concentrations. (The substrate concentration is always much higher than the enzyme concentration.) Use the hypothesis of the enzyme–substrate complex and the principles of equilibria and kinetics to explain why this occurs.

appendix A

Mathematics

Although this appendix may prove useful for review and reference, it is not a substitute for the necessary background in mathematics. You will not be able to follow the discussions in this book without a firm grasp of basic algebra, exponentials, and logarithms.

A.1 Exponentials

We can express any number in an exponential form, $N \times 10^n$.

Example 1: $750,000 = 7.5 \times 10^5$

Thus in exponential notation, 750,000 equals 7.5 times 100,000, and 100,000 is 10 multiplied by itself 5 times or 10 to the fifth power (10^5).

Numbers less than 1 involve negative powers of 10.

Example 2: $0.00018 = 1.8 \times 10^{-4}$

The rationale for this equation is

$$0.00018 = \frac{1.8}{10,000}$$
$$= \frac{1.8}{10^4}$$
$$= 1.8 \times 10^{-4}$$

Both examples involve a shift in the position of a decimal point. We must compensate for shifting a decimal point n positions to the *left* (as in Example 1) by multiplying by 10^n. A decimal shift of n positions to the *right* requires multiplication by 10^{-n}. The number, N, in the exponential form $N \times 10^n$ is usually chosen to be between 1 and 10.

A.2 Operations with Exponents

Addition and Subtraction

To add or subtract exponential numbers, n must be the same for all the numbers. If necessary, we must shift decimals to attain this equality.

Example 3: $(9.50 \times 10^{-3}) + (7.8 \times 10^{-4}) + (8 \times 10^{-5})$
$= (9.50 \times 10^{-3}) + (0.78 \times 10^{-3}) + (0.08 \times 10^{-3})$
$= 10.36 \times 10^{-3}$, or 1.036×10^{-2}

Multiplication

To multiply exponential numbers, the decimal parts of the number are multiplied, and the exponents are *added*.

$$a^n \cdot a^m = a^{(n+m)}$$

Example 4: $(3 \times 10^7)(5 \times 10^4)$

First group the decimal parts and exponential parts together, and then multiply the decimal parts and add the exponents of 10.

$$(3 \times 5)(10^7 \times 10^4) = 15(10^{7+4}) = 15 \times 10^{11} = 1.5 \times 10^{12}$$

Example 5: $e^{-(\Delta H/RT - \Delta S/R)} = e^{-\Delta H/RT} e^{\Delta S/R}$

Division

In division, the decimal parts of the number are divided, and the exponent of the denominator is *subtracted* from the exponent of the numerator.

$$\frac{a^n}{a^m} = a^{(n-m)}$$

Example 6:
$$\frac{1.0 \times 10^{-14}}{1.8 \times 10^{-5}} = \frac{1.0}{1.8} \times 10^{-14-(-5)}$$
$$= 0.56 \times 10^{-9} = 5.6 \times 10^{-10}$$

Powers

To raise an exponential number to a power, the exponent is multiplied by the power.

$$(a^n)^m = a^{nm}$$

Example 7: $(7 \times 10^{-8})^2 = 7^2 \times 10^{-8 \times 2} = 49 \times 10^{-16} = 4.9 \times 10^{-15}$

Roots

To take the root of an exponential number, the exponent is divided by the root.

$$\sqrt[m]{a^n} = (a^n)^{1/m} = a^{n/m}$$

When taking the root of an exponential number, shift the decimal point so that n is a whole-number multiple of m.

Example 8: $\sqrt[3]{2.7 \times 10^{-8}} = \sqrt[3]{27 \times 10^{-9}} = 3 \times 10^{-3}$

A.3 Significant Figures

The significant figures in a measured quantity include (the significant digits appear in boldface type):

1. All nonzero digits and zeros *between* nonzero digits, for example, **705**0 m, 0.00**402** kg, **1.008** amu.
2. All zeros following digits if the zeros are to the right of the decimal, for example, **40.0** cm, 0.00**50** Å.
3. Zeros that follow significant digits but appear to the left of the decimal point *may* be significant, for example, **7500.** kg, **286**,000. mi.

The ambiguity illustrated in rule 3 can be removed by using exponential notation. For example,

7500 kg is 7.5×10^3 kg if two significant digits are known
7.50×10^3 kg if three significant digits are known
7.500×10^3 kg if four significant digits are known

The purpose in writing the correct number of significant figures for an experimentally determined value is to present as much information as can be justified by the experimental results without claiming knowledge beyond the limits of experimental precision. For example, Millikan's estimates of the charge on an electron (Table 4.4) vary from experiment to experiment. These estimates agree well in their first two figures, but there is considerable fluctuation in the third figure. Given this large degree of uncertainty in the value of the third digit, it would be meaningless to quote the average value for these results to four significant figures.

Using too few significant figures is as wrong as using too many—it withholds valid information. If we weigh two grams of NaCl as precisely as possible on a balance reliable to the nearest milligram, we know the mass is 2.000 g. Ignoring significant figures and writing "2 g" would incorrectly imply that we have made a crude estimate of this mass.

The uncertainty in a measured quantity is implied to be ± 1 in the rightmost digit unless specifically stated.

Example 9: **1.05 Å** implies that the quantity is known to be between 1.06 and 1.04 Å but is closer to 1.05 Å.

Example 10: **2.056 ± 0.002 g** tells us that the uncertainty is 2 milligrams in 2.056 grams or 0.1%.

When *adding* or *subtracting* experimental values, we report an answer only to the least number of *decimal* places.

Example 11: $7.93 \times 10^{-3} + 4.38 \times 10^{-4} = ?$

Shifting the decimals to obtain a common exponential, we obtain

7.93×10^{-3}

$+0.438 \times 10^{-3}$

but the third digit (8) of 0.438×10^{-3} is not significant with respect to digits in the first number. We round it off before adding.

$$\begin{array}{r} 7.93 \times 10^{-3} \\ + \ 0.44 \times 10^{-3} \\ \hline 8.37 \times 10^{-3} \end{array}$$

The *product* of experimental quantities has the same number of significant digits as the *least precise individual quantity*.

Example 12: $(0.25 \text{ amp})(3.60 \times 10^3 \text{ sec}) = 9.0 \times 10^2$ coul

In this example, the product of two significant digits times three significant digits has two significant digits.

The same rule governing significant digits applies to division.

Example 13: $\dfrac{1.379 \text{ g}}{0.57 \text{ mL}} = 2.4$ g/mL

A.4 Logarithms

Base-Ten Logarithms

The base-ten logarithm (x) of a number (N) is the power to which 10 must be raised to equal that number, that is,

$10^x = N$ or $\log N = x$

Logarithms for powers of 10 such as 100 and 10^{-7} are whole numbers (2 and -7, respectively). Logarithms of numbers between 1 and 10 appear in Appendix B. For example, to find the logarithm of 3.76, look at the row marked 37 and the column headed 6. The value at this position is 5752, therefore,

$$\log 3.76 = 0.5752$$

To find logarithms of numbers larger than 10 or less than 1, apply the relationship

$$\log(N \times 10^n) = \log N + \log 10^n$$
$$= \log N + n$$

Example 14: $\log 376 = \log 3.76 \times 10^2$
$$= \log 3.76 + 2 = 2.5752$$

The 2 is called the *characteristic* (it positions the decimal) and 0.5752 is the *mantissa* (it gives the significant digits).

Example 15: $\log 0.00753 = \log 7.53 \times 10^{-3}$
$$= \log 7.53 - 3$$
$$= 0.8768 - 3$$
$$= -2.1232$$

Given a logarithm, x, you can find its antilogarithm—the number whose logarithm equals x—by reversing the procedure in Examples 14 and 15.

Example 16: $\quad x = 7.529$
$$\text{antilog } x = (\text{antilog } 0.529) \times (\text{antilog } 7)$$

The antilog of 7 is obviously 10^7, and we find the number whose logarithm is closest to 0.529 from the logarithm table.

$$\text{antilog } x = 3.38 \times 10^7$$

Example 17: $\log a = -2.7328$

The easiest way to find the antilogarithm of a negative number is to convert the logarithm to a positive fraction and a negative whole number because the mantissa must be a positive number between 0 and 1.

$$\log a = 0.2672 - 3$$
$$a = (\text{antilog } 0.2672) \times 10^{-3}$$
$$= 1.85 \times 10^{-3}$$

A.5 Natural Logarithms

Many scientific laws express natural exponential relationships between physical quantities. One example is the Arrhenius law (Section 12.4).

$$k = Ae^{-E_a/RT}$$

The number e is equal to about 2.72. Calculations involving natural exponents are easier if the equation is expressed in the form of the natural logarithm (ln). The natural logarithm (x) of a number (N) is the quantity such that

$$e^x = N$$

$$x = \ln N$$

It is not necessary to use a natural logarithm table to find $\ln N$ because we can estimate it from the value of the base-ten logarithm ($\log N$).

$$\ln N = (\ln 10)(\log N)$$
$$= 2.302(\log N)$$

Example 18: Find x if $\ln x = -2.000$.

$$-2.000 = 2.302 (\log x)$$
$$\log x = -0.8688$$
$$x = 0.135$$

A.6 Mathematical Operations with Logarithms

Because logarithms are exponents, mathematical operations with logarithms follow the same rules given for exponents.

$$\log a + \log b = \log ab$$

$$\log a - \log b = \log \frac{a}{b}$$

$$-\log a = \log \frac{1}{a}$$

$$n(\log a) = \log a^n$$

$$\frac{1}{n} (\log a) = \log \sqrt[n]{a}$$

The same relationships hold for natural logarithms.

A.7 Quadratic Equations

The solutions of a quadratic equation of the form

$$ax^2 + bx + c = 0$$

are

$$x = \frac{-b \pm \sqrt{b^2 - 4ac}}{2a}$$

Example 19: $x = 1.0 \times 10^{-7} + \dfrac{1.0 \times 10^{-14}}{x}$

Clear the fraction and collect terms.

$$x^2 - 1.0 \times 10^{-7}x - 1.0 \times 10^{-14} = 0$$

$$x = \frac{1.0 \times 10^{-7} \pm \sqrt{1.0 \times 10^{-14} + 4.0 \times 10^{-14}}}{2}$$

$$= \frac{1.0 \times 10^{-7} \pm 2.2 \times 10^{-7}}{2}$$

$$= 1.6 \times 10^{-7}, \text{ or } -0.6 \times 10^{-7}$$

Common sense tells us which solution is appropriate to our problem. If x represents a concentration, it must be positive.

A.8 Linear Equations

When two variable quantities (x and y) are related by a linear equation

$$y = mx + b$$

(m and b are constants), a graph of y versus x will be a straight line (Figure A.1) crossing the y-axis at the *intercept*, b. The *slope* of the line—the ratio of the increase in y to the increase in x—is m. The ease in interpreting linear graphs makes it worthwhile to graph quantities that are linearly related. For example, it is hard to estimate the heat of vaporization of a liquid from the curved graph of pressure versus temperature (Figure 9.32),

$$\log\left(\frac{P_2}{P_1}\right) = \frac{-\Delta H_{vap}}{2.303\,R}\left(\frac{1}{T_2} - \frac{1}{T_1}\right)$$

but we can estimate ΔH_{vap} from the slope of the linear graph of log P versus 1/T.

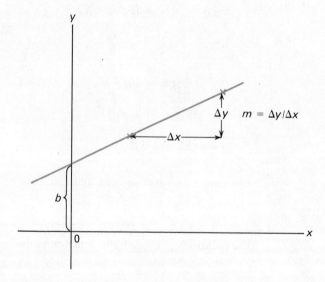

Figure A.1

A.9 Dimensional Analysis

The dimensions of the units of any quantity must be consistent with the way the quantity is computed. For example, the universal gas constant R (Chapter 3) equals PV/Tn, so it has dimensions of pressure times volume divided by absolute temperature times number of moles.

$$R = 0.08205 \frac{L\text{-atm}}{K\text{-mol}}$$

The dimensions of equal quantities must be equal. Keeping this principle in mind, you can avoid such common mathematical errors as inverted fractions. For example, suppose you carelessly write an incorrect formula for the number of moles in a gas sample

$$n = \frac{RT}{PV}$$

rather than the correct $n = PV/RT$. You can catch this error by dimensional analysis since n has the dimension of moles while the dimensions of RT/PV are

$$\frac{\left(\frac{\cancel{L}\cdot\cancel{atm}}{\cancel{K}\cdot mol}\right)\cdot \cancel{K}}{\cancel{atm}\cdot\cancel{L}} = mol^{-1}$$

The nature of the disagreement of the dimensions indicates the nature of the error—the inverted dimensions point to an inverted fraction. Carrying the dimensions *along* with the numerical values for the quantities in a mathematical calculation and checking to see that the answer has the correct dimensions as well, gives greater assurance that the numerical answer is correct.

appendix B

Logarithms

(Decimals have been omitted; numbers range from 0.0000 to 0.9996.)

	0	1	2	3	4	5	6	7	8	9
10	0000	0043	0086	0128	0170	0212	0253	0294	0334	0374
11	0414	0453	0492	0531	0569	0607	0645	0682	0719	0755
12	0792	0828	0864	0899	0934	0969	1004	1038	1072	1106
13	1139	1173	1206	1239	1271	1303	1335	1367	1399	1430
14	1461	1492	1523	1553	1584	1614	1644	1673	1703	1732
15	1761	1790	1818	1847	1875	1903	1931	1959	1987	2014
16	2041	2068	2095	2122	2148	2175	2201	2227	2253	2279
17	2304	2330	2355	2380	2405	2430	2455	2480	2504	2529
18	2553	2577	2601	2625	2648	2672	2695	2718	2742	2765
19	2788	2810	2833	2856	2878	2900	2923	2945	2967	2989
20	3010	3032	3054	3075	3096	3118	3139	3160	3181	3201
21	3222	3243	3263	3284	3304	3324	3345	3365	3385	3404
22	3424	3444	3464	3483	3502	3522	3541	3560	3579	3598
23	3617	3636	3655	3674	3692	3711	3729	3747	3766	3784
24	3802	3820	3838	3856	3874	3892	3909	3927	3945	3962
25	3979	3997	4014	4031	4048	4065	4082	4099	4116	4133
26	4150	4166	4183	4200	4216	4232	4249	4265	4281	4298
27	4314	4330	4346	4362	4378	4393	4409	4425	4440	4456
28	4472	4487	4502	4518	4533	4548	4564	4579	4594	4609
29	4624	4639	4654	4669	4683	4698	4713	4728	4742	4757
30	4771	4786	4800	4814	4829	4843	4857	4871	4886	4900
31	4914	4928	4942	4955	4969	4983	4997	5011	5024	5038
32	5051	5065	5079	5092	5105	5119	5132	5145	5159	5172
33	5185	5198	5211	5224	5237	5250	5263	5276	5289	5302
34	5315	5328	5340	5353	5366	5378	5391	5403	5416	5428
35	5441	5453	5465	5478	5490	5502	5514	5527	5539	5551
36	5563	5575	5587	5599	5611	5623	5635	5647	5658	5670
37	5682	5694	5705	5717	5729	5740	5752	5763	5775	5786
38	5798	5809	5821	5832	5843	5855	5866	5877	5888	5899
39	5911	5922	5933	5944	5955	5966	5977	5988	5999	6010
40	6021	6031	6042	6053	6064	6075	6085	6096	6107	6117
41	6128	6138	6149	6160	6170	6180	6191	6201	6212	6222
42	6232	6243	6253	6263	6274	6284	6294	6304	6314	6325
43	6335	6345	6355	6365	6375	6385	6395	6405	6415	6425
44	6435	6444	6454	6464	6474	6484	6493	6503	6513	6522
45	6532	6542	6551	6561	6571	6580	6590	6599	6609	6618
46	6628	6637	6646	6656	6665	6675	6684	6693	6702	6712
47	6721	6730	6739	6749	6758	6767	6776	6785	6794	6803
48	6812	6821	6830	6839	6848	6857	6866	6875	6884	6893
49	6902	6911	6920	6928	6937	6946	6955	6964	6972	6981
50	6990	6998	7007	7016	7024	7033	7042	7050	7059	7067
51	7076	7084	7093	7101	7110	7118	7126	7135	7143	7152
52	7160	7168	7177	7185	7193	7202	7210	7218	7226	7235
53	7243	7251	7259	7267	7275	7284	7292	7300	7308	7316
54	7324	7332	7340	7348	7356	7364	7372	7380	7388	7396

	0	1	2	3	4	5	6	7	8	9
55	7404	7412	7419	7427	7435	7443	7451	7459	7466	7474
56	7482	7490	7497	7505	7513	7520	7528	7536	7543	7551
57	7559	7566	7574	7582	7589	7597	7604	7612	7619	7627
58	7634	7642	7649	7657	7664	7672	7679	7686	7694	7701
59	7709	7716	7723	7731	7738	7745	7752	7760	7767	7774
60	7782	7789	7796	7803	7810	7818	7825	7832	7839	7846
61	7853	7860	7868	7875	7882	7889	7896	7903	7910	7917
62	7924	7931	7938	7945	7952	7959	7966	7973	7980	7987
63	7993	8000	8007	8014	8021	8028	8035	8041	8048	8055
64	8062	8069	8075	8082	8089	8096	8102	8109	8116	8122
65	8129	8136	8142	8149	8156	8162	8169	8176	8182	8189
66	8195	8202	8209	8215	8222	8228	8235	8241	8248	8254
67	8261	8267	8274	8280	8287	8293	8299	8306	8312	8319
68	8325	8331	8338	8344	8351	8357	8363	8370	8376	8382
69	8388	8395	8401	8407	8414	8420	8426	8432	8439	8445
70	8451	8457	8463	8470	8476	8482	8488	8494	8500	8506
71	8513	8519	8525	8531	8537	8543	8549	8555	8561	8567
72	8573	8579	8585	8591	8597	8603	8609	8615	8621	8627
73	8633	8639	8645	8651	8657	8663	8669	8675	8681	8686
74	8692	8698	8704	8710	8716	8722	8727	8733	8739	8745
75	8751	8756	8762	8768	8774	8779	8785	8791	8797	8802
76	8808	8814	8820	8825	8831	8837	8842	8848	8854	8859
77	8865	8871	8876	8882	8887	8893	8899	8904	8910	8915
78	8921	8927	8932	8938	8943	8949	8954	8960	8965	8971
79	8976	8982	8987	8993	8998	9004	9009	9015	9020	9025
80	9031	9036	9042	9047	9053	9058	9063	9069	9074	9079
81	9085	9090	9096	9101	9106	9112	9117	9122	9128	9133
82	9138	9143	9149	9154	9159	9165	9170	9175	9180	9186
83	9191	9196	9201	9206	9212	9217	9222	9227	9232	9238
84	9243	9248	9253	9258	9263	9269	9274	9279	9284	9289
85	9294	9299	9304	9309	9315	9320	9325	9330	9335	9340
86	9345	9350	9355	9360	9365	9370	9375	9380	9385	9390
87	9395	9400	9405	9410	9415	9420	9425	9430	9435	9440
88	9445	9450	9455	9460	9465	9469	9474	9479	9484	9489
89	9494	9499	9504	9509	9513	9518	9523	9528	9533	9538
90	9542	9547	9552	9557	9562	9566	9571	9576	9581	9586
91	9590	9595	9600	9605	9609	9614	9619	9624	9628	9633
92	9638	9643	9647	9652	9657	9661	9666	9671	9675	9680
93	9685	9689	9694	9699	9703	9708	9713	9717	9722	9727
94	9731	9736	9741	9745	9750	9754	9759	9763	9768	9773
95	9777	9782	9786	9791	9795	9800	9805	9809	9814	9818
96	9823	9827	9832	9836	9841	9845	9850	9854	9859	9863
97	9868	9872	9877	9881	9886	9890	9894	9899	9903	9908
98	9912	9917	9921	9926	9930	9934	9939	9943	9948	9952
99	9956	9961	9965	9969	9974	9978	9983	9987	9991	9996

appendix C

Commonly Used Units Derived from Basic SI Units

Quantity	Definition	Name of unit	Symbol
Density	Mass per volume	Kilogram/cubic meter	kg m^{-3}
Velocity	Distance per time	Meter/second	m sec^{-1}
Acceleration	$(v - v_0)/t$	Meter/second/second	m sec^{-2}
Force	Mass times acceleration	Newton	N (= kg m sec^{-2})
Pressure	Force per area	Pascal (newton/square meter)	Pa (= kg m^{-1} sec^{-2} or N m^{-2})
Energy	Force times distance or $\frac{1}{2}mv^2$	Joule	J (= kg m^2 sec^{-2})
Temperature		Degree Celsius	°C (= K − 273.15)
Electric charge	$\sqrt{Fd/K}$	Coulomb	coul (= amp · sec)
Frequency	$\nu = c/\lambda$	Hertz	Hz (= cycles sec^{-1})

appendix D

Conversion Factors

Mass	1 atomic mass unit (amu) = 1.6605×10^{-27} kg
Distance	1 angstrom (1 Å) = 10^{-8} cm = 10^{-10} m
Volume	1 liter (L) = 1000 cubic centimeters (cm^3) = 1.057 quarts (qt)
Force	1 newton (N) = 10^5 dyne
Energy	1 joule (J) = 10^7 erg
	1 cal = 4.184 J
	1 electron volt (1 eV) = 1.602×10^{-19} J
Pressure	1 atmosphere (1 atm) = 760 torr
	1 atm = 101,325 N m^{-2} = 101,325 Pa
Temperature	0 K = $-273.15°$C; 0°C = 273.15 K
Electric charge	1 coulomb (coul) = 2.99793×10^9 esu
	1 Faraday ($\mathscr{F}$) = 9.6487×10^4 coul

appendix E

Fundamental Physical Constants

Name	Symbol	Value
Universal gas constant	R	0.08205 L-atm/K-mol
		62.36 L-torr/K-mol
		8.3143 joule/K-mol
Boltzmann's constant	k	1.38062×10^{-23} J/K
Avogadro's number	N_{Avog}	6.02217×10^{23} molecules/mole
Molar volume of a gas		22.414 L at STP
Coulomb's law constant		8.99×10^9 N m^2/coul^2
Velocity of light	c	2.997925×10^8 m/sec
Planck's constant	h	6.62620×10^{-34} J-sec
Charge of electron	e	1.60210×10^{-19} coul
		4.86298×10^{-10} esu
Mass of electron	m_e	9.10956×10^{-31} kg
		5.48597×10^{-4} amu
Mass of proton	m_p	1.67261×10^{-27} kg
		1.00727 amu
Mass of neutron	m_n	1.67492×10^{-27} kg
		1.00867 amu
Rydberg constant (Hydrogen)	R_H	1.097373×10^7 m^{-1}

appendix F

Nomenclature of Inorganic Complexes (Coordination Compounds)*

F.1 Formulas

1. As usual cations are named before anions. The central atom is listed first, followed by the *negative* (anionic) and *positive* (cationic) ligands, and then the neutral ligands. Brackets are used to separate whole complex groups, for example,

    $[CoCl(NH_3)_5]^+$ $K[PtCl_3(C_2H_4)]$

2. The sequence of appearance *within* each of the above groups is in alphabetical order.

F.2 Nomenclature

1. In the nomenclature of complex ions the central atom is named *after* the ligands. Ligands are listed in alphabetical order regardless of the number of each. Greek prefixes designate the number of each ligand (mono is omitted) and the prefixes bis, tris, tetrakis, and so forth are used to indicate numbers of entire groups of atoms, for example

    $Na_3[Ag(S_2O_3)_2]$ sodium bis(thiosulfato)argentate(I)

    $[Co(en)_3]^{3+}$ tris(ethylenediamine)cobalt(III) ion

    In neutral molecules and complex cations the central atom is given no special designation; however, in complex anions the central atom name ends in *ate*. The oxidation state of the central atom is indicated (if necessary) by Roman numerals.

2. The names for anionic ligands end in *o*: F^-, fluoro; Cl^-, chloro; Br^-, bromo; I^-, iodo; H^-, hydrido; O^{2-}, oxo; OH^-, hydroxo; O_2^{2-}, peroxo; S^{2-}, thio; CN^-, cyano, and *ide, ite,* and *ate* endings become *ido, ito,* and *ato,* respectively. The names of neutral and cationic ligands are unchanged except for water (aqua) and ammonia (ammine). Nitrosyl (NO) and carbonyl (CO) are neutral ligands. Some examples are:

---

*For further reference, see "Preamble and Rules of the 1957 Report of the Commission on the Nomenclature of Inorganic Chemistry, International Union of Pure and Applied Chemistry," Butterworths, London (1959); and "Comments on American Usage," *Journal of the American Chemical Society*, **82**, 5523 (1960).

$K_4[Fe(CN)_6]$	potassium hexacyanoferrate(II)
$Al[(OH)(H_2O)_5]^{2+}$	pentaaquahydroxoaluminum ion (aluminum only forms +3 ions so the oxidation state is omitted)
$[CrCl_4(H_2O)_2]^-$	diaquatetrachlorochromate(III) ion
$[CoBr(NH_3)_5]^{2+}$	pentaamminebromocobalt(III) ion
$[CuCl_2(CH_3NH_2)_2]$	dichlorobis(methylamine)copper(II)
$Ni(NH_3)_4^{2+}$	tetraamminenickel(II) ion
$Pt[Cl_2(NH_3)_2]$	diamminedichloroplatinum(II)
$[Co(CN)_6]^{3-}$	hexacyanocobaltate(III) ion
$Fe(CO)_5$	pentacarbonyliron

appendix G

Vapor Pressure of Water from 0 to 100°C

Temperature (T) in °C, pressure (p) in torr

T	p	T	p	T	p	T	p
0	4.6	26	25.2	51	97.2	76	301.4
1	4.9	27	26.7	52	102.1	77	314.1
2	5.3	28	28.4	53	107.2	78	327.3
3	5.7	29	30.0	54	112.5	79	341.0
4	6.1	30	31.8	55	118.0	80	355.1
5	6.5	31	33.7	56	123.8	81	369.7
6	7.0	32	35.7	57	129.8	82	384.9
7	7.5	33	37.7	58	136.1	83	400.6
8	8.1	34	39.9	59	142.6	84	416.8
9	8.6	35	42.2	60	149.4	85	433.6
10	9.2	36	44.6	61	156.4	86	450.9
11	9.8	37	47.1	62	163.8	87	468.7
12	10.5	38	49.7	63	171.4	88	487.1
13	11.2	39	52.4	64	179.3	89	506.1
14	12.0	40	55.3	65	187.5	90	525.8
15	12.8	41	58.3	66	196.1	91	546.1
16	13.6	42	61.5	67	205.0	92	567.0
17	14.5	43	64.8	68	214.2	93	588.6
18	15.5	44	68.3	69	223.7	94	611.0
19	16.5	45	71.9	70	233.7	95	634.0
20	17.5	46	75.7	71	243.9	96	658.0
21	18.7	47	79.6	72	254.6	97	682.0
22	19.8	48	83.7	73	265.7	98	707.3
23	21.1	49	88.0	74	277.2	99	733.2
24	22.4	50	92.5	75	289.1	100	760.0
25	23.8						

appendix H

Equilibrium Constants (Room Temperature)

H.1 Ionization Constants for Acids

Name	Reaction	K_A (ionic strength = 0)
Acetic	$HC_2H_3O_2 \rightleftharpoons H^+ + C_2H_3O_2^-$	1.75×10^{-5}
Arsenic		
K_{A1}	$H_3AsO_4 \rightleftharpoons H^+ + H_2AsO_4^-$	6.5×10^{-3}
K_{A2}	$H_2AsO_4^- \rightleftharpoons H^+ + HAsO_4^{2-}$	1.3×10^{-7}
K_{A3}	$HAsO_4^{2-} \rightleftharpoons H^+ + AsO_4^{3-}$	3.2×10^{-12}
Boric	$HBO_2 \cdot H_2O(H_3BO_3) \rightleftharpoons H^+ + BO_2^- \cdot H_2O$	5.9×10^{-10}
Carbonic		
K_{A1}	$H_2CO_3 \rightleftharpoons H^+ + HCO_3^-$	4.5×10^{-7}
K_{A2}	$HCO_3^- \rightleftharpoons H^+ + CO_3^{2-}$	4.7×10^{-11}
Chloroacetic	$CH_2ClCOOH \rightleftharpoons H^+ + CH_2ClCOO^-$	1.54×10^{-3}
Formic	$HCOOH \rightleftharpoons H^+ + HCOO^-$	1.7×10^{-4}
Hydrocyanic	$HCN \rightleftharpoons H^+ + CN^-$	5.0×10^{-10}
Hydrofluoric	$HF \rightleftharpoons H^+ + F^-$	6.7×10^{-4}
Hydrogen peroxide	$H_2O_2 \rightleftharpoons H^+ + HO_2^-$	1.8×10^{-12}
Hydrosulfuric (hydrogen sulfide)		
K_{A1}	$H_2S \rightleftharpoons H^+ + HS^-$	1.0×10^{-7}
K_{A2}	$HS^- \rightleftharpoons H^+ + S^{2-}$	1.0×10^{-14}
Hypochlorous	$HOCl \rightleftharpoons H^+ + OCl^-$	3.0×10^{-8}
Maleic		
K_{A1}	$C_2H_2(COOH)_2 \rightleftharpoons H^+ + (COOH)C_2H_2COO^-$	1.2×10^{-2}
K_{A2}	$(COOH)C_2H_2COO^- \rightleftharpoons H^+ + C_2H_2(COO)_2^{2-}$	6.0×10^{-7}
Nitrous	$HNO_2 \rightleftharpoons H^+ + NO_2^-$	5.1×10^{-4}
Oxalic		
K_{A1}	$H_2C_2O_4 \rightleftharpoons H^+ + HC_2O_4^-$	6.5×10^{-2}
K_{A2}	$HC_2O_4^- \rightleftharpoons H^+ + C_2O_4^{2-}$	6.1×10^{-5}
Phenol	$C_6H_5OH \rightleftharpoons H^+ + C_6H_5O^-$	1.1×10^{-10}
Phosphoric		
K_{A1}	$H_3PO_4 \rightleftharpoons H^+ + H_2PO_4^-$	7.1×10^{-3}
K_{A2}	$H_2PO_4^- \rightleftharpoons H^+ + HPO_4^{2-}$	6.3×10^{-8}
K_{A3}	$HPO_4^{2-} \rightleftharpoons H^+ + PO_4^{3-}$	4.3×10^{-13}
Phosphorous		
K_{A1}	$H_2PHO_3 \rightleftharpoons H^+ + HPHO_3^-$	1×10^{-2}
K_{A2}	$HPHO_3^- \rightleftharpoons H^+ + PHO_3^{2-}$	2.6×10^{-7}
Sulfuric		
K_{A1}	$H_2SO_4 \rightarrow H^+ + HSO_4^-$	>10
K_{A2}	$HSO_4^- \rightleftharpoons H^+ + SO_4^{2-}$	1.02×10^{-2}
Sulfurous		
K_{A1}	$H_2SO_3 \rightleftharpoons H^+ + HSO_3^-$	1.7×10^{-2}
K_{A2}	$HSO_3^- \rightleftharpoons H^+ + SO_3^{2-}$	6.4×10^{-8}

H.2 Ionization Constants for Bases

Name	Reaction	K_B (ionic strength = 0)
Ammonia	$NH_3 + H_2O \rightleftharpoons NH_4^+ + OH^-$	1.80×10^{-5}
Aniline	$C_6H_5NH_2 + H_2O \rightleftharpoons C_6H_5NH_3^+ + OH^-$	4.0×10^{-10}
Hydrazine	$N_2H_4 + H_2O \rightleftharpoons N_2H_5^+ + OH^-$	1.0×10^{-6}
Hydroxylamine	$NH_2OH + H_2O \rightleftharpoons NH_2OH^+ + OH^-$	1.07×10^{-8}
Pyridine	$C_5H_5N + H_2O \rightleftharpoons C_5H_5NH^+ + OH^-$	1.5×10^{-9}

H.3 Solubility-Product Constants

Reactions are of the type $AB(s) \rightleftharpoons A^+ + B^-$

Compound	K_{sp}	Compound	K_{sp}	Compound	K_{sp}
AgBr	4.9×10^{-13}	$CaC_2O_4 \cdot H_2O$	4×10^{-9}	MgC_2O_4	4.2×10^{-5}
AgCN	1.6×10^{-14}	CaF_2	4.9×10^{-11}	$MgNH_4PO_4$	3×10^{-13}
Ag_2CO_3	9.5×10^{-12}	$Ca(OH)_2$	4×10^{-6}	MnS	6.3×10^{-12}
AgCl	1.8×10^{-10}	$CaSO_4$	2.4×10^{-5}	$Ni(OH)_2$	2.0×10^{-16}
Ag_2CrO_4	1.1×10^{-12}	CdS	1.0×10^{-28}	NiS	2×10^{-21}
AgI	8.3×10^{-17}	CoS	4×10^{-21}	$PbBr_2$	3.9×10^{-5}
$AgIO_3$	3.0×10^{-8}	$Cr(OH)_3$	6×10^{-31}	$PbCl_2$	1.6×10^{-5}
$AgOH(Ag_2O)$	2.6×10^{-8}	$Cu(OH)_2$	1.6×10^{-19}	$PbCrO_4$	1.8×10^{-14}
AgSCN	1.0×10^{-12}	CuI	1.1×10^{-12}	PbF_2	2.7×10^{-8}
Ag_2SO_4	1.6×10^{-5}	CuS	4×10^{-36}	PbI_2	6.5×10^{-9}
$Al(OH)_3$	5×10^{-33}	$Fe(OH)_3$	4×10^{-38}	$PbSO_4$	1.6×10^{-8}
$BaCO_3$	5.1×10^{-9}	Hg_2Br_2	5.8×10^{-23}	$SrCO_3$	1.1×10^{-10}
$BaCrO_4$	1.2×10^{-10}	Hg_2Cl_2	1.3×10^{-18}	SrF_2	2.5×10^{-9}
BaF_2	1.0×10^{-6}	Hg_2I_2	4.5×10^{-29}	$SrSO_4$	3.2×10^{-7}
BaS_2O_3	1.6×10^{-5}	HgS	1×10^{-53}	TlCl	1.7×10^{-24}
$BaSO_4$	1.3×10^{-10}	$MgCO_3$	3.5×10^{-8}	$Zn(OH)_2$	3×10^{-17}
Bi_2S_3	1×10^{-97}	$Mg(OH)_2$	1.2×10^{-11}	ZnS	1.6×10^{-24}
$CaCO_3$	4.5×10^{-9}				

H.4 Dissociation Constants for Complex Ions

Compound	Reaction	K_{diss}
$Ag(NH_3)_2^+$	$Ag(NH_3)_2^+ \rightleftharpoons Ag^+ + 2 NH_3$	7.2×10^{-8}
$Cu(NH_3)_4^{2+}$	$Cu(NH_3)_4^{2+} \rightleftharpoons Cu^{2+} + 4 NH_3$	9.2×10^{-13}
$Ni(NH_3)_4^{2+}$	$Ni(NH_3)_4^{2+} \rightleftharpoons Ni^{2+} + 4 NH_3$	3.3×10^{-8}
$Zn(NH_3)_4^{2+}$	$Zn(NH_3)_4^{2+} \rightleftharpoons Zn^{2+} + 4 NH_3$	2.0×10^{-9}
$Ag(CN)_2^-$	$Ag(CN)_2^- \rightleftharpoons Ag^+ + 2 CN^-$	1×10^{-21}
$Ni(CN)_4^{2-}$	$Ni(CN)_4^{2-} \rightleftharpoons Ni^{2+} + 4 CN^-$	1×10^{-22}
$Zn(CN)_3^-$	$Zn(CN)_3^- \rightleftharpoons Zn^{2+} + 3 CN^-$	5×10^{-18}
$Ag(OH)_2^-$	$Ag(OH)_2^- \rightleftharpoons Ag^+ + 2 OH^-$	3×10^{-4}
$Al(OH)_4^-$	$Al(OH)_4^- \rightleftharpoons Al^{3+} + 4 OH^-$	5×10^{-34}
$Zn(OH)_3^-$	$Zn(OH)_3^- \rightleftharpoons Zn^{2+} + 3 OH^-$	6×10^{-15}
$Ag(S_2O_3)_2^{3-}$	$Ag(S_2O_3)_2^{3-} \rightleftharpoons Ag^+ + 2 S_2O_3^{2-}$	3.5×10^{-14}

Sources of data: L. G. Sillen and A. E. Martell, *Stability Constants of Metal–Ion Complexes*, The Chemical Society, London, 1964; and L. Meites (ed.), *Handbook of Analytical Chemistry*, McGraw-Hill, New York, 1963.

appendix

Standard Reduction Potentials

Electrode	Half-cell reaction	E° volts
Acid solution		
Li^+/Li	$Li^+ + e^- \rightleftharpoons Li(s)$	−3.04
K^+/K	$K^+ + e^- \rightleftharpoons K(s)$	−2.92
Rb^+/Rb	$Rb^+ + e^- \rightleftharpoons Rb(s)$	−2.92
Cs^+/Cs	$Cs^+ + e^- \rightleftharpoons Cs(s)$	−2.92
Ba^{2+}/Ba	$Ba^{2+} + 2\ e^- \rightleftharpoons Ba(s)$	−2.91
Sr^{2+}/Sr	$Sr^{2+} + 2\ e^- \rightleftharpoons Sr(s)$	−2.89
Ca^{2+}/Ca	$Ca^{2+} + 2\ e^- \rightleftharpoons Ca(s)$	−2.87
Na^+/Na	$Na^+ + e^- \rightleftharpoons Na(s)$	−2.71
La^{3+}/La	$La^{3+} + 3\ e^- \rightleftharpoons La(s)$	−2.52
Ce^{3+}/Ce	$Ce^{3+} + 3\ e^- \rightleftharpoons Ce(s)$	−2.48
Eu^{3+}/Eu	$Eu^{3+} + 3\ e^- \rightleftharpoons Eu(s)$	−2.41
Mg^{2+}/Mg	$Mg^{2+} + 2\ e^- \rightleftharpoons Mg(s)$	−2.36
Lu^{3+}/Lu	$Lu^{3+} + 3\ e^- \rightleftharpoons Lu(s)$	−2.26
H_2/H^-	$H_2(g) + 2\ e^- \rightleftharpoons 2\ H^-$	−2.25
Sc^{3+}/Sc	$Sc^{3+} + 3\ e^- \rightleftharpoons Sc(s)$	−2.08
Be^{2+}/Be	$Be^{2+} + 2\ e^- \rightleftharpoons Be(s)$	−1.85
U^{3+}/U	$U^{3+} + 3\ e^- \rightleftharpoons U(s)$	−1.79
Al^{3+}/Al	$Al^{3+} + 3\ e^- \rightleftharpoons Al(s)$	−1.66
Ti^{2+}/Ti	$Ti^{2+} + 2\ e^- \rightleftharpoons Ti(s)$	−1.63
V^{2+}/V	$V^{2+} + 2\ e^- \rightleftharpoons V(s)$	−1.19
Mn^{2+}/Mn	$Mn^{2+} + 2\ e^- \rightleftharpoons Mn(s)$	−1.18
Zn^{2+}/Zn	$Zn^{2+} + 2\ e^- \rightleftharpoons Zn(s)$	−0.76
Cr^{3+}/Cr	$Cr^{3+} + 3\ e^- \rightleftharpoons Cr(s)$	−0.74
$TlCl/Tl$	$TlCl(s) + e^- \rightleftharpoons Tl(s) + Cl^-$	−0.56
Ga^{3+}/Ga	$Ga^{3+} + 3\ e^- \rightleftharpoons Ga(s)$	−0.53
Fe^{2+}/Fe	$Fe^{2+} + 2\ e^- \rightleftharpoons Fe(s)$	−0.44
Eu^{3+}/Eu^{2+}	$Eu^{3+} + e^- \rightleftharpoons Eu^{2+}$	−0.43
Cr^{3+}/Cr^{2+}	$Cr^{3+} + e^- \rightleftharpoons Cr^{2+}$	−0.41
Cd^{2+}/Cd	$Cd^{2+} + 2\ e^- \rightleftharpoons Cd(s)$	−0.40
Ti^{3+}/Ti^{2+}	$Ti^{3+} + e^- \rightleftharpoons Ti^{2+}$	−0.37
$PbSO_4/Pb$	$PbSO_4(s) + 2\ e^- \rightleftharpoons Pb(s) + SO_4^{2-}$	−0.36
Tl^+/Tl	$Tl^+ + e^- \rightleftharpoons Tl(s)$	−0.34
Co^{2+}/Co	$Co^{2+} + 2\ e^- \rightleftharpoons Co(s)$	−0.28
H_3PO_4/H_3PO_3	$H_3PO_4 + 2\ H^+ + 2\ e^- \rightleftharpoons H_3PO_3 + H_2O$	−0.28
Ni^{2+}/Ni	$Ni^{2+} + 2\ e^- \rightleftharpoons Ni(s)$	−0.25
AgI/Ag	$AgI(s) + e^- \rightleftharpoons Ag(s) + I^-$	−0.15
Sn^{2+}/Sn	$Sn^{2+} + 2\ e^- \rightleftharpoons Sn(s)$	−0.14
Pb^{2+}/Pb	$Pb^{2+} + 2\ e^- \rightleftharpoons Pb(s)$	−0.13
H^+/H_2	$2\ H^+ + 2\ e^- \rightleftharpoons H_2(g)$	0.000
$AgBr/Ag$	$AgBr(s) + e^- \rightleftharpoons Ag(s) + Br^-$	+0.07
Sn^{4+}/Sn^{2+}	$Sn^{4+} + 2\ e^- \rightleftharpoons Sn^{2+}$	+0.15
Cu^{2+}/Cu^+	$Cu^{2+} + e^- \rightleftharpoons Cu^+$	+0.15

Electrode	Half-cell reaction	$E°$ (volts)
SO_4^{2-}/H_2SO_3	$SO_4^{2-} + 4\,H^+ + 2\,e^- \rightleftharpoons H_2SO_3 + H_2O$	+0.17
$AgCl/Ag$	$AgCl(s) + e^- \rightleftharpoons Ag(s) + Cl^-$	+0.22
Hg_2Cl_2/Hg	$Hg_2Cl_2(s) + 2\,e^- \rightleftharpoons 2\,Hg(l) + 2\,Cl^-$	+0.27
Cu^{2+}/Cu	$Cu^{2+} + 2\,e^- \rightleftharpoons Cu(s)$	+0.34
$Fe(CN)_6^{3-}/Fe(CN)_6^{4-}$	$Fe(CN)_6^{3-} + e^- \rightleftharpoons Fe(CN)_6^{4-}$	+0.36
Ag_2CrO_4/Ag	$Ag_2CrO_4(s) + 2\,e^- \rightleftharpoons 2\,Ag(s) + CrO_4^{2-}$	+0.46
Cu^+/Cu	$Cu^+ + e^- \rightleftharpoons Cu(s)$	+0.52
I_2/I^-	$I_2(s) + 2\,e^- \rightleftharpoons 2\,I^-$	+0.54
I_3^-/I^-	$I_3^- + 2\,e^- \rightleftharpoons 3\,I^-$	+0.54
MnO_4^-/MnO_4^{2-}	$MnO_4^- + e^- \rightleftharpoons MnO_4^{2-}$	+0.56
O_2/H_2O_2	$O_2(g) + 2\,e^- + 2\,H^+ \rightleftharpoons H_2O_2$	+0.68
$PtCl_4^{2-}/Pt$	$PtCl_4^{2-} + 2\,e^- \rightleftharpoons Pt(s) + 4\,Cl^-$	+0.73
Fe^{3+}/Fe^{2+}	$Fe^{3+} + e^- \rightleftharpoons Fe^{2+}$	+0.77
Hg_2^{2+}/Hg	$Hg_2^{2+} + 2\,e^- \rightleftharpoons 2\,Hg(l)$	+0.79
Ag^+/Ag	$Ag^+ + e^- \rightleftharpoons Ag(s)$	+0.80
NO_3^-/N_2O_4	$2\,NO_3^- + 4\,H^+ + 2\,e^- \rightleftharpoons N_2O_4(g) + 2\,H_2O$	+0.80
Hg^{2+}/Hg	$Hg^{2+} + 2\,e^- \rightleftharpoons Hg(l)$	+0.85
NO_3^-/HNO_2	$NO_3^- + 3\,H^+ + 2\,e^- \rightleftharpoons HNO_2 + H_2O$	+0.94
NO_3^-/NO	$NO_3^- + 4\,H^+ + 3\,e^- \rightleftharpoons NO(g) + 2\,H_2O$	+0.96
N_2O_4/NO	$N_2O_4(g) + 4\,H^+ + 4\,e^- \rightleftharpoons 2\,NO(g) + 2\,H_2O$	+1.03
Br_2/Br^-	$Br_2(l) + 2\,e^- \rightleftharpoons 2\,Br^-$	+1.07
ClO_4^-/ClO_3^-	$ClO_4^- + 2\,H^+ + 2\,e^- \rightleftharpoons ClO_3^- + H_2O$	+1.19
IO_3^-/I_2	$2\,IO_3^- + 12\,H^+ + 10\,e^- \rightleftharpoons I_2(s) + 6\,H_2O$	+1.20
$ClO_3^-/HClO_2$	$ClO_3^- + 3\,H^+ + 2\,e^- \rightleftharpoons HClO_2 + H_2O$	+1.21
O_2/H_2O	$O_2(g) + 4\,H^+ + 4\,e^- \rightleftharpoons 2\,H_2O$	+1.23
MnO_2/Mn^{2+}	$MnO_2(s) + 4\,H^+ + 2\,e^- \rightleftharpoons Mn^{2+} + 2\,H_2O$	+1.23
Tl^{3+}/Tl	$Tl^{3+} + 3\,e^- \rightleftharpoons Tl(s)$	+1.25
$Cr_2O_7^{2-}/Cr^{3+}$	$Cr_2O_7^{2-} + 14\,H^+ + 6\,e^- \rightleftharpoons 2\,Cr^{3+} + 7\,H_2O$	+1.33
Cl_2/Cl^-	$Cl_2(g) + 2\,e^- \rightleftharpoons 2\,Cl^-$	+1.36
PbO_2/Pb^{2+}	$PbO_2(s) + 4\,H^+ + 2\,e^- \rightleftharpoons Pb^{2+} + 2\,H_2O$	+1.46
Au^{3+}/Au	$Au^{3+} + 3\,e^- \rightleftharpoons Au(s)$	+1.50
MnO_4^-/Mn^{2+}	$MnO_4^- + 8\,H^+ + 5\,e^- \rightleftharpoons Mn^{2+} + 4\,H_2O$	+1.51
BrO_3^-/Br_2	$2\,BrO_3^- + 12\,H^+ + 10\,e^- \rightleftharpoons Br_2(l) + 6\,H_2O$	+1.52
Ce^{4+}/Ce^{3+}	$Ce^{4+} + e^- \rightleftharpoons Ce^{3+}$	+1.61
$HClO/Cl_2$	$2\,HClO + 2\,H^+ + 2\,e^- \rightleftharpoons Cl_2(g) + 2\,H_2O$	+1.63
$HClO_2/HClO$	$HClO_2 + 2\,H^+ + 2\,e^- \rightleftharpoons HClO + H_2O$	+1.64
$PbO_2/PbSO_4$	$PbO_2(s) + SO_4^{2-} + 4\,H^+ + 2\,e^- \rightleftharpoons PbSO_4(s) + 2\,H_2O$	+1.68
Au^+/Au	$Au^+ + e^- \rightleftharpoons Au(s)$	+1.69
MnO_4^-/MnO_2	$MnO_4^- + 4\,H^+ + 3\,e^- \rightleftharpoons MnO_2(s) + 2\,H_2O$	+1.70
H_2O_2/H_2O	$H_2O_2 + 2\,H^+ + 2\,e^- \rightleftharpoons 2\,H_2O$	+1.78
Co^{3+}/Co^{2+}	$Co^{3+} + e^- \rightleftharpoons Co^{2+}$	+1.81
$S_2O_8^{2-}/SO_4^{2-}$	$S_2O_8^{2-} + 2\,e^- \rightleftharpoons 2\,SO_4^{2-}$	+2.01
O_3/O_2	$O_3(g) + 2\,H^+ + 2\,e^- \rightleftharpoons O_2(g) + H_2O$	+2.07
F_2/F^-	$F_2(g) + 2\,e^- \rightleftharpoons 2\,F^-$	+2.87
F_2/HF	$F_2(g) + 2\,H^+ + 2\,e^- \rightleftharpoons 2\,HF$	+3.06
Basic solution		
BeO/Be	$BeO(s) + H_2O + 2\,e^- \rightleftharpoons Be(s) + 2\,OH^-$	−2.61
$Al(OH)_4^-/Al$	$Al(OH)_4^- + 3\,e^- \rightleftharpoons Al(s) + 4\,OH^-$	−2.33
$Zn(OH)_4^{2-}/Zn$	$Zn(OH)_4^{2-} + 2\,e^- \rightleftharpoons Zn(s) + 4\,OH^-$	−1.22
SO_4^{2-}/SO_3^{2-}	$SO_4^{2-} + H_2O + 2\,e^- \rightleftharpoons SO_3^{2-} + 2\,OH^-$	−0.93
H_2O/H_2	$2\,H_2O + 2\,e^- \rightleftharpoons H_2(g) + 2\,OH^-$	−0.83
$Ni(OH)_2/Ni$	$Ni(OH)_2(s) + 2\,e^- \rightleftharpoons Ni(s) + 2\,OH^-$	−0.72
S/S^{2-}	$S(s) + 2\,e^- \rightleftharpoons S^{2-}$	−0.45
$CrO_4^{2-}/Cr(OH)_3$	$CrO_4^{2-} + 4\,H_2O + 3\,e^- \rightleftharpoons Cr(OH)_3(s) + 5\,OH^-$	−0.13
NO_3^-/NO_2^-	$NO_3^- + H_2O + 2\,e^- \rightleftharpoons NO_2^- + 2\,OH^-$	+0.01
$S_4O_6/S_2O_3^{2-}$	$S_4O_6^{2-} + 2\,e^- \rightleftharpoons 2\,S_2O_3^{2-}$	+0.08
IO_3^-/I^-	$IO_3^- + 3\,H_2O + 6\,e^- \rightleftharpoons I^- + 6\,OH^-$	+0.26

Electrode	Half-cell reaction	$E°$ (volts)
ClO_3^-/ClO_2^-	$ClO_3^- + H_2O + 2\ e^- \rightleftharpoons ClO_2^- + 2\ OH^-$	+0.33
ClO_4^-/ClO_3^-	$ClO_4^- + H_2O + 2\ e^- \rightleftharpoons ClO_3^- + 2\ OH^-$	+0.36
O_2/OH^-	$O_2(g) + 2\ H_2O + 4\ e^- \rightleftharpoons 4\ OH^-$	+0.40
MnO_4^-/MnO_2	$MnO_4^- + 2\ H_2O + 3\ e^- \rightleftharpoons MnO_2(s) + 4\ OH^-$	+0.59
MnO_4^{2-}/MnO_2	$MnO_4^{2-} + 2\ H_2O + 2\ e^- \rightleftharpoons MnO_2(s) + 4\ OH^-$	+0.60
BrO_3^-/Br^-	$BrO_3^- + 3\ H_2O + 6\ e^- \rightleftharpoons Br^- + 6\ OH^-$	+0.61
BrO^-/Br^-	$BrO^- + H_2O + 2\ e^- \rightleftharpoons Br^- + 2\ OH^-$	+0.76
HO_2^-/OH^-	$HO_2^- + H_2O + 2\ e^- \rightleftharpoons 3\ OH^-$	+0.88
ClO^-/Cl^-	$ClO^- + H_2O + 2\ e^- \rightleftharpoons Cl^- + 2\ OH^-$	+0.89
O_3/O_2	$O_3(g) + H_2O + 2\ e^- \rightleftharpoons O_2(g) + 2\ OH^-$	+1.24

Glossary

Absolute rate theory Theory of reaction rates based on the assumption that the formation of the activated complex is an equilibrium process.

Absolute zero Lowest attainable temperature: −273°C or 0 K.

Absorbance Negative base-ten logarithm of the ratio of light transmitted by a sample compared to the incident light.

Acceleration Rate of change of velocity.

Acid Substance that reacts in water to produce hydronium ion.

Acid solution Solution with a higher H^+ concentration and a lower OH^- concentration than pure water.

Actinides Series of fourteen elements following actinium in the periodic table.

Activated complex Molecular structure associated with the maximum potential energy during an elementary reaction.

Activation energy Minimum energy with which molecules must collide in order to react.

Active site Portion of an enzyme's structure responsible for its catalytic activity.

Activity Unitless quantity that compares the thermodynamic state of a component in a real system to an arbitrarily defined standard state.

Adenosine diphosphate (ADP) Substance formed when ATP reacts to release its free energy.

Adenosine triphosphate (ATP) Substance whose synthesis stores free energy in living systems.

Adiabatic Not involving heat exchange between a system and its surroundings.

Aerosol Colloidal dispersion of a liquid or a solid in the gas phase.

Aliphatic Not aromatic.

Alkali metal Element belonging to periodic group IA.

Alkaline earth metal Element belonging to periodic group IIA.

Alkane Hydrocarbon that contains only single bonds.

Alkene Hydrocarbon containing at least one carbon–carbon double bond.

Alkyne Hydrocarbon containing at least one carbon–carbon triple bond.

Allotropes Forms of the same element differing in the nature of their chemical bonds, their molecular mass, or their crystal structure.

Alloy Metallic mixture.

Alpha helix Spiral arrangement of a polypeptide chain.

Alpha particle ^4_2He nucleus.

Alpha rays Streams of alpha particles emitted by some radioactive substances.

Amalgam Liquid mercury alloy.

Amino acid Substance whose molecules contain a carboxylate group, a protonated amine group, and a hydrogen atom attached to the same carbon atom.

Amorphous solid Solid with a disordered internal structure.

Ampere Unit of electric current equal to a flow of 1 coulomb per second.

Amphoteric Capable of acting as an acid under sufficiently basic conditions and as a base under sufficiently acidic conditions.

Anhydride Oxide that is the dehydrated form of a particular hydroxyl compound.

Anion Negatively charged ion.

Anion exchange resin Synthetic ion exchanger for anions.

Anode Positive terminal of an electrolysis apparatus; the electrode at which oxidation occurs.

Anodizing Formation of an oxide film on a metal surface by electrolysis.

Antibonding molecular orbital Molecular orbital that is less stable than the atomic orbitals from which it is constructed.

Anticodon Three-unit sequence of a tRNA molecule that binds to mRNA.

Antifluorite lattice Ionic crystal lattice in which the cations occupy the tetrahedral holes of a face-centered cubic lattice of anions.

Antiparticles Two particles with the same mass but opposite charges of the same magnitude.

Aromatic Resembling benzene in many of its properties.

Arrhenius equation Equation describing the effect of temperature on a reaction rate.

Associated Containing molecules that cluster together due to hydrogen bonding.

Atom Quantum of mass of a particular element.

Atomic mass unit One-twelfth of the mass of the ^{12}C isotope.

Atomic number Position of an element in the periodic table. The number of protons in an atom's nucleus.

Aufbau principle Prediction of the ground-state electron configuration of an atom by filling the lowest-energy subshell first and then, as necessary, using the next higher-energy level.

Autoprotolysis Transfer of a proton from one water molecule to another, producing a hydronium ion and a hydroxide ion.

Avogadro's hypothesis Equal volumes of gases at the same temperature and pressure contain equal numbers of molecules, regardless of their identity.

Avogadro's number 6.023×10^{23}; number of particles in a mole.

Azeotrope Liquid solution that evaporates to produce vapor with the identical composition.

Azimuthal quantum number Quantum number describing the angular momentum of an electron.

Balmer's law Mathematical equation used to calculate the various wavelengths of the atomic emission spectrum of hydrogen.

Barometer Closed-end manometer used to measure atmospheric pressure.

Base Substance that produces hydroxide ion in aqueous solution.

Base group Nitrogen-containing organic group attached to each deoxyribose unit in DNA.

Basic oxygen process Industrial method of converting crude iron to steel.

Basic solution Solution with a higher OH^- concentration and lower H^+ concentration than pure water.

Beer–Lambert law Law describing the relationship between a sample's absorbance, path length, concentration, and absorptivity.

Beta particle High-energy electron emitted by an unstable nucleus.

Beta rays Streams of beta particles.

Beta structure Arrangement in which two polypeptide chains lie side by side.

Bidentate Able to occupy two coordination positions at the same time.

Bimolecular Involving collisions between two reactant molecules.

Blast furnace Industrial apparatus for producing iron by the reduction of iron oxides with carbon monoxide.

Body-centered lattice Lattice with lattice points at the corners and center of the unit cell.

Boiling point Temperature at which the equilibrium vapor pressure of a substance equals the external pressure.

Bond-dissociation energy Energy needed to break a covalent bond.

Bond length See Internuclear distance.

Bonding molecular orbital Molecular orbital that is more stable than the atomic orbitals from which it is constructed.

Boride Binary compound of boron and a metal.

Born–Haber cycle Method for analyzing the energy change involved in the formation of an ionic compound in terms of the energies of individual steps leading to the formation of the compound from its elements.

Boyle's law Volume of a gas sample is inversely proportional to its pressure.

Bragg's law Equation describing the relationships among the reflection angle of x rays, the wavelength of the x ray, and the distance between two parallel planes of atoms.

Branched-chain process Situation in which an energy-producing reaction increases the concentration of particles that initiate it.

Brass Copper–zinc alloy.

Bravais lattice One of the fourteen three-dimensional lattices that describe the internal structures of crystals.

Breeder reactor Nuclear reactor that creates more fissionable material than it consumes.

Brønsted–Lowry acid Proton donor.

Brønsted–Lowry base Proton acceptor.

Bronze Alloy of copper and tin.

Brownian motion The rapid, erratic motion of a small but visible particle due to its collisions with individual molecules.

Bubble chamber Instrument for observing the paths of radioactive particles by the formation of bubbles in a superheated liquid.

Buffer solution Solution containing a weak acid or base and its conjugate base or acid in comparable concentrations.

Calorimetry Measurement of heat changes due to chemical reactions.

Carbide Compound of carbon combined with elements of equal or lower electronegativity.

Carbohydrates Class of biochemically important substances including sugars, starches, and cellulose.

Carbon dating Method for measuring the age of organic matter by measuring intensity of carbon-14 radioactivity.

Carbon–nitrogen cycle Sequence of fusion reactions occurring in the sun.

Catalyst Substance that increases the rate of a reaction without itself being consumed in the reaction.

Catenation Formation of long-chain molecules.

Cathode Negative terminal of an electrolysis apparatus; the electrode at which reduction occurs.

Cathode ray Beam of electrons emanating from a cathode into a vacuum.

Catnode-ray tube Apparatus for generating and studying cathode rays; an evacuated tube with a large electric potential across it.

Cation Positively charged ion.

Cation exchange resin Synthetic ion exchanger for cations.

Cell notation Abbreviated method for displaying the structure of a galvanic cell.

Celsius scale Temperature scale in common use in chemical laboratories.

Charge-to-mass ratio Ratio of the electric charge on a particle to the particle's mass.

Charles' law Volume of a sample of gas at constant pressure is directly proportional to its temperature, expressed in K.

Chelate group Polydentate ligand.

Chelating agent See Chelate group.

Chelation Process of forming complexes of metal ions with polydentate ligands.

Chemical equation Method of summarizing the molecular changes occurring in a chemical reaction.

Chemical reaction Process that transforms one or more substances into a different set of substances.

Chemiluminescence Emission of light by an excited state of a product of a chemical reaction.

Chemisorb Adsorb molecular species on a surface by formation of chemical bonds.

Chirality Property of differing from one's mirror image reflection.

Chlorophyll Large organic molecule responsible for converting photons into chemical free energy during photosynthesis.

Chloroplasts Subcellular structures responsible for photosynthesis in green plants.

Cholesteric liquid crystal Liquid crystal in which the molecules lie lengthwise and parallel in planes.

Chromatography Separation of a mixture by selective distribution of the components between a stationary absorbing phase and a moving solvent phase.

Clathrates Solid substances in which small molecules are trapped in the holes of an open crystal structure.

Clausius–Clapeyron equation Equation describing the effect of temperature on the equilibrium vapor pressure of a substance.

Closed-end manometer Device for measuring gas pressure by balancing it against the pressure of a liquid column.

Closed system System that does not exchange matter with its surroundings.

Cloud chamber Instrument for observing the paths of radioactive particles by the formation of droplets from a supersaturated vapor.

Codon Three-unit sequence of mRNA that is complementary to a particular tRNA molecule.

Coenzyme Complicated organic molecule serving as a cofactor for an enzyme.

Cofactor Nonprotein molecule or ion necessary for an enzyme's activity.

Colligative properties Physical properties of solutions depending on solute concentration but not solute identity.

Collision frequency Number of collisions between reactant molecules per second when the reactant concentrations are each 1 M.

Collision theory Theory that explains the rates of elementary reactions on the basis of the conditions required for a productive collision between molecules.

Colloidal suspensions Dispersions of relatively large particles called colloids in a homogeneous medium.

Colloids Relatively large particles dispersed in a homogeneous medium.

Common-ion effect The high concentration of one of its ions suppresses the solubility of a salt.

Complementary sequence Sequence of one DNA chain that matches the sequence of the other chain in the double helix.

Complex See Complex ion.

Complex ion Polyatomic ion formed by the combination of a metal ion with anions or Lewis bases.

Compound Substance that can be chemically decomposed or synthesized.

Concentration Either a ratio of two components of a solution or a ratio of the amount of one component to the total amount of sample.

Concentration cell Cell composed of two half-cells using the same half-reaction but containing different concentrations.

Condensed phase State of matter in which the molecules are close together.

Conduction band Cluster of nearly identical energy levels for the outer-shell electrons in a metal crystal lattice.

Conjugate acid Species produced by proton addition to a base.

Conjugate base Species produced by removal of a proton from an acid.

Conservation of charge Charge is neither created nor destroyed in a chemical reaction.

Conservation of orbitals Number of hybrid orbitals must equal the number of atomic orbitals used to construct them.

Contact process Industrial synthesis of sulfuric acid by oxidation of sulfur dioxide.

Control rod Rod made of a substance that absorbs neutrons effectively that is used to control the rate of fission in a nuclear reactor.

Coordinate covalent bond Bond formed by electron-pair donation from a Lewis base to a Lewis acid.

Coordination complex See Complex ion.

Coordination number 1. Number of nearest neighbors of an atom in a crystal lattice. 2. Number of ligands around the central metal atom in a complex ion.

Cosmic rays Streams of small nuclei, mesons, and neutrons that bombard the earth from outer space.

Coulomb Unit of charge in practical electronics.

Coulomb's law Force acting between two charges is proportional to the charges and inversely proportional to the square of the distance between them.

Coulometric titration Titration in which the amount of titrant used is measured by the time and current needed to produce it.

Counts per minute (cpm) Common unit of relative radioactivity.

Covalent bond Chemical bond resulting from the sharing of electrons between two atoms.

Covalent crystal Crystal held together by a network of covalent bonds.

Covalent substance Substance whose atoms are held together by covalent bonds.

Critical mass Mass of fissionable material required to ensure a self-sustaining fission reaction.

Critical pressure Pressure that must be applied to a gas in order to liquefy it at its critical temperature.

Critical temperature Temperature above which a particular gas cannot be liquefied by compression.

Crystal Solid particle with well-defined planar surfaces.

Crystal field theory. Theory developed to describe the electronic states of transition metal ions in crystal lattices and later used to explain the colors of complex ions.

Crystal lattice energy (ionic) Energy released when cations and anions in the gas phase coalesce into an ionic crystal lattice.

Crystal systems Classification of crystals based on symmetry elements.

Crystallinity Measure of the internal order of a solid.

Cubic closest-packed Arranged with cubic symmetry and the smallest free space possible for a regular array of spheres of one size.

Curie (Ci) Quantity of radioactive material that produces 3.700×10^{10} nuclear disintegrations per second.

Cyclic compound Compound whose molecules contain rings of covalently bound atoms.

Cyclotron Circular particle accelerator.

Cytochromes Group of complicated molecules involved in the electron transport chain.

Dalton's law of partial pressures Total pressure of a gas mixture is the sum of the partial pressures of the individual components.

Daniell cell Galvanic cell consisting of a Zn-Zn^{2+} anode and a Cu^{2+}-Cu cathode.

Daughter nuclei Nuclei formed by radioactive decay of other nuclei.

Deacon process Commercial process for the preparation of chlorine by air oxidation of hydrogen chloride gas.

Decomposition potential Minimum electric potential required for a particular electrolysis.

Defect structures Various lattice imperfections that occur in solids.

Degenerate orbitals Two or more orbitals with the same energy.

Degree of dissociation Proportion of solute molecules that are dissociated at equilibrium.

Deliquesce Absorb so much water from the air that the substance turns into a moist paste.

Denaturation Reaction that destroys the shape of a protein molecule.

Denitrification Conversion of biologically useful nitrogen compounds into elemental nitrogen.
Deoxyribonucleic acid (DNA) Substance that contains the genetic code.
Deuterium Mass-2 isotope of hydrogen.
Diagonal similarity Similarity in behavior between elements lying in a diagonal band running from the upper left to the lower right of the periodic table.
Dialysis Removal of low-molecular-mass solutes from a colloidal dispersion by diffusion of the low-molecular-mass solutes through a membrane.
Diamagnetic Unable to act as a temporary magnet; being weakly repelled by magnetic fields.
Diffraction pattern Pattern of light and dark regions formed when light waves pass through two parallel slits.
Diffusion Migration of a substance from one region to another due to its random molecular motion.
Dipolar Neutral, but having equal positive and negative charges at opposite ends.
Dipole moment Measure of the charge separation between the two ends of a polar molecule.
Diradical Species containing two unpaired electron spins.
Disproportionation Reaction in which one particle of a particular species oxidizes an identical particle and is itself reduced.
Dissociation Reaction of an electrolyte in water to produce ions.
Dissociation constant Equilibrium constant for the dissociation of a weak acid or base into H^+ and an anion or OH^- and a cation.
Distillation Separation of a mixture by selective evaporation.
Double covalent bond Covalent bond formed by the sharing of two pairs of electrons.
Double helix Two intertwined DNA chains.
Dow process Commercial method for extracting magnesium from seawater.
Downs cell Apparatus for producing sodium and chlorine by electrolysis of molten sodium chloride.
Dry cell Galvanic cell based on the reduction of an electrolyte paste containing manganese dioxide and the oxidation of a zinc anode.

Effective reduction potential Reduction potential for a half-reaction minus the overpotential for that half-reaction.
Elastic collision Collision that conserves kinetic energy.
Electric charge One of two types of electric states, either positive or negative. Opposite charges attract each other, like charges repel each other.
Electric potential Potential energy of a charged object due to the electric forces acting on it.
Electrically neutral Carrying no net charge.
Electrolysis Use of electric potential energy to cause a chemical reaction that does not occur spontaneously.
Electrolyte Substance whose aqueous solution conducts electricity.
Electrolytic corrosion Accelerated corrosion of a metal when in contact with a less active metal.
Electromagnetism Formation of a magnetic field due to the movement of electric charge.
Electron Quantum of negative charge occurring in all atoms.
Electron affinity Decrease in potential energy as an atom in the gas phase attracts an electron.

Electron capture Type of nuclear decay in which a neutron-deficient nucleus absorbs one of its surrounding electrons.
Electron configuration Distribution of the electrons in an atom among its subshells.
Electron-deficient hydride Hydrogen compounds of beryllium, magnesium, group IIIA elements, and some transition metals in which some atoms lack octet electron configurations.
Electron-deficient species Molecule or ion in which at least one atom has fewer valence electrons than its nearby noble gas.
Electron-density function Equation describing the variations of electron density within a unit cell.
Electron-dot structure See Structural formula.
Electron spin Quantum number describing the magnetic state of an electron.
Electron transport chain Sequence of redox reactions occurring in mitochondria by which reduced nucleotides are oxidized by oxygen.
Electron volt Energy gained by an electron accelerated through a potential of 1 V.
Electronegativity Unitless number that expresses, on a relative scale, a covalently bonded atom's ability to attract the shared electrons.
Electrophoresis Migration of charged solute or colloid particles in an electric field.
Electroplating Plating of a metal on a surface by electrolysis.
Electrorefining Purification of a metal by electrolysis.
Electrowinning Production of an element from its compound by electrolysis.
Element Substance that cannot be decomposed, synthesized, or transformed into another element by a chemical reaction.
Elementary reaction Individual step in a reaction mechanism.
Emission spectrum Pattern of wavelengths of light emitted by atoms of a particular element when they are excited by heating or an electric arc.
Empirical formula See Simplest formula.
Emulsion Colloidal dispersion of a liquid in a liquid.
Enantiomers Pair of nonsuperimposable mirror images.
End point Stage of a titration at which the indicator changes color.
Endothermic reaction Chemical reaction that consumes heat.
Energy Ability to do work.
Energy level One of the discrete energies an electron in a particular atom may have.
Enthalpy Energy of a system plus the product of its pressure and volume.
Enthalpy of activation Excess heat a process must obtain over the ground state in order to occur.
Enthalpy of formation Enthalpy of the isothermal reaction needed to produce 1 mol of a substance from its elements in their most stable molecular and physical states.
Entropy Measure of the molecular randomness in a system.
Entropy of activation Excess entropy a process must obtain over the ground state in order to occur.
Enzyme Specific biochemical catalyst.
Enzyme–substrate complex Complex formed by the binding of a substrate molecule to the active site of an enzyme.
Equilibrium State at which no further net chemical change occurs.
Equilibrium constant Value of a reaction's mass-action ratio at equilibrium.

Equilibrium control Situation in which the major products of a reaction are the more stable substances.

Equilibrium vapor pressure Partial pressure of vapor in equilibrium with the liquid phase.

Equivalence point When equal amounts of equivalents of reactants have been combined in a titration.

Equivalent See Equivalent mass.

Equivalent mass Mass of a substance that supplies (1) in electrolysis, 1 mol of electrons, or (2) in acid–base reactions, 1 mol of H^+ or OH^-.

Excited state Less stable quantum state of an atom or molecule.

Exothermic reaction Chemical reaction that releases heat.

Extinction coefficient See Molar absorptivity.

Face-centered lattice Lattice with lattice points at the corners and centers of all the faces of the unit cells.

Fahrenheit scale Temperature scale commonly used in the United States.

Faraday Charge of 1 mol of electrons.

Faraday's law Charge needed to produce 1 mol of atoms of any element by electrolysis is a whole-number multiple of 9.65×10^4 coul.

Faraday–Tyndall effect Scattering of light by a colloidal dispersion.

Fat Substance whose molecules contain three ester bonds between the three alcohol groups of glycerol and three straight-chain carboxylic acids.

Ferromagnetism Ability to form permanent magnets.

First ionization potential Energy needed to remove an electron from an atom in the gas phase to form a +1 ion in the gas phase.

First law of thermodynamics See Law of conservation of energy.

Fission Splitting of a nucleus into two smaller fragments.

Flame spectrophotometer Device for measuring the intensities of various atomic emission lines.

Flame test Identification of an element by the characteristic color of its flame.

Flavin nucleotides Important class of biochemical oxidants.

Fluorescence Absorption of light at one wavelength and then rapid emission of longer-wavelength light.

Forbidden transition Change in quantized energy levels not readily stimulated by the absorption of light.

Force Any influence that alters the speed or direction of motion of an object.

Force constant Property related to a chemical bond's resistance to distortion.

Formality Moles of a component added to a solution (as opposed to present in a solution) per liter of solution.

Formula mass Total mass, expressed in atomic mass units, of all the atoms in a simplest formula.

Frasch process Method of extracting elemental sulfur from subterranean deposits.

Free radical Molecule, ion, or atom with one or more unpaired electron spins.

Frenkel defect Crystal defect in which positive or negative ions occupy interstitial locations or tetrahedral or octahedral holes instead of their regular lattice positions.

Frequency Number of wave cycles occurring at any point in a unit of time.

Frequency factor Rate at which a chemical reaction would occur if all molecular collisions had enough energy to react.

Fuel cell Galvanic cell in which the reactants flow into their respective electrodes as the cell operates.

Fuel elements Rods of subcritical amounts of fissionable material used in nuclear reactors.

Functional group Portion of a molecule responsible for the reactions that characterize it as a member of a particular family.

Fusion Merging of two nuclei to form a nucleus with a larger atomic number.

Galvanic cell Device for extracting electric potential energy from a redox reaction.

Gamma rays Light of very short wavelengths emitted by unstable nuclei.

Gas State of matter that has no definite volume but completely fills its container.

Gay-Lussac's law Pressure of a gas sample at constant volume is proportional to the temperature, expressed in degrees Kelvin.

Geiger–Müller counter Instrument for detecting radioactivity by ionization of a gas.

Geometric isomers Isomers differing in the geometric arrangements of groups attached to doubly bound carbon atoms or transition-metal ions.

Gibbs free energy Enthalpy of a system less the product of its temperature and entropy.

Glass electrode Special glass electrode that responds to changes in hydrogen-ion concentration.

Glycolysis Sequence of reactions by which glucose is metabolized to yield pyruvate.

Graham's law Relative rates of diffusion of two gases are inversely proportional to the square roots of their molecular masses.

Ground state Most stable quantum state of an atom or molecule.

Haber process Industrial synthesis of ammonia from nitrogen and hydrogen.

Half-cell Electrode of a galvanic cell.

Half-cell potential Potential of a half-cell measured against a standard hydrogen electrode.

Half-life Time required for the concentration of a reactant to decrease by half.

Half-reaction Reaction involving electrons, either as reactants or as products.

Halide Binary compound formed from any of the halogens (group VIIA elements) and a metallic element.

Hall process Commercial preparation of aluminum by electrolysis of its oxide.

Halogen Element belonging to periodic group VIIA.

Heat Energy of molecular motion.

Heat capacity Energy needed to raise the temperature of 1 g of a substance 1°C.

Heat of solution Enthalpy change occurring when a solution forms.

Heavy water Deuterium oxide, D_2O.

Heisenberg uncertainty principle It is impossible to determine both the momentum and the position of an object with absolute precision.

Heme Large organic group present in cytochromes.

Henry's law Vapor pressure of a solute in a dilute solution is proportional to its concentration.

Hess's law The sum of the enthalpies of two or more reactions equals the enthalpy of the net reaction that would result from these reactions occurring consecutively.

Heterocyclic compound Compound whose molecules contain rings made up of at least two different types of atoms.
Heterogeneous Consisting of distinct regions of differing composition.
Hexagonal closest-packed Arranged with hexagonal symmetry and the smallest free space possible for a regular array of spheres of one size.
Homogeneous Having the same composition throughout the sample.
Hund's rules Electrons occupy different orbitals of the same subshell, with parallel spins, before pairing.
Hybrid orbital Atomic orbital constructed by averaging two or more hydrogenlike atomic orbitals.
Hydrated hydrogen ion See Hydronium ion.
Hydrates Compounds in which water molecules occupy definite positions in a crystal lattice.
Hydration Solvation by water.
Hydration energy Energy released by hydration of a mole of ions.
Hydrocarbon Binary compound of carbon and hydrogen.
Hydrogen bond Moderately strong attraction between a hydrogen atom covalently bound to a small, strongly electronegative atom (N, O, or F) and an unshared electron pair on another of these atoms.
Hydrogen peroxide H_2O_2
Hydrolysis (basic) Reaction in which a weak base receives a proton from water and releases a hydroxide ion.
Hydronium ion H_3O^+
Hydroxide ion OH^-
Hydroxides Basic hydroxyl compounds.
Hydroxyl —OH

Icosahedron Three-dimensional figure with twenty triangular faces.
Ideal gas Mythical gas that would obey the gas laws perfectly under all conditions of temperature and pressure.
Ideal gas law Single equation, $PV = nRT$, that combines all the laws relating pressure, volume, number of moles, and temperature of a gas sample.
Ideal solution Solution that obeys Raoult's law.
Induced radioactivity Decay of nuclei produced by collisions of high-energy particles with stable nuclei.
Infrared Electromagnetic radiation whose frequency ranges from about 10^{14} to 10^{16} Hz.
Inner orbital complex Complex ion in which ligand bonding involves inner-shell d orbitals as opposed to outer-shell d orbitals.
Interference phenomena Interaction of two or more waves to either enhance or cancel the waves.
Intermediate Unstable species formed in one step of a mechanism and destroyed in a subsequent step.
Internuclear distance Distance between the two nuclei of covalently bonded atoms.
Interstice Space between three spheres in a plane.
Interstitial hydride Metallic hydride in which protons occupy interstitial positions of a metallic crystal lattice.
Iodometric titration Titration in which triiodide ion, formed in equivalent amounts by the oxidation of iodide ion by an oxidizing agent, is titrated with a standard thiosulfate solution.
Ion Charged atom or group of atoms.

Ion exchange Reaction in which ions in solution displace other types of ions bound to the surface of a water-insoluble substance.
Ion exchange chromatography Chromatographic separation of ions using an ion exchange resin as the stationary phase.
Ionic bond Type of chemical bonding that occurs when one type of atom gives one or more electrons to another type of atom.
Ionic carbide Carbide that is readily decomposed by water to form a metal hydroxide and a hydrocarbon.
Ionic compound Compound whose structure is an ionic crystal lattice.
Ionic crystal Crystal whose lattice points are occupied by ions.
Ionic hydride Ionic compound containing the H^- ion.
Ionic oxide Ionic compound containing the O^{2-} ion.
Ionization counter Instrument for detecting radioactivity by the ionization of a gas.
Ionization potential Energy required to remove an electron from a gaseous atom or ion.
Isoelectronic ions Ions with the same electron configuration.
Isomerism Existence of two compounds with the same molecular formula.
Isomers Compounds with the same molecular formula but differing structures.
Isothermal Occurring in a system maintained at constant temperature.
Isotope effect Difference in a chemical or physical property of two molecules of the same compound due to mass differences between two isotopes.
Isotopes Two or more atoms of the same element that differ from each other in mass.

Joule Unit of energy equal to 1 kg-m^2/sec^2.

Kelvin scale Temperature scale for which 0 K represents the lowest attainable temperature (equal to −273°C). The temperature change represented by a change of 1 K is the same as 1°C.
Kilogram Standard unit of mass in the SI system.
Kinetic characteristics Rates at which reactions approach equilibrium, and the factors influencing these rates.
Kinetic control Situation in which the major products of a reaction are those formed more rapidly, regardless of their relative stabilities.
Kinetic energy Energy an object possesses due to its own motion.
Kinetic isotope effect Difference in the rates of reaction of two molecules of the same compound due to a difference in mass between two isotopes.
Kinetic-molecular theory of gases Theory of gas behavior based on the random motion of molecules in a gas and their collisions with the container walls.

Lanthanide contraction Difference in atomic radius between lanthanum and lutetium.
Lanthanides (rare earths) Group of fourteen elements following lanthanum in the periodic table.
Lattice Periodic, repeating array of points with identical environments about each point in the lattice.
Lattice array Arrangement of molecules in a crystal.
Law of combining volumes Gases combine chemically in simple ratios of volume.
Law of conservation of energy The total energy of the universe remains constant.

Law of conservation of mass There is no change in the total mass of a system due to a chemical reaction.

Law of conservation of momentum The total momentum of interacting objects always remains the same.

Law of definite proportions A pure compound has a definite and characteristic composition by mass of its constituent elements.

Law of mass action The mass-action ratio of an equilibrium mixture equals the reaction's equilibrium constant, regardless of the initial concentrations.

Law of multiple proportions A ratio of whole numbers exists between the masses of one element that can combine with a certain mass of another element.

Lead storage battery Series of galvanic cells based on the reduction of lead dioxide and the oxidation of lead.

Le Chatelier's principle A system at equilibrium responds to an external stress so as to minimize the effects of that stress.

Leveling effect The strongest acid that can exist in a solution is the conjugate acid of the solvent, and the strongest base is the conjugate base of the solvent.

Lewis acid Electron pair acceptor in a reaction that forms a covalent bond.

Lewis base Electron pair donor in a reaction that forms a covalent bond.

Lewis octet rule Atoms, ions, and molecules tend to react so that each atom attains the valence electron configuration of a noble gas.

Lewis structure See Structural formula.

Ligand Molecule or ion that combines with a transition metal ion to form a complex.

Ligand field theory Extension of crystal field theory that deals with the electronic character of bonded ligands.

Light Unique form of energy that travels through empty space.

Limiting reactant Reactant that is completely consumed by an excess of the other reactants.

Linear accelerator Apparatus for accelerating bombarding particles so that they have enough energy to react with stable nuclei.

Lipids Class of water-insoluble organic compounds involved in biochemical systems.

Liquid State of matter with a definite volume but no definite shape.

Liquid crystal Phase that shows the mechanical properties of liquids and optical properties of crystals.

Liquid-drop model Theoretical model used to explain the upper limits in size and charge for stable nuclei.

Liquid–liquid juncture Boundary between two solutions in a galvanic cell.

Magic numbers Particular numbers of nucleons that confer added stability to nuclei.

Magnetic dipoles Indivisible combinations of north-seeking ends and south-seeking ends found in magnets.

Magnetic quantum number Quantum number describing the orientation of an orbital in space.

Mass Tendency of an object to resist a change in motion.

Mass-action ratio Ratio of the multiple of the concentrations of products to the multiple of the concentrations of reactants, each concentration being raised to an exponential power equal to the stoichiometric coefficient of the substance.

Mass number Sum of the protons and neutrons in an atomic nucleus.

Mass spectrometer Instrument for measuring the masses of atoms by deflecting beams of their ions in a perpendicular magnetic field.

Mechanism Sequence of molecular events producing an overall chemical change.

Melting point Temperature at which the solid and liquid phases of a substance are at equilibrium.

Mesons Group of subatomic particles whose masses are intermediate between those of electrons and those of nucleons.

Messenger RNA (mRNA) Ribonucleic acid; synthesized in the nucleus, and carries the genetic code to ribosomes.

Metal Element characterized by its luster, thermal and electric conductivity, and malleability.

Metal carbonyl Compound formed between a transition metal and carbon monoxide.

Metallic crystal Crystal in which the valence electrons are free to move throughout the crystal in response to an electrostatic force.

Metallic hydride Compound between hydrogen and a transition metal, lanthanide or actinide.

Metallic oxide Compound of oxygen and some transition metals in which oxide ions occupy interstitial positions in the metal crystal lattice.

Metastable Existing even though there are more stable states.

Microscopic reversibility Reverse reaction proceeds via the same intermediates and activated complexes as its forward reaction, but in reverse order.

Microwave Electromagnetic radiation whose frequency ranges from about 3×10^{10} to 3×10^{11} Hz.

Mischmetal Combustible alloy of the lighter lanthanide metals.

Mitochondria Subcellular structures in which many important metabolic oxidation reactions occur.

Mixed metal oxide Compound of two different metals with oxygen.

Mixture Sample containing two or more distinct substances that can be physically separated from each other.

Moderator Substance used in a nuclear reactor to slow neutrons to thermal speeds without capturing them.

Molal boiling-point elevation Increase in the boiling point of a solvent caused by 1 molal total solute concentration.

Molal freezing-point depression Decrease in the freezing point of a solvent due to 1 molal total solute concentration.

Molality Moles of a component of a solution per kilogram of solvent.

Molar absorptivity Constant equal to the absorbance of a sample at a particular wavelength divided by the product of the molar concentration and the path length.

Molar enthalpy of fusion Heat needed to melt 1 mol of a substance.

Molar enthalpy of vaporization Heat needed to evaporate 1 mol of a liquid.

Molar volume Volume occupied by 1 mol of an ideal gas (22.4 L at STP).

Molarity Moles of a component of a solution per liter of solution.

Mole Any collection of 6.023×10^{23} particles.

Mole fraction Ratio of the number of moles of a component of a solution to the total number of moles of all components in the solution.

Molecular crystal Crystal whose lattice points are occupied by discrete molecules.

Molecular electronic energy level Quantum state for an electron in a molecule.
Molecular formula Number and type of atoms of the various elements in a molecule of a compound.
Molecular hydride Covalent compound between hydrogen and an element of periodic groups IVA–VIIA.
Molecular mass Total mass, expressed in atomic mass units, of all the atoms in a molecule.
Molecular orbital Graph of the distribution in space of an electron occupying a particular energy level of a molecule.
Molecular spectroscopy Measurement of the wavelengths of electromagnetic radiation absorbed or emitted by molecules.
Molecule Cluster of atoms.
Momentum Mass of an object multiplied by its velocity.
Monatomic ion Charged atom.
Monomolecular Involving only one reactant molecule.
Monosaccharide Aldehyde or ketone containing several alcohol groups.

Natural radioactivity Radioactivity due to the presence of unstable nuclei on earth.
Near ultraviolet Wavelength range from 180 to 400 nm.
Negative charge Type of electric charge carried by electrons.
Negatron See Beta particle.
Nematic liquid crystal Liquid crystal in which the molecules are parallel to each other but are not restricted to planes.
Nernst equation Equation describing the effect of concentration on half-cell potentials.
Net ionic equation Chemical equation displaying only those species involved in the chemical change.
Neutral solution Solution whose molar concentration of H^+ equals the molar concentration of OH^-.
Neutralization reaction Reaction between hydronium ion and hydroxide ion to produce water.
Neutrino Small, neutral particle, having practically no rest mass, formed in radioactive decays.
Neutron Neutral particle with mass equal to 1 amu found in atomic nuclei.
Neutron activation analysis Elemental analysis by observing the radioactivities and decay particles produced when the sample is bombarded by thermal neutrons.
Newton's first law A body maintains a constant state of motion unless a force acts on it.
Newton's second law The force acting on an object equals the mass of the object multiplied by the acceleration it receives.
Newton's third law When one object exerts a force on another, the second object exerts an equal force acting in the opposite direction of the first.
Nicotinamide adenosine dinucleotide (NAD^+) Important biochemical oxidant.
Nicotinamide adenosine dinucleotide phosphate ($NADP^+$) Important biochemical oxidant.
Nitrogen cycle Natural cycle of nitrogen fixation, a chain of biological interconversions of nitrogen compounds and denitrification.
Nitrogen fixation Conversion of elemental nitrogen into biologically useful forms.
Noble gas Element belonging to the family in the right-hand column of the periodic table.

Nodal surface Surface on which an electron in a particular orbital has zero probability of being found.
Nonelectrolyte Substance whose aqueous solution does not conduct electricity.
Nonstoichiometric compound Compound whose elemental formula can vary and whose formula deviates from a simple stoichiometric ratio.
Normal boiling point Boiling point at 1 atm external pressure.
Normality Concentration unit equal to the number of equivalents of solute per liter of solution.
n–type semiconductor Semiconductor containing an impurity with an extra valence electron not needed for bonding.
Nuclear binding energy Difference in energy between a collection of nucleons and a nucleus formed from these nucleons.
Nuclear fission See Fission.
Nuclear fusion See Fusion.
Nuclear magnetic resonance (nmr) Absorption of electromagnetic radiation due to changes of nuclear spin states in a sample placed in a strong magnetic field.
Nuclear reaction Process that decomposes, synthesizes, or transforms elements; a change in the structure of an atomic nucleus.
Nuclear reactor Apparatus for the production of energy by controlled fission.
Nuclear spin states Quantized magnetic states of a nucleus.
Nucleation Start of crystal-lattice growth.
Nucleon Either a proton or a neutron.
Nucleus 1. Small, dense, positively charged center of an atom. 2. Portion of a cell containing its genetic material.

Octahedral hole Space in a close-packed lattice surrounded by six atoms arranged at the vertices of an octahedron.
Open-hearth process Industrial procedure for converting crude iron to steel.
Open system System that exchanges matter with its surroundings.
Optical activity Ability to change the plane of polarization of light.
Orbital Three-dimensional graph of the square of the wave function for a particular energy level.
Orbital splitting Energy difference between the e_g and t_{2g} orbitals of a transition metal ion in a crystal field.
Order Exponential power to which a particular concentration is raised in a rate law.
Osmotic pressure Pressure that must be applied to prevent the flow of solvent through a membrane into a solution.
Ostwald process Commercial process for production of nitric acid from air oxidation of ammonia.
Outer-orbital complex Complex ion in which ligand bonding involves outer-shell d orbitals as opposed to inner-shell d orbitals.
Overpotential Difference between the actual decomposition potential for an electrolysis and the theoretical decomposition potential.
Oxidant See Oxidizing agent.
Oxidation Increase in oxidation state.
Oxidation potential Potential of a cell in which the standard hydrogen electrode acts as a cathode.
Oxidation–reduction (redox) reaction Chemical reaction in which one species is oxidized and another is reduced.

Oxidation state Number of valence electrons in a free, neutral atom minus the number of valence electrons assigned to the atom in a particular chemical environment.

Oxidative phosphorylation Complex oxidative process that produces ATP.

Oxide Binary compound of oxygen that does not contain oxygen–oxygen covalent bonds.

Oxidizing agent Species that oxidizes another species and is itself reduced in the process.

Oxyacids Compounds containing acidic hydroxyl groups.

Paramagnetism Ability to act as a temporary magnet.

Partial pressure The pressure a component of a gas mixture would exert if it were the sole occupant of the container.

Pauli exclusion principle Two electrons in the same atom cannot be in identical quantum states.

Peptide bond Covalent bond joining the amino group of one amino acid and the carboxylate group of another.

Percent by weight Ratio of the mass of one component to the mass of the solution, expressed as a percentage.

Percent yield Ratio of the mass of product obtained compared to the theoretical yield, expressed as a percentage.

Period Horizontal row of elements in the periodic table.

Periodic family Group of elements with marked similarities in chemical and physical properties; elements lying in the same column of the periodic table.

Periodic law Elements of a particular chemical family appear at predictable places in the periodic table, which is arranged in order of atomic numbers.

Periodic table Table displaying the elements so that members of the same family lie in the same column.

Permanent hardness Presence in water of dissolved Ca^{2+}, Mg^{2+}, and Fe^{2+} salts other than the bicarbonates.

Permanent magnet Object that aligns itself with respect to another magnet so that their magnetic fields oppose each other.

Peroxide Ionic substance containing the O_2^{2-} ion.

pH Negative base-ten logarithm of hydrogen-ion activity.

Phase diagram Graph showing the conditions of temperature and pressure necessary for the existence of the various phases of a substance.

Phosphorescence Slow emission of long-wavelength light by an excited state.

Photochemical smog Air pollutants produced by the interaction of by-products of combustion of gasoline and fuel oil with intense sunlight.

Photoelectric effect Ejection of electrons from a metal surface by light with sufficiently high frequency.

Photon Quantum of light energy.

Photosynthesis Use of the sun's energy by green plants and algae to synthesize reduced carbon compounds and form oxygen.

Pi-meson Type of meson whose exchange between nucleons may be important in nuclear binding.

Pi-orbitals Molecular orbital constructed by the parallel overlap of p atomic orbitals.

Pig iron Crude form of iron produced in a blast furnace.

Pion See Pi-meson.

Planck's law The energy of a photon is proportional to the frequency of the light.

Planck's theory A light beam is a stream of quanta of energy, called photons.

Plane-polarized light Radiation whose electromagnetic waves all lie in the same plane.

Plasma Condensed state of nuclei and electrons.

Polar covalent bond Chemical bond involving the unequal sharing of electrons between two atoms.

Polar covalent substance Substance whose atoms are held together by polar covalent bonds.

Polyatomic ion Charged cluster of atoms.

Polydentate Capable of occupying several of a metal ion's coordination positions at the same time.

Polymeric Having a structure containing a repeating structural unit.

Polymeric covalent carbide Carbide formed by an element whose size and electronegativity are similar to carbon's.

Polymerize Form polymers from small molecules.

Polymorphic Having different crystal lattices for the same compound.

Polypeptide chain Polymeric chain of amino acids.

Polyprotic acid Acid containing more than one dissociable proton per molecule.

Polyprotic base Base capable of reacting with more than one proton per molecule.

Polysaccharide Polymer built from cyclic monosaccharide units.

Positive charge Type of electric charge carried by protons.

Positron Particle with the mass of an electron but a +1 charge.

Potential energy Energy of an object due to its position and the forces acting on it.

Pressure Force acting on a surface divided by the surface area.

Primary structure See Sequence.

Principal quantum number Quantum number describing the approximate average distance of an electron from the nucleus.

Product Substance formed in a chemical reaction.

Proportional counter Instrument for detecting radioactivity by the ionization of a gas.

Protein Biochemically important polymer of amino acids.

Protium Mass-1 isotope of hydrogen.

Proton Particle with mass equal to 1 amu and a +1 charge found in an atomic nucleus.

Proton–proton chain Sequence of fusion reactions occurring in the sun.

p-type semiconductor Semiconductor containing an impurity with too few valence electrons to bind to its nearest neighbors.

Pure substance Matter that cannot be separated into other substances by physical means.

Quantized Existing as quanta.

Quantum (plural: quanta) Indivisible unit of definite size.

Quantum mechanical liquid Liquid phase showing superfluidity and other unusual optical and acoustical properties.

Quantum mechanics Theory that the behavior of small particles can be explained in terms of their wave properties.

Quantum number Number describing a property of a particular energy level.

Quaternary structure Spatial arrangement of an aggregate of polypeptide chains.

Radioactivity Radiation from unstable nuclei.

Radiotherapy Destruction of tumors by radioactivity.

Random coil Three-dimensional arrangement of a polypeptide chain that cannot be classified as either alpha helix or beta structure.

Raoult's law The vapor pressure of a component of a solution is the vapor pressure of the pure component multiplied by the component's mole fraction.

Rate constant Proportionality constant in a rate equation.

Rate-determining step Slowest elementary reaction in a mechanism.

Rate law Equation showing the relationship between the rate of a particular reaction and its reactant concentrations.

Reactant Substance that is consumed in a chemical reaction.

Reaction rate Rate of change of the concentration of a reactant or product.

Reducing agent Species that reduces another species and is itself oxidized in the process.

Reductant See Reducing agent.

Reduction Decrease in oxidation state.

Reduction potential Potential of a cell in which the standard hydrogen electrode is defined as the anode.

Regulatory enzyme Enzyme that changes its catalytic activity in response to the cell's needs.

Replication Synthesis of identical molecules.

Resonance forms Two structural formulas with the same geometric arrangement of nuclei but differing in the positions of their electrons.

Resonance hybrid Description of a molecule or ion as the average of several Lewis structures.

Reversible reaction Reaction that ceases before the maximum amount of product, predicted by stoichiometry, forms.

Ribonucleic acid (RNA) Polymer of ribose units, joined by phosphate groups, in which each ribose unit has an attached base group.

Ribosome Subcellular structure in which protein synthesis occurs.

Rotational energy level Quantized energy of molecular rotation.

Salt bridge Tube filled with KCl solution that acts as a permeable barrier between two half-cells.

Saturated Having a concentration such that the solution is in equilibrium with the pure solute.

Schottky defect Crystal defect in which an equal number of positive and negative ions are missing from their lattice positions.

Schrödinger wave equation Equation used to calculate the energy levels of a system from the wave properties of its particles.

Scientific law Generalization based on experience.

Scintillation counter Device for detecting radioactivity by the formation of light flashes as the radioactivity passes through a crystal or liquid solution.

Second ionization potential Energy needed to remove an electron from a +1 ion in the gas phase to form a +2 ion in the gas phase.

Second law of thermodynamics The entropy of the universe always increases.

Secondary structure Relative amounts of alpha helix, beta structure, and random coil arrangements in a protein.

Selection rules Rules describing the probability that a photon with the correct energy will be absorbed due to a particular change in energy levels.

Semiconductor Poor electric conductor that is improved by increased temperature or the presence of crystal defects or impurities.

Semimetallic Having an electric conductivity that is low at room temperature but increases with increasing temperature.

Sequence Arrangement of amino acids in a polypeptide chain.

Shell Group of orbitals with the same principal quantum number.

Shells of energy levels Theory (often called the "shell model") that there are energy levels of limited capacity for the nucleons in a nucleus.

SI system Internationally recognized system of units.

Side-centered lattice Lattice with lattice points at the corners and centers of two opposite faces of the unit cell.

Sigma-orbital Molecular orbital constructed by head-on overlap of atomic orbitals.

Simplest formula Simplest whole-number ratio of atoms of the various elements that is consistent with the composition of a compound.

Single covalent bond Covalent bond formed by the sharing of an electron pair.

Slag Calcium silicate formed by the reaction of limestone with iron ore impurities in a blast furnace.

Smectic liquid crystal Liquid crystal in which the molecules are grouped in planes and lie perpendicular to these planes.

Smelting Melting and reduction of metal ores.

Sols See Colloidal suspensions.

Solubility Concentration of a saturated solution.

Solubility product Product of the equilibrium concentrations of the ions formed when a salt dissolves, each concentration being raised to a power equal to its stoichiometric coefficient.

Solute Component, other than the solvent, of a solution.

Solution Homogeneous mixture.

Solvation Insulation of an ion's charge by its interaction with solvent molecules.

Solvay process Production of sodium bicarbonate by the reaction of ammonia and carbon dioxide with brine.

Solvent Major liquid component of a liquid solution.

Specific heat Ratio of the heat capacity of a substance compared to the heat capacity of water.

Specificity Selection by an enzyme of a specific type of molecule whose particular reaction it catalyzes.

Spectator ion Ion comprising part of the structure of a reactant in a solution reaction but not involved in the net ionic reaction.

Spectrophotometer Device for measuring a substance's ability to absorb light at various frequencies.

Standard atmosphere Unit of pressure equal to 760 torr.

Standard enthalpy of formation Enthalpy of formation when the reactants and products are at the standard state of 1 atm pressure and 25°C.

Standard entropy Entropy increase that occurs when a mole of substance is heated from a perfectly crystalline state at 0 K to its standard state at 1 atm pressure and 25°C.

Standard free energy Free energy change of a reaction when the activity of each reactant and product equals one.

Standard hydrogen electrode Electrode in which hydrogen gas at 1 atm passes through a solution whose hydrogen-ion activity is one and over the surface of a catalytic platinum electrode.

Standard reaction potential Potential of a galvanic cell based on a particular reaction in which the activity of each reactant and product within its half-cell equals one.

Standard solution Solution whose concentration is accurately known.
Standard state Stable form of a substance at unit concentration and specific pressure and temperature.
Standard temperature and pressure (STP) 760 torr and 0°C.
Standardization Process whereby the exact concentration of a substance or solution is determined.
Standing wave Stable wave formed by a vibrating system.
State function Property that depends only on the system's chemical composition, physical state, temperature, and pressure but is independent of the past history of the system.
Steel Iron containing 0.05–2% carbon.
Stereospecificity Production by an enzyme-catalyzed reaction of one enantiomer in preference to the other or by the specificity of an enzyme for one of a pair of enantiomers.
Steric factor Ratio between the frequency factor of a reaction and the collision frequency.
Stern–Gerlach experiment Experiment showing that electron spin is quantized.
Stoichiometric coefficient Number of molecules of a species involved in a balanced chemical equation.
Stoichiometry Study of mass relationships in chemical reactions.
Strong electrolyte Electrolyte that is completely dissociated in aqueous solution.
Structural formula Symbolic method for displaying the types of chemical bonds and the arrangements of valence electrons in substances.
Sublimation energy Energy required to sublime a mole of solid.
Sublime Evaporate solid without melting it.
Subshell Group of orbitals with the same principal quantum number and the same azimuthal quantum number.
Substrate Specific reactant molecule for an enzyme.
Supercooling Cooling of a liquid below the melting point without freezing it.
Superfluidity Absence of viscosity found in certain phases of liquid helium.
Superoctet species Molecule or ion in which at least one atom has more than eight valence electrons.
Superoxide Ionic substance containing the O_2^- ion.
Supersaturated Having a concentration exceeding the solubility.
Symmetry element Imagined operation that relates one part of an object or group of objects to another part.

Temperature Intensity of heat energy in a sample.
Temporary hardness Presence of dissolved bicarbonates of Ca^{2+}, Mg^{2+}, and Fe^{2+} in water.
Temporary magnet Object that acts as a magnet in the presence of a magnetic field but does not retain its magnetism after removal from the field.
Termolecular Involving collisions between three reactant molecules.
Tertiary structure Characteristic three-dimensional shape of a protein molecule.
Tetradentate Able to occupy four ligand positions at the same time.
Tetrahedral hole Site in a close-packed lattice surrounded by four atoms arranged at the vertices of a tetrahedron.

Theoretical yield Maximum amount of product obtainable from a chemical reaction as calculated by a limiting reactant calculation.
Theory Proposed cause-and-effect relationship between hypothetical causes and experimental observations.
Thermal neutron Neutron with about as much kinetic energy as a molecule at room temperature.
Thermodynamic equilibrium constant Equilibrium constant based on activities rather than concentrations.
Thermodynamics Study of energy relationships in chemical reactions.
Thermonuclear reaction See Fusion.
Three-center bond Chemical bond, occurring in electron-deficient hydrides, in which a single molecular orbital containing one electron pair holds a hydrogen atom between two other atoms.
Titration Measurement of the amount of a substance by measuring the volume of standard reagent of known concentration needed to react with it.
Titration curve Graph of pH versus volume of added reagent for a titration.
Torr Unit of pressure equal to the pressure of a mercury column 1 mm high.
Transfer RNA (tRNA) Small RNA molecule that binds to mRNA during protein synthesis.
Transition-metal ion Ion formed by a transition-metal element.
Transition metals Group of 30 metallic elements lying in the short columns of the periodic table.
Transition state See Activated complex.
Translation Moving a unit cell exactly one edge length in a direction along the edge.
Transmittance Ratio of the light transmitted by a sample compared to the incident light.
Transuranium elements Unstable elements with atomic numbers 93 and above.
Tricarboxylic acid cycle Sequence of reactions that oxidizes pyruvate to carbon dioxide.
Tridentate Able to occupy three coordination positions at the same time.
Triple covalent bond Covalent bond formed by the sharing of three pairs of electrons.
Triple point Condition of temperature and pressure at which three different phases of the same substance are in equilibrium.
Tritium Mass-3 isotope of hydrogen.

Ultrafiltration Removal of colloids from a solution by filtering through a membrane.
Unit cell Smallest part of a lattice that contains all the properties of the lattice.
Universal gas constant Constant R in the ideal gas law.

Valence-bond method Approximate theoretical model for covalent bonding involving electron pairs.
Valence electrons Electrons in an atom's outer shell.
Valence-shell electron-pair repulsion (VSEPR) Method of predicting molecular shapes from the number of bound atoms and unshared electron pairs around a central atom.
van der Waals equation Equation that describes the pressure, volume, temperature relationships of real gases as opposed to ideal gases.

van der Waals forces Weak attractive forces between neutral, nonpolar molecules.
Velocity Direction of motion and speed of an object.
Vibrational energy level Quantized energy of molecular vibration.
Viscosity Resistance of a liquid to flow.
Visible–ultraviolet radiation Electromagnetic radiation whose frequency ranges from about 10^{14} to 10^{16} Hz.
Volt 1 J electric potential energy per coulomb charge.
Volumetric flask Flask used for accurate preparation of solutions of known molarity.

Water gas Mixture of hydrogen and carbon monoxide formed by the reaction of steam and coke.
Water of crystallization Intact water molecules combined with other atoms in a crystal.
Water of hydration See Water of crystallization.
Water softening Treatment of water to remove Ca^{2+}, Mg^{2+}, and Fe^{2+} salts.
Wave Periodically repeating disturbance.
Wave function Solution to the Schrödinger wave equation. The square of a realistic wave function describes the probability of an electron in a certain energy level being located in a particular region.
Wavelength Distance between two corresponding points of adjacent waves.
Weak electrolyte Electrolyte that is dissociated to only a slight extent at equilibrium.
Weight Gravitational force acting on an object.
Work Distance an object moves while under the influence of a force multiplied by the force.

x ray Light with a wavelength in the 0.1–1 nm range.
x-ray crystallography Determination of a crystal lattice structure by interpreting the diffraction patterns of x rays reflected from the crystal.

Zeolite Naturally occurring ion exchanger.
Zero-point energy Vibrational energy retained by a molecule at 0 K.
Zone refining Technique for purifying solids by selective melting and resolidification.

Answers to Selected Exercises

Chapter 1

1. b and c **3.** a, c, d are chemical reactions; b, e, f are nuclear reactions **5.** b and c **7.** 4.1×10^{-25} kg-m/sec
9. a. 6.3×10^{-19} J **b.** 6.3×10^{-19} J **10.** 2.67×10^{-2} m
12. a. 5.0 m; 3.5 m; 2.0×10^{-4} m **b.** 4.30×10^{-4} kg; 58.93 kg; 8.5×10^{-14} kg **c.** 1.200×10^{-9} sec; 3.825×10^{-8} sec; 5.35×10^{5} sec **14.** 0.339 to 1 **16.** 139 amu **18.** 2.37 g
20. 192 g **22.** 0.400 g oxygen **24. a.** 17.7% **b.** 4.31 g
26. 37°C **28.** 30.7°C **30.** 2.8°C **33.** None **35.** $N_A/O_A = 1.75$; $N_B/O_B = 0.876$ **37.** The ratio of chlorine atoms in compound 2 to chlorine atoms in compound 1 is 4:3. **39.** The old amu unit is 1.000038 times the new one.
41. a. 17.76% H **b.** 30.14% D **43.** 2.67×10^{8} m/sec^2
45. 1.5×10^{11} m/sec^2 **47.** 8.4×10^{-18} newtons

Chapter 2

1. 1.8×10^{22} **3.** 5.98×10^{-2} g **5.** 72 g **7.** 0.167 mol
9. 3:1 **10.** 5 carbons, 11 hydrogens, 1 nitrogen, 1 sulfur, 2 oxygens **12.** 4 potassiums, 1 iron, 6 carbons, 6 nitrogens, 6 hydrogens, 3 oxygens; 3 water molecules **14.** 20
16. a. 192 amu **b.** 92.0 amu **c.** 46.0 amu
18. a. 124 g **b.** 5.61 g **c.** 3.2×10^{-2} g
20. a. 2.74×10^{-2} mol **b.** 15.0 mol **22.** 0.50 mol
24. 1.20 mol **26.** 0.500 mol **28.** B_5H_9 **30.** B_2H_6
32. $Na_2S_2O_3$ **34. a.** $C_2H_3O_3$ **b.** $C_4H_6O_6$ **36.** 13.2%
38. 19.0% Na, 16.7% Al, 17.4% Si, 39.6% O, 7.3% Cl
40. a. $P_4(g) + 5\ O_2(g) \longrightarrow P_4O_{10}(s)$
  b. $2\ C_2H_6(g) + 7\ O_2(g) \longrightarrow 4\ CO_2(g) + 6\ H_2O(g)$
  c. $3\ CaCl_2(aq) + 2\ Na_2HPO_4(aq) \longrightarrow Ca_3(PO_4)_2(s) + 4\ NaCl(aq) + 2\ HCl(aq)$
  d. $H_2O + 2\ AgNO_3(aq) + Na_2SO_3(aq) \longrightarrow 2\ Ag(s) + Na_2SO_4(aq) + 2\ HNO_3(aq)$
  e. $2\ CrO_3(aq) + 12\ HCl(aq) \longrightarrow 2\ CrCl_3(aq) + 6\ H_2O + 3\ Cl_2(g)$
42. a. $2\ FeCl_3 + 2\ KI \longrightarrow 2\ FeCl_2 + I_2 + 2\ KCl$ **b.** KI
c. 0.764 g I_2
44. a. Co_2O_3 **b.** $3\ H_2 + Co_2O_3 \longrightarrow 3\ H_2O + 2\ Co$
46. a. 18.6 g **b.** 32.2% **48. a.** 0.5988 g **b.** 0.9758 g
50. a. 3.05% C, 0.26% H, 96.66% I **b.** CHI_3 **53.** $C_8H_{12}Br_2$
55. 10

Chapter 3

3. 1.25 atm **5.** 1.0×10^3 cm H_2O, or 34 ft **8. a.** 432 K
b. 205 K **c.** 192°C **d.** −93°C **10.** 764 torr
12. 18.3 mL **14.** 850 torr **16.** 35.5 mL **18.** 1000 torr
20. 109°C **22.** 132°C **24.** 22.5 mL **26.** 4.84 L
28. 671 torr **30.** 145 torr **32.** 6.28 L **34.** 7.28 atm
36. 38.8 mL **38. a.** $1\ S_8:8\ H_2$; $1\ N_2:2\ H_2$ **b.** 1:8; 1:1
40. 0.196 mol **42.** 2.71 g/L **44.** 4.50 L O_2; 3.00 L N_2O
46. 0.520 L **48.** 33.5 g/mol **50.** 0.050 L **52.** 0.131 atm
54. 53 g/mol **56.** Rate F_2/Rate $Cl_2 = 1.366$ **59.** 2.11 L N_2
61. 9.7 L STP **63.** 1.45×10^{-2} mol NO; 1.05×10^{-2} mol N_2
65. 3.03% **67.** Rate $D_2 = 1.4$ mL/min; rate HD = 1.6 mL/min; yes **69.** 1.64 torr **71. a.** 19.2 L **b.** 243 torr **c.** 6.6 L
73. Pressure decreases by ¼; 2.64 atm **75. a.** 3.33×10^{-3} sec
b. 3.33×10^{-3} sec **77.** 700 torr **79.** 34.5%

Chapter 4

1. O, S, Te, Po 3. b, d, f 5. b, g 7. Contains equal number of charges. 9. 0.00518 $\mathscr{F}$ 11. One chemical behavior for all isotopes of an element. 13. Charge is not conserved. 15. 59 17. 4+ net charge

19.
	Protons	Neutrons	Electrons
a.	43	55	43
b.	71	96	68
c.	85	123	85

21.
	Protons	Neutrons	Electrons
a.	16	16	18
b.	7	8	10
c.	35	46	36

23. W and Y 25. 6, 14, 32, 50, 82 27. Ra^{2+} must be the ion to be similar to Ba^{2+}, and mass of 225 is proper for next period after radon (222). 29. 7.0×10^3 coul 31. Result would be two separate beams. 33. Result would be a single electron beam, similar to what he observed. 35. $4.97 \approx 5$
37. a. 4.48×10^{-23} g b. 4.80×10^{-19} coul
c. 1.07×10^4 coul/g d. 1.07×10^4 coul/g
39. a. 1.66×10^{-24} g b. 2.66×10^{-23} g
c. 2.56×10^{-18} coul d. 6.02×10^3 coul/g
41. a. Attractive b. 3.0×10^{-9} newton
c. 2.4×10^{-8} newton d. Increase

Chapter 5

3. x ray
5. a. False. It is impossible to determine the distance of the electron from the nucleus at any time.
d. False. Degenerate implies only the same energy.
e. False. Each orbital can hold a maximum of two electrons.
f. False. Momentum is inversely proportional to wavelength.
h. False. The wave nature of a neutron is evidenced by its diffraction.
7. 6.0 cm 8. 7.41×10^{14} sec^{-1} 10. 231 m
12. 1.3×10^{-19} J 14. 6.52×10^{-19} J 16. 9.9×10^{-11} m
18. 1.03×10^{-7} m 20. $\approx 3.6 \times 10^7$ m/sec, or almost the speed of light 22. a. 50 b. s, p, d, f, g; 2, 6, 10, 14, 18, respectively c. 9; 2 24. a. 6 b. 4 c. 1
26. a. Impossible, $m_l \leq l$
c. Impossible, $l = n - 1$ maximum
f. Impossible, $m_s = \pm \frac{1}{2}$ only
28. Impossible: c. $3p^7$, p has maximum of 6 e. $3d^{11}$, maximum of 10 electrons in a d subshell Ground state: a, d Excited states: b and f 30. a. 0 b. 1 c. 0 d. 6
e. K and Cr 32. a. $1s^22s^22p^63s^23p^64s^2$
b. $1s^22s^22p^63s^23p^63d^64s^2$
c. $1s^22s^22p^63s^23p^63d^{10}4s^24p^64d^{10}5s^25p^66s^2$
34. Zr is $[Kr]5s^24d^2$, whereas Mo is $[Kr]5s^14d^5$ like Cr
36. No, electron spins are paired. 38. 5.52×10^{-26} kg-m/sec

45. n = 7 to n = 3
47. a. $[Xe]6s^24f^{1-14}$ for La–Yb; $[Xe]6s^24f^{14}5d^1$ for Lu
b. According to the building principle, all would form dichlorides, but Lu should also form a trichloride.
c. See Table 5.3. All form dichlorides, and other expected compounds are $LaCl_3$, $CeCl_3$, $GdCl_3$, and $LuCl_3$.
49. a. $\lambda_1 = 0.20 \times 10^{-9}$ m, $\lambda_2 = 0.10 \times 10^{-9}$ m, $\lambda_3 = 0.067 \times 10^{-9}$ m
Nodes: λ_1, 0; λ_2, 1; λ_3, 2
Momentum: $M_1 = 3.3 \times 10^{-24}$ kg-m/sec,
$M_2 = 6.6 \times 10^{-24}$ kg-m/sec,
$M_3 = 9.9 \times 10^{-24}$ kg-m/sec
$E_1 = 6.0 \times 10^{-18}$ J, $E_2 = 2.4 \times 10^{-17}$ J, $E_3 = 5.4 \times 10^{-17}$ J
$KE_1 = 6.0 \times 10^{-18}$ J, $KE_2 = 2.4 \times 10^{-17}$ J, $KE_3 = 5.4 \times 10^{-17}$ J
b. At $\frac{1}{6}$, $\frac{1}{2}$, $\frac{5}{6}$ times 0.10 nm the wave is a maximum; at $\frac{1}{3}$ and $\frac{2}{3}$ times 0.10 nm the nodes

41. a.
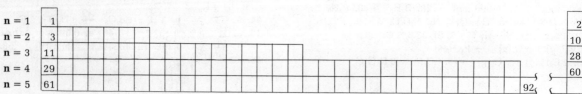

b. 2, 8, 18, 32, 50 c. 13, 31, 63; 3 electrons

Chapter 6

1. a. Cl b. He c. Ar 3. $I_2 < CH_4 < Cl_2O < HCl < H_2O$
5. a and d require removal of an inner-shell electron.
7. a. Antibonding orbitals are higher in energy. b. Bonding orbitals have more electron density between the nuclei and antibonding have less. c. Bonding orbitals increase molecular stability; antibonding decrease the stability.
9. Electric current through a solution results from the movement of ions. 11. A figure resembling Figure 6.4 with K^+ ions

migrating toward the cathode and F^- ions toward the anode. **13.** Rb is $[Kr]5s^1$ and can more easily lose its outer-shell electron. Kr has a completed octet. **15. a.** Less stable **b.** Less stable **c.** Less stable **d.** More stable **e.** More stable **17.** Toward N; neither; slightly toward H **19. a.** Trigonal planar **b.** Trigonal pyramidal **21.** No; polar bonds but they are equal and opposite and therefore cancel. **23.** Ionic **25. a.** $BaCl_2$ **b.** SrS **c.** K_2O **d.** LiF **27.** Be is $1s^22s^2$, whereas B is $1s^22s^22p^1$. The first ionization potential of boron involves removal of a p electron, which has been somewhat shielded by the $2s$ electrons and is farther out from the nucleus, hence, it is easier to remove. **31. a.** 11 valence electrons; $2s\sigma^22s\sigma^{*2}2p\sigma^22p\pi^42p\pi^{*1}$; 5 net bonding electrons; paramagnetic because there is one unpaired electron spin **b.** 12 valence electrons; $2s\sigma^22s\sigma^{*2}2p\sigma^22p\pi^42p\pi^{*2}$; there is one electron in each of the two $2p\pi^*$ orbitals; 4 net bonding electrons; paramagnetic, 2 unpaired electrons. **33.** $Li_2(g):2s\sigma^2$; two net bonding electrons, one covalent bond $Be_2(g):2s\sigma^22s\sigma^{*2}$; no net bonding electrons, no bond **35.** Yes. He_2^+ has the electron configuration $1s\sigma^21s\sigma^{*1}$, so it has one net bonding electron. **37. a.** B_2 has 6 valence electrons: $2s\sigma^22s\sigma^{*2}2p\pi^2$. There is one electron in each of the two $2p\pi$ orbitals and thus it is paramagnetic. C_2 has 8 valence electrons: $2s\sigma^22s\sigma^{*2}2p\pi^4$ and no unpaired electrons. **b.** C_2^{2+} has 6 valence electrons and it is just like B_2. **39.** -422 kJ/mol **41.** No **43.** The nearest neighbor ions are of the opposite sign so there is a smaller distance and larger coulombic attractive forces. **45.** Polar covalent
47.

___	$2p\sigma^*$	Two covalent bonds
↑ ___	$2p\pi^*$	
↑	$2p\sigma$	
↑↓ ↑↓	$2p\pi$	
↑↓	$2s\sigma^*$	
↑↓	$2s\sigma$	
↑↓	$1s\sigma^*$	
↑↓	$1s\sigma$	

49. a. Covalent; properties: melting point $-70°C$, boiling point $58°C$ **b.** Ionic; properties: melting point $1296°C$, boiling point $2239°C$ **c.** Covalent; properties: melting point $-133°C$, boiling point $-88°C$

Chapter 7

1. a. Gaining two electrons; gaining one electron and forming one covalent bond; forming two covalent bonds **b.** Na_2S; NaHS; H_2S **c.** SO_4^{2-}; SO_3^{2-} **d.** Four

3. :F:B: + :F:⁻ ⟶ :F:B:F:⁻ (with F atoms around B)

5. a. 1 unshared **b.** 4 shared, 4 unshared **c.** 0 **d.** 6 shared, 1 unshared **e.** 6 unshared

7. a. ·P̈· **b.** Ca^{2+} **c.** :Ö:²⁻ **d.** $[Sr^{2+}][:\ddot{C}l:^-]_2$ **e.** :Ö—H⁻

9. $2K\cdot + :\ddot{I}—\ddot{I}: \longrightarrow 2[K^+][:\ddot{I}:^-]$

11. a. 8 **b.** H—C(H)(H)—C(H)(H)—Ö—H and H—C(H)(H)(H)—Ö—C(H)(H)—H

13. a. :Ö—Cl—Ö: with :Ö: **b.** H—C(H)(H)—C(H)(H)—C(H)(H)—H **c.** H:S̈:H⁺ with H below **d.** Se with three O resonance structures

15. a. III, IV, VI, VII **b.** I **c.** II **d.** V
17. a. 0 **b.** 0 **c.** +5 **d.** +4 **e.** +3 **f.** −2
19. a. C(+4), S(−2) **b.** H(+1), C(0), O(−2) **c.** La(+3) **d.** O(−2) **e.** H(+1), N(−1), O(−2) **f.** P(+5), Cl(−1)
21. a. +4 **b.** +6 **c.** +3 **d.** +7 **23.** I
25. a. Trigonal planar **b.** Trigonal pyramidal **c.** Linear **d.** Trigonal pyramidal **e.** Trigonal bipyramidal
27. a and c **29.** Linear **31. a.** sp^3, trigonal pyramidal **b.** sp^2, trigonal planar **c.** sp, linear
33. $Mg^{2+}(aq) + 2\ OH^-(aq) \longrightarrow Mg(OH)_2(s)$
35. a. $Li^+(aq) + e^- \longrightarrow Li(s)$
b. $Ca(s) + 2\ OH^-(aq) \longrightarrow Ca(OH)_2(s) + 2\ e^-$
c. $Mg(s) \longrightarrow Mg^{2+}(aq) + 2\ e^-$
d. $Cd(s) + 4\ CN^-(aq) \longrightarrow Cd(CN)_4^{2-}(aq) + 2\ e^-$
e. $Zn(NH_3)_4^{2+}(aq) + 2\ e^- \longrightarrow Zn(s) + 4\ NH_3(aq)$
37. a. $4\ I^-(aq) + 2\ Cu^{2+}(aq) \longrightarrow I_2(aq) + 2\ CuI(s)$
b. Already balanced
c. $2\ HNO_2(aq) + 2\ H^+(aq) + 3\ I^-(aq) \longrightarrow I_3^-(aq) + 2\ NO(g) + 2\ H_2O(l)$
d. $2\ Hg^{2+}(aq) + 2\ CuCl(s) \longrightarrow Hg_2Cl_2(s) + 2\ Cu^{2+}(aq)$
e. $2\ Cr(OH)_4^-(aq) + 3\ HO_2^-(aq) \longrightarrow 2\ CrO_4^{2-}(aq) + OH^-(aq) + 5\ H_2O(l)$
39. a. $6\ H^+(aq) + 2\ MnO_4^-(aq) + 5\ H_2C_2O_4(aq) \longrightarrow 10\ CO_2(g) + 2\ Mn^{2+}(aq) + 8\ H_2O(l)$
b. $Cu(s) + 2\ Ag^+(aq) \longrightarrow Cu^{2+}(aq) + 2\ Ag(s)$
c. $8\ H^+(aq) + Cr_2O_7^{2-}(aq) + 3\ H_2CO(aq) \longrightarrow 3\ HCO_2H(aq) + 2\ Cr^{3+}(aq) + 4\ H_2O(l)$
d. $H_2O(l) + H_2SO_3(aq) + I_3^-(aq) \longrightarrow HSO_4^-(aq) + 3\ I^-(aq) + 3\ H^+(aq)$
e. $2\ H^+(aq) + MnO_2(s) + H_2O_2(aq) \longrightarrow Mn^{2+}(aq) + O_2(g) + 2\ H_2O(l)$
41. a. $BrO_3^-(aq) + 3\ I^-(aq) \longrightarrow Br^-(aq) + 3\ OI^-(aq)$
b. $3\ OH^-(aq) + H_2CO(aq) + 2\ Ag(NH_3)_2^+(aq) \longrightarrow HCO_2^-(aq) + 2\ Ag(s) + 4\ NH_3(aq) + 2\ H_2O(l)$
c. $2\ H_2O(l) + Mg(s) \longrightarrow Mg(OH)_2(s) + H_2(g)$
d. $4\ OH^-(aq) + 6\ H_2O(l) + 3\ O_2(g) + 4\ Cr(s) \longrightarrow 4\ Cr(OH)_4^-(aq)$
e. $2\ OH^-(aq) + 3\ H_2O(l) + S_2O_3^{2-}(aq) + 4\ MnO_2(s) \longrightarrow 2\ SO_4^{2-}(aq) + 4\ Mn(OH)_2(s)$
43. a. Hydrogen selenide **b.** Nitrogen triiodide **c.** Ammonium dichromate **d.** Strontium iodide **e.** Orthoarsenic acid **f.** Metaarsenic acid **g.** Copper(II) iodate or cupric iodate **h.** Zinc hydroxide **i.** Strontium nitrate **j.** Diantimony pentoxide or antimony(V) oxide or stibnic oxide **45. a.** Ba_3As_2 **b.** $Hg(ClO_3)_2$ **c.** $NiSO_4$ **d.** $NH_4CH_3CO_2$ **e.** $Mg(ClO_4)_2$ **f.** $Na_2C_2O_4$ **g.** HBr(aq) **h.** ZnS **i.** $BaCrO_4$ **j.** HIO_4 **47. a.** −1 **b.** −4 **c.** −3 **d.** −2 **49.** sp

Chapter 8

1. a. Into system   b. Positive   3. a and d   5. No net change   7. a. Open   b. Closed   c. Closed (neglecting cosmic rays and meteors)   d. Open   e. Open   f. Open   g. Closed   h. Open   9. a. $w = +4.56$ kJ; $\Delta E = +13.2$ kJ   b. $w = 0$; $\Delta E = +12.8$ kJ   c. $w = +9.3$ kJ; $\Delta E = +0.4$ kJ   10. 1.22 kJ   12. 1.19 kJ   14. 3.42 kJ   16. 0.690 kJ   18. -1197 kJ   20. -193 kJ   22. $+10.7$ kJ   24. -0.25 J/K   26. $+0.015$ J/K   28. a. -14.4 kJ   b. -14.6 kJ   30. a. -189.6 J/K   b. -116.8 J/K   c. -144.9 J/K   d. -200.4 J/K   e. -14.2 J/K   32. a. -140.0 kJ   b. -398.99 kJ   c. -894 kJ   d. -636.4 kJ   e. -222.26 kJ   35. c, d, f   37. 2.67 kJ/°C   39. a. 801 J   b. Increase by 1.27 L   c. $+479$ J   d. 801 J   41. The demon would allow O_2 molecules to pass through the hole from A to B, but close the hole to prevent O_2 molecules from moving from B to A.   43. a. -196.6 kJ   b. -433.9 kJ   c. -938 kJ   d. -696.0 kJ   e. -226.47 kJ   45. At equilibrium, $\Delta G = 0 = \Delta H - T\Delta S$. At constant pressure, $q = \Delta H$. Therefore, $0 = q - T\Delta S$ and $\Delta S = q/T$.   47. a. $+69$ kJ   b. -70 kJ   c. 672 K (399°C)   49. a. 0   b. 114 J   c. 114 J   d. 114 J

Chapter 9

3. c is true   7. a. Molecular crystal   b. van der Waals force   9. Both are disordered.   11. Solid benzene   15. Stretching produces an orderly arrangement of molecules.   17. False. This is true only if the two crystals are of the same lattice type or if they are both close-packed.

21. a., b., c. (diagrams)

23. Simple cubic   25. a. The solid sublimes to a gas as the temperature is raised at constant pressure.   b. An increase in pressure at constant temperature condenses the gas to the liquid.   c. The liquid evaporates as it is warmed at constant pressure.   d. The gas is compressed isothermally. No phase change.   27. 0   29. Solid sodium   31. a. 2.81 Å   b. 33.3° for $n = 2$; 55.4° for $n = 3$ 33. 19.7°   35. 2.72 g/cm³   37. 12.1 g/cm³   39. 1.51 g/cm³   41. 3.61 Å   43. 0.44 kJ   45. a. 11 J/K-mol   b. 203 J/K-mol   47. -0.3 kJ/mol   49. a. The gas condenses to the solid.   b. The solid sublimes.   51. A body-centered tetragonal cell can be drawn into the same lattice.   53. a. Yes   b. 36.2 torr   55. b. 32.1 kJ/mol   c. 82°C   57. 14°C   61. The tetrahedra of the C atoms surrounding an interior position point in opposite directions from the tetrahedra surrounding the lattice points.   63. A fourfold axis of rotation   65. It is wrong. According to the CRC *Handbook of Chemistry and Physics*, the entropies of fusion of benzene and toluene are nearly the same, whereas benzene has a much higher heat of fusion than toluene.   66. 110 torr   68. 34.0 kJ/mol

Chapter 10

1. Acetic acid is the solvent; water and methyl alcohol are solutes.   3. a. 30 g NaCl plus 170 g H_2O b. 16.0 g naphthalene and 484 g benzene   c. 0.233 g ethylene glycol plus water to make 25.0 mL   d. 7.80 g $CuSO_4 \cdot 6H_2O$ diluted to 500 mL with water   5. 34.5%   7. 0.00499 mol NaOH   9. 0.0328 M   11. a. 0.00486 M   b. 0.00972 M OH⁻   13. 0.591 or 59.1 mol %   15. a. Greater   b. Decrease   17. 6.8 torr   19. 205 torr   21. Graph is a straight line between 15.1 torr at 0 X_{CH_3OH} and 96.3 torr at 1.00 X_{CH_3OH}.   23. 0.788   25. 0.0273   27. 120 amu   29. 207 amu   31. 0.150 m   33. 1.22 atm   35. 115 amu   37. 5.2°C/m   39. Mole fraction of ethyl acetate is 0.926; mole fraction of ethyl propionate is 0.074.   41. 14.3%

Chapter 11

1. a. 3   b. 2   c. 3   d. 2   3. a. $K^+(aq) + Cd^{2+}(aq) + Fe^{3+}(aq) + Cl^-(aq) + SO_4^{2-}(aq)$ b. 1.70M Cl⁻, 0.300M Cd^{2+}, 0.200M K^+, 0.500M Fe^{3+}, 0.300M SO_4^{2-} 5. HI; 0.500M HI yields 1.00M concentration of ions in solution, whereas 0.500M HF yields only slightly more than 0.500M particles in solution.   7. 0.2278M   9. a. $NH_4OH(aq) + HI(aq) \longrightarrow NH_4^+(aq) + I^-(aq) + H_2O$ b. $NH_3(aq) + HI(aq) \longrightarrow NH_4^+(aq) + I^-(aq)$ c. $NH_3(aq) + H^+(aq) \longrightarrow NH_4^+(aq)$ 11. a. 3.67 mmol   b. 3.67×10^{-3} mol   13. a. H_2SO_3 b. $C_6H_7^+$ c. $H_2NO_3^+$ 15. a. AsH_2^- b. $HAsO_3^{2-}$ c. $HZnO_2^-$ 17. 0.157 N   19. 1.035 N   21. 0.1134 N   23. $HAsO_4^{2-} + H^+ \longrightarrow H_2AsO_4^-$
$HAsO_4^{2-} + OH^- \longrightarrow AsO_4^{3-} + H_2O$

25.

	Brønsted—Lowry acid	Brønsted—Lowry base
a.	HSO_4^-	$H_2PO_4^-$
b.	HCl	CH_3CO_2H
c.	H_2O	S^{2-}

27. a. Unfavorable   b. Favorable   c. Favorable   29. 34.3 g   31. 0.280   33. 17.0 g   35. a. 2.40×10^{-3}M   b. 4.80×10^{-3}N   37. 0.134N   39. -0.837°C   41. 1.54   43. a. NH_4^+ b. No   45. 39.3 mL   47. 392.2 g   49. 14.70%

Chapter 12

1. The reaction is very slow even though its equilibrium is favorable.   3. b. A catalyst is not consumed in a reaction.   d. A catalyst speeds up a reaction by altering the reaction mechanism.   5. Emulsification increases the surface area between the fat and the water.   7. a. Rate = $k[Ag^+][S_2O_8^{2-}]$ b. Rate = $k[C_6H_5NHNHC_6H_5][H^+]^2$ c. Rate = $k[NH_2CN][H_2S]$ 9. a and b   11. A one-step process   13. a. Rate = $k[Br_2]$ b. Rate = $k[H][Br_2]$ c. Rate = $k[CH_3CHO]$

15. Second order 17. a. Threefold increase b. Ninefold increase c. No change 19. Concentrations at various times: 0, 0.0193M; 10 min, 0.0171M; 20 min, 0.0148M; 60 min, 0.00500M; 67 min, 0.00350M; 70 min, 0.00300M; 75 min, 0.00160M; 80 min, 0.00065M; 83 min, 0.00005M
The graph is linear with a slope of -2.2×10^{-4} M/min.
21. Rate = $k[C_6H_{12}O_2][OH^-]$

Experiment	Estimated $k(M^{-1}sec^{-1})$
1	10.0×10^{-3}
2	9.1×10^{-3}
3	9.1×10^{-3}
4	8.9×10^{-3}
5	9.0×10^{-3}
6	9.2×10^{-3}
Average	9.2×10^{-3}

23. $1.44 \times 10^{-3} sec^{-1}$ 25. Concentrations at various times: 0, 0.100M; 1000 sec, 0.0862M; 2000 sec, 0.0744M; 3000 sec, 0.0641M; 4000 sec, 0.0553M; 5000 sec, 0.0477M; 6000 sec, 0.0411M; 7000 sec, 0.0355M; 8000 sec, 0.0306M; 9000 sec, 0.0264M; 10,000 sec, 0.0228M 27. The first step is slow, the second is fast. 29. One possible answer: $I^- + S_2O_8^{2-} \longrightarrow I + S_2O_8^{3-}$ (slow); $S_2O_8^{3-} + I^- \longrightarrow I + 2 SO_4^{2-}$ (fast); $2 I \longrightarrow I_2$ (fast); $I_2 + I^- \longrightarrow I_3^-$ (fast) 31. $ClCH_2CO_2^- \longrightarrow Cl^- + CH_2CO_2$ (slow); $NH_3 + CH_2CO_2 \longrightarrow NH_3CH_2CO_2$ (fast); $NH_3 + NH_3CH_2CO_2 \longrightarrow NH_2CH_2CO_2^- + NH_4^+$ (fast)
33. 4.2×10^{-8} M/sec 35. 35.3% 37. a. The β-isomer b. The one producing α-naphthalenesulfonic acid c. Kinetic control 39. $0.25 M^{-1} sec^{-1}$ 41. 2.5×10^{-33} sec^{-1} 43. 143 kJ/mol H_2 45. 0.045 47. The equilibrium $2 NO \rightleftharpoons N_2O_2$ shifts to the left at higher temperature, lowering the concentration of N_2O_2. Since the rate law for the rate-determining step is Rate = $k[N_2O_2][O_2]$, the rate falls as the temperature decreases $[N_2O_2]$.

Chapter 13

1. The mass-action ratio can have any value, depending on the concentrations of reactants and products. The equilibrium constant has a specific value for a given reaction and temperature and depends on the concentrations of reactants and products at equilibrium. 3. Use a large excess of N_2 and a high total pressure. 5. 2.03

7. a. $\dfrac{P_{C_2H_6}}{P_{C_2H_2}P_{H_2}^2}$ b. $\dfrac{P_{NO}^4}{P_{N_2O}^2 P_{O_2}}$ c. $\dfrac{P_{HF}^2 P_{Cl_2}}{P_{F_2}^2 P_{HCl}}$ d. $\dfrac{P_{HCl}^4 P_{O_2}}{P_{Cl_2}^2 P_{H_2O}^2}$

9. a. $\dfrac{[H^+]^6}{P_{H_2S}^3 [Fe^{3+}]^2} = K$ b. $\dfrac{[MgOH^+][H^+]}{[Mg^{2+}]} = K$ c. $\dfrac{P_{CO_2}}{[H^+][HCO_3^-]} = K$ d. $\dfrac{[Mn(OH)_3^-]}{[OH^-]} = K$

11. a. Left b. Right c. Left d. Right e. Same f. Right 13. 5.00 atm^{-1} 15. $3.7 \times 10^{-3} M^{-1}$
17. 0.212 atm^2 19. 0.116 atm 21. Reaction goes to right.
23. Reaction goes to left. 25. a. $7.36 \times 10^4 M^{-1}$
b. 5.60×10^3 c. 5.60×10^4 d. Yes 27. 2.5×10^{-6} atm
29. a. 0.122 atm b. 0.213 atm c. 0.122 atm
d. 0.213 atm 31. $P_{PCl_5} = 0.052$ atm
$P_{PCl_3} = P_{Cl_2} = 1.45$ atm

33. -98.7 kJ 35. 1.26×10^{18}
37. 2.3×10^5 J/mol 39. 10.0 atm 41. 0.018 atm
43. Reverse rate = $k_r[N_2][O_2]$; $k_r = 3.6 \times 10^{-3} M^{-1}$ sec^{-1}
45. Reverse rate = k_r[Isobutane][HAlBr$_4$]; $k_r = 1.2 \times 10^{-4} M^{-1}$ sec^{-1}

47. a. Rate$_{f1}$ = $k_{f1}[H^+][C_2H_5OH]$ Rate$_{r1}$ = $k_{r1}[C_2H_5OH_2^+]$
Rate$_{f2}$ = $k_{f2}[C_2H_5OH_2^+]$ Rate$_{r2}$ = $k_{r2}[C_2H_5^+][H_2O]$
Rate$_{f3}$ = $k_{f3}[C_2H_5^+]$ Rate$_{r3}$ = $k_{r3}[C_2H_4][H^+]$

b. $\dfrac{[C_2H_5OH_2^+]}{[H^+][C_2H_5OH]} = K_1$ $\dfrac{[C_2H_5^+][H_2O]}{[C_2H_5OH_2^+]} = K_2$ $\dfrac{[C_2H_4][H^+]}{[C_2H_5^+]} = K_3$

c. Yes

d. $K_1K_2K_3 = \dfrac{[\cancel{C_2H_5OH_2^+}]}{[H^+][C_2H_5OH]} \times \dfrac{[\cancel{C_2H_5^+}][H_2O]}{[\cancel{C_2H_5OH_2^+}]} \times \dfrac{[C_2H_4][H^+]}{[\cancel{C_2H_5^+}]}$
$= \dfrac{[H_2O][C_2H_4]}{[C_2H_5OH]}$

Chapter 14

3. Acidic: c; basic: a and d; neutral: b 5. a. 5.22 b. 10.9
7. 9.27 9. 12.63 11. a. $K_{sp} = s^2$ b. $K_{sp} = 4s^3$
c. $K_{sp} = 4s^3$ d. $K_{sp} = 27s^4$ e. $K_{sp} = s^3$ 12. 13.0
14. 2.04 16. 10.82 18. 2.1×10^{-9} 20. 1.0×10^{-4}
22. 7.80 24. 8.95 26. 6.59 28. a. 6.4×10^{-7}
b. 1.6×10^{-8} 30. a. 1.59×10^{-8} b. 2.59×10^{-4}
c. 1.4×10^{-18} 32. a. 9.1×10^{-9} b. 1.7×10^{-16}
c. 9.1×10^{-9} d. 8.3×10^{-16} 34. Yes 36. Yes
38. $2 \times 10^{-8} M$ 40. 0 mL, 0.60; 5.0 mL, 1.00; 9.0 mL, 1.76; 9.9 mL, 2.78; 10.0 mL, 7.00; 10.1 mL, 11.22; 11.0 mL, 12.21
42. a. 0 mL, 11.0; 1.0 mL, 10.3; 5.0 mL, 9.4; 10.0 mL, 8.7; 12.0 mL, 7.9; 12.5 mL, 5.4; 13.0 mL, 2.9; 15.0 mL, 2.2

b. Methyl red
45. a. 1.8×10^{-12} b. $1.8 \times 10^{-6} M$
47. a. $[Zn^{2+}] = 0.3 M$ $[Zn(OH)_3^-] = 5 \times 10^{-11} M$ $s = 0.3 M$
b. $[Zn^{2+}] = 3 \times 10^{-3} M$
$[Zn(OH)_3^-] = 5 \times 10^{-10} M$ $s = 3 \times 10^{-3} M$
c. $[Zn^{2+}] = 3 \times 10^{-15} M$
$[Zn(OH)_3^-] = 5 \times 10^{-4} M$ $s = 5 \times 10^{-4} M$
49. $NaH_2PO_4-Na_2HPO_4$; 0.050M NaH_2PO_4 and 0.097M Na_2HPO_4
51. 6.70 53. $[HSO_4^-] = 0.043 M$; $[SO_4^{2-}]$ 0.0073M
55. a. $1.0 \times 10^{-6} M$ b. 0.050M 57. a. $1.1 \times 10^{-8} M$
b. $7 \times 10^{-5} M$

Chapter 15

3. Copper, silver, and gold are hard to oxidize and so remain in their metallic states. Potassium, magnesium, and aluminum are readily oxidized and react in natural surroundings to form compounds.

5. Cathode (copper): $4\,e^- + O_2(g) + 4\,H^+(aq) \longrightarrow 2\,H_2O$
Anode (iron): $3\,H_2O + 2\,Fe(s) \longrightarrow Fe_2O_3(s) + 6\,H^+(aq) + 6\,e^-$
7. Overpotential at the smooth electrode prevents H_2 formation.
9. $8\,H^+(aq) + 3\,Cu(s) + 2\,NO_3^-(aq) \longrightarrow$
$\qquad\qquad\qquad 3\,Cu^{2+}(aq) + 2\,NO(g) + 4\,H_2O$

11. **a.** $Fe(s)\,|\,Fe(NO_3)_2(aq)\,\vdots\,AgNO_3(aq)\,|\,Ag(s)$
$\quad\quad\;$ Anode $\qquad\qquad\qquad\qquad\quad\;$ Cathode
Electrons flow through wire from left to right. Fe^{2+} and Ag^+ flow to right, NO_3^- to left.
b. $Zn(s)\,|\,ZnSO_4(aq)\,|\,Hg_2SO_4(s)\,|\,Hg(l)$
$\quad\;$ Anode $\qquad\qquad\qquad\qquad\quad\;$ Cathode
Electrons flow through wire from left to right. Zn^{2+} flows to right, SO_4^{2-} to left.
c. $Pt(s)\,|\,H_2(g)\,|\,HCl(aq)\,|\,AgCl(s)\,|\,Ag(s)$
$\quad\;$ Anode $\qquad\qquad\qquad\qquad\quad\;$ Cathode
Electrons flow through wire from left to right. H^+ flows to right, Cl^- to left.

Chapter 16

3. Because the major constituents of air are nonpolar and do not absorb microwaves.   5. 3.9×10^3 nm; infrared
7. **a.** 6.6×10^{-22} J; molecular rotation   **b.** 3.98×10^{-19} J; electron excitation, molecular vibration and rotation
9. 2.55×10^{-22} J   11. **a.** 0   **b.** 0.70   13. 20%
15. $4.79 \times 10^3\,M^{-1}cm^{-1}$ 17. 35.8%   19. **a.** 0.157
b. 1.41   21. **a.** Three   **b.** One   23. The intensities are in the ratio 3:2.   25. $4.08 \times 10^{-2}\,M$ 29. Both $H^{35}Cl$ and $H^{37}Cl$ are present in the sample.   31. Linear. A bent molecule would have two stretching bands (symmetric and asymmetric).
33. **a.** Conjugate base   **b.** The average value is 9.7×10^{-5}.
c. 1.83

Chapter 17

1. **a.** $C(s) + H_2O(g) \longrightarrow CO(g) + H_2(g)$
b. $3\,Fe(s) + 4\,H_2O(g) \longrightarrow Fe_3O_4(s) + 4\,H_2(g)$
c. $2\,H_2O + 2\,Cl^-(aq) \longrightarrow 2\,OH^-(aq) + H_2(g) + Cl_2(g)$
d. $2\,H_2O(l) \longrightarrow 2\,H_2(g) + O_2(g)$
e. $2\,H^+(aq) + Mg(s) \longrightarrow Mg^{2+}(aq) + H_2(g)$
3. **a.** $NiO(s) + H_2(g) \longrightarrow Ni(s) + H_2O(g)$
b. $2\,Ag^+(aq) + H_2(g) \longrightarrow 2\,H^+(aq) + 2\,Ag(s)$
c. $H_2(g) + 2\,Li(s) \longrightarrow 2\,LiH(s)$
7. **a.** Lewis acid   **b.** $BH_3 + H^- \longrightarrow BH_4^-$
9. **a.** RbH, PdH, LaH_2, BeH_2 **b.** HI, B_2H_6, SiH_4, H_2S
c. RbH **d.** HI, H_2S **e.** PdH **f.** PdH, LaH_2
g. B_2H_6, BeH_2 11. c and e   13. **a.** Expansion
b. Expansion   **c.** Expansion   14. **a.** CaO   **b.** K_2O

13. **a.** $3\,Sn(s) + 4\,Au^{3+}(aq) \longrightarrow 4\,Au(s) + 3\,Sn^{4+}(aq)$
b. $2\,H_2O + Fe(s) + NiO_2(s) \longrightarrow Fe(OH)_2(s) + Ni(OH)_2(s)$
c. $2\,H_2(g) + O_2(g) \longrightarrow 2\,H_2O$
15. -97.3 kJ   17. **a.** -283 kJ   **b.** -562 J/K
19. **a.** $+0.25$ V   **b.** -0.17 V   **c.** 0.00 V   21. $+0.483$ V
23. **a.** $+1.08$ V   **b.** -0.41 V   **c.** -0.68 V
25. **a.** $+1.39$ V   **b.** -0.80 V   **c.** $+0.89$ V
27. **a.** $+1.12$ V   **b.** $+0.83$ V   29. 3×10^{39} 31. 5.90
33. $+0.071$ V; electrons flow from left to right through the wire.
35. **a.** Cathode: $e^- + Na^+(l) \longrightarrow Na(l)$
Anode: $2\,Cl^-(l) \longrightarrow Cl_2(g) + 2\,e^-$
b. Cathode: $2\,e^- + Cu^{2+}(aq) \longrightarrow Cu(s)$
Anode: $2\,H_2O \longrightarrow O_2(g) + 4\,H^+(aq) + 4\,e^-$
37. PbO_2, yes; MnO_2, no   39. 4.83 g   41. **a.** Cu^{2+} and Fe^{2+}
b. No reaction   **c.** No reaction   43. **a.** -0.197 V
b. 1.56 V   45. $2\,e^- + Fe^{2+}(aq) \longrightarrow Fe(s)$ 47. **a.** $+85$ kJ; -74 kJ; $+11$ kJ   **b.** -0.038 V   **c.** No   49. 0.400 V
51. **a.** 15.5   **b.** 155

c. Cr_2O_3 **d.** SnO_2 16. **a.** $HClO_3$ **b.** HNO_2
c. H_2CO_3 18. **a.** $Sr(OH)_2$ **b.** $LiOH$ **c.** $Ga(OH)_3$
20. **a.** H_3PO_4 **b.** H_2SO_4 **c.** H_2SeO_4 22. **a.** $RbOH$
b. $Sr(OH)_2$ **c.** ScO 24. **a.** Li^+ **b.** Ba^{2+} **c.** Mg^{2+}
d. Fe^{3+} 27. **b.** No
29. $H_2O + SrCO_3(s) \rightleftharpoons Sr^{2+}(aq) + OH^-(aq) + HCO_3^-(aq)$
31. Hydrogen bonding in ethyl alcohol   35. **a.** $+0.83$ V
b. $+1.01$ V   **c.** $+1.10$ V   37. $F-H-F^-$ (hydrogen-bonded structure)   39. **a.** 237 kJ   **b.** 237 kJ   **c.** No
41. 3.00%   43. -2037 kJ   45. $1s\sigma^2\;1s\sigma^{*2}\;2s\sigma^2\,2s\sigma^{*2}\,2p\sigma^2$
$2p\pi^4\;2p\pi^{*2}\,(\;\uparrow\;\;$ or $\;\;\downarrow\;)$ 47. HBr   49. **a.** Two
b. One peak is twice as intense as the other.

Chapter 18

1. **a.** Na_2CO_3 **b.** $CaCO_3$ **c.** CaO   **d.** $CaSO_4 \cdot 2\,H_2O$
e. $CaSO_4 \cdot \tfrac{1}{2}\,H_2O$ **f.** $Na_2B_4O_7 \cdot 10\,H_2O$
3. **a.** $2\,Cs(s) + 2\,H_2O(l) \longrightarrow 2\,Cs^+(aq) + 2\,OH^-(aq) + H_2(g)$
b. $Be(s) + H_2O(l) \longrightarrow BeO(s) + H_2(g)$
c. $Ca(s) + 2\,H_2O(l) \longrightarrow Ca(OH)_2(s) + H_2(g)$
d. $2\,Li(s) + 2\,H_2O(l) \longrightarrow 2\,Li^+(aq) + 2\,OH^-(aq) + H_2(g)$
5. **a.** $3s^2 3p^1$ **b.** s^2 **c.** s^1 7. **a.** $Mg(HCO_3)_2$ **b.** $MgCl_2$ and $Ca(NO_3)_2$ 9. It contains sodium.   11. If an aqueous solution were used, hydrogen would form at the cathode instead of sodium.   13. **a.** Less   **b.** Less

15. $CaCO_3(s) + 2\,H^+ \longrightarrow Ca^{2+}(aq) + H_2O + CO_2(g)$
17. **a.** BeO; it is amphoteric.   **b.** Yes   19. **a.** Increase
b. Decrease   **c.** Decrease   21. By the flame test. NaCl would give a yellow flame.   23. Small size   27. 57.0%
29. 1.855 Å   33. 1.5 Å   35. Br^- radius is 1.94 Å; maximum radius for Li^+ is 0.81 Å.   37. Precipitate Ba^{2+} as $BaCrO_4$ by adding K_2CrO_4. After $BaCrO_4$ is removed, $CaCO_3$ can be precipitated by adding K_2CO_3.   39. 5.992 g/cm^3 41. 13.1% occupied space compared to 68.0% occupied space when atoms touch each other.

Chapter 19

1. **a.** $PbO_2(s) + 2\,OH^-(aq) \longrightarrow PbO_3^{2-}(aq) + H_2O$
b. $Sn(s) + 4\,Cl^-(aq) + 2\,H^+(aq) \longrightarrow SnCl_4^{2-}(aq) + H_2(g)$
c. $SnCl_4^{2-}(aq) + 4\,Cl^-(aq) + 2\,Hg^{2+}(aq) \longrightarrow$
$\qquad\qquad\qquad\qquad\qquad Hg_2Cl_2(s) + SnCl_6^{2-}(aq)$
d. $SiO_2(s) + 4\,OH^-(aq) \longrightarrow SiO_4^{4-}(aq) + 2\,H_2O$

3. The molecular formula of carbon dioxide is CO_2, whereas SiO_2 is the simplest formula of molecules with very high molecular masses. Carbon dioxide is a simple molecule because carbon, a second-row element, can form double bonds readily but silicon, a third-row element, does not.

5. a. Si(s) + 2 F$_2$(g) $\longrightarrow$ SiF$_4$(g)
   b. Li$_2$C$_2$(s) + 2 H$_2$O(l) $\longrightarrow$ 2 Li$^+$(aq) + 2 OH$^-$(aq) + C$_2$H$_2$(g)
   c. GeO$_2$(s) + 2 H$_2$(g) $\longrightarrow$ Ge(l) + 2 H$_2$O(g)

7. All the valence electrons in diamond are involved in localized covalent bonds. Graphite should conduct in the directions parallel to the planar hexagonal sheets but not in the direction perpendicular to them.

9. [Fischer projection: CO$_2$H at top, H on left, OH on right, CH$_3$ at bottom]

11. a. vi, vii, ix   b. i, iv, vii   c. vii   d. v   e. iii
13. 304.1 kg   15. Test for electric conductivity. Graphite conducts; amorphous carbon does not.
17. a. :C≡O: b. :Ö=C=Ö: c. :C≡N:$^-$
   d. [CO$_3^{2-}$ Lewis structure]

19. a. Amine   b. Ketone   c. Alkane   d. Alkyl bromide   e. Alcohol   f. Carboxylic acid

21. CH$_3$CHCH$_3$ or CH$_3$OCH$_2$CH$_3$
        |
        OH

23. Any of the following:
    CH$_3$CNHCH$_3$ CH$_3$CH$_2$CNH$_2$ HCNHCH$_2$CH$_3$
       ‖   ‖   ‖
       O   O   O

25. (Typical simple compounds are given for each functional group class.)
    a. CH$_3$CH$_2$CCH$_3$ b. CH$_3$CH$_2$OH
              ‖
              O
    c. CH$_3$CH=CH$_2$ d. CH$_3$OCH$_3$

27. b

29. CH$_3$CH$_2$CH$_2$CH$_2$Br   CH$_3$CH$_2$CHCH$_3$
                                        |
                                        Br

    CH$_3$CHCH$_2$Br   CH$_3$CBr
         | |
         CH$_3$ CH$_3$
                            |
                            CH$_3$

31. c   33. b and c   35. 3.53 g/cm^3

37. CH$_3$CH$_2$CH$_2$CH$_2$CH$_2$CH$_2$CH$_3$ CH$_3$CH$_2$CH$_2$CH$_2$CHCH$_3$
                                                                    |
                                                                    CH$_3$

    CH$_3$CH$_2$CH$_2$CHCH$_2$CH$_3$ CH$_3$CH$_2$CH$_2$CCH$_3$
                   | |
                   CH$_3$ CH$_3$
                                                |
                                                CH$_3$

    CH$_3$CH$_2$CHCHCH$_3$ CH$_3$CHCH$_2$CHCH$_3$
              | | | |
              CH$_3$ CH$_3$ CH$_3$ CH$_3$

    CH$_3$CH$_2$CCH$_2$CH$_3$ CH$_3$CH$_2$CHCH$_2$CH$_3$
              | |
              CH$_3$ CH$_2$
                                             |
                                             CH$_3$

    CH$_3$ CH$_3$
        \\ /
        CHCCH$_3$
        / \\
    CH$_3$ CH$_3$

39. 0.221 kg Na$_2$CO$_3$ and 0.751 kg SiO$_2$ 41. Aluminum and silicon are both abundant and they both form high molecular mass structures in which they occupy the centers of tetrahedra of oxygen atoms.   43. 1.41 × 10^4 atm   45. 12.3

Chapter 20

1. a. Bi   b. F   c. He   3. a. White   b. White
5. CH$_3$CO$_2$H + NH$_3$ $\rightleftharpoons$ CH$_3$CO$_2^-$ + NH$_4^+$ 7. a. NO$_2$
b. Oxygen is oxidized from the -2 to the 0 state.
   2 Cu(NO$_3$)$_2$ · 3 H$_2$O(s) $\longrightarrow$
        2 CuO(s) + 4 NO$_2$(g) + O$_2$(g) + 6 H$_2$O(g)
9. NH$_3$; hydrogen bonding   11. a. The formation of ammonia by reacting hydrogen and nitrogen at high temperature and pressure in the presence of a catalyst.   b. The use of superheated steam to melt sulfur and force it out from its deep subterranean deposits.   c. Air oxidation of ammonia to NO$_2$ in the presence of a platinum catalyst. Disproportionation of the NO$_2$ in water to form nitric acid and NO.   d. Oxidation of sulfur dioxide to sulfur trioxide by air in the presence of Pt or V$_2$O$_5$ catalyst and reaction of the sulfur trioxide with water to form sulfuric acid.
e. The Frasch process   13. A trigonal bipyramidal structure with the electron pair occupying an equatorial position.

[SF$_4$ structure with lone pair]

14. a. Sodium thiosulfate   b. Bromous acid   c. Selenious acid   d. Potassium dihydrogen phosphate
16. OH$^-$ + NH$_3$ $\rightleftharpoons$ H$_2$O + NH$_2^-$ 18. The outer two atoms are in the -1 state, and the middle one is in the $+1$ state. Yes.
20. 4 H$_2$O + NO$_3^-$(aq) + 3 SO$_2$(g) $\longrightarrow$
        H$_2$NOH(aq) + 3 SO$_4^{2-}$ + 5 H$^+$(aq)
22. a. 267.7 kg   b. 9.36 × 10^{4}L   24. 0.1815 g
26. 2 × 10^{-15}M   29. a. 15.5m   b. 11.6M   31. +0.33 V
33. 9.33%   35. a. 4   b. 8   37. 13.0   39. a. 6 × 10^5
b. +61 kJ   41. a. 235 amu   b. 58 torr S$_6$; 114 torr S$_8$
c. 7.6 torr or 0.010 atm   43. Orthorhombic is more dense because an increase in pressure converts the less dense phase (monoclinic) into the denser one (Le Chatelier's principle).

Chapter 21

3. Four and six are most common.   5. [Co(NH$_3$)$_5$SO$_4$]Br would give an immediate precipitate of AgBr when treated with AgNO$_3$.
7. Red-violet   9. a. [Rn]$7s^2 5f^{14} 6d^6$ b. +8   c. +3, +4, +6, +8   13. Cu(H$_2$O)$_4^{2+}$ is blue, uncomplexed Cu^{2+} is colorless, Cu(NH$_3$)$_4^{2+}$ is blue-violet.   15. Pd and Pt; lanthanide contraction
16. a. 4   b. 6   c. 4   18. a. +2   b. +2   c. +3
20. a. 2   b. 2   c. 1   22. No   24. a. Potassium tetracyanoaurate(III)   b. Hexaamminechromium(III) chloride   c. Potassium hexachloroplatinate(IV)   d. Hexacyanomolybdate(II) ion   e. Hexacyanoferrate(II) ion   f. Diiron nonacarbonyl   26. a. [VCl$_6$]$^{3-}$ b. [Ag(NH$_3$)$_2$]$^+$ c. [Co(C$_2$O$_4$)$_3$]$^{3-}$

28. a. [Kr] (4d^{10} filled) sp hybrids with CN$^-$, CN$^-$
    b. [Ar] 3d^8, dsp^2 hybrids with NH$_3$ (×4), 4p^1
    c. [Ar] 3d^6, d^{2}sp^3 hybrids with CO (×6)

30. cis and trans isomers of [Co(H$_2$O)$_3$Br$_2$(OH$_2$)] type complex with Br and OH$_2$ ligands.

32. a. 6, including 2 pairs of optical isomers   b. 2, no optical isomers   34. $\sigma_s^2 \sigma_x^2 \sigma_y^2 \sigma_z^2 \underbrace{\sigma_{z^2}^2 \sigma_{x^2-y^2}^2}_{e_g} \underbrace{\pi_{xy}^2 \pi_{xz}^2 \pi_{yz}^1}_{t_{2g}}$; one unpaired electron

36. Four large ligands
37. Measurement of dipole moment. The trans is not polar, the cis is.   39. The square-planar complex would be diamagnetic, whereas the tetrahedral complex would have two unpaired electrons.   41. [Co(NH$_3$)$_5$Br]$^{2+1}$
43. Rh: [Kr] 4d^8 (↑↓ ↑↓ ↑↓ ↑ ↑), 5s^1 (↑)
    Rh^{3+}: [Kr] 4d^6 (↑↓ ↑ ↑ ↑ ↑), 5s^0
    RhCl$_6^{3-}$: [Kr] 4d^6 filled with Cl$^-$ (×6) as d^{2}sp^3 hybrids

45. An octahedral field will not, a tetrahedral field will.
47. a. 2 Ce(s) + 6 H$_2$O(l) ⟶ 2 Ce^{3+}(aq) + 6 OH$^-$(aq) + 3 H$_2$(g)
    b. Ce^{4+}(aq) + Fe^{2+}(aq) ⟶ Ce^{3+}(aq) + Fe^{3+}(aq)
    c. Ti^{2+}(aq) + Fe^{3+}(aq) ⟶ Ti^{3+}(aq) + Fe^{2+}(aq)
49. It dissolves the unreacted AgBr by forming a silver complex.

AgBr(s) + 2 S$_2$O$_3^{2-}$(aq) ⟶ Ag(S$_2$O$_3$)$_2^{3-}$(aq) + Br$^-$(aq)

Chapter 22

3. An α particle has a smaller charge-to-mass ratio.
5. Magic numbers and discrete γ-ray emission spectra
9. Decay into smaller fragments. It is more massive than Fe-56, the most stable isotope.   10. a. 8 p, 10 n   b. 50 p, 68 n   c. 94 p, 145 n   12. a and c   14. a. $^4_2$He   b. $^{108}_{45}$Rh   c. $^4_2$He
16. a. $^{23}_{11}$Na + $^2_1$H ⟶ $^{21}_{10}$Ne + $^4_2$He
    b. $^{39}_{19}$K + $^1_1$H ⟶ $^{36}_{18}$Ar + $^4_2$He
    c. $^{115}_{49}$In + $^1_0$n ⟶ $^{116}_{49}$In
    d. $^{98}_{42}$Mo + $^2_1$H ⟶ $^{99}_{43}$Tc + $^1_0$n
18. a. $^{210}_{83}$Bi ⟶ $^{206}_{81}$Tl + $^4_2$He
    b. $^{222}_{86}$Rn ⟶ $^{218}_{84}$Po + $^4_2$He
    c. $^{194}_{81}$Tl ⟶ $^{190}_{79}$Au + $^4_2$He
    d. $^{232}_{90}$Th ⟶ $^{228}_{88}$Ra + $^4_2$He
20. a. $^{16}_8$O, $^{17}_8$O, $^{18}_8$O   b. $^{112}_{50}$Sn, $^{114}_{50}$Sn, $^{115}_{50}$Sn, $^{116}_{50}$Sn, $^{117}_{50}$Sn, $^{118}_{50}$Sn, $^{119}_{50}$Sn, $^{120}_{50}$Sn, $^{122}_{50}$Sn, $^{124}_{50}$Sn   c. $^{106}_{48}$Cd, $^{108}_{48}$Cd, $^{110}_{48}$Cd, $^{111}_{48}$Cd, $^{112}_{48}$Cd, $^{113}_{48}$Cd, $^{114}_{48}$Cd, $^{116}_{48}$Cd   22. a. $^{134}_{54}$Xe, $^{136}_{56}$Ba, $^{138}_{58}$Ce
b. $^{136}_{54}$Xe, $^{138}_{56}$Ba, $^{139}_{57}$La, $^{140}_{58}$Ce, $^{141}_{59}$Pr, $^{142}_{60}$Nd, $^{144}_{62}$Sm
c. $^{142}_{58}$Ce, $^{144}_{60}$Nd
24. a. 7.47 × 10^{11} J/mol (7.76 MeV); 2.59 MeV/nucleon
b. 8.1994 × 10^{13} J/mol (849.80 MeV); 8.5838 MeV/nucleon
c. 1.82661 × 10^{14} J/mol (1893.12 MeV); 7.45323 MeV/nucleon
26. 3.15 × 10^8 kJ/mol released   28. 9.77 × 10^{-4}
30. 1.10 billion yr   32. 1.7 × 10^4 yr   34. $t_{1/2}$ = 12 hr
37. 8.2 × 10^{14} atoms   39. 0.00673 amu; 1.01 × 10^{-12} J (6.31 MeV)   41. One possibility is an (n,p) reaction of Pb to form a Tl isotope that undergoes α decay to Au. The conversion of Pb to Tl could be brought about by a (p,α) reaction.   43. 8.30 × 10^{-12} J (51.9 MeV)   45. Both Rh-98 and Tc-98 decay to the same excited state of the Ru-98 nucleus. The same energy photon is released in both cases as the excited state changes to the nuclear ground state.

Chapter 23

3. a. iii   b. ii   c. vi   d. v   e. i   7. Proteins
9. False statements with errors corrected are: d. Electrostatic attraction and the tendency of nonpolar groups to cluster together also play a role in maintaining the three-dimensional structures.   e. Random coils appear in native proteins.   f. We still need to know the tertiary and quaternary structures.

11. $NH_3^+CHCNHCHCNHCHCO_2^-$
with structure:
- position 1: CH$_3$
- position 2: CH$_2$–OH
- position 3: CH$_2$–CO$_2$H

(showing two C=O groups in the backbone)

13. ACCTG   **15. a.** Adenine   **c.** AUC

17. Gly.Gly.Ala.Val   Gly.Gly.Val.Ala   Gly.Ala.Gly.Val
Gly.Val.Gly.Ala   Gly.Ala.Val.Gly   Gly.Val.Ala.Gly
Ala.Gly.Gly.Val   Val.Gly.Gly.Ala   Ala.Gly.Val.Gly
Val.Gly.Ala.Gly   Ala.Val.Gly.Gly   Val.Ala.Gly.Gly

19. Isoleucine and threonine

21. a. $NH_3^+CH_2CO_2^-(aq) + H^+(aq) \longrightarrow NH_3^+CH_2CO_2H(aq)$
   b. $NH_3^+CH_2CO_2^-(aq) + OH^-(aq) \longrightarrow NH_2CH_2CO_2^-(aq) + H_2O$

23. Buried within the interior   **25.** −31 kJ   **27.** c

29. It would be too difficult to synthesize.   **31. a.** No.

b. The enzyme for this reaction is a regulatory enzyme that is activated by ADP and inhibited by ATP.   **c.** It avoids wasting nutrients when the organism has enough available free energy as ATP and stimulates glycolysis when the organism is depleted of usable free energy.   **33. a.** AAA   **b.** GGG, GGU, GUG, UGG, GUU, UGU, UUG, UUU   **c.** Each codon would be present in the same amount.   **35.** The rate-determining step is the rate of decomposition of the enzyme–substrate complex, and is first order with respect to [ES]. At a low concentration of S, an increase in [S] shifts the binding equilibrium to the right, forming more complex and increasing the reaction rate. At high substrate concentrations, the binding equilibrium lies so far to the right that essentially all the enzyme is tied up as complex. Since a further increase in substrate concentration cannot raise the concentration of the complex, it does not increase the rate of the reaction.

Index

Absolute rate theory, 321-322
Absolute temperature, 52, 64-65
Absolute zero, 52, 64-65
Absorbance, 407-408
Acceleration 21, 64-66
Acetaldehyde, 316
Acetic acid, 162, 182, 355, 358, 363-365
Acetylene, 304, 476, 487
Acetylide ion, 476
Acid(s), 181, 287
   Arrhenius, 288
   Brønsted-Lowry, 289-290
   conjugate, 290-291
   Lewis, 292, 461, 467, 541, 550
   polymeric, 182
   polyprotic, 292, 365-366
Acid solutions, 179, 289, 352-353
Actinides, 466, 539, 541, 567
Actinium, 567
Activated complex, 318-319
Activation energy, 316, 318-320
Active site, 608-609
Activities, 337-338
Addition reactions, 487
Adenine, 609, 613
Adenosine diphosphate (ADP), 615
Adenosine triphosphate (ATP), 615, 617-619
Adiabatic systems, 191
Aerosol, 279
Albumin, 606-607
Alcohols, 479, 600
Aldehyde, 600
Aliphatic compounds, 489
Alkali metals, 79, 142, 449-455, 476
   compounds of, 453-455
   occurrence of, 449-450
   preparation of, 450-451
   properties of, 451-452
   reactions of, 452-453
Alkaline earth elements, 142, 455-462, 476
   carbonates and bicarbonates of, 458-460
   properties and occurrence of, 456
   stability of the +2 oxidation state of, 457-458
Alkanes, 480-481
Alkenes, 485-486, 487
Alkynes, 486-487
Allotropes, 433, 474, 509, 519-520, 560
Alloys, 452, 456, 463, 473, 511, 526, 539, 557, 558, 560, 563, 564
Alpha decay, 581, 585
Alpha helix, 605-606
Alpha particles, 92-93, 574, 580-582, 586
   scattering of, 92-94
Alpha rays, 574, 593
Alum, 217
Alumina, 316

Aluminum, 82, 392, 395, 397, 423, 453, 456, 462, 463-465, 467, 528, 556, 564
Aluminum hydroxide, 293, 370-371
Aluminum oxide, 395, 462-463
Amalgam, 455, 564
Amide ion, 289
Amino acids, 479, 507, 548, 602, 609, 613-614
Amino group, 602, 609
Ammonia, 163, 164, 167, 288, 291, 357-358, 371-372, 423, 454, 508-509, 512, 515, 543-544, 567
Ammonium cyanate, 479
Ammonium hydroxide, 288
Ammonium ion, 182, 512
Ammonium nitrate, 512
Ammonium nitrite, 509
Ammonium perchlorate, 512
Amorphous solid, 217, 235
Ampere, 396
Amphoterism, 293
Angstrom, 10
Anhydrides, 436
Anion(s), 136
Anion exchange resin, 460
Anode, 86, 380, 383, 395
Anodizing, 395, 463
Anticodon, 613
Antimonides, 517
Antimony, 81, 389-390, 507, 510-511
Antiparticle, 581-582
Aqua regia, 528
Argon, 82, 531-532, 574, 593
Aromatic compounds, 487-489
Arrhenius, Svante, 283-285, 316
Arrhenius equation, 316-318
Arsenic, 81, 507, 510-511
Arsenic acid, 518
Arsenic oxide, 517, 518
Arsenides, 517, 563
Arsine, 517
Asbestos, 456, 494
Astatine, 524
Atomic emission spectroscopy, 106, 531-532
Atomic mass unit, 15
Atomic masses
   average, 14
   formulas and, 15
Atomic number, 82, 94, 576, 580-582
   protons and, 95
   x-ray emission and, 108
Atomic theory. See Dalton's theory
Atoms, 3, 13
   elements and, 13
Aufbau principle, 123, 540-541
Austenite, 560

Autoprotolysis, 351-352, 354
Avogadro, Amadeo, 58-59
Avogadro's hypothesis, 58-60, 65-66
Avogadro's number, 32-33, 92
Axial positions, 169
Azeotropes, 266
Azide(s), 513
Azide ion, 513

Babbitt metal, 511
Baking soda, 478
Balmer's law, 108-110, 112
Barite, 461
Barium, 456
Barium oxide, 461
Barium sulfate, 442, 461
Barometer, 48
Bartlett, Neal, 532
Base(s), 287
   Arrhenius, 288
   Brønsted-Lowry, 289-290
   conjugate, 290-291
   Lewis, 292, 371, 478, 512, 548, 550
   polyprotic, 292-293
Base group, 609
Basic oxygen process, 560-561
Basic solutions, 179, 289, 352-353
Bauxite, 462
de Becquerel, Henri, 573
Beer-Lambert law, 407-408
Benzene, 143, 487-489
Beryl, 456
Beryllium, 453, 456, 458, 460-461, 462, 527
Beryllium chloride, 161, 460-461
Beta decay, 581, 585
Beta particles, 425, 475, 574, 581
Beta rays, 574, 593
Beta structure, 606
Bethe, Hans, 552
Bicarbonate(s), 458-460
Bicarbonate ion, 293, 454, 477-478
Bifluoride ion, 529
Bimolecular elementary reactions, 311
Bioenergetics, 615-619
Bis(ethylenediamine)dichlorocobalt(III) ion, 550
Bismuth, 81, 506-507, 565
Bismuthate ion, 517
Bismuthides, 517
Bismuthine, 517
Bisulfate, 523
Bisulfides, 521
Bisulfite, 523
Blast furnace, 560
Body-centered (I) lattices, 221
Bohr, Niels, 106-110
Bohr atomic model, 106-112, 114, 115

INDEX

Boiling, 245-246
Boiling point, 237, 245
  normal, 245
Bomb calorimeter, 194-195
Bond dissociation energy, 144-145
Bond length. See Internuclear distance
Borane, 429
Borax, 462, 466
Borazine, 466
Boric acid, 467
Borides, 466
Born-Haber cycle, 140-142, 201, 457
Boron, 453, 462, 464, 465-467
Boron carbide, 476
Boron hydrides, 466
Boron hydroxide. See Boric acid
Boron nitrides, 465-466
Boron trifluoride, 163-164
Boyle, Robert, 5, 7
Boyle's law, 50-54
Bragg's law, 223-225
Branched-chain process, 589-590
Brass, 397, 563, 565
Bravais lattices, 221
Breeder reactor, 591
Brines, 394, 450, 454
de Broglie, Louis, 112-114
Bromine, 79, 487, 526
Bronze, 473, 563
Brownian motion, 63-64
Bubble chamber, 574-575
Buffer solutions, 357-361
Bunsen, Robert William Eberhard, 106
Butanes, 143, 481
1-Butene, 485
cis-2-Butene, 485
trans-2-Butene, 485

Cadmium, 498, 558, 565, 590
Calcium, 82, 450
Calcium carbide, 476
Calcium silicate, 560
Calcium sulfate, 523
Calorie, 10, 21
Calorimeter, 194-195
Calorimetry, 194-197
Cannizzaro, Stanislao, 59, 76
Carbides, 476
Carbohydrates, 422, 600
Carbon, 81, 450, 465, 473-491, 523, 539, 560-561
  allotropes of, 474-475
  compounds of, 476-479
  isotopes of, 475-476
Carbon-14, 588, 593
Carbon cycle, 477
Carbon-dating method, 593
Carbon dioxide, 169, 293, 454, 458, 476-478, 497, 562
Carbon monoxide, 422, 478
Carbon-nitrogen cycle, 591
Carbon tetrachloride, 527, 529
Carbonate(s), 370, 432, 442, 450, 453, 458-460, 474, 477, 564
Carbonate ion, 169-171, 293, 477-478, 494
Carbonic acid, 288, 477-478
Carborundum, 476
Carboxyl group, 479, 490
Carboxylic acid, 479
Carboxypeptidase, 607
Cassiterite, 474, 494
Catalysis
  hetergeneous, 315-316, 563
  homogeneous, 313-315
  surface. See Catalysis, heterogeneous
Catalyst, 313-314
Catenation, 479
Cathode, 86, 380, 383, 395

Cathode ray(s), 90
Cathode ray tube, 89-90
Cation(s), 136
Cation exchange resin, 460
Caustic soda, 454
Cavendish, Henry, 531
Cell (biological), 599-600
Cell membranes, 599-600, 602
Cell notation, 381
Cellulose, 600
Celsius scale, 20
Cementite, 560
Cerium, 567
Cesium, 79, 439, 451, 452
Chalk, 456
Charge, 83-84
  conservation of, 83, 176
  of electron, 91-92
Charles' law, 52-54
Chelate(s), 548-550
Chelate group. See Ligand(s), polydentate
Chelating agents. See Ligand(s), polydentate
Chelation, 548
Chemical equations, 36-38
  balancing of, 37-38, 176-180
  stoichiometric calculations and, 38-42
Chemical families, 79
  periodic table and, 82
Chemical nomenclature, 180-183
  acids, 181-182
  anions, 182-183
  binary compounds of nonmetals, 180-181
  cations, 182
  inorganic complexes, 641-642
  salts, 183
Chemical reactions, 5
  distinction between nuclear reactions and, 6-7
  distinction between physical processes and, 5
  molecules and, 13
Chemiluminescence, 414-415, 516
Chemisorption, 315-316
Chile saltpeter, 507
Chiral center, 481-484
Chirality, 481-484
Chlorates, 530
Chlorides, 450, 456, 565
Chlorine, 79, 174, 342, 392, 394-395, 526, 530
Chlorine dioxide, 530
Chlorites, 530
2-Chlorobutane, 481
Chloroform, 529
Chlorophyll, 619-620
Chloroplasts, 600, 619
Chromate ion, 370, 558
Chromatography, 565
Chromium, 82, 394, 397, 456, 463, 495, 557-558, 563
Chromium trioxide, 558
Chymotrypsin, 609
Cinnabar, 564
Clathrates, 442-443
Clausius-Clapeyron equation, 242-245
Closed systems, 191
Cloud chamber, 574-575, 582
Cobalt, 82, 463, 495, 540, 543-544, 550, 559, 563
Codon, 613-614
Coenzyme, 607
Cofactor, 607-608
Coke, 422, 509, 560
Colligative properties, 267-274, 544
  of electrolytes, 284-285
Collision frequency, 320
Collision theory, 318-321
Colloid(s), 277-278

Colloidal suspensions, 277
Common-ion effect, 368-369
Complex ions, 371, 512, 541, 543-555
  colors of, 552-555
  magnetic properties of, 552
  octahedral, 544-546, 551
  square planar, 546, 551
  square pyramidal, 547
  tetrahedral, 547, 551
  trigonal bipyramidal, 547
Complexes. See Complex ions
Composition, 35-36
Compounds, 5
  molecules and, 13
Concentration(s), 258-260
Concentration cells, 300-309
Condensed phases, 236
Conduction band, 234, 451, 492
Conductivities, 287, 462, 466, 544
Conservation of orbitals, 166
Constructive interference, 103, 145
Contact process, 74, 523-524
Coordinate covalent bond, 550
Coordination complexes. See Complex ions
Coordination number
  for complex ions, 541, 544
  for crystals, 227
Copper, 82, 380, 393, 395, 397, 456, 463, 473, 539, 563-564
Copper(II) ion, 379, 383
Corundum, 462
Cosmic rays, 588, 593
Coulomb (unit), 84
Coulomb's law, 83, 107
Coulometric titrations, 395-397
Counts per minute, 575-576
Covalent bond, 135, 143-148
  double, 147
  single, 148
  triple, 146
Covalent substances, 142-143
Cracking, 422-423, 480
Crick, Sir Francis, 610
Cristobalite, 495
Critical mass, 590
Critical pressure, 248
Critical temperature, 248
Cryolite, 464, 525
Crystal(s), 217-235
  covalent, 234
  ionic, 234
  metallic, 234
  molecular, 234
Crystal-field theory, 552-555
Crystal imperfections, 235, 492
Crystal lattice(s), 220-221, 457
  antifluorite, 521
  body-centered cubic, 231, 451, 560
  calcium fluoride, 233, 521
  cesium chloride, 233
  diamond, 232
  face-centered cubic, 225, 227, 229-233, 451, 560
  simple cubic, 233, 254
  sodium chloride, 232-233
  wurtzite, 233
  zinc blende, 233
Crystal lattice energy, 141, 457, 462
Crystal systems, 221
Crystallinity, 236
Cubic closest packing, 226-231
Cuprite, 564
Curie (unit), 576
Curie, Marie and Pierre, 99, 573
Cyanate ion, 168
Cyanide ion, 183, 478-479
Cyclic compounds, 487
Cycloalkenes, 487

Cyclohexane, 487
Cyclopentane, 487
Cyclopropane, 487
Cyclotron, 586-587
Cytochromes, 616
Cytosine, 609

Dalton, John, 13, 55, 59
Dalton's law of partial pressures, 55-57
Dalton's theory, 13-15
Daniell cell, 380-382
Daughter nuclei, 585, 593
Davy, Sir Humphrey, 450
Deacon process, 526
Decomposition potential, 393
Defect structures, 492
Degenerate orbitals, 117, 125, 552-553
Deliquesce, 462, 515
Denaturations, 606-607, 609
Denitrification, 508
Density
  of close-packed crystals, 231
  of gases, 61-62
Deoxyribonucleic acid (DNA), 593, 599, 609-612
Deoxyribose, 609
Desiccants, 462
Destructive interference, 103, 145
Deuterium, 14, 425
Deuteron, 586
Dextrarotatory, 481
Diagonal similarities, 453
Dialysis, 276-277
Diamagnetism, 85
Diamond, 143, 232, 234, 440, 466, 474-475
Diatomic molecules, 58
Diborane, 429
Dichromate ion, 558, 561
Diethylenetriamine, 548
Differential equations, 116, 309
Diffraction pattern, 102-104
Diffusion, 67, 236
Dimensional analysis, 634-635
Dinitrogen pentoxide, 515
Dinitrogen tetroxide, 515
Dinitrogen trioxide, 515
Dipolar molecules, 149
Dipole moment, 151, 547
Diradicals, 164
Disproportionation, 457, 530, 558, 563
Dissociation, 283, 286
  constant, 355
  degree of, 287
  of strong acids, 354-355
  of strong bases, 356
  of weak acids, 288, 355-356
Distillation, 3-5, 264-265, 432, 509
Dolomite, 456, 478, 560
Double helix, 609-611
Dow process, 456
Downs cell, 450
Dry cell, 391
Dry ice, 247-248

Effective reduction potential, 394
Einstein, Albert, 105
Einstein's law of conservation of mass and energy, 576
Elastic collisions, 19, 65
Electric potential, 382
Electric potential energy, 84, 379
Electrodes. See Half-cells
Electrolysis, 86-89, 379, 391, 393, 422, 426, 438, 450, 454-455, 456, 464, 526, 530
  commercial, 395
  competing, 394
  concentration effects on, 394

Electrolytes, 283
  strong, 283-286
  weak, 286-287
Electrolytic corrosion, 397-398, 562
Electromagnetism, 85-86
Electron, 89-92, 574, 581-582
Electron affinity, 138-139
Electron capture, 582
Electron configuration, 123-125
  chemical similarity and, 128
Electron-deficient species, 163-164
Electron-density function, 226
Electron-dot structures. See Structural formulas
Electron spin, 121-122, 125, 147, 414, 542
Electron transport chain, 616-619
Electron volt, 10, 577
Electronegativity, 150-151, 172-173
Electronic spectra, 414, 552
Electrophoresis, 277
Electroplating, 392, 395
Electrorefining, 395, 494, 564
Electroscope, 574
Electrostatic forces, 83-84, 579
Electrowinning, 395
Elementary reactions, 311
Elements, 5-6
  atoms and, 13
  nuclear reactions and, 6
Emerald, 456
Emission spectra, 106, 454, 565
Empirical formula. See Simplest formula
Emulsion, 279
Enatiomers, 481-485, 495, 550
End point, 293, 362
Endothermic reaction, 189
Energy, 18-19, 191-194
Energy levels, 107-108
  electronic, 406, 413-414
  rotational, 406, 409
  vibrational, 406, 409, 410-411
Enthalpy, 197, 383
  of activation, 321
  of formation, 199
  of reaction, 197-199
Entropy, 203-210, 238-239, 245, 261, 383
  of activation, 321
  heat flow and, 205
  standard, 205-206
  useful work and, 208-210
Enzymes, 316, 484, 507, 539, 594, 602, 607-609, 613, 617, 619
Enzyme-substrate complex, 608
Epsom salts, 461
Equatorial positions, 169
Equilibrium, 208, 329
  effect of temperature on, 340, 341
  kinetics and, 342-343
Equilibrium constant, 331-332
  standard potential and, 388
  thermodynamic, 337
Equilibrium control, 304
Equilibrium vapor pressure, 241-242, 263-264, 267-270
  temperature effects on, 242-245
Equivalence point, 362
Equivalent(s), 295
  acid-base, 295-296
  redox, 296
Equivalent mass, 295-297
Ester bonds, 600
Ethane, 480
Ethanol, 316, 342
Ethyl acetate, 313-314
Ethyl group, 481
Ethylene, 316, 342, 485
Ethylene bromide, 526
Ethylenediamine, 548, 550

Ethylenediamine tetraacetate (EDTA), 548
Europium, 566-567
Evaporation, 241
  thermodynamics of, 245
Excited state, 107, 411, 414
Exothermic reaction, 189
Exponentials, 629-630
Extinction coefficient. See Molar absorptivity

Face-centered (F) lattices, 221
Fahrenheit scale, 20
Faraday (unit), 87, 92
Faraday, Michael, 87
Faraday's law, 87-89, 382
Faraday-Tyndall effect. See Light scattering
Fats, 422, 600-602
Feldspars, 462
Ferrite, 560
Ferromagnetism, 85, 542
First ionization potential, 136
First law of thermodynamics. See Law of conservation of energy
Flame spectrophotometer, 454
Flame test, 454
Flavin nucleotides, 615
Fluorapatite, 525
Fluorescence, 414
Flouride(s), 453, 464, 525-526, 527, 532-533
Fluoride ion, 369, 527, 528
Fluorine, 79, 173, 174, 525-526, 528, 532-533, 557
Fluorite, 525, 557
Fluorocarbons, 433
Fluorspar. See Fluorite
Forbidden transitions, 409
Force, 17, 22, 65-66
Force constant, 410
Formaldehyde, 167
Formality, 260
Formic acid, 182, 478
Formula mass, 34
Formulas, atomic masses and, 15
Francium, 452, 573
Frasch process, 519
Free radicals, 164, 513-514, 515
Frenkel defects, 492
Freon-12, 529
Frequency, 103-105
Frequency factor, 316
Fructose, 600
Fuel cell, 391-392
Fuel element, 590
Fumarate hydratase, 607
Functional groups, 479, 490, 548

Gadolin, Johan, 565
Gadolinite, 556, 565
Galena, 474, 494, 519
Gallium, 82, 462, 463, 465
Galvanic cells, 379-380
Galvanized iron, 398, 565
Gamma rays, 574, 575, 582, 593-594
Gas volume, stoichiometry of, 62-63
Gases, 47-70
  effect of pressure on, 50-51
  effect of temperature on, 52-54
  molar volume of, 59-60
Gay-Lussac, Joseph, 57-59
Gay-Lussac's law, 54
Geiger-Müller counters, 574-575
Genetic code, 609
Geologic-dating methods, 592-593
Geometric isomers
  of alkenes, 485-486, 487

Geometric isomers (continued)
 of complex ions, 544-547, 550
Germanium, 81, 474, 491-492, 493
Gibbs free energy, 206-210, 382-383
 of activation, 321
 equilibrium and, 208, 329, 339-340
 of formation, 207
 standard, 207, 339
 useful work and, 208-210
Glass electrode, 389
Glasses, 217, 440, 495-496
Glucose, 600, 617-619
Glycerol, 600
Glycolysis, 617-619
Gold, 395, 521, 563-564
Graham's law of diffusion, 67-68
Graphite, 391, 395, 450, 464, 474-475, 526, 590
Greenhouse effect, 418
Ground state, 123, 414
Group IA elements. See Alkali metals
Group IB elements, 540, 563-564
Group IIA elements. See Alkaline earth elements
Group IIB elements, 540, 564-565
Group IIIA elements, 462-467, 540
 compounds of, 465-467
 occurrence of, 462-463
 properties and preparation of, 463-465
Group IIIB elements, 540, 556
Group IVA elements, 473-498
 preparation of, 493-494
Group IVB elements, 556-557
Group VA elements, 505-518
 occurrence of, 507
 preparation of, 509
 properties of, 509-510
Group VB elements, 557
Group VIA elements, 518-524
 occurrence of, 519
Group VIB elements, 557-558
Group VIIA elements. See Halogens
Group VIIB elements, 540, 557-558
Guanine, 609
Gypsum, 461

Haber, Fritz, 508
Haber process, 508-509
Hafnium, 556-557
Half-cell(s), 380
Half-cell potentials, 383-386
 effect of concentration on, 386-388
Half-life, 309, 583-585, 588
Half-reactions, 177-178
Halides, 516, 527-529, 564, 565
Hall, Charles Martin, 464
Hall-Heroult process, 464
Halogens, 79, 142, 465, 516, 524-530
 occurrence of, 525
 properties of, 525-527
Heat, 16, 19-21, 64-66, 190-196
Heat capacity, 21, 195-196
Heat of solution, 261-263
Heating curves, 237
Heavy water, 12, 425, 590
Heisenberg uncertainty principle, 115, 144, 171
Helium, 82, 531-532, 576
 liquid, 532, 534
Hematite, 560
Heme group, 616
Hemoglobin, 478, 539, 548-550
Henry's law, 274-276
Heroult, Paul, 464
Hertz (unit), 104
Hess's law, 199-202
Heterocyclic compounds, 487
Heterogeneous mixture, 3

Hexagonal closest packing, 226-229, 451
Hexanes, 481
Histidine, 609
Hodgkins, Dorothy Crowfoot, 226
Homogeneous mixture, 3, 257
Hund's rules, 125-128, 147
Huygens, Christian, 101
Hybrid orbitals, 165-168
 dsp^2, 551
 d^2sp^3, 551-552
 sp, 167, 486, 494
 sp^2, 167, 474, 485, 489, 494
 sp^3, 166-167, 475, 494, 551
 sp^3d, 533
 sp^3d^2, 533, 552
Hydrated hydrogen ion. See Hydronium ion
Hydrates, 443
Hydration, 441, 457
Hydration energy, 452-453, 457, 458, 462
Hydrazine, 292-293, 512-513
Hydrazoic acid, 513
Hydrazonium ion, 513
Hydride ion, 159, 425, 427
Hydrogen, 82, 159, 173, 329, 381, 383, 391, 393, 394, 421-422, 452, 521
 compounds of, 426-429
 isotopes of, 14, 425-426
 molecular compounds of, 428
 monatomic, 424-425
 occurrence of, 421
 preparation of, 422-423
 properties of, 423-425
Hydrogen bomb, 592
Hydrogen bonds, 422, 429-431, 438, 439-440, 512, 528-529, 605, 609
Hydrogen chloride, 148, 291, 528-529
Hydrogen cyanide, 478-479
Hydrogen fluoride, 528-529
Hydrogen iodide, 329
Hydrogen peroxide, 173, 437-438, 512, 523
Hydrogen persulfide, 523
Hydrogen sulfide, 521
 precipitation of metal ions by, 521-522
Hydrolysis, 356-357, 436, 442
Hydrometer, 391
Hydronium ion, 160, 288, 351-352
Hydroxide(s), 436, 450, 452, 461, 467
Hydroxide ion, 159, 183, 288, 351-354, 370
Hydroxyl group, 436, 479, 612
Hydroxylamine, 513
Hypochlorous acid, 529

Ice, 237, 241, 440-441
Icosahedron, 464
Ideal gas, 51
Ideal gas law, 60-61
 derivation of, 65-67
Ideal solutions, 263
Ilmenite, 435, 557
Indicators, 293, 361
Indium, 82, 463, 465, 467
Infrared radiation, 406, 477
 molecular vibrations and, 410-413
Inner orbital complexes, 552
Insulin, 602
Interference, 102-104
 by electron beams, 113-114
Intermediate, 312
Intermolecular attraction, 65, 69, 234, 262
Internuclear distance, 144-145
Interstices, 227
Interstitial hydrides, 428, 466
Iodates, 525, 526
Iodimetric titrations, 527
Iodine, 79, 329, 525, 526-527
Iodometric titrations, 524, 527
Ion(s), 136-138
Ion exchange, 459-460

Ion exchange chromatography, 565
Ionic bond, 135-136
Ionic compounds, 139-142
 conductivities of, 139-140
 stability of, 140-142
Ionic hydrides, 427-428
Ionic lattice, 139, 232
Ionization, 574
Ionization counters, 574-575
Ionization potential, 136-138, 452-453, 457-458, 462, 532
Iron, 82, 397, 422, 463, 539, 540, 559-562
 rusting of, 397-398, 560
Iron-56, 578
Iron(III) hydroxide, 370
Iron(III) oxide, 397
Iron sulfide(pyrites), 519, 560
Isomer(s), 162, 171, 304
Isomerism, 161-163, 480-481
Isopropyl chloride, 416
Isothermal systems, 191
Isotope(s), 14, 82, 94-96, 425-426, 578, 585, 587, 593
 failure of the law of definite proportions and, 14
Isotope effects, 425
IUPAC nomenclature, 180-182

Joule (unit), 18, 193

Kelvin (unit), 52
Kelvin temperature scale, 52
Kernite, 462
Ketone, 600
Kilogram, 8-9
Kinetic characteristics, 303-304
Kinetic control, 304, 607
Kinetic energy, 19
 of molecules, 64-66, 190
Kinetic isotope effect, 425-426
Kinetic-molecular theory, 63-67
Kinetic orders, 307-310
 reaction mechanisms and, 311-313
Kinetics of radioactive decay, 582-585
Kjeldahl analysis, 301, 537-538
Krypton, 82, 532

Laboratory balance, 8
Lamellar silicates, 494
Lanthanide(s), 82, 466, 527, 539, 542-543, 556, 565-567
 electron configurations of, 128, 540-541
Lanthanide contraction, 542-543, 557
Lanthanum, 82, 540, 556
Lattice, 218
Lattice array, 139, 220
Lattice forces, 233-234
Law of combining volumes, 57
Law of conservation of energy, 19, 190
Law of conservation of mass, 10-11
Law of conservation of momentum, 17-18
Law of definite proportions, 11-12
Law of mass action, 331-332
Law of multiple proportions, 16
Le Chatelier's principle, 241, 245, 262, 287, 289, 332, 358, 368, 369, 425, 475
Lead, 81, 389-390, 452, 463, 473-474, 492-493, 497-498, 511, 558
Lead dioxide, 510
Lead storage battery, 389-390
Lead sulfate, 390
Levelling effect, 291-292
Levorotatory, 481
Lewis, Gilbert Newton, 143, 292
Lewis octet rule, 159-161
Lewis structures. See Structural formulas
Libby, Willard, 593
Ligand(s), 541, 548-550, 552-555

Ligand(s) *(continued)*
  bidentate, 548
  monodentate, 548
  polydentate, 548
  strong and weak, 553-555
Ligand field theory, 555
Light, 101-106
  photon model for, 105
Light scattering, 277
Lime, 180, 456, 461, 495
Limestone, 456, 459, 478, 560
Limiting reactants, 40-41
Linear accelerator, 586
Linear equations, 634
Lipids, 600-602
Liquid(s), 47, 236-237
Liquid crystals, 248-249
  cholesteric, 248
  nematic, 248
  smectic, 248
Liquid drop model for nuclei, 578, 589
Liquid-liquid juncture, 380
Liter, 10
Lithium, 79, 439, 449, 452-453
Lithium ion, 452-453
Logarithms
  base ten, 631-632, 636-637
  mathematical operations with, 633
  natural, 632-633
Lye. *See* Caustic soda

Magic numbers, 578-579
Magnesium, 453, 456, 458
Magnesium chloride, 456
Magnesium hydroxide, 370, 456
Magnesium oxide, 461
Magnetic dipoles, 85
Magnetite, 560
Malachite, 564
Maleic acid, 365
Manganate ion, 558
Manganese, 82, 397, 495, 558, 561
Manometers, 48
Marble, 456
Mass, 8-9
  quantization of, 12-13
Mass action ratio, 330-331, 386-388
Mass number, 96, 576, 578-579, 581-582
Mass spectrometer, 95
Matches, 510
Maxwell, James Clerk, 204
Maxwell's demon, 204, 343
Melting, 237-238
  pressure effects on, 240-241
  thermodynamics of, 238-239
Melting point, 237-239
Mendeleev, Dmitri Ivanovitch, 79-82, 531
Mercurous ion, 564
Mercury, 455, 465, 498, 521, 542, 558, 564-565
Mercury(I) chloride, 381-382
Mesons, 579-580, 588
Metal carbonyls, 478
Metallic bonding, 135, 451, 542
Metallic hydrides, 428, 563
Metaphosphoric acid, 516
Metastable, 263
Methane, 142, 165-167, 442, 480
Methanol, 422
Methyl formate, 162
Methyl group, 481
Methyl mercaptan, 521
Methyl radicals, 164
Methyl red, 365
Mica, 494
Microscopic reversibility, 342-343
Microwave(s), 406
  molecular rotations and, 409-410

Microwave oven, 410
Millikan, Robert A., 91
Millimoles, 293
Mirror images, 481
Mirror planes, 218
Mischmetal, 566
Mitochondria, 599, 616-617
Mixed metal oxides, 435
Mixtures, 3-5
Moderator, 590
Molal boiling point elevation, 270
Molal freezing point depression, 271
Molality, 259
Molar absorptivity, 407-408
Molar enthalpy of fusion, 237
Molar enthalpy of vaporization, 237
Molar heat of fusion. *See* Molar enthalpy of fusion
Molar heat of vaporization. *See* Molar enthalpy of vaporization
Molarity, 259-260
Mole, 32-34
Mole fraction, 258-259
Molecular formula, 29-31
Molecular mass, 30
  and gas density, 61-62
Molecular motion, 16-17, 68-69
Molecular orbital(s), 145-148
  antibonding, 145, 414, 555
  bonding, 145, 414, 555
  nonbonding, 555
  π, 146, 414, 474, 485, 486, 489, 494, 511
  σ, 146, 485, 486, 489, 492
Molecular orbital (MO) approach, 145-148, 555
Molecular shapes, 164-169, 533
  molecular polarities and, 151
Molecular spectroscopy, 405-417
Molecules, 3, 13
  volumes of, 69
Molybdenum, 557
Momentum, 17
Monatomic ions, 136
Monazite, 556
Monomolecular elementary reactions, 311
Monosaccharides, 600
Moseley, Henry Gwyn-Jeffreys, 94, 108
Muriatic acid, 528
Mustard gas, 526
Mutation, 593-594

Naphthalene, 143
Negatron. *See* Beta particles
Neon, 82, 532
Neptunium, 567
Nernst equation, 386-388
Net ionic equations, 175-176
Neurospora crassa, 594
Neutral solution, 289, 352-353
Neutrality (electrical), 83
Neutralization, 287, 288-289
Neutrino, 581
Neutron, 95, 576, 578-581, 587-591
  thermal, 588-589, 592
Neutron activation analysis, 592
Newton (unit), 17
Newton, Sir Isaac, 101
Newton's laws of motion, 21-22
Newtonian physics, 17, 112
Nickel, 82, 394, 397, 422, 428, 540, 559, 563
Nicotinamide adenosine dinucleotide (NAD$^+$), 615
Nicotinamide adenosine dinucleotide phosphate (NADP$^+$), 615
Niobium, 557
Nitrates, 450, 508-509, 513, 571
Nitric acid, 289, 514, 515, 528
Nitric oxide, 164, 513-514

Nitrites, 513, 515
Nitrogen, 81, 174, 453, 465, 505-509, 511, 531
Nitrogen compounds, 511-515
Nitrogen cycle, 507-508
Nitrogen dioxide, 515
Nitrogen fixation, 507-508, 514
Nitrous acid, 515
Nitrous oxide, 513
Noble gas(es), 82, 159, 174, 521, 526, 531-534
  discovery of, 531-532
Noble gas compounds, 532-533
Noble metals, 559, 563
Nodal surface, 117
Node, 114
Nonelectrolytes, 283
Nonideal solutions, 266, 285
Nonideality, 51, 69-70, 266, 337-338, 359, 369
Nonoctet structures, 163-164
Nonstoichiometric compounds, 428
Normality, 295-297
  for precipitation titrations, 297
Nuclear binding, 579-580
Nuclear binding energy, 576-578
Nuclear charge, 94
Nuclear fission, 578, 589-590
Nuclear fusion, 578, 591-592
Nuclear magnetic resonance, 415-417, 580
Nuclear model for the atom, 93-94, 107
Nuclear reactions, 6-7, 573-595
Nuclear reactor, 590-591, 592
Nuclear spin states, 415, 580
Nuclear stability, 578-579
Nuclear transformations, 580-582
Nucleation, 263
Nucleic acids, 422, 505, 600, 609-614
Nucleons, 576-577, 579-583
Nucleus (atom), 93-94, 576-593
Nucleus (cell), 599

Octahedral holes, 233
Octahedron, 169, 217, 221, 533, 544-546
Oleum, 523
Open-hearth process, 560
Open systems, 190
Optical activity, 481
Orbital(s), 116-118
  d, 118, 163, 511, 540, 541, 551, 552-553
  e_g, 554-555
  hydrogenlike, 123, 167
  p, 117-118, 511, 551
  s, 116-117, 511, 551
  t_{2g}, 554-555
Orbital splitting, 553-555
Organic chemistry, 479-491
Orthophosphoric acid, 292, 515, 516
Osmium, 559
Osmotic pressure, 273-274, 455
Ostwald process, 515
Outer orbial complexes, 552
Overpotential, 393-394, 526, 562
Oxalate ion, 294, 370
Oxalic acid, 182, 562
Oxidant. *See* Oxidizing agent
Oxidation, 174
Oxidation potential, 383-386, 452-453, 457
Oxidation-reduction reactions. *See* Redox reactions
Oxidation states, 171-174
  chemical similarity and, 174
Oxidative phosphorylation, 617
Oxide(s), 432, 434-435, 453, 456, 457, 507, 509, 510, 532-533, 560, 564, 565
  acid-base properties of, 436
Oxide ion, 159, 370, 432
Oxidizing agent, 174
Oxyacids, 181-182, 436
Oxygen, 159, 173-174, 392, 393, 395,

INDEX

Oxygen (continued)
397-398, 432, 453, 461, 465, 466, 506, 512, 518, 519, 523, 539, 548, 557, 561, 562, 615-619
  compounds of, 434-439
  preparation of, 432-433
  properties of, 432
Ozone, 433-434

Palladium, 428, 563
Paramagnetism, 85, 125, 147, 164, 432, 542
Partial pressure, 56-57
Pauli, Wolfgang, 581
Pauli exclusion principle, 123
Pauling, Linus, 150, 550-552
Pearlite, 560
Pentanes, 481
Percent by weight, 258
Percent yield, 41-42
Perchloric acid, 530
Period, 82
Periodic table, 81-82
  electron configuration and, 128
Permanent hardness, 459
Permanent magnet, 84-85
Permanganate ion, 294, 408, 558
Perovskites, 435, 557
Peroxide(s), 174, 437-438, 453, 461
Peroxide ion, 173, 183, 437
Peroxyacetyl nitrate (PAN), 514
Peroxydisulfate ion, 438
Perrin, Jean, 92
Petroleum, 473, 474, 480, 523
pH, 352-354
Phase diagrams, 246-248, 440
Phenolphthalein, 362, 365, 366
Phenylalanine, 614
Phosgene, 342, 526
Phosphate(s), 442, 453, 507, 509, 510, 609
Phosphate ion, 370
Phosphides, 517
Phosphine, 517
Phosphorescence, 414, 566
Phosphorous acid, 516
Phosphorus, 81, 163, 505-507, 509, 515, 516-517, 560-561
  black, 509-510
  red, 509-510, 515
  white, 509-510, 516
Phosphorus pentoxide, 515-517
Photochemical smog, 433
Photoelectric effect, 106, 451
Photon, 105-106, 405-406, 414, 582, 619
Photosynthesis, 619-620
Physical methods of separation, 3
Physical properties, 5
Pig iron, 560
Pi-meson, 580
Pion. See Pi-meson
Planck, Max, 105
Planck's constant, 105
Planck's law, 105, 107
Plane-polarized light, 248, 481
Plasma, 592
Plaster of Paris, 461
Platinum, 381, 383, 428, 521, 544, 563
Plutonium, 539, 567
Polar covalent bond, 135, 149
Polar covalent compounds, 148-149
Polonium, 518, 521, 573
Polyatomic ions, 136
Polyfunctional compounds, 479
Polymeric and electron-deficient hydrides, 429
Polymorphic structures, 233, 464, 465, 474-475, 493
Polynucleotide chains, 609
Polypeptide chains, 602-609, 612-613

Polyphosphate salts, 459
Polysaccharides, 600
Polyvinyl chloride, 529
Polyvinylidene chloride, 529
Porphyrin group, 548
Positron, 581-582
Positron-electron annihilation, 581-582, 591
Potash, 454
Potassium, 79, 82, 439, 449, 451, 454
Potassium chlorate, 432
Potassium hydrogen phthalate, 295
Potassium hydroxide, 422
Potassium nitrate, 369
Potential energy, 19, 190
Pressure, 47-51, 65
Primary structure, 602-605
Products, 5
Propane, 480
Propene, 485
n-Propyl chloride, 416
Protium, 14, 415, 425
Proportional counters, 574-575
Protein(s), 422, 505, 507, 548, 600, 602-607, 614
Protein synthesis, 599, 612-614
Protoactinium, 567
Proton, 95, 576, 578-581, 587
Proton-proton chain, 591-592
Prout's hypothesis, 95
Prussian blue, 562-563
Pure substance, 3-4
Pyridine, 489
Pyrophosphoric acid, 516
Pyroxenes, 494
Pyruvate, 619

Quadratic equations, 333, 334, 633-634
Quantization of energy, 107-108, 144, 409
Quantum (pl. quanta), 12-13
  of charge, 89
Quantum mechanics, 101, 112-122
  molecular, 144-148
Quantum number, 108
  azimuthal, 120-121
  magnetic, 121
  principal, 120
  spin, 122
Quantum theory, 105-106
Quartz, 143, 234, 474, 495
Quaternary structure, 606-607

Radar, 410
Radioactive decay, 456
Radioactivity, 573-576
  discovery of, 573-574
  induced, 586-589
  measurement of, 574-575
  natural, 574, 585
  uses of, 592-594
Radiotherapy, 461, 594
Radium, 456, 461, 573
Radon, 82, 531, 532, 573
Ramsay, Sir William, 99, 531-532
Random coils, 606
Randomness, 204, 236, 257
Raoult's law, 263-270
Rare earths. See Lanthanide(s)
Rate constant, 307
  effect of temperature on, 316-318
Rate-determining step, 311-313
Rate law, 307-310
Rayleigh, Lord, 99, 531-532
Reactants, 5
Reaction mechanisms, 140, 310-313
Reaction rates, 304-306
  concentration effects on, 306-307
  of heterogeneous reactions, 313
Red lead, 497-498

Redox reactions, 174, 379
  half-reaction method for balancing, 177-180
  oxidation number method for balancing, 176-177
Reducing agent, 174
Reductant. See Reducing agent
Reduction, 174
Reduction potential, 383-386, 453-454
Regulatory enzymes, 607
Replication, 609
Resonance, 169-171
Resonance forms, 171, 489
Resonance hybrid, 171
Reversible reactions, 329
Rhenium, 558
Ribonucleic acid (RNA), 612-614
  messenger (mRNA), 612-614
  transfer (tRNA), 613-614
Ribosomes, 599
Right-hand rule, 86
Ring strain, 487
Rotation axes, 218-219
Rotational-vibrational spectra, 409-413
Rubidium, 79, 439
Ruby, 463
Ruthenium, 559
Rutherford, Ernest, 92, 106-107, 573, 575
Rutile, 252, 557
Rydberg constant, 110

Salt bridge, 380
Salt domes, 450
Saltpeter, 507
Sand, 509
Sapphire, 462-463
Saturated molecules, 487
Saturated solution, 260-261
  equilibrium vapor pressure of, 267
Scandium, 82, 556
Sceptical Chymist, The, 5, 7
Schottky defects, 492
Schrödinger, Erwin, 115
Schrödinger wave equation, 116, 122, 145, 165
Scientific law, 2
Scientific method, 1-3
Scintillation counter, 575, 581
Second ionization potential, 137
Second law of thermodynamics, 204-205
Secondary structure, 606, 609
Selection rules, 408-409
Selenium, 520
Semiconductors, 491-492
  n-type, 492
  p-type, 492
Semimetallic, 462, 520
Serine, 609
Shell, 120
Shells of nuclear energy levels, 579, 582
SI system, 9-10, 638
Side-centered (C) lattices, 221
Significant figures, 630-631
Silica, 494-496
Silicates, 432, 440, 450, 474, 494, 560
Silicic acid, 496
Silicon, 81, 163, 453, 466, 473-474, 491-492, 493, 560-561
Silicon carbide, 476
Silicon compounds, 494-496
Silver, 395, 524, 563-564
Silver bromide, 526
Silver chloride, 396
Silver iodate, 368-369
Silver nitrate, 395
Simplest formula, 31-32
  method of determining, 34-35
Simultaneous equilibria, 371-372

INDEX

Singlet oxygen, 448
Slag, 560
Smelting, 494, 563, 564
Soda ash. See Washing soda
Sodium, 79, 392, 423, 439, 449, 450, 451-452, 454
Sodium acetate, 357
Sodium bicarbonate, 454
Sodium carbonate, 455
Sodium chloride, 139, 218, 232-233, 454, 527-528
Sodium hydroxide, 422, 454-455
Sodium hypochlorite, 530
Sodium thiosulfate, 524
Sol(s). See Colloidal suspensions
Solid(s), 47, 217
Solubility, 261
    effect of complex ion formation on, 371
    effect of gas pressure on, 275-276
    effect of pH on, 369-370
    effect of temperature on, 262-263
    of amphoteric hydroxides, 370-371
    thermodynamics of, 261-263
Solubility products, 367-369
Solute, 257
Solution, 257-258
    boiling point elevation of, 270-271
    freezing point depression of, 271-273
Solvation, 441-442
Solvation energy, 452
Solvay process, 454
Solvent, 257
    vapor pressure lowering, 267-270
Specific heat, 21
Specificity, 607
Spectator ions, 176
Spectrophotometer, 407
Speed of light, 101, 104-105
Sphalerite, 519, 564
Spinels, 435
Spontaneous change, 202-204
Square planar shaper, 169, 533
Stainless steel, 397, 557, 563
Stalactites, 459
Stalagmites, 459
Standard atmosphere, 48
Standard calomel electrode, 385
Standard enthapy of formation, 199
Standard hydrogen electrode, 383-385
Standard reaction potential, 382-383
Standard solution, 293
Standard state, 337
Standard temperature and pressure (STP), 55
Standardization, 294-295
Standing waves, 114
Starch, 527, 600
State function, 190, 193-194, 197
Statistical distribution, 69
Steel, 526, 539, 560-561, 565
Stereospecificity, 607
Steric factor, 320-321
Stern-Gerlach experiment, 121
Stibine, 517
Stock system. See IUPAC nomenclature
Stoichiometry, 29
Strontium, 456, 461
Structural formulas, 157-159
Structure proof, 490
Sublimation, 247
Sublimation energy, 452-453
Subshells, 120-121
    relative energies of, 123
Substrate, 608
Sucrose, 600
Sugars, 600

Sulfates, 432, 456, 461, 519, 523
Sulfides, 370, 442, 507, 509, 516, 519, 521, 523, 564
Sulfite ion, 523
Sulfur, 174, 465, 516, 519, 521, 523, 560-561
    monoclinic, 519-520
    orthorhombic, 519-520
    plastic, 520
    rhombohedral, 520
Sulfur dioxide, 519, 523
Sulfur trioxide, 523
Sulfuric acid, 292, 390-391, 519, 523-524
Sulfurous acid, 523
Supercooling, 239-240
Superfluidity, 534
Superheavy nuclei, 588
Superoctet species, 163, 511, 518, 528, 541
Superoxide ion, 438-439, 453
Supersaturated solution, 263
Symmetry elements, 218-219
Système International d'Unités. See SI system

Table salt, 454
Tantalum, 557
Technitium, 558
Teflon, 529
Tellurium, 520
Temperature, 19-20
Temporary hardness, 459
Temporary magnet, 85
Terbium, 567
Termolecular elementary reactions, 311
Tertiary structure, 606-607, 609
Tetraethyl lead, 452, 498, 526
Tetrahedral holes, 233
Tetrahedron, 165, 169, 232, 440, 466, 475, 481, 494, 509, 512
Thallium, 82, 463, 465, 467
Theories, 2-3
Thermodynamic systems, 190-191
Thermodynamics, 189-210
Thermometer, 19-20
Thermonuclear reaction. See Nuclear fusion
Thioacetamide, 522
Thomson, Joseph John, 90-91
Thomson's model for atomic structure, 93
Thorium, 567
Three-center bonds, 429
Thulium, 565
Thymine, 609, 613
Tin, 81, 473-474, 492-493, 497-498, 563, 578
    polymorphic forms of, 493
Tin disease, 493
Titanium, 82, 466, 556-557
Titration
    acid-base, 293-294
    precipitation, 294
    redox, 294
Titration curves, 362-366
Torr, 10, 48
Tourmaline, 462
Transition metal(s), 82, 128, 466, 521, 527, 539-565, 616
    electron configurations of, 128, 539-541
    properties of, 542
Transition metal ions, 163
Transition state. See Activated complex
Translation, 221
Transmittance, 407
Transuranium elements, 568, 588
Triangular planar arrangement, 169
Tricarboxylic acid cycle, 619
Trigonal bipyramidal, 169, 533

Triiodide ion, 183, 527
Triphosphoric acid, 516
Triple point, 246-247
Tris(ethylenediamine)cobalt(III) ion, 550
Tritium, 14, 425, 588
Trouton's rule, 255
Tungsten, 557
Tungsten carbide, 557

Ultracentrifugation, 278
Ultrafiltration, 276
Ultraviolet-visible spectrum, 406, 413-414
Unit cell, 221-222
Universal gas constant, 60, 67, 198
Uracil, 613
Uranium, 456, 495, 539, 541, 567, 573
Urea, 479

Valence(s), 543
Valence bond (VB) method, 145
    for complex ions, 550-552
Valence electrons, 128, 157-161, 163, 168-169
Valence-shell electron-pair repulsion method (VSEPR), 168-169
Vanadium, 82, 557
van der Waals equation, 69-70
van der Waals forces, 234, 531
Velocity, 17, 21, 65-68
Viscosity, 278
Vital forces, 479, 599
Vitamin B-12, 539
Volhard analysis, 302
Volt, 382, 386
Volumetric flask, 259
von Laue, Max, 221

Washing soda, 454, 459
Water, 164, 167, 237, 240-241, 289, 429, 439-443, 452, 516, 521, 523, 533, 566, 590
Water gas, 422
Water glass, 496
Water softening, 459-460
Watson, James, 610
Watson-Crick model, 609-613
Wave(s), 102-105
    particles and, 112-114
    phases of, 114
Wave functions, 116, 144
Wavelength, 102-105
    momentum and, 112-113
Weight, 8
Werner, Alfred, 543-544
Wöhler, Friedrich, 479
Wood's metal, 511
Work, 18, 191-194, 379, 382-383

Xenon, 82, 532-533
Xenon difluoride, 533
x-ray(s), 108, 221, 493, 582
x-ray crystallography, 221-226, 546
x-ray emission, 108

Ytterbium, 566, 567
Yttrium, 556
Yukawa, H., 579-580

Zeolites, 460
Zero-point energy, 411
Zinc, 82, 379-380, 383, 391, 394, 395, 398, 423, 542, 557, 563, 564-565
Zinc blende, 564
Zircon, 557
Zirconium, 556-557
Zone refining, 493-494

Fundamental Physical Constants

Name	Symbol	Value
Universal gas constant	R	0.08205 L-atm/K-mol
		62.36 L-torr/K-mol
		8.3143 joule/K-mol
Boltzmann's constant	k	1.38062×10^{-23} J/K
Avogadro's number	N_{Avog}	6.02217×10^{23} molecules/mole
Molar volume of a gas		22.414 L at STP
Coulomb's law constant		8.99×10^9 N m^2/coul^2
Velocity of light	c	2.997925×10^8 m/sec
Planck's constant	h	6.62620×10^{-34} J-sec
Charge of electron	e	1.60210×10^{-19} coul
		4.86298×10^{-10} esu
Mass of electron	m_e	9.10956×10^{-31} kg
		5.48597×10^{-4} amu
Mass of proton	m_p	1.67261×10^{-27} kg
		1.00727 amu
Mass of neutron	m_n	1.67492×10^{-27} kg
		1.00867 amu
Rydberg constant (Hydrogen)	R_H	1.097373×10^7 m^{-1}